Navigation: Science and Technology

Volume 19

This series *Navigation: Science and Technology (NST)* presents new developments and advances in various aspects of navigation - from land navigation, marine navigation, aeronautic navigation to space navigation; and from basic theories, mechanisms, to modern techniques. It publishes monographs, edited volumes, lecture notes and professional books on topics relevant to navigation - quickly, up to date and with a high quality. A special focus of the series is the technologies of the Global Navigation Satellite Systems (GNSSs), as well as the latest progress made in the existing systems (GPS, BDS, Galileo, GLONASS, etc.). To help readers keep abreast of the latest advances in the field, the key topics in NST include but are not limited to:

- Satellite Navigation Signal Systems
- GNSS Navigation Applications
- Position Determination
- Navigational instrument
- Atomic Clock Technique and Time-Frequency System
- X-ray pulsar-based navigation and timing
- Test and Evaluation
- User Terminal Technology
- Navigation in Space
- New theories and technologies of navigation
- Policies and Standards

This book series is indexed in **SCOPUS** and **EI Compendex** databases.

Kegen Yu · Nanshan Zheng
Editors

Advances in GNSS Reflectometry

Receiver Design, Signal Processing, Machine Learning-Based Model Construction, and Disaster Monitoring

Springer

Editors
Kegen Yu
School of Environmental Science
and Spatial Informatics
China University of Mining and Technology
Xuzhou, Jiangsu, China

Nanshan Zheng
School of Environmental Science
and Spatial Informatics
China University of Mining and Technology
Xuzhou, Jiangsu, China

ISSN 2522-0454 ISSN 2522-0462 (electronic)
Navigation: Science and Technology
ISBN 978-981-92-2324-4 ISBN 978-981-92-2325-1 (eBook)
https://doi.org/10.1007/978-981-92-2325-1

This Springer imprint is published by the registered company Springer Nature Singapore Pte Ltd.
The registered company address is: 152 Beach Road, #21-01/04 Gateway East, Singapore 189721, Singapore

Preface

Global navigation satellite systems reflectometry (GNSS-R), also termed as GNSS interferometric reflectometry (GNSS-IR) and GNSS multipath reflectometry (GNSS-MR) in some circumstances, is an emerging remote sensing technology. Although the idea of using navigation satellite signals for sensing the Earth surface especially the ocean surface was proposed in the late 1980s, Martin-Rainer conducted the in-depth study on GNSS-R the first time and published an article in *European Space Agency Journal* in 1993. Since then, great advances in GNSS-R have been made in both theory and practice, especially when a series of satellites with GNSS-R payloads have been successfully launched and huge datasets are available to public, including Technology Demonstration Satellite-1 by UK, Cyclone GNSS satellites by USA, and FengYun-3E/F/G and Tianmu-1 satellites by China. However, there are only a few books completely focusing on GNSS-R, while several other books just have chapters dealing with the technology. Nevertheless, it is worth noting that a significant number of Master's and Ph.D. theses worldwide focused on solving problems in the field of GNSS-R.

GNSS-R has been advancing quickly, so it is useful to report recent advances in this exciting emerging field. To the end, we edited this book and a good number of experts made contributions. Specifically, the book focuses on a range of GNSS-R topics, including receiver design (Chaps. 1 and 2), spaceborne data calibration (Chaps. 3 and 4), modeling and simulation tool (Chap. 5), GNSS synthetic aperture radar (Chap. 6), land applications (Chaps. 7–13), and ocean applications (Chaps. 14–22). In the case of land applications, the 7 chapters deal with ground-based determination of underground coal mining-induced ground subsidence (Chap. 7); soil moisture and flood inundation retrieval with spaceborne data (Chap. 8); ground-based measurement of snow depth and density and liquid water content (Chap. 9); ground-based snow depth retrieval under complex terrain conditions (Chap. 10); ground-based GNSS-MR sea level inversion (Chap. 11); mapping of surface water fraction (Chap. 12); and flood mapping and water level estimation (Chap. 13). On the other hand, the 9 chapters of ocean application cope with problems in static maritime target detection and location using ground-based data (Chap. 14); ground-based maritime moving target detection and localization (Chap. 15); oceanic eddy detection and shape reconstruction based on spaceborne data (Chap. 16); sea ice detection and classification using spaceborne data (Chap. 17); sea surface rainfall detection with spaceborne data (Chap. 18); sea surface wind speed retrieval using spaceborne Data (Chap. 19); sea surface wind direction inversion based on spaceborne delay-Doppler map's asymmetry (Chap. 20); sea surface rainfall intensity retrieval with spaceborne Data (Chap. 21); and significant wave height retrieval with spaceborne data (Chap. 22). All chapters are involved with model construction, using empirical mathematical modeling, machine learning-based modeling, or both. Disaster monitoring mainly focuses on the monitoring of flooding and drought through soil moisture retrieval, flooding detection, and surface water fraction mapping, on the estimation of sea surface wind speed and direction, and on the detection of ocean eddy and ocean ice. A range of signal and data processing techniques were employed in most of the chapters.

We sincerely thank all the chapter authors (see list of contributors) for their precious contributions. We would also like to thank the staff members of Springer Nature for their constant support over the entire process from proposal to final publication. Finally, we acknowledge the support by the National Key Research and Development Program of China under grant 2024YFB3910004, the Joint Key Program of the National Natural Science Foundation of China under grant U22A20569, the Priority Academic Program Development of Jiangsu Higher Education Institutions, and the National Natural Science Foundation of China under Grant 42174022.

Xuzhou, China

Kegen Yu
Nanshan Zheng

Contents

Contributors

Jinwei Bu Faculty of Land Resources Engineering, Kunming University of Science and Technology, Kunming, China;
Yunnan Key Laboratory of Intelligent Monitoring and Spatiotemporal Big Data Governance of Natural Resources, Kunming, China

Xu Chen College of Information Technology, Shanghai Ocean University, Shanghai, China

Hao Du Institute of Space Sciences (ICE-CSIC)/Institut d'Estudis Espacials de Catalunya (IEEC), Barcelona, Spain

Huan Feng School of Remote Sensing and Geomatics Engineering, Nanjing University of Information Science and Technology, Nanjing, China

Fan Gao School of Space Science and Technology, Shandong University, Weihai, China;
Institute of Space Science, Shandong University, Weihai, China;
Guangzhou Urban Planning & Design Survey Research Institute Co., Ltd., Guangzhou, China

Bofeng Guo School of Marine Science and Technology, Tianjin University, Tianjin, China

Wenfei Guo GNSS Research Center, Wuhan University, Wuhan, China

Yunqiao He School of Space Science and Technology, Shandong University, Weihai, China;
Institute of Space Science, Shandong University, Weihai, China;
Guangzhou Urban Planning & Design Survey Research Institute Co., Ltd., Guangzhou, China

Zhenyu He College of Information Science and Engineering, Hohai University, Changzhou, China

Yuxian Hu School of Cyber Science and Engineering, Wuhan University, Wuhan, China

Feixiong Huang National Space Science Center, Chinese Academy of Sciences, Beijing, China;
University of Chinese Academy of Sciences, Beijing, China

Cheng Jing National Key Laboratory of Space Microwave Communications, China Academy of Space Technology (Xi'an), Xi'an, China;
Xi'an Institute of Space Radio Technology, China Academy of Space Technology (Xi'an), Xi'an, China

Bochi Lei National Key Laboratory of Space Microwave Communications, China Academy of Space Technology (Xi'an), Xi'an, China;
Xi'an Institute of Space Radio Technology, China Academy of Space Technology (Xi'an), Xi'an, China

Yan Li Hubei University of Technology, Wuhan, China

Yunwei Li Shandong Provincial Lunan Geology and Exploration Institute (Shandong Provincial Bureau of Geology and Mineral Resources No. 2 Geological Brigade), Jining, China;
Technology Innovation Center of Restoration and Reclamation in Mining induced Subsidence Land, Ministry of Natural Resources of China, Jining, China

Gao Ma National Key Laboratory of Space Microwave Communications, China Academy of Space Technology (Xi'an), Xi'an, China;
Xi'an Institute of Space Radio Technology, China Academy of Space Technology (Xi'an), Xi'an, China

Zhongmin Ma College of Geological Engineering and Geomatics, Chang'an University, Xi'an, China

Xinyue Meng School of Space Science and Technology, Shandong University, Weihai, China;
Institute of Space Science, Shandong University, Weihai, China;
Guangzhou Urban Planning & Design Survey Research Institute Co., Ltd., Guangzhou, China

Yang Nan School of Marine Science and Technology, Tianjin University, Tianjin, China

Zhigang Sun CAS Engineering Laboratory for Yellow River Delta Modern Agriculture, Institute of Geographic Sciences and Natural Resources Research, Chinese Academy of Sciences, Beijing, China

Wei Wan School of Earth and Space Sciences, Peking University, Beijing, China

Changyang Wang National Satellite Meteorological Center, China Meteorological Administration,, Beijing, China

Feng Wang Department of Electronic Information Engineering, Beihang University, Beijing, China

Nazi Wang School of Space Science and Technology, Shandong University, Weihai, China;
Institute of Space Science, Shandong University, Weihai, China;
Guangzhou Urban Planning & Design Survey Research Institute Co., Ltd., Guangzhou, China

Xiaolei Wang School of Earth Sciences and Engineering, Hohai University, NanKing, Jiangsu, China

Xuerui Wu Shanghai Astronomical Observatory, Chinese Academy of Sciences, Shanghai, China

Jin Xing School of Artificial Intelligence, China University of Geosciences Beijing, Beijing, China

Nianfu Xu School of Environmental Science and Spatial Informatics, China University of Mining and Technology, Xuzhou, China

Tianhe Xu School of Space Science and Physics, Shandong University, Weihai, China

Yinghui Xue National Space Science Center, Chinese Academy of Sciences, Beijing, China;
University of Chinese Academy of Sciences, Beijing, China

Qingyun Yan School of Remote Sensing and Geomatics Engineering, Nanjing University of Information Science and Technology, Nanjing, China

Songhua Yan Wuhan University, Wuhan, Hubei Province, China

Shuhu Yang College of Information Technology, Shanghai Ocean University, Shanghai, China

Ting Yang CAS Engineering Laboratory for Yellow River Delta Modern Agriculture, Institute of Geographic Sciences and Natural Resources Research, Chinese Academy of Sciences, Beijing, China

Xiaotong Yang School of Marine Science and Technology, Tianjin University, Tianjin, China

Kegen Yu School of Environmental Science and Spatial Informatics, China University of Mining and Technology, Xuzhou, China

Jie Zhang School of Earth and Space Sciences, Peking University, Beijing, China

Shuangcheng Zhang College of Geological Engineering and Geomatics, Chang'an University, Xi'an, China

Yun Zhang College of Information Technology, Shanghai Ocean University, Shanghai, China

Naiquan Zheng College of Geodesy and Geomatics, Shandong University of Science and Technology, Qingdao, China

Nanshan Zheng School of Environmental Science and Spatial Informatics, China University of Mining and Technology, Xuzhou, China

Yu Zheng The School of Electronic Communication and Electrical Engineering, Changsha University, Changsha, Hunan, China

Yongchao Zhu School of Civil and Hydraulic Engineering, Hefei University of Technology, Hefei, China

Abbreviations

ACE-BOC	Asymmetric Constant-Envelope Binary Offset Carrier
ACL	Active Cold Loads
Adam	Adaptive Moment Estimation
ADC	Analog-to-Digital Converter
ADDMV	Allan Delay-Doppler Map Variance
ADT	Absolute Dynamic Topography
AE	Anticyclonic Eddy
AGC	Automatic Gain Control
AGM	Automatic Gain Mode
AIEM	Advanced Integral Equation Model
AIS	Automatic Identification System
AltBOC	Alternative Binary Offset Carrier
AMSR2	Advanced Microwave Scanning Radiometer 2
AMSR-E	Advanced Microwave Scanning Radiometer for EOS
ANN	Artificial neural network
APFA	Average Probability of False Alarm
API	Application Programming Interface
AS	Anti-Spoofing
ASCAT	Advanced Scatterometer
ASI	ARTIST Sea Ice
ATFII	Annual Threshold Flood Inundation Index
ATLAS	Advanced Terrain Laser Altimeter System
Ave	Average Value
AVISO	Archiving, Validation and Interpretation of Satellite Oceanographic data
AWGN	Additive White Gaussian Noise
azi	Azimuth Angle
BARTS	Bootstrap Aggregation of Regression Trees
BDS	BeiDou navigation satellite system
BDS2-IGSO	BeiDou Navigation Satellite System Phase 2 Inclined Geosynchronous Orbit

BDS2-MEO	BeiDou Navigation Satellite System Phase 2 Medium Earth Orbit
BDS-3	BeiDou Navigation Satellite System-3
BDS3-IGSO	BeiDou Navigation Satellite System Phase 3 Inclined Geosynchronous Orbit
BDS3-MEO	BeiDou Navigation Satellite System Phase 3 Medium Earth Orbit
BEA	Base elevation angle
BeiDou	Beidou Navigation Satellite System
BF-1	BuFeng-1
Bfemp	BF-1 SM derived from an empirical method
Bfphy	BF-1 SM retrieved using the proposed physics-based approach
BOC	Binary Offset Carrier
BP	Back-projection
BP-ANN	Back-Propagation Artificial Neural Network
BPSK	Binary Phase Shift Keying
BSA	Backscatter Alignment
BT	Bagging Tree
C/A	Coarse/Acquisition
C3S	Copernicus Climate Change Service
CA	Cell Averaging
CAP	Coherent accumulation process
CART	Classification and Regression Tree
CAST	China Academy of Space Technology
CC	Correlation Coefficient
CCI	Climate Change Initiative
CDMA	Code Division Multiple Access
CDW	Central Delay Waveform
CE	Cyclonic Eddy
CEMS	Copernicus Emergency Management Service
CFAR	Constant False Alarm Rate
CHIRPS	Climate Hazards Group Infrared Precipitation with Station
CIT	Coherent integration time
CNES	Centre National d'Études Spatiales
CNN	Convolutional Neural Networks
CNR	Contrast-to-noise ratio Carrier-to-Noise Ratio
CP	Carrier Phase
CSWF	CYGNSS Surface Water Fraction
CYG-05	Cyclone Global Navigation Satellite System satellite 05
CYGNSS	Cyclone Global Navigation Satellite System
CYGv3.0	CYGNSS level 1 data product version 3.0
CYGv3.1	CYGNSS level 1 data product version 3.1
dB	Decibel
DD	Delay Doppler

DDM	Delay-Doppler Map
DDMA	Delay-Doppler Map Average
DDMI	Delay Doppler Mapping Instrument
DDMV	Delay-Doppler Map Variance
DDW	Delay-Doppler Waveform
DEM	Digital elevation model
DFM	Doppler Frequency Migration
DFR	Doppler Frequency Rate
DFT	Discrete Fourier Transform
DMC	Disaster Monitoring Constellation
DMR	Delay Mapping Receiver
DOY	Day of Year
DSSS	Direct Sequence Spread Spectrum
DT	Decision tree
DTM	Digital terrain model
EASE-Grid	Equal-Area Scalable Earth Grid
ECEF	Earth-Centered Earth-Fixed
ECMWF	European Center for Medium-Range Weather Forecasts
EIRP	Effective Isotropically Radiated Power
EKE	Eddy Kinetic Energy
ele	Elevation Angle
EM	Engineering Model
ENOB	Effective Number of Bits
EOS	Earth Observing System
ERA5-Land	European Centre for Medium Range Weather Forecasts Reanalysis v5 Land
ERO	Exact Repeat Orbit
ESA	Effective Scattering Area/European Space Agency
F2C	Dual-Frequency Carrier Phase and Code Combination
F3	Triple-Frequency Carrier Phase
FC	Fully Connected (Layer)
FCN	Fully Connected Network
FDMA	Frequency Division Multiple Access
FFNN	Feed-Forward Neural Network
FFT	Fast Fourier Transform
FFZ	First Fresnel Zone
FN	False Negatives
FOM	Figure of Merit
FORMOSAT-7R	FORMOSAT-7R (TRITON)
FP	False Positives
FPGA	Field-Programmable Gate Array
FrFT	Fractional Fourier transform
FSA	Forward Scatter Alignment
FT	Fourier Transform
FY-3E	FengYun-3E

FYI	First-year ice
GAB	Great Artesian Basin
GAL-FOC	Galileo Full Operational Capability
Galileo	Galileo Satellite Navigation System
GAL-IOV	Galileo In-Orbit Validation
GCPM	Ground-based Constellation Power Monitor
GEDI	Global Ecosystem Dynamics Investigation
GEE	Google Earth Engine
GEO	Geostationary Earth Orbit
GEROS-ISS	GNSS Reflectometry, Radio Occultation and Scatterometry on the International Space Station
GFM	Global Flood Monitoring
GFZ	German Research Centre for Geosciences
GLAD	Global Land Analysis and Discovery
GLOBE	Global Land One-kilometer Base Elevation
GLONASS	Global Navigation Satellite System
GMF	Geophysical Model Function
GNOS	Gnss Occultation Sounder
GNSS	Global Navigation Satellite Systems
GNSS+IR	GNSS positioning and GNSS-IR
GNSS-IR	GNSS interferometric reflectometry
GNSS-MR	Global Navigation Satellite System Multipath Reflectometry
GNSS-R	Global Navigation Satellite System Reflectometry
GNSS-RO	GNSS Radio Occultation
GO	Geometrical Optics
GOLD-RTR	GNSS-Ocean Reflection Delay and Ranging Receiver
GORS	GNSS Occultation, Reflectometry and Scatterometry
GPM	Global Precipitation Measurement
GPM-IMERG	Global Precipitation Measurements from Integrated Multi-Satellite Retrievals
GPS	Global Positioning System
GPSII-F	Global Positioning System Block IIF
GPSIII-A	Global Positioning System IIIA
GPSII-R	Global Positioning System Block IIR
GPSIIR-M	Global Positioning System Block IIR-M
GPU	Graphics Processing Unit
GSARM	GNSS Signal Amplitude Ratio Model
GSMaP	Global Satellite Mapping of Precipitation
GSVM	Gaussian support vector machine
GSW	Global Surface Water
HMDL	Heterogeneous Multimodal Deep Learning
HOD	Hour of the day
HVCE	Helmert variance component estimation
HydroGNSS	The HydroGNSS Scout mission
IC	Intensity Contrast

ICESat-2	Ice, Cloud and land Elevation Satellite-2
ICF	Interferometric complex field
IDW	Integrated delay waveform
IEEC	Institute of Space Sciences (Spain)
IEM	Integral Equation Model
IF	Intermediate Frequency
IGBP	International Geosphere–Biosphere Programme
IGSO	Inclined Geosynchronous Orbit
IMERG	Integrated Multi-Satellite Retrievals for Global Precipitation Measurement
IMERG-F	IMERG Final
IoT	Internet of Things
IPT	Interference pattern technique
IR	Improvement Rate
IRE	Integrated Relative Error
ISMN	International Soil Moisture Network
ISS	International Space Station
JPL	Jet Propulsion Laboratory
KA-GO	Kirchhoff Approximation-Geometric Optics
KPIM	Probability integral model combined with the Knothe time
KT	Keystone Transform
L4	Geometry-Free Linear Combination
LAGRS	Land Surface GNSS Reflection Simulator
LAGRS-Soil	Land Surface GNSS Reflection Simulator for Soil
LAGRS Veg	Land Surface GNSS Reflection Simulator for Vegetation
LE	Left edge
LEO	Low Earth Orbit
LES	Leading-edge Slope
LES-NIDW	Leading Edge Slope of NIDW
LEWS	Leading-Edge Waveform Summation
LEWS-NIDW	Leading-Edge Waveform Summation of NIDW
LHCP	Left-Hand Circular Polarization/left-Hand Circularly Polarized
LiDAR	Light Detection and Ranging
LNA	Low Noise Amplifiers
LO	Local Oscillator
LR	Left-Hand Circular Polarization Received, Right-Hand Circular Polarization Transmitted
LSC	Land Surface Clustering
LSF	Least squares fitting
LSP	Lomb–Scargle periodogram
LUT	Look-Up Table
LWC	Liquid Water Content
MAE	Mean Absolute Error
MAPE	Mean Absolute Percentage Error
ME	Mean error

MEO	Medium Earth Orbit
MF	Matched Filtering
ML	Machine learning
MODIS	Moderate Resolution Imaging Spectroradiometer
MORP	Mean of Relative Percentage
MSE	Mean square error
MSL	Mean Sea Level
MSS	Mean Square Slope
MSWEP	Multi-Source Weighted-Ensemble Precipitation
MTD	Moving Target Detection
MV	Minimum Variation
MYI	Multi-year ice
NASA	National Aeronautics and Space Administration
NBRCS	Normalized Bistatic Radar Cross Section
NCDW	Normalized Center Delay Waveform
NDBC	National Data Buoy Center
NDDM	Normalized DDM
NDVI	Normalized Difference Vegetation Index
NetCDF	Network Common Data Form
NIDW	Normalized Integral Delay Waveform
NOAA	National Oceanic and Atmospheric Administration
NOHRSC	National Operational Hydrologic Remote Sensing Center
NP	Notice point
NRCS	Normalized Radar Cross Section
NRCSNWCC	Natural Resources Conservation Service National Water and Climate Center
NSIDC	National Snow and Ice Data Center
NSSC	National Space Science Center
OCXO	Oven-Controlled Crystal Oscillator
OSISAF	Ocean and Sea Ice Satellite Application Facility
OVA	One-versus-all
OW	Open water
OWE	Observed Water Extent
PARIS	Passive Reflectometry and Interferometry System
PBO	Plate Boundary Observatory
PD	Probability of Detection Phase difference
PDF	Probability Density Functions
PFAr	Probability of false alarm detection of rainfall
PGYSER	PGYSER: physics-based spaceborne GNSS-R SM retrieval framework
PIRA	PARIS Interferometric Receiver
PLL	Phase-Locked Loop
PMR	Probability of Missing in Rainfall

PN	Polynomial function
	Pixel Number Observable
PNR	Peak-to-Noise Ratio
POBI	Previously-Observed Behavior Interpolation
POD	Precision Orbit Determination
POE	Probability of Overall Error
PR	Power Ratio
PRN	Pseudo Random Noise
QC	Quality Control
QMBOC	Quadrature Multiplexed Binary Offset Carrier
QZSS	Quasi-Zenith Satellite System
RAG	Range-Corrected Average Gain
RANSAC	RANdom SAmple Consensus
RBF	Radial Basis Function
RC	Range compression
RCG	Range Corrected Gain
RCM	Range Cell Migration
RCS	Radar Cross Section
RD	Range-Doppler
RDD	Range-Doppler-DFR
RE	Right edge
ReLU	Rectified Linear Unit
RESC	RE slope of CDW
RESD	RE slope of DDW
REWD	RE waveform summation of DDW
REWI	RE waveform summation of IDW
REWS	Regional Ensemble Weighted Smoothing
RF	Random Forest
	Radio Frequency
RFI	Radio Frequency Interference
RFT	Radon-Fourier Transform
RH	Reflector Height
RHCP	Right-Hand Circular Polarization
	Right-Hand Circularly Polarized
RI	Rainfall Intensity
RINEX	Receiver Independent Exchange Format
RM	Range migration
RMS	Root Mean Square
RMSD	Root Mean Square Difference
RMSE	Root Mean Square Error
RS	Remote sensing
RSS	Root Sum Square
RT	Regression Tree
RTK	Real time kinematic
RZE	Ratio of Zero Intensity to Echo

SAR	Synthetic aperture radar
SAST	Shanghai Academy of Spaceflight Technology
SAVERS	Soil and Vegetation Reflection Simulator
SBAS	Satellite-Based Augmentation System
SC	Similarity Coefficient
SCoBi-Veg	Signals of Opportunity Coherent Bistatic Scattering Model for Vegetated Terrains
SDR	Software-defined radio
SGR-ReSI	SGR-ReSI (Surrey GNSS Remote Sensing Instrument)
SHTS	Significant height of total swells
SHWW	Significant height of wind waves
SIC	Sea Ice Concentration
SIE	Sea Ice Edge
SIT	Sea Ice Type
SLR	Stepwise linear regression
SM	Soil moisture Surface moisture
SM2RAIN	Soil Moisture to Rain
SMAP	Soil Moisture Active Passive
SMAP-R	Soil Moisture Active Passive-Reflectometry
SMcali	Calibrated soil moisture
SMOS	Soil Moisture and Ocean Salinity
SMUCAR	UCAR/CU soil moisture product
SNR	Signal-to-Noise Ratio
SNR_COM	SNR combination
SoOp	Signals of Opportunity
SoR-COR	Soil Reflectivity CORrecting
SP	Specular Point
SPA	Snow Pack Analysing System
SPM	Small Perturbation Method
SQNR	Signal-to-Quantization-Noise Ratio
SR	Surface reflectivity
SRcali	Calibrated Surface Reflectivity
SRTM	Shuttle Radar Topography Mission
SSA	Small Slope Approximation
SSA-ELM	Sparrow Search Algorithm—Extreme Learning Machine
SSD	Solid-State Drive
SSH	Sea Surface Height
SSMIS	Special Sensor Microwave Imager/Sounder
SST	Sea Surface Temperature
SuR-CAL	Surface Reflectivity CALibrating
SVM	Support Vector Machine
SVR	Support Vector Regression
SWAMPS	Surface Water Microwave Product Series
SWE	Snow Water Equivalent

SWF	Surface Water Fraction
SWH	Significant Wave Height
TCXO	Temperature-Compensated Crystal Oscillator
TDS-1	Techdemosat-1
TES	Trailing Edge Slope
TMC	Target Motion Compensation
TN	True Negatives
TP	True Positives
TPR	True Positive Rate
TRMM TMPA	Tropical Rainfall Measurement Mission Multi-Satellite Precipitation Analysis
TSS	Tilted Surface Strategy
TTC	Terrain tilt correction
TWS	Total Water Storage
UAV	Unmanned aerial vehicle
ubRMSD	Unbiased Root Mean Square Difference
ubRMSE	Unbiased Root Mean Square Error
UK-DMC	United Kingdom Disaster Monitoring Constellation
ULA	Uniform Linear Array
UNAVCO	University NAVSTAR Consortium
USDA	United States Department of Agriculture
VC	Vapnik–Chervonenkis
VIIRS	Visible Infrared Imaging Radiometer Suite
VMD	Variational mode decomposition
VWC	Vegetation Water Content
WAF	Woodward Ambiguity Function
WD	Wind Direction
WF	Water fraction
WL	Water level
WS	Wind Speed
WW3	WaveWatch 3
ZPP	Zero-Pixel Percentage
ZSR	Zenith-to-Specular Ratio
Z-V	Zenith-Vertical

Chapter 1
Hardware System Design for BeiDou/GNSS Reflectometry Sensors: From the Bistatic Radar Equation to Error Analysis

Cheng Jing, Gao Ma, and Bochi Lei

Abstract GNSS Reflectometry (GNSS-R) has emerged as an effective passive bistatic radar technique for observing ocean surface roughness, soil moisture, sea ice conditions, and atmospheric parameters. With the continuous enhancement of BeiDou signals, hardware-based optimization has become a decisive factor for enabling high-precision scattering measurements. This chapter provides a hardware-centric perspective on BeiDou/GNSS-R system design, starting from the classical bistatic radar equation and extending to error propagation across all receiver subsystems. We establish a unified analytical framework linking bistatic geometry, antenna radiation characteristics, RF front-end impairments, sampling and quantization constraints, and digital baseband processing errors to the final observable parameters such as received power, delay–Doppler maps, and normalized scattering cross-section σ^0. Particular emphasis is placed on low-SNR regimes associated with spaceborne and airborne GNSS-R missions, where system noise figure, oscillator phase stability, and beam-pointing accuracy critically determine the achievable coherent and incoherent integration gains. By integrating theoretical derivations with practical engineering considerations, this chapter outlines key limitations, calibration requirements, and design recommendations for next-generation BeiDou/GNSS-R hardware, forming the foundation for high-quality geophysical retrievals in subsequent chapters.

C. Jing (✉) · G. Ma · B. Lei
National Key Laboratory of Space Microwave Communications, China Academy of Space Technology (Xi'an), Xi'an, China
e-mail: jingcheng@radi.edu.cn

Xi'an Institute of Space Radio Technology, China Academy of Space Technology (Xi'an), Xi'an, China

B. Lei
e-mail: leibochi@cast504.com

K. Yu and N. Zheng (eds.), *Advances in GNSS Reflectometry*, Navigation: Science and Technology 19, https://doi.org/10.1007/978-981-92-2325-1_1

1 Introduction

1.1 Background of GNSS Bistatic Remote Sensing

Global Navigation Satellite System Reflectometry (GNSS-R) has transformed from an experimental bistatic radar concept into a rapidly expanding remote-sensing technique supported by a diverse ecosystem of navigation, communication, and Earth-observation satellite constellations [1, 2]. Early demonstrations—such as ESA's GNSS-R experiments on the UK TechDemoSat-1 (TDS-1) mission [3] and NASA's Cyclone Global Navigation Satellite System (CYGNSS) [4]—provided the first global datasets of ocean surface wind speed derived exclusively from reflected GNSS signals. Since then, an increasing number of spaceborne, airborne, and stratospheric missions have embraced GNSS-R as a lightweight, power-efficient, and weather-penetrating sensor modality [5–10].

In recent years, several new constellations have significantly broadened the opportunities for GNSS-R. China's BuFeng-1 (BF-1) twin satellites demonstrated one of the earliest dedicated GNSS-R missions outside the United States [11–14]. The Chinese Tianmu meteorological constellation, launched beginning in 2023, also incorporates GNSS-R payloads for atmospheric and oceanic monitoring, forming an emerging large-scale low Earth orbit (LEO) platform for operational reflectometry [15]. Similarly, ESA's GEROS-ISS experiment [16], Taiwan's FORMOSAT-7R (also known as TRITON) [17], and Spire's commercial GNSS-RO/reflectometry nanosatellite fleet further expand both the scientific and commercial ecosystems [18, 19].

Beyond classical navigation systems, the proliferation of broadband LEO mega-constellations provides new signal opportunities for bistatic and opportunistic reflectometry. Systems such as Starlink (SpaceX), OneWeb, and the Chinese GW constellation broadcast high-power Ku/Ka-band communication waveforms, which are potentially useful for ocean and land-surface bistatic scattering [20, 21]. Other Chinese constellations, though primarily designed for communications or Internet of Things (IoT), create additional downlink signals that can be exploited for reflection measurements, particularly in Ku/Ka-bands. These developments signal a shift toward multi-frequency, multi-source reflectometry, where GNSS signals represent only one class of illuminating sources.

Within the traditional GNSS domain, the modernization of BeiDou (BDS-3) is especially impactful for GNSS-R. The B1C and B2a signals incorporate advanced multiplexed BOC modulations, improved correlation properties, higher equivalent isotropically radiated power (EIRP), and dedicated pilot components suitable for long coherent integrations—capabilities that significantly enhance observable power levels and retrieval sensitivity [22]. Other GNSS constellations have also advanced: GPS III increases transmit power and adds L1C; Galileo offers exceptionally high EIRP and wideband E5 AltBOC signals; and GLONASS-K2 introduces CDMA signals compatible with modern receivers. Together, these constellations create a dense global illumination environment ideal for high-temporal-coverage bistatic remote sensing.

Meanwhile, new experimental missions are pushing GNSS-R into new domains. ESA's HydroGNSS [23, 24], China's spaceborne dual-frequency reflectometry demonstrators, airborne wide-swath GNSS-R/SAR hybrid sensors, and nanosatellite swarms demonstrate a trend toward higher spatial resolution, multi-parameter retrieval, and real-time environmental monitoring. The field is also expanding toward non-traditional applications such as soil freeze–thaw state detection, vegetation biomass, sea-ice thickness, inland water mapping, and ionospheric disturbance sensing.

This evolving landscape highlights a crucial technological implication: as more constellations with stronger signals, higher bandwidth, and greater diversity emerge, the limiting factor for GNSS-R performance is shifting away from illumination availability and toward receiver hardware design—including antenna pattern stability, RF front-end noise figure, oscillator coherence, sampling precision, and digital processing robustness. These hardware constraints directly determine the achievable coherent integration time, the effective radiometric sensitivity, and the overall fidelity of Delay–Doppler maps (DDMs) and scattering coefficients.

1.2 *Motivation for a Hardware-Centered Perspective*

As GNSS-R evolves into a multi-constellation, multi-frequency, and multi-mission remote sensing modality, the fundamental limiting factor for measurement quality has increasingly shifted from signal availability to receiver hardware performance. Earlier generations of GNSS-R instruments operated in environments where the scarcity and weakness of illuminating signals were the dominant constraints. However, with the proliferation of modernized GNSS signals (e.g., BeiDou B1C/B2a, GPS III L1C, Galileo E5 AltBOC) and the emergence of powerful non-navigation sources—including Ku/Ka-band downlinks from GW, Starlink, and other broadband LEO systems—the bottleneck has moved toward how effectively the receiver can exploit these abundant electromagnetic opportunities.

A central challenge is that the reflected signal remains extremely weak, even when the illuminating source is strong. Surface scattering, geometric spreading loss, and glistening-zone averaging generally reduce the received power for L-band GNSS-R to the range of –160 to –196 dBW (decibels relative to a watt). In higher-frequency reflections (e.g., Ku/Ka-band opportunistic reflectometry from broadband LEO), the situation can be equally severe due to increased path loss and surface decorrelation. Consequently, the achievable Signal-to-Noise Ratio (SNR) and coherent integration time are dominated not by the transmitter but by the thermal noise, phase stability, and calibration accuracy of the receiver hardware.

The growing diversity of missions amplifies these hardware challenges.

- Spaceborne wide-swath systems (e.g., BF-1, Tianmu GNSS-R, HydroGNSS) experience significant platform dynamics, requiring oscillators and phase Locked loops (PLLs) with exceptional phase stability to sustain even millisecond-level coherent integration.
- Airborne platforms must suppress vibration-induced local-oscillator (LO) phase modulation and maintain stable antenna patterns across rapidly varying incident angles.
- Nanosatellite constellations rely on ultra-low-power front ends whose noise figure, quantization resolution, and clock jitter directly limit their ability to detect diffuse reflections.
- Hybrid reflectometry–radar or reflectometry–SAR concepts place additional constraints on calibration, linearity, and timing synchronization.

Moreover, GNSS-R measurement quality depends on the end-to-end consistency of the signal chain—from the antenna to the FPGA—because the DDM is effectively a radiometric product reconstructed through precise correlation and compensation operations. Errors in antenna gain stability, polarization purity, IF filtering, ADC nonlinearities, and digital tracking loop parameters propagate nonlinearly into DDM amplitude, peak shape, delay resolution, Doppler bandwidth, and subsequently the retrieved surface geophysical parameters (e.g., σ^0, wind speed, wave slope distribution, soil moisture).

As reflectometry moves toward long coherent integration using modern pilot channels (e.g., BeiDou B1C pilot, Galileo E1C/E5), previously negligible hardware effects become mission-defining factors. LO phase noise limits the maximum coherent integration time; ADC quantization determines effective radiometric sensitivity; clock jitter introduces decorrelation; and antenna thermal drift biases the absolute calibrated σ^0 scale. These effects are further coupled with bistatic geometry, making a hardware-centered error-propagation framework indispensable for any mission seeking geophysical accuracy beyond 1–2 dB in σ^0 or 0.5–1 m/s in ocean wind speed.

In this context, the need for a dedicated, comprehensive analysis of GNSS-R hardware is clear. Modern reflectometry cannot be viewed solely as a signal-processing or geophysical inversion problem; rather, it is a hardware–software–geometry integrated remote sensing system, where hardware design directly determines the attainable information content. This chapter provides the foundation for such an integrated understanding, linking bistatic radar physics, hardware-induced uncertainties, and final observable performance.

2 Classical Bistatic Radar Equation for GNSS-R

GNSS-R can be rigorously interpreted within the framework of classical bistatic radar theory. In this configuration, GNSS satellites act as non-cooperative transmitters of opportunity, while a separate receiver collects both the direct signal and the signal scattered from the Earth's surface. The transmitter and receiver are spatially separated, and the scattering geometry is inherently bistatic.

2.1 General Form of the Bistatic Radar Equation

For a bistatic radar system, the average received power from a distributed scattering surface can be expressed as [7]

$$P_r = \frac{P_t G_t G_r \lambda^2}{(4\pi)^3 R_t^2 R_r^2} \int_A \sigma^0(\mathbf{r}) dA, \tag{1}$$

where

- P_t is the transmitted power of the GNSS satellite,
- G_t and G_r are the gains of the transmitting and receiving antennas in the directions of the scattering point,
- λ is the signal wavelength,
- R_t and R_r are the distances from the transmitter to the scattering point and from the scattering point to the receiver, respectively,
- σ^0 is the bistatic normalized radar cross section (NRCS) of the surface, and
- A denotes the illuminated scattering area.

This formulation highlights the fundamental dependence of the received scattered power on the bistatic geometry, surface scattering properties, and antenna characteristics.

Special Characteristics in GNSS-R Applications

Compared with conventional bistatic radar systems, GNSS-R exhibits several distinctive features:

1. Low effective EIRP and wideband signals

GNSS satellites transmit spread-spectrum signals with relatively low EIRP, typically on the order of tens of dBW. Consequently, GNSS-R systems operate in an extremely low SNR regime, requiring coherent integration and correlation processing.

2. Specular-dominant scattering geometry

For smooth surfaces such as calm oceans, lakes, or gentle land surfaces, the dominant contribution to the scattered power arises from the vicinity of the specular reflection point. In such cases, the surface integral can be approximated by a localized glistening zone, whose size depends on surface roughness, wavelength, and geometry.

3. Bistatic NRCS dependence

The bistatic NRCS σ^0 in GNSS-R depends on both the incidence angle and scattering angle, as well as surface parameters such as roughness, dielectric constant, and vegetation structure. This contrasts with monostatic radar, where the scattering angle is fixed by the incidence angle.

2.2 GNSS-R Power Expression in Practice

For practical GNSS-R analyses, the bistatic radar equation is often rewritten in terms of the GNSS satellite EIRP and receiver system parameters:

$$P_r = \frac{EIRP \cdot G_r \lambda^2}{(4\pi)^3 R_t^2 R_r^2} \int_A \sigma^0(\mathbf{r}) dA, \tag{2}$$

This expression forms the theoretical basis for link budget analysis, receiver sensitivity estimation, and system design in GNSS-R. It also provides the physical link between measured observables—such as DDMs—and geophysical parameters of the reflecting surface.

2.3 Relevance to GNSS-R System Design

The classical bistatic radar equation clarifies several key design trade-offs in GNSS-R systems, including antenna gain versus coverage, receiver noise figure versus integration time, and platform altitude versus spatial resolution. As such, it serves as the foundational model upon which more advanced formulations—incorporating correlation processing, ambiguity functions, and delay–Doppler mapping—are developed in subsequent sections.

2.4 From Bistatic Radar Equation to GNSS-R Observables

While the classical bistatic radar equation provides the physical relationship between the scattered power and the surface bistatic normalized radar cross section (NRCS), GNSS-R receivers do not directly measure instantaneous scattered power from an individual surface element. Instead, the received signals are processed through correlation with a locally generated replica of the GNSS spreading code, followed by coherent and incoherent integration. As a result, the fundamental GNSS-R observable is not power itself, but a two-dimensional distribution in delay and Doppler frequency, commonly referred to as the DDM.

The transition from the bistatic radar equation to the DDM formulation is achieved by incorporating the GNSS signal ambiguity function and explicitly accounting for the mapping between surface scattering points and their corresponding delay–Doppler coordinates. In this sense, the DDM can be interpreted as a delay–Doppler-resolved bistatic radar equation, where the surface integral is weighted by the GNSS signal correlation properties.

2.5 General DDM Expression

Under the assumption of a stationary scattering surface during the coherent integration interval, the mean power of a GNSS-R DDM cell centered at delay τ and Doppler frequency f can be expressed as [7]

$$P(\tau,f) = \frac{EIRP \cdot G_r\lambda^2}{(4\pi)^3} \iint_A \frac{\sigma^0(\mathbf{r})}{R_t^2(\mathbf{r})R_r^2(\mathbf{r})} |\chi(\tau - \tau(\mathbf{r}), f - f(\mathbf{r}))|^2 dA, \qquad (3)$$

where

- $\chi(\cdot,\cdot)$ is the GNSS signal ambiguity function determined by the spreading code, coherent integration time, and receiver processing strategy,
- $\tau(\mathbf{r})$ and $f(\mathbf{r})$ are the geometric delay and Doppler shift associated with surface point $\mathbf{r}$, and
- other symbols follow the definitions of the classical bistatic radar equation.

This formulation explicitly links the DDM to the bistatic scattering process, with each surface element contributing energy to a specific region in delay–Doppler space.

2.6 Physical Interpretation

The DDM can be physically interpreted as a projection of the glistening zone onto delay–Doppler coordinates:

- Delay dimension reflects the excess path length relative to the specular point and is primarily related to surface height and geometry.
- Doppler dimension captures the relative motion between transmitter, receiver, and scattering surface, as well as surface roughness–induced Doppler spreading.

For smooth surfaces, such as calm oceans, most of the scattered power is concentrated near the specular delay–Doppler bin, resulting in a compact DDM. Increasing surface roughness broadens the effective glistening zone, leading to power spreading over multiple delay and Doppler bins.

2.7 *Relation to Classical Radar Cross Section*

In contrast to conventional radar, where the radar cross section is often inferred from a single range bin, GNSS-R retrieves geophysical information from the shape, width, and peak power of the DDM. Parameters such as ocean surface roughness, wind speed, soil moisture, or vegetation structure are thus indirectly inferred through their impact on σ^0 and its spatial distribution within the glistening zone.

Accordingly, the DDM formulation represents a natural extension of the classical bistatic radar equation, embedding surface scattering physics into a signal-processing-aware framework that directly corresponds to GNSS-R measurements.

2.8 *Implications for System Design and Inversion*

The DDM-based formulation highlights several key aspects of GNSS-R system design and data inversion:

- Receiver sensitivity and antenna gain directly affect the absolute DDM power level.
- Coherent and incoherent integration times control the delay–Doppler resolution and noise statistics.
- The inversion of geophysical parameters relies on forward models that map surface properties to DDM observables rather than to point-wise radar cross sections.

These considerations motivate the adoption of DDM-centric modeling and analysis in modern GNSS-R systems.

3 GNSS-R Receiver Hardware Architecture

A modern GNSS-R receiver is a highly integrated passive bistatic sensing system designed to extract extremely weak scattered signals from a dense electromagnetic environment. Unlike conventional GNSS positioning receivers, GNSS-R instruments must preserve radiometric fidelity, phase stability, and geometric consistency across the entire signal chain. Consequently, the hardware architecture plays a decisive role in determining achievable coherent integration time, DDM quality, and the accuracy of derived geophysical observables.

Figure 1 illustrates a generic hardware architecture for a spaceborne or airborne GNSS-R receiver, which can be decomposed into five principal subsystems: (1) antenna system, (2) RF front end, (3) intermediate frequency (IF) and sampling subsystem, (4) digital baseband processing unit, and (5) timing and control subsystem.

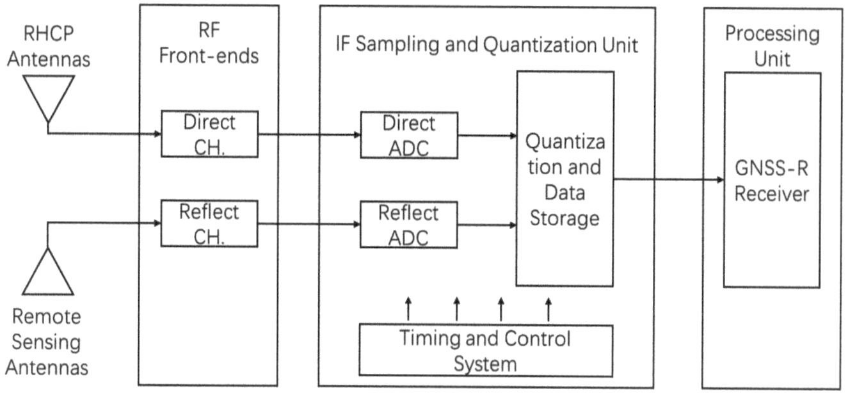

Fig. 1 Block diagram of a GNSS-R receiver hardware architecture with explicit timing and control support. The receiver employs parallel RF front-end channels for the direct and reflected GNSS signals. After down-conversion, both channels are independently digitized by the IF sampling and quantization unit. The quantized data are temporarily stored and subsequently forwarded to the GNSS-R processing unit for baseband correlation and reflectometry processing. A dedicated timing and control system provides synchronized clocking, control, and timing references to the RF front-ends and the sampling unit, ensuring coherent operation of the entire receiver chain

3.1 Antenna Subsystem

The antenna subsystem serves as the first and often most critical interface between the bistatic scattering environment and the receiver. Spaceborne GNSS-R antennas are typically designed for wide field-of-view coverage, moderate gain (10–20 dB), and dual polarization to accommodate the polarization change induced by surface scattering. Depending on mission requirements, configurations range from single wide-beam antennas to multi-beam or electronically steerable arrays.

From a system perspective, the antenna directly modulates the bistatic radar equation through its gain pattern, polarization response, and pointing accuracy. Small deviations in beam alignment or gain stability translate linearly into received power uncertainty. Moreover, antenna thermal deformation, mutual coupling in array configurations, and polarization mismatch between incident and scattered fields introduce slowly varying biases that cannot be fully mitigated by post-processing. For this reason, antenna design and calibration are foundational to GNSS-R radiometric performance.

Antenna Element Design

The antenna element serves as the basic building block of the array and is responsible for the conversion between electromagnetic waves and radio-frequency signals. Key design considerations include operating frequency, polarization, radiation pattern, efficiency, and phase stability.

To support wide-angle scanning and large-area coverage, the antenna element is designed with a broad beamwidth and low sidelobe characteristics. The radiation pattern of an individual element can be expressed as

$$F_e(\theta, \phi), \tag{4}$$

and its gain is given by

$$G_e = \eta_e \frac{4\pi A_e}{\lambda^2}, \tag{5}$$

where η_e denotes the radiation efficiency, A_e is the effective aperture area, and λ is the operating wavelength. To enhance robustness under complex propagation and scattering environments, dual-polarization or circular-polarization configurations can be employed.

Array Antenna Fundamentals

An antenna array is formed by arranging multiple antenna elements according to a predefined geometry and combining their outputs with appropriate amplitude and phase weighting. The overall array radiation pattern can be expressed as the product of the element pattern and the array factor:

$$F(\theta, \phi) = F(\theta, \phi) \cdot AF(\theta, \phi), \tag{6}$$

The array factor is defined as

$$AF(\theta, \phi) = \sum_{n=1}^{N} w_n \exp(j\mathbf{k} \cdot \mathbf{r}_n), \tag{7}$$

where w_n is the complex weight applied to the n th element, $\mathbf{r}_n$ denotes its position vector, and $\mathbf{k}$ is the wave vector. For an ideal uniform array, the peak array gain can be approximated by

$$G_{array} \approx G_e + 10 \log_{10} N(dB), \tag{8}$$

with N being the number of elements.

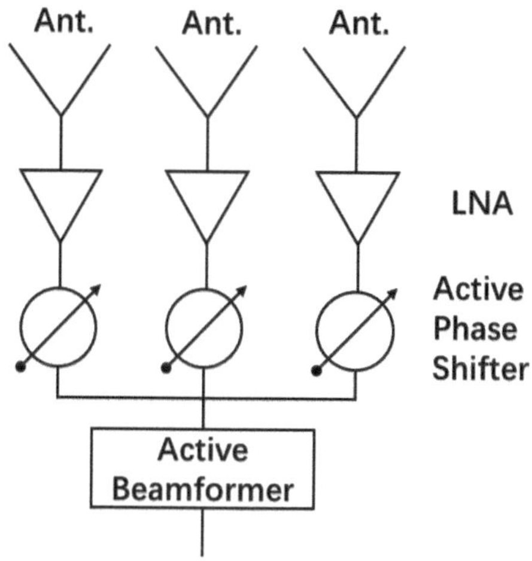

Fig. 2 Conceptual diagram of an active receive-only beamformer with digitally controlled phase shifters

Active Phased Array Configuration

In the active phased array configuration, each antenna element or subarray is equipped with active components such as low-noise amplifiers (LNAs), phase shifters, and variable gain modules. Electronic beam steering and multi-beam generation are achieved by independently controlling the amplitude and phase of each channel (Fig. 2).

The received signal at the array output can be modeled as

$$y(t) = \sum_{n=1}^{N} g_n s(t - \tau_n) e^{j\phi_n} + n(t) \tag{9}$$

where g_n and ϕ_n represent the gain and phase control of the n-th channel, τ_n is the corresponding propagation delay, and $n(t)$ denotes noise.

Active phased arrays offer high flexibility, fast beam reconfiguration, and strong adaptability to dynamic scenarios. However, the introduction of numerous active components increases system complexity, power consumption, and sensitivity to temperature variations and component aging, which can degrade long-term phase coherence if not carefully calibrated.

Passive Front-End Array Configuration

The passive front-end array configuration avoids active components at the antenna aperture and performs signal combination through a passive beamforming network. Signals are then routed to a limited number of highly consistent receiver channels (Fig. 3).

The array output can be expressed in matrix form as

Fig. 3 Conceptual diagram of a receive-only antenna array with phase-controlled elements for beamforming

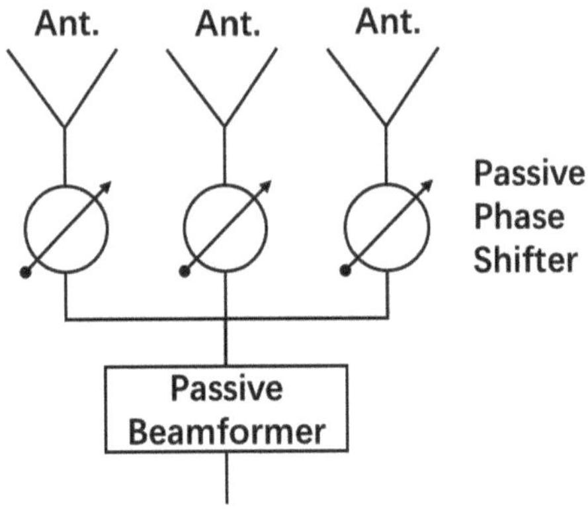

$$y(t) = W_{passive}x(t), \tag{10}$$

where $W_{passive}$ denotes the passive combining matrix and $x(t)$ is the vector of element signals. Because passive networks exhibit intrinsically stable phase responses, the relative phase relationships among array channels are directly measurable and traceable. When combined with temperature-dependent gain compensation models applied at the receiver stage, this architecture provides superior long-term amplitude and phase consistency.

Comparison and Design Considerations

A qualitative comparison between the two array configurations is summarized as follows (Table 1):

Considering the requirements for measurement stability, system simplicity, and low power consumption, the passive front-end array is adopted as the baseline solution, while the active phased array remains a viable option for scenarios requiring enhanced beam agility.

Array Element Layout and Dual-Frequency Considerations

The spatial layout of antenna elements is a critical factor in array performance, directly affecting beamwidth, sidelobe characteristics, grating lobe suppression, and multi-frequency compatibility. For GNSS-R and navigation signal reception, the array geometry must simultaneously accommodate multiple carrier frequencies, most notably the L1 band at 1.575 GHz and the L5 band at 1.176 GHz. This section discusses the fundamental layout principles and the design trade-offs associated with dual-frequency operation.

A. General Principles of Array Element Spacing

Table 1 Comparison between active phased array and passive front-end array configurations

Aspect	Active phased array	Passive front-end array
Beam steering capability	Electronic beam steering with high flexibility and rapid reconfiguration	Fixed or quasi-fixed beamforming determined by passive network
Phase stability	Dependent on active component calibration and thermal compensation	Intrinsically stable due to passive feed network
System complexity	High, owing to per-element active RF chains	Low, with simplified front-end architecture
Power consumption	Relatively high due to distributed active components	Low, as no active devices are placed at the antenna aperture
Long-term consistency	Sensitive to temperature variation and component aging	Highly consistent over long-term operation
Typical application scenarios	Dynamic multi-beam and agile tracking applications	High-precision and stability-critical measurement applications

For a planar or linear antenna array, the inter-element spacing $\boldsymbol{d}$ is typically constrained by the requirement to suppress grating lobes over the desired field of view. For a uniformly spaced array, the grating-lobe-free condition can be expressed as

$$d \leq \frac{\lambda}{1 + \sin\theta_{\max}}, \tag{11}$$

where λ is the wavelength and $\theta_{\max}$ denotes the maximum scan angle from broadside. For broadside or limited-angle reception, a commonly adopted practical guideline is

$$d \leq \frac{\lambda}{2}, \tag{12}$$

This condition ensures that no grating lobes appear within the visible region for a single-frequency operation.

B. Wavelength Difference Between L1 and L5 Bands

The dual-frequency nature of modern GNSS signals introduces additional constraints on array layout. The wavelengths corresponding to the two primary navigation bands are

$$\lambda_{L1} = \frac{c}{f_{L1}} \approx \frac{3 \times 10^8}{1.575 \times 10^9} \approx 0.190m \tag{13}$$

$$\lambda_{L5} = \frac{c}{f_{L5}} \approx \frac{3 \times 10^8}{1.176 \times 10^9} \approx 0.255m \tag{14}$$

The wavelength ratio between L5 and L1 is therefore

$$\frac{\lambda_{L5}}{\lambda_{L1}} \approx 1.34 \tag{15}$$

indicating that an array spacing optimized for L1 will be electrically denser at L5, whereas a spacing optimized for L5 may lead to grating lobes at L1.

C. Dual-Frequency Layout Strategies

To ensure simultaneous compatibility with both frequency bands, several layout strategies can be adopted.

(1) Spacing Determined by the Highest Frequency

A conservative and widely used approach is to determine the inter-element spacing based on the highest operating frequency (shortest wavelength), i.e.,

$$d \leq \frac{\lambda_{L1}}{2} \approx 0.095m \tag{16}$$

Under this configuration:

- Grating lobes are avoided at both L1 and L5 bands;
- The array appears electrically compact at L5, resulting in a wider main beam;
- Beamforming performance is primarily governed by the physical aperture size rather than the electrical spacing at L5.

This approach is particularly suitable for GNSS-R receiving arrays where robustness and simplicity are prioritized.

It is worth noting that, in GNSS-R configurations characterized by fixed-beam operation and constrained observation angles, the inter-element spacing is not strictly limited to 0.5λ and can be moderately increased without degrading the effective reception geometry.

(2) Multi-Scale or Subarray-Based Layout

An alternative strategy employs a multi-scale layout, where the array is conceptually divided into subarrays optimized for different frequencies. For example:

- A dense subarray satisfying the L1 spacing constraint;
- A larger effective aperture formed by grouping multiple elements for L5 processing.

Such configurations can be described as

$$y_{L1} = W_{L1}x \ , \ y_{L5} = W_{L5}x \,, \tag{17}$$

where W_{L1} and W_{L5} denote frequency-dependent beamforming weight matrices. While offering greater flexibility, this approach increases system complexity and calibration requirements.

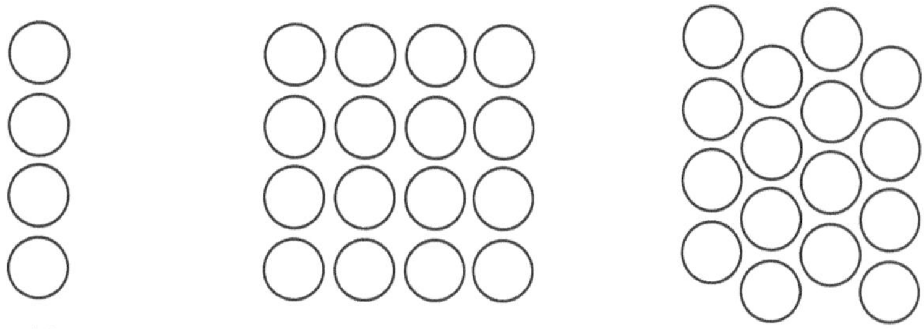

Fig. 4 Conceptual layouts of three antenna array geometries—uniform linear, rectangular planar, and triangular lattice—illustrated using circular elements. The figure is schematic and not to scale, intended to highlight differences in spatial sampling patterns rather than specific design parameters

D. Array Geometry Considerations

Beyond element spacing, the geometric arrangement of elements also plays an important role. Common geometries include:

- Uniform linear arrays (ULA);
- Rectangular planar arrays;
- Triangular or staggered lattices.

For dual-frequency GNSS reception, planar arrays with staggered or triangular layouts are often preferred, as they provide improved spatial sampling efficiency and reduced grating-lobe sensitivity compared to simple rectangular grids, especially when operating over wide angular sectors (Fig. 4).

E. Design Implications for GNSS-R Applications

In GNSS-R systems, the primary objective is stable and coherent reception rather than aggressive beam scanning. Consequently:

- Array spacing is typically dictated by the L1 band to ensure grating-lobe-free operation;
- L5 performance benefits from increased wavelength, leading to inherently smoother spatial responses;
- A single, frequency-agnostic physical layout can be employed, with frequency-dependent beamforming weights applied in the receiver.

This design philosophy enables a unified antenna array to support multi-frequency GNSS observations while maintaining simplicity, phase stability, and long-term consistency.

3.2 RF Front-End Subsystem with Power Calibration

The RF front-end subsystem provides the critical interface between the antenna array and the digital receiver, directly determining the sensitivity, amplitude stability, and long-term consistency of the received signals. For GNSS-R and other weak-signal navigation applications, the received signal power is typically far below the thermal noise floor, making accurate and stable power characterization a key design requirement.

The primary functions of the RF front-end subsystem include:

4. Low-noise amplification of weak navigation signals;
5. Band-limited filtering to suppress out-of-band interference;
6. Signal conditioning for subsequent beamforming and correlation;
7. Stable and traceable power measurement for calibration purposes.

Unlike conventional communication receivers, GNSS-R front ends emphasize relative amplitude consistency and long-term stability rather than instantaneous dynamic range, motivating the inclusion of dedicated calibration paths.

A conceptual block diagram of the RF front-end subsystem can be described as

$$x_{RF}(t) \xrightarrow{LNA} x_{amp}(t) \xrightarrow{BPF} x_{filt}(t) \xrightarrow{Downconversion} x_{IF/BB}(t) \tag{18}$$

where $x_{\mathrm{RF}}(t)$ denotes the received RF signal at the antenna port.

In addition to the main signal path, a power calibration branch is introduced at the front-end stage, enabling continuous or periodic monitoring of signal amplitude without disrupting normal reception.

The power calibration mechanism is designed to establish a stable reference between the RF input power and the digital output level. A representative calibration signal model can be expressed as

$$P_{meas} = G_{RF}(T)P_{in} + P_{offset}, \tag{19}$$

where P_{in} is the input signal power, $G_{\mathrm{RF}}(T)$ denotes the temperature-dependent front-end gain, and P_{offset} accounts for noise and measurement bias.

To mitigate gain drift, the front-end incorporates one or more of the following calibration elements:

- Internal reference signal injection;
- Power detection or envelope sensing circuits;
- Switchable calibration paths sharing the same RF chain as the main signal.

By periodically measuring the response to a known reference level, the effective gain can be estimated as

$$\hat{G}_{RF} = \frac{P_{ref,meas}}{P_{ref}} \tag{20}$$

which is then used to compensate the received signal amplitude in the digital domain.

Temperature variation is a dominant source of gain instability in RF components, particularly LNAs and mixers. To address this issue, the RF front-end adopts a temperature-aware calibration strategy, in which the gain is modeled as

$$G_{RF}(T) = G_0 + \alpha(T - T_0) \tag{21}$$

where G_0 is the nominal gain at reference temperature T_0, and α is the gain temperature coefficient.

By combining temperature measurements with periodic power calibration, the front-end achieves:

- Stable amplitude scaling across long observation periods;
- Improved inter-channel consistency in array configurations;
- Reduced reliance on frequent external recalibration.

When integrated with an antenna array, the calibrated RF front-end provides several advantages:

- Consistent amplitude weighting for beamforming and correlation;
- Improved comparability between channels and observation epochs;
- Enhanced robustness against environmental and hardware-induced variations.

Importantly, the calibration mechanism operates independently of the beamforming architecture, making it compatible with both passive analog beamformers and active receive-only phased arrays.

The proposed RF front-end design follows a measurement-driven philosophy: rather than attempting to eliminate gain variation at the hardware level, it emphasizes accurate characterization and compensation. This approach is particularly well suited for GNSS-R systems, where relative power stability and long-term repeatability are often more critical than absolute radiometric accuracy.

3.3 *IF Chain and Sampling Subsystem*

The intermediate-frequency (IF) chain and sampling subsystem bridges the analog RF front end and the digital signal processing unit. Its primary role is to preserve the amplitude and phase integrity of weak GNSS signals while enabling reliable digital processing, including correlation, beamforming, and long-term averaging. For GNSS-R and navigation signal reception, the design of the IF chain is primarily driven by SNR considerations rather than throughput requirements.

GNSS signals received at the antenna output are typically well below the thermal noise floor. After RF amplification and downconversion, the IF signal can be modeled as

$$x_{IF}(t) = s(t) + n(t), \tag{22}$$

where $s(t)$ is the navigation signal component and $n(t)$ represents additive thermal noise.In most practical GNSS receivers, the instantaneous SNR before correlation is negative, often on the order of – 20 to – 30 dB. Under such conditions, the sampling subsystem must ensure that quantization noise does not dominate thermal noise, as this would irreversibly degrade subsequent processing gain.

The sampling rate f_s is primarily determined by the signal bandwidth B. For GNSS signals, the effective bandwidth depends on the modulation scheme and filtering strategy. A conservative requirement is to satisfy the Nyquist criterion. In practice, modest oversampling is often adopted:

$$f_s \geq 2B, \tag{23}$$

$$f_s = \kappa B, \kappa \in [2, 4], \tag{24}$$

which simplifies digital filtering and relaxes analog anti-aliasing constraints. For L-band GNSS signals with typical bandwidths of 10–20 MHz, sampling rates on the order of 20–80 MSPS are sufficient for faithful signal representation and flexible digital processing.

The quantization noise power introduced by an N-bit analog-to-digital converter (ADC) can be approximated as

$$P_q = \frac{\Delta^2}{12}, \tag{25}$$

where Δ is the quantization step size. The corresponding signal-to-quantization-noise ratio (SQNR) for a full-scale sinusoidal input is [25]

$$SQNR \approx 6.02N + 1.76dB \tag{26}$$

For GNSS applications, the critical design criterion is not the absolute SQNR, but whether quantization noise remains below the thermal noise floor after IF amplification. This requirement can be expressed as

$$P_q \ll P_n, \tag{27}$$

where $P_n = kTB$ denotes the thermal noise power within bandwidth B.

Analysis and extensive empirical evidence indicate that:

- 2–3 bit quantization is sufficient for basic GNSS correlation when large processing gain is available;
- 4–6 bit quantization ensures that quantization loss is negligible (< 0.2 dB) for coherent integration;
- 8-bit or higher quantization provides ample dynamic range for multi-channel array systems and stable amplitude estimation.

Consequently, IF sampling resolutions in the range of 6–8 bits represent a practical and well-balanced choice for GNSS-R receivers.

The effective number of bits (ENOB) determines the usable dynamic range of the sampling subsystem [25]:

$$DR \approx 6.02ENOB\ dB \tag{28}$$

For an 8-bit ADC with an ENOB of approximately 7 bits, the achievable dynamic range is on the order of 40–45 dB, which is sufficient to accommodate:

- Thermal noise fluctuations;
- Moderate interference levels;
- Calibration signal injection without saturation.

Importantly, excessive dynamic range does not significantly improve GNSS performance, as the signal detection relies primarily on long-term correlation rather than instantaneous amplitude fidelity.

For typical GNSS integration times, processing gains exceeding 30–40 dB are common. As long as quantization noise remains well below thermal noise prior to correlation, its impact after integration becomes negligible.

Therefore, increasing ADC resolution beyond a certain point yields diminishing returns, while increasing power consumption and system complexity.

Based on the above considerations, the following design guidelines can be established for the IF chain and sampling subsystem:

- Sampling rate: 2–4 times the signal bandwidth;
- Quantization resolution: 6–8 bits, ensuring quantization noise remains below thermal noise;
- Dynamic range: 40 dB-class, sufficient for GNSS-R reception and calibration operations.

These parameters provide a robust balance between performance, power efficiency, and implementation complexity, and are well aligned with the requirements of array-based GNSS-R systems.

In summary, the IF chain and sampling subsystem should be designed to preserve thermal-noise-limited signal characteristics rather than maximize instantaneous precision. Quantization and sampling parameters must be selected such that digital processing gain can be fully exploited without introducing unnecessary hardware complexity.

3.4 Digital Baseband Processing Unit

The digital baseband processing unit constitutes the core computational subsystem of the receiver, transforming sampled IF data into navigation observables and geophysical information products. In contrast to the RF and IF subsystems, which primarily preserve signal fidelity, the baseband unit focuses on signal extraction, integration, and parameter estimation under extremely low SNR conditions.

In GNSS-R systems, the baseband processing unit must accommodate both conventional navigation signal processing and reflection-specific operations, placing higher demands on flexibility and long-term numerical stability.

At a high level, the digital baseband processing unit performs the following functions:

1. Digital downconversion and channelization;
2. Code and carrier correlation;
3. Coherent and non-coherent integration;
4. Parameter estimation and observable generation.

These functions are applied to the sampled IF signals obtained from the preceding subsystem and form the foundation for both navigation and reflectometry processing chains.

The sampled signal at the digital baseband input can be expressed as

$$x[n] = s[n] + n[n], \tag{29}$$

where $s[n]$ represents the desired GNSS signal component and $n[n]$ denotes additive noise. The baseband processing aims to enhance the detectability of $s[n]$ by exploiting known signal structures, including spreading codes and carrier properties.

A typical processing flow includes:

- Carrier wipe-off and frequency compensation;
- Code replica generation and alignment;
- Correlation and accumulation over predefined integration intervals.

Through these operations, significant processing gain is achieved, enabling reliable signal detection and parameter estimation even when the instantaneous SNR is negative.

In array-based GNSS-R receivers, the digital baseband processing unit interfaces closely with both the antenna array and the RF front-end calibration mechanisms. Frequency-dependent beamforming weights and amplitude compensation factors obtained from upstream subsystems are applied prior to or within the baseband processing chain.

This architecture ensures that:

- Inter-channel amplitude and phase consistency are preserved;
- Beamformed signals remain compatible with standard GNSS correlation algorithms;
- Calibration effects are transparently integrated into digital processing.

While the RF and IF subsystems determine the instantaneous signal quality, the digital baseband processing unit largely governs the effective sensitivity and resolution of the overall system through long-term integration and estimation strategies. Its design therefore emphasizes algorithmic robustness and numerical accuracy rather than raw computational throughput.

Importantly, the baseband unit is designed to remain largely agnostic to specific hardware implementations, allowing different processing strategies to be explored without modifying the analog front-end.

The description provided in this section is intentionally limited to a system-level overview. Detailed discussions of baseband processing algorithms, including correlation structures, delay–Doppler map generation, and advanced estimation techniques, are presented in subsequent chapters.

In summary, the digital baseband processing unit serves as the computational core of the GNSS-R receiver, bridging the gap between sampled IF data and high-level observables. Its detailed algorithmic design is discussed in later chapters, while this section establishes its role within the overall system architecture.

3.5 Timing, Frequency, and Control Subsystem

The timing and control subsystem provides the frequency references, synchronization signals, and configuration management required for coherent operation. High-stability reference oscillators, such as oven-controlled crystal oscillators (OCXO) or temperature-compensated crystal oscillators (TCXO), are commonly employed. In some high-performance systems, external references or inter-channel synchronization networks are used to support multi-beam or multi-frequency operation.

Timing errors manifest as both phase and amplitude degradation in the received signal, influencing correlation alignment and DDM stability. As GNSS-R moves toward longer coherent integration and multi-signal fusion, the demands on timing accuracy and long-term frequency stability continue to increase.

3.6 Architectural Implications for Error Propagation

System architecture determines not only the functional organization of subsystems but also the way hardware-induced errors propagate, accumulate, or are suppressed throughout the receiver chain. Unlike component-level errors, which can be characterized locally, architectural effects manifest through error coupling and amplification mechanisms that are inherently system-level phenomena. This section provides a quantitative interpretation of how architectural choices influence error propagation prior to the detailed component-level analysis in Sect. 4.

Let the ideal baseband signal after beamforming be denoted as

$$y(t) = \sum_{n=1}^{N} w_n s(t) \tag{30}$$

where w_n is the complex weight applied to the n-th array element and N is the number of elements. In the presence of hardware-induced amplitude and phase errors, the actual output becomes

$$\tilde{y}(t) = \sum_{n=1}^{N} w_n\left[1 + \varepsilon_{A,n}\right]s(t)\exp\left(j\varepsilon_{\phi,n}\right). \tag{31}$$

Assuming small errors ($\left|_{A,n}\right| \ll 1$, $\left|_{\phi,n}\right| \ll 1$), a first-order approximation yields

$$\tilde{y}(t) = y(t) + \sum_{n=1}^{N} w_n s(t)\left(\varepsilon_{A,n} + j\varepsilon_{\phi,n}\right). \tag{32}$$

This expression shows that architectural choices directly affect how element-level errors are spatially combined.

In fully distributed architectures, amplitude and phase errors are statistically independent across channels. Let the variances be

$$Var\left(\varepsilon_{A,n}\right) = \sigma_A^2,\ Var\left(\varepsilon_{\phi,n}\right) = \sigma_\phi^2. \tag{33}$$

The expected relative power loss due to phase errors after beamforming can be approximated as

$$\mathrm{E}\left[\frac{|\tilde{y}|^2}{|y|^2}\right] \approx \exp\left(-\sigma_\phi^2\right), \tag{34}$$

which is independent of N but becomes increasingly significant as coherence requirements tighten. In contrast, architectures with shared reference paths introduce correlated errors, for which

$$\varepsilon_{\phi,n} = \varepsilon_{\phi,common} + \delta\varepsilon_{\phi,n}. \tag{35}$$

In this case, the common-mode term can be compensated globally, leaving only the residual variance $\mathrm{Var}\left(\delta_{\phi,n}\right)$, thereby reducing effective error propagation.

Let beamforming be performed either before or after calibration. If beamforming precedes calibration, the effective error variance at the output becomes

$$\sigma_{eff}^2 \approx \sigma_A^2 + \sigma_\phi^2, \tag{36}$$

since individual errors are spatially combined without compensation.

If calibration is applied prior to beamforming, residual errors are reduced to demonstrating that architectural sequencing directly determines the magnitude of propagated errors.

$$\sigma_{eff}^2 \approx \sigma_{A,res}^2 + \sigma_{\phi,res}^2, \sigma_{res}^2 \ll \sigma^2, \tag{37}$$

For analog RF phase shifters, phase errors are typically continuous and slowly varying, which can be modeled as

$$\varepsilon_\phi(t) = \alpha(T(t) - T), \tag{38}$$

where α is the temperature coefficient. Such errors primarily occupy low-frequency spectral components and can be mitigated through long-term calibration.

For digitally controlled phase shifters with b-bit resolution, the quantization-induced phase error is bounded by

$$|\varepsilon_\phi| \leq \frac{\pi}{2^b}. \tag{39}$$

The corresponding variance for uniformly distributed phase quantization error is

$$\sigma_\phi^2 \approx \frac{\pi^2}{3 \cdot 2^{2b}} \tag{40}$$

providing a deterministic upper bound on phase-induced beamforming loss.

GNSS-R receivers rely on large processing gains achieved through long-term integration. As long as the effective phase error variance satisfies

$$\sigma_\phi^2 \ll \frac{1}{BT} \tag{41}$$

where B is signal bandwidth and T is coherent integration time, hardware-induced errors do not significantly degrade correlation gain.

Architectures that limit error propagation prior to integration therefore preserve the full benefit of processing gain, whereas architectures that amplify inter-channel inconsistencies reduce effective integration efficiency.

From a quantitative perspective, architectural design should aim to:

1. Minimize uncorrelated inter-channel errors;
2. Confine dominant error sources to observable calibration points;
3. Ensure that residual error variances remain below integration-limited thresholds.

These principles provide measurable criteria for evaluating architectural robustness prior to detailed hardware implementation.

To translate these architectural principles into a concrete and verifiable design framework, it is necessary to explicitly associate each architectural choice with its corresponding hardware error sources and quantitative metrics. Rather than treating error mechanisms in isolation, a structured mapping between subsystem-level design decisions and their dominant error characteristics enables a systematic assessment of architectural robustness.

From an architectural standpoint, residual hardware errors in a calibrated GNSS-R receiver are concentrated in a small number of dominant subsystems, rather than being uniformly distributed along the signal chain. The antenna–feed assembly, RF front-end, clock and local oscillator distribution, and calibration path typically govern the overall coherence and radiometric stability, while errors introduced by digital processing stages are largely deterministic and can be suppressed below integration-limited levels through proper design.

After calibration and temperature compensation, practical multi-channel receivers commonly exhibit residual per-stage amplitude instabilities on the order of several hundredths of a decibel and phase uncertainties of a few degrees RMS. These values reflect post-calibration engineering residuals, consistent with widely adopted assumptions in phased-array, SAR, and GNSS-R receiver modeling, rather than worst-case component specifications. Clock-related timing skew and phase noise further impose a global constraint on coherent integration, whereas ADC and digital baseband errors generally remain secondary when a common reference and stable numerical implementation are employed.

Architectural robustness is therefore achieved by deliberately confining dominant error sources to observable and calibratable subsystems, ensuring that cumulative residual errors remain below thresholds at which coherent integration and averaging become error-limited.

The cumulative impact of hardware-induced errors can be estimated by considering the dominant amplitude and phase error contributions across the signal chain. Assuming that residual errors after calibration are largely uncorrelated, the effective variance of the aggregated error can be approximated as the sum of individual error variances:

$$\sigma_{total}^2 \approx \sum_i \sigma_{A,i}^2 + \sum_j \sigma_{\phi,j}^2, \tag{42}$$

where $\sigma_{A,i}$ and $\sigma_{\phi,j}$ denote the residual amplitude and phase error variances associated with the i-th and j-th subsystems, respectively.

For a calibrated GNSS-R receiver architecture, typical residual error magnitudes are on the order of:

amplitude errors: $\sigma_A \approx 0.05-0.1$ dB per dominant stage;

phase errors: $\sigma_\phi \approx 1^\circ-3^\circ$ per dominant stage.

Under conservative assumptions involving several contributing subsystems, the cumulative phase error standard deviation can therefore be bounded as $\sigma_{\phi,\text{total}} \lesssim 5^\circ-8^\circ$, while the cumulative amplitude error typically remains below 0.5 dB.

The resulting coherence loss due to phase errors can be approximated by

$$G_{loss} \approx \exp\left(-\sigma_{\phi,total}^2\right) \tag{43}$$

where $\sigma_{\phi,\text{total}}$ is expressed in radians. For $\sigma_{\phi,\text{total}} \leq 8^\circ$, the corresponding beamforming gain reduction remains below approximately 1 dB, which is negligible relative to the processing gains achieved through long-term coherent integration in GNSS-R systems.

Similarly, amplitude errors within the estimated bounds primarily affect sidelobe structure rather than main-beam gain and do not significantly degrade correlation efficiency when proper normalization is applied.

This quantitative assessment indicates that, under a calibration-aware architecture, the cumulative hardware-induced errors remain well below integration-limited thresholds. Consequently, the dominant performance limitations shift from hardware imperfections to signal environment factors and processing strategy choices. The architecture therefore provides sufficient robustness to support high-coherence beamforming and long-term GNSS-R observations without requiring excessive hardware precision.

4 Error Analysis of Hardware Components

4.1 Error Propagation Framework Based on the Bistatic Radar Equation

GNSS reflectometry is a passive bistatic remote sensing technique in which geophysical parameters are retrieved from the power and structure of reflected navigation signals. The received power of a GNSS-R system can be described using the classical bistatic radar equation:

$$P_r = \frac{P_t G_t G_r \lambda^2 \sigma^0}{(4\pi)^3 R_t^2 R_r^2} \cdot L_{sys} \tag{44}$$

where P_t and G_t denote the equivalent isotropically radiated power and antenna gain of the GNSS transmitter, G_r is the receiving antenna gain, λ is the signal wavelength, σ^0 represents the bistatic normalized radar cross section of the reflecting surface, R_t and R_r are the transmitter–surface and surface–receiver distances, respectively, and L_{sys} accounts for the total system loss of the receiving chain.

For error analysis, the relative uncertainty of the received power can be expressed in differential form as

$$\frac{\Delta P_r}{P_r} = \frac{\Delta P_t}{P_t} + \frac{\Delta G_t}{G_t} + \frac{\Delta G_r}{G_r} + \frac{\Delta \sigma^0}{\sigma^0} + \frac{\Delta L_{sys}}{L_{sys}} \tag{45}$$

In GNSS-R remote sensing, transmitter-related terms (P_t and G_t) are generally well characterized by navigation system specifications and vary slowly with time. Geometric terms are determined by precise orbit and specular reflection modeling. Consequently, uncertainties associated with the receiver-side hardware components constitute the dominant and most critical error sources affecting the accuracy and stability of GNSS-R observables and derived geophysical parameters.

In practice, GNSS-R remote sensing does not use P_r as the final product; instead, geophysical retrievals are based on the inferred bistatic reflectivity σ^0(or closely related observables) obtained by rearranging the bistatic radar equation:

$$\sigma^0 = P_r \cdot \frac{(4\pi)^3 R_t^2 R_r^2}{P_t G_t G_r \lambda^2} \cdot \frac{1}{L_{sys}} \tag{46}$$

Taking differentials yields the key error-propagation relation for remote sensing retrievals:

$$\frac{\Delta\sigma^0}{\sigma^0} = \frac{\Delta P_r}{P_r} - \frac{\Delta P_t}{P_t} - \frac{\Delta G_t}{G_t} - \frac{\Delta G_r}{G_r} - \frac{\Delta L_{sys}}{L_{sys}} + 2\frac{\Delta R_t}{R_t} + 2\frac{\Delta R_r}{R_r} - 2\frac{\Delta\lambda}{\lambda} \tag{47}$$

where $\Delta\lambda/\lambda$ is typically negligible given the stability of GNSS carrier frequencies. This formulation makes explicit how receiver-side uncertainties—especially $\Delta G_r/G_r$, $\Delta L_{\mathrm{sys}}/L_{\mathrm{sys}}$, and receiver-induced contributions to $\Delta P_r/P_r$—directly translate into biases and variance in σ^0, and consequently in derived geophysical parameters.

To facilitate a structured analysis, the terms are mapped to system elements as follows:

- Transmitter terms: $\Delta P_t/P_t$, $\Delta G_t/G_t$(typically treated as known or slowly varying inputs in GNSS-R).
- Geometry terms: $2\Delta R_t/R_t$, $2\Delta R_r/R_r$(dominated by orbit/attitude, specular-point modeling, and timing).
- Receiver antenna term: $\Delta G_r/G_r$(pattern calibration, polarization response, orientation/attitude coupling).
- Receiver chain term: $\Delta L_{\mathrm{sys}}/L_{\mathrm{sys}}$(RF gain drift, filter insertion loss variation, temperature dependence).
- Measurement/processing term: $\Delta P_r/P_r$(noise floor estimation, quantization, correlation losses, integration efficiency).

4.2 Receiving Antenna–related Error Contributions

This subsection quantifies receiver antenna contributions, which enter the retrieval as the effective gain uncertainty $\Delta G_r/G_r$ in the propagation of $\Delta\sigma^0/\sigma^0$.

The receiving antenna plays a central role in determining the effective gain G_r toward the specular reflection region. In GNSS-R, surface reflectivity and related geophysical parameters are retrieved through power-based observables, such as delay–Doppler maps (DDMs). As a result, uncertainties in the effective antenna gain propagate directly and linearly into the retrieved normalized radar cross section σ^0.

Unlike active radar systems, GNSS-R does not rely on beam steering to improve spatial resolution. Instead, wide angular coverage is required to ensure continuous illumination of the specular region under varying bistatic geometries. Any uncertainty in the antenna radiation pattern, including gain nonuniformity across the field of view or inaccuracies in pattern calibration, leads to systematic biases in the estimated reflected power.

Polarization effects further amplify antenna-related errors. GNSS signals are transmitted in right-hand circular polarization, while the reflected signals generally exhibit polarization mixing depending on surface roughness and incidence angle. Imperfect polarization purity or uncharacterized cross-polarization responses of the receiving antenna introduce additional uncertainties in the effective gain, which directly affect the absolute accuracy of reflectivity retrievals.

Therefore, antenna-related errors in GNSS-R primarily manifest as biases in geophysical parameter estimation rather than as random measurement noise, emphasizing the importance of stable and well-characterized antenna responses.

Uncertainty in the angular gain model $G_r(\theta, \phi)$, including polarization-dependent responses, induces a systematic multiplicative error in σ^0 that is often indistinguishable from true surface reflectivity changes unless explicitly calibrated.

4.3 RF Front-End Gain and Noise Figure Uncertainties

RF front-end non-idealities contribute primarily through the system loss term $\Delta L_{sys}/L_{sys}$, and secondarily through receiver-induced variations in the measured power $\Delta P_r/P_r$ via noise floor and SNR-dependent estimation errors.

The RF front-end, including low-noise amplifiers, filters, and analog signal paths, contributes to the system loss term L_{sys} in the bistatic radar equation. In GNSS-R applications, reflected signals typically exhibit extremely low power levels, often below − 190 dBW at the antenna output, making the system particularly sensitive to gain and noise figure variations.

Slow temporal variations in RF front-end gain, caused by temperature changes or component aging, do not merely alter the instantaneous signal amplitude. Instead, they distort the baseline and relative power distribution of the delay–Doppler map, which degrades the robustness of geophysical parameter retrieval. Such variations effectively introduce uncertainty in the estimation of L_{sys}, leading to biased or unstable reflectivity estimates.

Uncertainty in the noise figure of the receiving chain affects the effective signal-to-noise ratio, thereby influencing the detectability of weak reflected signals and the precision of power-based observables. From a remote sensing perspective, these effects are equivalent to an uncertainty in the received power term of the radar equation and must be treated as a systematic component of the error budget rather than as purely random noise.

Slow gain drift and temperature-dependent insertion loss can be modeled as a time-varying $L_{sys}(t)$, leading to low-frequency biases in σ^0 and distortions in DDM baseline estimation.

4.4 Phase Instability and Its Impact on Coherent GNSS-R Observables

Phase instability does not appear explicitly in the scalar radar equation, but in GNSS-R it manifests as a correlation/coherent integration efficiency loss, which enters the retrieval as a multiplicative factor on P_r(i.e., contributes to $\Delta P_r/P_r$).

Let $\eta_{\text{coh}} \in (0, 1]$ denote the effective coherent integration factor accounting for oscillator phase noise, residual Doppler, and tracking errors. Then the measured power-like observable can be written as $P_r^{(\text{meas})} = \eta_{\text{coh}} P_r$, yielding

$$\frac{\Delta\sigma^0}{\sigma^0} \supset \frac{\Delta P_r}{P_r} \approx \frac{\Delta\eta_{coh}}{\eta_{coh}} \tag{48}$$

This term is negligible for purely incoherent DDM products but becomes dominant for coherent GNSS-R modes.

While many GNSS-R applications rely on incoherent power measurements, an increasing number of techniques exploit coherent processing, such as coherent integration and GNSS-based synthetic aperture approaches. In these cases, phase stability along the receiving chain becomes a critical hardware requirement.

Phase instability introduced by oscillators, mixers, or analog signal paths does not directly alter the instantaneous received power. However, it reduces the achievable coherent integration time and degrades the effective coherence of the reflected signal. From the perspective of the bistatic radar equation, this loss of coherence can be interpreted as a reduction in the effective scattering efficiency, leading to an underestimation of σ^0.

Consequently, phase instability represents a hardware-induced error source that is negligible for purely incoherent GNSS-R processing but becomes dominant in coherent or partially coherent remote sensing applications.

4.5 Error Budget Dominance and Remote Sensing Implications

For completeness, the remaining terms in $\Delta P_t/P_t$, $\Delta G_t/G_t$, and the geometric contributions $2\Delta R_t/R_t$ and $2\Delta R_r/R_r$—are included in the error budget, although they are typically treated as external inputs rather than receiver hardware design parameters.

Table 2 Mapping between bistatic radar equation error terms, receiver hardware subsystems, and their impacts on GNSS-R observables and geophysical parameter retrievals

Propagation term	Primary source	Typical manifestation in GNSS-R products
$\Delta G_r/G_r$	Antenna pattern & polarization	Multiplicative bias in σ^0, angle-dependent artifacts
$\Delta L_{\text{sys}}/L_{\text{sys}}$	RF gain/loss drift, temperature	DDM baseline drift, long-term bias
$\Delta P_r/P_r$	Noise-floor estimation, quantization, correlation loss	Increased variance, reduced sensitivity
$\Delta P_t/P_t$, $\Delta G_t/G_t$	Transmitter power/attitude	Slowly varying bias (often treated as known)
$2\Delta R_t/R_t$, $2\Delta R_r/R_r$	Geometry and timing	Systematic scaling error if mis-modeled

In most GNSS-R implementations, these terms are constrained by navigation system specifications and precise orbit determination; consequently, receiver hardware stability ($\Delta G_r/G_r$, $\Delta L_{\text{sys}}/L_{\text{sys}}$, and receiver-induced $\Delta P_r/P_r$) remains the dominant controllable contributor to σ^0 uncertainty.

Based on the above analysis, the dominant hardware-related error sources in GNSS-R remote sensing can be summarized by their corresponding radar equation terms and impacts on observables. Receiver antenna gain uncertainty and RF front-end stability are the primary contributors to systematic errors in reflectivity retrieval, while phase instability governs the performance limits of coherent GNSS-R techniques.

Overall, the analysis indicates that receiver-side hardware stability, rather than absolute peak performance, dominates the error budget in GNSS-R remote sensing systems. Ensuring stable, well-characterized, and repeatable hardware responses is therefore essential for achieving reliable and physically meaningful geophysical parameter retrievals (Table 2).

5 End-to-End Error Propagation Framework

The end-to-end framework highlights that, for GNSS-R remote sensing, controlling the propagation of errors is more critical than maximizing isolated hardware performance metrics. In particular, multiplicative errors dominate the bias characteristics of geophysical retrievals and must be tightly constrained through stable architecture and calibration.

Effective system design should therefore aim to (i) minimize uncorrelated errors across channels, (ii) confine dominant error sources to observable and calibratable points in the signal chain, and (iii) ensure that residual error variances remain below

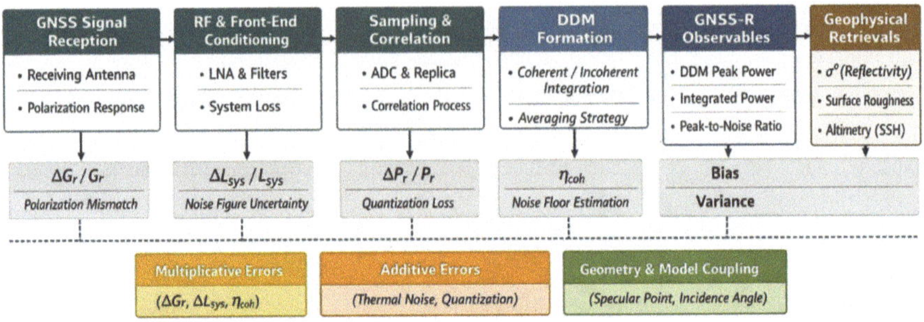

Fig. 5 End-to-end error propagation framework for GNSS reflectometry, illustrating how receiver-side hardware uncertainties propagate through the signal reception, processing, and inversion chain to affect GNSS-R observables and geophysical parameter retrievals. Multiplicative errors (e.g., antenna gain uncertainty, system loss variation, and coherent efficiency), additive errors (e.g., thermal noise and quantization), and geometry- and model-coupled effects are mapped to their corresponding stages in the bistatic remote sensing workflow, highlighting their distinct impacts on bias and variance of retrieved parameters

integration-limited thresholds. These principles provide quantitative criteria for evaluating architectural robustness and guide the design of GNSS-R receivers optimized for reliable remote sensing applications (Fig. 5).

6 Hardware Design Recommendations

6.1 Application-Driven Design Philosophy

The hardware design of GNSS reflectometry receivers should be guided by the error sensitivity of the targeted remote sensing application rather than by generic receiver classifications. As demonstrated in Sects. 4 and 5, the dominant contributors to retrieval uncertainty are receiver-side multiplicative errors, including antenna gain uncertainty, system loss variation, and coherence-related efficiency, while additive noise terms primarily affect measurement variance.

Different GNSS-R applications emphasize different observables and processing modes, which results in distinct dominant error terms and design priorities. Consequently, hardware architectures and calibration strategies should be selected and optimized based on application-level requirements, ensuring that system complexity is commensurate with the information content and robustness demands of the retrieved geophysical parameters.

6.2 Power-Based GNSS-R Applications

Power-based GNSS-R applications, such as surface reflectivity, surface roughness, and soil moisture retrieval, rely primarily on integrated delay–Doppler map power or closely related observables. For these applications, the dominant uncertainty arises from multiplicative receiver-side errors, most notably the effective receiving antenna gain G_r and the system loss term L_{sys}.

Hardware design for power-based GNSS-R should therefore prioritize stability and calibratability over peak performance metrics. The receiving antenna should provide sufficiently wide angular coverage to continuously illuminate the specular reflection region under varying bistatic geometries, while maintaining a stable and well-characterized gain response within this region. Polarization characteristics must be explicitly measured and modeled, as polarization mismatch propagates directly into systematic bias in retrieved reflectivity.

The RF front-end should be designed to minimize gain drift and temperature-dependent variations. Rather than maximizing gain, emphasis should be placed on repeatability and traceable calibration, enabling residual system loss uncertainty to be constrained through direct-signal referencing or temperature-based compensation. Sampling and correlation stages should preserve linearity across the dynamic range of both direct and reflected signals, ensuring that quantization and correlation losses remain below the noise floor.

For these applications, single-element receivers or array receivers operated in power-only modes are generally sufficient, provided that the dominant multiplicative error sources are confined to observable calibration points. Increasing architectural complexity without corresponding calibration control yields limited benefit for power-based retrieval accuracy.

6.3 Spatially Enhanced GNSS-R Applications

Spatially enhanced GNSS-R applications aim to improve spatial sampling or coverage through multi-beam or multi-channel observations. Examples include mesoscale surface mapping and observations requiring broader swath coverage. In these cases, the dominant error terms shift from absolute gain uncertainty to relative inter-channel consistency.

Hardware design for spatially enhanced applications must therefore emphasize channel-to-channel stability and symmetry. Multi-beam or array architectures should ensure that gain, phase, and polarization responses across channels are traceable to a common reference. Uncorrelated inter-channel variations directly manifest as artificial spatial variability in GNSS-R observables and can be easily misinterpreted as geophysical signals.

Beamforming in GNSS-R should not be treated analogously to active radar systems. The primary objective is not agile scanning or spatial resolution enhancement, but continuous and consistent coverage of the specular reflection region. Effective beam-level gain models, including frequency- and polarization-dependent responses, must be available to support relative normalization across beams.

Array architectures are justified for spatially enhanced GNSS-R applications only when accompanied by explicit inter-channel calibration mechanisms, such as dedicated direct-signal reference channels and synchronized RF and timing distributions. Without these measures, the additional degrees of freedom introduced by array receivers may increase, rather than reduce, overall retrieval uncertainty.

6.4 Coherent GNSS-R Applications

Coherent GNSS-R applications, including coherent integration, GNSS-based synthetic aperture techniques, and certain altimetric modes, impose fundamentally different constraints on hardware design. In these applications, phase stability and correlation efficiency dominate the error budget, and coherence loss becomes the primary limiting factor.

Hardware design for coherent GNSS-R should therefore prioritize stable frequency and timing reference distribution, minimizing phase noise and differential phase drift across the receiving chain. Phase stability requirements must be explicitly matched to the intended coherent integration time; imposing unnecessarily stringent phase constraints on power-based modes should be avoided.

Coherent efficiency should be treated as a measurable system parameter rather than an assumed constant. Continuous monitoring using direct-signal observations provides a practical means of estimating effective correlation efficiency and bounding coherence-related uncertainty. Where possible, coherent and incoherent processing paths should be logically separated, allowing each mode to operate under hardware constraints appropriate to its error sensitivity.

Coherent GNSS-R modes should be regarded as specialized operating configurations, activated only when their additional information content justifies the increased hardware and calibration complexity.

6.5 *Multi-Frequency and Multi-Polarization GNSS-R Applications*

Multi-frequency and multi-polarization GNSS-R observations offer enhanced sensitivity to surface properties and scattering mechanisms, but they also introduce additional systematic error pathways. Cross-frequency and cross-polarization inconsistencies primarily manifest as multiplicative biases that can obscure genuine geophysical signatures.

Hardware design for these applications should treat each frequency band as an independent signal chain with separately characterized antenna responses and system loss models. Assumptions of shared calibration across frequencies are rarely justified in practice. Similarly, polarization responses should be represented explicitly, even when only co-polarized observables are operationally exploited.

Consistency checks across frequencies and polarizations, using stable targets or repeat observations, provide an effective means of bounding residual systematic errors. The added information content of multi-dimensional GNSS-R observations is realized only when cross-domain calibration uncertainties are controlled to levels below the expected geophysical variability.

6.6 *Synthesis and Recommended Design Principles*

Across all GNSS-R application categories, the dominant design objective is the control of receiver-side multiplicative uncertainties rather than the maximization of isolated hardware specifications. Hardware complexity should increase only when it is supported by corresponding calibration observability and stability guarantees.

Effective GNSS-R hardware design should therefore aim to minimize uncorrelated inter-channel errors, confine dominant error sources to observable calibration points, and ensure that residual error variances remain below integration-limited thresholds. These principles provide a practical and application-driven basis for selecting and evaluating GNSS-R receiver architectures.

7 Summary

This chapter has presented a systematic analysis of GNSS-R receiver hardware from an application-driven and remote sensing–oriented perspective. Rather than treating hardware components as isolated engineering elements, the discussion has focused on how receiver-side uncertainties propagate through the bistatic radar equation and the GNSS-R processing chain to ultimately affect geophysical parameter retrieval.

Section 3 introduced the overall GNSS-R receiver hardware architecture, emphasizing the distinct characteristics of passive bistatic reception and the role of direct and reflected signal paths. Building on this architecture, Sect. 4 analyzed the dominant hardware-induced error sources by explicitly mapping receiver components to the corresponding uncertainty terms in the bistatic radar equation. A quantitative error budget demonstrated that receiver-side multiplicative errors—particularly antenna gain uncertainty and RF front-end stability—dominate the bias characteristics of GNSS-R reflectivity retrieval, while additive noise primarily affects measurement variance.

In Sect. 5, an end-to-end error propagation framework was established, linking hardware uncertainties to GNSS-R observables and geophysical retrievals through the complete signal reception and processing chain. This framework clarified the distinct propagation mechanisms of multiplicative, additive, and geometry-coupled errors, and highlighted the application-dependent sensitivity of different GNSS-R products to these uncertainties.

Based on these analyses, Sect. 6 formulated hardware design recommendations driven by GNSS-R application requirements rather than by generic receiver classifications. Power-based, spatially enhanced, coherent, and multi-frequency or multi-polarization GNSS-R applications were shown to impose fundamentally different constraints on hardware architecture and calibration strategy. Across all application categories, controlling receiver-side multiplicative uncertainties and confining dominant error sources to observable calibration points emerged as the primary design principle.

Overall, this chapter establishes a coherent methodological link between GNSS-R hardware design, error propagation theory, and remote sensing performance. The presented framework provides a quantitative basis for evaluating GNSS-R receiver architectures and serves as a foundation for subsequent discussions on implementation, calibration, and experimental validation.

References

1. Hall C, Cordey R (1988) Multistatic scatterometry. International geoscience and remote sensing symposium,' remote sensing: moving toward the 21st century 1:561–562. https://doi.org/10.1109/igarss.1988.570200
2. Martin-Neira M (1993) A passive reflectometry and interferometry system (PARIS): application to ocean altimetry. ESA J 17(4):331–355
3. Unwin M, Van Steenwijk RDV, Gommenginger C et al (2010) The SGR-ReSI-a new generation of space GNSS receiver for remote sensing. In: 23rd international technical meeting of the satellite division of the institute of navigation 2010 (ION GNSS 2010). Institute of Navigation: Portland, OR, pp 1061–1067
4. Ruf C, Unwin M, Dickinson J et al (2013) CYGNSS: enabling the future of hurricane prediction [remote sensing satellites]. IEEE Geosci Remote Sens Magaz 1(2):52–67
5. Garrison JL, Katzberg SJ, Hill MI (1998) Effect of sea roughness on bistatically scattered range coded signals from the global positioning system. Geophys Res Lett 25(13):2257–2260
6. Komjathy A, Zavorotny VU, Axelrad P et al (2000) GPS signal scattering from sea surface: wind speed retrieval using experimental data and theoretical model. Remote Sens Environ 73(2):162–174

7. Zavorotny VU, Voronovich AG (2000) Scattering of GPS signals from the ocean with wind remote sensing application. IEEE Trans Geosci Remote Sens 38(2):951–964
8. Egbert GD, Erofeeva SY (2002) Efficient inverse modeling of barotropic ocean tides. J Atmos Oceanic Tech 19(2):183–204
9. Garrison JL, Komjathy A, Zavorotny UV et al (2002) Wind speed measurement using forward scattered GPS signals. IEEE Trans Geosci Remote Sens 40(1):50–65. https://doi.org/10.1109/36.981349
10. Onrubia R, Pascual D, Querol J et al (2019) The global navigation satellite systems reflectometry (GNSS-R) microwave interferometric reflectometer: Hardware, calibration, and validation experiments. Sensors 19(5):1019
11. Jing C, Niu X, Duan C et al (2019) Sea surface wind speed retrieval from the first Chinese GNSS-R mission: technique and preliminary results. Remote Sens 11(24):3013. https://doi.org/10.3390/rs11243013
12. Jing C, Niu X, Lu F et al (2021) GNSS-R from the Bufeng-1 twin satellites for sea surface winds under hurricane condition. In: 2021 IEEE international geoscience and remote sensing symposium IGARSS, pp 1299–1302. https://doi.org/10.1109/igarss47720.2021.9554797
13. Guo Z, Liu B, Wan W et al (2022) Soil moisture retrieval using BuFeng-1 A/B based on land surface clustering algorithm. IEEE J Selected Topics App Earth Observat Remote Sens 15:4680–4689. https://doi.org/10.1109/jstars.2022.3179325
14. Lu F, Shou Y, Chen H et al (2023) BF-1 GNSS-R sea surface wind speed retrieval and observing experiments. In: The proceeding of 16th international wind workshop, Montreal, Canada
15. Huang F, Yin C, Liu Y et al (2025) Tianmu-1 constellation GNSS-R in-orbit performance: spatiotemporal characteristics, product applications and polarimetric features. IEEE Trans Geosci Remote Sens 63:5802020
16. Wickert J, Cardellach E, Martín-Neira M et al (2016) GEROS-ISS: GNSS reflectometry, radio occultation, and scatterometry onboard the international space station. IEEE J Selected Topics App Earth Observat Remote Sens 9(10):4552–4581
17. Juang, JC, Tsai FY, Lin CT (2019) FORMOSAT-7R mission for GNSS reflectometry. In: 2019 IEEE international geoscience and remote sensing symposium
18. Freeman V, Masters D, Jales P et al (2020) Earth surface monitoring with spire's new GNSS reflectometry (GNSS-R) CubeSats. EGU General Assembly Conference Abstracts, p 13766 https://doi.org/10.5194/egusphere-egu2020-13766
19. Jales P, Esterhuizen S, Masters D et al (2020) The new Spire GNSS-R satellite missions and products. Image Signal Process Remote Sens 26(11533):1153316. https://doi.org/10.1117/12.2574127
20. Michel F, Trevisan M, Giordano D et al (2022) A first look at starlink performance. In: Proceedings of the 22nd ACM internet measurement conference, Nice, France
21. Kozhaya S, Kassas ZM (2024) A first look at the OneWeb LEO constellation: beacons, beams, and positioning. IEEE Trans Aerosp Electron Syst 60(5):7528–7534
22. Zhang Q, Y. Zhu Y, Chen Z (2021) An In-depth assessment of the new BDS-3 B1C and B2a Signals. Remote Sensing 13(4):788
23. Unwin MJ, Pierdicca N, Cardellach E et al (2021) An introduction to the HydroGNSS GNSS reflectometry remote sensing mission. IEEE J Selected Topics App Earth Observat Remote Sens 14:6987–6999
24. Peng J, Li W, Cardellach E et al (2024) Signal coherence and water detection algorithms for the ESA HydroGNSS mission. IEEE Trans Geosci Remote Sens 62:5801218
25. Balagopal S (2014) High-speed delta-sigma data converters for next-generation wireless communication. Boise State University Theses and Dissertations. 822. https://scholarworks.boisestate.edu/td/822

Chapter 2
Design of GNSS-R Software Defined Receiver and Its Application in Altimetry

Fan Gao, Yunqiao He, Xinyue Meng, and Nazi Wang

Abstract This chapter systematically presents the theoretical methodology for designing a GNSS-R altimetry software defined receiver (SDR) and the corresponding test results. Currently, the accuracy of sea surface altimetry using GNSS-R technology remains insufficient, limiting its utility for applications in geodesy and marine environmental dynamics. To enhance the accuracy of retrieval results, it is essential to obtain higher-precision observations of the path delay between the direct and reflected signals. Fortunately, next-generation signals with higher measurement accuracy are being or will be broadcast by various GNSS satellites. Accordingly, we have developed a GNSS-R software receiver capable of processing next-generation signals, with the objective of acquiring stable, high-precision path delay observations. To evaluate the performance, we conducted shore-based, shipborne, and airborne experiments to collect raw GNSS-R intermediate frequency (IF) data, which were subsequently processed using the software receiver. The results demonstrate that centimeter-level sea surface altimetry retrievals can be achieved using both code-phase and carrier-phase observations.

1 Introduction

Among the various parameters retrievable from GNSS-R, sea surface height holds particular significance for the fields of geodesy and marine environmental dynamics. By enabling high-precision observations of sea surface height, this technology can significantly advance research efforts such as the retrieval of high spatiotemporal resolution marine gravity fields, the monitoring of mesoscale and sub-mesoscale

F. Gao (✉) · Y. He · X. Meng · N. Wang
School of Space Science and Technology, Shandong University, Weihai, China
e-mail: gaofan@sdu.edu.cn

Institute of Space Science, Shandong University, Weihai, China

Guangzhou Urban Planning & Design Survey Research Institute Co., Ltd., Guangzhou, China

N. Wang
e-mail: wnz@sdu.edu.cn

K. Yu and N. Zheng (eds.), *Advances in GNSS Reflectometry*, Navigation: Science and Technology 19, https://doi.org/10.1007/978-981-92-2325-1_2

oceanic processes, and the development of real-time sea state forecasting models. The core observation is the path delay between the direct and reflected signals, where its precision directly dictates the accuracy of the retrieval results. However, GNSS signals reflected from the sea surface suffer from significant power attenuation and pronounced scattering. This poses substantial challenges for receivers in tracking the carrier phase of the reflected signals, which ultimately prevents the continuous and stable acquisition of high-precision GNSS-R path delay observations. While tracking the ranging code is relatively straightforward for receivers, the meter-level accuracy of traditional GNSS ranging codes is insufficient for sea surface altimetry. Consequently, extracting high-precision path delay observations requires dedicated research into baseband signal processing methods to achieve GNSS-R altimetry results with high accuracy and spatiotemporal resolution. The implementation of baseband signal algorithms is dependent on the GNSS-R receiver.

Unlike conventional GNSS receivers, a dedicated dual-antenna GNSS-R receiver must provide measurements of the path delay between the direct and reflected signals. Furthermore, since reflected signals are typically weak and suffer from significant scattering, the relevant algorithms must be directly integrated into the receiver's baseband signal processing. Therefore, receiver development constitutes a core technology in GNSS-R altimetry.

Based on the implementation form of signal processing, GNSS-R receivers can be categorized into software defined receivers (SDRs) and hardware receivers. Hardware receivers are capable of processing signals and outputting observations in real time. However, their baseband processing algorithms are solidified into chips or burned into FPGAs, making debugging, upgrading, and modification complex and costly. In contrast, SDRs start from the digitally sampled intermediate frequency (IF) signal, with all baseband signal processing performed by computer software. This approach allows for easier code modification, debugging, and upgrading, and enables repeated processing of the sampled IF signals, making it suitable for researchers to test and develop algorithms. Nevertheless, due to the substantial computational load, existing GNSS-R software receivers typically require longer processing times and cannot output observations in real time.

Representative GNSS-R altimetry hardware receivers developed internationally include Spain's IEEC GOLD-RTR and PIRA, as well as Germany's GFZ GORS. Among them, GOLD-RTR and PIRA are cGNSS-R and iGNSS-R group-delay receivers, respectively, both capable of outputting a Delay-Doppler Map (DDM) every millisecond [1, 2]. GORS was developed based on an existing GNSS radio occultation instrument and can output interferometric phase information at a frequency of 200 Hz while supporting dual-frequency GPS L1/L2 signals [3]; its spaceborne version is named GROS-ISS. In addition, key receivers adopted in relevant planned or ongoing spaceborne GNSS-R missions include: ESA's PARIS-IOD, the University of Surrey's SGR-ReSI, and the PYCARO receiver developed by the Universitat Politècnica de Catalunya [4–6]. Several institutions in China also possess the capability to develop GNSS-R hardware receivers. Representative organizations include Beihang University, the National Space Science Center (NSSC) of the Chinese Academy of Sciences, the China Academy of Space Technology (CAST), and the Shanghai Academy of Spaceflight Technology (SAST) [7–11]. Among these, the "Buffeng-1" receiver developed by CAST and the GNOS-II receiver developed by NSSC have been successfully launched and are both capable of receiving reflected signals from China's BeiDou and GPS satellites [12, 13].

An early GNSS-R software receiver was developed by NASA's Jet Propulsion Laboratory (JPL) in the United States for altimetry applications. This receiver is capable of outputting both Delay-Doppler Maps (DDMs) and interferometric phase information based on GPS C/A and P(Y) codes [14]. In 2017, Spain's IEEC developed a corresponding interferometric software receiver for the European Space Agency's (ESA) PARIS GNSS-R altimetry mission, which is capable of supporting wide-bandwidth, multi-frequency signals [15]. Additionally, Spain's Starlab developed the OCEANPAL GNSS-R software receiver [11]. The French space agency (CNES) also developed a GNSS-R software receiver based on an open-source GNSS software receiver platform [16]. In this domain within China, Beihang University has developed a general-purpose GNSS-R software receiver [17]. A research team at the National Space Science Center (NSSC) of the Chinese Academy of Sciences developed the MRPS software receiver for processing raw intermediate frequency data from the GNOS-II instrument [18]. Furthermore, the Satellite Navigation and Remote Sensing team at Shandong University has developed group-delay GNSS-R altimetry software receivers based on the GPS C/A code, BDS B1I code, and GPS L5 code [12, 19, 20]. GNSS-R altimetry utilizes freely available, abundant, and seamless GNSS L-band ranging signals. Compared to existing similar techniques, it offers inherent advantages such as high revisit rates, high spatial resolution, and low cost. In the coming years, major GNSS systems will continue to be deployed and modernized, leading to significant improvements in both the quantity and quality of their signals, which will further enhance the advantages of GNSS-R altimetry. Examining the development history and current research status of GNSS-R altimetry reveals a bottleneck: group-delay observations offer good stability but low accuracy, while phase-delay observations provide high accuracy but suffer from low availability.

This chapter addresses the aforementioned issues by establishing a comprehensive theoretical and methodological framework for sea surface altimetry. This framework encompasses the entire process from baseband signal processing to retrieval, specifically designed for next-generation GNSS signals. It is worth noting that GNSS-R altimetry technology is suitable for various observation platforms. As the technology matures, its application prospects are extensive. On shore-based platforms, it can be used for tide level monitoring. Combined with precise positioning information from direct signals, it can determine absolute sea surface height variations. On shipborne platforms, it can serve as a mobile tide gauge. Integrated with the ship's attitude and positioning data, it can effectively address the current lack of tide observations in open oceans. On airborne platforms, it can be employed to obtain high-precision, high spatial resolution regional sea surface height information. On spaceborne platforms, it can serve as a valuable complement to existing radar altimetry techniques, mitigating their limitations in spatiotemporal resolution. It will provide crucial data support for research on mesoscale and sub-mesoscale ocean circulation, the development of real-time sea state forecasting models, and the retrieval of high spatiotemporal resolution marine gravity fields.

2 GNSS-R Altimetry: Theory and SDR Design

GNSS-R altimetry systems are primarily composed of four key components: a signal acquisition system, a code-phase-based software receiver, a carrier-phase-based software receiver, and an altimetry module. The basic theory of each method is briefly described in the remainder of this section.

2.1 GNSS-R Signal Acquisition System

The hardware of our GNSS-R receiver system mainly comprises two dual-circularly polarized crossed dipole antennae and one GNSS IF raw data recorder with four radio frequency signal input ports. The two antennae are connected with the recorder. The collected raw IF data are written into an internal hard drive.

Figure 1 presents photographs of the GNSS-R antennae. Left-hand circularly polarized antennas and right-hand circularly polarized antennas have the same appearance. Each antenna has dimensions of 130 × 130 × 35 mm and a mass of about 500 g. We connected the down-looking LHCP antenna with the IF data collector to receive GNSS signals reflected from the sea surface. The zenith-looking RHCP antenna was used to receive direct signals. The two antennae have the same parameter settings, including the settings of the operating band and gain. The operating band of the antennae ranged from 1.16 to 1.62 GHz, which covers GNSS L-band signals. The antenna gain was 5.5 dBi (dB isotropic) and the precision of the antenna phase center was better than 2 mm. The noise in the signals output by the antennae was less than 2.0 dB.

Figure 2 shows the photograph of our GNSS digital IF signal collector, which has four channels. Of the four channels, two operate at a center frequency of 1575.42 MHz, and the other two at 1176.45 MHz. In terms of functionality, two channels are dedicated to receiving the direct signals, while the remaining two are designated for the reflected signals. The data collector has a bandwidth of 20.46 MHz and an IF sampling rate of 62 MHz. Consequently, all signals operating within the shared GNSS frequency bands can be captured and archived. The data collector is equipped with a built-in 4 TB solid-state drive (SSD) for continuous storage of the IF data.

The digital IF data recorded by the receiver are a sequence of real-valued samples expressed as $S_d(n)$ and $S_r(n)$. S_d and S_r respectively are the direct and reflected digital IF real signals. n ranges from 1 to 62,000,000 because the system works at 62 MHz for 1-s duration of data. The A/D converter has a resolution of 2 bits.

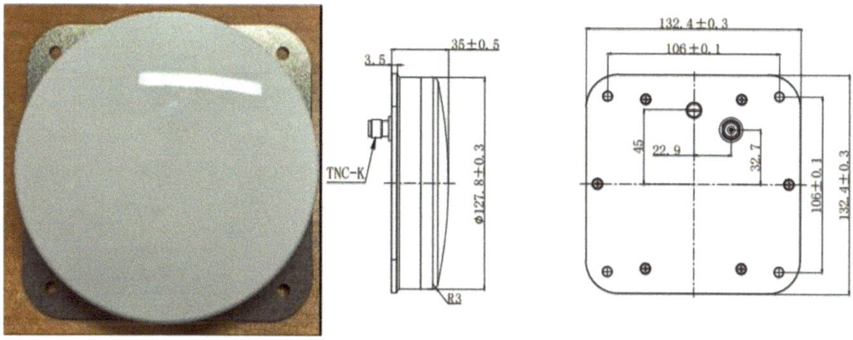

Fig. 1 Photograph and the size of GNSS-R antenna

Fig. 2 Photograph and the size of the GNSS IF signal collector with four radio frequency signal input ports

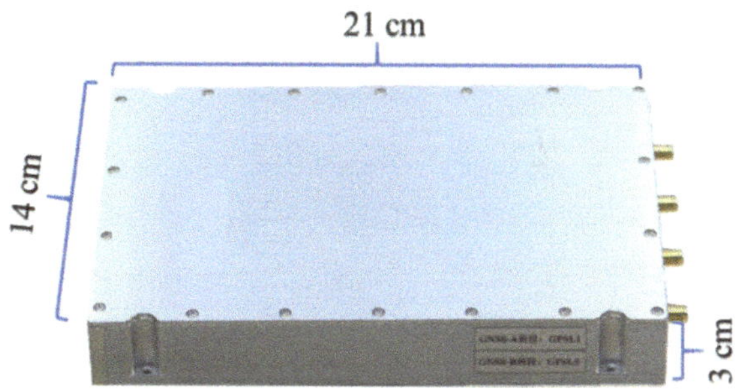

2.2 GNSS-R SDR Based on Ranging Code

We developed an SDR for GNSS-R altimetry to process the collected digital IF data. The SDR separately provides the code chip delays between direct and reflected signals from GNSS shared frequencies. The code chip delays can be converted into path delays and used to derive the reflector heights.

The SDR acquires reflected signals using openly available code firstly. After acquisition of the reflected signals, the SDR generates the local code sequences and up-converts them to the sampling frequency of the digital IF signals provided by the collector. The clean replicas of the openly available code are cross correlated with direct and reflected signals separately to generate correlation waveforms. The code delays are obtained from the positions of the waveform peaks as shown in Fig. 3. In the following part of this subsection, the specific software receiver will be introduced, focusing on its processing of the BDS-3 B1C and B2a signals.

BDS-3 B1C and B2a signals for fundamental positioning, navigation, and timing service are broadcast from 24 medium circular orbit (MEO) satellites and three inclined geostationary orbit (IGSO) satellites. Their pseudo random noise code (PRN) numbers range from 19 to 46, in which 38, 39, and 40 are the PRNs of IGSO. The two kinds of new civil ranging codes have the same length of 10,230, while the

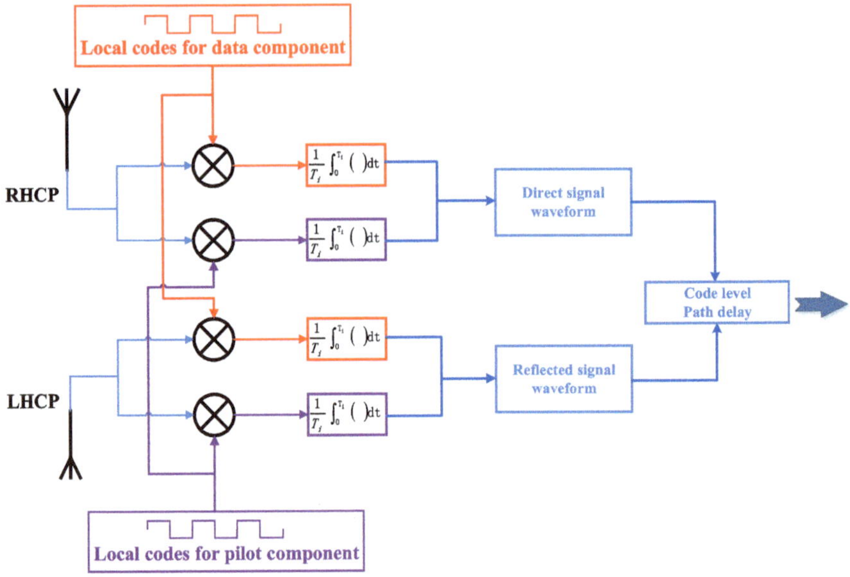

Fig. 3 Basic concept of deriving code-level path delay from waveforms of direct and reflected signals

chip rates of B1C and B2a are 1.023Mbps and 10.23Mbps respectively. These are separately modulated on carrier signals with centre frequencies of 1575.42 MHz and 1176.45 MHz, and different ranging codes are modulated on data component and pilot component. The power ratio of the data component to the pilot component is 1:1 for B2a while the ratio is 1:3 for B1C. In addition, both the components of B2a adopt BPSK(10) modulation while those for B1C adopt more advanced BOC (binary offset carrier) and QMBOC (quadrature multiplexed BOC) modulation.

The modulated B2a signal can be described as:

$$S_{B2a}^{(i)}(t) = \sqrt{2P_{B2a}}\left[D_{B2a\_d}^{(i)}(t)C_{B2a\_d}^{(i)}(t)\cos(2\pi ft) - C_{B2a\_p}^{(i)}(t)\sin(2\pi ft)\right] \quad (1)$$

where $S_{B2a}^{(i)}$ stands for the signals from satellite i; $D_{B2a\_d}^{(i)}$ is modulation data; $C_{B2a\_d}^{(i)}$ stands for the ranging code on data component; $C_{B2a\_p}^{(i)}$ is the ranging code on pilot component. In Eq. (1), the pilot channel is data-less, so it can be used for estimating the distance better without the problem of sign transitions [17].

The modulated B1C signal can be described as:

$$S_{B1C}^{(i)}(t) = \sqrt{2P_{B1C}}\left[\frac{1}{2}D_{B1C}^{(i)}(t)C_{B1C\_d}^{(i)}(t)\cos(2\pi ft) - \frac{\sqrt{3}}{2}C_{B1C\_p}^{(i)}(t)\sin(2\pi ft)\right] \quad (2)$$

where $S_{B1C}^{(i)}$ stands for the signals from satellite i; $D_{B1C}^{(i)}$ is modulation data; $C_{B1C\_d}^{(i)}$ stands for the ranging code on data component; $C_{B1C\_p}^{(i)}$ is the ranging code on pilot component, which employs QMBOC(6,1,4/33) modulation [18]. In Eq. (2), we can see that not only the pilot channel is free of data information, but that the modulation is also more complicated. These improvements will enhance the ranging ability of B1C code.

Since our altimetry solutions are derived from the differential measurement of the direct and reflected code delays, we investigated the cross-correlated waveforms of the two new BDS-3 civil signals for further exploration. Figures 4 and 5 respectively show the waveforms of B1C and B2a codes for direct and reflected signals in shore-based and drone-borne experiment. During the experiment, the sea surface had no appreciable roughness, so that the path delays can be calculated from the peak positions of the waveforms [6]. From Figs. 4 and 5, we can see that the waveforms are smooth and cross-correlation peaks of B2a are much sharper than those of B1C. As the path delays between direct and reflected signals are derived from the peak positions of their waveforms, cubic spline interpolations were used to find the positions of the peaks.

Raw IF data of direct and reflected signals were processed to obtain the code-level path delay measurements using a SDR modified from an open-source code using MATLAB. Figure 3 shows a brief flow chart of this GNSS-R SDR. Firstly, both the direct and reflected signals are cross-correlated separately with the pilot and data local replicas. The coherent span for both B1C and B2a codes is 10 ms; each signal produces two waveforms because of its pilot and data components. In order to increase the SNR, the direct signal waveform and the reflected signal waveform are added incoherently with power ratios. Then, the code or path delays are computed from the positions of waveform peaks by applying cubic spline interpolations. Furthermore, in order to increase the stability and precision, we computed 21 path delays in 1 s at intervals of 50 ms and selected their median value for further processing. At last, the

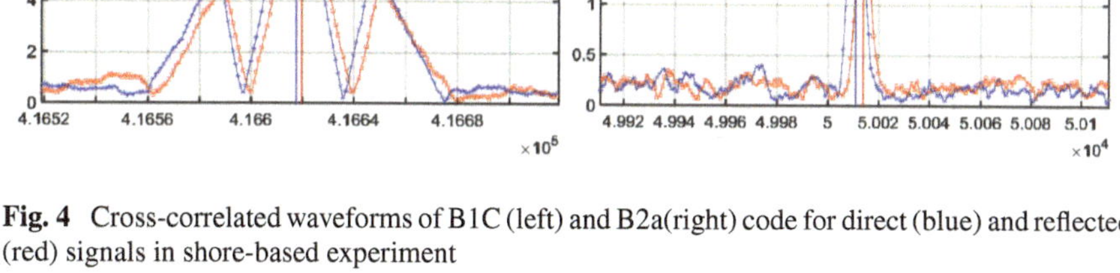

Fig. 4 Cross-correlated waveforms of B1C (left) and B2a(right) code for direct (blue) and reflected (red) signals in shore-based experiment

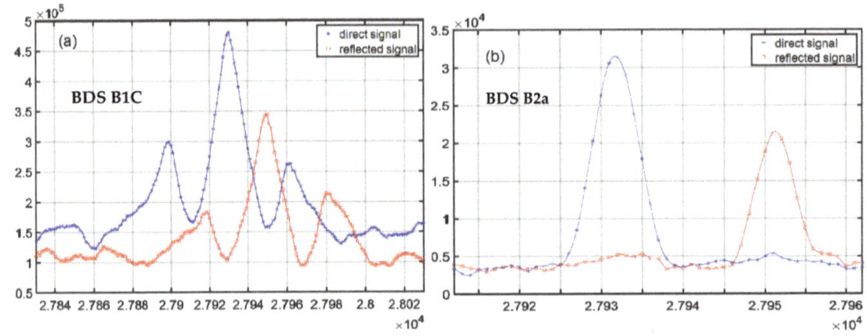

Fig. 5 Cross-correlated waveforms for the B1C **a** and B2a **b** codes for the direct and reflected signals acquired in the drone-borne experiment

reflector heights from the sea surface are calculated in accordance with the geometry of ground-based GNSS-R altimetry.

2.3 GNSS-R SDR Based on Carrier Phase

Considering the GNSS signal transmitted by a single satellite, the signal reaching the direct antenna at epoch t can be expressed as:

$$S_D(t) = A_D D(t - \tau(t))C(t - \tau(t))\sin\left(2\pi f_D - \phi_{Ini}^D\right) + \epsilon_D(t) \tag{3}$$

In the coastal condition, the signals would also reach the down-looking antenna through an extra path delay $\Delta\rho$ after reflecting off the sea surface, which can be simply expressed as:

$$S_R(t) = A_R D(t - \tau(t) - \Delta\rho/c)C(t - \tau(t) - \Delta\rho/c)\sin\left(2\pi f_D - \phi_{Ini}^D - \phi(t)\right) + \epsilon_R(t) \tag{4}$$

where S_D and S_R are the direct and reflect signals. A_D and A_R are the amplitude of the direct and reflected GNSS signals, respectively. $D(\cdot)$ is the navigation message. $C(\cdot)$ is the ranging code sequences. $\tau(t)$ is the transmitting time in the space. $\Delta\rho$ is the path delay between the direct and reflect signals. f_D is the Doppler frequency. ϕ_{Ini}^D is the initial phase. $\phi(t)$ is the extra phase and is related to the path delay. c is the light speed. $\epsilon_D(t)$ and $\epsilon_R(t)$ are the noises. Therefore, the reflected signal can be regarded as the delay form of the direct signals with most of the same parameters. The code-phase, Doppler frequency, the carrier phase measurements, and the navigation data bits obtained by the direct signal tracking loop are used to construct a continuously updating reference signal in coastal conditions:

$$S_R(t) = D(t - \tau(t))C(t - \tau(t))\exp\left(2\pi f_D - \phi_{Ini}^D\right) + \epsilon_R \tag{5}$$

The reflected signals are then correlated with the reference signals. The in-phase and quad-phase correlation values are:

$$\begin{cases} I(t) = \frac{A_R}{2}D(t)R\left(-\frac{\Delta\rho}{c}\right)\sin c\left(-\frac{\Delta\rho}{c}\right)\cos(-\phi(t)) + \epsilon_I \\ Q(t) = \frac{A_R}{2}D(t)R\left(-\frac{\Delta\rho}{c}\right)\sin c\left(-\frac{\Delta\rho}{c}\right)\sin(-\phi(t)) + \epsilon_Q \end{cases} \tag{6}$$

where $R(\cdot)$ is the correlation function of the code sequence. $\Delta\rho$ is the path delay. c is the light speed. $\phi(t)$ is the interferometric phase between the direct and reflected signals. ϵ_I and ϵ_Q are the noises in the in-phase and quad-phase components. The complex correlation field of the L1 reflected signals is:

$$U_{R_{L1}} = I_{L1} + jQ_{L1} \tag{7}$$

The amplitude and the interferometric phase of the L1 reflected signal are:

$$A_{R_{L1}} = 2\sqrt{I_{L1}^2 + Q_{L1}^2} \tag{8}$$

$$\phi_{L1}(t) = -a\tan 2\left(\frac{Q_{L1} * sign(I_{DL1}(t))}{I_{L1} * sign(I_{D_{L1}}(t)}\right) \tag{9}$$

where $I_{D_{L1}}(t)$ is the in-phase component of the direct signals, which contain the navigation message. The phase reversal caused by the navigation message is then eliminated by (9). In single frequency carrier phase altimetry, the interferometric phase observations have to be accumulated by unwrapping the carrier phase when the difference between consecutive values exceeds $\pm\pi$. Furthermore, the interferometric phase is related to the path delay $\Delta\rho$ rather than the propagation path of the reflected signals. As a result, the phase cycles will be greatly reduced by the open-loop tracking strategy and lead to a longer coherent time in coastal conditions. For example, 60-s coherent integration can be adopted to raise the SNR.

A similar configuration of GNSS-R software has been described in previous studies [1, 17]. To facilitate the analysis of the GNSS-R altimetry performance of L5 signals, the joint tracking approach of the data and pilot components is always adopted. The complex correlation field of the L5 reflected signals is:

$$U_{R_{L5}} = I_{L5} + jQ_{L5} \tag{10}$$

The amplitude and the interferometric phase of the L5 reflected signal are:

$$A_{R_{L5}} = 2\sqrt{I_{L5}^2 + Q_{L5}^2} \tag{11}$$

$$\phi_{L5}(t) = -a\tan 2\left(\frac{Q_{L5} * sign(I_{DL5}(t))}{I_{L5} * sign(I_{D_{L5}}(t)}\right) \tag{12}$$

In the one-dimension Fraunhofer region, the complex cross-correlation between fields at L1 and L5 in a gentle ocean is [21]:

$$U_1 U_5^* = e^{j\frac{4\pi}{\lambda_{15}}H} e^{-8\pi^2 \frac{\sigma^2}{\lambda_{15}}} \tag{13}$$

where λ_{15} is the synthetic wavelength, σ is the standard deviation of the sea surface height.

Then the complex cross-correlation between correlation values at L1 and L5 can also be given by:

$$\mathrm{U}_{15} = U_{R_{L1}} U_{R_{L5}}^* \tag{14}$$

Using (7), (9), (10), and (12), the in-phase and quad-phase components can be expressed below:

$$\begin{cases} \mathrm{I}_{\mathrm{L}15} = I_{L1} I_{L5} + Q_{L1} Q_{L5} \\ \mathrm{Q}_{\mathrm{L}15} = I_{L5} Q_{L1} - I_{L1} Q_{L5} \end{cases} \tag{15}$$

The amplitude and the phase measurements of the L1 and L5 combination can be further expressed as:

$$\mathrm{A}_{15} = \frac{1}{4} \mathrm{A}_{\mathrm{R}_{\mathrm{L}1}} \mathrm{A}_{\mathrm{R}_{\mathrm{L}5}} \tag{16}$$

$$\phi_{15}(t) = -atan2\left(\frac{Q_{L15}}{I_{L15}}\right) \tag{17}$$

When the phases of L1 and L5 are available, the phase of the combined signal can be expressed as:

$$\phi_{15}(t) = \phi_1(t) - \phi_5(t) \tag{18}$$

Therefore, the combination between L1 and L5 can be implemented by the cross-correlation between the two fields, without the information of the phase measurements of L1 and L5. The amplitude of the combined signal is greatly enlarged, which implies a higher SNR. Thus, the phase measurements of the combination can be robustly achieved under stormy sea states.

The synthetic wavelength of the L1 and L5 combination is 75.1 cm, which is not significantly short than the L1 and L2 combination wavelength of 86.2 cm. Under calm sea states, the in-phase (I) and quad-phase (Q) distribution of the combined signal from QZSS J07 is demonstrated in Fig. 6, which is from 00:01 to 14:21 (Local Time) on Nov. 6, 2020. The running filter of 10-min is used to reduce the gross errors

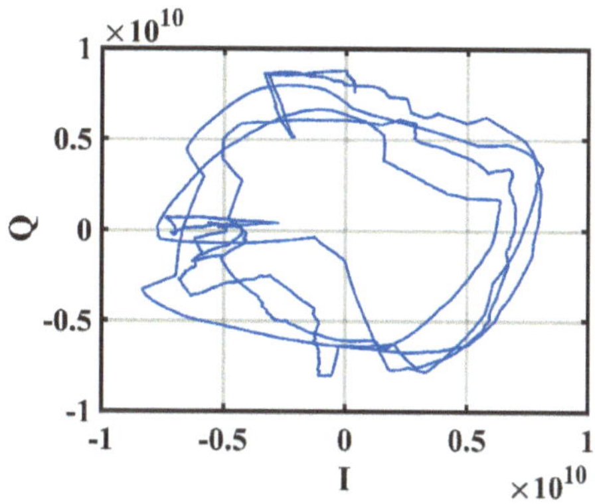

Fig. 6 Distribution of I and Q for the L1 and L5 combination from J07 (GEO) in low sea states

after 1-s coherent integration. The interferometric phase delay measurements of the combined signals can be seen from the phase angles, which proves that the method is practical to extract the phase measurements of the combined signals. And the amplitude information can be seen from the radius of the distribution circle, which is stable because the signals are transmitted from the GEO satellite.

Once the interferometric phase is obtained. The sea surface height can be calculated using least-squares method with a period of observations.

2.4 *The Geometry of GNSS-R Altimetry*

Code-level path delay measurements between direct and reflected signals are the core observations required for deriving the height of the reflector above the sea surface in this study. Thus, SSH can be deduced according to the geometry of GNSS-R altimetry based on the path delay measurements. The drone was able to fly up to the height of approximately 60 m in this experiment because it was limited by the cable length of the tethered power supply; therefore, the effect of Earth's curvature on the observed geometry could be disregarded. A geometric schematic of the drone-borne GNSS-R altimetry experiment is depicted in Fig. 7.

ΔR represents the geometric path delay between the direct and reflected signals, θ is the elevation angle, h is the reflector height above the sea surface, and d is the vertical distance between the phase centers of the direct (RHCP: right-hand circular polarization) and reflector (LHCP: left-hand circular polarization) antennas.

According to the geometry shown in Fig. 7, the following equation can be derived:

$$\Delta R = (2h + d) \sin \theta \tag{19}$$

There are various error factors related to the actual code-level path delay, such as atmospheric refraction, hardware bias, sea state bias, platform attitude errors, measurement errors, and other unmodeled errors. Thus, the path delay can be

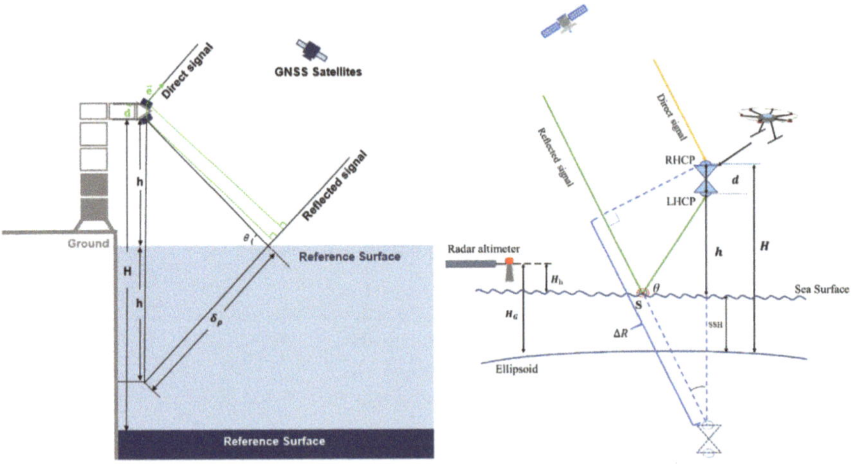

Fig. 7 Schematic of the drone-borne GNSS-R altimetry experiment

expressed more fully as follows:

$$\Delta\rho = \Delta R + \delta_{atm} + \delta_{hw} + \delta_{ss} + \delta_{att} + \delta_m + \epsilon \tag{20}$$

where $\Delta\rho$ represents the delay between the direct and reflected signals from the satellite to the receiver, δ_{atm} represents the tropospheric delay error, δ_{hw} represents the bias between the two channels attributable to hardware error, δ_{ss} represents the sea state bias, δ_{att} represents the attitude error of the hovering drone, δ_m represents the measurement error, and ϵ represents other unmodeled errors. The sea state bias is caused by the roughness of the sea surface, which will result in the first Fresnel zone deviating from the specular point. The attitude of the drone will have a certain tilt that will lead to reduction in the vertical distance between the two antennas, i.e., d will not remain constant. Because the reflected signal will travel over an extra section of the sea compared with the direct signal, tropospheric refraction will be induced. One of the main errors contained in the path delay measurements is the ranging error of the code.

The main data processing principle adopted in this study was as follows. The direct and reflected signals were cross-correlated with the same locally replicated codes, including pilot and data components. At 50-ms intervals, 10-ms coherent integration was conducted to obtain the waveforms of the direct and reflected signals. Cubic spline interpolation was applied to compute the path delays from the positions of the waveform peaks. Thus, 21 path delays were resolved over a 1-s period and their median values were selected as the final $\Delta\rho$ used for further processing.

Because the sea surface had small roughness during the flight, the path delay was computed from the positions of the waveform peaks by applying cubic spline interpolation. Therefore, the reflector height could be deduced by:

$$h = \frac{\Delta\rho - \delta_{atm} - \delta_{hw} - \delta_{ss} - \delta_{att} - \delta_m - \epsilon}{2sin\theta} - \frac{d}{2} \tag{21}$$

The above-mentioned errors can be divided into two categories: random errors and systematic errors. The random errors can be mitigated by applying a moving average.

In the experiment, the vertical distance d between the phase centers of the upward and downward antennas was measured in advance. The geodetic height H of the upward antenna was obtained from the Real-Time Kinematic (RTK) terminal with centimeter precision during the experiment. Therefore, the SSH was finally obtained from the airborne GNSS-R using the following equation:

$$SSH = H - h - d \tag{22}$$

To improve the accuracy of the solutions, a moving average was used in this study. Different sliding average windows were applied to the SSH solutions in 1-s intervals of the different signals from the three satellites and the results were compared with in situ measurements. Figure 8 shows the root mean square error (RMSE) values of the differences between the results after application of different sliding average windows and the in-situ observations. The RMSE values tended to become stable after 5 min; therefore, a 5-min sliding average window was selected in this work.

$$\overline{h} = \frac{1}{n}\sum_{i=1}^{n} h_i \tag{23}$$

The accuracy of GNSS-R code delay altimetry can only reach decimeter level due to the limitation of ranging code width. Besides, signal quality varies with sea state, so that the GNSS-R SDR cannot output continues altimetry results all the time. Therefore, it is necessary to choose appropriate filtering to improve the accuracy and ensure the continuity.

Fourier series is widely used in signal processing and analysis. It can be seen as a special form of trigonometric series, which can be decomposed into a series of sine or cosine functions as follows:

$$f(t) = \frac{a_0}{2} + \sum_{k=1}^{\infty}(a_k\cos(k\omega t) + b_k\sin(k\omega t)) \tag{24}$$

where a_0, a_k, and b_k are Fourier coefficients; k is the order of Fourier, ω is the angular frequency, t is the fitted data. By solving for these coefficients, it is possible to fit the data with a finite Fourier series.

In tidal harmonic analysis, SSH can also be described as the accumulation of several cosine functions:

$$Z(t) = H_0 + \sum_{i}^{j}\left[h_i\cos(\omega_i t - g_i)\right] \tag{25}$$

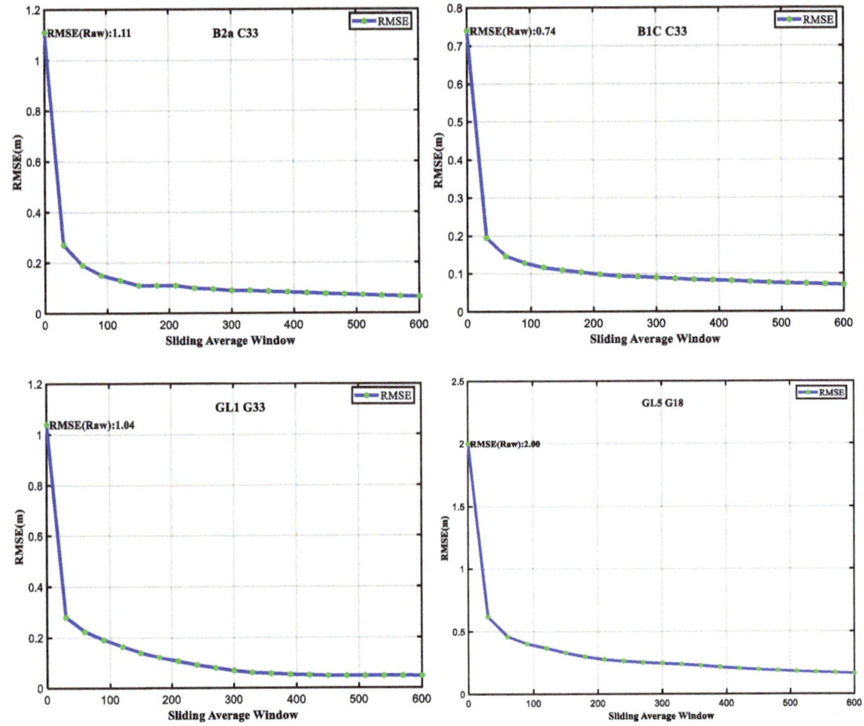

Fig. 8 Relationship between the sliding average window ranges and the RMSE of the smoothed solutions compared with in situ observations based on typical BDS and GPS signals

where each cosine function represents a tidal constituent, $Z(t)$ is the water level; j is the total number of tidal constituents, S_0 is the mean sea level, h, ω, and g are the amplitude, frequency and epoch of partial tide, respectively. Linear transformation on (3) generates:

$$Z(t) = H_0 + \sum_{i}^{j} [a_i \cos(\omega_i t) + b_i \sin(\omega_i t)]$$

$$a_i = h_i \cos g_i \tag{26}$$

$$b_i = h_i \sin g_i$$

Comparing Eq. (3) with (4), it can be seen that Fourier series is suitable to fliting GNSS-R altimetry results. Accordingly, we chose this method to optimize the altimetry results that output by SDR.

3 Experiments for GNSS-R Altimetry

3.1 Shore-Based GNSS-R Altimetry Experiment

We performed the experiment on a shore trestle bridge located at Weihai in Shandong Province with latitude and longitude coordinates (37°32′ 2.62″ N, 122°2′44.11″E) on 5 November 2020. An upward RHCP antenna was used for receiving direct signals while a down-looking LHCP one is for reflected signals. As the operating band of the antennas ranges from 1.16 to 1.62 GHz, it can cover B1C and B2a signals. Before entering the recorder, both direct and reflected signals are spilt into two channels by two two-way power dividers. One is for B1C and another is for B2a. The configuration of antennas and the satellite image of the trestle bridge are showed in Fig. 9. The tilt angles of the antennas are both 30° to the horizontal and their phase centres are in a single plumb line. The height of the antennas above the sea surface ranged about from 3 to 5 m during our experiment. In addition, as the beam width angle of the antennas is 60°, the available satellite elevation ranges from 30° to 90° for this experiment. The two antennas faced south to receive more reflected signals, and, so, the available satellite azimuth ranged from 150° to 210°.

Apart from GNSS-R setups, a geodetic GNSS receiver, a chock-ring antenna, and an electronic total station were used to obtain the geodetic height of the GNSS-R antennas. An independent 26 GHz radar altimeter was installed on the trestle, which can provide vertical distance from the sea surface to its phase centre with 3 mm accuracy every 1 s. We measured the precise height differences among the phase centres of the chock-ring GNSS antenna, GNSSS-R antenna, and radar altimeter using the electronic total station. Figure 10 shows the relevant photos.

In addition to temporary shore-based experiments, we have also established a GNSS-R altimetry station designed for long-term observations. The stationary GNSS-R tide gauge was established on a breakwater located in Weihai City, Shandong Province, China, as shown in Fig. 11a and b.There is a conventional tide gauge nearby, which can provide the tide height with 1-min interval for in-situ data. And

(a) (b) (c)

Fig. 9 The configuration of antennas for our GNSS-R altimetry experiment **a**, the photograph of the antennas' arrangement **b** and top view of the trestle bridge **c** during the experiment

Fig. 10 Photos of the geodetic GNSS chock-ring antenna, electronic total station, and the monostatic radar altimeter

the GNSS-R altimetry data were collected and processed for about 31 days from July 28, 2023 to August 27, 2023.

As shown in Fig. 11c and d, this platform composed with two antennas, one IF data collector and a computer. Two antennas are fixed by an equilateral triangle frame. The up-looking right- RHCP was used to receive GNSS direct signals, while the LHCP antenna was used to receive the signals that reflected by sea surface. This IF collector can cover both L1 and L5 frequency. We choose only L5 signals for its better performance in code delay altimetry. The sampling rate of IF data collector is 80 MHz and the GNSS L5 band real signals were quantified to 2 bits. In order to process huge amounts of IF data more efficiently, the computer with Nvidia GPU A2000 was used to run SDR and output altimetry results in real time.

Besides the above-mentioned hardware, this experiment used a self-development SDR to generate the GNSS-R altimetry results. This SDR was developed based on CPU and GPU [20] and designed to process L5 ranging? Code which is modulated with data component and pilot component. With appropriate parameters, it can realize real-time code delay altimetry. In order to obtain high accurate results, it was designed to process several data segments per second and calculate corresponding code delays. The mean value of these measurements is used to invert SSH. Additionally, the coherent time was set to 2 ms and the number of code delay per second was set to 40 in this experiment.

3.2 *Drone-Borne GNSS-R Altimetry Experiment*

The drone-borne GNSS-R altimetry experiment was conducted during 15:58:38–16:56:50 (local time) on July 15, 2023, at a pier on the Yellow Sea coast (37°32′2.60”N, 122°2′44.06”E) near the Weihai Campus of Shandong University (Fig. 12a). The pier extends westward into the sea for approximately 100 m, and the drone flew and hovered over the sea to the west of the pier so that the reflected

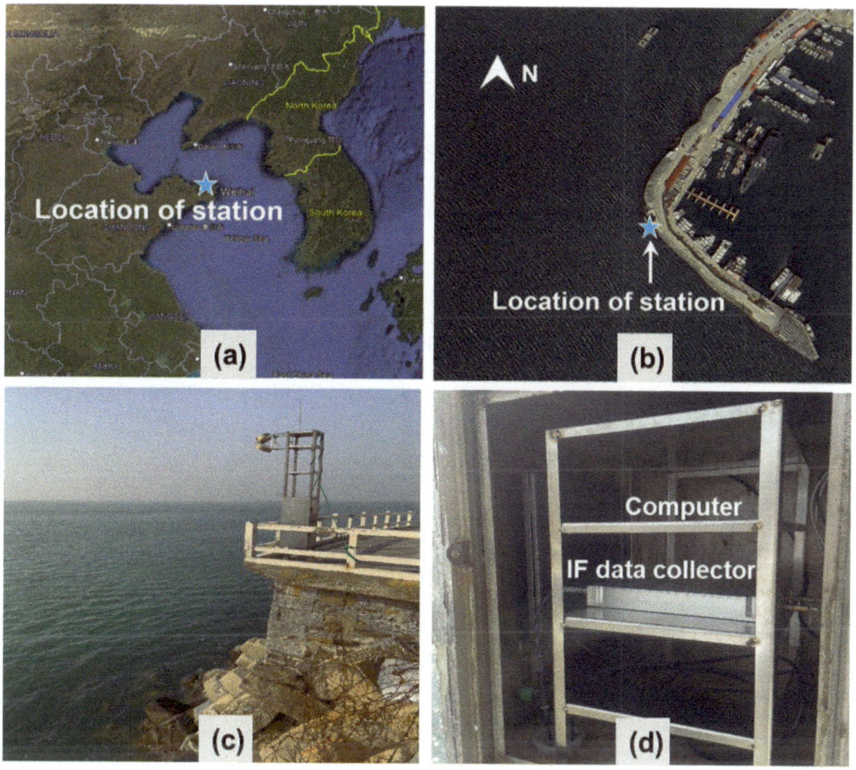

Fig. 11 (**a**) and (**b**) the location of stationary GNSS-R tide gauge; **c** the on-site photo of this experiment; **d** experiment equipment including the IF data collector and the computer equipped with SDR

signals came from the open sea surface, thereby ensuring that more effective observations could be obtained. The experiment used a six-rotor DJI M600 pro drone, which has a maximum effective payload of 5 kg. The flight time supported by its battery is only 15 min, which was insufficient to meet the needs of the experiment. Therefore, an external tethered power supply and a gasoline generator were used to overcome this limitation (Fig. 12b). The weather conditions during the experiment were favorable (average wind speed: 2–3 m/s) and the sea state was goodcalm? A raw IF data recorder with four radio frequency channels was used to receive the direct and reflected signals at two frequencies. A high-precision quartz clock provided the time base for the four channels, while the two channels used to receive signals at the set frequency shared a common channel. The two antennas with 5-dBi gain were the same except for their polarization and the position in which they were mounted on the drone, i.e., pointing vertically upward and downward. Therefore, bias on the path delay measurement due to hardware could be avoided. Additionally, the raw IF data obtained with a 62 MHz sampling rate were stored in a solid-state hard drive, which was integrated into the recorder. A 26 GHz radar altimeter mounted at the end of the pier was used to obtain in-situ SSH observations (Fig. 12c). A GNSS RTK terminal mounted on the drone, together with a nearby reference station located approximately 500 m from the end of the pier (Fig. 4d), provided highly precise coordinates of the position of the drone at 1-s intervals.

The drone selected for use in this study has very advanced flight control systems that could ensure that the tilt of the drone remained within 5°. Because the vertical distance between the two antennas was 0.62 m, the maximum error attributable to tilt was 0.0024 m, i.e., sufficiently small to be ignored.

To verify the performance of the drone-born GNSS-R altimetry at different heights, the drone hovered at heights of approximately 20, 40, and 60 m above the sea for periods of approximately 15 min. The drone relied on its internal battery to fly to the height of 100 m and to move in the horizontal direction. Table 1 lists the details of the three flights.

The signal bandwidth of the IF data recorder was 20.46 MHz, and the analog signals were quantized by an analog-to-digital converter with 2-bit quantization at a

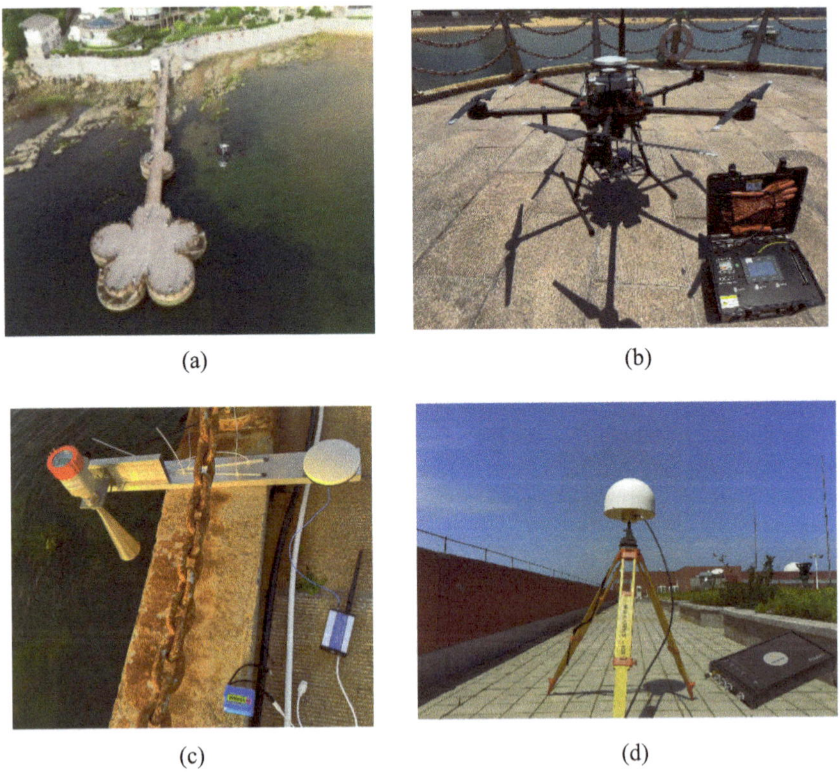

Fig. 12 Photographs of the experimental site and related equipment: **a** the drone carrying the GNSS-R payload above the experiment site, **b** the drone and the tethered power supply on the pier, **c** the 26-GHz radar altimeter and a GNSS RTK terminal, and **d** the nearby GNSS reference station

Table 1 Records of experimental drone observations

Experimentation	Flight height/m	Period/min	Hover duration/min
Hovered	20	16:03–16:18	15
	40	16:20–16:35	15
	60	16:37–16:52	15

sampling rate of 62 MHz. The raw IF data stored in the solid-state drive covered a duration of 58 min (file size: 201 GB).

The visibility of various GNSS satellites during each drone flight is illustrated in Fig. 13. It is evident that the elevation of the six GPS satellites, six BDS-3 satellites, and two QZSS satellites considered in this study varied over the period of the experiment. Because the beamwidth of the antennas is 120°, signals from satellite with elevation > 30° were used in this study.

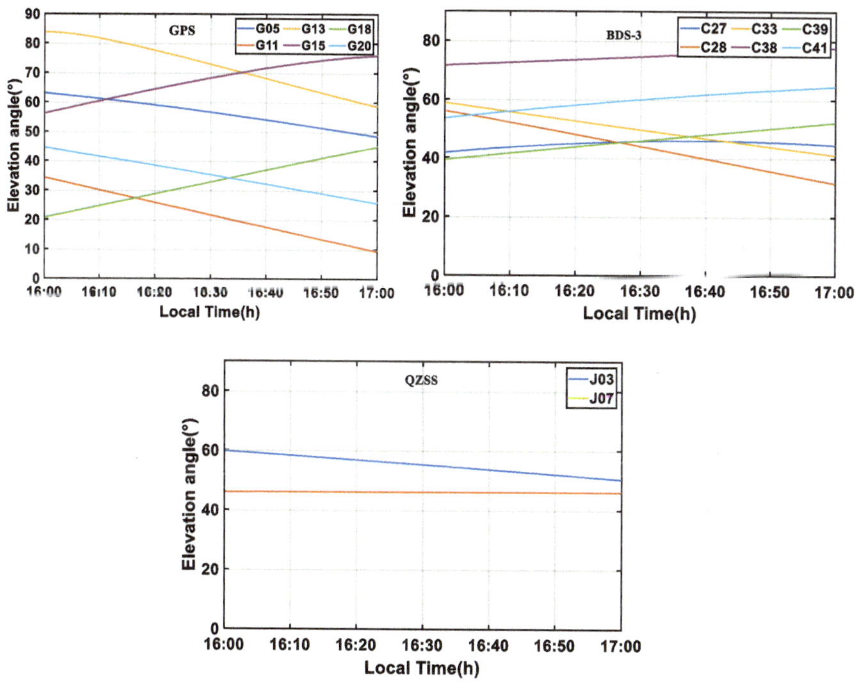

Fig. 13 Variation in elevation of the GPS, BDS-3, and QZSS satellites visible during the drone flights

4 Experimental Analyses

4.1 *Ground-Based Experimental Results for GNSS-R Code-Level Altimetry*

In order to receive continuous and stable GNSS signals, we selected the QZSS GEO satellite for this experiment. The data acquisition lasted for 31 days, during which the data stream was briefly disrupted on three occasions. The satellite elevation and azimuth are about 46.2° and 171.7° respectively. During the first 4 days, the SDR output frequency was 1 Hz. Subsequently, due to the high temperature, the output frequency for the remaining days was set to 0.5 Hz. After obtaining the original altimetry results from SDR, Fourier fitting method was chosen to smooth the results of each day. The altimetry results are shown in Fig. 14.

As shown in Fig. 14a, the navy-blue scattered points represent the original results output by SDR. The magenta line is the in-situ data and the light cyan line is the altimetry result after Fourier fitting. It can be seen that the accuracy has been significantly improved after smoothing. The overall RMSE of the raw results is 78.56 cm, and it is 9.78 cm after Fourier fitting. Because of the poor isolation between up-looking and down-looking antennas, direct signals may enter the down-looking antenna, resulting in overall bias in the results [19]. Besides, cross-talk will also lead to obviously dithering of results. Fourier fitting can smooth the results while the commonly moving average can hardly remove this effect.

The residuals of altimetry results are shown as Fig. 14b and c. The residuals of original raw points mostly concentrated between -2.5 m and 2.5 m, while the residuals are reduced to $\pm$ 0.25 m by Fourier fitting. It is obviously that the residuals at the peaks and valleys of the tide are relatively high than other stages. In addition, since this chapter uses single-day data to fitting separately, the fitted curves may not align, especially when the link points are located at the peaks or valleys of the tide. It should be noted that the reference surface of our results and tide station data are not the same, thus we removed bias from each day results before precision statistics are calculated.

The maximum RMSE of raw results is 116.72 cm and minimum one is 63.59 cm before fitting. As Fig. 15c shows, the accuracy of each day is improved efficiently after Fourier fitting. The dates for the best and worst results are August 4th and 12th, with RMSEs of 5.01 cm and 21.55 cm, respectively. It is worth noting that when using first-order Fourier fitting, the RMSE of the August 12th can be reduced to 12.71 cm, but other data are more suitable for second-order Fourier fitting. The influence of wind on the altimetry accuracy is not intuitive and the correlation between them is low. The ocean waves analyised by this chapter including wind wave, swell and mixed wave formed by a mixture of multiple factors. In general, there is a certain correlation between the change of waves and the altimetry accuracy. Also, the accuracy of August 3rd, 4th and 5th seems to have no obvious correlation with waves. After excluding these three days results, the correlation coefficients between the single-day GNSS-R

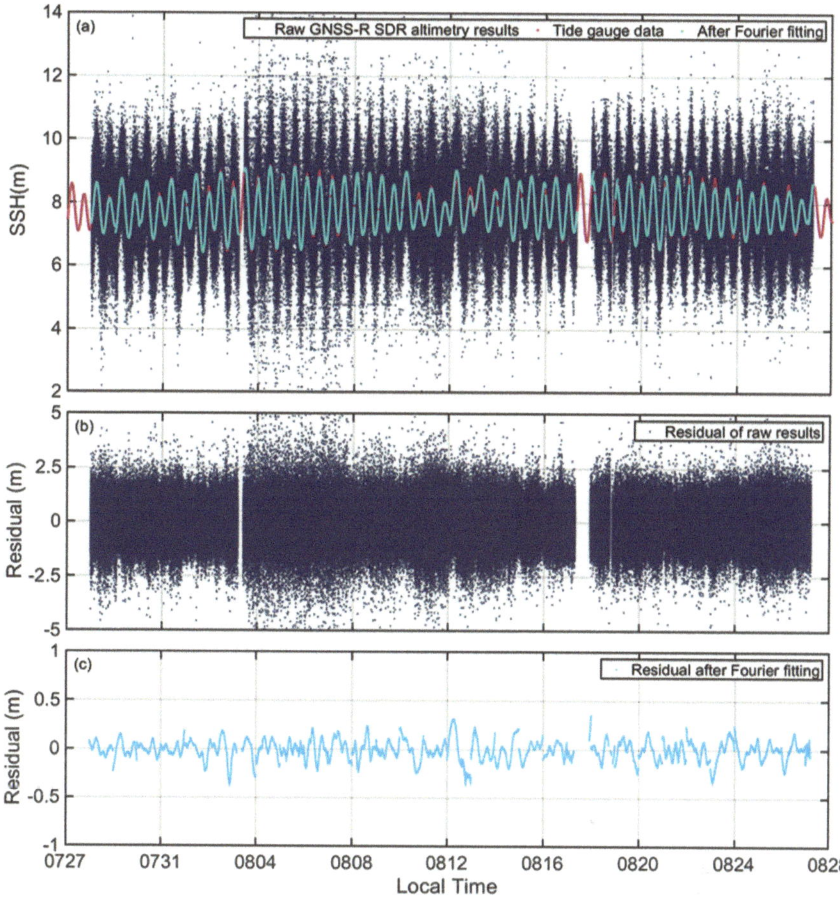

Fig. 14 **a** Raw GNSS-R altimetry results and smoothed GNSS-R results; **b** the residuals of raw altimetry results; **c** the residuals after Fourier fitting

altimetry accuracy and the mixed wave, wind wave and swell are 0.6018, 0.4571 and 0.8093 respectively.

The most obvious reason of these variances shown in Fig. 16 is the fact that these days are during spring tides. With the significant rise and fall of the water level, the roughness of the sea surface will undergo remarkable changes, especially that of the nearshore surface. It should also be noted that during spring tides, although the accuracy is poor at low tide levels, the performance of altimetry is relatively good at high tide levels. This is because when the tide level descends to a specific point, the inshore stones will become exposed above the sea surface. This occurrence triggers an augmented error in the reflected signals captured by the down-looking antenna, causing a sudden decrease in SNR of reflected signals. The combination of spring

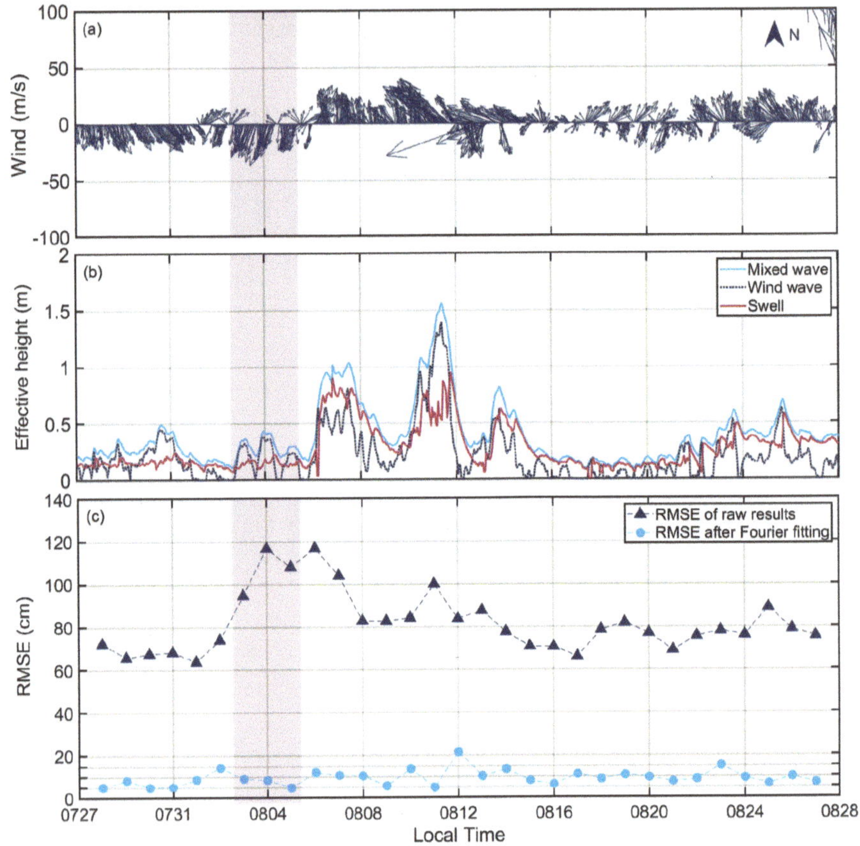

Fig. 15 **a** the wind speed and wind direction during the experiments; **b** effective height of swell, wind-wave and mixed-wave; **c** RMSE (cm) of daily altimetry results and daily results after Fourier fitting

tides, winds and waves led to the poor SNR of the reflected signals during August 3rd to 5th, ultimately leading to a unsatisfied precision.

4.2 Drone-Borne Experimental Results for GNSS-R Code-Level Altimetry

In this study for drone-borne experiment, raw IF signal data at the BDS B1 and B2 frequencies were collected and processed to resolve SSH based on GPS L1 C/A, L5, BDS B1C/B2a, and QZSS L5 code-level path delay measurements. The recorder started to store raw IF data approximately 1-min before the drone took off. Because the drone did not fully fly to the open sea at first, the data obtained during the first

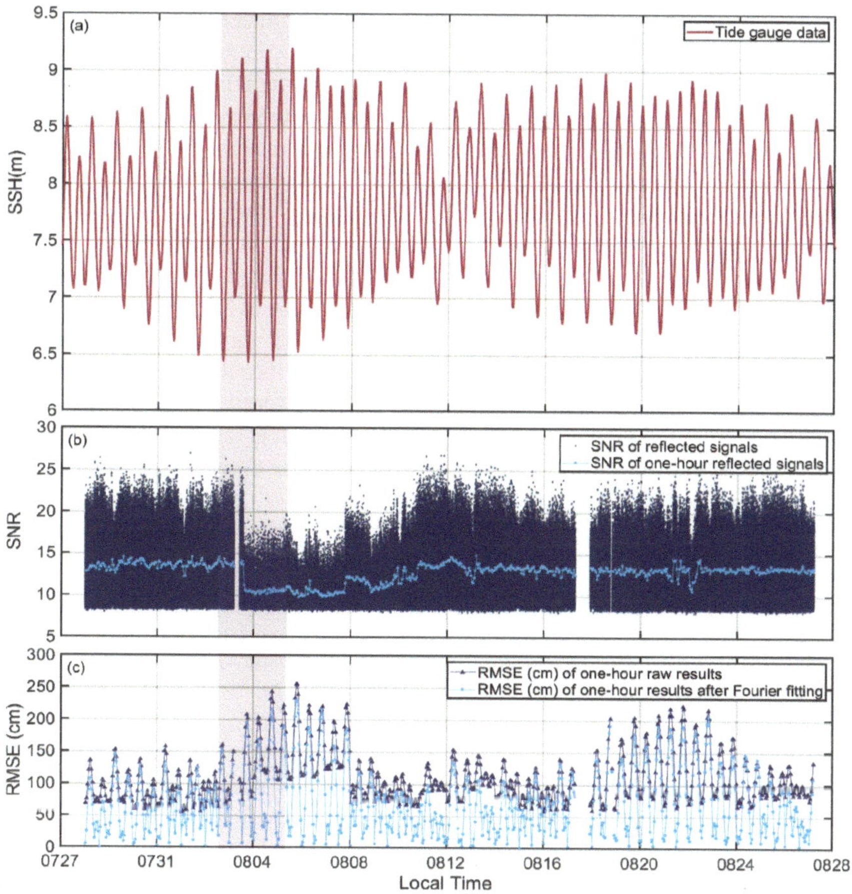

Fig. 16 **a** SSH data from tide gauge; **b** SNR of reflected signals and mean SNR of one hour; **c** RMSE of one-hour results including raw results and Fourier fitting results

100 s were discarded to ensure the quality and precision of the observations. The SSH retrievals were resolved according to the method described in Sect. 12.2, and the solutions obtained from the BDS, GPS, and QZSS platforms are detailed separately in the remainder of this section.

The SSHs derived from B1C/B2a signals at different elevations and the in-situ observations are depicted in Fig. 17. The scattered points represent the initial altimetric results derived from the raw data of the BDS satellites at three different flight heights: 20 m (cyan), 40 m (blue), and 60 m (green). The black scatter plot shows the results after application of the 5-min moving average, and the red dotted line represents the in-situ observations derived from the radar altimeter. The RMSEs and the mean values of the differences between the retrievals from the different heights and the in-situ observations are listed in each panel. The numbers of effective GNSS-R code-level path delay measurements of B1C and B2a at different heights are shown in Fig. 18.

The RMSE values of the original estimations from the B1C and B2a path delay measurements are approximately 1 m, whereas those of the results derived following application of the moving average are within 10 cm in most cases. The RMSE values of the original solutions from C27, C28, and C39 are greater than1 m because of their lower elevations. The RMSE value of the moving-averaged solution from the C28 B2a signal is 14 cm, which is attributable to the lower number of effective measurements. Furthermore, the inversion outcomes for satellites C33 and C41, sharing two bands, demonstrate the superior performance of the B1C band over that of B2a at the drone height of 20 m. This superiority is attributed to the broader bandwidth of the B1C signal, the more distinct peak of its autocorrelation function, and the enhanced precision in code delay estimation. Overall, the accuracy is better at higher elevations, and the accuracy of smoothed result is related to the number of effective measurements. There is no notable difference in the accuracy of the results at different drone flight heights.

The SSHs derived from the GPS L1 C/A and L5 signals at various elevations are shown in Fig. 19, together with the in-situ observations. The quantities of effective code-level path delay measurements for these signals at different heights are shown in Fig. 20.

The RMSE values of the original estimations from the GPS L1 (C/A) and L5 path delay measurements are approximately 1.0 and 2.5 m, respectively, whereas the moving average results are 10 and 30 cm, respectively, in most cases. The RMSE value of the original G20 solution was 1.7 m owing to the low elevations, which prevented the satellite from being captured later in the experiment and resulted in fewer effective measurements. Furthermore, the two satellites at the L5 band had generally low elevation angles and poor inversion results throughout the experiment. Specifically, G11 had the smallest elevation angle and the lowest amount of effective data, resulting in the highest RMSE value (32 cm) among the GPS satellites.

Regarding the QZSS satellite system, its orbital position in the L1 band is mainly distributed near Earth's equator, which means that it could not be captured in this experiment. However, using the same method as described above, we obtained results for the data of two satellites in the QZSS L5 band, namely J03 and J07, as shown in Fig. 21.

From the information presented in Figs. 22, it is evident that the overall inversion results of the two QZSS satellites in the L5 band are more accurate than those obtained using the BDS and GPS systems, and the residuals can be maintained within the range of $\pm$ 1 m. The J07 satellite achieves the best inversion accuracy, with the minimum RMSE value (0.06 m) among the QZSS satellites. The reason for the lower accuracy of the inversion results of the J03 satellite with a higher elevation angle, compared with that of the results of the J07 satellite with a lower elevation angle, is that the amount of effective data for J03 is lower, as shown in Fig. 22. Overall, the QZSS satellites provide a better inversion and more accurate results.

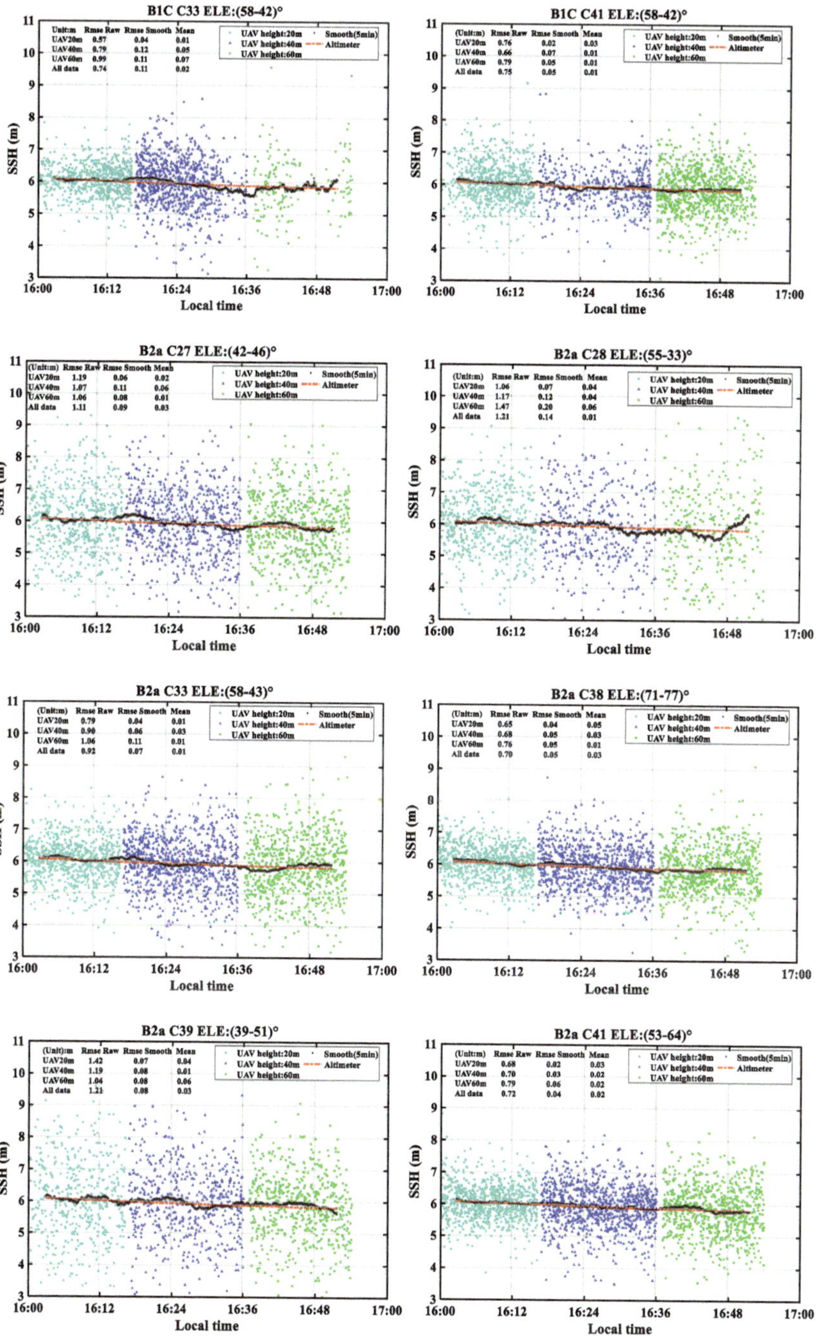

Fig. 17 SSHs derived from BDS B1C/B2a signals at different elevations and the in situ observations

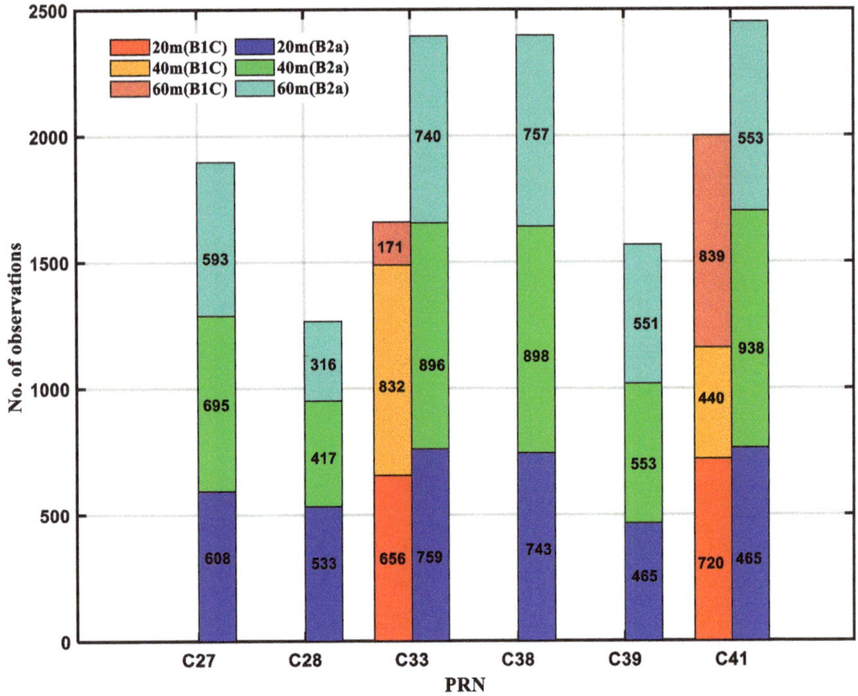

Fig. 18 Numbers of effective GNSS-R code-level path delay measurements of B1C and B2a at different heights

4.3 Ground-Based Experimental Results for GNSS-R Phase Altimetry

The GNSS-R altimetry SDR is developed to process the collected raw digital IF data based on an open-source GNSS SDR [17]. Delay waveforms are computed by correlating the reference signal and reflected signal, and then the correlation sums of both in- and quad-phase components at their peaks are obtained. In order to eliminate the noise and errors of the measurements, we apply a running mean filter with 5-s window to each in-phase and quad-phase component. The interferometric phase measurements are derived from the filtered sequences as shown in Fig. 23.

As the reflected signals are affected by sea roughness, multipath effects, noise floor and so on, the interferometric phase values are not continuous in one cycle sometimes as shown in Fig. 24. These discontinuities will result in phase unwrapping errors. However, since the sea level change is a gradual process, a second-time phase unwrapping is adopted by adjusting $\pm 2n\pi$ to advoid big jumps of the interferometric phase.

Furthermore, we find that there are many interferometric phase jumps, and they are repaired by adding or subtracting a bias to form smooth curves. As the observations

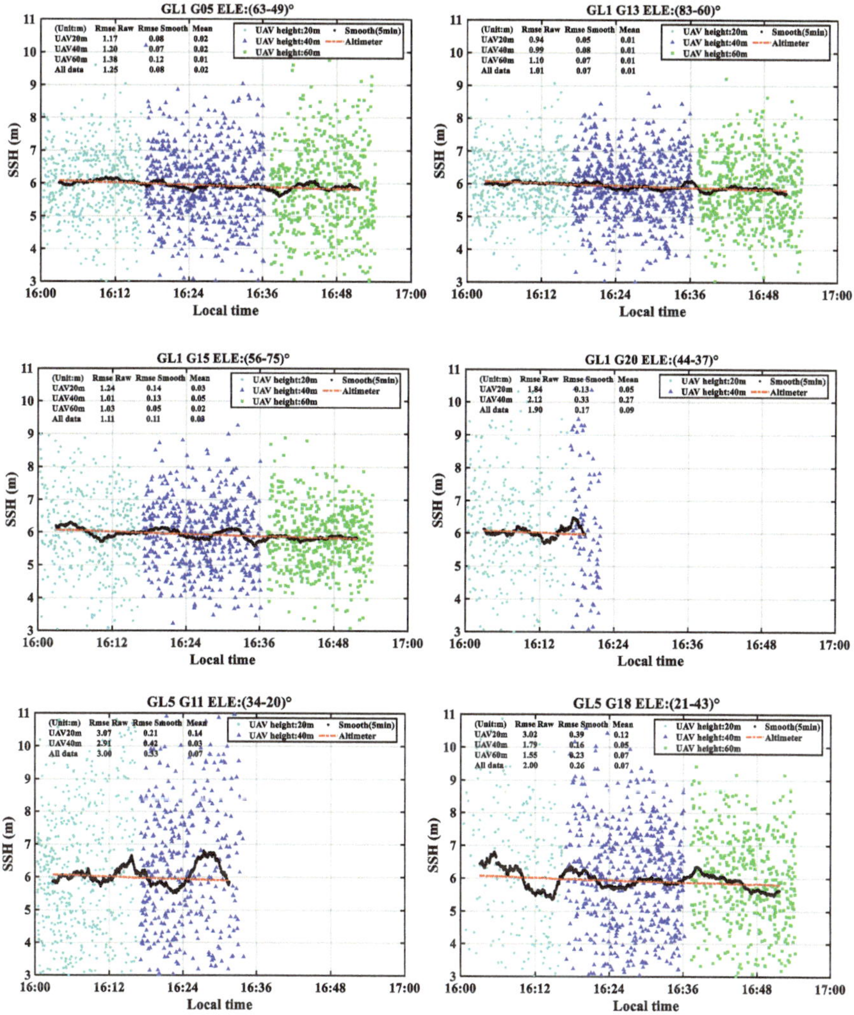

Fig. 19 SSHs derived from the GPS L1 C/A and L5 signals at different elevations and the in situ observations

are not continuous, interferometric phase measurements are obtained using a third-order polynomial interpolation.

As the elevation angle of QZSS GEO satellite does not change almost, the path delay between direct and reflected signals is mainly related to the height of sea surface. For example, only about 10 cycles of change in interferometric phase is observed during 7 h, which are much less than that for the MEO case [18].

In addition, we compute the SNR values of L1C/A, L1Sb and L1S of direct signals from QZSS GEO satellite and they are 37.1 dB-Hz,39.9 dB-Hz,43.7 dB-Hz, respectively. As the SNR of L1S is highest and higher SNR leads to better

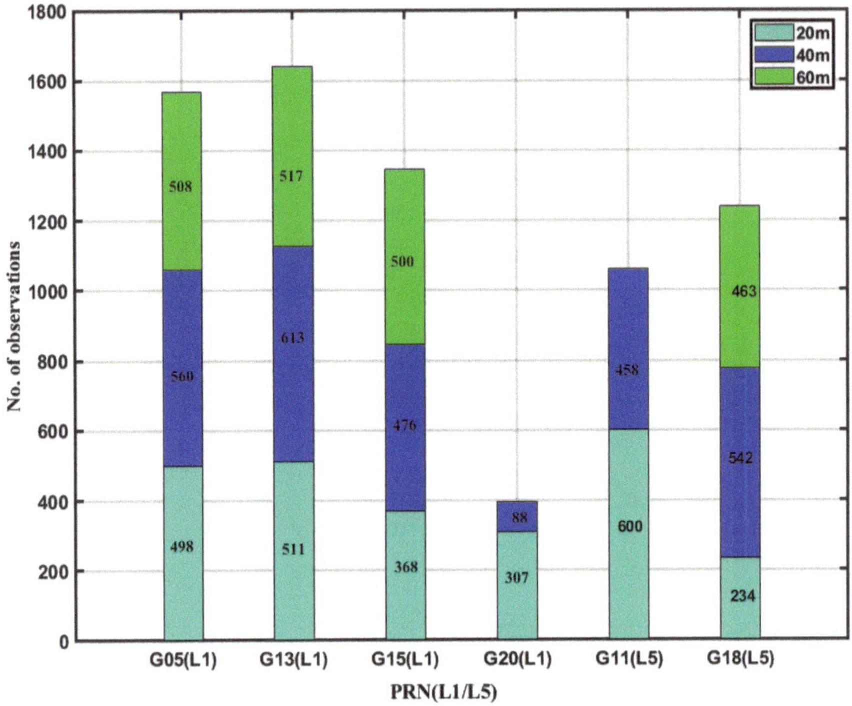

Fig. 20 Numbers of available observations for each GPS satellite at different heights

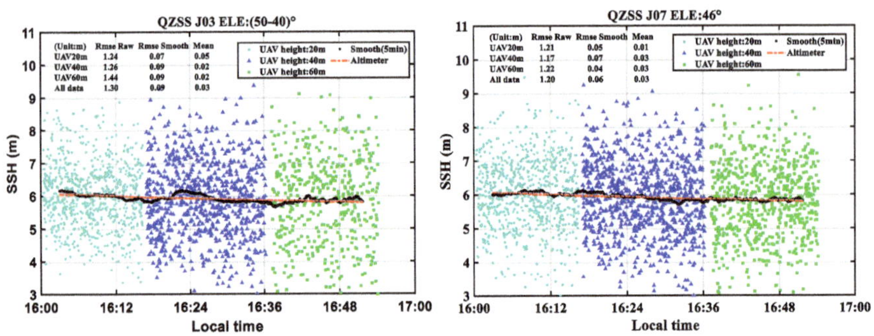

Fig. 21 Solutions resolved from the QZSS L5 satellites

performance, L1S is selected for further processing in the experiment. The QZSS GEO L1S signal is used to extract the interferometric phase.

Variation of the reflector height is estimated based on interferometric phase measurements in our experiment according to Eq. (9) at 1 s interval. The QZSS GEO satellite elevation angle is assumed to be 46.20° at the experiment site, while the actual GEO satellite elevation angle has small change ranges from 46.13° to

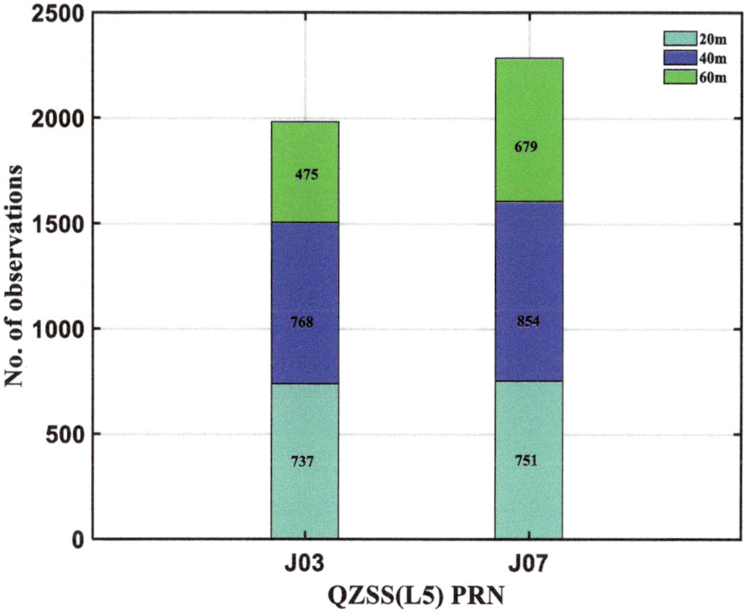

Fig. 22 Numbers of available observations for each QZSS satellite at different heights

Fig. 23 The interferometric phase measurement sequence using the mean filter with 5-s length

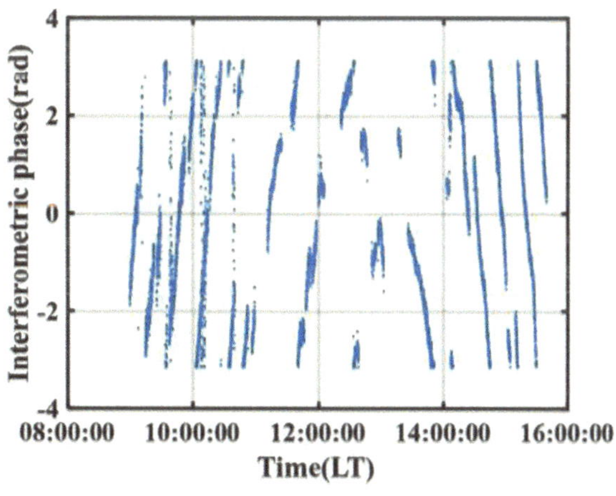

46.27° during the observation period, which could cause errors of a few millimeters. Then, we evaluate the accuracy of our solutions by comparing with the observations of the radar altimeter. Figure 25a shows the variations of the reflector height derived from GNSS-R interferometric phase measurements (green dots) are consistent with those derived from the radar altimeter measurements (red dots). We can see that the subtle sea level changes also can be detected using this method. Figure 25b shows the residuals and their RMSE is 1.3 cm.

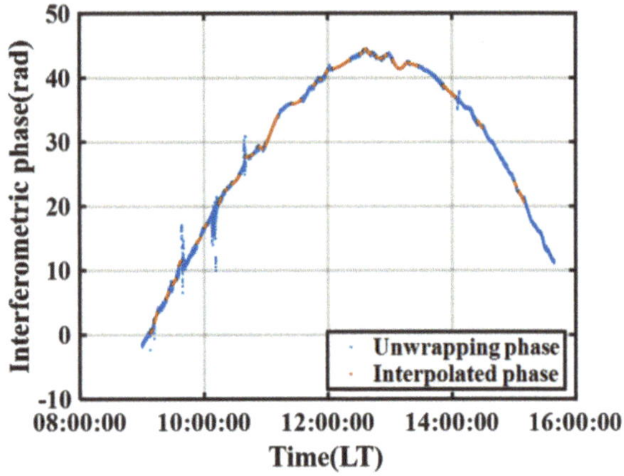

Fig. 24 The unwrapping interferometric phase measurement and interpolated phase

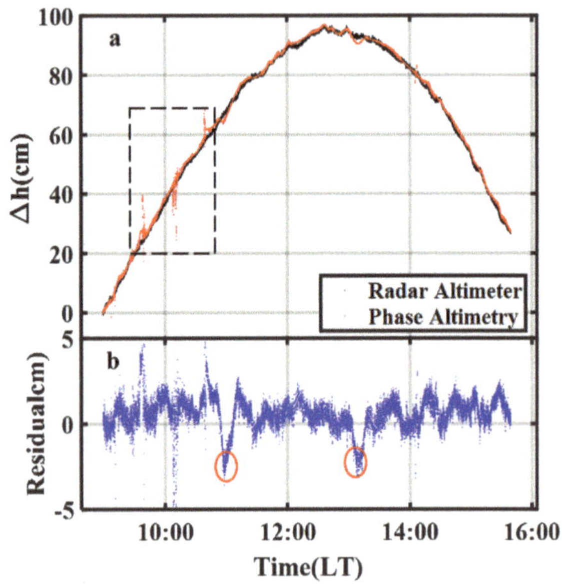

Fig. 25 Sea surface phase altimetry results. The black box is the effect of the passing boat; The red circles are the large error caused by the improper interpolation

In the black dotted box, the height difference measured by the phase altimetry changes dramatically. Considering the experimental environment during this time, it can be inferred that this is caused by the passing of a ship on the sea. Most of the altimetry errors are less than ± 2 cm. The two large peaks of the errors in the red circles are caused by improper interpolation due to the irregular changes in the sea surface over a short period.

The process of the sea level rising for a long period and some subtle sea-level changes are accurately detected, as shown in Fig. 25. From the residual error distribution, most of the altimetry errors are less than ± 2 cm. The other errors in the altimeter results are mainly related to the model, the experimental environment, and the equipment.

According to the "Rayleigh criterion", a relative sea roughness is introduced to evaluate the signals' altimetry performance which is related to the signal's wavelength and the incident angle. For the same sea roughness condition, when the signals' wavelength became larger, the interferometric phase between the direct and reflected signals are less affected. As known, the wavelength of GNSS L1 and L5 is 19.03 cm and 25.48 cm, respectively. Based on the satellite elevation angles, we can roughly judge that if the sea surface is smooth for L1 and L5. Figure 26 shows the elevation angle variations of the four QZSS satellites on November 7 during the experiment. Among them, J07 is a GEO satellite with a constant elevation angle value, which is about 46° with less variation. With these information, it can be seen that that the sea is relatively rough for L1. Consequently, we regard it as the stormy sea state after 13:00 (LT), as shown in Fig. 27. In addition, the sea surface would become rougher for both L1 and L5 signals transmitted by J01 and J02 when the elevation angles are higher than 46°. In the contrast, the relative sea surface would become smoother for both L1 and L5 signals transmitted by J03 due to the lower elevation angles than J07. The altimetry performance of the phase combination will then be analyzed under different elevation angles.

The carrier-to-noise ratio (CNR) of the direct signal from the J07 GEO satellite is used to illustrate the tracking performance of the direct signals and the quality of the reference signals. The pilot channel of L5 has the same level CNR as that of L1. Since the IF data is packedin 2-min, the CNR is low at the beginning of per data packet due to the initialization of the tracking loop in the left panel of Fig. 28. In addition, the data is processed by the SDR afterward, and the CNR decreases to 40 dB-Hz possibly due to the inappropriate loop parameters. Even so, the direct signals are continuously tracked from the distribution of in-phase (I) and quad-phase (Q) components in the right panel of Fig. 28. As a result, the reference signal can be well built due to the accurate tracking results and then correlated with the reflected

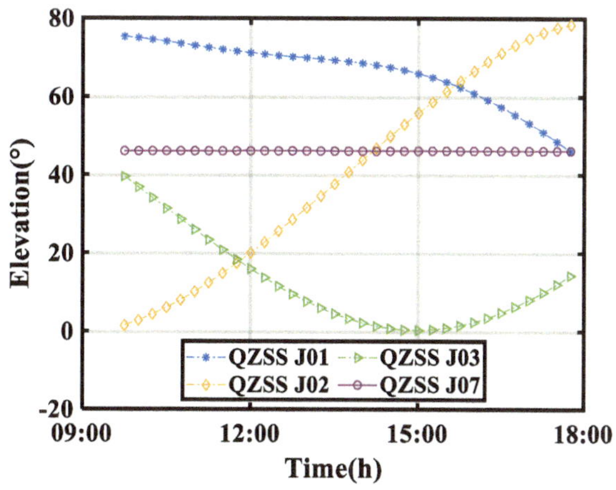

Fig. 26 The elevation angle variation of the QZSS during the observation on 7 November 2020

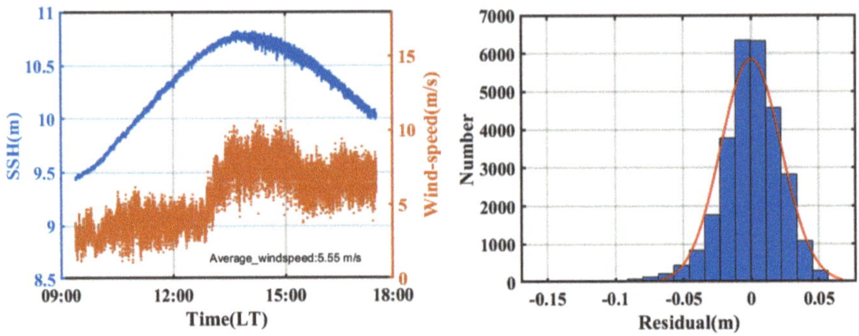

Fig. 27 The SSH and wind speed recorded in the experiment on November 7, 2020 (left) and the histogram of the residual between the SSH and the fitting value (right)

signals. The amplitude values and the carrier phase combination measurements of L1 and L5 can be obtained by (16) and (17).

The amplitudes of the reflected signals are shown to analyze the quality of the reflected signals in Fig. 29. With the increase of the sea surface roughness, the amplitudes of L1 and L5 GNSS reflected signals decrease rapidly as shown in the left panel of Fig. 29. The amplitude of the L5 reflected signal is larger than that of L1 due to the higher transmission power and longer wavelength. After 12:00, both the amplitudes of L1 and L5 reflected signals become very small and have no significant change compared with the initial of the sequences, which indicates that the reflected signals are dominated by the non-coherent components. However, the coherent components of the reflected signals still exist and can be combined with proper method. As shown in the right panel of Fig. 29, the amplitude of the combined signal is demonstrated, which is much larger than any of L1 and L5. The amplitude of the combined signal is smaller than the expected value in (16), due to that the

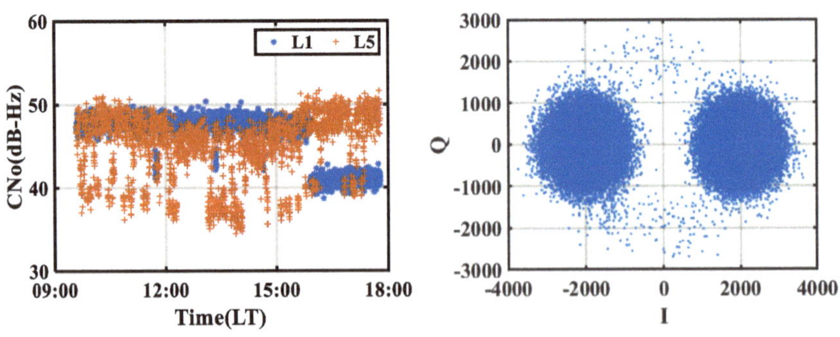

Fig. 28 The tracking performance of the direct signals from the J07 (GEO) satellite during the experiment: (left) the direct signal CNR of the L1 and L5 (pilot channel); (right) the distribution of the in-phase (I) and quad-phase components (Q) of L1 when the CNR is 40 dB-Hz

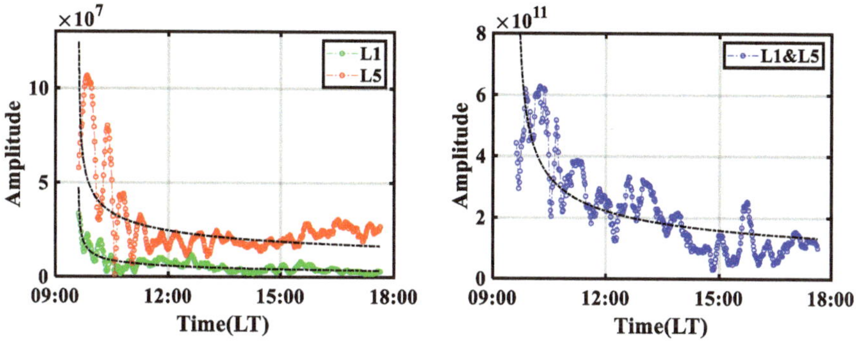

Fig. 29 The reflected signal amplitude of the L1, L5 (left), and the L1&L5 combination (right) from the J07 (GEO) satellite during the experiment

amplitude of the combined signal is the combination of the coherent components of L1 and L5, rather than the whole signals. Also, the noise would be enlarged by the combination. The amplitude of the combined signal decreases gradually with the increase of the sea surface roughness, but the value is still large under stormy sea states.

For reflected signals of the GNSS GEO satellite, the relative roughness signals differ from each other due to the signal wavelength for the same sea state. Figure 30 shows the interferometric phase delay measurements of L1, L5, and the combined signal from the GEO satellite before unwrapping. Also, the phase delays are calculated according to (17). The phase delay models are different due to the different wavelengths. The real interferomctic phase are calculated to compare with the measured interferometric phase shown in Fig. 30. The results show that the phase measurements of L1 and L5 are in good agreement with the models under calm sea states before 12:00. High-precision phase altimetry results can be obtained in this period. With the sea roughness increasing, the phase delay measurements have major differences with the delay models, which implies that the single-frequency phase measurements cannot be well extracted from the reflected signals under stormy sea states, and are incapable of phase altimetry. While the phase measurements of the L1 and L5 combination have been in good agreement with the phase model under different sea conditions, the phase altimetry using the phase measurements of the combined signals can be robustly implemented even under high sea states.

The phase altimetry based on L1 and L5 is greatly affected by the sea states owning to their short wavelengths and low power. What is more, the sea surface is relatively smooth for L5. But the phase measurements cannot be well extracted due to the rapid signal power decreasing with the sea surface roughness increasing. Due to the combination of the coherent components of L1 and L5, the decreasing of the combined signal is tolerable compared with the single frequency signals shown in Fig. 9b. In addition, the longer coherent integration can be implemented to improve the SNR and reduce errors. Therefore, the phase altimetry performance using the

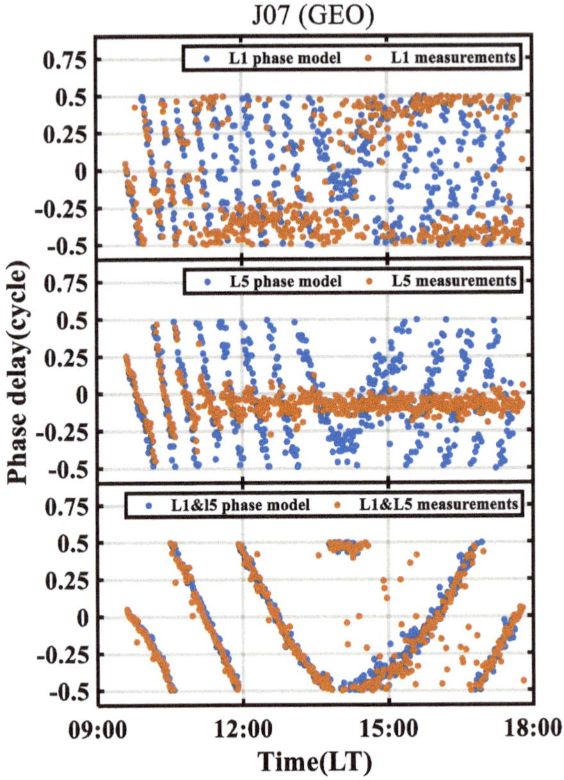

Fig. 30 The comparison between the phase delay measurements and the phase delay models of the L1 (top), L5 (middle), and the L1&L5 combination (bottom) from the J07 (Elev. 46°) satellite during the experiment

combination of L1 and L5 can be significantly improved than any of L1 and L5 under severe ocean conditions.

To carry out GNSS-R carrier phase altimetry, low elevation angle observations are usually selected. In the shore-based experiment, the reflected signals from J03 are blocked seriously by the land buildings owning to the low elevation angles, so it is difficult to obtain good observations under stormy sea states. Therefore, the phase delay measurements from J01 and J02 are exhibited in Fig. 31. The elevation angle range is from 75.5° to 46.1° for J01 and from 8.2° to 78.5° for J02 shown in Fig. 26. The phase altimetry performance under different elevation angles will be analysed.

For J01 satellites, the phase delay measurements of L1 and L5 are very different from the phase delay models, indicating that the interferometric phase measurements are not well extracted. The reflected signals of L1 and L5 are incoherent. The reason is that the elevation angles are higher than that of the GEO satellite, the sea surface is regarded as a rougher surface and the coherent components become smaller. For J02 satellites, when the elevation angles are small at the start of the observation before 12:00, the phase delay measurements of L1 and L5 are in good agreement with the phase delay models. With the elevation angle increasing, phase delay measurements are rapidly unavailable. The results show that the single-frequency carrier phase

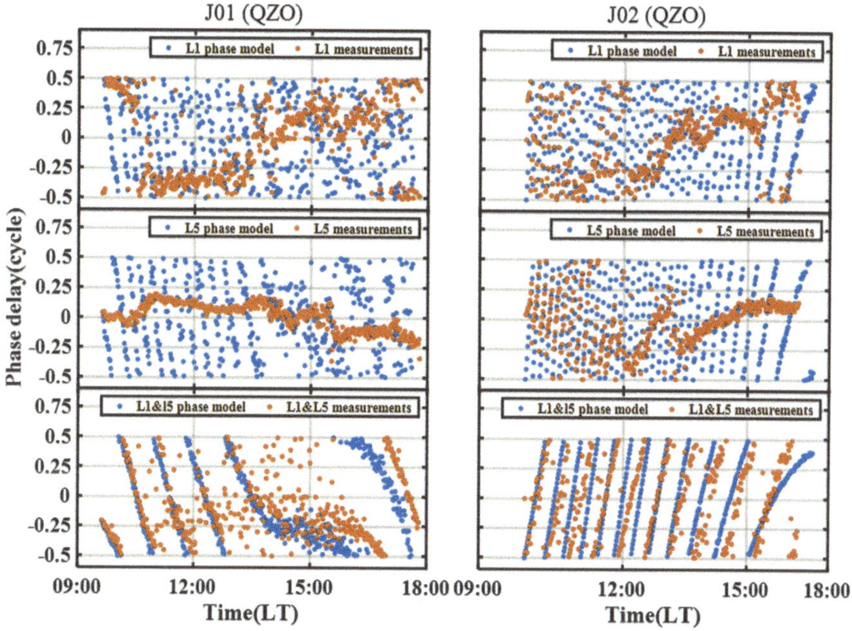

Fig. 31 The comparison between the phase delay measurements and the phase delay models from the J01 (Elev. 75.5° ~ 46.1°) and J02 (Elev. 8.2° ~ 78.5°) satellites during the experiment

altimetry can be implemented when the GNSS satellites' elevation angle is low and the sea surface is relatively smooth in the coastal condition. The limited elevation angle range constrains the application of the GNSS-R carrier phase altimetry.

On the contrary, the phase measurements of the L1 and L5 combination have been in good agreement with the combination phase delay models from the observations at any elevation angles. The main reason for the small mismatch is that the model cannot be established completely according to the tidal changes. According to the analysis, the carrier phase altimetry can be implemented with the combination of L1 and L5 with a higher elevation angle even though the sea surface is rough. The available elevation angles range for phase altimetry can be increased using the combination of L1 and L5.

Because the carrier phase altimetry of L1 and L5 is impossible under stormy sea states, the statistics based on the L1 and L5 combination phase measurements from four QZSS satellites and the in-situ SSH are listed in Table 2. We can see that the general accuracies are good except for the J01 satellite. Its accuracy is the worst among the four satellites, and more than 15 cm on account of its high elevation angle and the bad performance of the L1 under severe sea conditions shown in Fig. 31. According to the error propagation law and (18), the errors of the L1 and L5 combination measurements are larger than the respective errors of L1 and L5, the altimetry precision of the L1 and L5 combination is thus worse than any of L1 and L5. After the elimination of the gross errors, centimeter-level altimetry accuracy can

Table 2 The mean, STD, and RMSE between the l1 & l5 combination measurements from different satellites and the in-situ SSH under high sea states

Satellite	Mean (cm)	Std (cm)	RMSE (cm)
J07	− 0.64	3.44	3.49
J01	1.91	16.05	16.14
J02	0.19	4.41	4.40
J03	− 0.16	4.82	4.81

be achieved under severe sea states using the phase measurements of the L1 and L5 combination.

5 Conclusion

The utilization of next-generation GNSS signals enables centimeter-level GNSS-R altimetry retrieval on both ground-based and airborne platforms. This also demonstrates the effectiveness of the algorithms presented in this chapter. It will significantly promote the application of the technology in the field of water level monitoring.

For GNSS-R code level altimetry, the effectiveness and accuracy of drone-based GNSS-R multi-frequency sea surface altimetry were verified through experiments, and the results demonstrated that the inversion accuracy of most of the signals could reach the centimeter level. The results showed that greater accuracy can be achieved using satellites at higher elevation, and that there are differences in the satellite inversion effects of different satellite orbits. Specifically, the accuracy achievable using a medium Earth orbit satellite is higher than that achievable using a satellite in geosynchronous Earth orbit. Higher orbital altitude makes the signal propagation path longer and therefore more susceptible to interference. However, the accuracy of the satellite signal inversion results is not correlated with either the modulation of the signal or the altitude of the drone. Most of the errors can be eliminated through application of sliding average processing, and better smoothing results are obtained with a sliding average duration of 5 min. This approach can improve the results of the inversion from meter-level accuracy to centimeter-level accuracy. During the smoothing process, the numbers of effective measurements have substantial impact on the smoothing results; that is, greater numbers of effective data ensure more robust statistical processing and better smoothing results.

For GNSS-R phase altimetry, we improve the configuration of SDR by reconstructing ‘clean’ direct signals to compute phase differences between direct and reflected signals. The high frequency complex correlated values of L1 and L5 are obtained within the antenna coverage. The proposed method can reduce the correlated errors between the phase measurements of L1 and L5 under the proposed open-loop tracking. Thus, the carrier phase altimetry can be robustly implemented with dual-frequency signals, especially under stormy sea states. In addition, 60-s coherent integration is allowed to raise the SNR and reduce the statistic errors owning to a wide wavelength of the combined signal. Under stormy sea states, the phase measurements cannot be well defined from L1 and L5 separately, especially for the observations with high elevation angles. Yet, the solutions can be achieved on the basis of the L1 and L5 combination in severe sea conditions. The elevation angle range for phase altimetry is also increased based on the combination of L1 and L5. The results show that centimeter-level altimetry accuracy can be achieved in stormy sea states. Our method is also suitable for the new blocks of GPS satellites, which transmit L1 and L5 signals.

It should be noted that the centimeter level GNSS-R altimetry solutions presented in this chapter were obtained through post-processing using the software defined receiver. However, the real-time and accurate acquisition of water level data is critical for monitoring applications including floods and storm surges. Therefore, the relevant baseband signal processing algorithms must be refined, and the software upgraded accordingly.

References

1. Bai W, Xia J, Wan W et al (2015) A first comprehensive evaluation of China's GNSS-R airborne campaign: part II—River remote sensing. Sci Bull 17:61–68
2. Camps A, Park H, Domenech EV et al (2014) Optimization and performance analysis of interferometric GNSS-R altimeters: application to the PARIS IoD mission. IEEE J Sel Top Appl Earth Obs Remote Sens 7(5):1436–1451
3. Camps A, Park H et al (2017) GNSS-R altimetry performance analysis for the GEROS experiment on board the international space station. Sensors 17(7):1583. https://doi.org/10.3390/s17071583
4. Cardellach E, Rius A, Martin-Neira M et al (2014) Consolidating the precision of interferometric GNSS-R ocean altimetry using airborne experimental data. IEEE Trans Geosci Remote Sens 52(8):4992–5004
5. Cardellach E, Li W, Rius A et al (2019) First precise spaceborne sea surface altimetry with GNSS reflected signals. IEEE J Sel Top Appl Earth Obs Remote Sens 13:102–112
6. Cardellach E (2004) Carrier phase delay altimetry with GPS-reflection/occultation interferometry from low Earth orbiters. Geophys Res Lett 31(10):L10402
7. Carreno-Luengo H, Camps A, Ramos-Perez I et al (2014) Experimental evaluation of GNSS-reflectometry altimetric precision using the P(Y) and C/A signals. IEEE J Sel Top Appl Earth Obs Remote Sens 7(5):1493–1500
8. Clarizia MP, Ruf C, Cipollini P et al (2016) First spaceborne observation of sea surface height using GPS-reflectometry. Geophys Res Lett 43(2)
9. Cotton PD, Menard Y (2006) Future requirements for satellite altimetry: recommendations from the GAMBLE project for future mission and research programs. In: Proceedings of the symposium on 15 years of progress in radar altimetry, ESA SP-614
10. D'Addio S, Martin-Neira M (2013) Comparison of processing techniques for remote sensing of earth-exploiting reflected radio-navigation signals. Electron Lett 49(4):292–293
11. Egido A, Ruffini G, Caparrini M et al (2008) Soil moisture monitorization using GNSS reflected signals. arXiv:0805.1881. https://arxiv.org/abs/0805.1881
12. Gao F, Xu TH, Wang NZ et al (2020) A shipborne experiment using a dual-antenna reflectometry system for GPS/BDS code delay measurements. J Geod 94:88
13. Helm A, et al (2007) GORS—A GNSS occultation, reflectometry and scatterometry space receiver. In: Proceedings of the 20th international technical meeting of the Satellite Division of The Institute of Navigation, Fort Worth, TX, USA, Sep 2007, pp 2011–2021

14. Lowe ST, Zuffada C, Chao Y, Kroger P, Young LE, LaBrecque JL (2002) 5-cm-Precision aircraft ocean altimetry using GPS reflections. Geophys Res Lett 29:13. https://doi.org/10.1029/2002GL014759
15. Ribó S, Arco-Fernández JC, Cardellach E, Fabra F, Li W, Nogués-Correig O, Rius A, Martín-Neira M (2017) A software-defined GNSS reflectometry recording receiver with wide-bandwidth, multi-band capability and digital beam-forming. Remote Sens 9:450. https://doi.org/10.3390/rs9050450
16. Lestarquit L, Peyrezabes M, Darrozes J et al (2016) Reflectometry with an open-source software GNSS receiver: use case with carrier phase altimetry. IEEE J Sel Top Appl Earth Obs Remote Sens 9(10):4843–4853
17. Yang DK, Zhang YQ, Zhang QS, Li ZW (2006) Airborne ocean wind-field retrieval system based on GPS scattering signals. Acta Aeronaut Astronaut Sin 27(2):310–313
18. Tian Y et al (2019) Multifunctional GNSS-R processing software for GNOS II. In: IGARSS 2019–2019 IEEE international geoscience and remote sensing symposium, pp 8669–8671. https://doi.org/10.1109/IGARSS.2019.8897997
19. He Y, Gao F, Xu T, Meng X, Wang N, Ning B (2023) Coastal GNSS-R phase altimetry based on the combination of L1 and L5 signals under high sea states. J Geod 97:19
20. Meng X, Gao F, Xu TH et al (2024) Design of real-time GNSS-R software-defined receiver for coastal altimetry using GPS/BDS/QZSS signals. GPS Solut 28:20. https://doi.org/10.1007/s10291-023-01563-w
21. Martin-Neira M et al (2002) Altimetry precision of 1 cm over a pond using the wide-lane carrier phase of GPS reflected signals. Can J Remote Sens 2002

Chapter 3
Calibration of Spaceborne GNSS-R Measurements

Feixiong Huang and Yinghui Xue

Abstract Calibration is an important step in the spaceborne GNSS-R data processing. It converts the received signals from the receiver in raw digital counts to physical observables such as Normalized Bistatic Radar Cross Section (NBRCS) and reflectivity by removing their dependency on instrument, geometry and environmental changes. The accuracy of calibration determines the accuracy of follow-on geophysical parameter retrievals. This chapter presents the calibration of parameters for spaceborne GNSS-R measurements including receiver gain, GNSS Effective Isotropic Radiated Power (EIRP), antenna pattern, effective scattering area, and polarimetric factors. The calibration assessment methods are also introduced.

Keywords GNSS-R · Calibration · Spaceborne mission · EIRP · Antenna pattern

1 Introduction

1.1 Principle of Calibration

The Global Navigation Satellite System Reflectometry (GNSS-R) technology realizes remote sensing of the Earth's surface parameters (such as sea surface wind speed, soil moisture, sea ice parameters, etc.) by receiving Global Navigation Satellite System (GNSS) signals reflected from the surface. Spaceborne GNSS-R utilizes receivers on low-Earth orbit satellites, forming a bistatic radar system. The receiver's zenith antenna captures direct signals from GNSS satellites, while the nadir antenna receives signals reflected from the Earth's surface. Raw signals acquired by the receiver do not directly reflect surface physical information and are subject to spatial

F. Huang (✉) · Y. Xue
National Space Science Center, Chinese Academy of Sciences, Beijing, China
e-mail: huangfeixiong@nssc.ac.cn

Y. Xue
e-mail: xueyinghui217@163.com

University of Chinese Academy of Sciences, Beijing, China

K. Yu and N. Zheng (eds.), *Advances in GNSS Reflectometry*, Navigation: Science and Technology 19, https://doi.org/10.1007/978-981-92-2325-1_3

environment interference and instrument parameters. Therefore, in data processing, it is necessary to convert these raw signals into observables with physical meanings such as NBRCS or reflectivity. This process eliminates systematic instrument errors and external environmental disturbances, establishing quantitative relationships between raw signals and observables. The calibrated observables are then used to retrieve geophysical parameters such as sea surface wind speed and land soil moisture.

Therefore, calibration is the process of accurately converting the signal radiation intensity obtained by the receiver into quantitative observables with clear physical meanings through the establishment of models and algorithms. The received signal contains not only the raw signal power but also instrument thermal noise power and antenna thermal noise power [1]. Errors in the antenna pattern will affect the accuracy of signal power calibration. Therefore, calibrating data errors enables the establishment of a conversion relationship between the received raw count values and signal power. This allows the raw data to be converted into a unified unit of NBRCS or reflectivity, facilitating the analysis and comparison of observation data from different satellites and time periods. The accuracy of calibration determines the retrieval precision of surface physical parameters such as sea surface wind speed, soil moisture, and vegetation biomass.

A GNSS-R receiver aboard the low-earth orbit satellite tracks signals reflected from the Earth's surface, generating Delay-Doppler maps (DDM) in raw digital counts, namely GNSS-R Level 0 (L0) data. Subsequent Level 1 (L1) data calibration computes observables including NBRCS and reflectivity. Level 2 (L2) data refers to geophysical parameters derived from the L1 data [2].

1.2 Calibration Process

The L1 data calibration consists of Level 1a (L1) power calibration and Level 1b (L1b) observable calculation. L1a calibration converts raw L0 DDM counts into signal power in watts [3]:

$$P(\tau, f) = G \cdot C_r(\tau, f) \tag{1}$$

where $C_r(\tau, f)$ is the DDM in raw counts after subtracting the noise floor, $P(\tau, f)$ is the DDM power, and G is the total instrument gain.

For GNSS-R observations over the ocean, sea surface exhibits smoothness and uniformity, with a uniquely solvable specular point. After L1a calibration, the position of specular point is determined using the DTU10 mean sea surface model. A forward scattering power model is then employed to calculate the NBRCS. In contrast, land terrain is complex with non-unique specular points, and different surface types exhibit distinct scattering behaviors. For locating specular points over land, the WGS84 ellipsoidal Earth model is used to estimate initial values. These initial values are then projected onto the land surface using the 1 km resolution STRM digital elevation map [4].

The Level 1b (L1b) calibration converts signal power into NBRCS or reflectivity. When the measured specular point is over the sea, scattering is generally incoherent. Therefore, for incoherent scattering regions, the NBRCS at the specular point is calculated [5]:

$$\sigma_0 = \overline{P}_{sp} \left[\frac{\lambda^2 P_t G_t G_r}{(4\pi)^3 R_t^2 R_r^2} \right]^{-1} \overline{A}^{-1} \tag{2}$$

where $\overline{P}_{sp}$ is the average observed DDM power at the specular point, $\overline{A}$ is the Effective Scattering Area (ESA), R_t and R_r are the distances from the specular point to the transmitter and receiver, respectively, λ is the signal wavelength, $P_t G_t$ is the transmitter Effective Isotropic Radiated Power (EIRP) at the specular point, and G_r is the receiver antenna gain at the specular point.

When the specular point is over land, assuming the reflection is coherent, the major observable is the reflectivity Γ:

$$\Gamma = \frac{P_{peak} \cdot (4\pi)^2 (R_t + R_r)^2}{\lambda^2 P_t G_t G_r} \tag{3}$$

where P_{peak} is the DDM peak power.

The Power Ratio (PR), defined as the ratio of power in the inner and outer regions of the DDM, is commonly used to distinguish between coherent and incoherent scattering types:

$$PR = \frac{C_{in}}{C_{out}} \tag{4}$$

where, C_{in} is the total power in the small region near the specular point, and C_{out} is the total power in the remaining area of the DDM. For a smooth surface (lakes, rivers, low-roughness desserts), signal power is concentrated, resulting in a high PR value indicating coherent scattering; for rough surfaces (rough sea, forests, farmland), power is dispersed, leading to a low PR value indicating incoherent scattering [4].

To accurately calculate NBRCS and reflectivity, the GNSS-R calibration process requires good estimation of instrument gain, noise floor, transmitter EIRP, receiver antenna pattern, and effective scattering area. The overall calibration process for spaceborne GNSS-R measurements is illustrated in Fig. 1.

1.3 Calibration Methods of Different GNSS-R Missions

TechDemoSat-1 (TDS-1), an early GNSS-R technology demonstration satellite, relies primarily on pre-launch static measurements for calibration. Antenna patterns are measured in an anechoic chamber, instrument gain is set to a fixed value, and the transmitter's EIRP is assumed to be a constant value [6, 7]. The science team of CYGNSS spent a lot of efforts in the accurate calibration of GPS L1 C/A signals. For

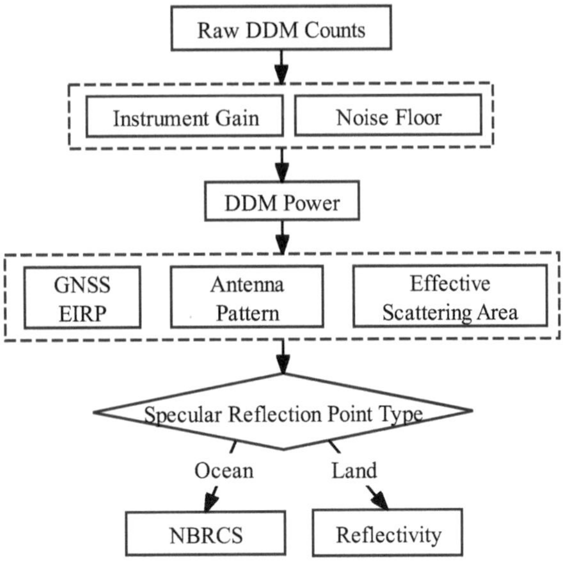

Fig. 1 Flowchart of spaceborne GNSS-R calibration process

the L1a calibration, it switches to a blackbody load every 60 s to calibrate instrument gain. For the L1b calibration, the zenith antenna receiving direct GPS signals was used for real-time dynamic GPS EIRP calibration, correcting errors in the ground-based static lookup-table (LUT). Additionally, it introduced a "bin ratio" method to correct digital sampling issues, achieving a total calibration error within 0.59–0.74 dB [2, 5, 8]. The GNOS-II payload onboard FY-3E is the first multi-GNSS operational payload. The Receiver gain was calibrated using a LUT established through prelaunch thermal cycling experiment. Antenna patterns for BDS, GPS, and GAL were measured separately, and GNSS EIRP was derived from a lookup table constructed using two static power monitors [3]. BuFeng-1 uses blackbody loads and microwave switch for power calibration, correcting the impact of LNA temperature drift on DDM power. It also plans to utilize ground-based BDS EIRP observations to improve multi-system calibration [9].

Spire satellite calibration is adapted to the characteristics of low-orbit constellations. The effective scattering is based on delay, Doppler, and altitude. EIRP is calibrated through joint observations by multiple satellites. Although it supports multi-GNSS signals, its calibration accuracy is slightly lower than that of CYGNSS [10, 11]. HydroGNSS, an upcoming European Space Agency Scout mission, innovatively incorporates dual polarization (LHCP/RHCP) and coherent channels in its instrument design. Pre-launch anechoic chamber measurements will characterize dual-frequency (L1/E1, L5/E5) antenna patterns. On-orbit calibration will use blackbody loads and ground targets to calibrate antenna noise, and it plans to calculate multi-GNSS EIRP biases based on sea surface observation data [12, 13]. FSSCAT, equipped with the FMPL-2 payload, calibrates instrument gain by measuring and correcting noise temperature and system gain drift in real time using internal thermal loads and Active Cold Loads (ACL). Antenna patterns were measured in a prelaunch anechoic chamber, with on-orbit correction of pattern biases through observations over uniform scenes such as sea surfaces and sea ice. The effective scattering area is dynamically adjusted based on the orbital altitude and incidence angle [14, 15].

2 Receiver Gain Calibration

When converting DDM digital quantities measured by the receiver into DDM power, accurate calibration of the receiver gain is required. The total gain of a GNSS-R receiver is influenced by the amplification of various components in the receiver chain and is related to the temperature of components like the Low-Noise Amplifier (LNA).

2.1 Prelaunch Ground Calibration

For the receiver gain calibration of FY-3/GNOS-II, prelaunch thermal cycling experiments were first conducted to measure the total receiver gain at different instrument temperatures, establishing a lookup table (LUT) of the total instrument gain in the function of instrument temperature and Digital Automatic Gain Control (AGC) attenuation [16]. During on-orbit operation, the real-time instrument temperature is measured to query the pre-established LUT and obtain the corresponding receiver gain G. This gain is then used to convert raw DDM counts $C(\tau, f)$ into DDM power $P(\tau, f)$. Figure 2 shows the LUT of the GNOS-II instrument gain as a function of temperature and AGC attenuation levels through prelaunch thermal cycling experiments.

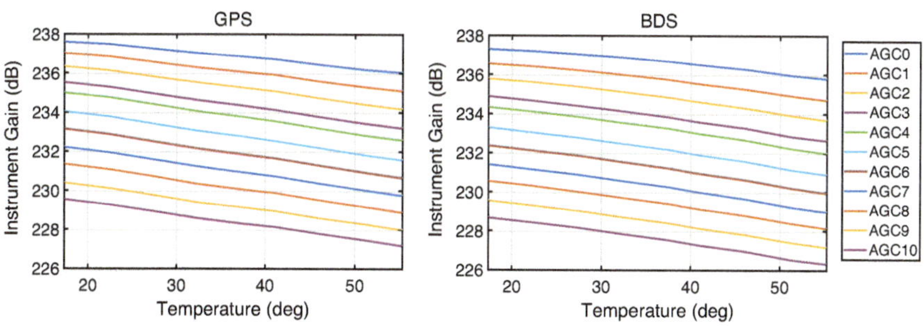

Fig. 2 The instrument gain of FY-3E/GNOS-II as a function of temperature and AGC attenuation levels measured from the prelaunch experiment

The Digital Automatic Gain Control (AGC) of the GNOS-II receiver operates in a fixed-gain mode, whose front-end gain does not change with input signal power. When encountering Radio Frequency Interference (RFI) on orbit, this can lead to quantization errors. To address this, an analog AGC operating in Automatic Gain Mode (AGM) is designed as a monitor for input signal power. When strong RFI signals appear, the analog AGC values can be used to mitigate interference from ground signals and Satellite-Based Augmentation Systems (SBAS) [16].

2.2 *On-Orbit Calibration*

The CYGNSS mission employs an onboard blackbody load for the receiver gain calibration, further reducing calibration errors [5]. Each receiver channel of the satellite incorporates a blackbody load at the LNA input. A switch controls whether the receiver input port connects to the antenna signal or the load signal, enabling monitoring and calibration of receiver gain during L1a calibration. A thermistor is used to accurately measure the temperature of the blackbody load resistor.

CYGNSS combines blackbody calibration and open-ocean calibration to achieve precise on-orbit estimation of the receiver gain and instrument noise power. The total signal power received by the receiver includes not only GPS signal power scattered from the sea surface but also thermal noise power generated by the antenna and receiver. The raw DDM counts generated by the Delay Doppler Instrument (DDMI) are linearly related to the total signal power:

$$C = G(P_a + P_r + P_g) \tag{5}$$

where C is the raw DDM count, G is the total instrument gain, P_a is the thermal noise power from the antenna, P_r is the thermal noise power from the instrument, and P_g is the scattered GPS signal power.

Signal processing first requires removing the influence of thermal noise power on the total signal. The raw noise signal count is expressed as:

$$C_N = G(P_a + P_r) \tag{6}$$

A microwave switch is used to receive signals from either the nadir antenna or the blackbody. When the switch is connected to the nadir antenna, the antenna thermal noise power can be expressed as:

$$P_a = kT_a B_W \tag{7}$$

where k is the Boltzmann constant, T_a is the brightness temperature of the upper atmosphere obtained by integrating over the receiving antenna pattern, and $B_W =$ 1000 Hz is the bandwidth for receiving GPS signals by the antenna.

When the switch is connected to the blackbody, there is no GPS signal input to the DDMI, only the blackbody load noise power:

$$P_B = kT_B B_W \tag{8}$$

where T_B is the effective temperature of the blackbody.

When the CYGNSS satellite is over open ocean, the received signal contains no scattered signals from land. The noise power from the ocean can be expressed as:

$$P_O = kT_O B_W \tag{9}$$

where T_O is the integrated beam brightness temperature of the antenna, which can be estimated using a microwave radiative transfer model based on environmental conditions.

The combination of measurements from the open ocean and blackbody allows for estimation of the instrument gain and thermal noise. The raw DDM noise counts under open ocean and blackbody states can be expressed as:

$$C_O = G(P_O + P_r) \tag{10}$$

$$C_B = G(P_B + P_r) \tag{11}$$

Combining Eqs. (10) and (11), the receiver gain G and the instrument noise power P_r can then be solved as:

$$G = \frac{C_O - C_B}{P_O - P_B} \tag{12}$$

$$P_r = \frac{P_O C_B - P_B C_O}{C_O - C_B} \tag{13}$$

By routinely switching to the blackbody, this process updates the gain in orbit, optimizing the receiver gain calibration.

Other than using a blackbody for on-orbit calibration, it can also be achieved by using a reference signal. The Spire mission improved its payload by designing a radio frequency coupler with multiple input channels. A known GNSS signal is input into the channels, and the gain and signal phase characteristics of each channel are inverted by measuring the response of this known signal in each channel [17].

3 GNSS EIRP Calibration

Effective Isotropic Radiated Power (EIRP) of GNSS satellites, defined as the product of satellite transmitting power and antenna gain, is a critical parameter determining the L1b calibration accuracy. Current calibration methods for GNSS EIRP include estimation based on ground power monitor and estimation based on real-time zenith antenna measurements. The ground-based method establishes a ground power monitoring system to establish LUTs of EIRP for each GNSS satellite. This method is relatively simple but yields incomplete GNSS antenna patterns, only obtaining data for off-boresight angles within 13°, and cannot account for dynamic changes of GNSS transmitting power such as the flex power event [18]. On-orbit real-time EIRP measurement using zenith antennas calibrates the EIRP at the specular point based on the power measurements at the receiver's zenith antenna. This method effectively adapts to real-time variations in the GNSS transmitting power but is affected by errors in the calibration of the direct channel and the zenith antenna pattern.

3.1 Estimation Based Ground Power Monitor

Wang et al. proposed the construction of a Ground-based Constellation Power Monitor (GCPM) system for GPS constellation power [19]. This system consists of two parts: outdoors, that receives zenith GPS signals using a passive choke ring antenna and integrates components like a thermal control chamber, hot and cold loads, noise source, and Low-Noise Amplifier (LNA) to ensure signal amplification and calibration stability; Indoors, a GPS receiver collects raw GPS signal counts, a data processing computer, load switching, and temperature control.

Figure 3 illustrate the process of the power calibration for a single GPS satellite, respectively. Radiometric calibration is first performed: a liquid nitrogen load is used to calibrate the brightness temperatures of the internal cold load and noise source of the system. The system is switched between six different states, and the base band output of the receiver is recorded to establish a linear relationship between the base

Fig. 3 Flowchart of single GPS satellite calibration process

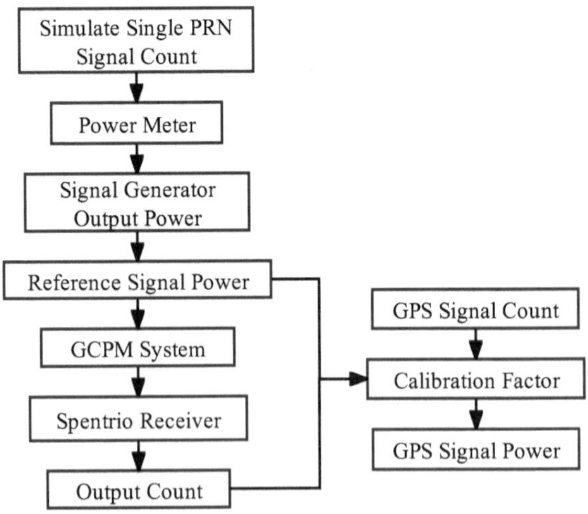

band output and system gain, determining the dynamic range of the system gain and calibrating the system gain. For single PRN power calibration, a GPS simulator generates a simulated single PRN signal, which is input into the GCPM thermally controlled box after passing through an attenuator. Parameters such as receiver gain, LNA gain, and temperature must be consistent with those when receiving real GPS signals. The input signal power at the GCPM input port is P_{sim}, and the output is the raw signal count C_{sim}. The input power P_{sim} is varied, and the corresponding output count C_{sim} is recorded. Within the linear range of the system, the ratio of the input power P_{sim} to the output count C_{sim} is constant, defined as the calibration factor G.

After calculating the calibration factor G, the raw count C_{GPS} of the real GPS signal can be converted to power P_R in Watts:

$$P_R = GC_{GPS} \tag{14}$$

A forward model is established based on the Friis transmission equation:

$$P_R = \frac{P_T G_T(\theta_T, \phi_T)}{4\pi R^2 L_a} \frac{\lambda^2}{4\pi} G_R(\theta_R, \phi_R) \tag{15}$$

where P_T is the signal transmit power, G_T is the transmit antenna gain, R is the distance between the transmitter and receiver, L_a is the atmospheric loss due to oxygen attenuation at the GPS L1 frequency, G_R is the receive antenna gain, and λ is the GPS L1 signal wavelength. According to Eq. (15), GNSS EIRP ($P_T G_T$) can be calculated. To determine the optimal transmit power P_T, a cost function is defined to minimize the difference between the forward model's predicted value and the GCPM's actual measured value:

$$\mathrm{cost}(P_T) = \sum_{t=t_1}^{t_N} \left[P_R^{Model}(P_T, t) - P_R^{Meas}(t)\right]^2 \tag{16}$$

where $P_R^{\mathrm{Meas}}(t)$ is the actual received power measured and obtained by the single PRN calibrated GCPM, $P_R^{\mathrm{Model}}(P_T, t)$ is the simulated received power based on P_T. Among all possible P_T values, the one that minimizes the cost function is selected as the optimal estimate of the transmit power of GPS satellite.

Using around 1 month of data, the transmit power P_T of the same GPS satellite was estimated multiple times and averaged to obtain the final transmit power value of the satellite. The estimation results indicate that the transmit power P_T varies among different types of GPS transmitters, with a difference of approximately 4 dB. The CYGNSS Cal/Val and Science team applied the estimated transmit power P_T to CYGNSS L1 calibration, successfully demonstrating that this method reduces the dependence of L1 NBRCS and L2 wind speed calibration on the PRN numbers of different GPS satellites [19].

A similar method has also been used to obtain EIRP LUTs of multiple constellations including GPS, Beidou (BDS), Galileo (GAL) and GLONASS for FY-3 and

Tianmu-1 missions [3, 20]. Figure 4 shows the EIRP profiles of each satellite from GPS, BDS, GAL and GLONASS.

3.2 On-Orbit Real-Time Calibration Using the Zenith Antenna

Calibration of GNSS satellite EIRP is primarily challenged by power variations among different GPS satellites, limitations on antenna gain direction, and asymmetry of the antenna pattern [21]. To address the issue caused by GPS satellite flex power mode (GPS satellites dynamically allocate transmit power to different frequencies to effectively reduce inter-system signal interference [22]), two methods were proposed based on on-orbit dynamic EIRP calibration using the zenith antenna of the GNSS-R

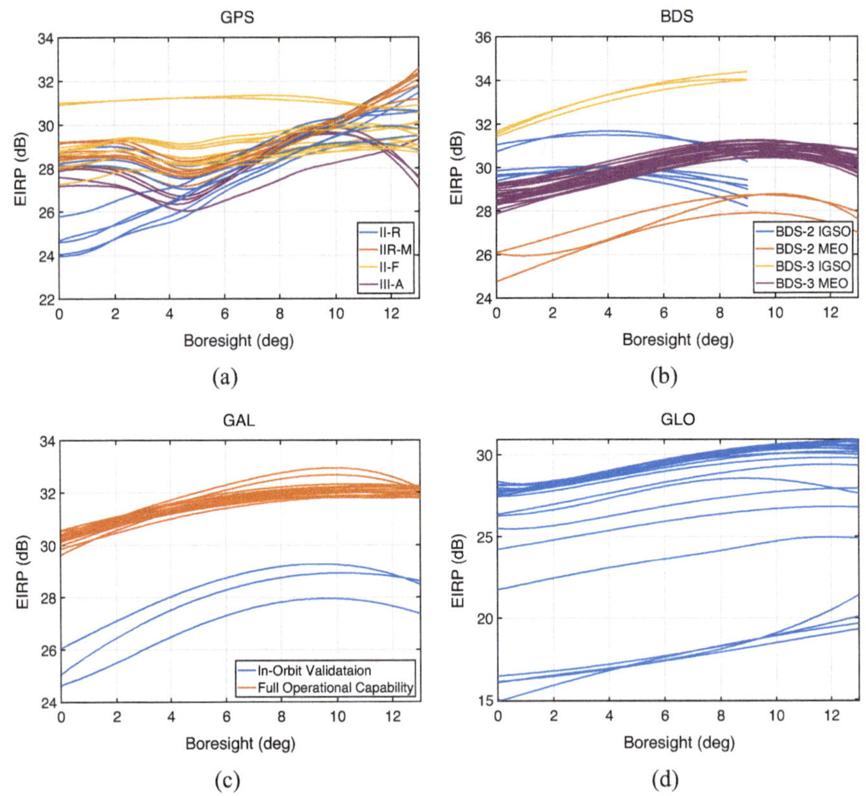

Fig. 4 The EIRP in a function of boresight angle for satellites from each GNSS constellation: **a** GPS, **b** BeiDou, **c** Galileo and **d** GLONASS. Each line stands for the profile of one satellite

receiver: the Zenith-to-Specular Ratio Method [18] and Normalized Antenna Pattern Method [23].

Zenith-to-Specular Ratio Method

The Zenith-to-Specular Ratio (ZSR) method was proposed to estimate GNSS EIRP for the CYGNSS mission [23]. The raw counts from the zenith GNSS signal and the LNA gain are used to obtain the output power P_R at the zenith antenna. Using the Friis equation, the EIRP in the zenith direction is calculated. Zenith-to-Specular Ratio (ZSR) is defined as the ratio of the antenna gain in the zenith direction to that in the specular point direction, i.e., the ratio of EIRP in the zenith direction to that in the specular direction. The EIRP at the specular point can then be calculated using ZSR and the zenith direction EIRP, achieving dynamic EIRP calibration. Figures 5 and 6 show the diagram and flowchart for calibrating the EIRP at the specular point using the ZSR method, respectively. The detailed steps are as follows.

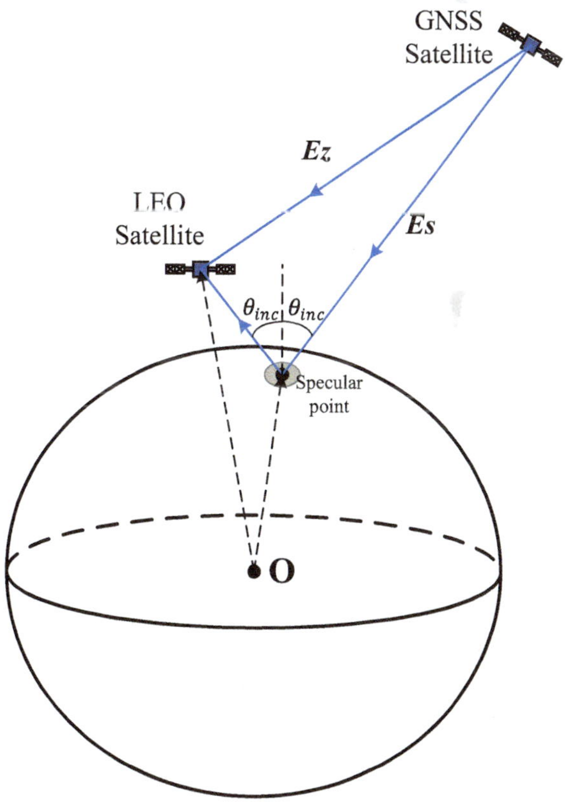

Fig. 5 Diagram of Zenith-to-specular ratio method for GNSS EIRP calibration

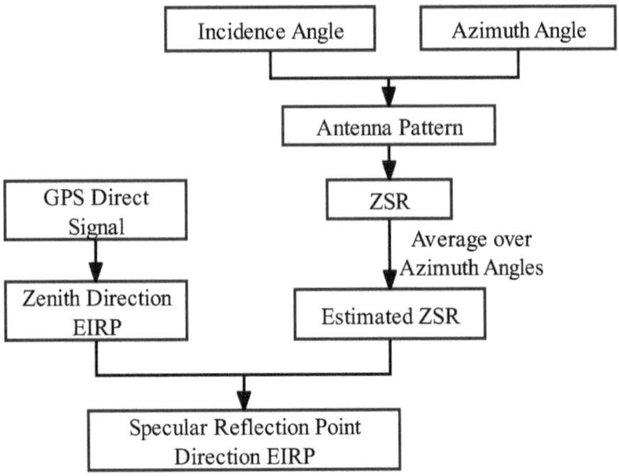

Fig. 6 Flowchart of Zenith-to-specular ratio method for GNSS EIRP calibration

(1) Calibration of the zenith channel: An expression was first developed to convert the raw signal counts C_Z from the receiver's zenith channel into the signal power P_Z at the receiver input port:

$$P_Z = f(C_Z) \tag{17}$$

A LUT for the LNA gain versus temperature was also established by prelaunch experiments. Then, the signal power P_R at the zenith antenna output port can be calculated based on the signal power P_Z at the receiver input port:

$$P_R = \frac{P_Z}{G_{LNA}(T_{LNA})} \tag{18}$$

where T_{LNA} is the real-time temperature of the LNA, and G_{LNA} is the corresponding LNA gain obtained from the prelaunch determined LUT.

(2) Estimation of zenith direction EIRP: The CYGNSS antenna pattern was measured in an anechoic chamber before satellite launch. Then, using in-orbit GPS signals with constant transmit power over time, the actual antenna gain pattern is determined. According to the Friis transmission equation, the EIRP at the receiver of the GPS satellite received by the CYGNSS satellite's zenith antenna is calculated as:

$$E_Z = \frac{(4\pi R)^2}{\lambda^2} \frac{P_R}{G_R(\theta_R, \phi_R)} \tag{19}$$

where R is the distance from the GPS satellite to the CYGNSS satellite, $\lambda = 0.19$ m is the GPS signal wavelength, P_R is the GPS signal power calculated from Eq. (18), and $G_R(\theta_R, \phi_R)$ is the zenith antenna gain obtained by looking up the antenna pattern based on θ_R (off-boresight angle) and ϕ_R (azimuth angle).

(3) Definition of ZSR function: ZSR is defined as the ratio of the EIRP in the zenith direction to that in the specular point direction:

$$ZSR \triangleq \frac{E_Z}{E_S} = \frac{E(t, \theta_Z, \phi_Z)}{E(t, \theta_S, \phi_S)} \tag{20}$$

where E_Z is the zenith direction EIRP, E_S is the specular point EIRP, θ_Z and ϕ_Z are the off-boresight and azimuth angles of the receiver relative to the CYGNSS satellite, and θ_S and ϕ_S are the off-boresight and azimuth angles of the receiver relative to the specular point.

EIRP of the transmitting satellite is defined as the product of transmit power P_T and antenna gain G_T:

$$E(t, \theta, \phi) = P_T(t)G_T(\theta, \phi) \tag{21}$$

For the same GPS satellite, the transmit power $P_T(t)$ is the same in both the zenith and specular point directions at time t, allowing ZSR to be simplified to the ratio of antenna gains, eliminating dependence on transmit power:

$$\text{ZSR} = \frac{P_T(t)G_T(\theta_Z, \phi_Z)}{P_T(t)G_T(\theta_S, \phi_S)} = \frac{G_T(\theta_Z, \phi_Z)}{G_T(\theta_S, \phi_S)} \tag{22}$$

Based on geometric relationships, the azimuth angles (ϕ_Z, ϕ_S) for the zenith and specular directions are approximately equal, and the elevation angles (θ_Z, θ_S) have a mapping relationship with the specular point incidence angle θ_{inc}. Therefore, ZSR can be approximated as a function depending only on θ_{inc} and the azimuth angle ϕ:

$$\text{ZSR}(\theta_Z, \theta_S, \phi_Z, \phi_S) \simeq \frac{G_T(\theta_Z(\theta_{\text{inc}}), \phi)}{G_T(\theta_S(\theta_{\text{inc}}), \phi)} \tag{23}$$

To further eliminate dependence on the GPS satellite's yaw state, ZSR is averaged over all azimuth angles, resulting in an estimate $\widehat{\text{ZSR}}$ that depends only on the specular incidence angle, $\text{ZSR}(\theta_{\text{inc}})$.

(4) Specular point EIRP estimation: Based on the GPS satellite's EIRP in the zenith antenna direction and the zenith-to-specular power ratio $\text{ZSR}(\theta_{\text{inc}})$, the GPS satellite's EIRP in the specular point direction is estimated:

$$E_S = \frac{E_Z}{\text{ZSR}(\theta_{\text{inc}})} \tag{24}$$

Substituting Eqs. (19) into (24) yields:

$$E_S = \frac{(4\pi R)^2}{\lambda^2} \frac{P_Z}{G_{\text{LNA}}(T_{\text{LNA}})G_R(\theta_R, \phi_R)\text{ZSR}(\theta_{\text{inc}})} \tag{25}$$

By calculating the Root Sum Square (RSS) of various error sources using CYGNSS data, the mean squared error of the estimated GNSS satellite EIRP at the specular point was calculated to be around 0.3 dB. Monte Carlo simulation also estimated a similar error, validating that the dynamic estimation method for GNSS satellite EIRP significantly improves calibration accuracy [19].

Normalized Antenna Pattern Method

The Normalized Antenna Pattern method was firstly proposed for the FY-3E mission [23]. By dynamically estimating the EIRP at the specular point in orbit using GNSS satellite direct signals [23], it overcomes the limitation of static GNSS satellite EIRP measurements being restricted to limited off-boresight angles. This enables the construction of normalized antenna patterns covering the full off-boresight angle range, thereby utilizing all GNSS-R observation data. Figures 7 and 8 present the diagram and flowchart for the Normalized Antenna Pattern method, respectively. The detailed steps are as follows.

(1) Direct signal power calibration: The GNSS direct signal is tracked using the Precision Orbit Determination (POD) module to obtain its raw count value C_d, After subtracting the noise floor, it is converted to the actual power value P_d

Fig. 7 Diagram of normalized antenna pattern method for GNSS EIRP calibration

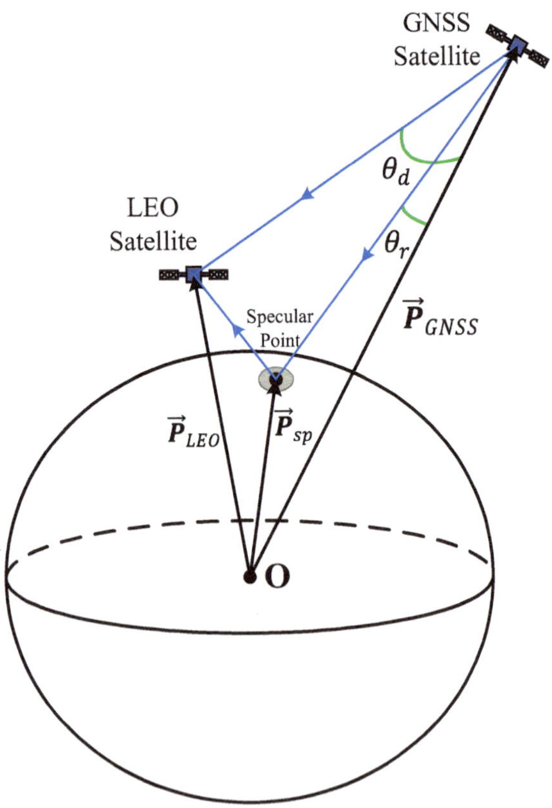

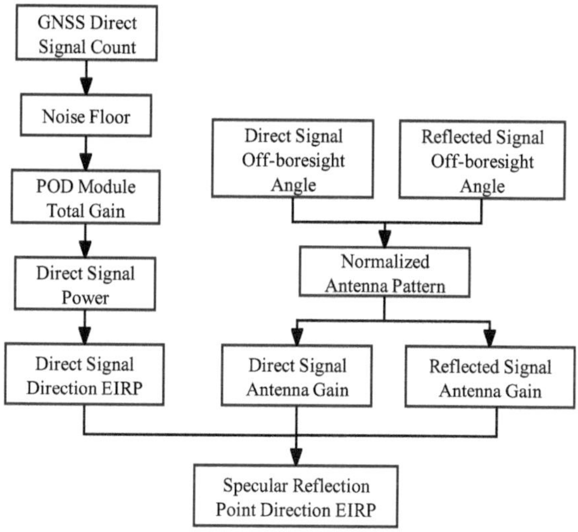

Fig. 8 Flowchart of normalized antenna pattern method for GNSS EIRP calibration

using the POD module's total gain G_{POD}:

$$P_d = G_{\text{POD}}(T_{\text{POD}}) \cdot (C_d - \eta_d) \tag{26}$$

where T_{POD} is the temperature of the POD module's RF front-end, and G_{POD} is obtained from a look-up table based on the empirical relationship between G_{POD} and T_{POD} established through prelaunch thermal cycling tests.

(2) Calculation of GNSS EIRP in the direct signal direction:

$$P_t G_{t,d} = P_d \cdot \left(\frac{\lambda^2 G_d}{(4\pi R_d)^2} \right)^{-1} \tag{27}$$

where λ is the GNSS signal wavelength, G_d is the zenith antenna gain in the direct signal direction, and R_d is the distance from the GNSS satellite to the LEO satellite.

(3) Calculation of GNSS EIRP in the specular point direction: Using the established normalized GNSS antenna pattern, convert the EIRP in the direct direction to the EIRP in the specular point direction:

$$P_t G_{t,r} = P_t G_{t,d} \cdot \frac{\overline{G_t}(\theta_r)}{\overline{G_t}(\theta_d)} \tag{28}$$

where $\overline{G_t}$ is the normalized antenna pattern of the corresponding GNSS satellite. θ_r is the off-boresight angles of the reflected in the GNSS antenna coordinate system,

$$\theta_r = \text{acos}\left(\frac{-P_{GNSS}}{|P_{GNSS}|} \cdot \frac{P_{SP} - P_{GNSS}}{|P_{SP} - P_{GNSS}|}\right) \tag{29}$$

θ_d is the off-boresight angles of the direct signals in the GNSS antenna coordinate system,

$$\theta_d = \text{acos}\left(\frac{-P_{GNSS}}{|P_{GNSS}|} \cdot \frac{P_{LEO} - P_{GNSS}}{|P_{LEO} - P_{GNSS}|}\right) \tag{30}$$

where P_{GNSS} is the position vector of the GNSS satellite, P_{SP} is the position vector of the specular point, and P_{LEO} is the position vector of the LEO satellite as illustrated by Fig. 7.

4 Antenna Pattern Calibration

An antenna pattern describes the variation of antenna gain with azimuth angle and off-boresight angle. In GNSS-R missions, calibration of the incident signal power largely depends on the accuracy of the patterns of the transmitting and receiving antennas. Antenna patterns are mainly obtained through ground and on-orbit measurements: initial pattern is measured in an anechoic chamber, and then corrected using on-orbit data.

The antenna pattern of TechDemoSat-1 was determined through prelaunch measurements, but its accuracy was limited [24]. The antenna patterns for GPS (1575 MHz) and BDS (1561 MHz) of FY-3 series were measured separately in an anechoic chamber and are shown in Fig. 9.

Gleason et al. obtained the initial pattern through electromagnetic simulation and measurements of the CYGNSS Engineering Model (EM) installed in an anechoic

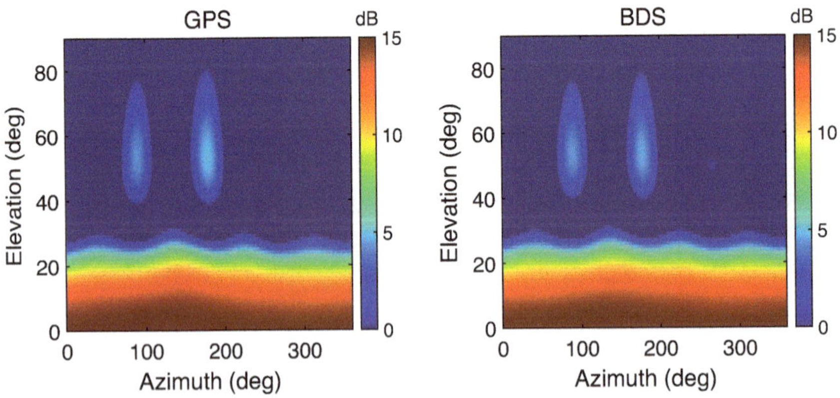

Fig. 9 The receiver antenna patterns of FY-3F/GNOS-II

chamber before launch. Considering the impact of solar panels on the antenna pattern, further optimization of the antenna pattern estimation using on-orbit data is required [2]. The anomaly of the NBRCS in azimuth ($\tilde{\sigma}_0$) is analyzed:

$$\tilde{\sigma}_0 = \frac{\sigma_0 - \langle\sigma_0\rangle}{\langle\sigma_0\rangle} \tag{31}$$

where σ_0 is the measured NBRCS, and $\langle\sigma_0\rangle$ is the average value of all σ_0 observations.

Averaging the anomalies by 1° azimuth angle revealed obvious systematic deviations, with the maximum deviation reaching 40%. Therefore, it is assumed that the deviations are mainly caused by inaccurate antenna patterns, and on-orbit correction of the antenna patterns is necessary. For each CYGNSS antenna, the variation of anomalies with azimuth angle and off-boresight angle is calculated, the data is smoothed using a 5° sliding window, and a 0.1° resolution pattern correction map is obtained through nearest neighbor interpolation. The corrected pattern is superimposed on the original pattern in logarithmic form (dB) to obtain the modified antenna pattern. After correction, the initial antenna pattern anomalies greater than 40% are reduced to less than 20%, and the dependence on azimuth angle is decreased [2]. This method uses ECMWF reanalysis wind speed as a reference, introducing uncertainties. The corrected antenna pattern still has an error of approximately 10%, with certain limitations. The antenna patterns measured on the ground have asymmetry, and the yaw attitude of GPS satellites in on-orbit operation will affect the gain of the transmitting antenna, leading to inaccurate gain calibration [25, 26].

Wang et al. proposed a joint calibration method for the GPS transmit antenna and CYGNSS receive antenna patterns based on the precise orbit and attitude data of GPS and CYGNSS satellites, which allows for the measurement of geometric relationships [27]. The orbit altitude is higher than the troposphere, so only ionospheric scintillation needs to be considered, forming an ideal measurement environment. Both the GPS transmitting antenna and the CYGNSS receiving antenna use the Right-Hand Circular Polarization (RHCP) component for calibration. The eight CYGNSS satellites provide more sufficient sampling of the GPS antenna pattern angles; the sampling density from eight satellites in one day is higher than that from ground stations in 100d. Finally, the sampling resolution of the GPS antenna pattern is 0.5° × 0.5°, and the angular resolution of the CYGNSS antenna pattern is 1° × 1°, with sampling concentrated near azimuth angles of 90° and 270°. Therefore, a spaceborne antenna range can be constructed to jointly measure the GPS transmitting antenna and CYGNSS receiving antenna patterns on orbit [27]. Using a pre-calibrated LNA LUT, the power P_Z at the receiver input port is converted to the received power P_R at the antenna output port. Using precision orbit determination software such as GipsyX, the geometric relationship between each pair of GPS satellites and CYGNSS satellites is calculated, mapping each power measurement to the corresponding direction in the GPS antenna coordinate system and CYGNSS antenna coordinate system to obtain the antenna pattern.

For the receiving antennas of eight CYGNSS satellites, first select seven GPS satellites with stable transmit power and calculate corresponding antenna pattern for each GPS satellite using the Friis transmission equation:

$$G_R(\theta_R, \phi_R) = \frac{(4\pi R)^2}{\lambda^2} \frac{P_R}{P_T G_T(\theta_T, \phi_T)} \tag{32}$$

For each measured signal power P_R, the corresponding antenna gain value G_R is calculated. Antenna patterns of GPS satellites were then developed based on a large amount of data.

To quantify the consistency between G_R patterns estimated from different GPS satellites, the normalized integral gain G_{NI} was then calculated:

$$\begin{aligned} G_{NI} = \sum_{\theta} \Bigg[& \sum_{\phi} G_R(\theta_i, \phi_i) \sin\theta_i \\ & + \sum_{\phi} G_R(\theta_i, \phi_i) \sin\theta_i \Bigg] \Delta\theta \Delta\phi \end{aligned} \tag{33}$$

The consistency of the calculation results from different GPS satellites is evaluated through the normalized integrated gain G_{NI}. If the deviation of G_{NI} between satellites exceeds 0.2 dB, it is considered to be caused by the inaccuracy of the initially assumed power P_T. By Adjusting the GPS transmit power, the G_{NI} estimated by all seven GPS satellites converge to a consistent level. Finally, the calculation results of the 7 GPS satellites are weighted averaged to generate a high-resolution on-orbit antenna pattern for a single CYGNSS satellite.

Using the obtained CYGNSS antenna pattern as a known quantity, for each GPS satellite, the transmit antenna pattern of the GPS satellite is inversely estimated through the Friis equation using measurement data from all 8 CYGNSS satellites:

$$G_T(\theta_T, \phi_T) = \frac{(4\pi R)^2}{\lambda^2} \frac{P_R}{P_T G_R(\theta_R, \phi_R)} \tag{34}$$

After obtaining the GPS antenna pattern corresponding to a single CYGNSS satellite, the estimation results of the eight CYGNSS satellites are weighted averaged to generate the final GPS antenna pattern. The on-orbit calculated antenna pattern is basically consistent with the prelaunch pattern, meeting the CYGNSS L1 calibration requirements.

5 Effective Scattering Area Estimation

5.1 Definition of Effective Scattering Area

Under incoherent scattering condition, for each time delay and Doppler frequency, the signal received by the GNSS-R receiver originates from scattered signals over a specific area of the ocean surface. When calculating NBRCS as Equation (2), it is necessary to accurately estimate the size of this scattering area, which is also affected by the ambiguity function of the GNSS signal. The accuracy of the effective scattering area (ESA) directly determines the accuracy of the NBRCS (σ_0) [4].

The ESA is different from the physical scattering area. Physical scattering area refers to the geometric area of the surface region that actually contributes to the received power within a DDM unit:

$$A_{\tau,f} = \iint_A dx\,dy \tag{35}$$

where the integration region A is the physical surface area corresponding to the DDM unit.

From the physical scattering area, the ESA further considers the effect of the GNSS spreading function. Therefore, it is obtained by a weighted integration of the delay and Doppler spreading function [5, 28]:

$$A_{\tau,f} = \iint_A \Lambda^2_{\tau;x,y} S^2_{f;x,y} dx\,dy \tag{36}$$

where $\Lambda_{\tau;x,y}$ is the delay spreading function, $S_{f;x,y}$ is the Doppler spreading function, and $\Lambda^2_{\tau;x,y} S^2_{f;x,y}$ is the Woodward ambiguity function.

5.2 Estimation Methods

In the CYGNSS mission, LUT for the ESA is generated through an end-to-end simulator during the prelaunch phase to provide an initial estimation for on-orbit calibration [4]. A high-resolution grid is constructed on the surface near the specular point. Based on the precise orbit and geometric attitude of the satellite and receiver, the path delay and Doppler frequency are calculated for each point in the grid, and the ESA of each DDM bin is obtained through the integration.

The LUT is a function of incidence angle and azimuth angle. Due to the elliptical orbit of CYGNSS satellites, which causes fluctuations in orbital altitude, CYGNSS version 2.1 added the orbital altitude dimension to the ESA LUT to correct errors caused by the satellite's elliptical orbit.

Initially, the forward binning scheme was used to calculate the ESA, which overestimated the ESA by 22%. The improved version 3.1 adopted central binning scheme, which halved the sampling error and achieved more accurate calculation of the ESA. The forward binning defines a DDM bin as the surface region with delay and Doppler frequency within the ranges $[\tau, \tau + \Delta\tau]$ and $[f, f + \Delta f]$, while the central binning defines each DDM bin as the range $[\tau - \Delta\tau/2, \tau + \Delta\tau/2]$ and $[f - \Delta f/2, f + \Delta f/2]$. This improvement reduces the maximum delay-Doppler sampling error from $\Delta\tau$ and Δf to $\Delta\tau/2$ and $\Delta f/2$, significantly improving the accuracy of ESA estimation [4].

5.3 Comparison and Analysis of ESA

Since the effective scattering area ESA is related to the chipping rate and modulation of GNSS signals, and each system has different signal chipping rates and modulation methods, there are differences in the calculation of ESA for different systems. Using the same ESA calibration parameters for different GNSS systems or different satellite types will introduce certain errors. Therefore, relationship diagrams of ESA versus incident angle at the DDM specular point for different GNSS signals needs to be established to generate empirical LUT for real-time signal calculation.

From the FY-3E/GNOS-II mission, a grid with 1 km resolution is constructed. According to the orbit information of FY-3E and GNSS satellites, delay, Doppler frequency, and ambiguity function value of each grid point are calculated, and the ESA of each DDM bin is obtained through Eq. (36). The ESA distributions of BDS MEO, BDS IGSO, and GPS under different incident angles are calculated within a specific range. The chip length of the BDS B1I signal (146.6 m) is only half of that of the GPS L1 C/A signal (293.3 m).

Figure 10 shows the ESA at the specular bin as a function of incidence angle from GPS, BDS and GAL for FY-3E. Figure 11 shows the ESA at different delay and Doppler bin as incidence angle of 30°. At the same incidence angle, the ESA of BDS is significantly smaller than that of GPS, and the difference between the two gradually increases with the increase of incident angle. Comparing the MEO and IGSO satellites of BDS-2/BDS-3, it is found that although IGSO satellites have higher orbits, their ESA is similar to that of MEO satellites, indicating that the influence of GNSS transmitter altitude on ESA is negligible [3, 29]. Furthermore, analysis shows that for high chipping rate systems such as BDS, the effective scattering area is relatively low; the GAL system uses BOC modulation, and its ESA is lower than that of GPS and BDS systems using BPSK modulation [29]. The size of ESA directly affects the signal-to-noise ratio (SNR) and observation quality of DDM. A smaller ESA means higher spatial resolution but may also lead to lower SNR.

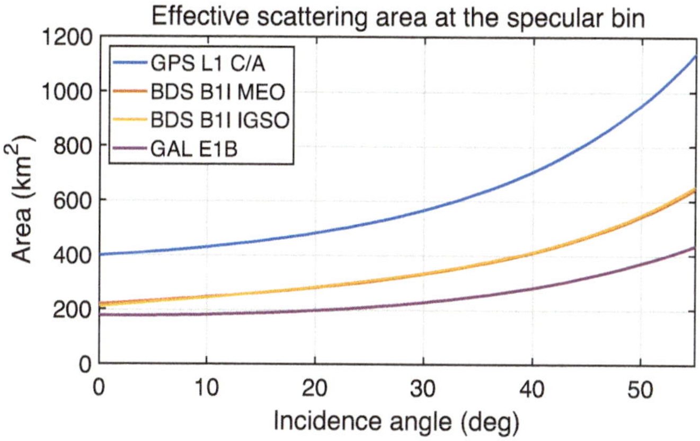

Fig. 10 The effective scattering area at the specular bin as a function of incidence angle from GPS, BDS and GAL for FY-3E

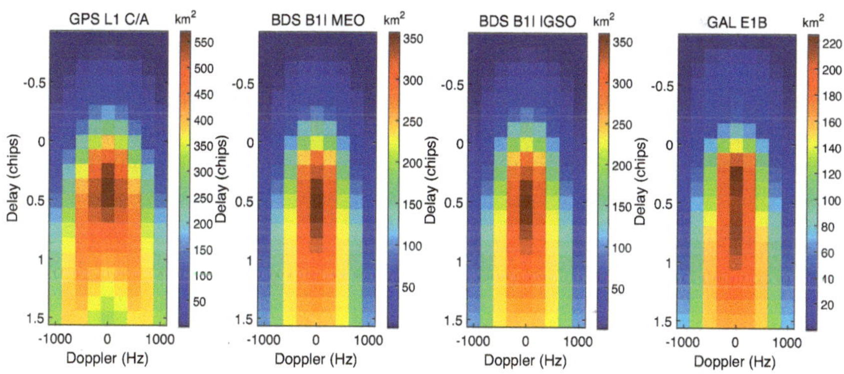

Fig. 11 The effective scattering area at different delay and Doppler bin as incidence angle of 30° from GPS, BDS and GAL for FY-3E

6 Polarization Calibration

Polarization calibration in spaceborne GNSS-R aims to eliminate the effect of system non-idealities (such as transmitter elliptical polarization, receiver channel crosstalk, and Faraday rotation) on polarization observations. It converts raw measurements into Stokes parameters and scattering matrix elements that are directly related to surface scattering characteristics. Based on the electromagnetic wave polarization transmission model and matrix transformation theory, systematic errors are inverted and compensated by establishing a mathematical relationship between observations and the true surface polarization response [30–32].

The Stokes parameters are a four-dimensional vector describing the polarization state of electromagnetic waves, defined as:

$$\mathbf{S}=\begin{pmatrix}S_0\\S_1\\S_2\\S_3\end{pmatrix}=\begin{pmatrix}\langle|E_X|^2+|E_Y|^2\rangle\\\langle|E_X|^2-|E_Y|^2\rangle\\2\langle\mathrm{Re}\left(E_XE_Y^*\right)\rangle\\2\langle\mathrm{Im}\left(E_XE_Y^*\right)\rangle\end{pmatrix}\tag{37}$$

where E_X and E_Y are the complex signal voltages received from two orthogonally polarized antennas, respectively; $\langle\cdot\rangle$ denotes time averaging or statistical averaging to remove random noise and obtain stable polarization characteristics. S_0 is the total power, S_1 is the linear polarization difference, S_2 is the diagonal or anti-diagonal linear polarization difference, and S_3 is the circular polarization difference.

The scattering matrix characterizes the surface polarization scattering properties. For the forward scattering scenario of GNSS-R, the reciprocity $S_{hv}=-S_{vh}$ is satisfied, and the matrix form is:

$$S=\begin{pmatrix}S_{hh}&S_{hv}\\-S_{hv}&S_{vv}\end{pmatrix}\tag{38}$$

where S_{hh} and S_{vv} are the co-polarization components, and S_{hv} is the cross-polarization component.

Due to the non-idealities of the transmitter caused by the axial ratio and non-ideal phase of GPS signals, Faraday rotation and polarization alignment errors will cause coupling between Stokes parameters, as well as antenna gain imbalance and cross-polarization response. Mueller matrices are used to correct these non-idealities:

$$\mathbf{S}^c=M_{corr}\cdot\mathbf{S}\tag{39}$$

$$M_{corr}=M_{tx}^{-1}\cdot M_{rot}^{-1}\cdot M_{rx}^{-1}\tag{40}$$

where M_{tx}, M_{rot}, and M_{rx} are the Mueller matrices for the transmitter, rotation, and receiver, respectively.

Moisture Active Passive-Reflectometry (SMAP-R) collects bistatic radar measurement data by switching its radar receiver to the GPS L2C frequency band. It is the world's first dual-polarimetric GNSS-R mission [33]. The least squares fitting between the Stokes parameters measured by SMAP-R and the reference SM data is used to invert the cross-polarization component S_{hv}, with the objective function:

$$J=\sum_{x=1,2,3}\left|\tilde{S}_x^c-\tilde{S}_x^M\right|^2\tag{41}$$

where $\tilde{S}_x^M$ are the Stokes parameters simulated by the scattering model.

By correcting systematic errors using Mueller matrices and combining measured Stokes parameters with the surface scattering model, the global estimation of the cross-polarization component in the forward scattering direction is realized. This method significantly improves the inversion accuracy of parameters such as soil moisture, but it also has problems such as dependence on reference data and high model complexity.

The calibration of polarimetric GNSS-R observations is complicated and is still an ongoing research area [34]. It is important to accurately calibrate measurements for polarimetric systems such as Tianmu-1 [20] and HydroGNSS [12].

7 Calibration Assessment

7.1 *Significance of Calibration Assessment*

Assessment of calibration quality is important for Level 2 (L2) parameter retrieval. Sufficiently high Level 1 (L1) calibration accuracy ensures accurate inputs for L2 retrieval. Insufficient calibration accuracy will cause systematic biases and random errors in L1 data which will propagate to the L2 process, leading to deviations in the retrieved geophysical parameters. Therefore, it is necessary to quantify the error of each component in the calibration process and the total calibration error [35–37].

7.2 *Methods*

Assessment Based on Error Propagation Model

Assessment based on the error propagation model first estimates the error of each calibration parameter individually. Assuming all errors are independent, the contribution of each error to the total error is computed using the partial derivative method, and the total error is obtained via the Root Sum Square (RSS) method [2, 5, 38].

For a calibration formula $Z = f(q_1, q_2, \ldots, q_n)$, the error caused by each error source q_i is defined as:

$$E(q_i) = \left| \frac{\partial Z}{\partial q_i} \right| \Delta q_i \tag{42}$$

And the total error is then calculated as:

$$\Delta Z = \sqrt{\sum_{i=1}^{n} [E(q_i)]^2} \tag{43}$$

Based on the calibration formulas (42) and (43), the main error sources are analyzed. Error components in the Level 1a (L1a) mainly include

- Instrument noise power estimation error (ΔP_r)
- Blackbody load power error (ΔP_B)
- Noise floor error (ΔC_N).

These errors directly affect the accuracy of extracting effective power from raw signals. Error components in the Level 1b (L1b) includes

- Uncertainty in GPS EIRP ($\Delta P_t G_t$): one of the largest error sources, mainly arising from imprecise calibration of GPS satellite transmit power and antenna pattern.
- Antenna gain error (ΔG): jointly caused by on-orbit antenna pattern errors, attitude determination deviations, and ground measurement errors.
- Effective scattering area error (ΔA): depends on the accuracy of the simulator and is related to the uncertainty of Earth models and scattering models.
- Specular point positioning error: caused by inaccurate orbit, timing, and Earth models, leads to incorrect extraction of the DDMA region and introduces significant errors.

Gleason et al. assessed the overall calibration accuracy of CYGNSS NBRCS to be 0.39 dB [2]. The analysis indicates that GPS transmit EIRP and receive antenna gain are the primary sources of error, and the impact of calibration errors varies with wind speed. At low wind speeds (wind speed < 5 m/s): NBRCS is highly sensitive to wind speed, and the Geophysical Model Function (GMF) has a large slope. The same calibration error results in a wind speed retrieval error of only 0.5 m/s. At high wind speeds (wind speed > 20 m/s): The GMF slope decreases, and the same calibration error leads to a wind speed retrieval error increasing to 3 m/s.

Assessment Based on On-Orbit Data

Assessment based on on-orbit data leverages the fact that measurements from adjacent GNSS-R satellites should exhibit minimal differences. By comparing data from adjacent satellites and analyzing their discrepancies, the overall performance is evaluated. For CYGNSS satellites, data pairs are selected that meet the following criteria: measurements from two satellites of the same GPS satellite, with a time difference of 10 min, a specular point difference of 5 km, and an observation incidence angle difference of 1°. A total of 200,000 measurement pairs are selected, and the root mean square (RMS) of data differences is calculated. The results show that the RMS error of NBRCS caused by all uncertain measurement factors is 0.34 dB, and the RMS error of the Leading Edge Slope (LES) is 0.50 dB [39].

This method can clearly identify and quantify individual error sources, facilitating improvements in instrument design and algorithms. For example, GNSS EIRP and receive antenna gain are the primary targets for improvement in L1b calibration. However, this method relies on accurate modeling of error terms and the assumption of independence. If hardware parameters drift on-orbit or correlations exist between error sources, the total error may be underestimated. The on-orbit data-based assessment method is directly based on actual observations and can reflect the true system performance, including unmodeled errors. Its limitations include the need for high-quality and independent reference data, limited sample size under certain conditions, and an inability to resolve specific error-contributing links, providing only overall error estimate [2, 39].

8 Conclusion

This chapter gives an overview for the calibration of spaceborne GNSS-R measurements. This includes the calibration of receiver gain, GNSS EIRP, antenna pattern, effective scattering area and polarimetric factors.

For receiver gain calibration, it can be performed by pre-launch ground measurements or on-orbit using blackbody and open-ocean measurements. The latter one needs the receiver to routinely switch to a blackbody but can be adapted to the degradation of the instrument over time.

GNSS EIRP can be estimated using static look-up tables developed from a ground power monitor but the result suffers from the dynamic change of the transmitting power and limited observation angle on the ground. These errors can be mitigated by on-orbit real-time estimation using the zenith antenna. Two real-time estimation methods, the Zenith-to-Specular Ratio Method and Normalized Antenna Pattern Method are presented.

The antenna patterns of GNSS-R zenith and nadir antennas are important for calibration. They are usually measured in the anechoic chamber before launch, but should be improved using on-orbit data.

The effective scattering area is an important parameter when computing the NBRCS. It should be modeled as a function of incidence angle, azimuth angle, satellite altitude, and signal type. The effective scattering areas of different GNSS signal types are compared and analyzed.

Polarization calibration is crucial to polarimetric GNSS-R systems. System nonidealities such as transmitter elliptical polarization, receiver channel crosstalk, and Faraday rotation need to be correct to accurately compute polarimetric observables such NBRCS and reflectivity. The computation of Stokes parameters in a compact polarimetric GNSS-R system is presented.

Finally, the calibration assessment methods are overviewed. It includes the one based on an error propagation model and the one based on on-orbit data by matching with winds from numerical weather prediction models or between different satellites.

References

1. Gleason S, Al-Khaldi M, McKague D (2023) Level 1a DDM calibration algorithm theoretical basis document. CYGNSS, Ann Arbor, MI, USA, pp 136–148
2. Gleason S, Ruf CS, O'Brien AJ, McKague DS (2018) The CYGNSS level 1 calibration algorithm and error analysis based on on-orbit measurements. IEEE J Sel Top Appl Earth Obs Remote Sens 12(1):37–49
3. Sun Y et al (2023) GNOS-II on Fengyun-3 satellite series: exploration of multi-GNSS reflection signals for operational applications. Remote Sens 15(24):5756
4. Gleason S, Al-Khaldi M, McKague D (2023) Level 1b DDM calibration algorithm theoretical basis document. CYGNSS, Ann Arbor, MI, USA, pp 137–148
5. Gleason S, Ruf CS, Clarizia MP, O'Brien AJ (2016) Calibration and unwrapping of the normalized scattering cross section for the cyclone global navigation satellite system. IEEE Trans Geosci Remote Sens 54(5):2495–2509

6. Foti G, Gommenginger C, Jales P, Unwin M, Shaw A, Robertson C et al (2015) Spaceborne GNSS reflectometry for ocean winds: first results from the UK TechDemoSat-1 mission. Geophys Res Lett 42(13):5435–5441
7. Lin W, Portabella M, Foti G, Stoffelen A, Gommenginger C, He Y (2019) Toward the generation of a wind geophysical model function for spaceborne GNSS-R. IEEE Trans Geosci Remote Sens 57(2):655–666
8. Gleason S, Al-Khaldi MM, Ruf CS, McKague DS, Wang T, Russel A (2021) Characterizing and mitigating digital sampling effects on the CYGNSS level 1 calibration. IEEE Trans Geosci Remote Sens 60:1–12
9. Jing C et al (2024) A review of the BuFeng-1 GNSS-R mission: calibration and validation results of sea surface and land surface. Geo-spatial Information Science 27(3):638–652
10. Setti PT, Tabibi S (2023) Evaluation of spire GNSS-R reflectivity from multiple GNSS constellations for soil moisture estimation. Int J Remote Sens 44(20):6422–6441
11. Al-Khaldi MM et al (2024) An analysis of a commercial GNSS-R soil moisture dataset. IEEE J Sel Top Appl Earth Obs Remote Sens 17:15480–15493
12. Unwin MJ et al (2021) An introduction to the HydroGNSS GNSS reflectometry remote sensing mission. IEEE J Sel Top Appl Earth Obs Remote Sens 14:6987–6999
13. Peng J, Cardellach E, Li W, Ribó S, Rius A (2025) Toward the exploitation of HydroGNSS coherent channel: high sampling rate products. IEEE Trans Geosci Remote Sens 63:1–13
14. Munoz-Martin JF, Llaveria D, Herbert C, Pablos M, Camps A (2021) Soil moisture retrieval using the FMPL-2/FSSCat GNSS-R and microwave radiometry data. In: 2021 IEEE international geoscience and remote sensing symposium (IGARSS), pp 7638–7641
15. Munoz-Martin JF, Fernandez L, Perez A, Ruiz-de AJA, Park H, Camps A et al (2020) In-orbit validation of the FMPL-2 instrument—the GNSS-R and L-band microwave radiometer payload of the fsscat mission. Remote Sens 13(1):121
16. Huang F et al (2023) Analysis and mitigation of radio frequency interference in spaceborne GNSS ocean reflectometry data. IEEE Trans Geosci Remote Sens 61:1–12
17. Jales PY, Guruprasad S, Nogués-Correig O, Cartwright J (2024) Antenna digital beamforming on spire's GNSS-reflectometry cubesat constellation. In: 18th European conference on antennas and propagation (EuCAP), pp 1–5
18. Wang T et al (2022) Dynamic calibration of GPS effective isotropic radiated power for GNSS-Reflectometry earth remote sensing. IEEE Trans Geosci Remote Sens 60:1–12
19. Wang T, Ruf CS, Block B, McKague DS, Gleason S (2019) Design and performance of a GPS constellation power monitor system for improved CYGNSS L1B calibration. IEEE J Sel Top Appl Earth Obs Remote Sens 12(1):26–36
20. Huang F, Yin C, Liu Y, Sun Y, Xia J, Bai W et al (2025) Tianmu-1 constellation GNSS-R in-orbit performance: spatiotemporal characteristics, product applications, and polarimetric features. IEEE Trans Geosci Remote Sens 63:1–20
21. Wang T, Ruf C, Block B, McKague D, Gleason S (2018) Characterization of GPS L1 EIRP: transmit power and antenna gain pattern. In: Proceedings of the 31st international technical meeting of the satellite division of the institute of navigation (ION GNSS+2018), pp 2879–2890
22. Jiménez-Baños D, Perelló-Gisbert JV, Crisci M (2010) The measured effects of GPS flex power capability collected on sensor station data. In: 2010 5th ESA workshop on satellite navigation technologies and European workshop on GNSS signals and signal processing (NAVITEC) Noordwijk, Netherlands, pp 1–6
23. Qiu T et al (2024) Retrieving ocean surface wind speeds in real time on spaceborne GNSS-R receivers: algorithm and validation. IEEE J Sel Top Appl Earth Obs Remote Sens 17:2201–2212
24. Asgarimehr M, Wickert J, Reich S (2018) TDS-1 GNSS reflectometry: development and validation of forward scattering winds. IEEE J Sel Top Appl Earth Obs Remote Sens 11(11):4534–4541
25. Reigh WAM et al (2015) The GPS block IIR and IIR-M broadcast L-band antenna panel: its pattern and performance. J Inst Navig 62(4):329–347
26. Wang T, Ruf C, Gleason S, Block B, McKague D, Provost D (2017) Development of GPS constellation power monitor system for high accuracy calibration/validation of the CYGNSS

L1B data. In: 2017 IEEE international geoscience and remote sensing symposium (IGARSS). IEEE, pp 1008–1011
27. Wang T et al (2024) Measurement of GPS antenna gain patterns for CYGNSS using a spaceborne antenna range. IEEE Trans Aerosp Electron Syst 60(5):6883–6893
28. Zavorotny VU, Voronovich AG (2000) Scattering of GPS signals from the ocean with wind remote sensing application. IEEE Trans Geosci Remote Sens 38(2):951–964
29. Huang F, Xia J, Yin C, Bai W, Sun Y, Du Q et al (2022) Characterization and calibration of spaceborne GNSS-R observations over the ocean from different BeiDou satellite types. IEEE Trans Geosci Remote Sens 60:1–11
30. Bai D, Ruf CS, Moller D (2025) Calibration of the polarimetric GNSS-R sensor in the Rongowai mission. IEEE Trans Geosci Remote Sens 63:1–18
31. Munoz-Martin JF, Rodriguez-Alvarez N, Bosch-Lluis X, Oudrhiri K (2024) Scattering matrix retrieval using full-polarimetric GNSS-R. IEEE Trans Geosci Remote Sens 62:1–15
32. Munoz-Martin JF, Bosch-Lluis X, Rodriguez-Alvarez N, Oudrhiri K (2023) Calibration strategy for compact polarimetric GNSS-R instruments. IEEE Trans Geosci Remote Sens 61:1–13
33. Rodriguez-Alvarez N, Munoz-Martin JF, Bosch-Lluis X, Oudrhiri K (2022) Introduction to the new SMAP-Reflectometry (SMAP-R) dataset: status and science capabilities. In: IGARSS 2022—2022 IEEE international geoscience and remote sensing symposium, pp 7624–7627
34. Ruf C, Gleason S (2025) Spaceborne GNSS-R bistatic radar remote sensing, CYGNSS, and future missions. In: Proceedings of the IEEE
35. Hammond ML, Foti G, Gommenginger C, Srokosz M (2021) An assessment of CyGNSS v3.0 level 1 observables over the ocean. Remote Sen 13(17):3500
36. Said F, Jelenak Z, Chang PS, Soisuvarn S (2018) An assessment of CYGNSS normalized bistatic radar cross section calibration. IEEE J Sel Top Appl Earth Obs Remote Sens 12(1):50–65
37. Yang X, Nan Y, Guo B, Du H, Zhu J, Huang F (2025) Level 1 products calibration assessment of FENGYUN-3E GNOS-II GNSS-R. IEEE Trans Geosci Remote Sens
38. Jansen MA, Ruf CS, Keihm SJ (1995) TOPEX/Poseidon microwave radiometer (TMR): II. Antenna pattern correction and brightness temperature algorithm. IEEE Trans Geosci Remote Sens 33(1):138–146
39. Ruf CS, Gleason S, McKague DS (2019) Assessment of CYGNSS wind speed retrieval uncertainty. IEEE J Sel Top Appl Earth Obs Remote Sens 12(1):87–97

Chapter 4
CYGNSS and FY-3E Level 1 Product Data Calibration

Yang Nan, Xiaotong Yang, and Bofeng Guo

Abstract This chapter outlines the Level 1 data calibration process for GNSS-R missions, with a focus on the methodologies adopted in CYGNSS and FY-3E/GNOS-II. Beyond introducing the calibration procedures, a comparative analysis is conducted to evaluate their distinct characteristics in terms of observation geometry, normalized bistatic radar cross section (NBRCS), and key parameters in the Level 1B calibration procedure. Results show that FY-3E and CYGNSS achieve a high degree of consistency in NBRCS values, although a systematic bias of about 1.9 dB exists. Further assessment across multiple GNSS constellations (GPS, BDS, and Galileo) highlights differences in calibration performance associated with transmitter power, modulation schemes, and antenna gain models. These findings underscore the critical role of Level 1 calibration in ensuring the reliability of GNSS-R data products for wind retrieval, and suggest the need for standardized calibration protocols to support future operational GNSS-R missions.

1 Introduction

GNSS-R missions generally distribute their observations in the form of multi-level data products. Taking the CYGNSS mission as an example, its publicly released products are organized into three levels. The Level 1 dataset mainly consists of Delay–Doppler Maps (DDMs) generated from raw Level 0 measurements, together with the corresponding normalized bistatic radar cross section (NBRCS) and ancillary parameters required for geophysical retrieval. These Level 1 parameters serve as the fundamental inputs for subsequent wind retrieval algorithms. At Level 2, CYGNSS provides geophysical products derived through spatial averaging. The mean square slope (MSS) product represents the averaged sea-surface slope statistics within a 25 $\times$ 25 km^2 footprint, accompanied by uncertainty information. In parallel, the Level 2 wind speed product reports the mean surface wind speed over the same spatial scale with corresponding uncertainty estimates. The Level 3 product further aggregates

Y. Nan (✉) · X. Yang · B. Guo
School of Marine Science and Technology, Tianjin University, Tianjin, China
e-mail: nanyang@tju.edu.cn

K. Yu and N. Zheng (eds.), *Advances in GNSS Reflectometry*, Navigation: Science and Technology 19, https://doi.org/10.1007/978-981-92-2325-1_4

the retrieved winds in both space and time, delivering gridded surface wind fields on a 0.2° latitude–longitude grid. In addition to ocean wind observations, CYGNSS also supplies gridded soil moisture products for land surface applications.

The FY-3E/GNOS-II GNSS-R mission provides L1 data products and orbital sea surface wind products to the public. The Level 1 data products are derived from raw observations and preprocessed static parameters, and are generated through a series of steps including decoding, quality control, precise orbit determination, additional phase calculation, and sea surface reflection signal processing. These products include additional phase and GNSS reflected signal data, which are made available to users for applications such as data assimilation and geophysical parameters retrievals. The GNOS-II sea surface wind product refers to the wind speed at 10 m above the sea surface, retrieved from GNSS signals (including GPS, BDS, and GALILEO) reflected off the ocean surface. The spatial resolution of this product is 25 km.

This chapter focuses on the differences in Level 1 data product calibration between the CYGNSS mission and the FY-3E GNSS-R mission. Specifically, it compares the NBRCS and key calibration parameters between the CYGNSS Version 3.1 Science Data Record Level 1 products and the FY-3E/GNOS-II GNSS-R Version 1.0 Level 1 products.

2 Procedure of Level 1 Product Calibration

Figure 1 provides an overview of the Level 1 calibration workflow commonly adopted in GNSS-R processing. In a Delay–Doppler Map (DDM), each bin contains raw digital measurements recorded by the receiver, which are expressed in uncalibrated units known as counts. These measurements are proportional to the total received signal power after receiver processing.

The recorded signal is not solely composed of the reflected GNSS contribution from the Earth's surface, but also includes thermal noise originating from both the Earth and the receiver electronics. As a result, the measured power reflects the combined contribution of all input signals modulated by the receiver gain. Through Level 1A calibration, the raw DDM counts are transformed into absolute power values expressed in physical units of watts.

Following Level 1A, Level 1B calibration converts the mapped power into NBRCS values by incorporating auxiliary information. This conversion is performed on a pixel-by-pixel basis in the DDM and accounts for factors such as the transmitted signal's effective isotropically radiated power (EIRP), the bistatic geometry linking the transmitter, specular reflection point, and receiver, as well as the directional gain characteristics of the receiving antenna.

This section first outlines the Level 1 calibration framework implemented for the CYGNSS mission, and subsequently examines the differences in algorithmic implementation between FY-3E/GNOS-II and CYGNSS.

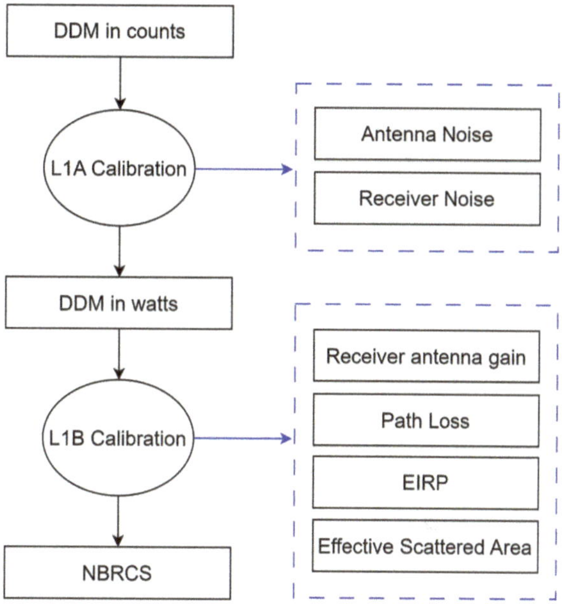

Fig. 1 Routine GNSS-R Level 1 data calibration process

2.1 CYGNSS L1 Calibration

The DDM count values output by the CYGNSS receiver are the result of the reflected signal being processed through various internal stages of the instrument, effectively applying an instrument gain to the received signal. Therefore, the count value in any bin of the DDM can be related to the received signal power, which can be simply expressed by the following equation:

$$C = G(N_a + N_r + P_g) \tag{1}$$

where C is the DDM count value output by the receiver; G is the total receiver gain accounting for both noise and scattered signal during the instrument processing; N_a is the power of thermal noise generated by the antenna in watts; N_r is the power of thermal noise generated by the receiver in watts; and P_g represents the power of the scattered signal by the receiver in watts.

The quantities C and P_g vary with both delay and Doppler coordinates, whereas the noise contributions N_a and N_r are assumed to remain invariant across the Delay–Doppler Map. These noise components correspond to DDM bins that contain no scattered signal energy and therefore represent pure noise samples. Such noise-only bins can be described in terms of digital counts as

$$C_N = G(N_a + N_r) \tag{2}$$

where G denotes the receiver gain.

The antenna noise power N_a and receiver noise power N_r are parameterized using empirical relationships. Specifically, the antenna noise contribution can be written as

$$N_a = kT_a B_W \tag{3}$$

where k s Boltzmann's constant, T_a denotes the antenna temperature—defined as the effective brightness temperature obtained by weighting the atmospheric brightness temperature with the receive antenna pattern—and B_W represents the effective signal bandwidth. For GPS signals, this bandwidth is governed by the coherent integration interval applied at the receiver.

The receiver thermal noise contribution is expressed as

$$N_r = kT_r B_W = k[(NF - 1)290]B_W \tag{4}$$

where T_r is the instrument temperature expressed in degrees Celsius. The receiver noise figure NF characterizes the relationship between the equivalent noise temperature and the physical temperature of the receiver front end. Prior to launch, the dependence of NF on temperature was characterized for all low-noise amplifiers (LNAs), enabling the instrument noise power to be estimated using Eq. (4).

Starting from the general expression of the instrument DDM in digital counts given in Eq. (1), which accounts for both the scattered signal contribution and the noise floor defined by Eq. (2), the signal-only DDM can be obtained by removing the noise component as described below.

$$C_g = C - C_N = GP_g \tag{5}$$

Using Eq. (5), we obtain the signal-only DDM in units of watts, denoted as P_g. However, the instrument gain G remains unknown. To accurately estimate the instrument's thermal noise gain, the receiver is equipped with a black body calibration load mode. A black body load resistor is placed on the thermally coupled LNA mainboard, with a temperature sensor located; thus, it is assumed that the black body load temperature is equivalent to the LNA section temperature [1]. When the instrument switches to the black body calibration mode, the expression for the DDM noise counts is given by Eq. (6).

$$C_B = G(N_B + N_r) \tag{6}$$

$$N_B = kT_I B_W \tag{7}$$

where N_B and N_r represent the black body resistor noise power and the instrument noise power, respectively. N_B can be expressed by Eq. (7), where T_I denotes the physical temperature of the instrument's LNA and the black body load resistor.

Finally, using Eq. (1) as the base formulation and removing the noise component defined in Eq. (2), where the instrument noise gain is accurately obtained through the black body load mode as described in Eq. (6), the final Level 1A calibration is derived as follows:

$$P_g = \frac{(C - N_a - N_r)(N_B + N_r)}{C_B} \tag{8}$$

Level 1B calibration is performed following Level 1A calibration and involves converting the DDM in watts obtained from L1A into the NBRCS, a key parameter used for retrieval of ocean surface geophysical parameters. This conversion is carried out on a per-pixel basis within the DDM and requires the following information:

1. The reception geometry of the CYGNSS platform, characterized by the receiver's GPS time, position, and velocity, all expressed in the WGS-84 Earth-Centered Earth-Fixed (ECEF) reference frame.
2. The corresponding orbital state of the transmitting GPS satellite, including its position and velocity, also represented in the WGS-84 ECEF coordinate system.
3. $P^T G^T$: EIRP of the transmitted GPS signal, defined in the transmitter's own reference frame.
4. G^R: The gain pattern of the CYGNSS receiving antenna, together with the satellite attitude information required to map the antenna response into the Earth-fixed geometry.
5. $R^R R^T$: The bistatic scattering geometry, obtained through precise determination of the specular reflection point, which enables accurate calculation of the propagation path lengths between the transmitter, the specular point, and the receiver.
6. The mapping of the specular point into the delay–Doppler domain, allowing its location to be identified within the corresponding delay and Doppler bins of the DDM.
7. A: The effective scattering area associated with the vicinity of the specular point, evaluated over all delay and Doppler bins contributing to the DDM.

$$P^g_{\tau f} = \frac{P^T \lambda^2}{(4\pi)^3} \iint\limits_A \frac{G^T_{x,y} \sigma^0_{x,y} G^R_{x,y}}{(R^R_{x,y})^2 (R^T_{x,y})^2} \Lambda^2_{\tau;x,y} S^2_{f;x,y} dxdy \tag{9}$$

$$P^g_{\tau f} = \frac{\lambda^2 P^T G^T_{\tau f} \left\langle \sigma^0_{\tau f} \right\rangle \overline{G}^R_{\tau f} \overline{A}_{\tau f}}{(4\pi)^3 (\overline{R}^R_{\tau f})^2 (\overline{R}^T_{\tau f})^2} \tag{10}$$

The bistatic scattering power model for GPS signals was originally formulated by Zavorotny and Voronovich (2000) and is summarized in Eq. (9). In this formulation, the functions $\Lambda_{\tau;x,y} S_{f;x,y}$ describe the spreading of the GPS signal in the delay and frequency domains, respectively. To facilitate practical implementation, the integral appearing in Eq. (9) is approximated by introducing effective values for several parameters over the corresponding delay and Doppler integration intervals. The impact of the delay and frequency spreading functions is thus embedded into these effective quantities, allowing $\Lambda_{\tau;x,y}S_{f;x,y}$ to be factored out of the integral and represented using overbar notation. Under this approximation, the original formulation expressed in the physical (x, y) coordinate system is subsequently mapped into the delay–Doppler domain, which forms the basis for the forward scattering representation used in L1B calibration.

Equation (10) represents the simplified forward scattering model, in which the received power in each delay–Doppler bin is related to NBRCS. This expression is obtained by substituting the effective values of the spreading functions into Eq. (9) and transforming the geometry into the delay–Doppler coordinate system.

2.2 FY-3E/GNOS-II GNSS-R L1 Calibration

FY-3E, which was launched on 5 July 2021, operates as an early-morning polar-orbiting meteorological satellite and carries the GNOS-II payload for GNSS signal reception and reflection processing [2]. A summary of the key orbital characteristics and instrument configurations of FY-3E in comparison with CYGNSS is provided in Table 1 [3].

The GNOS-II payload has a mass of 24.5 kg and a nominal power consumption of 55 W. Its antenna subsystem consists of multiple elements, including two zenith-pointing antennas for navigation purposes, two side-looking antennas dedicated to

Table 1 In-orbit instrument parameters for FY-3E/GNOS-II and CYGNSS/DDMI [4]

	FY-3E/GNOS-II	CYGNSS/DDMI
Altitude (km)	836	520
Inclination angle (°)	98.5	35
GNSS frequency	GPS L1 C/A, BDS B1I and GAL E1B	GPS L1 C/A
Coherent integration time (ms)	1	1
Non-coherent integration time (ms)	1000	500/1000
DDM dimension	122 delays × 20 dopplers	17 delays × 11 dopplers
DDM delay resolution	Non-uniform with 0.125 and 0.25 chip	0.25 chip
DDM doppler resolution (Hz)	500	500
Number of nadir reflection antenna	1	2
Antenna tilt angle (°)	20	± 28 (port and starboard)
Antenna gain (dBi)	15	14

GNSS radio occultation observations, and a nadir-pointing antenna designed for GNSS reflectometry measurements.

In terms of signal processing capability, GNOS-II is equipped with eight independent channels for reflected signal acquisition, allowing simultaneous tracking of up to eight GNSS transmitters. The instrument supports multiple GNSS constellations, including GPS L1, BeiDou B1, and Galileo E1. Within this configuration, four channels are assigned to GPS signals, three channels to BeiDou, and one channel to Galileo.

Similar to the CYGNSS constellation, the FY-3E/GNOS-II GNSS-R payload is designed to support ocean surface wind speed observations. Reliable wind retrieval relies on accurately calibrated Level 1 observables, including the NBRCS, leading-edge slope, and other radiometric parameters. Accordingly, this study focuses on the calibration characteristics of GNSS-R products derived from the FY-3E/GNOS-II multi-GNSS system.

Differences in Level 1 GNSS-R calibration between FY-3E and CYGNSS arise mainly from their distinct satellite platforms and instrument configurations. FY-3E operates as a large, operational meteorological satellite hosting multiple payloads, such as MERSI-LL, HIRAS-II, WindRAD, and GNOS-II. By contrast, CYGNSS is implemented on a constellation of small satellites, whose compact platforms are subject to relatively rapid thermal variations.

As a result of these platform differences, the L1A calibration strategy adopted for FY-3E differs from the blackbody radiation–based gain calibration scheme employed by CYGNSS. For FY-3E, pre-launch thermal cycling experiments were conducted to characterize the dependence of receiver gain on instrument temperature [5]. The resulting gain–temperature relationship was stored in a lookup table, enabling the in-orbit instrument gain to be determined using onboard temperature measurements, as summarized in Eq. (11).

$$P_g = G(T)C \tag{11}$$

where C denotes the DDM expressed in raw counts after incoherent averaging, and P_g is the corresponding DDM power converted into watts. The receiver gain G is treated as temperature dependent, i.e., $G = G(T)$.

For both FY-3E and CYGNSS, the conversion from DDM power to NBRCS follows the same bistatic scattering formulation (Eq. 10). Their main divergence arises from how the transmitter EIRP is obtained and applied. CYGNSS estimates the EIRP in a time-varying manner by tracking the transmitting GNSS signals, whereas FY-3E relies on pre-established EIRP characterizations. Specifically, EIRP values for GPS, BDS, and Galileo are taken from look-up tables derived using two ground-based power monitoring stations in Beijing following the procedure in [6]. For each GNSS satellite, an EIRP–boresight-angle look-up table (referenced to the receiver line-of-sight) is generated. An example of the monitoring setup is shown in Fig. 2 [4].

Fig. 2 The rooftop GNSS EIRP power monitoring facility at the NSSC building [4]

3 Level 1 Products Calibration Assessment of FY-3E GNOS-II GNSS-R

Level 1 datasets from FY-3E/GNOS-II and CYGNSS were used to assess the calibration performance. FY-3E/GNOS-II L1 data were accessed via the Feng Yun Satellite Remote Sensing Data Service, while CYGNSS L1 data were retrieved through NASA's Earthdata Search portal. As an independent wind reference, ERA5 reanalysis products provided by ECMWF under the Copernicus Climate Change Service were employed, offering hourly fields on a $0.25° \times 0.25°$ latitude–longitude grid. The ERA5 wind speeds were interpolated to the GNSS-R specular point locations to generate collocated reference values. The analysis covers the period from July 2022 to June 2023.

3.1 Processing of Data

In addition to the final calibrated parameter NBRCS obtained through Level 1 data calibration, several other parameters are also required to evaluate the quality and effectiveness of the calibration. These include the NBRCS ratio, EIRP, range correction gain, and the peak signal-to-noise ratio of the DDM.

(1) NBRCS Ratio: NBRCS of GNSS-R is dependent on the incidence angle and wind speed, which can be obtained not only through GNSS-R calibration but also through simulation. To evaluate the calibration accuracy, a similar parameter to [7], this study introduces the concept of NBRCS ratio:

$$NBRCS_r = \frac{\sigma_0}{\sigma_{Model}(\theta, \varepsilon, U_{10})} \quad (12)$$

where σ_0 is the NBRCS calibrated by L1 and σ_{Model} is the simulated value, generated using the geometrical optics (GO) approximation and an ocean surface spectrum described by the Elfouhaily model [8] with a cutoff number from [9], defined as:

$$\sigma_{Model}(\theta, \varepsilon, U_{10}) = \frac{|R(\theta, \varepsilon)|^2}{2\sigma_u\sigma_c} \quad (13)$$

where R is the complex Fresnel reflection coefficient for vertical or horizontal polarization [10], a function related to the angle of incidence (θ) and the seawater dielectric constant (ε). $2\sigma_u\sigma_c$ is the square of the surface MSS, 'u' and 'c' refer to upwind and crosswind, U_{10} denotes the wind speed at 10 m above the sea surface under neutral atmospheric stability conditions. Physically, MSS can be modeled as a weighted integration over the sea surface spectrum,

$$MSS_{Model}^2 = 2\sigma_u\sigma_c = 2\left[\int_0^{k_{cut}} d\vec{\kappa}\,\Psi(\vec{\kappa})\kappa^2\cos^2\varphi\right]^{\frac{1}{2}}\left[\int_0^{k_{cut}} d\vec{\kappa}\,\Psi(\vec{\kappa})\kappa^2\sin^2\varphi\right]^{\frac{1}{2}} \quad (14)$$

the cutoff frequency $k_{cut} = 2\pi\cos(\theta_{SP})/3\lambda$, with λ the incident wavelength and θ_{SP} the specular incidence angel.

(2) EIRP: The transmitter EIRP was calculated for each GNSS satellite by using two ground-based static power monitoring stations for FY-3E/GNOS-II [11]. A LUT (look up table) of EIRP versus boresight angle was created for each GNSS satellite. Unlike being provided directly in the L1 data of CYGNSS, the EIRP can only be back-calculated from other parameters using the following Eq. (15):

$$EIRP = \frac{(4\pi)^3 R_t^2 R_r^2}{\lambda^2 G_r} \cdot G_{BRCS} \quad (15)$$

(3) RCG: RCG is a parameter of the electromagnetic wave propagation distance together with the receiver antenna gain. According to [8], RCG can be described by the Eq. (16):

$$RCG = \frac{G_r}{R_t^2 R_r^2} \times 10^{27} \quad (16)$$

(4) DDM Peak SNR: It is worth noting that "Ddm_peak_snr" in FY-3E and "ddm_snr" in CYGNSS are not computed in the same way, so we have uniformly adopted the "Ddm_peak_snr" calculation method used in FY-3E. The calculation of the peak SNR in this work is obtained uniformly by the calculation of Eq. (17):

$$SNR_{peak} = 10 \times \log_{10}(\frac{S_{\max} - Noise}{Noise}) \quad (17)$$

where SNR_{peak} is peak SNR in dB, $S_{\max}$ is the peak value of the raw DDM counts and *Noise* is the mean noise floor of the raw DDM.

3.2 Grouping Methods

In this study, we categorize the possible effects on NBRCS into five categories (here the categorization refers to some generalized classifications): wind speed, incidence angle, different GNSS systems, different satellite types under the same GNSS system, and antennas:

(1) To control the influence of sea surface roughness on the calibration, we opted for a specific range of wind speeds for the study. Figure 3a illustrates the probability density function of wind speeds in the dataset. Utilizing this wind speed probability density function, we selected a range of 6.5–7 m/s for our analysis.
(2) To control the impact of incidence angle, we selected the interval with the highest data density for our study. Figure 3b illustrates the probability density function

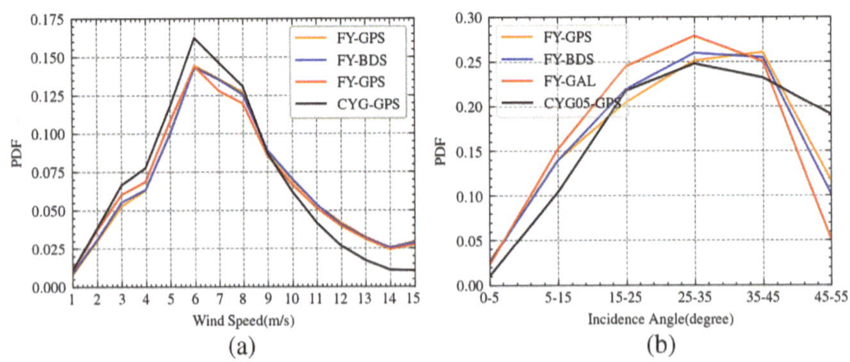

Fig. 3 The probability density function of wind speed (**a**) or incidence angle (**b**) in the dataset

of the incidence angles in the dataset. Utilizing this function, we opted for an incidence angle range of 25°–35°.

(3) To evaluate the calibration performance of different GNSS systems, the FY-3E dataset was partitioned into three groups: GPS, BDS, and GAL. CYGNSS relies only on the GPS system.

(4) Even within the same system, variations exist among individual satellites. Consequently, within the GPS system, GPS satellites were classified into four groups: GPSII-R, GPSII-F, GPSIII-A and GPSIIR-M. Similarly, within the BDS system, BDS satellites were divided into four groups: BDS2-IGSO, BDS2-MEO, BDS3-IGSO and BDS3-MEO. GALILEO satellites can be categorized based on their stage of development into IOV (In-Orbit Validation) and FOC (Full Operational Capability). For the CYGNSS GPS system, GPSII-R, GPSII-F and GPSII-R-M are available.

(5) The CYGNSS constellation consists of eight satellites, simplifying the reference data and listing the statistical figures of CYG-05, which has the largest amount of data. Also considering the possible influence of different antennas on the observed volume, we differentiate between the two antennas of CYGNSS. For the FY-3E reflectometry mission there is only one satellite and one antenna.

3.3 *Performance of NBRCS*

GNSS-R is used to perform wind speed inversion by establishing a relationship between wind speed and NBRCS. In this study, we analyzed this relationship by segmenting the matched data into windows with a step size of 0.15 m/s, computing the mean and standard deviation of NBRCS for each window. The standard deviation is employed to assess the degree of dispersion of the NBRCS.

Figure 4 illustrates the variations in NBRCS and wind speed for FY-3E/GNOS-II and CYGNSS. From Fig. 4a, b, it is evident that NBRCS values of FY-3E are generally higher than those of CYGNSS under the same wind speed conditions. The numerical deviations are believed to be due to differences arising from the iteration of the CYGNSS L1 version itself, from 3.0 to 3.1. As a previous study [12] has demonstrated, there is good numerical agreement between FY-3E and CYGv3.0 data. For FY-3E, there is consistency in NBRCS values across different GNSS systems, with no apparent bias with wind speed. In contrast, CYGNSS exhibits a systematic bias between its two antennas, the systematic bias of observations that exists between antennas can be eliminated.

The mean standard deviations of the NBRCS values for different systems are shown in Table 2 Compared with CYGNSS, FY-3E exhibits a similar mean standard deviation in NBRCS values from GPS reflections. However, FY-3E shows larger standard deviations when processing reflected signals from the BDS and GAL systems for NBRCS calibration. The poorer SNR of the BDS reflections is primarily due to the increased noise power from BDS-IGSO signals, which is caused by interference from regional GNSS and Satellite-Based Augmentation System (SBAS) systems, as

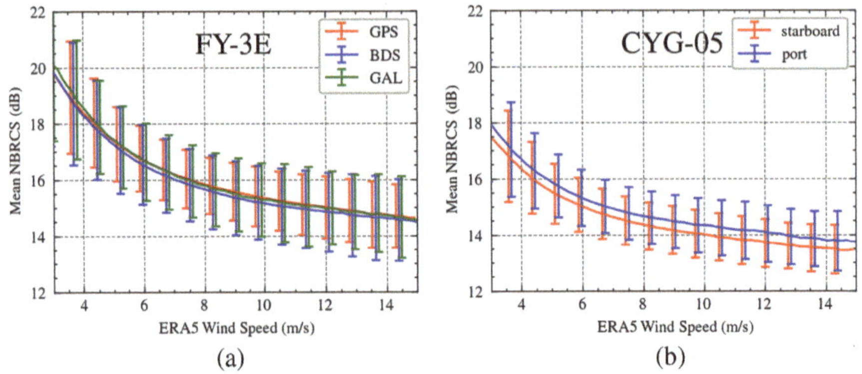

Fig. 4 The correlation between NBRCS and ERA5 reference wind speed over incidence angle 25°–35° between FY-3E GNSS system (**a**) and CYG-05 antenna (**b**)

cited in [11]. The lower SNR of the GAL reflected signals is attributed to the Binary Offset Carrier (BOC) modulation, as verified in [13, 14]. Additionally, Fig. 5 and Table 2 indicate that the CYGNSS starboard antenna produces a smaller standard deviation in NBRCS compared to the port antenna.

The relevant analyses for each satellite system are presented in Fig. 5 as follows:

(1) Figure 5a shows the NBRCS values for the four types of GPS satellites. Specifically, the NBRCS values for GPSII-R and GPSIIR-M are consistently similar,

Table 2 Mean standard deviation of NBRCS in different wind bins of FY-3E and CYG-05

	FY-3E/GPS	FY-3E/BDS	FY-3E/GAL	CYG-05/PORT	CYG-05/STARBOARD
3.75 m/s	2.003	2.184	2.117	1.690	1.618
4.50 m/s	1.586	1.773	1.666	1.339	1.271
5.25 m/s	1.323	1.553	1.471	1.121	1.065
6.00 m/s	1.157	1.418	1.381	1.005	0.955
6.75 m/s	1.070	1.307	1.323	0.946	0.895
7.50 m/s	1.046	1.292	1.277	0.939	0.854
8.25 m/s	1.001	1.310	1.306	0.944	0.841
9.00 m/s	1.025	1.313	1.302	0.941	0.858
9.75 m/s	1.055	1.323	1.304	0.972	0.859
10.05 m/s	1.057	1.358	1.401	0.989	0.844
11.25 m/s	1.085	1.386	1.402	1.021	0.857
12.00 m/s	1.100	1.411	1.300	1.014	0.863
12.75 m/s	1.119	1.464	1.334	0.981	0.818
13.5 m/s	1.188	1.503	1.394	0.994	0.840
14.25 m/s	1.130	1.460	1.456	1.066	0.876

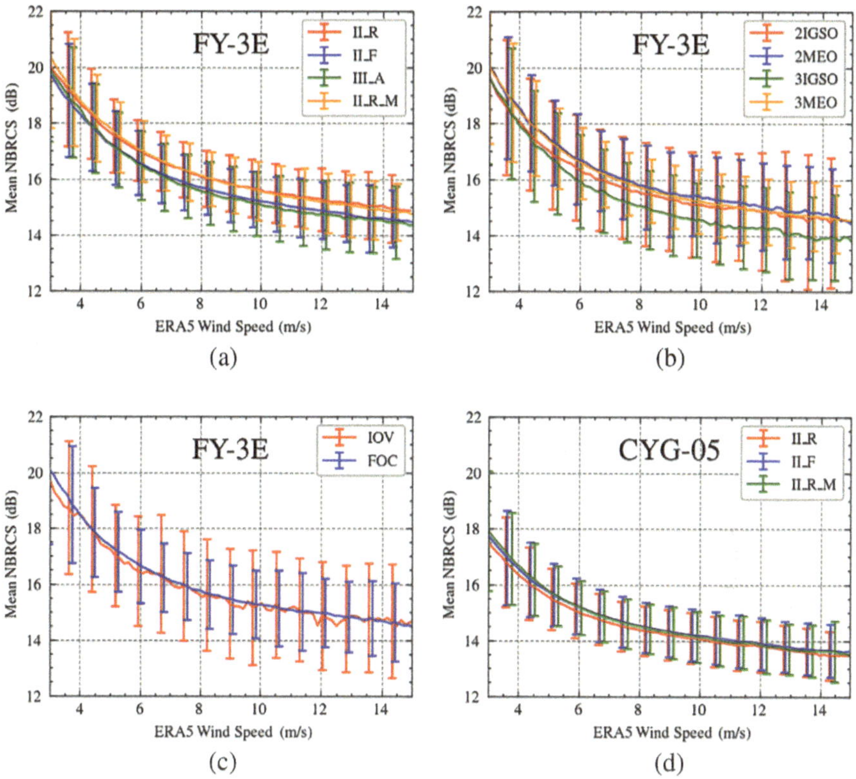

Fig. 5 The relationship between NBRCS and ERA5 reference wind speed over incidence angle 25°–35° for FY-3E and CYG-05 under different satellite types is shown. **a**–**c** are the results of FY-3E from GPS, BDS, and GAL, respectively. **d** is the result of CYG-05 from GPS

while those for GPSIII-A and GPSII-F are consistently lower. This discrepancy is due to differences in the transmit signal power among the various types of GPS satellites. The calibration process did not fully eliminate these differences in transmitter power. In fact, for FY-3E, the calibration of such offsets can be easily achieved through empirical corrections, as the offset between GPS blocks in NBRCS aligns consistently with the ERA5 wind speed. This also explains the constant bias in FY-3E's estimation of the transmit power across the different GPS satellites. In contrast, as shown in Fig. 5d, there is no deviation between different GPS types for CYGNSS. This NBRCS bias between different systems may stem from discrepancies in the EIRP monitoring of the various GNSS satellites. A method that dynamically monitors the EIRP of the satellite is more accurate in correcting for the transmit power variations across different satellites.

(2) Figure 5b illustrates the NBRCS values for the four types of BDS satellites. Within the wind speed range of 3–6 m/s, the sequence of NBRCS values is

BDS2-MEO, BDS3-MEO, BDS3-IGSO, and BDS2-IGSO. Beyond 6 m/s, the sequence changes to BDS3-MEO, BDS2-IGSO, BDS3-IGSO, and BDS2-MEO. These variations are attributed to differences in transmit signal power and orbital altitudes among the BDS satellite types. In addition, the calibration process did not completely eliminate the power differences between transmitters, leading to systematic deviations in NBRCS values for various BDS satellites across different orbits. Notably, the standard deviation for BDS3-MEO and BDS3-IGSO is smaller than that for BDS2-MEO and BDS2-IGSO, which may reflect differences in stability between BDS2 and BDS3 satellites. This issue primarily arose from the increased noise power of the BDS-IGSO satellites due to interference from regional GNSS and SBAS systems [11]. However, it has been resolved in the subsequent update of the FY-3E/GNOS-II system algorithm.

(3) Figure 5c illustrates the calibration of the NBRCS for the GAL system. The two types of Galileo satellites (IOV and FOC) exhibit a stable NBRCS trend with wind speed. Notably, the GAL-IOV curve shows a significant deviation. This may be due to the limited number of IOV satellites as test satellites and their own low signal-to-noise ratio.

Restricting the wind speed range to 6.5–7 m/s ensures that the calibration of NBRCS is analyzed under comparable ocean surface conditions (Fig. 6). CYGv3.1, serving as a reference, shows relatively stable and consistent NBRCS variations with incidence angle across different GPS systems. For all GNSS systems processed by FY-3E, significant NBRCS fluctuations are observed at low incidence angles ($< 10°$) and high incidence angles ($> 40°$). Within the 20°–30° range, the NBRCS values generally decrease under the GPS system, with II-R and II-R-M exhibiting similar patterns. In contrast, II-F shows a systematic deviation relative to these GPS types, and III-A demonstrates a pronounced jump. For the BDS system, results from 2-IGSO, 2-MEO, and 3-MEO are largely consistent, while 3-IGSO shows a notable systematic bias. The GAL system results are generally comparable, but IOV measurements display relatively higher noise, likely due to the limited dataset size.

The discrepancies in NBRCS values observed by FY-3E across different systems and incidence angles are likely due to biases in the EIRP calibration for various satellites. These variations can be minimized through empirical debiasing or by refining the EIRP lookup table for improved accuracy. Since the NBRCS trends with incidence angle are consistent within the same GNSS system but vary across different GNSS systems, it is recommended to perform wind speed retrievals independently for each system to ensure higher accuracy.

The normalized NBRCS is employed as a standard metric to assess the calibration accuracy of antenna gain levels, following the methodology outlined in [7]. The GNSS-R reflectometry technique operates across various angles of incidence and azimuth, making it challenging to isolate the impact of different incidence angles on the NBRCS due to the influence of sea surface roughness. To address this, the σ_{Model} represents a simplified model that can effectively filter out sea surface roughness effects. It is important to emphasize that the interpretation of results based on the numerical magnitude of the normalized NBRCS is inherently constrained by the

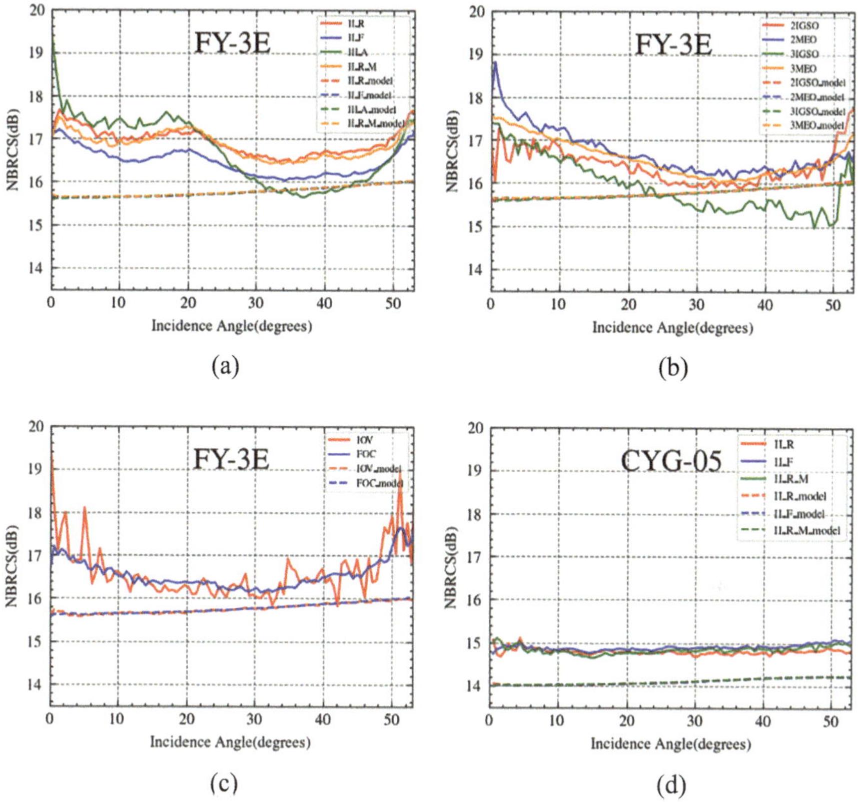

Fig. 6 The variation of the NBRCS with incidence angle over wind speed 6.5–7 m/s for FY-3E and CYGNSS. **a–d** are FY-3E GPS system, FY-3E BDS system, FY-3E GAL system and CYG-05 GPS system respectively

modeled NBRCS values and other ocean surface information tied to the actual observations. Consequently, the focus of this section is on the variation of the NBRCS ratio with respect to the angle of incidence. This variation serves as a more representative metric for assessing errors in estimating the gain level in the directional map of the receiving antenna as a function of incidence angle.

Figure 7 depicts the fluctuations in NBRCS ratio in terms of incidence angle for FY-3E and CYGNSS over 7–15 m/s. From these figures, several observations can be made:

(1) It is important to note that the modeled NBRCS values are not significantly influenced by the angle of incidence. The use of the NBRCS ratio mitigates some effects of sea surface roughness while amplifying the influence of antenna gain pattern evaluation. A notable outcome is that CYGNSS has undergone corrections for this form of antenna gain pattern evaluation [15], whereas FY-3E has not, resulting in the better performance illustrated in Fig. 7.

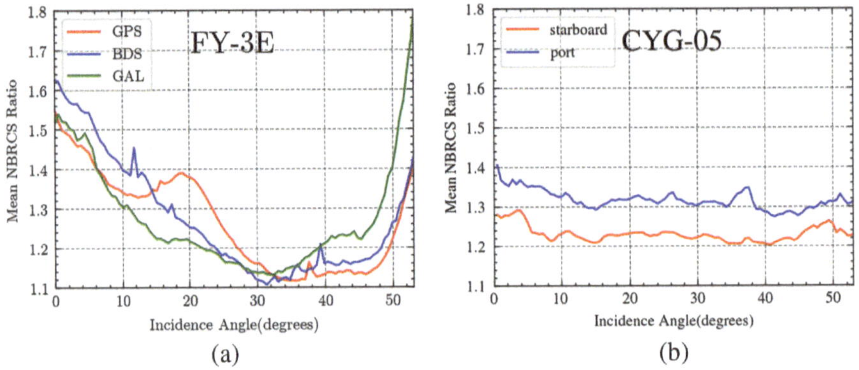

Fig. 7 The relationship between NBRCS ratio and the incidence angle over wind speed 7–15 m/s. **a, b** are the result of FY-3E and CYG-05

(2) The NBRCS ratio for FY-3E exhibits significant fluctuation. The curve of the NBRCS ratio displays a parabolic shape, with higher values observed at both extreme incidence angles. This suggests that the antenna gain is underestimated at both low and high incidence angles. The deviation in the antenna gain pattern varies more gradually at low incidence angles but becomes more pronounced at high incidence angles. Notably, we observed unusual fluctuations in the GPS curve between 10° and 30°, which subsequent analysis revealed to be caused by variations in its EIRP within this range.

(3) Additionally, the variation of the NBRCS ratio with incidence angle for CYGv3.0 is not entirely uniform. Later updates to CYGv3.1 incorporated a correction for the dependence of NBRCS on sea surface roughness in the antenna gain pattern, as shown in Fig. 7b.

3.4 Performance of Calibration Parameters

L1 calibration parameters offer valuable insights into the calibration process of FY-3E. Much statistical work has been done on the calibration parameters of the FY-3E in [12]. However, there are relatively few studies that directly compare these parameters with CYGNSS data. During the conversion from power unit DDM to BRCS, G_{BRCS} is a composite parameter that includes electromagnetic wave wavelength, EIRP, and range-corrected gain (RCG). CYGNSS performs on-orbit EIRP calibration, while FY-3E relies on two static GNSS EIRP power monitors located in Beijing. Additionally, the peak signal-to-noise ratio is used to evaluate the quality of the collected raw DDMs. The following section provides a statistical analysis of these calibration parameters.

(1) EIRP: The variation of the calibration parameter EIRP with offboresight angle for FY-3E and CYGNSS is illustrated in Fig. 8. Figure 8a–c show that EIRP

values obtained from ground-based EIRP power monitors for different GNSS satellites are relatively stable. Figure 8a, d demonstrate that, despite differences from the on-orbit EIRP power monitoring method used by CYGNSS, the ground-based EIRP power monitoring method employed by FY-3E is similarly effective. Among the GPS satellites, the EIRP power monitoring results for GPSII-R are closest to those of CYGNSS. However, the average EIRP values for GPSII-R-M and GPSII-F show consistent trends but are smaller compared to those of CYGNSS.

(2) RCG: Figure 9 illustrates the variation of RCG values with incidence angle for FY-3E and CYGNSS. Compared to CYGNSS, FY-3E satellites operate at a higher altitude, resulting in smaller RCG values. Additionally, the RCG values for different GNSS satellite systems correspond to their respective orbital altitudes. The satellites are ordered by descending altitude as follows: BDS-IGSO, GAL, BDS-MEO, and GPS, which aligns with the results presented in the figure.

Further analysis involves examining the relationship between the angle of incidence and the received antenna gain, as shown in Fig. 10. We find that both the antenna

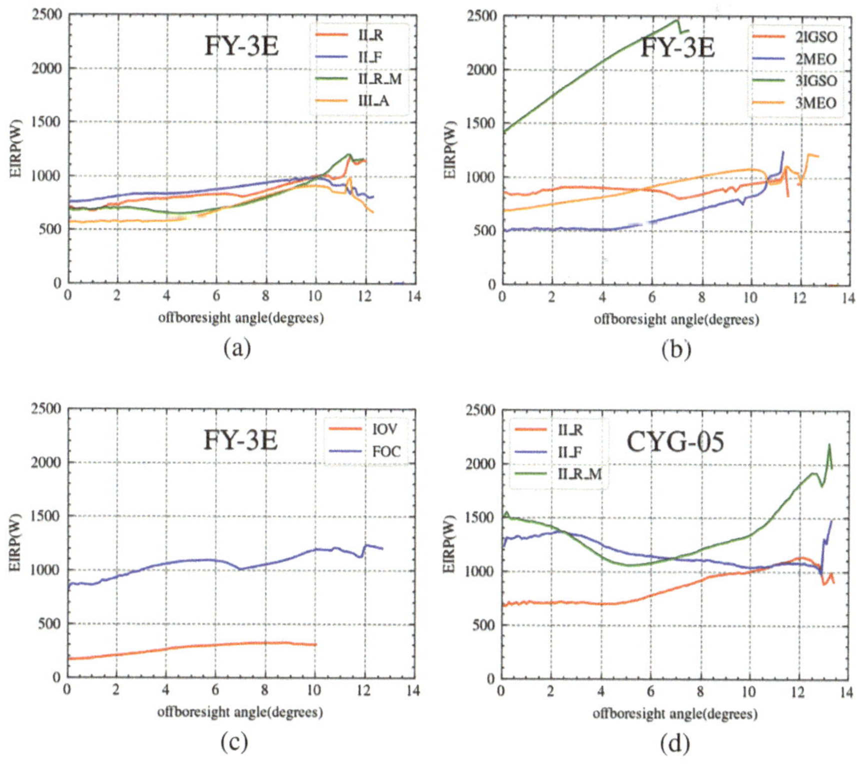

Fig. 8 The variation of the calibration parameter EIRP with offboresight angle over wind speed 6.5–7 m/s for FY-3E and CYGNSS. **a**–**d** are FY-3E GPS system, FY-3E BDS system, FY-3E GAL system and CYG-05 GPS system respectively

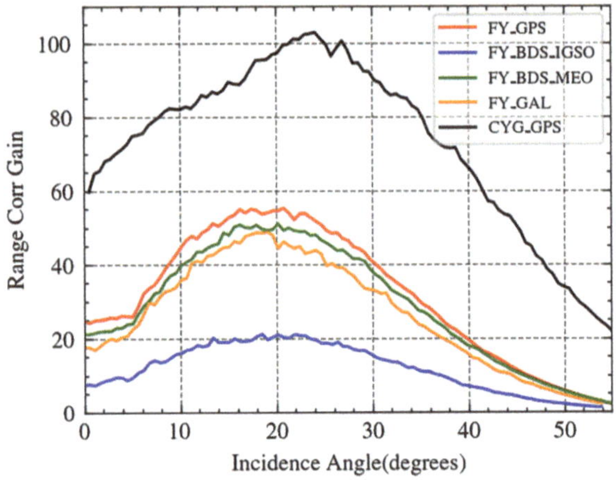

Fig. 9 The variation of the calibration parameter RCG with incidence angle for FY-3E and CYG-05

gain value and the range of antenna gain variation for FY-3E are larger compared to those for CYGNSS. This result is consistent with the observations presented in Fig. 7a in Sect. 3.3. It is important to note that the CYGNSS science antenna gain patterns have been adjusted to enhance the accuracy of the ocean surface NBRCS calibration in version 3.1 [15]. These adjustments ensure that the annual average observed NBRCS aligns with the model-predicted average, which is derived from Wavewatch-3 estimates of surface roughness, including the necessary spectral tail extension. This accounts for the observed stability in CYGNSS's antenna pattern across varying angles of incidence.

(1) Peak SNR: Figure 11 illustrates the variation of FY-3E's peak SNR with antenna gain and wind speed. The peak SNR of CYGNSS is numerically greater than that of FY-3E over the same range of receive antenna gain and wind speed variation. The peak SNR is influenced by both the noise floor and the peak of

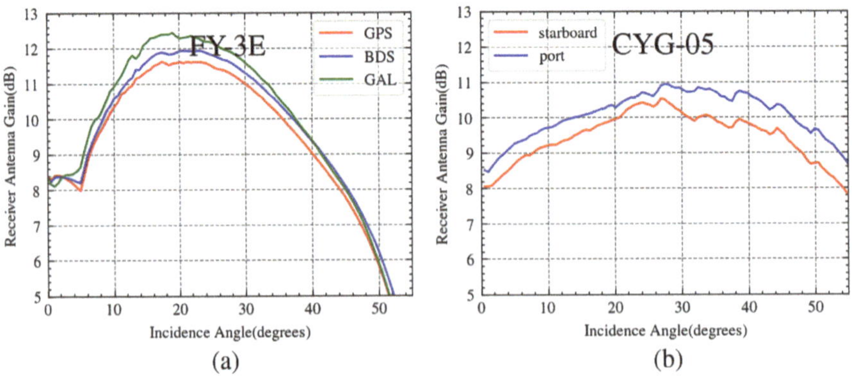

Fig. 10 The variation of receiver antenna gain with incidence angle. **a**, **b** are the statistics of FY-3E and CYG-05

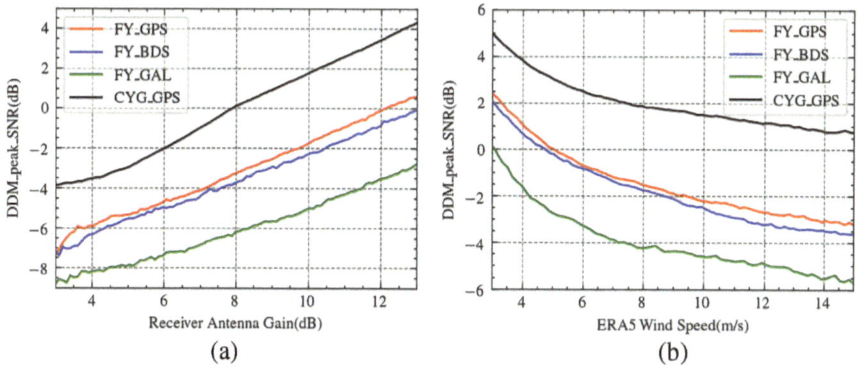

Fig. 11 The variation of peak SNR calibration parameters with respect to the receiver antenna gain and the wind speed of FY-3E and CYG-05. **a** over the wind speed 6.5–7 m/s and **b** over the incidence angle 25°–35°

the DDM, which typically reflects the quality of DDM acquisition in GNSS-R data calibration. Due to the signal modulation and signal transmission powers, Fig. 11a, b show that the peak SNR for GPS and BDS systems is higher than that for the GAL system. According to the signal-to-noise ratio formula, the signal processed by FY-3E maintains a similar noise floor. However, the lower EIRP of the GAL-IOV satellites leads to a significantly reduced signal-to-noise ratio. Therefore, special attention should be given to the signal source of GAL-IOV when analyzing the data. Compared to CYGNSS data shown in Fig. 11a, b, the peak SNR of the FY-3E is affected by the altitude of the orbit in which it is located, with a systematic deviation of − 2 dB in GPS system at the same receive antenna gain.

Additionally, we compared the SNR across different channels of the FY-3E satellite. The variations in signal-to-noise ratios between channels primarily stem from engineering design choices made during the receiver development process. The deviation in signal processing between different GNSS systems, based on their channel ordering, can be attributed to the antenna gain allocation principle in the design of FY-3E. This design prioritizes channels by assigning signals with higher antenna gain levels to top-ranked channels within the system. Once these channels reach capacity, signals are then allocated to lower-ranked channels. Consequently, the peak signal-to-noise ratio within the same system is ordered from largest to smallest according to the channel ranking.

Fig. 12 NBRCS scatter density distribution plots of FY-3E and CYGv3.1

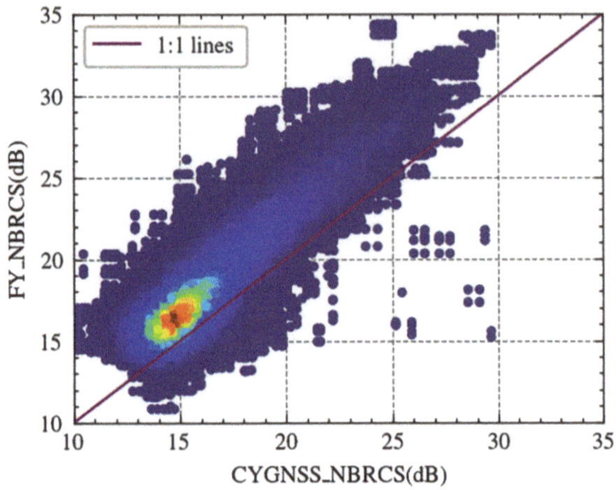

3.5 *Impacts of Data Version Evolution on GNSS-R Observations*

CYGNSS and FY-3E utilize similar calibration processes for NBRCS. To evaluate the differences in NBRCS values across various satellite-based GNSS-R payloads, we matched the GPS systems of CYGNSS and FY-3E using the data matching method described in Sect. 3.2. Additionally, the collection of colocated measurements are limited to only those having SNRs greater than 1.5 dB (CYGNSS) and reflectivity greater than − 3 dB (FY-3E). This process resulted in a total of 1,564,253 spatiotemporal matching points covering the period from July 2022 to June 2023. A scatter density distribution plot was subsequently generated from this data (see Fig. 12).

In previous study, the NBRCS from FY-3E and CYGv3.0 showed strong agreement [12]. As shown in Fig. 12, FY-3E and CYGv3.1 exhibit similar consistency, though a fixed deviation of 1.924 dB is present. This discrepancy arises from the numerical magnitude of the NBRCS which varies with iterations of the CYGNSS Level 1 data product. Nevertheless, the high correlation between FY-3E and CYGv3.1 confirms that both datasets are valid for wind speed inversion, with the only difference being their respective signal-to-noise ratios.

4 Conclusion

In this chapter, we introduced the general Level 1 data calibration methods for spaceborne GNSS-R receivers, detailing the calibration approaches used in the CYGNSS and FY-3E/GNOS-II GNSS-R missions. We highlighted their key differences and conducted a comparative analysis. Since FY-3E was launched after CYGNSS, we used CYGNSS as a reference to assess the calibration performance of FY-3E.

The results indicate that FY-3E demonstrates a high correlation across different GNSS systems and exhibits minimal overall bias in NBRCS estimation. A comparative analysis of NBRCS between FY-3E and CYGNSS was conducted, revealing a strong correlation between FY-3E and 1,564,253 co-located Level 1 data products from CYGNSS, with an overall NBRCS correlation coefficient of 0.882. However, the analysis also identified a bias between FY-3E and CYGv3.1, which differs from the findings in [12], where FY-3E was compared with CYGv3.0. This discrepancy is likely due to differences in the calibration methods used in CYGv3.0 and CYGv3.1. Additionally, the NBRCS ratio of FY-3E/GNOS-II varies with the incidence angle, possibly caused by inaccuracies in antenna gain estimation. CYGNSS addressed this issue with improvements introduced in version 3.1 [15]. Furthermore, for the GAL-IOV satellite series, the relatively low SNR is attributed to their lower EIRP.

Overall, it is difficult to definitively determine which GNSS-R mission achieves better calibration, as such an evaluation must take into account a range of integrated factors, including the characteristics of the host satellite platform, flight altitude, and the design of the antenna gain pattern. Given that spaceborne GNSS-R missions are progressing toward operational applications, we recommend the planning and implementation of standardized GNSS-R calibration procedures. This would benefit international GNSS-R scientific research and promote the advancement of its operational use.

References

1. Janssen MA, Ruf CS, Keihm SJ (2002) TOPEX/Poseidon Microwave Radiometer (TMR). II. Antenna pattern correction and brightness temperature algorithm. IEEE Trans Geosci Remote Sens 33(1):138–146. https://doi.org/10.1109/36.114568
2. Huang F, Xia J, Yin C, Zhai X, Xu N, Yang G, Bai W, Sun Y, Du Q, Liao M et al (2022) Assessment of FY-3E GNOS-II GNSS-R global wind product. IEEE J Sel Top Appl Earth Obs Remote Sens 15:7899–7912. https://doi.org/10.1109/JSTARS.2022.3205331
3. Ruf C, Unwin M, Dickinson J, Rose R, Rose D, Vincent M, Lyons A (2013) CYGNSS: enabling the future of hurricane prediction [remote sensing satellites]. IEEE Geosci Remote Sens Mag 1(2):52–67. https://doi.org/10.1109/MGRS.2013.2260911
4. Sun Y, Huang F, Xia J, Yin C, Bai W, Du Q, Wang X, Cai Y, Li W, Yang G, Zhai X, Xu N, Hu X, Liu Y, Liu C, Wang D, Qiu T, Tian Y, Duan L, Li F, Meng X, Liu C, Tan G, Hu P, Wu R, Song D (2023) GNOS-II on Fengyun-3 satellite series: exploration of multi-GNSS reflection signals for operational applications. Remote Sens 15(24):5756. https://www.mdpi.com/2072-4292/15/24/5756
5. Yang G, Bai W, Wang J et al (2022) FY-3E GNOS II GNSS reflectometry: mission review and first results. Remote Sens 14(4):988. https://doi.org/10.3390/rs14040988
6. Wang T, Ruf CS, Block B et al (2018) Design and performance of a GPS constellation power monitor system for improved CYGNSS L1B calibration. IEEE J Sel Top Appl Earth Obs Remote Sens 12(1):26–36. https://doi.org/10.1109/JSTARS.2018.2867773
7. Al-Khaldi MM, Gleason S, Linnabary R, Policelli FS (2023) Level-1 calibration assessment of Spire's LEMUR-2 GNSS-R ocean normalized bistatic radar cross section estimates. IEEE Geosci Remote Sens Lett 20:1–5. https://doi.org/10.1109/LGRS.2023.3249405

8. Elfouhaily T, Chapron B, Katsaros K, Vandemark D (1997) A unified directional spectrum for long and short wind-driven waves. J Geophys Res Oceans 102(C7):15781–15796. https://doi.org/10.1029/97JC00467
9. Garrison J, Komjathy A, Zavorotny V, Katzberg S (2002) Wind speed measurement using forward scattered GPS signals. IEEE Trans Geosci Remote Sens 40(1):50–65. https://doi.org/10.1109/36.981349
10. Cardellach E, Fabra F, Rius A, Pettinato S, D'Addio S (2012) Characterization of dry-snow sub-structure using GNSS reflected signals. Remote Sens Environ 124:122–134. https://doi.org/10.1016/j.rse.2012.05.012
11. Huang F, Yin C, Xia J, Wang X, Sun Y, Bai W, Qiu T, Du Q, Yang G, Zheng Q (2023) Analysis and mitigation of radio frequency interference in spaceborne GNSS ocean reflectometry data. IEEE Trans Geosci Remote Sens 61:1–12. https://doi.org/10.1109/TGRS.2023.3289998
12. Huang F, Xia J, Yin C, Bai W, Sun Y, Du Q, Wang X, Cai Y, Duan L (2022) Characterization and calibration of spaceborne GNSS-R observations over the ocean from different BeiDou satellite types. IEEE Trans Geosci Remote Sens 60:1–11. https://doi.org/10.1109/TGRS.2022.3224844
13. Nan Y, Ye S, Liu J, Guo B, Zhang S, Li W (2021) Signal-to-noise ratio analyses of spaceborne GNSS-reflectometry from Galileo and BeiDou satellites. Remote Sens 14(1):35. https://doi.org/10.3390/rs14010035
14. Huang F, Xia J, Yin C, Zhai X, Yang G, Bai W, Sun Y, Du Q, Wang X, Qiu T, Cai Y, Duan L, Xu N, Liao M, Hu X, Zhang P (2023) Spaceborne GNSS reflectometry with Galileo signals on FY-3E/GNOS-II: measurements, calibration, and wind speed retrieval. IEEE Geosci Remote Sens Lett 20:1–5. https://doi.org/10.1109/LGRS.2023.3241358
15. Ruf C (2022) CYGNSS handbook. Michigan Publishing Services, Ann Arbor. ISBN 978-1-60785-787-7 (open access); ISBN 978-1-60785-549-1 (print)

Chapter 5
Scattering Models in Signals of Opportunity Reflectometry for Land Surface Remote Sensing: Focus on GNSS-R

Xuerui Wu

Abstract Global Navigation Satellite System Reflectometry (GNSS-R) enables the retrieval of critical land surface parameters, such as soil moisture, vegetation biomass, and soil freeze–thaw states, through analysis of signals reflected from the Earth's surface. This chapter delineates the fundamental bistatic scattering principles and the pivotal role of the Delay-Doppler Map (DDM) as the primary observable. It further presents the Land Surface GNSS Reflection Simulator (LAGRS) framework, comprising the LAGRS-Soil and LAGRS-Veg modules, which incorporate full-polarization simulations, incoherent scattering processes, and temperature-dependent dielectric constants to enhance the physical understanding of surface scattering mechanisms, particularly for soil freeze–thaw detection (LAGRS-FT). The sensitivity of GNSS-R signals to diverse land surface characteristics is explored, underscoring the advantages of multi-polarization and multi-frequency observations in disentangling contributing factors. Future advancements, including multi-constellation signal fusion, hybrid statistical-physical inversion techniques, and integration with complementary remote sensing datasets, are expected to strengthen GNSS-R's role in large-scale environmental monitoring. Ultimately, the development and refinement of physically-based scattering models represent essential mechanistic tools for interpreting and analyzing satellite observations, simulating reflected signals, facilitating data assimilation, developing quantitative retrieval algorithms for land surface parameters, and guiding the design of novel sensors.

1 Introduction

Global Navigation Satellite System Reflectometry (GNSS-R) represents an emerging bistatic radar remote sensing technique that employs L-band signals transmitted by GNSS satellites as sources of illumination. By analyzing the signals reflected from

X. Wu (✉)
Shanghai Astronomical Observatory, Chinese Academy of Sciences, Shanghai, China
e-mail: xrwu@shao.ac.cn

K. Yu and N. Zheng (eds.), *Advances in GNSS Reflectometry*, Navigation: Science and Technology 19, https://doi.org/10.1007/978-981-92-2325-1_5

the Earth's surface, GNSS-R enables the retrieval of key geophysical parameters. In comparison with conventional polar-orbiting satellite remote sensing systems, GNSS-R offers distinct advantages, including an abundance of signal sources, high spatiotemporal resolution, diverse observation geometries, and relatively low operational costs.

Initial GNSS-R investigations were predominantly focused on oceanographic applications, such as retrieval of sea surface wind fields and estimation of significant wave heights [1]. The successful deployment of spaceborne missions, including UK-DMC, TechDemoSat-1, and especially NASA's Cyclone Global Navigation Satellite System (CYGNSS) constellation, has provided global-scale validation of GNSS-R technology and facilitated its transition toward land surface applications [2]. Studies have shown that GNSS-R signals exhibit high sensitivity to surface dielectric properties and roughness characteristics, thereby demonstrating substantial potential for soil moisture monitoring, vegetation biomass estimation, flood inundation mapping, and detection of near-surface soil freeze–thaw states [3].

In recent years, Signals of Opportunity (SoOp) reflectometry has attracted growing research interest [4]. This promising remote sensing paradigm utilizes transmissions from a variety of radio-frequency systems as opportunistic illumination sources, extending beyond the L-band navigation signals employed in GNSS-R to encompass P-band, C-band, X-band, and other frequency bands [5]. The differing wavelengths and penetration depths associated with these bands provide complementary capabilities for sensing specific geophysical parameters; for instance, P-band signals are particularly effective for soil moisture and forest biomass retrieval, whereas higher-frequency bands are better suited for monitoring snow properties. The increasing availability of diverse spectral resources opens new avenues for land parameter estimation using SoOp reflectometry. A key challenge in this field lies in the effective exploitation and integration of these multifaceted signals. In this regard, in-depth investigations of scattering mechanisms and polarization characteristics are essential to fully elucidate the remote sensing capabilities of these opportunistic sources and to accelerate the broader adoption of the technology.

The development of physically-based bistatic scattering models plays a pivotal role in advancing GNSS-R and SoOp remote sensing of land surfaces. These mechanistic models provide a rigorous theoretical foundation for interpreting the complex interactions between electromagnetic waves and heterogeneous land surfaces, enabling accurate forward simulation of reflected signals under varying surface conditions, polarization states, and observation geometries. By bridging observable quantities (such as Delay-Doppler Maps) with underlying geophysical parameters, such models are indispensable for analyzing and validating satellite measurements, supporting data assimilation into land surface models, developing robust quantitative retrieval algorithms, and evaluating the performance of existing and future sensors. Ultimately, continued refinement of these scattering models will be critical to unlocking the full potential of GNSS-R and broader SoOp techniques for operational large-scale environmental monitoring.

2 A Full-Polarization GNSS-Reflectometry Model for Bare Soil Applications

2.1 Physical Basis of GNSS-R Bare Soil Scattering

Observation Geometry

The observation geometry in GNSS-R is determined by the relative positions and velocities of the transmitter, receiver, and specular reflection point on the surface. This geometry governs the signal propagation path, incidence angle, scattering angle, and associated Doppler shift. In processing, the reflected signal is cross-correlated with a replica of the direct signal to generate a two-dimensional waveform known as the Delay-Doppler Map (DDM), with axes corresponding to relative delay and Doppler frequency. The DDM serves as the primary observable in GNSS-R, where its shape and power distribution directly reflect surface scattering properties and provide the basis for geophysical parameter retrieval [1, 6].

Scattering Mechanisms

Figure 1 illustrates the primary scattering mechanisms involved in GNSS-R signals interacting with land surfaces of varying roughness. In the case of a perfectly smooth surface (panel a), scattering is predominantly specular and coherent, with the reflected signal concentrated in the specular direction according to the law of reflection, preserving phase information and resulting in a strong, focused peak in the Delay-Doppler Map (DDM).

As surface roughness increases (panel b), diffuse (incoherent) scattering begins to dominate, where electromagnetic energy is redistributed in multiple directions away from the specular path due to random phase interference from surface facets. This incoherent component broadens the scattering distribution and reduces the peak intensity of the coherent reflection.

For surfaces with intermediate roughness (panel c), both coherent and incoherent scattering coexist: a residual coherent component remains along the specular direction, while diffuse scattering contributes energy across a wider angular range. This mixed regime is typical of many natural land surfaces, such as bare soil or sparsely vegetated areas.

In highly rough surfaces (panel d), coherent scattering becomes negligible, and diffuse scattering prevails, with energy scattered broadly in various directions. Under such conditions, the DDM exhibits a more extended and flattened shape, reflecting the loss of phase coherence. Understanding the transition from coherent to incoherent scattering with increasing roughness is fundamental to interpreting GNSS-R observations and developing accurate forward scattering models for land surface parameter retrieval.

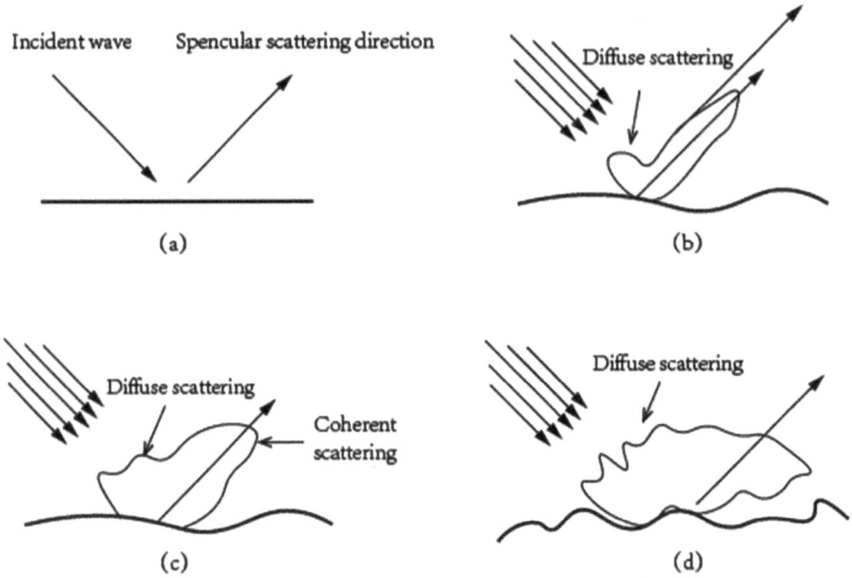

Fig. 1 Schematic illustration of scattering mechanisms in GNSS-R land surface remote sensing. **a** Specular (coherent) scattering from a smooth surface, where the reflected signal follows the law of reflection. **b** Diffuse (incoherent) scattering from a rough surface, resulting in scattered energy distributed in multiple directions away from the specular direction. **c** Mixed scattering from an intermediate roughness surface, with both coherent specular reflection and incoherent diffuse scattering components present. **d** Predominantly diffuse scattering from a very rough surface, where the coherent component is negligible and most energy is scattered diffusely in various directions

In GNSS-R, the received signal at the receiver is thus a superposition of these coherent and incoherent contributions originating from the surface within the glistening zone—the area on the Earth's surface from which reflected signals can reach the receiver. GNSS-R systems are inherently bistatic. Signals transmitted from GNSS satellites are captured by a GNSS-R receiver after being reflected from the Earth's surface. Reflection occurs within the glistening zone, with the strongest energy at the specular point. The total received power comprises coherent and incoherent scattering components. The total received power P_r comprises coherent P_r^{coh} and incoherent P_r^{inc} scattering components [7]:

$$P_r = P_r^{coh} + P_r^{inc} \tag{2.1}$$

(1) Coherent Scattering

Coherent scattering predominates near the specular direction, with energy concentrated in a small area. It dominates over very smooth surfaces. The coherent power is expressed as:

$$P_r^{coh} = \frac{P_T \lambda^2 G_T G_R}{(4\pi)^2 (R_t + R_r)^2} \cdot \left|\Gamma_{pq}\right|^2 \tag{2.2}$$

where P_T is the transmitted power, λ is the wavelength, G_T and G_R are the transmitter and receiver antenna gains, R_t and R_r are the distances from the transmitter and receiver to the specular point, and Γ_{pq} is the Fresnel reflection coefficient, with subscripts p and q denoting the receive and transmit polarizations.

(2) Incoherent Scattering

As surface roughness increases, diffuse scattering becomes significant, dispersing energy over a wider area (the glistening zone). Its extent and shape depend on the incidence angle, surface roughness, and antenna gains. The incoherent scattering power is approximated as:

$$P_r^{inc} \approx \frac{P_T \lambda^2 G_T G_R \sigma_{pq}^o A_{eff}}{(4\pi)^3 R_{t1}^2 R_{r1}^2} \tag{2.3}$$

where σ_{pq}^0 is the bistatic scattering coefficient and A_{eff} is the effective scattering area. Incoherent scattering contains richer spatial information about the surface and is particularly important for parameter inversion over complex terrain.

2.2 Modeling Approaches in GNSS-R Simulators

Early model development relied on the Z-V (Zenith-Vertical) model based on the KA-GO (Kirchhoff Approximation—Geometric Optics) approach [1]. This model established physical links between observation geometry, environmental parameters, and the received signal through a bistatic radar integral formulation, successfully extending from ocean surface scattering to soil moisture retrieval applications, using the LR polarization (left-hand circular polarization received and right hand circular polarization transmitted) scheme.

Building on the Z-V framework, Pierdicca et al. developed the SAVERS (Soil and Vegetation Reflection Simulator) [7]. This model consists of geometric computation and surface reflectivity calculation modules. Its innovation lies in applying Fung and Eom's coherent scattering coefficient and the AIEM (Advanced Integral Equation Model) [8, 9] to compute coherent and incoherent scattering properties from bare soil, respectively, while integrating the Tor Vergata model for vegetation scattering [10]. Inputting transmitter/receiver positions and velocities into the geometry module yields parameters such as the specular point, antenna pattern, and Doppler shift, which are then coupled with the scattering coefficient module to generate the DDM. The model supports both RR (right-hand circular polarization received and left-hand circular polarization transmitted) and LR polarizations.

For vegetation remote sensing, the SCoBi-Veg (Signals of opportunity Coherent Bistatic scattering model for Vegetated terrains) model uses a Monte Carlo framework to achieve full-polarization bistatic coherent scattering modeling over vegetated terrain [11]. It is suitable for P/S-band signals from navigation and communication satellites, accurately simulating polarized reflection characteristics of vegetation canopies.

The LAGRS (Land Surface GNSS Reflection Simulator)-Veg model is a spaceborne DDM simulation and analysis system for vegetation based on a modified first-order radiative transfer equation, capable of full-polarization scattering calculations [12]. By establishing physical links between navigation signals and vegetation, it quantitatively analyzes scattering contributions and their impact on DDM waveforms for various vegetation types, from tall forests to low shrubs.

The LAGRS-Soil model is a GNSS-R soil simulation and analysis system featuring several novel aspects [13]: DDM waveform computation incorporating both specular and diffuse scattering characteristics; a full-polarization calculation capability designed as an extensible module for future spaceborne payloads; and a dedicated DDM module for surface freeze–thaw characterization, aimed at soil moisture retrieval in cold regions and monitoring near-surface freeze/thaw states. A comparison among the models mentioned is summarized in Table 1.

On 5 July 2021, China launched its Fengyun-3E (FY-3E) meteorological satellite. One of the payloads onboard this satellite is the GNSS Occultation Sounder (GNOS)-II, which combines GNSS-R and GNSS radio occultation techniques. To support scientific applications of FY-3E GNOS-R data, particularly for retrieving soil moisture and freeze–thaw states over bare soil, the LAGRS-Soil was developed. This model introduces two major improvements: it comprehensively incorporates both coherent and incoherent scattering from bare soil based on precise observation geometry, achieved through optimized coherent scattering and rough surface scattering models; and it features full-polarization calculation capability, which enhances soil moisture retrieval accuracy and provides technical support for future GNOS-R payload upgrades, implemented via polarization synthesis techniques.

Table 1 Comparison between space-borne DDM simulators

Abbreviation	Application	Scattering model	Polarization
Z-V	Ocean	KA-GO	LR pol
Z-V-M	Bare soil	KA-GO	LR pol
SAVERS	Bare soil	Fung/Eom and AIEM	RR/LR pol
	Veg	Tor Vergata model	RR/LR pol
SCoBi-Veg	Bare soil	Monte Carlo model	Full pol
	Veg	Monte Carlo	Full pol
LAGRS-Veg	Veg	1st order RT model	Full pol
LAGRS-Soil	Bare soil	Fung/Eom, GO, PO, SPM, IEM, and AIEM	Full pol

2.3 *Core Architecture of the LAGRS-Soil Model*

The LAGRS-Soil model builds on established microwave scattering theory and incorporates key enhancements tailored to GNSS-R system characteristics. Its core architecture is illustrated in Fig. 2.

Comprehensive Modeling of Incoherent Scattering

Unlike many traditional models that emphasize coherent scattering, a defining feature of LAGRS-Soil is its concurrent treatment of both coherent and incoherent scattering mechanisms. It incorporates a suite of classical rough surface scattering models (e.g., GO, PO, SPM, IEM, AIEM) [14]. By intelligently selecting the most suitable model based on surface roughness conditions, LAGRS-Soil accurately computes diffuse scattering contributions from various geometries within the glistening zone, thus providing a more realistic representation of the complete DDM waveforms.

Full-Polarization Calculation Capability

Conventional GNSS-R applications have often been restricted to receiving the Left-Hand Circular Polarization (LHCP) reflected component. GNSS satellites transmit Right-Hand Circular Polarization (RHCP) signals, whose polarization state changes upon surface reflection, typically producing a strong LHCP component and a weaker RHCP component that carry vital information on surface dielectric properties and roughness. LAGRS-Soil employs polarization synthesis to achieve full-polarization

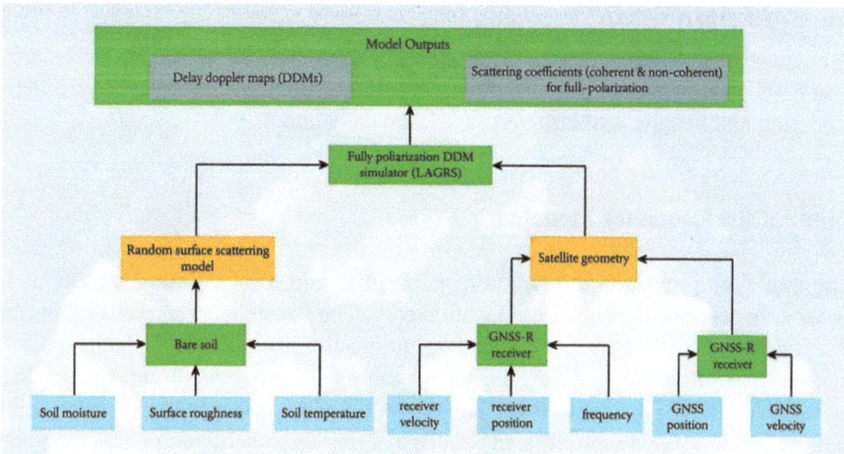

Fig. 2 Flowchart of the LAGRS-Soil

simulation. By constructing and solving the Mueller Matrix that describes the scattering process, it can simulate signals under RHCP incidence for various polarization states (e.g., LR, RR, VR, HR, ± 45° linear) [15, 16]. That is, the bistatic scattering coefficient is described as:

$$\sigma_{pq}(\psi_p, \chi_p, \psi_q, \chi_q) = 4\pi \tilde{Y}_m^T(\psi_p, \chi_p) M_m Y_m(\psi_q, \chi_q) \tag{2.4}$$

where $\tilde{Y}_m^T$ and Y_m are the modified Stokes vectors for the receive and transmit polarizations, respectively, defined by the orientation angle ψ and ellipticity angle χ of their polarization states. This capability provides a crucial theoretical tool for designing future polarized GNSS-R payloads and leveraging polarization information to improve retrieval accuracy.

Model of Surface Freeze–Thaw Processes

To address cold region monitoring needs, LAGRS-Soil integrates a frozen soil dielectric model (e.g., the Zhang-Zhao model). This model accurately captures the dramatic dielectric constant variations during soil freezing and thawing caused by water phase transitions. Incorporating temperature-dependent dielectric constants into scattering calculations allows LAGRS-Soil to generate distinct DDMs for frozen and thawed soils, paving the way for large-scale, frequent monitoring of surface freeze–thaw states and offering new data sources for climate change studies, carbon cycle estimation, and cold region engineering.

2.4 *Theoretical Framework and Key Formulas of LAGRS-Soil*

The LAGRS-Soil model primarily consists of two modules: observation geometry modeling and bistatic scattering cross-section calculation.

Observation Geometry Modeling

The model employs a combination of geocentric and local coordinate systems to accurately describe the signal propagation path. The key geometric relationships are illustrated in Fig. 3, with parameters including: the position and velocity vectors of the transmitter and receiver, the local radius of Earth's curvature, the satellite elevation angle, the distances from the transmitter and receiver to the specular point, and the incidence and scattering angles. By solving a series of geometric equations, the model determines the specific geometric configuration corresponding to each

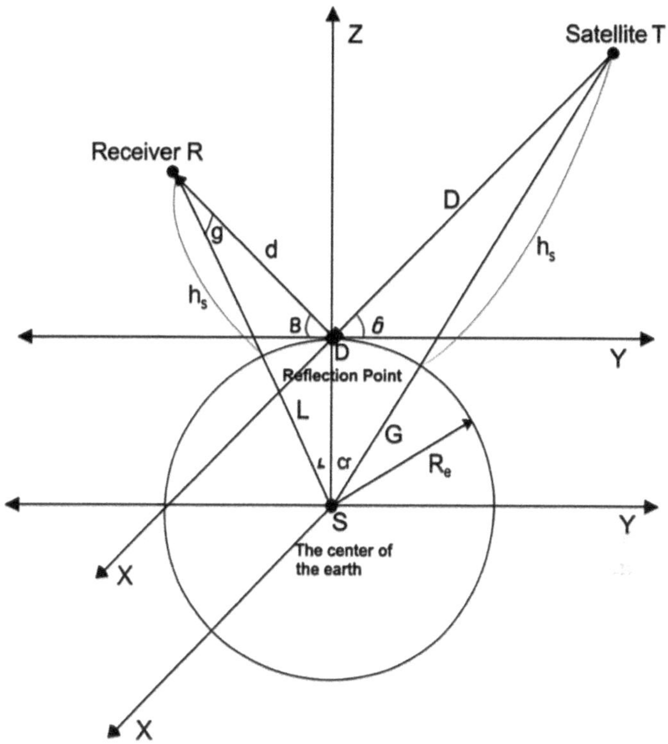

Fig. 3 Observational geometry of LAGRS

observation, which forms the basis for calculating the time delay, Doppler shift, and scattering angle.

Polarization Synthesis

Polarization synthesis is central to LAGRS-Soil's full-polarization capability. Since GNSS signals are circularly polarized while many traditional rough surface scattering models are formulated for linear polarization, a robust polarization conversion framework is essential.

When considering scattering from an object, two primary polarization basis conventions exist within GNSS-R systems.

Forward Scatter Alignment (FSA): Primarily used in bistatic scattering scenarios, its polarization basis is defined according to the wave propagation direction, making it suitable for describing polarization transformation during the scattering process.

Backscatter Alignment (BSA): Commonly employed in traditional monostatic radar systems, its polarization basis for the received wave aligns with the direction of the polarization basis for the transmitted wave.

LAGRS-Soil adopts the FSA convention. For the conversion from circular to linear polarization, the model related to the standard Stokes vectors by a simple transformation matrix. A 4 × 4 real matrix is introduced to relate the incident and scattered Stokes vectors. This formulation enables the calculation of the bistatic scattering cross section for any combination of transmitted and received polarizations through wave synthesis techniques, as defined in Eq. (2.4).

For coherent scattering in the specular direction, the polarization characteristics are primarily described by the Fresnel reflection matrix. In accordance with the above-mentioned models and the wave synthesis technique, we adopted the specular reflectivity:

$$\Re = \begin{bmatrix} |R_v|^2 & 0 & 0 & 0 \\ 0 & |R_h|^2 & 0 & 0 \\ 0 & 0 & \mathrm{Re}(R_v R_h^*) & -\mathrm{Im}(R_v R_h^*) \\ 0 & 0 & \mathrm{Im}(R_v R_h^*) & \mathrm{Re}(R_v R_h^*) \end{bmatrix} \tag{2.5}$$

where R_e and I_m represent the real part and the imaginary part of the corresponding equations, respectively, R is the Fresnel reflectivity, subscripts v and h are the vertical and horizontal polarizations, respectively, and $*$ denotes the conjunctive values. Based on the complex formulas, the random surface scattering models are utilized to derive the corresponding Mueller matrix, thereby enabling the acquisition of scattering properties across various polarizations. For random rough surface scattering, its general form is given by:

$$\mathbf{M} = \begin{bmatrix} \langle |S_{vv}|^2 \rangle & \langle |S_{vh}|^2 \rangle & \mathrm{Re}\langle S_{vv} S_{vh}^* \rangle & -\mathrm{Im}\langle S_{vv} S_{vh}^* \rangle \\ \langle |S_{hv}|^2 \rangle & \langle |S_{hh}|^2 \rangle & \mathrm{Re}\langle S_{hv} S_{hh}^* \rangle & -\mathrm{Im}\langle S_{hv} S_{hh}^* \rangle \\ 2\mathrm{Re}\langle S_{vv} S_{hv}^* \rangle & 2\mathrm{Re}\langle S_{vh} S_{hh}^* \rangle & \mathrm{Re}\langle S_{vv} S_{hh}^* + S_{vh} S_{hv}^* \rangle & \mathrm{Im}\langle S_{vh} S_{hv}^* - S_{vv} S_{hh}^* \rangle \\ 2\mathrm{Im}\langle S_{vv} S_{hv}^* \rangle & 2\mathrm{Im}\langle S_{vh} S_{hh}^* \rangle & \mathrm{Im}\langle S_{vv} S_{hh}^* + S_{vh} S_{hv}^* \rangle & \mathrm{Re}\langle S_{vv} S_{hh}^* - S_{vh} S_{hv}^* \rangle \end{bmatrix} \tag{2.6}$$

where S is the average scattering amplitude coefficient at pq polarization, where pq can be either vertical (v) or horizontal (h) polarization. R_e and I_m are used to obtain the real and imaginary parts of the dielectric constant, respectively, while $*$ is the conjugate.

The full-polarization synthesis capability enables LAGRS-Soil not only to simulate existing GNOS-R LR observations but also to assess the potential performance of future polarized GNSS-R payloads. It facilitates systematic investigation of different polarization combinations' sensitivity to soil parameters, advancing the development of joint inversion algorithms for soil moisture and roughness using polarization features, and supports polarization scattering analysis in vegetated scenarios.

2.5 Response Mechanisms of Soil Physical Parameters to GNSS-R Signals

Understanding how soil parameters influence GNSS-R signals, manifested as DDM variations, is essential for successful parameter retrieval.

Soil Moisture

Soil moisture is the dominant factor controlling the soil dielectric constant. The dielectric constant of liquid water (~ 80) far exceeds that of dry soil (~ 3–5) and air (~ 1). Consequently, increased volumetric water content significantly elevates the soil's bulk dielectric constant.

Impact on Coherent Scattering: A higher dielectric constant increases the Fresnel reflection coefficient Γ_{pq}, thereby enhancing the coherent scattering power P_r^{coh} near the specular point and raising the DDM peak power.

Impact on Incoherent Scattering: As soil wetness increases, incoherent scattering intensity across the glistening zone generally strengthens. The DDM's delay spreading may also vary with moisture content.

Figure 4 illustrates the effect of soil moisture on the final DDMs as the soil moisture content is changed from 0.1 to 0.6. It can be observed that the soil moisture content not only affects the scattering at the center of the coherent direction but also affects the edges and pattern of the scattering in the diffuse direction. This phenomenon becomes more pronounced as the delay (Chips) increases, as shown in Fig. 5. The accuracy of spaceborne GNSS-R data for soil moisture retrieval can be enhanced through a deeper understanding of its impact on signal scattering behavior.

Surface Roughness

Surface roughness is a key factor determining scattering directivity, directly influencing the energy distribution between coherent and incoherent scattering. As roughness increases, scattering energy diffuses from the specular direction, weakening the coherent component and strengthening the incoherent part. Accordingly, the DDM peak decreases and broadens, while the glistening zone expands. Roughness is thus a critical confounding factor in soil moisture retrieval that must be addressed. LAGRS-Soil's incorporation of multiple rough surface scattering models enables quantitative analysis of roughness effects, aiding in separating the contributions of moisture and roughness.

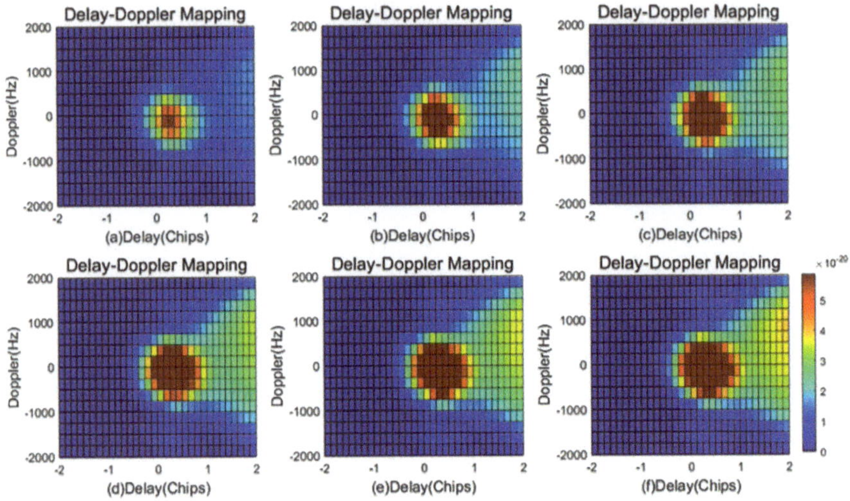

Fig. 4 Effect of soil moisture on the final DDMs: **a**–**f** volumetric soil moisture content from 0.1 to 0.6 in intervals of 0.1

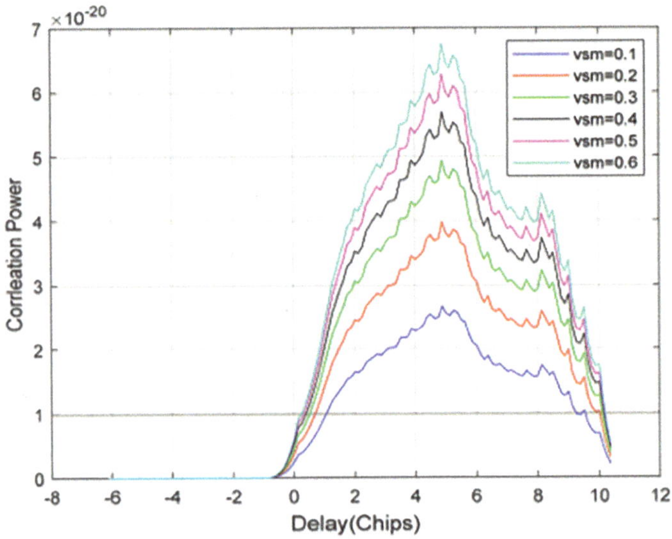

Fig. 5 Simulated correlation power versus delay (Chips) for different soil moisture contents with vsm from 0.1 to 0.6 from subfigure (**a**) to (**f**)

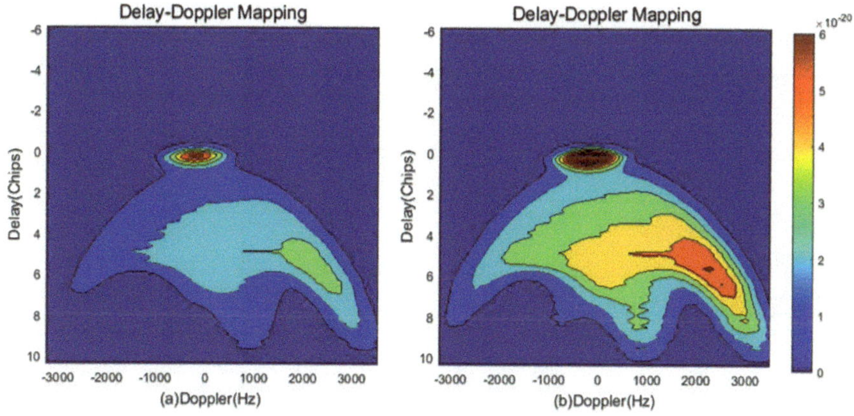

Fig. 6 Simulated correlation power versus delay (Chips) for **a** frozen and **b** thawed soil

Temperature and Freeze/Thaw State

Soil freeze–thaw cycles induce dramatic changes in dielectric properties, fundamentally affecting reflected GNSS signals. When soil temperature drops below freezing, liquid water transitions to ice, causing the dielectric constant real part to plunge from ~ 80 to ~ 3.2, significantly reducing surface reflectivity. This dielectric shift directly impacts the DDM: compared to thawed soil, frozen soil typically exhibits a 40–60% lower DDM peak power, a flatter waveform, and wider delay distribution. By integrating a temperature-dependent frozen soil dielectric model, LAGRS-Soil successfully simulates this contrast, providing key theoretical support for large-scale freeze–thaw monitoring using GNSS-R. The DDM patterns at the center point of the delay and Doppler frequency, shown in Fig. 6, are evidently different and show that frozen soil has higher values.

Influence of Observation Geometry

The observation geometry including the incidence angle, scattering angle, and scattering azimuth angle, serves as a key external factor governing the morphology and power distribution of the Delay-Doppler Map (DDM). The simulations corresponding to the setting listed in Table 2 are shown in Fig. 7, variations in geometric configuration alter the relative signal propagation path and scattering surface relationship, thereby directly influencing both the proportion and spatial distribution of coherent and incoherent scattering components within the DDM. Moreover, as the definition of the coherent scattering direction is relaxed—for instance, by increasing the azimuth angle tolerance—the incoherent component in the DDM is significantly enhanced, resulting in a corresponding shift in its energy distribution.

Table 2 Settings of the observational geometry

	Coherent calculations	Noncoherent calculations
Case 1	All the scatterings are coherent	No consideration
Case 2	$\theta_i = \theta_s$ no consideration of scattering azimuth angle	Other observation geometry
Case 3 Veg	$\theta_i = \theta_s, \varphi_s < 1^{\circ}$	Other observation geometry
Case 4 Bare soil	$\theta_i = \theta_s, \varphi_s < 2^{\circ}$	Other observation geometry
Case 5	$\theta_i = \theta_s, \varphi_s < 5^{\circ}$	Other observation geometry
Case 6 6	$\theta_i = \theta_s, \varphi_i = \varphi_s = 0^{\circ}$	Other observation geometry

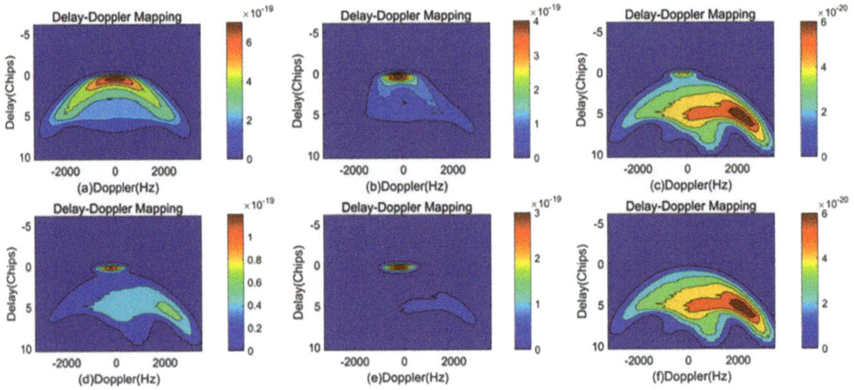

Fig. 7 DDMs illustrating noncoherent components. The observational geometry settings are shown in Table 2, while the setting for **a** is case 1, **b** is case 2, **c** is case 3, **d** is case 4, **e** is case 5, and **f** is case 6

Polarization Response

The full-polarization capability of LAGRS-Soil enables the investigation of the sensitivity of different polarization channels to soil parameters. Figure 8 presents the simulated DDM results under three different transmit-receive polarizations: LR, VR, and HR. It can be observed that the impact of different polarizations on the DDM waveform exhibits differences not only in the specular direction but also in the diffuse scattering direction. By comprehensively utilizing multi-polarization information, it is anticipated that the effects of surface roughness and vegetation cover can be more effectively separated during soil moisture retrieval, thereby improving the inversion accuracy.

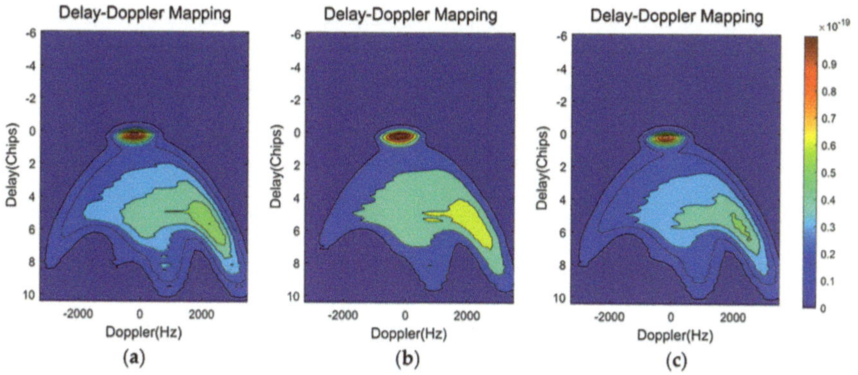

Fig. 8 DDMs at LR polarization (**a**), VR polarization (**b**), and HR polarization (**c**)

GNSS Reflectometry provides a transformative approach for monitoring key soil physical parameters globally. As one of the most advanced GNSS-R bare soil simulators, the LAGRS-Soil model, through its integration of incoherent scattering calculation, full-polarization simulation, and freeze–thaw response analysis, significantly deepens the understanding of GNSS-R signal interactions with soil.

Future directions for GNSS-R soil sensing include: First, utilizing multi-GNSS (GPS, BeiDou, Galileo) and multi-frequency (e.g., L1, L5) signals to enhance observational dimensions and spatiotemporal density. Second, developing more robust and efficient semi-empirical and machine learning inversion algorithms based on physical models like LAGRS-Soil to advance operational high-precision soil moisture and freeze–thaw products. Third, strengthening space-ground synergy and multi-source data fusion by integrating GNSS-R with active/passive microwave (e.g., SMAP, SMOS), optical, and thermal infrared data for complementary advantages and improved retrieval accuracy and application scope. Finally, expanding into emerging applications beyond soil moisture and freeze–thaw, such as vegetation optical depth, snow water equivalent, and inland water level estimation.

3 GNSS-R Vegetation Model

Initially focused on ocean, soil, and other applications, GNSS-R's potential for vegetation monitoring has increasingly been recognized. Compared to traditional monostatic radar, GNSS-R exhibits less saturation in its response to biomass, maintaining sensitivity even over dense, high-biomass vegetation [17, 18]. Advancing this application requires physical models that accurately characterize vegetation-GNSS signal interaction. However, prior research often emphasized statistical analysis of experimental data, lacking depth in physical mechanisms, particularly concerning polarization properties and scattering geometry modeling. Addressing this gap, researchers developed the Land surface GNSS-R Simulator—Vegetation (LAGRS-Veg), a physically-based, full-polarization capable spaceborne GNSS-R vegetation simulator that combines vegetation radiative transfer models with GNSS scattering models, providing end-to-end simulation from vegetation parameters to GNSS-R observables.

3.1 Overview of the LAGRS-Veg Model

With growing application of GNSS-R technology in vegetation monitoring, a pressing need emerged for accurate physical models simulating vegetation-GNSS signal interaction. Traditional statistical models, while useful under specific conditions, lack generalizability and a strong physical basis, limiting their adaptability across different vegetation types and observation scenarios. To address this, the LAGRS-Veg model was developed to establish a physically-based, full-polarization simulation platform for spaceborne GNSS-R vegetation scenarios.

LAGRS-Veg supports full-polarization simulation (RHCP, LHCP, H, V, $\pm$ 45° linear), is multi-frequency and multi-GNSS compatible (BeiDou, GPS, Galileo, GLONASS), and simulates scattering processes under various incidence, scattering, and azimuth angles using physics-based microwave scattering theory to describe vegetation volume and surface scattering. Its modular architecture also supports extension to other surface types like soil, snow, and wetlands.

Structurally, LAGRS-Veg comprises three core modules: an observation geometry module computing transmitter/receiver positions, velocities, and scattering geometry; a vegetation scattering module solving for the vegetation layer's bistatic scattering coefficient based on radiative transfer equations; and a DDM generation module synthesizing coherent and incoherent scattering components to produce the Delay-Doppler Map. Auxiliary modules for noise generation and Fresnel zone calculation complete the integrated simulation system.

As shown in Fig. 9, the workflow begins with input parameters including soil layer parameters (moisture, roughness), vegetation layer parameters (water content, biomass, temperature, size), and GNSS transmitter/receiver parameters (position, velocity). The vegetation scattering module then computes vegetation scattering properties stepwise via dielectric constant models, single-scatterer models, and radiative transfer equation models. The geometry module subsequently determines key parameters like incidence angle, scattering angle, the first Fresnel zone, and the glistening zone. Finally, the DDM generation module computes coherent and incoherent scattering components separately, combining them with a noise model to output DDM power, delay waveform power, and bistatic scattering coefficients.

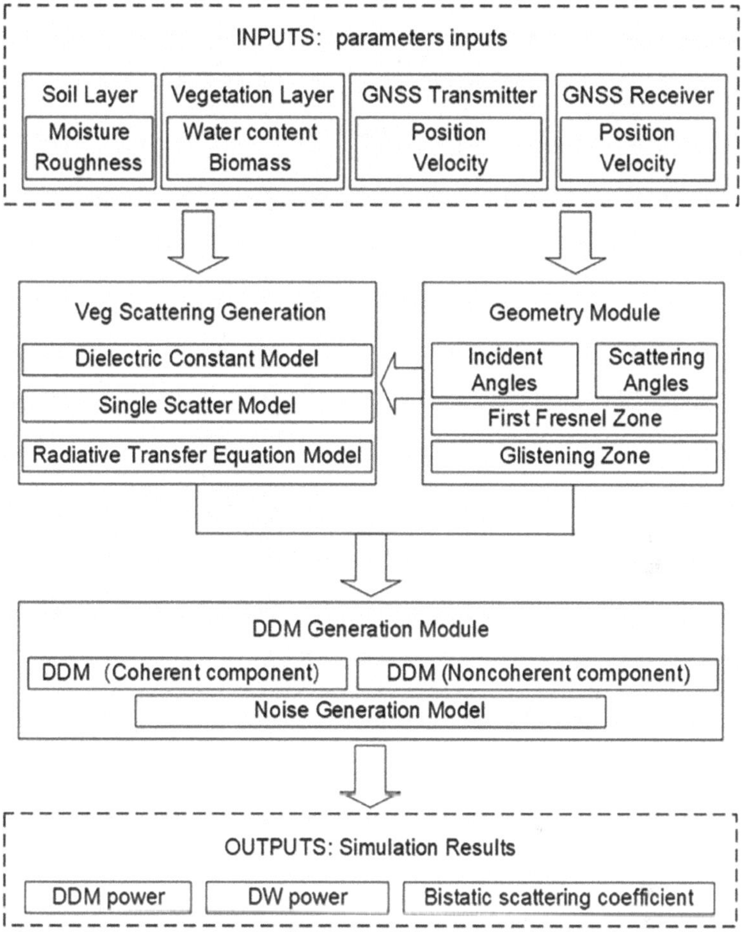

Fig. 9 Flowchart of the LAGRS-Veg model, which is mainly composed of three components: vegetation scattering generation module, GNSS-R observation geometry module and DDM generation module. Model inputs include vegetation physical parameters and transmitter/receiver parameters. Simulation output includes DDM/DW power and the vegetation bistatic scattering coefficient

3.2 Theoretical Basis of Vegetation Scattering Models

Basic Principles of Electromagnetic Wave-Vegetation Interaction

The interaction between GNSS signals and vegetation involves complex physical processes, including scattering, transmission, and absorption. Incident GNSS signals undergo multiple interactions with vegetation components (leaves, branches, trunks, etc.). These interactions manifest primarily in three forms: volume scattering (signals scattered multiple times within the vegetation volume before returning to

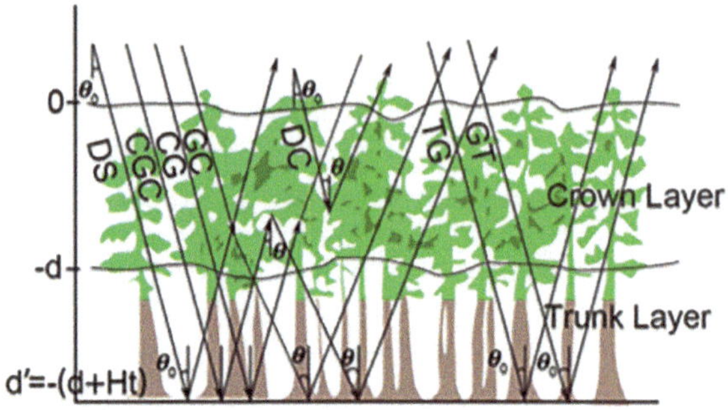

Fig. 10 Scattering mechanisms considered in the vegetation radiative transfer equation model: G–C–G, C–G, D–C, G–C, G–T, D–S, and T–G (see text for definitions)

the receiver); surface scattering (signals reflected directly by the ground surface); and volume-surface interactions (combined paths involving vegetation scattering and ground reflection).

From a mechanistic perspective, vegetation affects GNSS signals in two primary ways: through attenuation, where the vegetation layer reduces signal strength via absorption and scattering, and through alteration of polarization properties, where the anisotropic structure of vegetation changes the signal's polarization state, causing significant depolarization. Understanding and quantifying these physical processes is crucial for developing accurate scattering models.

In the LAGRS-Veg model, for forests and other tall vegetation, the vegetation layer is subdivided into canopy and trunk layers. The canopy includes leaves and branches, while the trunk layer is represented by vertically distributed cylinders. When the trunk layer is excluded from simulations, the model adapts to short vegetation like crops. Scattering mechanisms in forests encompass seven primary types, which collectively constitute the total scattering response. As annotated in the Fig. 10, the scattering mechanisms include ground reflection and crown scattering and ground reflection (G–C–G), crown scattering and ground reflection (C–G), direct crown bistatic scattering (D–C), ground reflection and crown scattering (G–C), ground reflection and trunk scattering (G–T), direct ground specular scattering (D–S), and trunk scattering and ground reflection (T–G) [16]. These mechanisms can be effectively described using ray-tracing methods, providing a solid theoretical foundation for the model.

Radiative Transfer Theory

Traditional GNSS-R vegetation scattering models often employed simplified formulations, for linear polarizations [19]:

$$\sigma_{\mathrm{vv}} = \frac{\varepsilon \cos\theta_i - \sqrt{\varepsilon - \sin^2\theta_i}}{\varepsilon \cos\theta_i + \sqrt{\varepsilon - \sin^2\theta_i}} \tag{3.1}$$

For circular polarization:

$$\sigma_{\text{hh}} = \frac{\cos\theta_i - \sqrt{\varepsilon - \sin^2\theta_i}}{\varepsilon\cos\theta_i + \sqrt{\varepsilon - \sin^2\theta_i}} \tag{3.2}$$

where σ is the reflection coefficient (with subscripts vv and hh denoting the polarization states), ε is the dielectric constant, and θ_i is the incidence angle.

For circular polarization, it is typically expressed in terms of a combination of linear polarizations:

$$\sigma_{\text{RL}} = \sigma_{\text{LR}} = \frac{1}{2}\left[(\sigma_{\text{vv}} - \sigma_{\text{hh}}) + j(\sigma_{\text{vh}} + \sigma_{\text{hv}})\right] \tag{3.3}$$

$$\sigma_{\text{RR}} = \sigma_{\text{LL}} = \frac{1}{2}\left[(\sigma_{\text{vv}} + \sigma_{\text{hh}}) + j(\sigma_{\text{vh}} + \sigma_{\text{hv}})\right] \tag{3.4}$$

where the LR and RR scattering coefficients represent linear combinations of the VV and HH scattering coefficients. Considering the effects of surface roughness, the final calculation is given by:

$$\Gamma = |\sigma_{\text{LR}}| e^{-4k^2 z^2 \cos^2\theta_i} \tag{3.5}$$

where k is the wavenumber, and z is the root mean square (RMS) height of the surface roughness. In GNSS-R, the LR signal is the dominant observable due to effective depolarization by land surfaces. Thus, the Eq. (2.5) focuses on LR to link measurements to geophysical parameters like soil moisture and roughness. The LL signal is much weaker and more susceptible to noise, so it's often omitted in practical retrieval models.

These traditional approaches potentially lose information in two key areas: detailed polarization effects, and the full influence of scattering azimuth and zenith angles, impacting the consideration of polarization and scattering geometry. Therefore, LAGRS-Veg employs more comprehensive microwave scattering models to account for these effects.

LAGRS-Veg utilizes an incoherent scattering model based on Radiative Transfer (RT) theory. The radiative transfer approach starts from the scattered energy of individual scatterers, treating the vegetation layer as a collection of scatterers with specific water content, geometric size, and orientation distribution. This incoherent scattering model, based on RT theory, is implemented in the LAGRS-Veg model.

Consistent with the principles outlined in Sect. 1.4, analyzing the Delay-Doppler Map (DDM) and its corresponding delay waveform enables effective retrieval of surface vegetation parameters, such as vegetation water content and biomass. In LAGRS-Veg, key input parameters for vegetation include vegetation water content and biomass, while soil moisture and surface roughness are input as soil parameters. Based on the radiative transfer equation model, the scattering coefficient of the vegetation layer is calculated and fed directly into the DDM generation module. This mechanism establishes a physical connection between vegetation/soil parameters and GNSS-R observables (DDM), allowing systematic analysis of their combined influence on DDM observations using the LAGRS-Veg simulator.

3.3 Radiative Transfer Formulas in the Model

LAGRS-Veg is a physical model based on RT theory, simulating microwave transmission through vegetation via the processes of extinction and emission. The model employs three core matrices to fully characterize this process: the scattering matrix, describing the amplitude and phase relationship between incident and scattered waves; the phase matrix, characterizing the angular distribution of scattered energy; and the extinction matrix, quantifying energy attenuation along the propagation path [14].

The scattering matrix defines the relationship between the incident field E_v^s and the scattered field E_h^s.

$$\begin{bmatrix} E_v^s \\ E_h^s \end{bmatrix} = \frac{e^{ikr}}{r} \begin{bmatrix} S_{vv} & S_{vh} \\ S_{hv} & S_{hh} \end{bmatrix} \begin{bmatrix} E_v^i \\ E_h^i \end{bmatrix} \tag{3.6}$$

The coefficient $\frac{e^{ikr}}{r}$ indicates that the scattered wave approximates a spherical wave at a certain distance. The scattering amplitude is a function of the polarization (v, h), the incident distance$r$, the wavenumber$k$, the zenith angleθ, and the azimuth angles φ of both the incident and scattered waves.

The phase matrix is calculated from the average modified Mueller matrix of composite scatterers within the canopy. It relates the Stokes vectors of the incident and scattered waves. For a medium containing one type of scatterer, it is given by a specific formula:

$$P = 4\pi \begin{bmatrix} \langle |S_{vv}|^2 \rangle & \langle |S_{vh}|^2 \rangle & \langle \mathrm{Re}(S_{vh}S_{vv}^*) \rangle & \langle \mathrm{Im}(S_{vh}S_{vv}^*) \rangle \\ \langle |S_{hv}|^2 \rangle & \langle |S_{hh}|^2 \rangle & \langle \mathrm{Re}(S_{hh}S_{hv}^*) \rangle & \langle \mathrm{Im}(S_{hh}S_{hv}^*) \rangle \\ 2\langle \mathrm{Re}(S_{vv}S_{hv}^*) \rangle & 2\langle \mathrm{Re}(S_{vh}S_{hh}^*) \rangle & \langle \mathrm{Re}(S_{vv}S_{hh}^* + S_{vh}S_{hv}^*) \rangle & \langle \mathrm{Im}(S_{vv}S_{hh}^* + S_{vh}S_{hv}^*) \rangle \\ 2\langle \mathrm{Im}(S_{vv}S_{hv}^*) \rangle & 2\langle \mathrm{Im}(S_{vh}S_{hh}^*) \rangle & \langle \mathrm{Im}(S_{vv}S_{hh}^* + S_{vh}S_{hv}^*) \rangle & \langle \mathrm{Re}(S_{vv}S_{hh}^* - S_{vh}S_{hv}^*) \rangle \end{bmatrix} \tag{3.7}$$

It is defined as 4π times the Mueller matrix, describing the transformation of polarization states during scattering.

For the GNSS-R configuration, four angular combinations are used to compute the extinction matrix: two upward and two downward directions. The same applies to the phase matrix. The extinction matrix relates the incident Stokes vector to the attenuated Stokes vector:

$$K_e = n_0 \begin{bmatrix} -2\mathrm{Im}(M_{vv}) & 0 & -\mathrm{Im}(M_{vh}) & \mathrm{Re}(M_{vh}) \\ 0 & -2\mathrm{Im}(M_{hh}) & -\mathrm{Im}(M_{hv}) & -\mathrm{Re}(M_{hv}) \\ -2\mathrm{Im}(M_{hv}) & -2\mathrm{Im}(M_{vh}) & -\mathrm{Im}(M_{vv}+M_{hh}) & \mathrm{Re}(M_{vv}-M_{hh}) \\ 2\mathrm{Re}(M_{hv}) & -2\mathrm{Re}(M_{vh}) & -\mathrm{Re}(M_{vv}-M_{hh}) & -\mathrm{Im}(M_{vv}+M_{hh}) \end{bmatrix} \tag{3.8}$$

When scatterers are uniformly distributed in azimuth, M_{vh}=M_{hv}= 0.For a single particle (such as a leaf, branch, or trunk element considered in isolation), the modified Mueller matrix is typically used, converting the incident intensity into the scattered intensity by treating both as modified Stokes vectors.

Looking ahead, GNSS-R vegetation remote sensing holds broad prospects. Increasing numbers of spaceborne GNSS-R missions, such as the HydroGNSS mission, will provide richer data resources. Advancements in full-polarization technology will further enhance vegetation parameter monitoring capabilities, supporting applications like precision agriculture and forest resource management. Multi-source data fusion and assimilation will become a key trend, combining GNSS-R with optical, microwave, and other remote sensing data for complementary advantages.

However, several areas require further development. First, vegetation parameterization schemes need refinement, particularly regarding generalizability across different vegetation types (e.g., crops, forests). Second, topographic effects are not yet fully considered, and model accuracy under complex terrain conditions requires validation. Furthermore, model verification efforts need strengthening with more ground measurement support. As models continue to improve and application experience deepens, GNSS-R technology is poised to play an increasingly vital role in global vegetation monitoring, providing new technical means for ecological environment monitoring and sustainable development.

4 GNSS-R Freeze–Thaw Model

This section focuses on the application of GNSS-R for monitoring the critical geophysical process of bare soil freeze–thaw cycles. Soil freezing and thawing are crucial processes in cold region environments, directly influencing surface-atmosphere energy exchange (sensible and latent heat fluxes), water migration, carbon cycling, and ecosystem productivity. Seasonal frost or permafrost affects roughly 35% of the global land area. Traditionally, monitoring freeze–thaw processes has relied on limited ground station observations, which are costly and often unrepresentative at regional scales. While microwave remote sensing (including radar and radiometers) has been employed for this purpose, its spatiotemporal resolution frequently falls short of requirements for detailed scientific studies [20–22]. Consequently, GNSS-R technology, with its high spatiotemporal resolution, offers a promising new solution for dynamic, large-scale monitoring of soil freeze–thaw states.

Although GNSS-R has demonstrated promising potential for large-scale soil freeze–thaw monitoring, the underlying physical mechanisms governing the response of reflected signals to freeze–thaw transitions remain incompletely understood. A deeper insight into these mechanisms is essential not only for interpreting existing observations but also for informing the design of future satellite missions and the development of robust data processing algorithms. To this end, the development of a physically based forward model capable of accurately simulating full-polarization GNSS-R observables is of particular importance.

In this section, we present the application and extension of the LAGRS-Soil model to bare soil freeze–thaw cycle detection, e.g., LAGRS-FT. By integrating temperature-dependent dielectric mixing models with the core full-polarization scattering framework described in Sect. 2, this configuration enables quantitative analysis of the distinct signal responses associated with frozen and thawed states.

4.1 Electromagnetic Properties and Modeling of Bare Soil Freeze–Thaw Processes

The dielectric constant, a core parameter determining electromagnetic scattering characteristics, describes a material's ability to polarize in response to an external electric field. For soil, the dielectric constant depends primarily on the dielectric properties and volume fractions of its constituents. Dry soil typically has a low dielectric constant real part (around 2–5), while liquid water's is very high (approximately 80 at L-band), and ice's is low (about 3.2). Consequently, even small changes in liquid water content cause significant variations in the soil's bulk dielectric constant.

The most fundamental physical change during soil freeze–thaw cycles is the phase transition of water: liquid in the thawed state and solid ice in the frozen state. The substantial dielectric contrast between liquid water and ice causes a sharp discontinuity in the soil's bulk dielectric constant during the phase transition. This forms the fundamental physical basis for using microwave remote sensing, including GNSS-R, to detect soil freeze–thaw states.

To quantitatively compute the dielectric constant of frozen/thawed soil, several dielectric mixing models have been developed:

Thawed Soil Models: Often employ four-component models (e.g., Dobson, Hallikainen) [23], treating soil as a mixture of soil solids, bound water, free water, and air. These models, fitted to experimental data, effectively describe the dielectric behavior of soils with different textures and moisture levels.

Frozen Soil Models: Extend thawed soil models by incorporating the ice component. A critical aspect is determining the unfrozen water content at sub-zero temperatures, which is a function of soil temperature, texture, and salinity. It is commonly estimated using an empirical formula, for example:

$$W_u = A \cdot |T - 273.2|^{(-B)} \tag{4.1}$$

where W_u is the unfrozen water content, T is the soil temperature (K), and A & B are empirical parameters related to soil texture. The general form, applicable to both thawed and frozen models, can be expressed as:

$$\varepsilon_m^\alpha = V_s\varepsilon_s^\alpha + V_{bw}\varepsilon_{bw}^\alpha + V_{fw}\varepsilon_{fw}^\alpha + V_a\varepsilon_a^\alpha + V_i\varepsilon_i^\alpha \tag{4.2}$$

where ε_{m} is the effective complex dielectric constant of the mixture, {α (a superscript index, not a power exponent) is the shape factor, V and ε represent the volume fraction and complex dielectric constant of each component, respectively, and the subscripts s, bw, fw, a, i denote soil solids, bound water, free water, air, and ice, respectively.

Figure 11 presents typical curves of soil dielectric constant (real and imaginary parts) versus temperature simulated using the aforementioned models. A clear discontinuity is evident near 0 °C, directly reflecting the freeze–thaw transition. The figure also compares scenarios with different soil moisture levels, showing that higher moisture leads to a larger dielectric contrast between frozen and thawed states, implying greater sensitivity of GNSS-R signals to freeze–thaw in wetter soils.

An increase in the dielectric constant real part enhances soil reflectivity. According to Fresnel reflection formulas, for a smooth surface, a higher dielectric constant increases the power reflection coefficient, particularly at vertical polarization. Therefore, when soil transitions from frozen (low dielectric) to thawed (high dielectric) state, its reflectivity to GNSS signals strengthens significantly, leading to increased reflected power received by the GNSS-R receiver. Although natural surfaces are

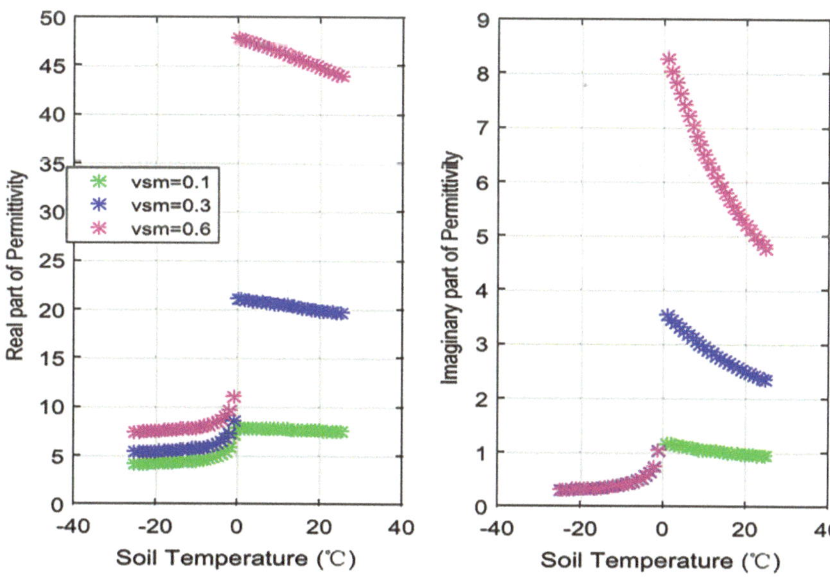

Fig. 11 Real part (left) and imaginary part (right) of the permittivity versus soil temperature at different soil moisture

rough and GNSS-R is bistatic, this fundamental trend persists. Changes in the dielectric constant imaginary part primarily affect signal penetration depth and energy loss within the soil.

4.2 *Construction of a Full-Polarization GNSS-R Scattering Model*

Accurately simulating the signal power reflected from the land surface and received by a GNSS-R instrument requires a scattering model that precisely calculates the bistatic scattering coefficient σ^0, a key parameter in the Z-V model [1].

Surface Scattering Models: From IEM to AIEM

Scattering from a bare soil surface is a classical random rough surface scattering problem. Traditional models like the Perturbation Method are suitable for surfaces with small height variations, while the Kirchhoff Approximation works best for surfaces with gentle undulations. The Integral Equation Model (IEM) was developed to overcome the limitations of these models under specific roughness conditions, covering a broader range of natural surface roughness.

However, the IEM has shortcomings in certain assumptions and its representation of the roughness spectrum. Consequently, Wu, Chen, and others introduced improvements, leading to the Advanced Integral Equation Model (AIEM) [9]. The AIEM offers higher accuracy and wider applicability and is widely used for land surface scattering simulation. In the AIEM, the bistatic scattering coefficient σ_{pq}^0 can be expressed as the sum of three terms:

$$\sigma_{\mathrm{qp}}^c = \sigma_{\mathrm{qp}}^k + \sigma_{\mathrm{qp}}^{kc} + \sigma_{\mathrm{qp}}^c \tag{4.3}$$

Here, σ_{qp}^k is the Kirchhoff term, describing coherent scattering from large-scale roughness; σ_{qp}^c is the complementary term, describing incoherent scattering from small-scale roughness; and $\sigma_{\mathrm{qp}}^{kc}$ is the cross term, describing coupling between different roughness scales.

Input parameters for the AIEM include the incident and scattered wave geometry, surface dielectric constant, and surface roughness parameters (typically described by root mean square height and correlation length).

Full-Polarization Simulation: Polarization Synthesis Technique

GNSS satellites transmit circularly polarized waves, but surface reflection alters the polarization state. To fully exploit polarization information, our simulator must compute scattering coefficients for arbitrary polarization combinations. However, traditional scattering models typically provide results for linear polarizations (VV, HH, HV). To address this, polarization synthesis technology is introduced. This technique utilizes the Mueller matrix and Stokes vector. Using polarization synthesis, the scattering coefficient for any transmit/receive polarization combination can be calculated via specific formulas involving the azimuth and ellipticity angles of the polarization ellipse. The calculation follows the same method as in Eqs. (2.4).

The full-polarization GNSS-R DDM simulator systematically integrates observation geometry and surface scattering models to achieve forward simulation from land surface parameters to observables. Its workflow involves three main stages: during parameter initialization, geometric parameters (spatial positions and velocity vectors of GNSS satellite and receiver) and physical parameters (soil temperature, volumetric water content, roughness parameters, soil texture) are defined; a parallel computation stage runs the geometry calculation module and the scattering calculation module simultaneously, with the former resolving geometric relationships for scattering points within the specular zone and the latter performing a three-stage calculation (dielectric constant solution, linear scattering simulation, and full-polarization synthesis) to obtain scattering coefficients for any polarization combination; finally, in the DDM generation stage, these results are substituted into the Z-V model for integration, incorporating factors like antenna gain and path loss, to output the Delay-Doppler Map for a specific surface state and polarization combination. This integrated workflow provides a vital quantitative analysis tool for studying freeze–thaw detection mechanisms and developing retrieval algorithms. The flowchart of the fully polarization DDM simulator for soil freeze/thaw process detection is presented in Fig. 12.

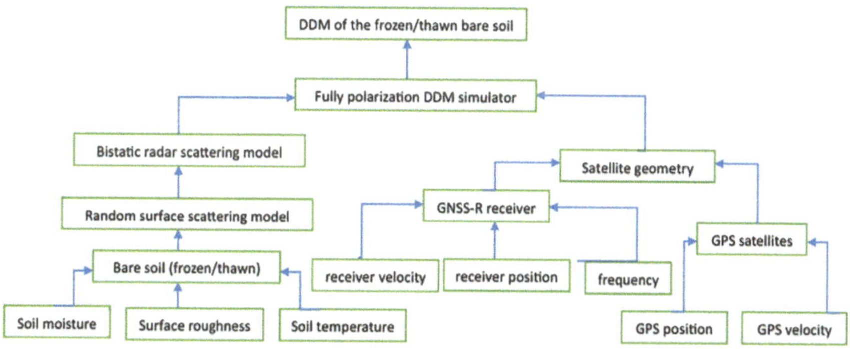

Fig. 12 Flowchart of AIEM-based Z–V model for bare soil freeze/thaw process detection

4.3 Discussion on the LAGRS-FT Model

Although this simulator offers an effective theoretical tool for understanding GNSS-R freeze–thaw detection mechanisms, its current version has several limitations that point toward key future research directions. A primary limitation is the bare soil assumption, which often diverges from real-world conditions. Most frozen ground regions feature seasonal snow cover or vegetation, which significantly modify signal propagation and reflection characteristics. Therefore, integrating scattering models for vegetation and snow layers into the existing framework is a crucial step toward enhancing model practicality. Second, model validation remains challenging. Due to the scarcity of dedicated field campaigns providing synchronous GNSS-R observations during freeze–thaw transitions, current validation relies heavily on inter-model comparison. Future efforts urgently need specialized observational experiments across different climate zones and surface types to optimize model parameters and structure using measured data. Finally, important steps remain in transitioning from theoretical models to operational applications. Based on patterns revealed by simulations, robust freeze–thaw state identification algorithms need further development. These algorithms must reliably extract freeze–thaw information from real GNSS-R DDM observations while effectively separating the influences of confounding factors like soil moisture and roughness. Progress in these research areas will enable GNSS-R technology to play a more significant role in global change studies and cold region environment monitoring.

5 Conclusions and Future Perspectives

GNSS Reflectometry (GNSS-R) has evolved from its original ocean-focused applications into a versatile Earth observation technique capable of providing valuable information on land surface parameters, particularly soil moisture, surface roughness, vegetation characteristics, and freeze–thaw state. This chapter systematically presented the theoretical foundations, model development, and application-oriented advancements of the **LAGRS family of simulators** (LAGRS-Soil, LAGRS-Veg, and the full-polarization freeze–thaw extension), which represent a coherent set of physically-based forward modeling tools tailored for spaceborne GNSS-R observations over complex land surfaces.

The **LAGRS-Soil** model constitutes one of the most comprehensive GNSS-R bare soil simulators currently available. By integrating optimized coherent scattering formulations, a wide range of classical and advanced rough surface scattering theories (GO, PO, SPM, IEM, AIEM), full-polarization synthesis capability via Mueller matrix formalism, and a temperature-dependent frozen soil dielectric model, it enables realistic simulation of Delay-Doppler Maps (DDMs) under diverse soil conditions—from arid to cold-region environments. The model successfully reproduces the strong sensitivity of GNSS-R observables to volumetric soil moisture, surface roughness, incidence/scattering geometry, and especially the pronounced dielectric discontinuity during soil freeze–thaw transitions. These capabilities provide a solid theoretical foundation for interpreting current FY-3E GNOS-R data and for guiding the design of future multi-polarization GNSS-R payloads.

Building upon similar full-polarization and multi-GNSS architecture, the **LAGRS-Veg** model extends the simulation capability to vegetated terrains. By coupling advanced radiative transfer theory with detailed vegetation structural parameterization (canopy + trunk layers), multiple scattering mechanisms (G–C–G, C–G, D–C, G–C, G–T, D–S, T–G), and polarization-dependent volume/surface interactions, it offers a physically consistent end-to-end simulation framework from biophysical parameters to spaceborne DDMs. The model addresses a long-standing gap in GNSS-R vegetation studies—the lack of polarization-resolved, geometry-aware forward models—and lays the groundwork for quantitative biomass, vegetation water content, and vegetation optical depth retrieval, even in high-biomass regimes where conventional backscatter techniques tend to saturate.

The development of these models also highlights several key scientific and technical advances achieved in recent years:

- recognition of the critical role of **incoherent (diffuse) scattering** over land surfaces and its proper integration into DDM simulation;
- systematic utilization of **full-polarization information** (LR, RR, VR, HR, $\pm 45°$ linear, etc.) to enhance parameter sensitivity and reduce ambiguity in retrieval;
- quantitative modeling of **dielectric phase transitions** in cold-region soils, providing a mechanistic basis for freeze–thaw state detection using GNSS-R;
- establishment of modular, extensible simulation platforms compatible with current (GPS, BeiDou, Galileo, GLONASS) and emerging multi-frequency signals (L1/L2/L5).

Despite these advances, several important challenges and research directions remain open for the coming 5–10 years:

1. **Model validation and calibration with in-situ and airborne datasets**

Comprehensive, synchronous ground/airborne GNSS-R experiments across different climate zones, soil textures, vegetation types, and freeze–thaw stages are urgently needed to quantitatively evaluate model performance and refine key parameterization schemes.

2. **Extension to more realistic land surface conditions**

Current versions mainly focus on bare soil or simplified vegetation structures. Future efforts should incorporate seasonal snow cover, complex topography, layered soil profiles, and dynamic soil moisture–temperature coupling.

3. **Development of physically consistent retrieval algorithms**

Moving from forward simulation to operational inversion requires robust semi-empirical, machine learning-assisted, or hybrid (physics-informed neural network) retrieval frameworks that can effectively disentangle the confounding effects of moisture, roughness, vegetation, and geometry.

4. **Multi-source data fusion and assimilation**

Combining GNSS-R with active/passive microwave (SMAP, SMOS, Sentinel-1), optical (Sentinel-2, Landsat), and thermal infrared observations is expected to significantly improve retrieval accuracy, spatial continuity, and temporal resolution.

5. **Support for next-generation GNSS-R missions**

With missions such as HydroGNSS, GEROS-ISS follow-on, and potential Chinese FY-3 series upgrades on the horizon, full-polarization, multi-frequency, multi-constellation GNSS-R observations will become available. The LAGRS family of models should continue to evolve to serve as reference forward operators and sensitivity analysis tools for these new observation capabilities.

In summary, the LAGRS-Soil and LAGRS-Veg models, together with their extensions for freeze–thaw processes, represent a significant step forward in establishing a physically sound, polarization-resolving, and multi-scenario GNSS-R land surface simulation capability. They not only deepen our understanding of GNSS signal interaction mechanisms over complex terrestrial environments but also provide essential theoretical support for transforming GNSS-R from a promising technique into a reliable operational data source for global soil moisture, vegetation, and cryosphere monitoring in the coming decade.

References

1. Zavorotny VU, Voronovich AG (2000) Scattering of GPS signals from the ocean with wind remote sensing application. IEEE Trans Geosci Remote Sens 38:951–964. https://doi.org/10.1109/36.841977
2. Pierdicca N, Comite D, Camps A et al (2022) The potential of spaceborne GNSS reflectometry for soil moisture, biomass, and freeze–thaw monitoring: summary of a European Space Agency-funded study. IEEE Geosci Remote Sens Mag 10(2):8–38. https://doi.org/10.1109/MGRS.2021.3115448
3. Chew C, Shah R, Zuffada C et al (2018) Vegetation sensing using GPS-interferometric reflectometry: theoretical effects of canopy parameters on signal coherence. IEEE Trans Geosci Remote Sens 56(2):838–851. https://doi.org/10.1109/TGRS.2017.2759023
4. Cardellach E, Rius A, Ribó S et al (2022) Consolidation of GNSS and radio occultation reflections: theory and applications. Surv Geophys 43:335–366. https://doi.org/10.1007/s10712-021-09653-5
5. Kurum M, O'Neill PE, Lang RH, Joseph AT, Cosh MH (2019) The status of soil moisture retrievals using P-band signals of opportunity. In: *IGARSS 2019—2019 IEEE international geoscience and remote sensing symposium*, Yokohama, Japan, pp 6924–6927. https://doi.org/10.1109/IGARSS.2019.8899196
6. Clarizia MP, Pierdicca N, Costantini F, Floury N (2019) Analysis of CYGNSS data for soil moisture retrieval. IEEE J Sel Top Appl Earth Obs Remote Sens 12(1):2227–2235. https://doi.org/10.1109/JSTARS.2019.2895510

7. Pierdicca N, Guerriero L, Giusto R, Brogioni M, Egido A (2014) SAVERS: a simulator of GNSS reflections from bare and vegetated soils. IEEE Trans Geosci Remote Sens 52:6542–6554. https://doi.org/10.1109/TGRS.2014.2299811
8. Fung AK, Eom HJ (1983) Coherent scattering of a spherical wave from an irregular surface. IEEE Trans Antennas Propag 31(1):68–72. https://doi.org/10.1109/TAP.1983.1142984
9. Chen KS, Wu TD, Tsang L, Li Q, Shi JC, Fung AK (2003) Emission of rough surfaces calculated by the integral equation method with comparison to three-dimensional moment method simulations. IEEE Trans Geosci Remote Sens 41(1):90–101. https://doi.org/10.1109/TGRS.2002.807587
10. Ferrazzoli P, Guerriero L (1996) A numerical model to simulate the emission from forest canopies. Int J Remote Sens 17(4):797–809. https://doi.org/10.1080/01431169608949046
11. Kurum M, Deshpande M, Joseph AT, O'Neill PE, Lang RH, Eroglu O (2019) SCoBi-Veg: a generalized bistatic scattering model of reflectometry from vegetation for signals of opportunity applications. IEEE Trans Geosci Remote Sens 57(2):1049–1068. https://doi.org/10.1109/TGRS.2018.2864634
12. Wu X, Wang F (2023) LAGRS-Veg: a spaceborne vegetation simulator for full polarization GNSS-reflectometry. GPS Solut 27(3):107. https://doi.org/10.1007/s10291-023-01441-5
13. Wu X, Xia J, Yan Z, Wang F (2023) LAGRS-soil: a full-polarization GNSS-reflectometry model for bare soil applications in FY-3E GNOS-R payload. Remote Sens 15(22):5296. https://doi.org/10.3390/rs15225296
14. Ulaby FT, Long DG (2014) Microwave radar and radiometric remote sensing. University of Michigan Press
15. Wu X, Jin S (2020) Models and theoretical analysis of SoOp circular polarization bistatic scattering for random rough surface. Remote Sens 12:1506. https://doi.org/10.3390/rs12091506
16. Ulaby FT, Sarabandi K, Mcdonald K, Whitt M (1990) Michigan microwave canopy scattering model. Int J Remote Sens 11:1223–1253. https://doi.org/10.1080/01431169008955090
17. Carreno-Luengo H, Camps A (2021) Global navigation satellite system-reflectometry: fundamentals, status, and trends. IEEE Geosci Remote Sens Mag 9(3):10–35. https://doi.org/10.1109/MGRS.2021.3055923
18. Carreno-Luengo H, Luzi G, Crosetto M (2020) Above-ground biomass retrieval over tropical forests: a novel GNSS-R approach with CyGNSS. Remote Sens 12(9):1368. https://doi.org/10.3390/rs12091368
19. Park H, Camps A, Pablos M, Foti G, Gommenginger C (2020) Analytical modelling of ocean-reflected GNSS-R delay-Doppler maps using the Kirchhoff Approximation. IEEE Trans Geosci Remote Sens 58(10):6790–6800. https://doi.org/10.1109/TGRS.2020.2977646
20. Zhang T (2005) Influence of the seasonal snow cover on the ground thermal regime: an overview. Rev Geophys 43:RG4002. https://doi.org/10.1029/2004RG000157
21. Zhang T, Barry RG, Knowles K, Ling F, Armstrong RL (2003) Distribution of seasonally and perennially frozen ground in the Northern Hemisphere. In: Proceedings of the 8th international conference on Permafrost, Zurich, Switzerland, 21–25 July 2003, pp 1289–1294
22. Zhao T, Zhang L, Jiang L, Zhao S, Chai L, Jin R (2021) A review of remote sensing for permafrost mapping: progress, challenges, and opportunities. Earth-Sci Rev 218:103684. https://doi.org/10.1016/j.earscirev.2021.103684
23. Dobson MC, Ulaby FT, Hallikainen MT, El-Rayes MA (1985) Microwave dielectric behavior of wet soil-part II: dielectric mixing models. IEEE Trans Geosci Remote Sens GE-23(1):35–46. https://doi.org/10.1109/TGRS.1985.289498

Chapter 6
Iterative Range Compressions for Object Enhancement of Passive GNSS-Based SAR Under Weak Signal Circumstance

Yu Zheng

Abstract Passive global navigation satellite system-based synthetic aperture radar (GNSS-SAR) has attracted numbers of research interests in recent decade. However, due to the distance between GNSS satellites and the earth surface, the respective power flux density is extremely low, where object detection can be unfavourable. To address this drawback, this chapter proposes an iterative range compression scheme based imaging method. In the proposed method, firstly, through the correlation of the reflected signal with the local direct signal replica, range compression is performed. Meanwhile, using auto-correlation of the local replica, an iterative range-matched filter is generated. Thereafter, the correlation of the initial range-compressed pulse with the iterative range-matched filter is carried out for the first round iteration. A noise threshold is settled. In the first round of iteration, if the noise of the compressed signal is higher than the settled threshold, the signal is correlated with the iterative range-matched filter for performing another round of iteration. Otherwise, azimuth compression for generating full image is carried out. To examine the proposed imaging method, simulation tests and field experimental validation are conducted using the BeiDou B3I signal. The results shows that a significantly higher image contrast-to-noise ratio (CNR) can be achieved by the proposed imaging method, compared to the state-of-the-art imaging method for imaging enhancement.

Keywords Global navigation satellite system · Synthetic aperture radar · Passive GNSS-based SAR · Iterative range compressions · Object detection

1 Introduction

Synthetic aperture radar (SAR) plays a vital role for object detection [1]. However, because of low system flexibility and low concealment for conventional active SAR, in recent decades, the technology that utilizes other sources of opportunity for SAR

Y. Zheng (✉)
The School of Electronic Communication and Electrical Engineering, Changsha University, Changsha, Hunan, China
e-mail: y_zheng170@sina.com

K. Yu and N. Zheng (eds.), *Advances in GNSS Reflectometry*, Navigation: Science and Technology 19, https://doi.org/10.1007/978-981-92-2325-1_6

receivers has been proposed, which is widely known as passive SAR [2]. As for passive SAR applications, Digital Video Broadcasting-Terrestrial (DVB-T) signals [2] and global navigation satellite system (GNSS) signals [3, 4] are the mostly welcomed sources. Specifically, due to the fact that GNSS signals have the advantages of global coverage and uninterrupted transmission, passive GNSS-based SAR (GNSS-SAR) can perform object detection under all-day, all-weather conditions.

Preliminarily, the feasibility with respect to GNSS-SAR imaging [5–13] was tested by several research groups. Image generation is mainly based on the back-projection (BP) algorithm, including two domain compressions i.e. range and azimuth compressions. Later on, methods to improve imaging gain has been proposed, as the power flux density of the GNSS at the earth's surface is very low. For instance, through joint coherent and non-coherent integration for azimuth compression, imaging gain for bi-static models is improved by Zheng et al. [14]. Meanwhile, multiple GNSS-SAR images fusion scheme were firstly considered in [15], while multi-static system model and coherently combined signals from multiple GNSS satellites were presented in [16]. Moreover, some researchers have investigated imaging methods under non-linear trace circumstances. For example, a sub-aperture combination scheme to address the azimuth defocusing problem was proposed in [17]. A two-stage imaging processing method relying on the translational motion of object over a long observation time was proposed in [18].

Low range resolution is another vital problem. To deal with this issue, methods for improving range resolution has been studied by some scholars. For example, in both [19, 20], a higher range resolution was obtained by high-bandwidth Galileo signals, compared to those provided by GPS and GLONASS signals. Typically, a range resolution at the 3 m level was obtained by Ma et al. [19] employing joint Galileo E5a and E5b signals. To distinguish objects within one pseudo-random noise (PRN) chip, range compression schemes based on intermediate frequency GNSS signals and the Diff2 operator have been proposed in [21, 22], respectively. Furthermore, the improved Diff2 operator for reducing side-lobe level and the incremental Wiener filter-based scheme for improving range resolution in situations with non-obvious correlation peaks was proposed by Fang et al. [23, 24], respectively.

Some scholars have investigated the applications of GNSS-SAR technique, including maritime surveillance, environmental surveillance, and surface deformation monitoring. The methods for detecting maritime ships, such as the imaging and localization of ships, has been examined in Refs. [25–27]. To improve maritime ship detection accuracy, a hybrid integration method and a spawning particle filter-based method for correcting movement-related errors have been proposed in Refs. [28, 29], respectively. In [30], parameter estimation of a moving object was used by a short-time integration scheme. Moreover, several studies have investigated the feasibility of BeiDou satellite system (BDS) in GNSS-SAR applications. The related studies in the latest five years can be seen as follows: On the basis of new-generation GNSS signal-BeiDou B3I signal, the feasibility of surface deformation monitoring was verified in [31]. Literature [32] presented a phase error compensation method for BDS GEO satellite application in GNSS-R deformation retrieval. A decimetre level range resolution for GNSS-SAR was achieved by the work [33]. At the same time, some literatures carried out sea ice coverage estimation, forest coverage estimation, soil moisture estimation, and snow depth estimation [34–37], respectively.

Among current literatures, a bottleneck problem for GNSS-SAR implementations is remained in weak surveillance signal power. Although extending the coherent integration time can do some helps for improving imaging gain, when dwell time is limited, this method may not be effective. For instance, an object in motion may have already disappeared before the integration can provide a satisfactory imaging gain for indication. To deal with this situation, a scheme for increasing imaging gain within limited dwell times is investigated in this chapter, and an imaging method based on iterative range compressions is proposed. In the proposed method, according to the auto-correlation of the local direct signal replica, an iterative range-matched filter is generated. Thereafter, the initial range-compressed pulse (generated through the correlation of the reflected signal with the local replica) is corrected with the iterative range-matched filter for one or several rounds until the noise level is below the expected threshold. Finally, for generating full GNSS-SAR image, azimuth compression is conducted in the manner presented in the previous work of the author in this chapter [14]. Through both simulations and experiments, the feasibility with respect to the proposed imaging method is demonstrated successfully. A higher image contrast-to-noise ratio (CNR) can be obtained for object detection, compared to the state-of-art algorithm in [14].

The rest of this chapter is organized as follows: Sect. 2 shows the principal of GNSS-SAR. The proposed imaging method is given and analyzed in Sect. 3. The simulation demonstrations and experimental validation are provided in Sect. 4 and Sect. 5, respectively. Section 6 concludes the entire chapter, and discusses the further research direction.

2 Principal of GNSS-SAR

Using a bi-static GNSS-SAR mode as an example, the respective schematic is provided in Fig. 1

From Fig. 1, we can have that GNSS-SAR system contains the following modular: (1) GNSS satellites, serves as the transmission source of opportunity; (2) Sensing area, where a targeted region for object detection; (3) GNSS signal receiver, where both reference signals from the satellites and surveillance signals from sensing area are collected simultaneously. Currently, the GNSS systems i.e. GPS, Galileo, GLOnass and BeiDou has been demonstrated for SAR imaging successfully.

Reference signal is used for synchronization. At the GNSS-SAR receiver, the base-band direct signal can be expressed as:

$$S_d = A_d \cdot C(t, u) \cdot D(u) \cdot \exp\left(j\left(2\pi f_d t + \phi_d(u)\right)\right) + n_d \tag{1}$$

where A_d denotes the magnitude; C denotes the pseudo-random noise (PRN) code; D denotes the navigation bit; f_d denotes the Doppler frequency; ϕ_d denotes the

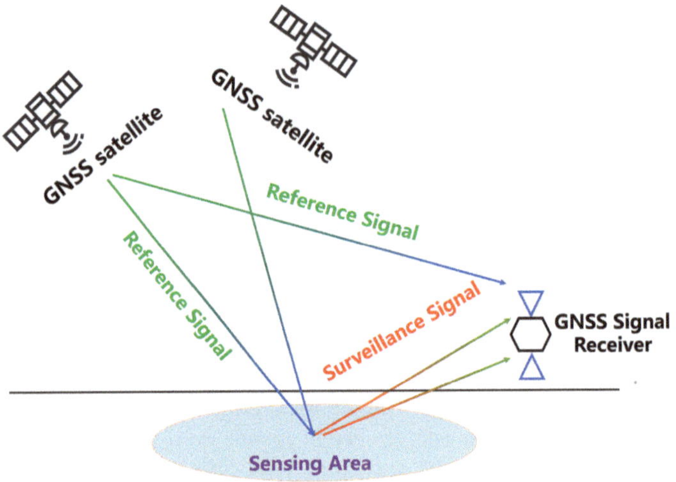

Fig. 1 The schematic diagram of the system model

carrier phase; n_d denotes the background noise at the reference channel; t denotes the range domain, which is perpendicular to the moving trace of the receiver; and u denotes the azimuth domain, which is parallel to the receiver's moving trace.

Meanwhile, the surveillance signal received from the sensing region can be considered as a delayed version of (1), which is expressed as:

$$S_r = A_r \cdot C\,(t - \tau, u) \cdot D\,(u) \cdot \exp\,(j\,(2\pi f_r t + \phi_r\,(u))) + n_r \tag{2}$$

where A_r represents the magnitude of the reflected signal, τ represents the signal delay at the range domain, f_r represents the Doppler frequency of the reflected signal, ϕ_r represents the carrier phase of the reflected signal, and n_r represents the background noise at the surveillance channel.

Before performing the imaging procedure, reference signal synchronization is carried out. The procedure of signal synchronization can be seen in Fig. 2.

According to Fig. 2, we can have that after carrying out reference signal synchronization, the parameters carrier phase, code delay, navigation bits are outputted for

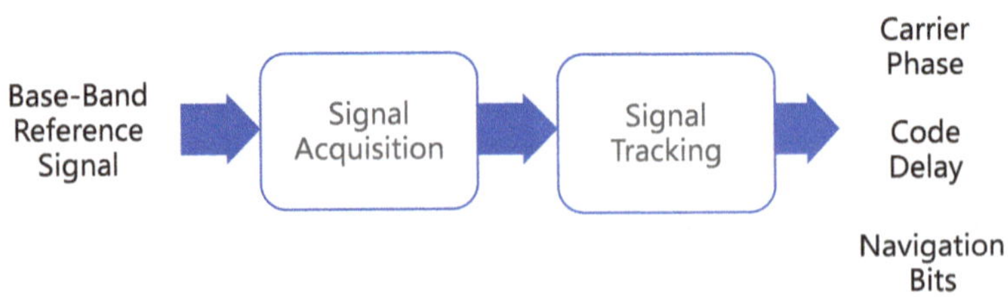

Fig. 2 The flow of signal synchronization

generating local reference signal. The local reference signal is a noise-free replica, which serves as a range domain matched filter, can be expressed as follows:

$$S_m = C(t, u) \cdot D(u) \cdot \exp(j(2\pi f_d t + \phi_d(u))). \tag{3}$$

Thereafter, starting on range compression by the correlating and signal (2) with (3) at range domain t, imaging procedure is conducting. The range compressed signal can be expressed as:

$$\begin{aligned} R_1(t, u) &= S_r(t, u) \circledast (S_m(t, u))^* \\ &= A_r \wedge (t - \tau, u) \\ &\quad \cdot \exp(j(2\pi(f_r - f_d)t + \phi_r - \phi_d)) + n_r \end{aligned} \tag{4}$$

in which, $\circledast$ represents correlation operation, $\wedge$ represents code correlation function, $*$ represents conjugate operation.

On the basis of the results as (4), azimuth compression is carried out for generating full image by integration the signals over azimuth domain u, which can be expressed as:

$$I_{GNSS-SAR} = \int_0^{T-1} R_1(t, u)\,\mathrm{d}u \tag{5}$$

where T denotes the length for performing aperture synthesising.

The overall flow for GNSS-SAR imaging can be seen in Fig. 3.

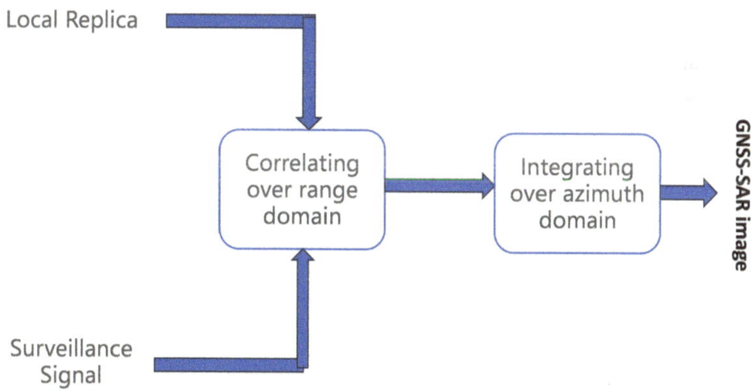

Fig. 3 GNSS-SAR imaging flow chart

3 The Iterative Range Compressions Scheme Based Imaging Method

3.1 The Algorithm Flow

The initial range compressed result is obtained in a similar manner as back-projection (BP) imaging method. The respective results can be expressed in the same manner as (4).

Then, using auto-correlation of local replica (3), iterative range matched filter is obtained, which can be expressed as:

$$\begin{aligned} R_m(t,u) &= S_m(t,u) \circledast (S_m(t,u))^* \\ &= \wedge(t,u)\exp(j0) \\ &= \wedge(t,u) \end{aligned} \tag{6}$$

On the basis of correlating initial range compressed result in (4) with (6), the first round iterative range compression is carried out, which can be expressed as:

$$\begin{aligned} R_2(t,u) &= R_1(t,u) \circledast (R_m(t,u))^* \\ &= A_r \cdot \wedge^2(t-\tau,u) \\ &\quad \cdot \exp\left(j\left(2\pi\left(f_r - f_d\right)t + \phi_r - \phi_d\right)\right) + n_r \end{aligned} \tag{7}$$

The peak value of the pulse and noise variance are extracted based on (7), and a threshold is settled for determining whether to carry out the next round iteration. The rule for setting threshold at each range domain can be theoretically modelled as a binary hypothesis problem as follows:

$$\begin{aligned} H_0 &: T = n_r; \\ H_1 &: T = R_2(t,u). \end{aligned} \tag{8}$$

Assuming in current range domain, the noise variance is σ_v, the peak value of the compressed pulse is μ_s, because the sample quantity in each range domain is very large, according to central limit theory [41], the distribution of (8) can be modelled as Gaussian distribution. Thus, the false alarm and detection probability of (8) can be derived as

$$P_f = Q\left(\frac{\varepsilon}{\sigma_v}\right) \tag{9}$$

and

$$P_d = Q\left(\frac{\varepsilon - \mu_s}{\sigma_v}\right) \tag{10}$$

respectively, where $Q(\cdot)$ represents cumulative probability density function of Gaussian distribution, ε represents noise threshold.

An optimization problem is modelled as follows for selecting a proper threshold:

$$\begin{aligned} &\max_{\varepsilon} 1 - P_f(\varepsilon) + P_d(\varepsilon) \\ &\text{s.t.} \quad 0 < \varepsilon \le \mu_s \end{aligned} \tag{11}$$

The convexity of the optimization problem (11) is studied. The respective second order differentiation of $1 - P_f(\varepsilon)$ and $P_d(\varepsilon)$ can be seen as follows:

$$\frac{\partial^2}{\partial \varepsilon^2}\left(1 - P_f\right) = \frac{1}{\sqrt{2\pi}\sigma_v} \cdot \frac{\varepsilon}{\sigma_v^2} \cdot \exp\left(-\frac{\varepsilon^2}{2\sigma_v^2}\right) \tag{12}$$

$$\frac{\partial^2}{\partial \varepsilon^2} P_d = -\frac{1}{\sqrt{2\pi}\sigma_v} \cdot \frac{\varepsilon - \mu_s}{\sigma_v^2} \cdot \exp\left(-\frac{(\varepsilon - \mu_s)^2}{2\sigma_v^2}\right) \tag{13}$$

As ε is larger than zero, (12) is larger than zero; as ε is smaller than μ_s, (13) is larger than zero as well. Thus $\frac{\partial^2}{\partial \varepsilon^2}\left(1 - P_f(\varepsilon) + P_d(\varepsilon)\right)$ is definitely larger than zero. This indicates that the problem (11) is not a convex optimization problem, a global optimal value of ε cannot be obtained. Then incremental algorithm [42] is employed to find out the local optimal value of ε. As incremental algorithm for the optimization problem in (11) is well documented, the respective description is omitted. By solving the optimization problem in (11), the proper threshold in the current round can be found out.

After choosing the proper noise threshold ε, if the noise level is below the threshold, then the next step is carried out; if the noise level is higher than the threshold, then another round of iteration is carried out by correlating the compressed result (7) with (6). Assuming n rounds of iteration are performed, the range compressed result can be expressed as:

$$\begin{aligned} R_s(t,u) &= R_1(t,u) \circledast \underset{1}{(R_m(t,u))^*} \circledast \underset{2}{(R_m(t,u))^*} \\ &\quad \circledast \cdots \circledast \underset{n-1}{(R_m(t,u))^*} \\ &= A_r \cdot \wedge^n(t-\tau,u) \\ &\quad \cdot \exp\left(j\left(2\pi\left(f_r - f_d\right)t + \phi_r - \phi_d\right)\right) + n_r \end{aligned} \tag{14}$$

Comparing (14) with (4), it can be seen that the differential phase information $\phi_r - \phi_d$ is still preserved for the lateral azimuth compression.

Followed by, to obtain a higher azimuth imaging again, different from the conventional imaging method illustrated in the Sect. 2, azimuth compression is carried out based on the work [14] of the author this chapter with joint coherent and non-coherent integration. Assuming T_a is the length of one azimuth resolution cell, the compressed result can be expressed as:

$$I_i = \int_{-\frac{T_a}{2}}^{\frac{T_a}{2}} R_s\left(t/m_s, u\right) \circledast \left(R_s\left(t/m_s, u\right)\right)^* du \tag{15}$$

where m_s denotes the interval for non-coherent integration.

The full GNSS-SAR image is obtained based on consecutively merging the results in (15), which can be expressed as:

$$I = \left[I_1, I_2, \ldots, I_{M-1}\right] \tag{16}$$

where M represents the quantity of azimuth resolution cell.

All in all, Algorithm 1 [44] summaries the steps of the proposed imaging method.

Algorithm 1 The steps of the proposed iterative imaging method

1. Performing initial range compression as (4), on the basis of output of signal synchronized results;
2. Generating the iterative range matched filter as (6);
3. Performing first round iterative range compression as (7);
4. Extracting peak and noise variance of the current range domain, and setting a proper noise threshold ε on the basis of the rule in (11);
5. Judging whether the noise level of the current range domain is lower than the threshold ε?
   - If yes, go to next step;
   - if no, performing another round of iteration by correlating with (6) until the noise level is below ε.
6. Performing azimuth compression as (15);
7. Consecutively merging azimuth compressed result in each resolution cell as (16);
8. Outputting final GNSS-SAR image.

3.2 The Theoretical Performance Analysis

Performance with respect to the proposed imaging method is analyzed theoretically. Assuming the mean of reflected signal is μ_s, the variance of Guassian white noise is σ_s, the variance of other noise is Var_m, the quantity of samples at each range domain is N, the mean of initial local replica (3) at range domain is $\alpha\mu_s$, then after performing initial range compression as (4), the mean of signal s_r^1, the variance of Guassian white noise σ_{n_1}, the variance of other noise Var_{m_1} can be derived as:

$$s_r^1 = \mu_s \cdot \alpha\mu_s = \alpha\mu_s^2 \tag{17}$$

$$\sigma_{n_1} = \frac{\alpha}{N} \cdot \mu_s \cdot \sigma_s \tag{18}$$

$$Var_{m_1} = \alpha\mu_s Var_m \tag{19}$$

Thus, the signal to noise ratio (SNR) after performing initial range compression as (4) can be derived as:

$$SNR_1 = \frac{s_r^1}{\sigma_{n_1} + Var_{m_1}} = \frac{\mu_s}{\frac{\sigma_s}{N} + Var_m} \tag{20}$$

After performing the first round iteration as (7), the power of white Guassian Noise has been averaged, thus the SNR will be higher than (20). The respective signal mean s_r^2, variance of white Guassian Noise σ_{n_2}, variance of other noise Var_{m_2} can be expressed as:

$$s_r^2 = \alpha^2\mu_s^2 \cdot \alpha\mu_s^2 = \alpha^3\mu_s^4 \tag{21}$$

$$\sigma_{n_2} = \frac{\alpha^2}{N} \cdot \mu_s^2 \cdot \left(\frac{\alpha}{N} \cdot \mu_s \cdot \sigma_s\right) \tag{22}$$

$$Var_{m_2} = \alpha^2\mu_s^2\alpha\mu_s Var_m \tag{23}$$

Thus after first round iteration, the SNR can be expressed as:

$$SNR_2 = \frac{s_r^2}{\sigma_{n_2} + Var_{m_2}} = \frac{\mu_s}{\frac{\sigma_s}{N^2} + Var_m} \tag{24}$$

The SNR after $n + 1$ rounds of iteration can be derived as follows, based on recursion rule:

$$SNR_n = \frac{\mu_s}{\frac{\sigma_s}{N^n} + Var_m} \tag{25}$$

The SNR of proposed imaging method (25) is compared with conventional BP algorithm based method. Since the range compression of conventional BP algorithm is carried out as (4), the SNR should be the same as (20) theoretically. Because $N > 1$, we can have that $\frac{\sigma_s}{N^n} << \frac{\sigma_s}{N}$, therefore equation (25) should be much larger than equation (4). This indicates that the proposed imaging method can provide a significant higher SNR for object detection compared to conventional BP algorithm based imaging method.

Meanwhile, according to the principle of the state-of-art imaging method in [14], the respective SNR can be derived as:

$$SNR_s = \frac{\mu_s}{\frac{\sigma_s}{N \cdot m_s} + Var_m}, \quad \text{where} \quad 0 < m_s < N \tag{26}$$

Comparing (25) with (26), due to $0 < m_s < N$, we can have $\frac{\sigma_s}{N^n} < \frac{\sigma_s}{N \cdot m_s}$. Thus (25) is larger than (26). This reveals that the proposed method can provide a significantly higher SNR for object detection than the state-of-art method in [14].

4 The Simulation Tests

Simulation tests are carried out to examine the proposed iterative range compression scheme. The model of simulation scenario is provided in Fig. 4.

The signal from Beidou B3I satellite is selected as the source of opportunity. The dual channel Beidou signal receiver is employed, in which, reference channel is used for collecting direct signal for synchronization, surveillance channel is used for collecting reflected signal from sensing area. For obtaining angular trace for synthetic aperture radar imaging, the circular movement of Beidou B3I signal receiver is adopted, where the length of trace is 100°. The objects are the two strong scattered areas located at the range distance 3 m from the receiver. We assume that the background noise is Additive White Gaussian Noise (AWGN). The parameter values of the simulation are shown in Table 1 [44].

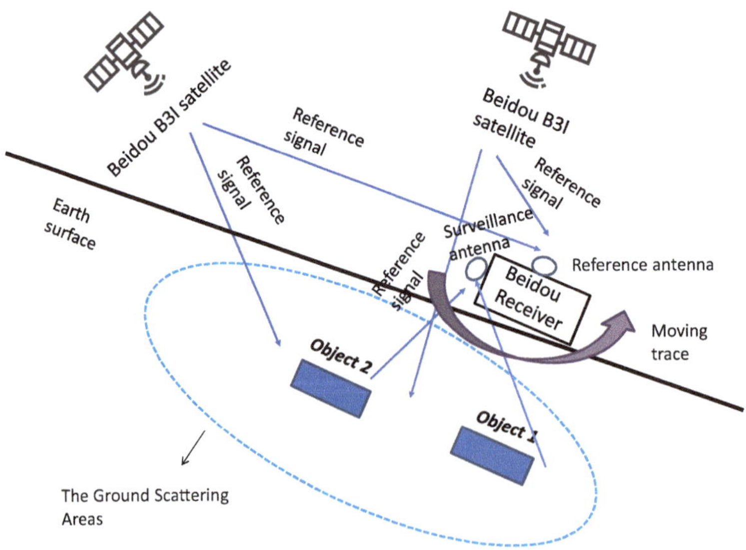

Fig. 4 The model of simulation scenario

Table 1 Parameter values with respect to the simulation tests [44]

Parameters	Values
Operating frequency	1268.52 MHz (L1 band)
Distance between satellite orbit and Earth surface	21, 528 km
Signal transmission velocity c	3×10^8 m/s
PRN code period T	1 ms
Signal bandwidth B	10.23 MHz
Sampling frequency	1.5×10^9 Hz
Boltzmann constant k	1.38×10^{-23} J/K
Scenario temperature	290 K

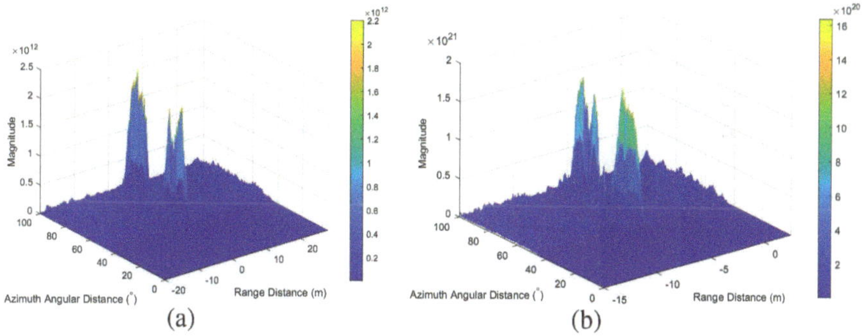

Fig. 5 Imaging results under the SNR −45 dB: **a** The imaging method in [14], **b** the proposed method with iterative range compressions [44]

Based on Fig. 4 and Table 1, the simulations were carried out under the SNR conditions as −45 dB, −50 dB and −55 dB with respect to both the proposed imaging method and the method in [14]. The results can be seen from Figs. 5, 6, and 7 [44].

Figures 5b and 6b are generated by two round of iteration, while Fig. 7b is generated by three rounds of iteration. From Figs. 5, 6, and 7, it can be seen that when SNR is −45 dB, both the imaging method in [14] and the proposed method with iterative range compressions can illuminate the two objects. However due to the impacts of iterative convolution, the magnitude of the proposed method is significantly higher than the imaging method in [14]. This indicates that the proposed method can give a higher imaging gain. When the input SNRs become worse, a faster falling rate with respect to detectability of the imaging method in [14] can be seen, compared to the proposed method. Specifically, the objects can still be illuminated clearly by the proposed imaging method, however, on the basis of the imaging method in [14], the objects are very difficult to be identified, when SNR falls to −55 dB.

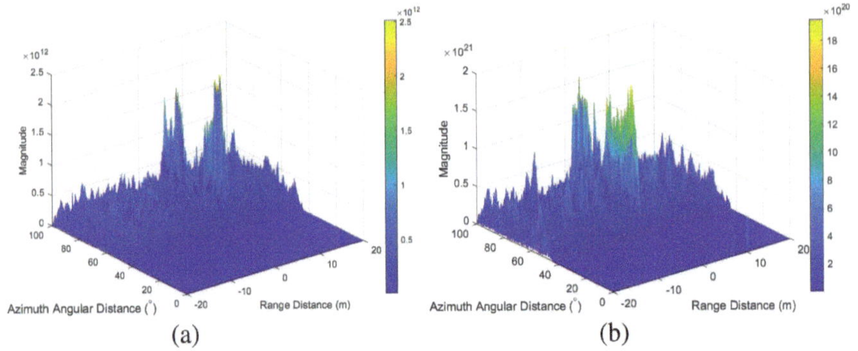

Fig. 6 Imaging results under the SNR −50 dB: **a** The imaging method in [14], **b** the proposed method with iterative range compressions [44]

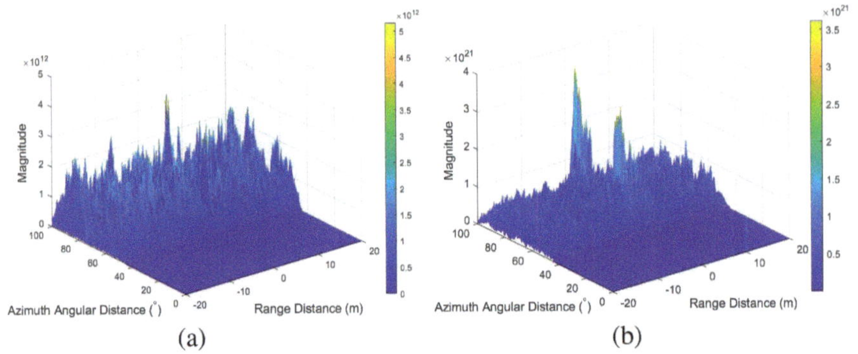

Fig. 7 Imaging results under the SNR −55 dB: **a** The imaging method in [14], **b** the proposed method with iterative range compressions [44]

Table 2 The CNRs for the simulation tests [44]

Input SNR	The CNR for the proposed method	The CNR for the method in [14]
−45 dB	20.10 dB	26.50 dB
−50 dB	11.28 dB	18.53 dB
−55 dB	2.36 dB	11.34 dB

The results from Fig. 5, 6, and 7 are quantitatively analyzed by examining the contrast-to-noise ratio (CNR) of the images on the basis of the scheme in [38, 39]. The respective CNRs can be seen as in Table 2 [44].

From Table 2, it can be seen that when the input SNR is −45 dB, −50 dB, −55 dB, the proposed imaging method can improve CNR by 6.40 dB, 7.25 dB and 8.98 dB, respectively. This reveals that the proposed imaging method can give a higher object detectability compared to the method in [14]. According to Fig. 7, Table 1,

Table 2 and propagation theory [43], it can be seen that the minimum object can be detected in the simulation scenario in Fig. 4 is with the Radar Cross Section (RCS) 0.04 m^2.

5 Experimental Confirmation with BeiDou B3I Signals

The experiments with BeiDou B3I signals were carried out for further confirm feasibility of the proposed imaging method. The same experimental parameter values as Table 1 are used. The software defined radio (SDR) receiver produced by Beijing CNSENS Technology Company is employed for collecting signal data, which is shown in Fig. 8 [44].

In Fig. 8, both reference channel and surveillance channel are triggered by the same clock for collecting the respective signals simultaneously. On the basis of the parameter values and the equipment in Fig. 8, indoor proof-of-concept experiment and outdoor field experiment were carried out.

5.1 *Indoor Proof-of-Concept Experiment*

Fig. 9 shows an optical image with respect to indoor proof-of-concept experiment.

In Fig. 9, BeiDou B3I signal was generated by a signal simulator and transmitted by the transmitting antenna. The surveillance channel was collecting the echo signal from the wall surface, which distance is 2 m. The detailed models of both reference and surveillance antennas in Fig. 9 can be seen in Fig. 10.

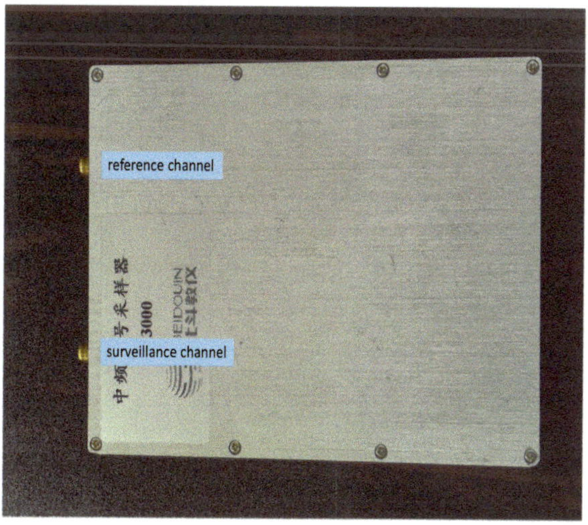

Fig. 8 The SDR receiver front end for collecting field B3I signal [44]

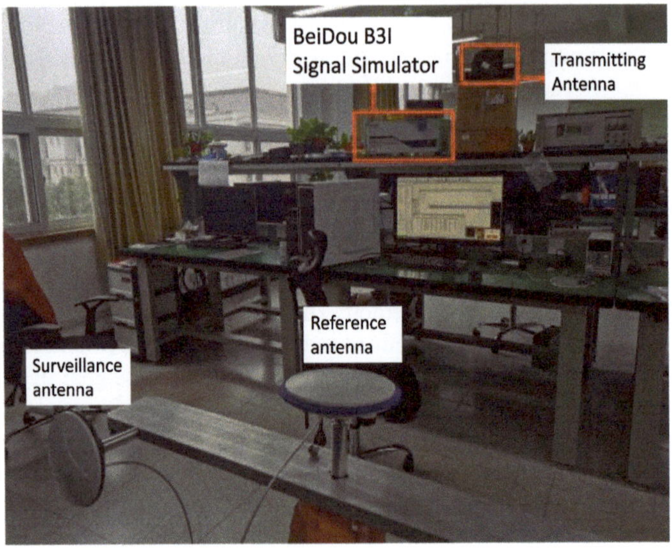

Fig. 9 Optical image of indoor proof-of-concept experiment

Table 3 CNRs for the imaging results in Fig. 12 [44]

The method in [14]	9.73 dB
The proposed method	16.25 dB

For generating an angular trace for synthetic aperture imaging, the shaft was moved rotationally, where the azimuth angular length is 90°. The flow chart for imaging in this experiment can be seen in Fig. 11.

Both the proposed imaging method and the state-of-art imaging method in [14] are used for processing the raw signal data, where the round of iterations for the proposed method is set as two. The respective experimental results presented in 3-D image manner are given in Fig. 12.

From Fig. 12, we can see that the imaging gain in Fig. 12b is significantly higher than (a). This is because the range compression is iteratively carried out. Meanwhile, due to the fact that threshold is settled, the magnitudes below the threshold are zoomed out by iterative range compressions, the CNR of Fig. 12b is higher than Fig. 12a visually. To be accurate, Table 3 [44] provides the quantitatively analysis with respect to CNRs in Fig. 12.

According to Table 3, compared to the imaging method in [14], we can have that under the first scenario, the proposed method can improve image CNR by 6.52 dB.

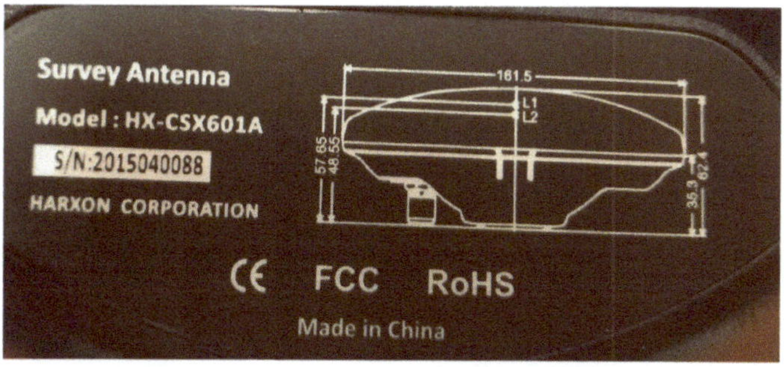

(a)

(b)

Fig. 10 The models of both reference and surveillance antennas

5.2 *Outdoor Field Experiment*

Meanwhile, outdoor field experiment was carried out at a playground in Changsha University. Figure 13 illustrates the respective optical image.

In Fig. 13, in order to eliminate multipath signals, the reflecting board was placed several tens meters from background. The receiver platform and the models of both reference and surveillance antennas are the same as the indoor proof-of-concept experiment scenario. The distance between the reflecting boards and surveillance antenna is 2 meters. Synthetic aperture image is still generating based on angular trace, where length of the trace is 70°. To satisfy the backscatter geometry for optimally avoiding direct interference, the satellite C27 is used as source of opportunity. The imaging procedure is the same as Fig. 11. The rounds of iterations for the proposed imaging method is two. Fig. 14 shows the respective imaging results.

From Fig. 14, indeed, the proposed imaging method can provide a significant higher gain in signal strength visually according to magnitude. The respective quantitative analysis for the results in Fig. 14 can be seen in Table 4 [44].

Based on Table 4, we can have that the proposed imaging method enhances CNR by 5.51 dB.

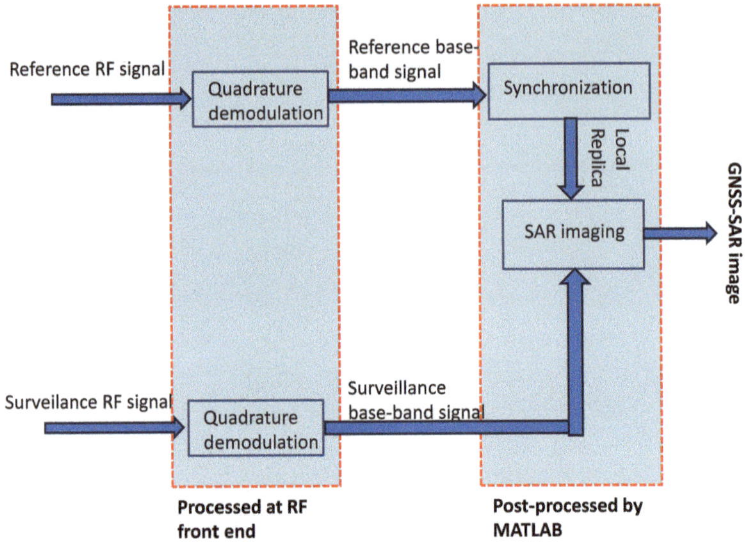

Fig. 11 The flow chart for imaging in experimental confirmation

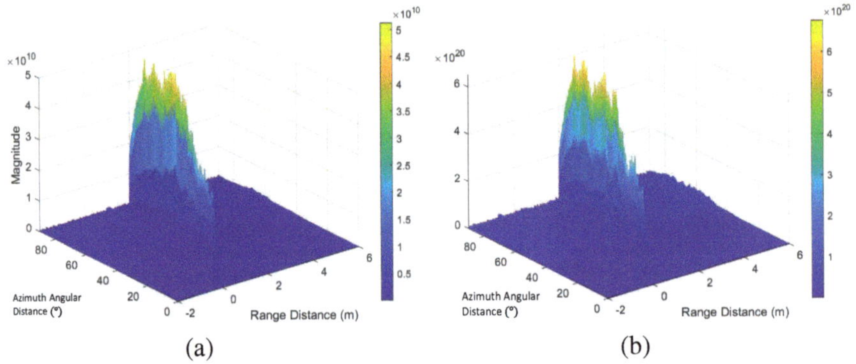

Fig. 12 Field imaging results for object imaging at the lawn in front of school library: **a** The imaging method in [14], **b** the proposed method with iterative range compressions

Table 4 CNRs for the imaging results in Fig. 14

The method in [14]	4.90 dB
The proposed method	10.41 dB

To further validate the proposed imaging method, another outdoor field experiment with two corner reflectors was conducted at the roof garden at Engineering Building in Changsha University. The optical image is shown in Fig. 15.

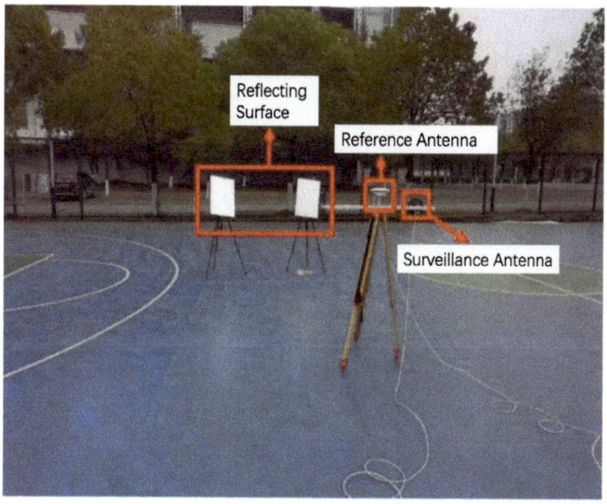

Fig. 13 Optical image for the playground in Changsha University with respect to outdoor experimental demonstration

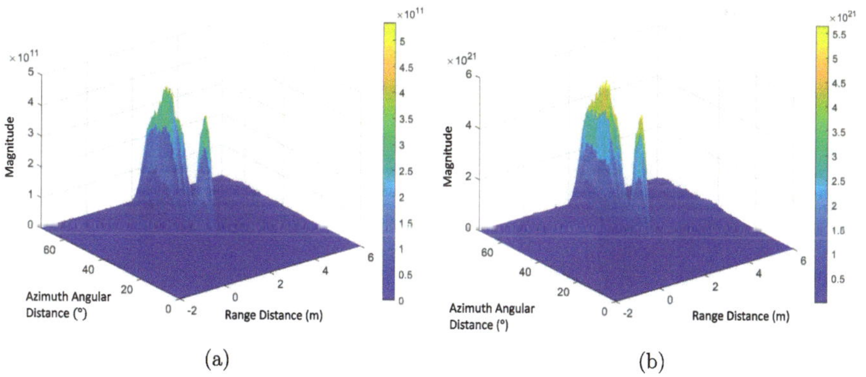

Fig. 14 Field imaging results for the second scenario: **a** The imaging method in [14], **b** the proposed method with iterative range compressions [44]

Under the scenario in Fig. 15, the objects for imaging are the two corner reflectors, which located at 3 m from the surveillance antenna. The circular movement length for performing synthetic aperture is also 3 m. To optimally resist direct signal interference at surveillance channel, the satellite C12 was chosen as transmitter, as it satisfied back scattering geometry. The parameter values were the same as Table 1. Based on the experimental set-up. Still the imaging procedure in Fig. 11 was employed. The collected signals were processed by both the state-of-art imaging method [14] and the proposed imaging method in this chapter. The respective imaging results can be seen in Fig. 16.

From Fig. 16, intuitively we can have the same verdict as the experiment at the playground in Changsha University that a higher imaging magnitude can be offered by the proposed imaging method in this chapter. This indicates that a higher imaging

Fig. 15 Optical image for the roof garden at Engineering Building with two corner reflectors

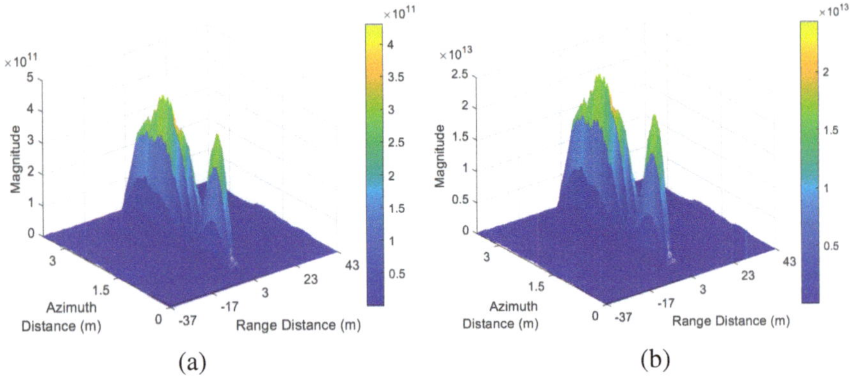

Fig. 16 Field imaging results for the scenario at roof garden of engineering building: **a** The imaging method in [14], **b** the proposed method with iterative range compressions [44]

Table 5 CNRs for the imaging results in Fig. 16

The method in [14]	11.96 dB
The proposed method	15.78 dB

gain can be obtained. To be more accurate, the CNRs with respect to Fig. 16 was extracted, which is shown in Table 5.

From Table 5, it can be seen that under the roof garden signal receiving circumstance, the proposed imaging method can have the CNR 3.82 dB higher than the state-of-art method in [14].

All in all, according to the experimental results in Figs. 12, 14, 16 and the quantitative analyzed results in Table 3, 4, 5, the feasibility of the proposed imaging method has been demonstrated successfully. Moreover, as the multi-static system model can be considered as the combination of multiple bi-static model, theoretically the imaging method proposed can also be applied to multi-static system model, such as the author's previous work [16].

6 Conclusions

To address object detection under weak echo GNSS signal circumstance, this chapter proposes an imaging method on the basis of iterative range compression. In the proposed imaging method, after performing the initial range compression, using the auto-correlation of the local direct signal replica, the iterative matched filter is generated. To improve the CNR, range compression is iteratively performed through the correlation of the compressed results in the former state with the iterative matched filter until the noise level falls below the expected threshold. Both simulation tests and field experimental results indicate that the proposed method can provide a significantly higher CNR for GNSS-SAR imaging than the state-of-the-art method [14] for imaging enhancement. This reveals that a higher object detectability can be achieved by the proposed imaging method. Furthermore, the proposed imaging method can also be applied in the range compression state of the coherently integrated scheme of multi-satellite signals in the first author's previous work [16] and the multi-image fusion scheme [15]. In summery, this research should provide helpful insights for imaging enhancement with respect to GNSS-SAR system.

In this chapter, when the iteration rounds increase, the number of operations will also increase, which will add the computational burdens for GNSS-SAR imaging processing. However, the balance with respect to object detectability and the number of iterations has not been discussed in detail, as the main goal of this chapter is to demonstrate the feasibility of the proposed imaging method based on iterative range compressions. In the future, we will aim at developing a scheme to achieve a trade-off between object detectability and numbers of processing. A helpful tool for the further research will be deep learning [45].

Acknowledgements The authors would like to thank researchers from College of Electronic Science, National University of Defense Technology, China, for providing technical guidance, and AsiaEdit (www.asiaedit.com) for English polishing of this chapter.

Declarations Funding: This research was substantially funded by the National Natural Science Foundation of China under Projects 42471384 and 62402066; the Natural Science Foundation of Hunan Province under Projects 2024JJ9190, 2022JJ40522, and 2024JJ7617; the Research Foundation of Education Department of Hunan Province under Projects 22B0821, 24B0798 and 2026J81624; the Research Foundation of Education Department of Hunan Province under Project 23A0609; National Key Laboratory of Unmanned Aerial Vehicle Technology in NPU (Grant WR202403); and the Aid Program for Science and Technology Innovative Research Team in Higher Educational Institutions of Hunan Province.

References

1. Curlander J, McDonough R (1991) Synthetic aperture radar. Wiley, New York, NY, USA
2. Attalah M, Laroussi T, Gini F, Greco MS (2019) Range-Doppler fast block LMS algorithm for a DVB-T-based passive bistatic radar. Signal Image Video Process 13(1):27–34
3. Borre K, Akos DM, Bertelsen N et al (2007) A software-defined GPS and Galileo receiver: a single-frequency approach. Springer Science & Business Media
4. Yang Y, Gao W, Guo S, Mao Y, Yang Y (2019) Introduction to BeiDou? 3 navigation satellite system. Navigation 66(1):7–18
5. Zuo R (2012) Bistatic synthetic aperture radar using GNSS as transmitters of opportunity. Ph.D. Thesis, University of Birmingham
6. Zeng Z (2013) Passive bistatic SAR with GNSS transmitter and a stationary receiver. University of Birmingham,PhD Thesis
7. Antoniou M, Cherniakov M (2013) GNSS-based bistatic SAR: a signal processing view. EURASIP J Adv Signal Process 1:1–16
8. Antoniou M, Zeng Z, Feifeng L, Cherniakov M (2012) Experimental demonstration of passive BSAR imaging using navigation satellites and a fixed receiver. IEEE Geosci Remote Sens Lett 9(3):477–481
9. Antoniou M, Hong Z, Zhangfan Z, Zuo R, Zhang Q, Cherniakov M (2013) Passive bistatic synthetic aperture radar imaging with Galileo transmitters and a moving receiver: experimental demonstration. IET Radar Sonar Navig 7(9):985–993
10. Santi F, Pastina D, Bucciarelli M, Antoniou M, Tzagkas D, Cherniakov M (2014) Passive multistatic SAR with GNSS transmitters: preliminary experimental study. In: IEEE 11th European radar conference (EuRAD), pp 129–132
11. Shi S, Liu J, Li T, Tian W (2017) Basic performance of space-surface bistatic SAR using BeiDou satellites as transmitters of opportunity. GPS Solut 21(2):727–737
12. Wang P, Zhou X, Fang Y et al (2022) GNSS-based passive inverse SAR imaging. IEEE J Sel Top Appl Earth Observ Remote Sens 16:508–521
13. Huang C, Li Z, An H et al (2022) Optimal GNSS-based passive SAR large field-of-view imaging via multistatic configuration: method and experimental validation. IEEE J Sel Top Appl Earth Observ Remote Sens 15:9873–9884
14. Zheng Y, Yang Y, Chen W (2019) Object detectability enhancement under weak signals for passive GNSS-based SAR. Signal Image Video Process 13(8):1549–1557
15. Zeng T, Zhang T, Tian W, Hu C (2016) Space-surface Bistatic SAR image enhancement based on repeat-pass coherent fusion with Beidou-2/Compass-2 as illuminators. IEEE Geosci Remote Sens Lett 13(12):1832–1836
16. Zheng Y, Zhang Z, Feng L, Zhu P, Zhou F (2020) Enhanced passive GNSS-based radar imaging based on coherent integrated multi-satellite signals. Sensors 20(3):842
17. Zheng Y, Zhang Z, Zhu P et al (2023) Passive GNSS-based SAR imaging with sub-apertures combination. Signal Image Video Process 17(1):2177–2184
18. He Z, Chen W, Yang Y, Weng D, Cao N (2023) Maritime ship target imaging with GNSS-based passive multistatic radar. IEEE Trans Geosci Remote Sens 61:1–18
19. Ma H, Antoniou M, Cherniakov M (2015) Passive GNSS-based SAR resolution improvement using Joint Galileo E5 signals. IEEE Geosci Remote Sens Lett 12(8):1640–1644
20. Santi F, Bucciarelli M, Pastina D, Antoniou M, Cherniakov M (2016) Spatial resolution improvement in GNSS-Based SAR using multistatic acquisitions and feature extraction. IEEE Trans Geo-sci Remote Sens 54(10):6217–6231
21. Zheng Y, Yang Y, Chen W (2017) A novel range compression algorithm for resolution enhancement in GNSS-SARs'. Sensors 17(7):1496
22. Zheng Y, Yang Y, Chen W (2019) New imaging algorithm for range resolution improvement in passive global navigation satellite system-based synthetic aperture radar. IET Radar Sonar Navig 13(12):2166–2173
23. Fang Y, Chen J, Wang P, Zhou X (2020) An image formation algorithm for bistatic SAR using GNSS signal with improved range resolution. IEEE Access 8:80333–80346

24. He Z, Yang Y, Chen W, Weng D (2022) Range resolution improvement of GNSS-based passive radar via incremental Wiener filter. IEEE Geosci Remote Sens Lett 19:1–5
25. Ma H, Antoniou M, Pastina D, Santi F, Pieralice F, Bucciarelli M, Cherniakov M (2018) Maritime moving target indication using passive GNSS-based bistatic radar. IEEE Trans Aerosp Electron Syst 54(1):115–130
26. Santi F, Pastina D, Bucciarelli M (2017) Maritime moving target detection technique for passive bistatic radar with GNSS transmitters. In: 2017 18th IEEE International radar symposium (IRS), pp 1–10
27. Ma H, Antoniou M, Stove AG, Winkel J, Cherniakov M (2018) Maritime moving target localization using passive GNSS-based multistatic radar. IEEE Trans Geosci Remote Sens 56(8):4808–4819
28. He Z, Yang Y, Chen W (2020) A hybrid integration method for moving target detection with GNSS-based passive radar. IEEE J Sel Top Appl Earth Observ Remote Sens 14:1184–1193
29. Zeng H et al (2023) A spawning particle filter for defocused moving target detection in GNSS-based passive radar. J Syst Eng Electron 34(5):1085–1100
30. Zhang C, Shi S, Yan S et al (2023) Moving target detection and parameter estimation using BeiDou GEO satellites-based passive radar with short-time integration. IEEE J Sel Top Appl Earth Observ Remote Sens 16:3959–3972
31. Zhang Z, Zheng Y, Zheng L et al (2022) Deformation monitoring using passive BeiDou B3I signal-based radar: a proof of concept experimental demonstration. Acta Geodyn Geophys 57(4):625–638
32. Chen Y, Yan S, Gong J (2022) Phase error analysis and compensation of GEO-satellite-based GNSS-R deformation retrieval. IEEE Geosci Remote Sens Lett 19:1–5
33. Zheng Y, Zhang Z, Wu P et al (2021) Decimetre level range resolution for BeiDou B3I signal-based passive SAR imaging: proof-of-concept experimental demonstrations. IEEE Access 9:22927–22938
34. Roesler CJ, Morton YJ, Wang Y et al (2021) Coherent GNSS-reflections characterization over ocean and sea ice based on spire global CubeSat data. IEEE Trans Geosci Remote Sens 60:1–18
35. Zribi M, Guyon D, Motte E et al (2019) Performance of GNSS-R GLORI data for biomass estimation over the Landes forest. Int J Appl Earth Observ Geoinf 74:150–158
36. Wu X, Ma W, Xia J et al (2020) Spaceborne GNSS-R soil moisture retrieval: status, development opportunities, and challenges. Remote Sens 13(1):45
37. Zhang Z, Guo F, Zhang X et al (2023) Physical modeling and compensation for systematic negative errors in GNSS-R snow depth retrieval. IEEE Trans Geosci Remote Sens 61:1–12
38. Buhari MD, Tian GY, Tiwari R et al (2018) Multicarrier SAR image reconstruction using integrated MUSIC-LSE algorithm. IEEE Access 6:22827–22838
39. Buhari MD, Tian GY, Tiwari R (2018) Microwave-based SAR technique for pipeline inspection using autofocus range-Doppler algorithm. IEEE Sens J 19(5):1777–1787
40. Zhang Z, Dong Y, Wen Y et al (2023) Modeling, refinement and evaluation of multipath mitigation based on the hemispherical map in BDS2/BDS3 relative precise positioning. Measurement 213:112722
41. Johnson O (2004) Information theory and the central limit theorem. World Scientific
42. Floudas CA, Pardalos PM (eds) (2008) Encyclopedia of optimization. Springer Science & Business Media
43. Souganidis PE (2006) Front propagation: theory and applications, viscosity solutions and applications: lectures given at the 2nd Session of the Centro Internazionale Matematico Estivo (CIME) held in Montecatini Terme, Italy, June 12?20, 1995. Springer, Berlin Heidelberg, Berlin, Heidelberg, pp 186–242
44. Zheng Y, Zhang Z, Zhu P et al (2024) GNSS-based SAR imaging for object detection based on iterative range compressions. IEEE Sens J 24(6):8493–8502
45. Goodfellow I, Bengio Y, Courville A (2016) Deep learning. MIT Press

Chapter 7
Retrieval of Underground Coal Mining Induced Ground Subsidence Surface with a Single GNSS Receiver Collected Observations

Yunwei Li and Tianhe Xu

Abstract Monitoring of ground subsidence induced by underground coal mining over a large area is essential for evaluating the impacts of subsidence on environment. This chapter firstly presents a ground subsidence retrieval method using a single station collected Global Navigation Satellite System (GNSS) carrier-phase and signal-to-noise (SNR) observations. The proposed method can be utilized to monitor the underground coal mining induced ground subsidence over a large area (~8000 m^2) continuously. The geometry models related to the subsidence retrieval are developed based on GNSS positioning technique and GNSS reflectometry. The former observes GNSS antenna movement and the latter estimates phase difference of the reflected SNR signals, both of which are utilized to calculate plane coordinates and subsidence on the ground reflection point. Using semi-empirical formulae instead of in-situ measurement data, an analytical exponential model is established in advance to describe relationship between plane coordinates and subsidence of the ground point. When the model coefficients are obtained based on the plane coordinates and subsidence observations at the ground reflection points by exploiting the weighted least-squares method, the model is used to determine ground subsidence surface around the GNSS station. Three experimental data sets collected by three different measuring techniques during an experimental campaign were used to test the proposed method. The results show that the root-mean-square error of the

Y. Li
Shandong Provincial Lunan Geology and Exploration Institute (Shandong Provincial Bureau of Geology and Mineral Resources No. 2 Geological Brigade), Jining 272100, China

Technology Innovation Center of Restoration and Reclamation in Mining induced Subsidence Land, Ministry of Natural Resources of China, Jining 272100, China

Y. Li
e-mail: yunweili@whu.edu.cn

T. Xu (✉)
School of Space Science and Physics, Shandong University, Weihai 264209, China
e-mail: thxu@sdu.edu.cn

K. Yu and N. Zheng (eds.), *Advances in GNSS Reflectometry*, Navigation: Science and Technology 19, https://doi.org/10.1007/978-981-92-2325-1_7

proposed method-based ground subsidence estimations is less than 40 mm when the subsidence is between 0 and 3000 mm.

1 Introduction

Coal is the most abundant source of electricity worldwide, currently providing more than 36% of global electricity. Due to the limitation of geology conditions, surface mining is impractical or uneconomical in most circumstances where the coal seam is buried deeply. Hence, most of the coal comes from underground mining especially in China [1]. On the other hand, underground coal mining can also induce subsidence over a large area on the ground surface, resulting in obvious impacts on the geo-ecological environment of the ground (e.g., land degradation, water table drawdown, and building deformation) [2–4]. Continuously monitoring underground coal mining induced ground subsidence in a large scale is crucial for assessment of environmental impacts arising from the subsidence above the coal seam and prevention of the possible damage of buildings on the ground [5].

However, it is impractical to monitor underground coal mining induced ground subsidence with a high temporal and spatial resolution using the traditional techniques due to limitations of observing capability or economic cost. For instance, the airborne based remote sensing methods (e.g. LiDAR or photogrammetry) are insufficiency in temporal resolution and are easily affected by weather conditions [6–8]. The monitoring performance of satellite based InSAR technique is affected wavelength of the electromagnetic wave transmitted by the satellite, orbit repeat period, and ground subsidence velocity. The InSAR technique is generally incapable of monitoring of large-magnitude and high-velocity subsidence induced by excavation of thick coal seam directly, which is the common mining pattern worldwide [9–12].

Global navigation satellite system (GNSS) has been widely used for positioning, navigation, and timing applications. In the last 30 years, the GNSS positioning techniques (e.g., GNSS Real time kinematic, GNSS RTK) has also been used to monitor ground subsidence caused by underground coal mining [13–15]. Although the traditional GNSS positioning techniques can be used to measure the ground subsidence with centimeter-grade (or, even millimeter-grade) precision, the techniques can only provide subsidence observations in few ground points. In addition to these typical applications, the GNSS signal reflected from a specific surface can be utilized to remotely sense a series of geophysical parameters of the surface. This remote sensing technique is termed as GNSS interferometric reflectometry (GNSS-IR) when the direct and reflected signals are received by the same antenna such as a zenith-looking GNSS antenna. The technique can be applied to estimate the land's surface characteristics (e.g., snowpack depth, soil moisture, and vegetation height) around the GNSS station [16–21].

In this chapter, a combined technique of GNSS positioning and GNSS-IR (or, may be termed as GNSS + IR) is firstly proposed to retrieve ground subsidence surface induced by underground coal mining. The proposed method can be applied to monitor the ground subsidence over an area of ~ 8000 m^2 accurately and continuously by making use of a single GNSS receiver collected carrier-phase and SNR observations, which is a much larger and more useful scale than the traditional GNSS positioning measurement in terms of assessment of environmental impacts due to the coal mining. Furthermore, because of the use of the consumer-grade GNSS instrument, the method presented in the chapter is also promising for retrieval of ground subsidence surface at larger scale in a cost-effective way.

The remainder of this chapter is organized as follows. In Sect. 2, the fundamentals of the proposed method are provided. Section 3 presents details of the proposed ground subsidence surface retrieval method. Section 4 evaluates the proposed method using three data sets collected by three different measuring techniques. Section 5 concludes this chapter.

2 Fundamentals of Subsidence Surface Retrieval Based on GNSS Positioning and GNSS-IR Technology

2.1 Underground Coal Mining Induced Ground Subsidence

Before underground coal mining, the pressures of rocks and soils are carried by the coal seam. After the coal seam is mined out, the pressures are transferred to the solid coal sides. With the increase of the mined-out area (or goaf), the goaf progressively becomes too wide to be self-supporting. As a result, rocks above the coal seam sag and separate along bedding planes and then falls into the goaf. When the influence of the rock collapse arrives to the ground surface, ground subsidence commences. Basically, shape of the subsided ground surface is very similar to a bowl as shown in Fig. 1, which is also termed as subsidence bowl [22].

In general, area of the subsidence bowl is considerably larger than that of the mined-out area; subsidence magnitudes are also different for different ground points in the bowl. Both the subsidence area and magnitude depend on geological and mining conditions of the coal seam (e.g. coal seam depth, mining thickness, and mining speed), and are spatial- and temporal-related [23, 24]. Based on a large amount of ground truth observations in the past several decades, many physical or mathematical models have been developed to describe the underground coal mining induced ground subsidence. Among these models, the probability integral model combined with the Knothe time function model (or, KPIM model) has been widely used in many countries especially in China [25]. For a given ground point with plane coordinates of (X, Y), the KPIM model-based subsidence estimation at the time of t_1 can be predicted by [26–28]:

$$W(X,Y)_{t=t_1} = W_{\max} \cdot \iint\limits_D \frac{1}{r^2} \cdot e^{-\pi \frac{(X-\xi)^2+(Y-\eta)^2}{r^2}} dXdY \cdot T(t_1) \tag{1}$$

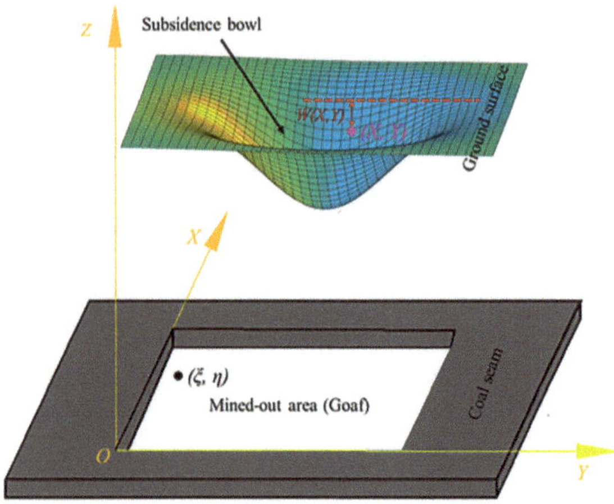

Fig. 1 Underground coal mining induced subsidence bowl over the ground surface

where W_{max} is the maximum subsidence of the bowl; r is the main influence radius; (ξ, η) are plane coordinates of a mining point within the mined-out area of D; T(t) is the Knothe time function, which is given by:

$$T(t) = 1 - e^{-ct} \tag{2}$$

where c is the model coefficient. Note that, all these three model parameters (i.e., W_{max}, r, c) are related to geological and mining parameters of the coal mining working face. Interested readers are referred to [29] for more information of the underground coal mining induced ground subsidence prediction models.

2.2 GNSS Reflection Model for the Subsided Ground Surface

As shown in Fig. 2, a GNSS antenna is connected on a pole that is vertically fixed on the ground above a coal seam. The raw GNSS SNR observation series collected by the antenna and receiver is an interfered signal composed by the direct signal and the signal reflected from the ground specular reflection point, if the ground is smooth and mirrorlike. The low-frequency component of the interfered signal can be obtained by making use of low-order polynomial fitting method; after removing the low-frequency component from the raw SNR series, the high-frequency component of SNR signal, i.e. the detrended SNR signal, can be obtained [30]. Assuming that the ground surface is flat enough when the ground is subsidence-free, the high-frequency component of SNR signal can be expressed as [31–33]:

$$SNR_m = A_m \cos \Phi_0(\theta) \tag{3}$$

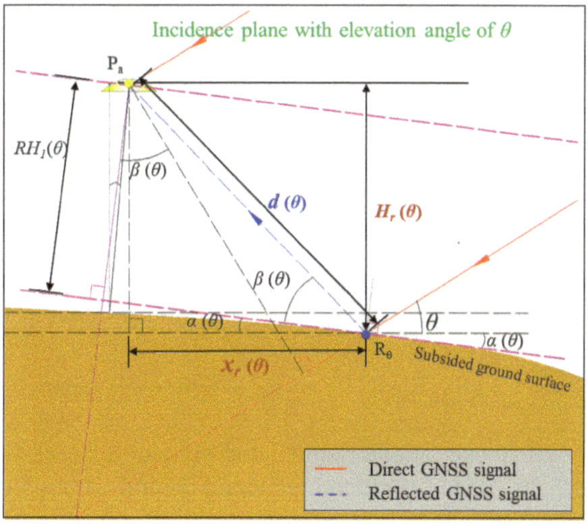

Fig. 2 GNSS reflection model for subsided ground surface induced by underground coal mining

where A_m is the amplitude of the signal; $\Phi_0(\theta)$ is the phase of the signal at the GNSS satellite elevation angle of θ, which is given by:

$$\Phi_0(\theta)=\frac{4\pi \sin\theta}{\lambda}RH_0(\theta) \tag{4}$$

where $RH_0(\theta)$ is the reflector height (RH) of the ground specular reflection point at elevation angle of θ; and the RH satisfies:

$$RH_0(*)=H_0 \tag{5}$$

where H_0 is initial GNSS antenna height, or vertical distance from the antenna to the ground in the case of ground subsidence-free.

After the ground subsides due to underground coal mining, the RH and thus phase of the reflected signal would be changed. Based on the geometry model shown in Fig. 2, phase of the detrended SNR signal at elevation angle of θ in the case of subsided ground surface can be written as [34]:

$$\begin{cases} \Phi_1(\theta)=\frac{4\pi \sin\beta(\theta)}{\lambda}RH_1(\theta) \\ \beta(\theta)=\theta+\alpha(\theta) \end{cases} \tag{6}$$

where α(θ) is the ground tilt angle of the ground reflection point within the incidence plane; $RH_1(\theta)$ is the RH of ground reflection point at elevation angle of θ. Subtracting (4) from the first equation of (6) and with some manipulation, the phase difference (PD) between the detrended signals collected when the ground is subsidence-free and when the ground subsides due to underground coal mining can be obtained by:

$$\Delta\Phi(\theta) = \Phi_1(\theta) - \Phi_0(\theta) \\ = 4\pi[\sin\beta(\theta) \cdot RH_1(\theta) - \sin\theta \cdot H_0]/\lambda \tag{7}$$

Then, the RH in the case of subsided ground can be derived from (7) as:

$$RH_1(\theta) = \left[\frac{\lambda}{4\pi}\Delta\Phi(\theta) + \sin\theta \cdot H_0\right] / \sin\beta(\theta) \tag{8}$$

It can be observed from Fig. 2 that, distance from the reflection point to the GNSS antenna, i.e. d(θ), can be calculated by:

$$d(\theta) = RH_1(\theta) / \sin\beta(\theta) \tag{9}$$

Obviously, the horizontal and vertical distance from the GNSS antenna phase center (Pa) to the ground reflection point (R_θ), i.e. Hr(θ) and xr(θ), can be obtained by:

$$\begin{cases} H_r(\theta) = d(\theta) \cdot \sin[\beta(\theta) + \alpha(\theta)] \\ x_r(\theta) = d(\theta) \cdot \cos[\beta(\theta) + \alpha(\theta)] \end{cases} \tag{10}$$

Note that, the realistic ground surface is usually not perfectly mirror-like but rough, so the reflection is diffuse and the GNSS antenna will capture signals reflected from many points of the ground. However, among these reflected signals, the signal reflected from the specular point has the shortest propagation path and is the first one to arrive at the GNSS antenna. The signals energy captured by the antenna mostly comes from the first Fresnel zone, and the signals reflected around the specular point contribute the most [30]. Thus, it would be reasonable that one only considers the specular reflection for simplicity of the theoretical modeling.

3 Proposed Method for Retrieval of Underground Coal Mining-Induced Ground Subsidence Surface

In this section, details of the proposed method are presented. Geometry model of GNSS + IR based ground subsidence estimation are derived at first. Then, an enhanced estimation method of phase difference (PD) of the detrended SNR signals is provided; and the typical variation characteristics of the PD time series were also studied. Based on the KPIM model simulated results, a mathematic function is established to describe the relationship between plane coordinates and subsidence of the ground point. Finally, an iterative algorithm is developed to calculate optimal parameters of the function.

3.1 Geometry of the Proposed Method Based Ground Subsidence Surface Retrieval

Underground coal mining could induce movement of ground and thus the GNSS antenna fixed on ground in both vertical and horizontal directions. As illustrated in Fig. 3, a GNSS antenna is fixed on a pole at the position of P_0 when the ground is subsidence-free; and the pole is vertically fixed the ground surface. However, the antenna moves to Pa due to ground movement caused by underground coal mining. Vertical component of displacement of the GNSS antenna from P_0 to Pa can be measured accurately by using GNSS carrier-phase observations with traditional positioning technique (e.g. GNSS RTK). If the initial antenna height H_0 is known in advance and the vertical distance Hr(θ) is estimated by exploiting the reflected SNR observations, the underground coal mining induced ground subsidence on the reflection point R_θ can be obtained by:

$$W_\theta = H_a + H_r(\theta) - H_0 \tag{11}$$

where Ha is the vertical displacement of the GNSS antenna.

In addition to displacement in the vertical direction, position of the GNSS antenna in the horizontal plane would also be changed due to the ground movement. As shown in Fig. 4, taking the raw position of the GNSS antenna as the reference point (or, the origin of the coordinate system shown in the figure), plane coordinates of the raw GNSS antenna position P_0 can be written as (0, 0). After the ground subsides due to underground coal mining, the antenna moves to P_a and position of the antenna is (x_a, y_a) in the case. Obviously, x_a and y_a equals to horizontal displacement of the antenna in the east and north directions, respectively; and the displacements can also be measured by using the GNSS RTK technique. If horizontal distance from the ground reflection point R_θ to the GNSS antenna P_a, i.e. $x_r(\theta)$, is estimated with

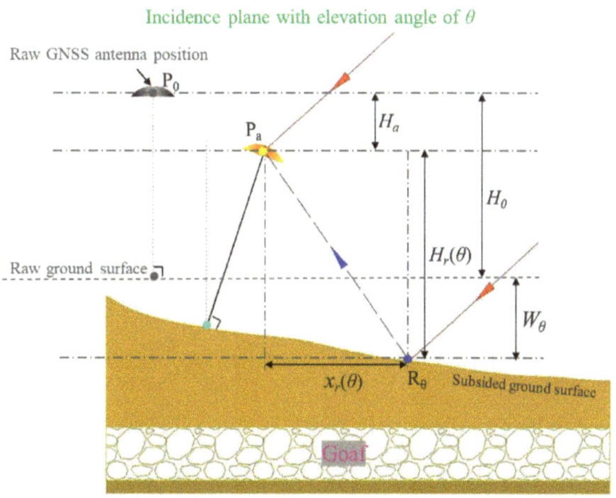

Fig. 3 Geometry of underground coal mining induced ground subsidence estimation with combined technique of GNSS positioning and GNSS-IR

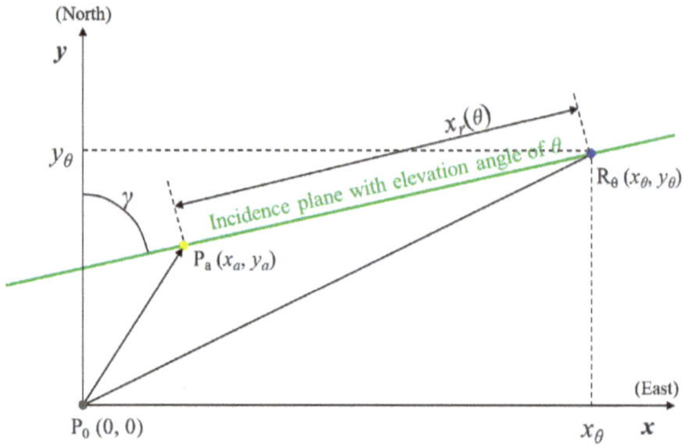

Fig. 4 Geometry of plane position estimation for ground specular reflection point

GNSS-IR, plane coordinates of the ground specular reflection point can be calculated by:

$$\begin{cases} x_\theta = x_a + x_r(\theta) \cdot \sin \gamma \\ y_\theta = y_a + x_r(\theta) \cdot \cos \gamma \end{cases} \tag{12}$$

where γ is azimuth angle of the GNSS satellite when the satellite elevation angle is θ.

It can be seen from (11) and (12) that, for each ground reflection point around the GNSS station, plane coordinates and subsidence of the point can be determined, if displacements of the antenna (i.e., H_a, x_a, and y_a) are measured by using the GNSS RTK technique and the distances (i.e., $H_r(\theta)$ and $x_r(\theta)$) are estimated with GNSS-IR. Further, an analytical function can be developed to describe subsidence surface around the GNSS station, i.e.:

$$W = f(x, y) \tag{13}$$

After obtaining coefficients of the function based on the GNSS + IR derived plane coordinates and subsidence observations on the ground reflection points around the GNSS station (i.e. x_θ, y_θ, W_θ), ground subsidence surface induced by underground coal mining can be readily determined. This is the principle of the proposed method for retrieval of underground coal mining induced ground subsidence surface.

3.2 Estimating Phase Difference of the Reflected SNR Signals

It can be derived from (9–12) that the RH on the ground reflection point, i.e. $RH_1(\theta)$, is the key to calculate plane coordinates and subsidence of the point. The RH can be obtained by substituting PD of the reflected SNR signals into (8), if the tilt angle at the point is known. That is, it is necessary to estimate PD of the reflected SNR signals for calculation of plane coordinates and subsidence on the ground reflection points.

When the GNSS satellite elevation angle is lower than a certain value such as 15°, the reflected GNSS SNR observation series is a quasi-sinusoidal signal, periodically oscillating with respect to sine of elevation angle, as presented in Fig. 5 which shows a typical example of the reflected SNR series collected on four different time periods. Note that the series are received by the low-cost GNSS chip of u-blox M8N and u-blox ANN-MB antenna; and the antenna is fixed on the ground surface above a coal seam. Besides, for improving signal quality, the raw SNR observations with elevation angle difference less than 0.1° are treated as one observation at a given elevation angle; then, the mean filtering method is used to reduce measurement noise, obtaining the filtered SNR series presented in the figure. Clearly, for the quasi-sinusoidal signal, crests and troughs of the signal are noticeable and easily to be identified. As studied in [35], crests and troughs of the signal can be determined by fitting the SNR series over each semi-cycle using a parabolic function; and the determined crest or trough can also be termed as "notice point" (NP).

Selecting a reflected SNR signal collected when the ground surface is subsidence-free as the base signal, NP derived from the signal can be taken as the base point, and the elevation angles at NPs are taken as the base elevation angles (BEAs). Phase of the base signal at each BEA can be configured as 0°, i.e.:

$$\Phi_0(\theta_{\text{BEA}}) = 0^\circ \tag{14}$$

After the ground subsides due to underground coal mining, RH and thus phase of the SNR series at the BEA will be changed. In the case, phase of the SNR series at the BEA can be calculated through the linear interpolation method:

$$\Phi_1(\theta_{\text{BEA}}) = \Phi_1^{\text{consistent}}(\theta_{\text{BEA}}) = 180^\circ \cdot \frac{\sin\theta_{\text{BEA}} - \sin\theta_{\text{left\_NP}}}{\sin\theta_{\text{right\_NP}} - \sin\theta_{\text{left\_NP}}} \tag{15}$$

where $\theta_{\text{left\_NP}}$ is the elevation angle of the left NP nearest to the BEA; $\theta_{\text{right\_NP}}$ is the elevation angle of the right NP nearest to the BEA. For example, as shown in Fig. 5, the reflected SNR series collected on 15 October 2021 is taken as the base signal; and elevation angle of 9.782°, which corresponds to the NP of PB_1, is identified as one of a BEA. On 20 December 2021, the ground subsided due to underground coal mining, and phase of the SNR series also changed. The left and right NP nearest to the BEA are PL_1 and PR_1 respectively for the reflected SNR series collected on 20 December 2021; and the elevation angles of the left and right NP (i.e. $\theta_{\text{left\_NP}}$ and $\theta_{\text{right\_NP}}$) are 9.663° and 10.136°, respectively. Substituting these two elevation angles into (15), phase of the reflected SNR series at the BEA of 9.782° on the day can be obtained, which is 45.098°. Subsequently, PD of the reflected signal at the BEA when the ground subsides can be easily obtained by substituting (14) and (15) into (7).

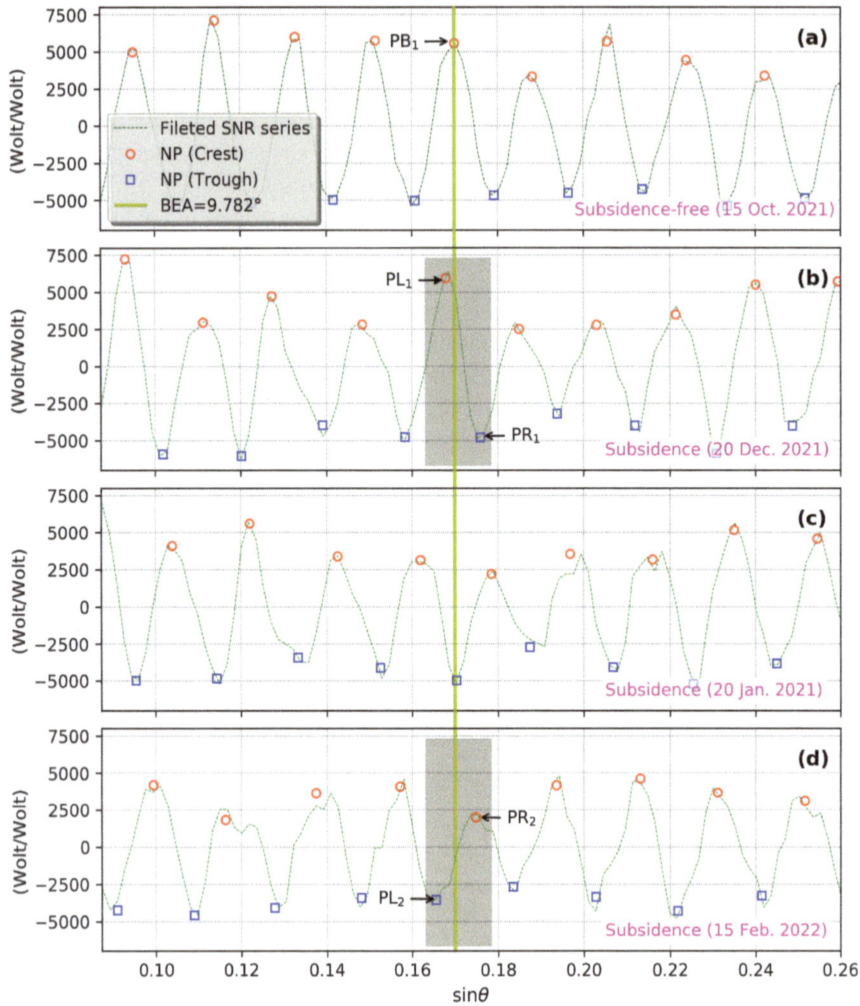

Fig. 5 The reflected SNR series for GPS G03 satellite with L1 band signal. The signals were collected by using the low-cost u-blox M8N chip and ANN-MB antenna fixed on the ground above a longwall coal mining working face

Notice that, Eq. (15) can be used to calculate phase of the reflected signal at the BEA only when the NP closest to the BEA of the base signal is consistent with the type of left NP nearest to the BEA. For instance, the BEA of 9.782° associated NP (i.e. PB_1) is crest for the base signal; for the reflected SNR series collected on 20 Dec. 2021, the left NP nearest to the BEA (i.e. PL_1) is also crest. Thus, phase of the reflected SNR series at the BEA can be calculated by using (15). However, for the reflected signal collected on 15 Feb. 2022, the left NP nearest to the BEA (i.e. PL_2 in Fig. 5) is trough, which is inconsistent with the BEA-associated NP type. In such case, a new formula should be utilized to calculate the phase at the BEA:

$$\Phi_1(\theta_{\text{BEA}}) = \Phi_1^{\text{inconsistent}}(\theta_{\text{BEA}}) = 180^\circ \cdot \frac{\sin\theta_{\text{BEA}} - \sin\theta_{\text{right\_NP}}}{\sin\theta_{\text{right\_NP}} - \sin\theta_{\text{left\_NP}}} \tag{16}$$

3.3 *Variation Characteristics of the PD Time Series*

For the continuously observed SNR signals on a given GNSS satellite ground reflection track, the raw PD time series at a given BEA can be obtained by making use of the method above to process each signal in sequence. Figure 6a shows two examples of the raw PD time series at BEA of 6.024° and 9.782° respectively. Note that the GNSS satellite track of the SNR series used to derive the PD is the same as that of the series displayed in Fig. 5. It can be seen from Fig. 6a that, before 30 November 2021, the observed PD is very close to 0° for both two series. This is because excavation position of the underground coal mining working face is far enough away from the GNSS station in the period; and the ground surface at the reflection point is not affected by the coal mining. Since then, impact of the underground coal mining began to reach the ground surface around reflection point. Thus, ground surface at the reflected point begins to subside. Because impact of the mining on reflection point is greater than that on the GNSS station, ground subsidence on the former is larger than that on the latter. As a result, RH of the ground reflection point at the BEA increases, causing the increase of the PD at the BEA. With the intensification of the mining impacts, the RH and thus corresponding PD of the reflection point at the BEA continues to increase.

When the PD exceeds 180°, the "phase-jump" with a suddenly negative angle shift of 360° happens for the increased PD time series, as clearly display in the grey zone of Fig. 6a. In the case, the PD exception induced by the phase-jump should be corrected by:

$$\Delta\Phi_{\text{correct}}^{\text{increase}} = \Delta\Phi(\theta_{\text{BEA}}) + 360^\circ \tag{17}$$

Figure 6b shows two examples of the corrected PD time series, which are derived from the series presented in Fig. 6a. It can be seen from Fig. 6b that, the corrected PD series changes continuously, which is in accordance with the continuous variation of ground subsidence induced by underground coal mining. Note that, because influence degree of the underground coal mining on the ground reflection point at BEA of 6.024° is greater than that at BEA of 9.782°, ground subsidence at the former reflection point is larger than that at the latter reflection point. Consequently, RH and thus PD of the reflected signal at the BEA of 6.024° is larger than that at the BEA of 9.782° in general, when the ground subsided due to the underground coal mining. In addition, to improve accuracy of PD estimations, the moving average method with

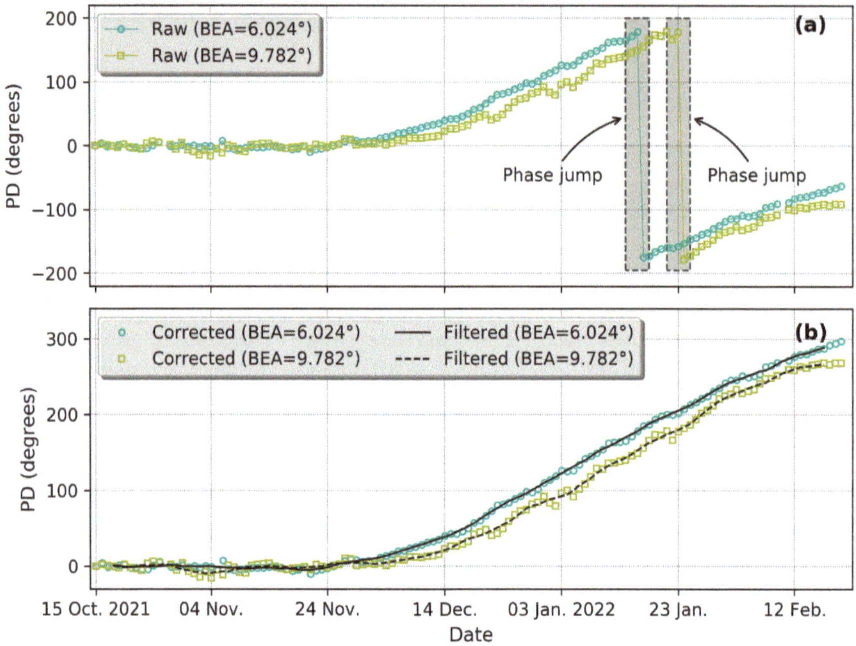

Fig. 6 The PD time series with increase pattern at BEA of 6.024° and 9.782° for the reflected SNR signal of GPS G03 satellite with L1 band signal

window length of 6 days is used to mitigate measurement noise, resulting in the filtered PD observations as shown in Fig. 6b.

Impacts of underground coal mining on the ground subsidence are rather complex. The PD time series shown in Fig. 6 presents a typical increase pattern of the PD series, which corresponds to the circumstance that subsidence at the ground reflection point is larger than that at the GNSS station, and the absolute difference between the subsidence at the former and that at the latter increases with time. In fact, in addition to the increase pattern, the amplitude of PD time series can also be decreased; this corresponds to the realistic circumstance that subsidence at the ground reflection point is always less than that at the GNSS station, and the absolute difference also increases with time.

A typical example of the raw PD time series with decrease pattern is given in Fig. 7a. Before 4 Nov. 2021, difference of ground subsidence at the GNSS station and that at the reflection point corresponding to the BEA of 6.082° is marginal basically; thus, PD at the BEA is very close to 0°. After then, the PD time series decreases with time, indicating the decrease of RH at the reflection point. Notice that the negative value of PD suggests that RH observed on a day is less than the initial RH of the GNSS antenna as demonstrated in (7). The decrease of RH is mainly caused by the uneven ground subsidence; specifically, the coal mining induced ground subsidence at the reflection point is less than that at the GNSS station. As a result, relative to the ground surface at the GNSS station, the ground surface at the reflection point rises; subsequently, RH and the corresponding PD at the reflection point decreases. When the PD is lower than $-180°$, phase-jump also happens for the decreased PD time series as clearly illustrated in the grey zone of Fig. 7a, but the angle shift of the series is opposite to that of the increased PD series. To maintain continuity of the PD series, a correction should also be performed for the decreased PD series, i.e.:

$$\Delta\Phi_{\text{correct}}^{\text{decrease}} = \Delta\Phi(\theta_{\text{BEA}}) - 360° \tag{18}$$

By making use of the moving average method to the corrected PD time series, the fileted PD estimations can be obtained, as shown Fig. 7b.

Substituting the filtered PD instead of the raw one into (8), RH of the ground reflection point at the BEA can be obtained. Then, by making use of (9–12), plane coordinates and corresponding subsidence of the reflection point can be calculated. Note that, tile angle at the reflection point is the other parameter to be determined for the calculation; and the retrieval method will be detailed in the later section.

Figure 8a–c show three examples of plane coordinates and corresponding subsidence observations of the ground reflection points at the BEAs, which are derived from the reflected SNR series displayed in Fig. 5b–d respectively. Notice that, to simplify analysis, subsidence and planar movement of the GNSS antenna (i.e. H_a,

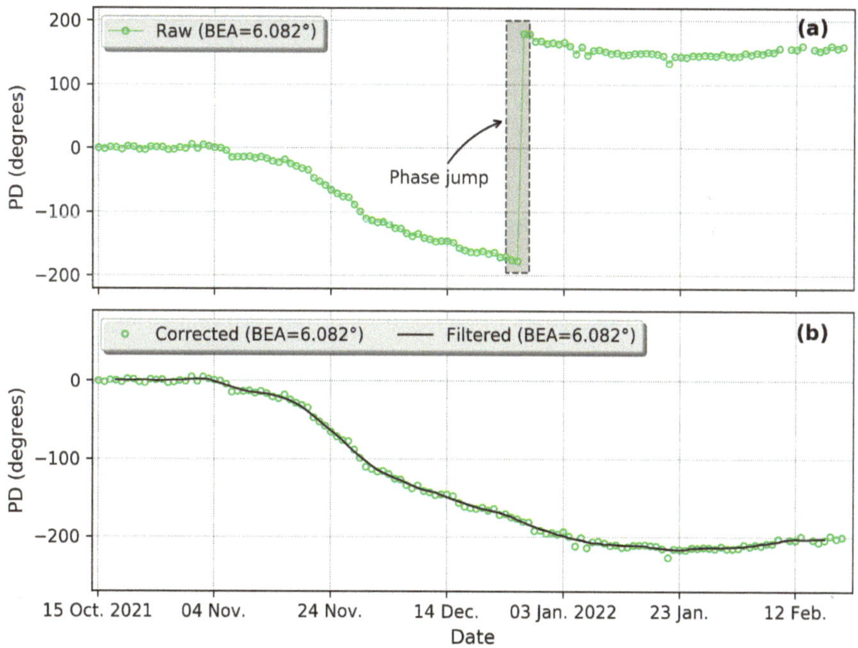

Fig. 7 The PD time series with decrease pattern at BEA of 6.082° for the reflected SNR signal of GPS G08 satellite with L1 band signal

x_a, and y_a), and all tilt angles at the reflection points are assumed as 0 for calculation of the observations shown in Fig. 8. It can be seen from the figure that, the observed subsidence on the reflection point increases along the southeast direction basically. In general, magnitude of subsidence at a ground point is inversely proportional to the distance from the point to the goaf center [29]. Since distance between the reflection point and the goaf center decreases along the southeast direction, ground subsidence on the point increases along the direction. In addition, underground coal mining induced ground subsidence develops progressively; and subsidence at the ground point within the subsidence bowl increases with time [36]. This can also be seen clearly in Fig. 8. For example, the observed subsidence around the ground point of (20, −35) is ~ 160 mm on 20 December 2021; while the subsidence increases to ~ 320 mm on 20 January 2022, and to ~ 500 mm on 15 February 2022. Basically, both spatial-and-temporal variation patterns of the ground subsidence derived by the proposed method agree with the previous study results.

Besides, it can be inferred from (12) that, the ground subsidence caused variation of RH can also induce variation of plane position of the reflection point at a given BEA. As shown in Fig. 8, the observed plane coordinates on the reflection point with BEA of 9.782° is (14.9, −25.3) on 20 December 2021; while the point moves to (15.2, −26.5) on 20 January 2021, and moves to (15.7, −27.2) on 15 February 2022, although difference of sky position of the GNSS satellite is marginal on these three days as demonstrated in Fig. 8d.

3.4 Ground Subsidence Surface Modeling

The plane coordinates and corresponding subsidence observations on the ground reflection points can be calculated by exploiting the GNSS RTK and GNSS-IR; however, it is a challenging problem to derive a mathematic model describing subsidence surface around the GNSS station with the discrete observations on the reflection points.

Suppose that width and length of a typical long wall coal mining working face is 350 m and 1000 m respectively; coal seam depth, mining thickness, and mining speed of the working face are 600 m, 8500 mm, and 2.5 m/d, respectively. A GNSS station is fixed on the ground above upper boundary of the working face; and distance between the station and the mining start position is 500 m, as shown in Fig. 9. One takes a square area centered at the raw GNSS station position with edge length of 160 m as the modeling area; then the KPIM can be utilized to simulate the daily ground subsidence in the modeling area. Note that, in Fig. 9, d_i is the distance from the mining position to the GNSS station along the mining direction at a certain time; and the negative sign of the distance denotes the mining position is behind the station (i.e., coal seam under the GNSS station has not been excavated). Besides, four different mining positions displayed in Fig. 9 (i.e. d_1, d_2, d_3, and d_4) correspond to four different time instants.

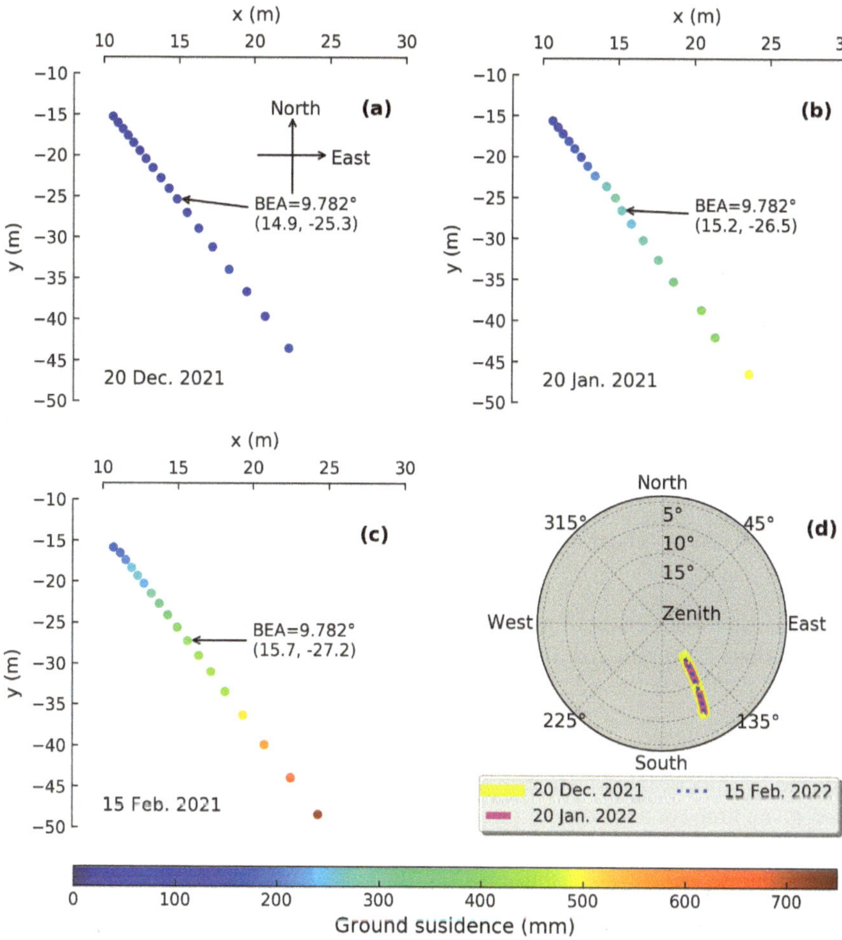

Fig. 8 **a**, **b**, and **c** are plane coordinates and corresponding subsidence observations on the ground reflection points at the BEAs for the reflected GPS G03 L1 band SNR series which are collected on 20 Dec. 2021, 20 Jan. 2022, and 15 Feb. 2022, respectively; **d** is sky plot of the GPS satellite track on these three days

When the working face is mined to the position with distances of −150, 0, 150 and 300 m away from the GNSS station, the simulated ground subsidence surfaces of the modeling area are shown in Fig. 10a–d respectively. It can be observed from the figure that, when the mining position is behind the GNSS station (i.e., $d = -150$ m), the ground surface in the monitoring area already subsides with the maximum subsidence of about 400 mm, although the coal seam under the area has not been excavated. This is mainly related to early influence of the underground coal mining. As the mining position approaches to the GNSS station, subsidence of the modeling area increases. After the mining position exceeds the GNSS station (i.e., the coal seam under the modeling area has been excavated), ground subsidence in the modeling area

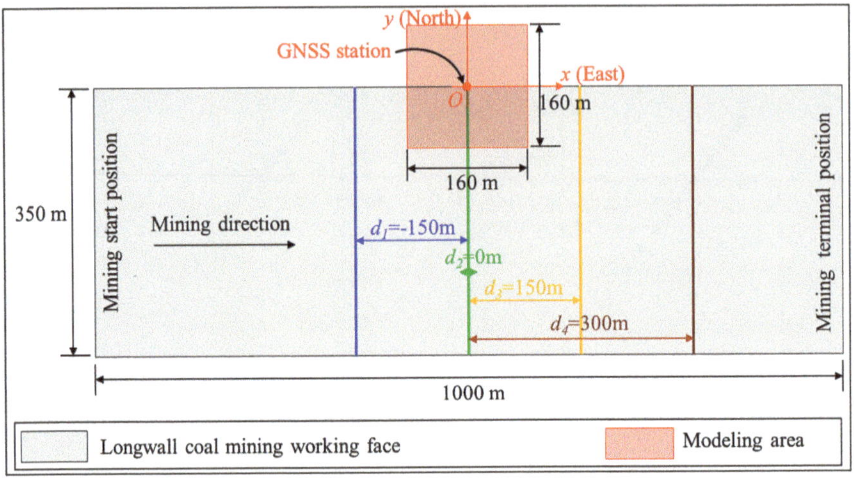

Fig. 9 Relative plane position between the GNSS station and the mining line for the simulated long wall coal mining working face

still increases significantly; for example, the maximum subsidence of the modeling area is about 2800 mm and 3500 mm at the mining position of 150 m and 300 m, respectively. This is caused by the lag influence of the underground coal mining as studied in [23].

In addition, it can also be observed from Fig. 10 that, the ground subsidence induced by underground coal mining varies continuously in the spatial-dimension; for a given time (or a given mining position), magnitude of the subsidence depends on plane position of the ground point. Thus, it is possible to develop an analytical function to describe relationship between the subsidence and corresponding plane coordinates.

Based on the ground surface subsidence pattern simulated by KPIM, the exponential models of 1st-, 2nd- and 3rd-order polynomial are developed to describe the subsidence surface induced by the underground coal mining in the monitoring area on a given time period (e.g. a day):

$$\begin{cases} f(x,y)_{1\text{st}} = \exp(a_0 + a_1x + a_2y) \\ f(x,y)_{2\text{nd}} = \exp(a_0 + a_1x + a_2y + a_3xy + a_4x^2 + a_5y^2) \\ f(x,y)_{3\text{rd}} = \exp(a_0 + a_1x + a_2y + a_3xy + a_4x^2 + a_5y^2 \\ \qquad + a_6x^2y + a_7xy^2 + a_8x^3 + a_9y^3) \end{cases} \tag{19}$$

where (x, y) are the plane coordinates of the ground point in the modeling area; $f(x, y)_{1\text{st}}$, $f(x, y)_{2\text{nd}}$, $f(x, y)_{3\text{rd}}$ are the 1st-, 2nd-, and 3rd-order model based subsidence on the point respectively; $\{a_i\}$ are model coefficients.

To evaluate performance of the models, a series of ground points in the modeling area is selected; distance interval of these points in both the x and y direction is

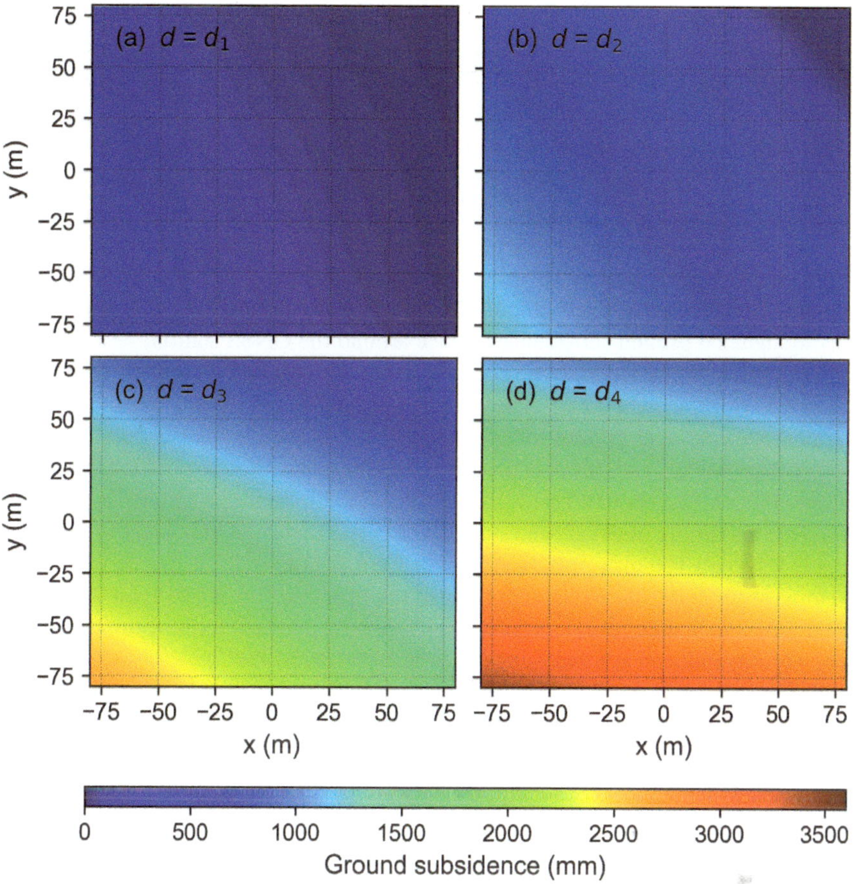

Fig. 10 The KPIM simulated ground subsidence surface in the modeling area when the mining position is − 150 m (**a**), 0 m (**b**), 150 m (**c**), and 300 m (**d**) away from the GNSS station

2 m, yielding 1600 simulated ground points. Substituting the KPIM-simulated plane coordinates and subsidence of these points into (19), coefficients of the 1st-, 2nd-, and 3rd-order polynomial model can be obtained by exploiting the nonlinear least-squares method respectively. Then, by substituting plane coordinates of the point and the model coefficients into corresponding polynomial model, the model-based subsidence can be determined.

Scatterplot of the model-based subsidence on the ground point versus the KPIM simulated one in the cases of four different mining positions is shown in Fig. 11. It can be seen from the figure that, there exists an obvious divergence for the subsidence results fitted by the 1st-order model, indicating an inferior performance of the model in description of the ground subsidence surface. Meanwhile, the divergence can be well corrected by using the higher-order models; accordingly, correlation between the fitted subsidence and the simulated one, and the fitting error also improve obviously.

For example, when distance is d_3, correlation coefficient between the model-based subsidence and the simulated one improves to 0.999 for the 2nd- and 3rd-order model; also, RMS of the fitting error decreases to 3.0 and 0.3 mm for the 2nd- and 3rd-order model respectively, as shown in Table 1. Although the model performance in terms of both correlation coefficient and RMSE improves with higher-order polynomial models, the improvement is relatively marginal from the 2nd- to 3rd-order model as evidenced in Fig. 11 and Table 1. It would be more sensible to opt for the simpler model with a lower-order polynomial at the cost of negligible performance degradation. Therefore, the 2nd-order polynomial model, i.e. the second equation of (19), can be selected as the mathematical function describing the underground coal mining induced ground subsidence surface around the GNSS station.

Notice that, more simulation results indicate that, for the typical long wall coal mining working face, the 2nd-order polynomial model can also be used to describe the underground coal mining induced ground subsidence surface accurately under different geological and mining conditions and the GNSS station positions, although the model coefficients can be much different for different conditions. In addition, when the maximum of the ground subsidence is less than 5000 mm and the modeling area is less than 160 m × 160 m, RMSE of the model-based subsidence is less than 10 mm; and the maximum relative error of the fitting results is less than 0.5% basically. All these results clearly suggest that it is practicable to describe the underground coal mining induced ground subsidence surface using the 2nd-order polynomial model.

In the realistic data processing stage, if there exists k GNSS + IR derived plane positions and subsidence observations on the reflection points at a given time period (e.g. a day), k observation equations with 6 unknown model coefficients can be obtained by:

$$\begin{cases} \ln \hat{W}_{\theta 1} = a_0 + a_1 \cdot \hat{x}_{\theta 1} + a_2 \cdot \hat{y}_{\theta 1} + a_3 \cdot \hat{x}_{\theta 1}\hat{y}_{\theta 1} + a_4\hat{x}_{\theta 1}^2 + a_5 \cdot \hat{y}_{\theta 1}^2 \\ \ln \hat{W}_{\theta 2} = a_0 + a_1 \cdot \hat{x}_{\theta 2} + a_2 \cdot \hat{y}_{\theta 2} + a_3 \cdot \hat{x}_{\theta 2}\hat{y}_{\theta 2} + a_4\hat{x}_{\theta 2}^2 + a_5 \cdot \hat{y}_{\theta 2}^2 \\ \cdots \\ \ln \hat{W}_{\theta k} = a_0 + a_1 \cdot \hat{x}_{\theta k} + a_2 \cdot \hat{y}_{\theta k} + a_3 \cdot \hat{x}_{\theta k}\hat{y}_{\theta k} + a_4\hat{x}_{\theta k}^2 + a_5 \cdot \hat{y}_{\theta k}^2 \end{cases} \tag{20}$$

where $(\hat{x}_{\theta i}, \hat{y}_{\theta i})$ and $\hat{W}_{\theta i}$ are the GNSS + IR derived plane position and subsidence on the ground reflection point corresponding to the i-th BEA. Besides, an additional equation can also be obtained to describe the ground movement at the GNSS station observed by the GNSS positioning technique:

$$\ln \hat{W}_a = a_0 + a_1 \cdot \hat{x}_a + a_2 \cdot \hat{y}_a + a_3 \cdot \hat{x}_a\hat{y}_a + a_4\hat{x}_a^2 + a_5 \cdot \hat{y}_a^2 \tag{21}$$

where $(\hat{x}_a, \hat{y}_a)$ and $\hat{W}_a$ are the plane position and subsidence at the GNSS station on a day, observed by the GNSS positioning technique.

Based on (20) and (21), the subsidence surface model coefficients can be determined by making use of the weighted least-squares method:

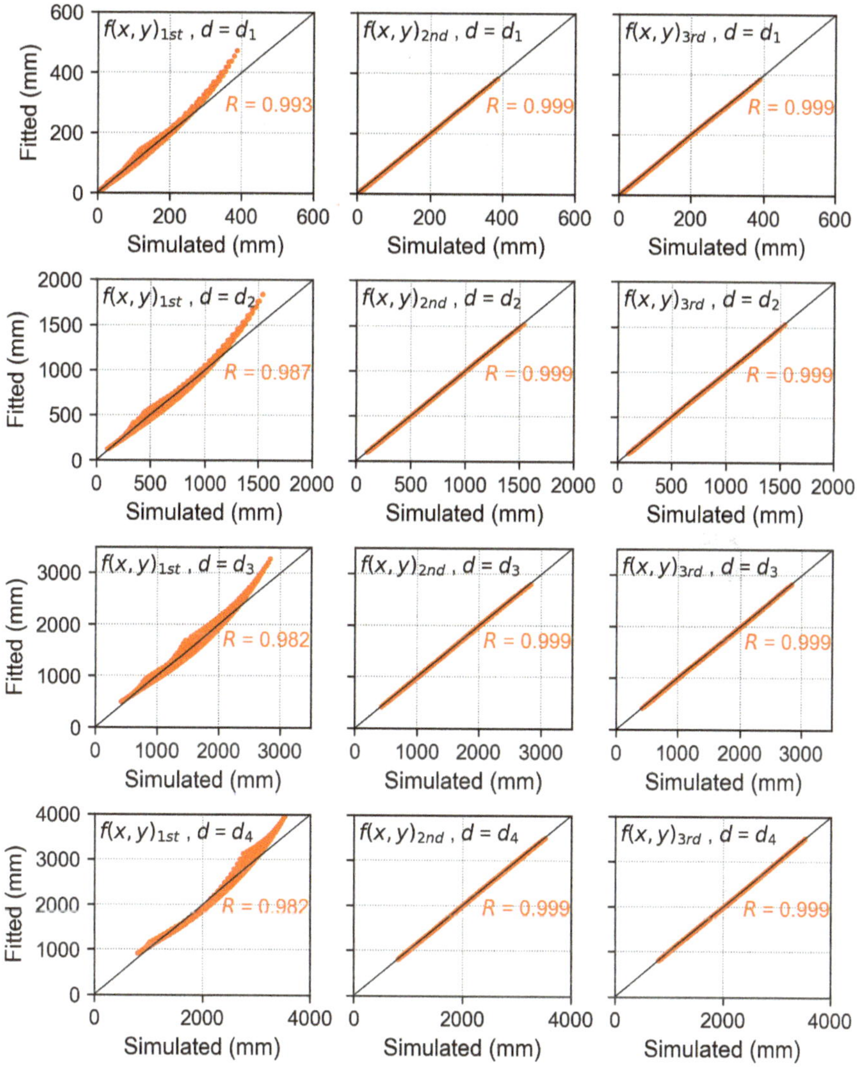

Fig. 11 Scatterplot of the KPIM-simulated ground subsidence versus the model-fitted one for 1st- (left column), 2nd- (center column), and 3rd- (right column) order polynomial model under four different mining positions

$$\mathbf{A} = (\mathbf{X}^T\mathbf{G}\mathbf{X})^{-1}\mathbf{X}^T\mathbf{G}\mathbf{Y} \tag{22}$$

where $\mathbf{A}$ is the coefficients matrix of the subsidence surface model, which given by:

$$\mathbf{A} = \left[a_0\ a_1\ a_2\ a_3\ a_4\ a_5 \right]^T, \tag{23}$$

Table 1 Mean, STD and RMS of the model-based ground subsidence errors under four different mining positions

Position	Polynomial	Mean (mm)	STD (mm)	RMSE (mm)
d_1	1st	1.2	10.7	10.8
	2nd	0.0	0.3	0.3
	3rd	0.0	0.3	0.3
d_2	1st	2.3	44.4	44.5
	2nd	0.0	1.2	1.2
	3rd	0.0	0.3	0.3
d_3	1st	2.4	86.1	86.2
	2nd	0.0	3.0	3.0
	3rd	0.0	0.3	0.3
d_4	1st	3.2	113.5	113.6
	2nd	0.0	3.4	3.4
	3rd	0.0	0.3	0.3

X is the plane coordinates observations related matrix:

$$\mathbf{X} = \begin{bmatrix} 1 & \hat{x}_{\theta 1} & \hat{y}_{\theta 1} & \hat{x}_{\theta 1}\hat{y}_{\theta 1} & \hat{x}_{\theta 1}^2 & \hat{y}_{\theta 1}^2 \\ 1 & \hat{x}_{\theta 2} & \hat{y}_{\theta 2} & \hat{x}_{\theta 2}\hat{y}_{\theta 2} & \hat{x}_{\theta 2}^2 & \hat{y}_{\theta 2}^2 \\ \vdots & \vdots & \vdots & \vdots & \vdots & \vdots \\ 1 & \hat{x}_{\theta k} & \hat{y}_{\theta k} & \hat{x}_{\theta k}\hat{y}_{\theta k} & \hat{x}_{\theta k}^2 & \hat{y}_{\theta k}^2 \\ 1 & \hat{x}_a & \hat{y}_a & \hat{x}_a\hat{y}_a & \hat{x}_a^2 & \hat{y}_a^2 \end{bmatrix}, \tag{24}$$

Y is the subsidence observation related matrix:

$$\mathbf{Y} = \begin{bmatrix} \ln \hat{W}_{\theta 1} & \ln \hat{W}_{\theta 2} & \ldots & \ln \hat{W}_{\theta k} & \ln \hat{W}_a \end{bmatrix}^T, \tag{25}$$

and **G** is the weight matrix and is given by:

$$\mathbf{G} = \begin{bmatrix} g_\theta & 0 & \cdots & 0 & 0 \\ 0 & g_\theta & 0 & 0 & 0 \\ \vdots & \vdots & \vdots & \vdots & \vdots \\ 0 & 0 & 0 & g_\theta & 0 \\ 0 & 0 & 0 & 0 & g_a \end{bmatrix} \tag{26}$$

Here g_θ is weight of the GNSS + IR based observation at the reflection point; and g_a is weight of the GNSS positioning technique-based observation at the GNSS station. {{The former can be set as 1; and the data processing results indicate that, the proposed method calculated model coefficients and ground subsidence have a good accordance with the in-situ one, when the latter (i.e. g_a) is in the range of 3~6 basically.

3.5 Calculation of Ground Tilt Angle on the Reflection Point

A massive number of previous studies have proved that the underground coal mining induced subsidence on the ground points with different plane positions are usually different especially for the points around border of the goaf, resulting in tilt of the ground surface. Because tilt angle at the reflection point is the other key parameter when calculating plane coordinates and subsidence observation on the reflection point as indicated in (10–12), it is also necessary to retrieve the parameter accurately for the proposed method. It can be deduced that, slope at the reflection point in the incidence plane of the GNSS signal equals directional derivative of the subsidence surface function along the direction pointing from the reflection point R_θ to the GNSS station P_a, i.e.:

$$k(x_\theta, y_\theta) = \frac{\partial f(x, y)_{2\text{nd}}}{\partial x}\mu_1 + \frac{\partial f(x, y)_{2\text{nd}}}{\partial y}\mu_2\bigg|_{x=x_\theta, y=y_\theta} \tag{27}$$

where $k(x_\theta, y_\theta)$ is the slope on the ground reflection point with BEA of θ; (μ_1, μ_2) is unit vector, and the two components are given by:

$$\begin{cases} \mu_1 = \cos\dfrac{x_a - x_\theta}{\sqrt{(x_a - x_\theta)^2 + (y_a - y_\theta)^2}} \\ \mu_2 = \cos\dfrac{y_a - y_\theta}{\sqrt{(x_a - x_\theta)^2 + (y_a - y_\theta)^2}} \end{cases} \tag{28}$$

Then, tilt angle on the ground reflection point can be easily determined by:

$$\alpha(\theta) = \alpha(x_\theta, y_\theta) = \text{atan}\big[k(x_\theta, y_\theta)\big] \tag{29}$$

However, in fact, it is impossible to calculate the tilt angle with equations of (27–29) directly, unless the subsidence surface function (coefficients of the function, specifically) can be determined in advance. However, determination of the function coefficients depends on plane coordinates and subsidence observations of the reflection points; and tile angle is the necessary parameter to calculate these observations. That is, if one wants to obtain the tilt angle on the reflection point with (29), subsidence surface function must be determined at first; meanwhile, to determine the subsidence surface function, the tilt angle on the reflection point must be known in advance.

To solve the contradictory problem, an iterative algorithm is developed to calculate the tilt angle and coefficients of the subsidence surface function simultaneously.

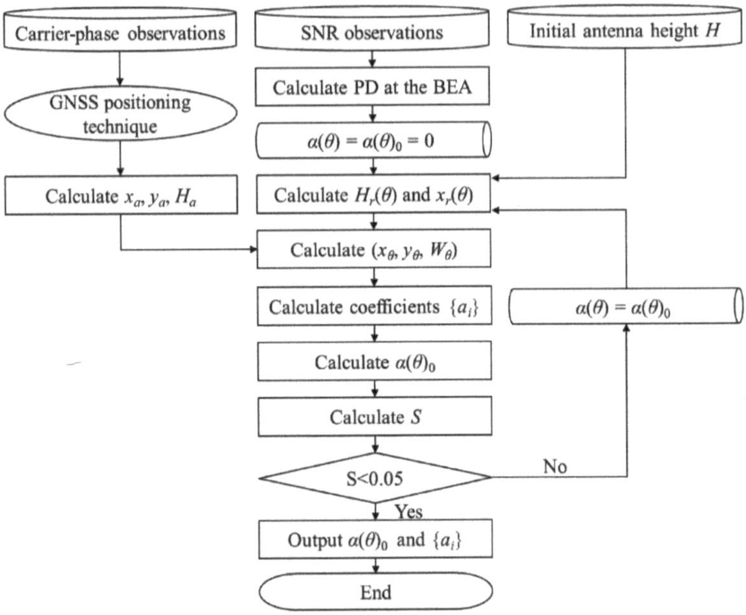

Fig. 12 Algorithm flowchart of GNSS + IR based ground subsidence surface retrieval

Figure 12 shows the flowchart of the algorithm, and the algorithm procedure can be described as follows:

Step 1: Calculate plane coordinates (x_a, y_a) and subsidence (H_a) of the GNSS station on a day using GNSS positioning technique based on carrier-phase observations.
Step 2: Calculate PD of the reflected SNR signal at the BEA.
Step 3: Set initial value of tilt angle as to 0° for each BEA.
Step 4: Calculate horizontal and vertical distance from the GNSS antenna phase center to the ground reflection point, i.e. $H_r(\theta)$, $x_r(\theta)$ with (10).
Step 5: Calculate plane coordinates (x_θ, y_θ) and subsidence (W_θ) on the ground reflection point with (11) and (12).
Step 6: Calculate the coefficients of the ground subsidence surface model $\{a_i\}$ using (22).
Step 7: Calculate ground tilt angle at each BEA using (29) respectively.
Step 8: Calculate RMSE of the tilt angle estimation S, which is given by:

$$S = \sqrt{\frac{\sum_{i=1}^{k} [\alpha(\theta_i) - \alpha(\theta_i)_0]}{k-1}} \tag{30}$$

Step 9: If S is greater than the threshold error, assign the tilt angle derived from Step 7 to the initial tile angle and go to Step 4; otherwise, output the tilt angles and coefficients of the subsidence surface function.

Note that, the data processing results indicate that the tilt angle and function coefficients can be accurately determined when the threshold error is less than 0.05° in general. By substituting the estimated optimal function coefficients into the second equation of (19), the underground coal mining induced ground subsidence surface around the GNSS station can be determined finally.

4 Experimental Results

In this section, the proposed subsidence surface determination method is analyzed and evaluated by making use of the data set collected in Jining mining area, China, over about 120 days.

4.1 Experiment Setup

The Jining mining area is one of the largest coal-producing areas in China, which locates in the southwest of Shandong Province, China, as indicated in Fig. 13. The underground mining has caused large-scale ground subsidence, resulting in a serious impact on the local ecological environment. To monitor the impact of underground coal mining on the ground surface, a GNSS station was established within the mining area on 7 September 2020. Latitude and longitude of the GNSS station are about 35°29′21″ and 117°00′40″. The station is above upper boundary of a long wall coal mining working face, as shown in Fig. 14; distance between the station and open-off cut of the working face is about 450 m. Mining width and length of the working face are about 358 m and 1500 m respectively; and the averaged coal seam thickness, mining depth, and mining speed of the work face are about 8500 mm, 600 m, and 2.5 m/d, respectively. The mining direction of the working face is parallel to strike direction of the coal seam (i.e., southeast, approximately). The existing bore holes show that stratums of the working face mainly include Quaternary, Jurassic, Permian, Carboniferous, and Ordovician; and the coal seam distributes on the top of the stratum of Carboniferous basically. Geological structure of the working face is relatively simple without any large fault or fold. As a typical Quaternary alluvial plain, ground surface above the coal mining working face is rather flat without obvious slope or terrain undulation when the surface is subsidence-free, although the ground is slightly rough.

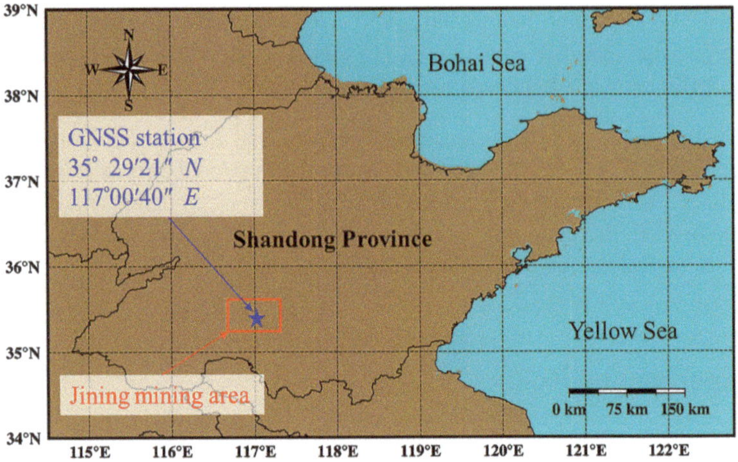

Fig. 13 Location of the Jining mining area and the GNSS station

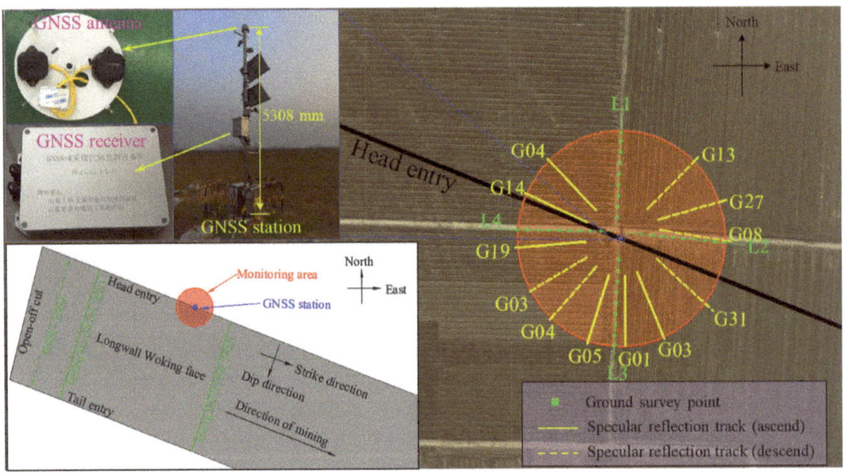

Fig. 14 Local environment around the GNSS station and specular reflection tracks of GPS satellites under elevation angle of 5°–15°

The GNSS station is equipped with a receiver which is developed based on two GNSS chips including a u-blox F9P chip and a u-blox M8N chip. Two u-blox ANN-MB GNSS antennas fixed on a metal plate are connected with the two chips respectively, as shown in left upper of Fig. 14. The u-blox M8N chip is a navigational grade single-frequency GNSS chip, which has 32 signal channels configured to receive and process the GPS L1 band signals in the experimental campaign. Each u-blox F9P chips is a high precision GNSS positioning chip; the chip can make use of dual-band signals of all GNSS constellations to realize fast and reliable RTK positioning, and

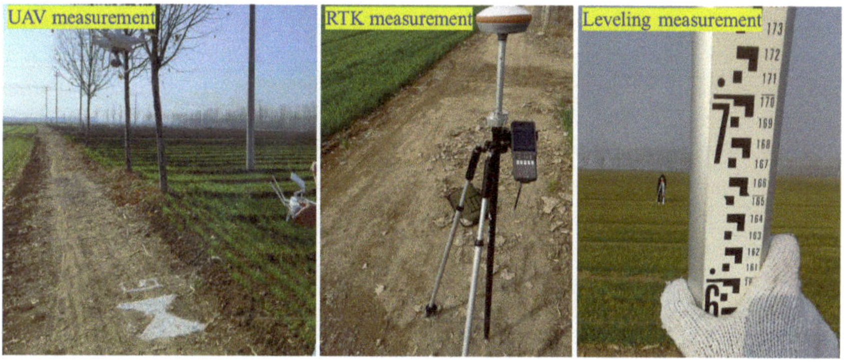

Fig. 15 Measurement of ground subsidence with the UAV photogrammetry (left), GNSS RTK (center) and leveling survey (right) during the experimental campaign

measure ground movement at the station. The F9P chip outputs SNR observation as the integer value, which is inadequate for calculating PD of the reflected SNR series accurately; therefore, the high-resolution SNR observations collected by the M8N chip are utilized to calculate PD of the reflected signal [37]. The observation interval is set as 1 s for both two types of chips. Before ground subsides due to underground coal mining, vertical height from the GNSS antenna to the ground surface (or, the initial antenna height) is 5308 mm. Two solar cells with 300AH battery capacity are utilized to provide power supply for the GNSS receiver, ensuring continuous operation of the receiver sufficiently.

In addition, an unmanned aerial vehicle (UAV) of DJI Phantom 4 was applied to measure digital elevation model (DEM) around the GNSS station based on the photogrammetry, as shown in the left figure of Fig. 15. By calculating the difference between the DEM before and after the surface subsides, the UAV-based subsidence surface can be obtained. Further, four ground survey lines (L1, L2, L3, and L4) including 36 ground survey points with horizontal interval of about 5 m were deployed on the ground surface, as shown in Fig. 14. Ground subsidence on these survey points were measured by using GNSS RTK method and the leveling survey method, as shown in the center and right figure of Fig. 15; the instruments utilized by these two methods are Hi-Target iRTK10 GNSS receiver and KOLIDA KL-03 level, respectively. The UAV, GNSS RTK, and leveling survey-based subsidence observations are taken as the reference data to compare with the results generated by the proposed method in the experimental campaign.

4.2 Data

The experimental period is from 15 October 2020 to 15 February 2021, over 124 days. There are more than 20 thousand plane movement and subsidence observations with fixed RTK resolution on each day, which were recorded by the GNSS receiver with the u-blox F9P chip. By averaging all the plane movement and subsidence observations on a day respectively, the daily subsidence and plane movement of the GNSS antenna can be obtained. The GNSS receiver with the u-blox M8N chip recorded GPS L1 band SNR observations of 10 satellites with 12 different ground reflection tracks under satellite elevation angle range from 5° to 15°, to retrieve PD of the reflection signal on each day. The ground reflection tracks of the GPS satellites are also displayed in Fig. 14; and these tracks include about 240 BEA corresponded ground reflection points on a day.

Because most of the BEA corresponded reflection points are scattered in a circular area centered at the GNSS station with radius of 50 m, the area is taken as the monitoring area, as shown in Fig. 14. Taking the initial GNSS antenna position as the origin point, a local coordinate system related to the monitoring area can be established; the positive direction of x and y axis of the coordinate system is parallel to the east and north direction, respectively. On 15 October 2021, distance from the mining position to the GNSS station (i.e., d) was about −280 m. Due to the distant mining position, the monitoring area cannot be impacted by the underground coal mining in general. Therefore, the reflected SNR signals collected on the day are taken as the base signals to calculate PD of the reflected signal at the BEA. Combined with the F9P chip recorded movement observations of the GNSS antenna and the M8N chip provided PD estimations at the ground reflection point on a day, 124 ground subsidence surface measurements in the monitoring area can be obtained by making use of the proposed method (or, GNSS + IR). Note that, because the experiment was conducted in late autumn and winter, vegetation on the ground is rather sparse; impact of the ground vegetation on the reflected signal could be marginal and was neglected during the experiment.

The UAV, GNSS RTK, and leveling measurement were firstly performed on 10 October 2021. Another 6 UAV-based photogrammetry measurements were conducted on 20 November 2021, 2 December 2021, 20 December 2021, 15 January 2022, 6 February 2022, and 11 February 2022 respectively, producing 6 UAV-based subsidence surfaces of the motoring area. Measurement interval of subsidence on the ground survey lines is about 4 days and 9 days for GNSS RTK and leveling survey method, respectively; the total number of subsidence observations on the ground survey points is 828 and 468 for GNSS RTK and leveling survey respectively. Details of the observation data type and number during the experimental campaign can also be seen clearly in Table 2.

Table 2 The observed data type and number during the experimental campaign

Method	Observation type	Observation number
GNSS + IR	Surface	124
UAV Photogrammetry	Surface	6
GNSS RTK	Point	828
Leveling survey	Point	468

4.3 The Proposed Method Derived Ground Subsidence Surface

By making use of all the plane coordinates and subsidence observations at the ground reflection points of 12 GPS satellite trajectories on a day with the proposed iterative algorithm, coefficients of the ground subsidence surface model and thus the model-based ground subsidence at the point can be obtained. Scatterplot of subsidence observed by GNSS + IR on the BEA-associated reflection point versus the model-based one during the experimental campaign is shown in Fig. 16. Basically, all the correlation coefficients between the observed subsidence and the model-based ones are greater than 0.95 especially when the maximum subsidence is larger than about 800 mm, indicating a good accordance between these two subsidence estimations. These is a slightly negative deviation with mean error of about −1 mm for the model-based subsidence. This might attribute to the error of the developed exponential surface model; that is, the developed model derived subsidence surface could underestimate the GNSS + IR based subsidence estimation to some extent. Nevertheless, the maximum RMS of the model-based subsidence error is less than 50 mm; and the averaged relative error is less than 1% in general. This also indicates the feasibility to describing the underground coal mining induced ground subsidence surface with the developed exponential model.

The ground subsidence surface derived by the proposed method during the experiment campaign is shown in Fig. 17. Basically, when the coal mining working face excavates to the position behind a ground point with horizontal distance in range of 1/5 to 1/3 of mining depth, the ground point begins to be impacted by the coal mining, inducing ground subsidence at the point [23]. With the excavation of the working face, area impacted by the underground coal mining and ground subsidence in the area increases; and the area impacted by coal mining is much larger than the excavating area [24]. These can also be seen clearly from Fig. 17. On 10 November 2021, excavation position of the working face is about 130 m behind of the monitoring area; while the ground in southwest of the area has been significantly affected by the underground coal mining with ground subsidence larger than 150 mm. When the excavating position approaches to the GNSS station, area affected by the coal mining increases; on 25 November 2021, the contour line with subsidence of 150 mm moved to northeast of the monitoring area, indicating that the notable impact induced by the underground coal mining has spread to almost all the monitoring area. In addition to the increase of subsidence area, subsidence magnitude on the ground point also increases when the excavating position approaches. For example, on 15 November 2021, subsidence of the contour line nearest to southwest of the monitoring area is 300 mm, as displayed in Fig. 17; however, subsidence of the ground point on the contour line increases to about 450 mm on 20 November 2021, and to about 600 mm on 25 November 2021, respectively. Note that, because southwest side of the monitoring area has the shortest distance to the goaf, the side is the most seriously affected by the underground coal mining.

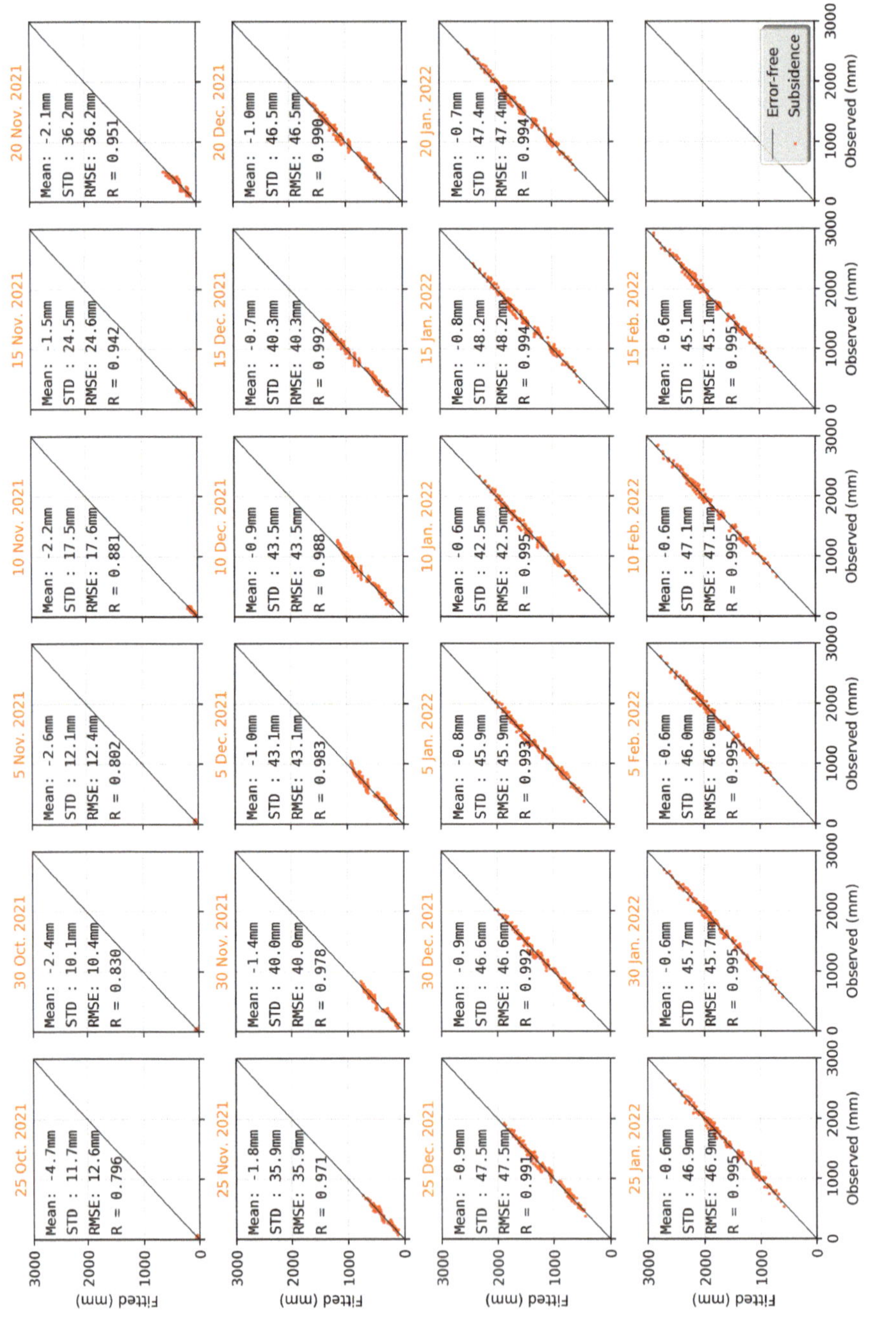

Fig. 16 Scatterplot of the GNSS + IR observed subsidence versus the surface model-based one on the ground reflection point during the experimental campaign

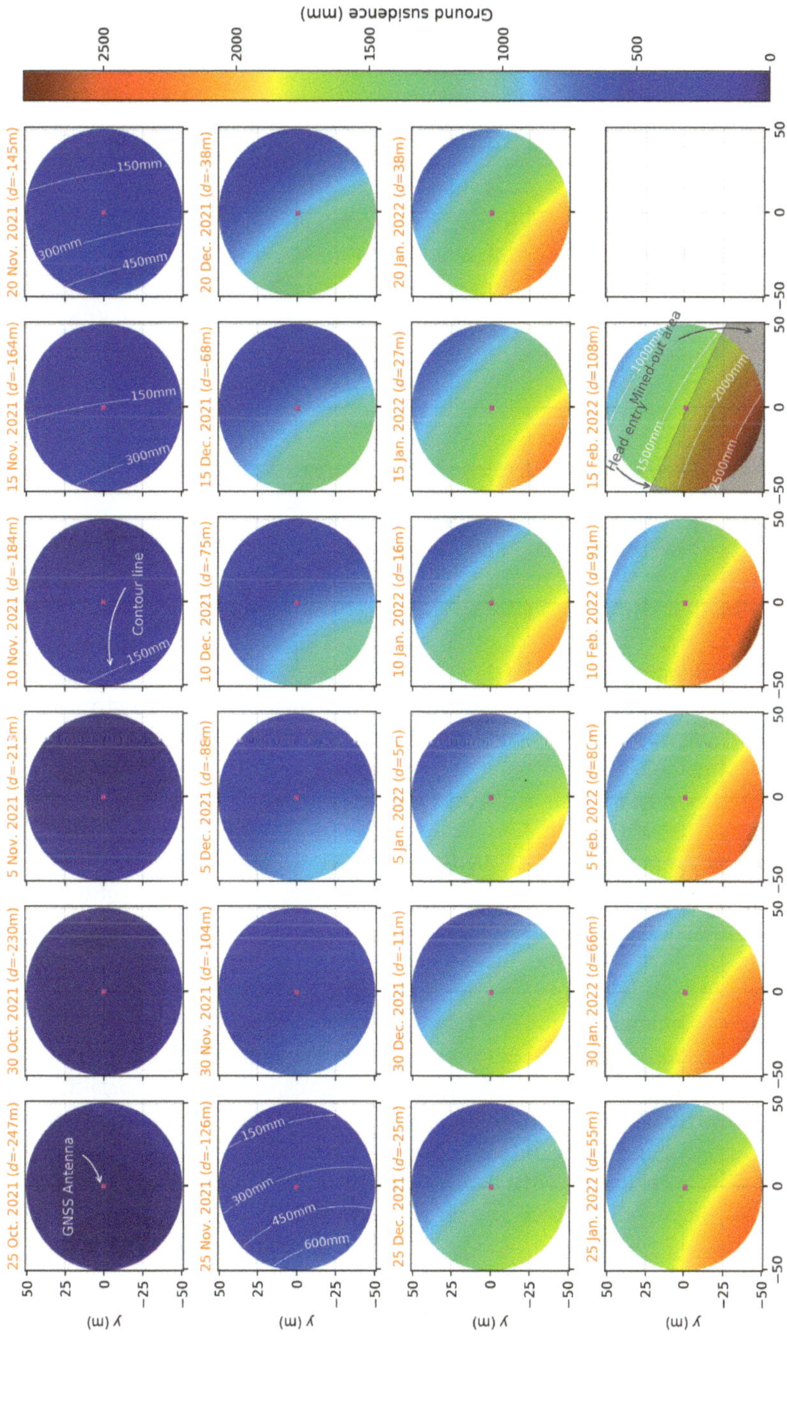

Fig. 17 The proposed method derived subsidence surface during the experimental campaign

On 25 January 2022, the mining position is in front of the monitoring area about 5 m away; and the coal seam underneath the monitoring area has been excavated. However, ground subsidence in the monitoring area still increases obviously; e.g., the maximum subsidence of the area is 2583 mm, 2651 mm, and 2748 mm on 30 January 2022, 5 February 2022, and 10 February 2022 respectively. In addition, for the coal seam mined out area, the ground subsidence magnitude gradually decreases outward along the mining boundary. This can also be observed from the proposed method-derived subsidence surface on 15 February 2022. All the contour lines are approximately parallel to the upper boundary of the working face (i.e. the head entry) on the day; and subsidence value of these contour lines decrease along the northeast direction, which is perpendicular to the mining boundary. Basically, the temporal and spatial variation pattern of the ground subsidence can be well captured by the proposed subsidence surface model.

4.4 Comparison with the Reference Observations

For evaluating subsidence estimation performance of the proposed method with the UAV-based subsidence surface, a series of ground points within the monitoring area is selected. Both x and y axis coordinate range of these points are set from -50 to 50 m with an interval of 2 m, producing about 2000 ground points. The proposed method-based ground subsidence on the point can be calculated by substituting plane coordinates of the point into the developed subsidence surface model directly; while the UAV-based subsidence on the point is obtained by exploiting Kriging interpolation method to the UAV-based subsidence observations on the ground feature points. Note that the UAV-based DEMs and thus subsidence observations are obtained by using photogrammetry; the elevation accuracy of UAV Photogrammetry typically ranges from 5 to 30 cm. Taking the UAV-based subsidence observations as the reference data, mean, STD, and RMS of the proposed method-based ground subsidence estimation errors on these 2000 points can be obtained, which is given in Table 3. It can be seen from the table that, there exists an obvious systematic bias between the proposed method-based ground subsidence and the UAV-based one; and the minimum and maximum bias are -236.7 mm and 212.6 mm respectively. The in-situ maximum subsidence over the monitoring area are 606 mm, 885 mm, 1362 mm, 2049 mm, 2431 mm, and 2553 mm for these six days, respectively. The bias should be related to measurement error of the reference data (position measurement error of the UAV-carried camera, specifically). The UAV makes use of GNSS RTK technique to measure coordinates of the camera; and the position is taken as one of the inputs to determine coordinates of ground points by exploiting aerial triangulation in the photogrammetry. Because of high dynamic variation of the UAV position and influence of the atmospheric environment (e.g. wind) on the UAV, the GNSS RTK measured camera position could produce a significant error, resulting in the systematic offset on the photogrammetry derived ground point position (about 100–800 mm) especially for elevation observation of the ground point [38]. Nevertheless, all the correlation coefficients between the proposed method-based ground subsidence estimation and the UAV-based one are greater than 0.99, which implies a good accordance between the two estimations in terms of underground coal mining induced subsidence variation on the ground surface.

By subtracting the UAV-based ground subsidence surface from the proposed method derived one and removing the systematic bias, difference between the UAV-based subsidence surface and the proposed method based one can be obtained, as shown in Fig. 18. It can be seen from the figure that the maximum subsidence difference is less than 80 mm in the monitoring area for all six measurements. Furthermore, estimating error of the proposed method-based subsidence surface in the north of the monitoring area is relatively larger than that in other areas, especially for the subsidence surface collected on 20 December 2021, 15 January 2022, and 11 February 2022. This should be caused by the lack of subsidence observation of the ground reflection point in the north of the monitoring area. Due to the reason of design of the GPS satellite orbits, almost all the GPS satellite reflection tracks distribute in area centered on a GNSS station with azimuth angle range from about 30° to 330° for the Asian region [39]. Therefore, in the north of the monitoring area, none of GPS satellite reflection tracks and thus the ground subsidence observations on the reflection points can be utilized to participate in the subsidence surface estimation by the proposed method. As a result, performance of the proposed method is insufficient in retrieval of the ground subsidence in the north area of the GNSS station to some extent. Monitoring the ground subsidence induced by underground coal mining with multiple GNSS constellation reflection signals (e.g. GPS and BDS satellite reflection signal) based on the proposed method should be able to improve the insufficiency significantly.

Figure 19 shows the comparison between the proposed method derived ground subsidence on the 36 ground survey points versus the GNSS RTK and leveling survey observed one during the experimental campaign. Taking the GNSS RTK method and leveling survey method observed ground subsidence as the reference data respectively, mean, STD, and RMS of the proposed method-based ground subsidence estimation error can be obtained, which are given in Table 4. It can be seen from Fig. 19 and Table 4 that there exists a slightly positive mean error for the proposed method derived subsidence estimates in the cases of two different reference data

Table 3 Mean, STD and RMS of errors in the proposed method derived ground subsidence estimation for the reference data of the UAV-based subsidence surface

Date	Mean (mm)	STD (mm)	RMSE (mm)	*R*
20 Nov. 2021	− 156.0	9.1	156.3	0.998
2 Dec. 2021	− 236.7	14.6	237.3	0.997
20 Dec. 2021	114.5	13.7	115.3	0.999
15 Jan. 2022	212.6	21.5	213.7	0.998
6 Feb. 2022	− 233.9	10.3	234.1	0.999
11 Feb. 2022	− 220.5	24.5	221.9	0.998

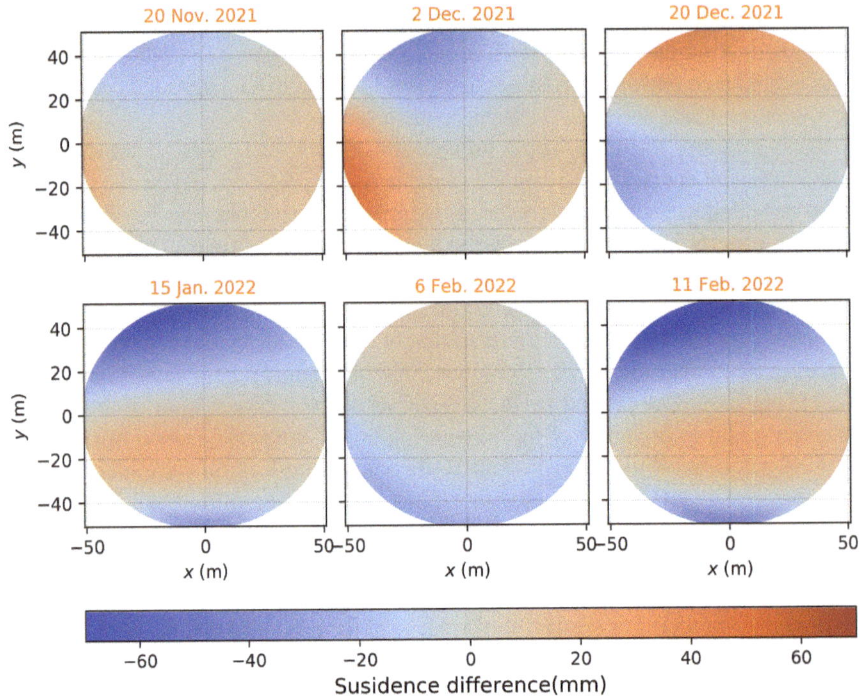

Fig. 18 Difference between the proposed method derived ground subsidence surface and the UAV-based one

sets. This is probably caused by penetration of the GNSS signal. As studied in [40, 41], the GNSS signal can penetrate the ground surface with a certain of depth and can be scattered again from underneath of the surface; and the penetrating depth is inversely proportional to soil moisture of the ground surface. The {base date} of the experimental campaign is selected in mid-autumn (i.e. 15 October 2021); soil moisture of the ground surface was relatively large in the period due to the frequent precipitations. Thus, the reflected base signal captured by the GNSS receiver should mainly come from the ground surface with marginal penetrating depth. After entering winter, precipitation of the experimental area decreases significantly. Due to seldom precipitation, ground soil moisture decreases obviously, resulting in the increase of penetrating depth of the reflected GNSS signal. Therefore, reflector height and thus ground subsidence on the ground reflection point could be overestimated by the proposed method.

Because measurement precision of the GNSS RTK is different from that of the leveling survey, RMS of the proposed method-based ground subsidence estimation errors are also different, which are 54.7 mm and 35.9 mm for the two techniques, respectively as shown in Table 4. In general, the measurement precision of the GNSS RTK method-based elevation observation is inferior to that of the leveling survey method [42, 43]; the latter-based ground subsidence measurement should be closer

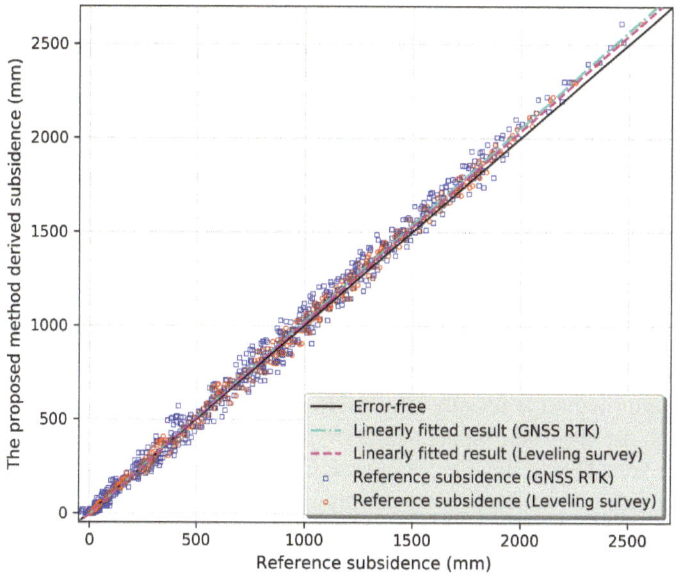

Fig. 19 Scatterplot of the proposed method derived ground subsidence observation versus the GNSS RTK based one, and versus the leveling survey method based one during the experimental campaign

Table 4 Mean, STD and RMS of errors in the proposed method derived ground subsidence estimation for the reference data of GNSS RTK and leveling survey based observations

Reference	Mean (mm)	STD (mm)	RMSE (mm)	*R*
GNSS RTK	18.1	51.6	54.7	0.997
Leveling survey	12.2	33.8	35.9	0.998

to the in-situ one. Thereby, the comparison between the leveling survey-based subsidence observations and the GNSS + IR-based ones could better reflect the realistic accuracy of the proposed method (i.e. 35.9 mm). In addition to the estimation error induced by the penetration of the GNSS signal, the PD retrieval error of the reflected signal should be the main reason causing the ground subsidence estimation error for the proposed method. Besides, the leveling survey-based ground subsidence measurements were observed once a day at some specific time point, whereas the GPS-based subsidence estimates were produced by making use of the GNSS observations over irregular periods on the day. Such a temporal mismatch may also contribute to the estimation error, particularly when the ground subsidence velocity is significant. Nevertheless, the correction coefficient between the proposed method-based subsidence observation and the reference observation is greater than 0.99 for these two different reference data sets. Moreover, the significant variation of ground subsidence on the survey points under different temporal and spatial circumstances can also be well estimated by the proposed method as proved in Fig. 19. All these results clearly exhibit the accuracy of the proposed method in retrieval of the ground subsidence surface induced by underground coal mining.

5 Conclusions

In this chapter, a new method has been proposed for retrieval of underground coal mining induced ground subsidence surface, which uses combination of GNSS positioning technique and GNSS-IR. The key advantage of the proposed method is that it can make use of a single station collected GNSS observations to monitor underground coal mining induced subsidence over a large area (~8000 m^2) continuously. The proposed method is usually applicable since carrier-phase and SNR observations, which are the main inputs of the proposed method, are recorded by typical GNSS receivers. Due to the use of low-cost GNSS instrument, the proposed method can also provide a cost-effective way for ground subsidence monitoring in the mining area.

The geometry models and PD estimation formulas are derived to obtain estimations of plane coordinates and subsidence on the BEA corresponded ground reflection point. The 2nd-order exponential model has been developed to describe ground subsidence surface for the typical long wall coal mining working faces. The model is independent of GNSS station location and the mining position as long as the modeling area is less than 160 m $\times$ 160 m, and the maximum subsidence in the area is less than 5000 mm. Coefficients of the developed model, which are much different in different mining conditions, can be obtained by using the plane coordinates and subsidence observations on the ground reflection points with the weighted least-squares method. Due to lack of GPS signal-based subsidence observation in the north of the monitoring area, error of the proposed method derived subsidence surface in the north area is relatively larger. Because of penetration of the GNSS signal, the ground subsidence can be slightly overestimated by the proposed method.

Future research will focus on using multiple GNSS constellations combinations to improve estimation accuracy of the proposed method. Retrieval of ground subsidence surface under the obvious interference of vegetation is another future research focus.

References

1. Shi Y, Chen J, Hao J, Bi J, Qi M, Wang X (2019) Statistical analysis of coal mine accidents of China in 2018. In: Paper presented at the 2019 prognostics and system health management conference (PHM-Qingdao), Qingdao, China, pp 1–6
2. Qian T, Bagan H, Kinoshita T, Yamagata Y (2014) Spatial–temporal analyses of surface coal mining dominated land degradation in Holingol, Inner Mongolia. IEEE J Sel Top Appl Earth Obs Remote Sens 7(5):1675–1687
3. Bayer P, Duran E, Baumann R, Finkel M (2009) Optimized groundwater drawdown in a subsiding urban mining area. J Hydrol 365(1):95–104

4. Anantrasirichai N et al (2021) Detecting ground deformation in the built environment using sparse satellite InSAR data with a convolutional neural network. IEEE Trans Geosci Remote Sens 59(4):2940–2950
5. Edmonds C (2018) Five decades of settlement and subsidence. Q J Eng Geol Hydrogeol 51(4):403–416
6. Tang F, Lu J, Wei S, Li X, He K, Yang Q (2020) Improvement of mining subsidence modeling method based on UAV LiDAR in Yushen mining area. J China Coal Soc 45(7):2655–2666
7. Park S, Choi Y (2020) Applications of unmanned aerial vehicles in mining from exploration to reclamation: a review. Minerals 10(8):663–675
8. Akturk E, Altunel AO (2019) Accuracy assessment of a low-cost UAV derived digital elevation model (DEM) in a highly broken and vegetated terrain. Measurement 136(1):382–386
9. Wang T et al (2023) An algorithm for locating subcritical underground goaf based on InSAR technique and improved probability integral model. IEEE Trans Geosci Remote Sens 61:5214914
10. Karamvasis K, Karathanassi V (2020) Performance analysis of open source time series InSAR methods for deformation monitoring over a broader mining region. Remote Sens 12(9):1380–1398
11. Chen Y et al (2021) Accuracy verification and correction of D-InSAR and SBAS-InSAR in monitoring mining surface subsidence. Remote Sens 13(21):4365–4380
12. Ferretti A, Fumagalli A, Novali F, Prati C, Rocca F, Rucci A (2011) A new algorithm for processing interferometric data-stacks: SqueeSAR. IEEE Trans Geosci Remote Sens 49(9):3460–3470
13. Muntean A, Mocanu V, Ambrosius B (2015) A GPS study of land subsidence in the Petrosani (Romania) coal mining area. Nat Hazards 80(2):797–810
14. Suresh D, Yarrakula K (2018) Subsidence monitoring techniques in coal mining: Indian scenario. J Geo-Mar Sci 47(10):1918–1933
15. Tao T, Liu J, Qu X, Gao F (2018) Real-time monitoring rapid ground subsidence using GNSS and Vondrak filter. Acta Geophys 67(1):133–140
16. Larson KM, Nievinski FG (2013) GPS snow sensing: results from the earth scope plate boundary observatory. GPS Solut 17(1):41–52
17. Roussel N et al (2015) Sea level monitoring and sea state estimate using a single geodetic receiver. Remote Sens Environ 171:261–277
18. Rodriguez-Alvarez N et al (2012) Vegetation water content estimation using GNSS measurements. IEEE Geosci Remote Sens Lett 9(2):282–286
19. Chew CC, Small EE, Larson KM (2015) An algorithm for soil moisture estimation using GPS-interferometric reflectometry for bare and vegetated soil. GPS Solut 20(3):525–537
20. Small EE, Larson KM, Braun JJ (2010) Sensing vegetation growth with reflected GPS signals. Geophys Res Lett 37(12):1–5
21. Yu K et al (2014) An overview of GNSS remote sensing. EURASIP J Adv Signal Process 134:1–14
22. Kratzsch H (1983) Mining subsidence engineering. Springer-Verlag, Berlin
23. Khanal M, Hodgkinson JH (2021) Subsidence prediction versus observation in Australia: a short comment. Environ Impact Assess Rev 86:106479
24. Malinowska A, Hejmanowski R, Dai H (2020) Ground movements modeling applying adjusted influence function. Int J Min Sci Technol 30(2):243–249
25. Jiang Y, Misa R, Li P, Yuan X, Sroka A, Jiang Y (2019) Summary and development of mining subsidence theory. Metal Mine 48(10):1–7
26. Knothe S (2005) Asymmetric function of distribution of mining exploatation influences in the medium with changing properties. Arch Min Sci 50(4):401–415
27. Saeidi A, Deck O, Al Heib M, Verdel T, Rouleau A (2013) Adjusting the influence function method for subsidence prediction. Key Eng Mater 553:59–66
28. Tan X et al (2020) Extraction of irregularly shaped coal mining area induced ground subsidence prediction based on probability integral method. Appl Sci 10(5523):1–18

29. Bahuguna PP, Srivastava A, Saxena NC (1991) A critical review of mine subsidence prediction methods. Min Sci Technol 13(3):369–382
30. Yu K, Li Y, Chang X (2019) Snow depth estimation based on combination of pseudorange and carrier phase of GNSS dual-frequency signals. IEEE Trans Geosci Remote Sens 57(3):1817–1828
31. Larson KM, Ray RD, Nievinski FG, Freymueller JT (2013) The accidental tide gauge: a GPS reflection case study from Kachemak Bay, Alaska. IEEE Geosci Remote Sens Lett 10(5):1200–1204
32. Wang X, He X, Zhang Q (2019) Evaluation and combination of quad-constellation multi-GNSS multipath reflectometry applied to sea level retrieval. Remote Sens Environ 231:111229
33. Larson KM et al (2009) Can we measure snow depth with GPS receivers? Geophys Res Lett 36(17):L17502
34. Bo H et al (2023) Estimation of ground subsidence deformation induced by underground coal mining with GNSS-IR. Remote Sens 15(96):1–17
35. Li Y et al (2022) Measuring soil moisture with refracted GPS signals. IEEE Geosci Remote Sens Lett 19:2504205
36. Yan W, Chen J, Tan Y, He R, Yan S (2022) Surface dynamic damage prediction model of horizontal coal seam based on the idea of wave lossless propagation. Int J Environ Res Public Health 19(68):1–15
37. Li Y, Yu K, Jin T, Chang X, Wang Q, Li J (2021) Development of a GNSS-IR instrument based on low-cost positioning chips and its performance evaluation for estimating the reflector height. GPS Solut 25(4):1–12
38. Ekaso D, Nex F, Kerle N (2020) Accuracy assessment of real-time kinematics (RTK) measurements on unmanned aerial vehicles (UAV) for direct geo-referencing. Geo Spat Inf Sci 23(2):165–181
39. Betz JW (2015) Engineering satellite-based navigation and timing: global navigation satellite systems, signals, and receivers. Wiley, Hoboken, NJ
40. Ha MC et al (2022) GNSS-R monitoring of soil moisture dynamics in areas of severe drought: example of Dahra in the Sahelian climatic zone (Senegal). Eur J Remote Sens. https://doi.org/10.1080/22797254.2022.2156931
41. Roussel N et al (2016) Detection of soil moisture variations using GPS and GLONASS SNR data for elevation angles ranging from 2° to 70°. IEEE J Sel Top Appl Earth Obs Remote Sens 9(10):4781–4794
42. Kim KB, Lee CK, An S (2014) Evaluation of network-RTK survey accuracy for applying to ground control points survey. J Korean Soc Geospatial Inf Sci 22(4):127–133
43. Cherciu C, Croitoru AN, Svasta PM, Nastac DI (2019) Analysis of RTK corrections for GNSS ground stations. In: Paper presented at the 2019 IEEE 25th international symposium for design and technology in electronic packaging (SIITME), Cluj-Napoca, Romania, pp 304–307

Chapter 8
Soil Moisture and Flood Inundation Retrieval with Spaceborne GNSS Reflectometry

Ting Yang, Zhigang Sun, and Wei Wan

Abstract This chapter presents an integrated overview of soil moisture retrieval and flood inundation monitoring using multi-mission spaceborne GNSS-R observations, including TDS-1, CYGNSS, and BF-1. Several complementary retrieval strategies are introduced, including empirical modeling, back-propagation artificial neural network (BP-ANN) retrieval, a CYGNSS–SMAP physics-coupled approach, and the physics-based PHYSER framework. For flood applications, a CYGNSS-based Annual Threshold Flood Inundation Index (ATFII) is used to detect inundation and quantify inundation grades. Results show that GNSS-R observations provide useful hydrological information across multiple scales. Over Mainland China, the monthly soil moisture derived from TDS-1 and CYGNSS agrees well with SMAP and in situ measurements, with correlation coefficients of 0.676 and 0.798 against SMAP, and 0.687 and 0.724 against in situ observations, respectively. The CYGNSS–SMAP coupled method further improves retrieval accuracy, while the PHYSER framework demonstrates that independent physics-based GNSS-R soil moisture retrieval is feasible through accurate soil reflectivity correction. For flood monitoring, the ATFII captures daily inundation dynamics and shows monthly correlations of 0.51–0.64 with the VIIRS flood product. These results highlight the strong potential of GNSS-R for integrated soil moisture and flood studies.

T. Yang (✉) · Z. Sun
CAS Engineering Laboratory for Yellow River Delta Modern Agriculture, Institute of Geographic Sciences and Natural Resources Research, Chinese Academy of Sciences, Beijing, China
e-mail: yangt@igsnrr.ac.cn

Z. Sun
e-mail: sun.zhigang@igsnrr.ac.cn

W. Wan
School of Earth and Space Sciences, Institute of Remote Sensing and GIS, Peking University, Beijing, China
e-mail: w.wan@pku.edu.cn

K. Yu and N. Zheng (eds.), *Advances in GNSS Reflectometry*, Navigation: Science and Technology 19, https://doi.org/10.1007/978-981-92-2325-1_8

1 Introduction

Soil moisture (SM) is a key variable in land–atmosphere interactions and plays an important role in hydrological processes, evapotranspiration, water migration, and the carbon cycle [1]. Accurate SM information with high spatial and temporal resolution is therefore essential for a wide range of applications, including drought monitoring, agricultural management, climate modeling, and hydrological prediction [2]. In parallel, flood inundation is another critical land hydrological variable, particularly during extreme precipitation events. Floods are among the most destructive natural disasters worldwide because of their rapid onset, large spatial extent, and severe impacts on ecosystems, infrastructure, and human society. Timely and reliable retrieval of both SM and flood inundation is thus of great significance for understanding land surface hydrological dynamics and assessing environmental hazards [3, 4].

Remote sensing has become one of the most effective approaches for monitoring land hydrological conditions at regional to global scales. Optical remote sensing provides observations with relatively high spatial resolution, but its applicability is often constrained by cloud cover and atmospheric conditions. In contrast, microwave remote sensing has the advantage of all-weather capability and is more directly related to surface dielectric properties. In recent decades, several microwave missions have been developed for SM monitoring, such as AMSR-E, ASCAT, SMOS, and SMAP. However, because of orbital sampling limitations, spatial gaps and insufficient temporal coverage still exist in daily SM products derived from conventional microwave sensors. Similar challenges also affect flood monitoring, where optical data are limited by cloud contamination and radar data, although accurate, often have relatively low temporal resolution for tracking rapidly evolving inundation processes [5, 6].

Global Navigation Satellite System Reflectometry (GNSS-R) is a bistatic remote sensing technique that uses navigation satellites as transmitters and low-Earth-orbit receivers to measure the reflected L-band signals from the Earth's surface [7]. Because GNSS-R operates at L-band, similar to SMAP and SMOS, it is well suited for sensing near-surface SM and surface wetness conditions [8]. The first spaceborne GNSS-R experiments demonstrated the feasibility of retrieving environmental parameters over both ocean and land surfaces, and subsequent missions further expanded its applications. TechDemoSat-1 (TDS-1) provided one of the earliest publicly available spaceborne GNSS-R datasets, while the Cyclone Global Navigation Satellite System (CYGNSS), launched in 2016 as an eight-satellite constellation, greatly improved temporal sampling. Although CYGNSS was originally designed for tropical ocean wind measurements, its reflected signals have been shown to be sensitive to several land surface variables, including SM, flood inundation, wetland extent, and other hydrological conditions. More recently, Chinese GNSS-R missions such as BuFeng-1 (BF-1) have provided new opportunities for extending GNSS-R land applications and for testing retrieval frameworks on independent platforms [9, 10].

The physical basis for using GNSS-R in land hydrology is that L-band reflections are highly sensitive to changes in the dielectric properties of the land surface [11]. For SM retrieval, changes in soil water content alter the soil dielectric constant and consequently modify the reflected GNSS signal. For flood inundation monitoring, the presence of standing water and strong surface wetting significantly enhances surface reflectivity (SR) and changes the scattering behavior observed by GNSS-R [12]. However, the relationship between GNSS-R observables and hydrological variables is often influenced by vegetation attenuation, surface roughness, topography, observation geometry, and scale mismatch with reference datasets. Therefore, reducing the effects of these confounding factors remains a central issue in both SM retrieval and flood inundation retrieval [13].

To address these challenges, different GNSS-R retrieval strategies have been developed. Empirical and machine-learning approaches have been used to estimate SM by linking reflectivity and ancillary land surface variables to reference products [14]. For example, back-propagation artificial neural network (BP-ANN) models can incorporate reflectivity, vegetation, terrain, and precipitation variables to estimate monthly SM. Physics-based and semi-empirical methods further attempt to represent vegetation attenuation and surface roughness more explicitly through radiative transfer theory and reflectivity calibration [15]. In parallel, flood inundation studies have shown that calibrated CYGNSS reflectivity can be transformed into inundation-related indices and further classified into multiple flood grades when combined with reference flood products such as VIIRS. Despite this progress, it remains challenging to develop retrieval frameworks that are physically interpretable, transferable across regions, and capable of supporting both SM and flood applications [16].

This chapter focuses on GNSS-R retrieval of two key land hydrological variables, namely SM and flood inundation. Several complementary retrieval strategies are integrated. First, a BP-ANN-based model is introduced to generate continuous monthly SM estimates using GNSS-R reflectivity and environmental predictors. Second, a physics-based CYGNSS–SMAP coupled retrieval method is presented, in which SMAP brightness temperature is used to derive a vegetation–roughness combined correction parameter for calibrating CYGNSS reflectivity. Third, the PHYSER framework is introduced as an independent physics-based approach for SM retrieval and is validated using BF-1 observations. Finally, a GNSS-R-based flood inundation retrieval method is presented through the construction of the Annual Threshold Flood Inundation Index, which enables daily flood mapping and inundation grading at rapid timescales. Through these methods, the chapter demonstrates the potential of multi-mission GNSS-R observations for integrated land hydrological retrieval.

The remainder of this chapter is organized as follows. Section 2 introduces the spaceborne GNSS-R observations, the reference and ancillary datasets, and the preprocessing and quality-control procedures. Section 3 presents the GNSS-R retrieval methodologies, including SR calculation, SM retrieval models, and the flood inundation retrieval method. Section 4 provides the corresponding results and analysis for SM and flood inundation retrievals. Finally, conclusions and outlook are given in Sect. 5.

2 Data Acquisition and Preprocessing

This chapter integrates multiple spaceborne GNSS-R observations, satellite reference products, ancillary environmental datasets, and ground-based measurements to support SM retrieval and flood inundation retrieval. Because the datasets used in this chapter differ in sensor type, spatial resolution, temporal sampling, and physical meaning, a unified data acquisition and preprocessing strategy is required before retrieval and validation. This section first introduces the spaceborne GNSS-R missions used in this chapter, including TDS-1, CYGNSS, and BF-1. It then describes the reference and ancillary datasets employed for calibration, model construction, and independent validation, such as SMAP SM and brightness temperature, MODIS vegetation products, GPM precipitation, VIIRS floodwater fraction, land cover data, and in situ observations. Finally, the preprocessing and quality-control procedures applied to these multi-source datasets are summarized, including observation filtering, spatial and temporal collocation, gridding, and sample screening, in order to provide a consistent data basis for the retrieval methods presented in the following sections.

2.1 Spaceborne GNSS-R Observations

This chapter uses spaceborne Global Navigation Satellite System Reflectometry (GNSS-R) observations from three satellite missions, namely the UK TechDemoSat-1 (TDS-1), the Cyclone Global Navigation Satellite System (CYGNSS), and the Chinese BuFeng-1 (BF-1) satellites [17, 18]. These missions employ reflected Global Positioning System (GPS) L-band signals in a bistatic radar configuration to sense surface scattering properties, and their observations have been shown to be sensitive to land hydrological variables such as SM and flood inundation [19]. In this chapter, TDS-1 and CYGNSS data are mainly used for SM retrieval studies, whereas BF-1 data are used to provide an independent dataset for validating the PHYSER framework. TDS-1 and CYGNSS coexisted from May 2017 to April 2018, which makes it possible to compare the performance of the two publicly available spaceborne GNSS-R missions for SM estimation [20].

The main observable in spaceborne GNSS-R measurements is the delay–Doppler map (DDM), which describes the reflected signal power as a function of delay and Doppler frequency relative to the specular point [21]. A DDM is generated by correlating the received reflected signal with locally produced replicas of the transmitted GNSS signals under different delay and Doppler offsets. From the DDM, the signal-to-noise ratio (SNR) can be derived and subsequently used to estimate surface reflectivity or bistatic radar cross section, which provide the fundamental observables for the SM and flood inundation retrieval methods introduced in this chapter [22].

TDS-1 and CYGNSS provide the two earliest publicly available spaceborne GNSS-R datasets used in this chapter [8]. TDS-1 Level-1B data were obtained from the MERRByS archive, and CYGNSS Level-1 Version 2.1 data were downloaded from the NASA PO.DAAC archive for the SM studies. For the TDS-1/CYGNSS comparison over Mainland China, the two missions share an overlapping period from May 2017 to April 2018. CYGNSS consists of eight microsatellites and provides substantially denser temporal sampling than TDS-1. Although CYGNSS was originally designed for observing ocean surface winds, previous studies have demonstrated that its reflected signals are also sensitive to land surface conditions [11–13]. For retrievals dominated by coherent reflections, the effective spatial footprint of CYGNSS over land can be approximated by the along-track first Fresnel zone, which is about 6 × 0.5 km before July 2019 and about 3 × 0.5 km afterward, depending on the signal integration time [18].

In addition to the CYGNSS data used for SM studies, CYGNSS Level-1 Version 3.0 observations from 2020 to 2021 are used in this chapter for flood inundation retrieval [21]. In the flood application, CYGNSS observations are gridded to the 9 km EASE-Grid 2.0 after quality screening, and the calibrated CYGNSS surface reflectivity is further used to construct the annual threshold flood inundation index. This allows the same GNSS-R observation system to be used for both SM retrieval and flood inundation monitoring [22].

BF-1 observations are introduced to provide an independent dataset for the realization and validation of the PHYSER framework. One year of BF-1 Level-1C data, spanning from August 2019 to July 2020, is used in this chapter. The BF-1 A/B twin satellites operate in low Earth orbit at an altitude of about 579 km and receive GPS L1C/A signals, with coverage extending to approximately ± 45° latitude. The GNSS-R receiver onboard BF-1 produces DDMs of surface scattered power near the specular point; the downlinked DDM has dimensions of 128 × 21, and the L1C product provides 17 × 11 bistatic radar cross sections (BRCSs). Owing to the observation geometry and surface scattering complexity, the effective spatial resolution of BF-1 is typically assumed to be about 10–25 km in practical SM retrieval applications [23].

Overall, these three GNSS-R missions provide complementary observational capabilities for this chapter: TDS-1 offers an early demonstration dataset, CYGNSS provides dense temporal sampling for both SM and flood studies, and BF-1 offers an independent platform for testing the feasibility of a physics-based SM retrieval framework. Together, they form the observational basis for the multi-method land hydrological retrieval experiments presented in the following sections.

3 Reference and Ancillary Datasets

In addition to the spaceborne GNSS-R observations, several reference and ancillary datasets are used in this chapter to support SM retrieval, flood inundation retrieval, and independent validation. These datasets include satellite-based SM and brightness temperature products, precipitation and topographic data, vegetation- and roughness-related variables, land cover information, optical flood products, and ground-based in situ observations. Together, they provide the environmental constraints, calibration inputs, and validation references required by the different retrieval approaches discussed in this chapter [18–21].

For SM retrieval, the SMAP Level-3 products are used as the primary reference datasets in several parts of this chapter. In the TDS-1/CYGNSS study over Mainland China, SMAP SM is used both as a benchmark for sensitivity analysis and as the target variable for training the monthly BP-ANN model. In the CYGNSS–SMAP coupled retrieval algorithm, the 36 km gridded SMAP SM and horizontally and vertically polarized brightness temperatures are further used to derive the vegetation–roughness combined correction parameter within the radiative transfer framework. In this way, SMAP provides not only reference SM but also the auxiliary microwave observations needed to calibrate CYGNSS reflectivity [24].

A number of environmental ancillary datasets are introduced to characterize land surface conditions affecting GNSS-R retrievals. For the BP-ANN SM model, vegetation and terrain descriptors are included as explanatory variables, such as normalized difference vegetation index (NDVI), vegetation water content (VWC), elevation, slope, and surface roughness. In the original study, NDVI and VWC were derived from MODIS surface reflectance products, while elevation, slope, and roughness were derived from the SRTM 90 m digital elevation model. Precipitation was taken from the GPM Level-3 product to account for rainfall-driven SM variability. These variables were used together with GNSS-R reflectivity to construct the monthly SM retrieval model [25, 26].

For the PHYSER framework, additional auxiliary datasets are used to support the physically based estimation of land surface parameters. The SMAP SM product is used to derive soil dielectric constant through the Topp dielectric model, while SMAP brightness temperature provides the basis for estimating the combined effects of vegetation attenuation and surface roughness. In addition, MODIS land cover data based on the International Geosphere–Biosphere Programme (IGBP) classification are used to examine the applicability of the retrieval framework across different land cover types. These datasets are essential for establishing the physical linkage between GNSS-R observables and SM [27–29].

For flood inundation retrieval, both meteorological and optical reference datasets are employed. GPM precipitation is used to identify the timing and intensity of extreme rainfall events and to compare the temporal evolution of precipitation with the CYGNSS-derived inundation grades. SMAP SM is also used as a supplementary reference to evaluate changes in near-surface wetness before and after flood events. To calibrate and validate the inundation classification, the VIIRS floodwater fraction product is used as the benchmark dataset for defining inundation thresholds and assessing the performance of the ATFII. These datasets provide complementary information for linking GNSS-R reflectivity changes with rainfall forcing and floodwater extent [30–32].

Ground-based measurements are further used for independent validation of the SM retrieval results. In the Mainland China study, monthly averaged in situ SM observations from dense meteorological networks are used to evaluate the SM derived from TDS-1 and CYGNSS. In the PHYSER study, in situ SM measurements from the International SM Network (ISMN), as well as dense sites from the Hebi network in Henan Province, are used to assess the temporal dynamics and spatial representativeness of the BF-1-derived SM product. These ground observations provide an important reference for evaluating the retrieval accuracy beyond satellite-to-satellite comparison [18, 20].

Overall, the reference and ancillary datasets used in this chapter play three complementary roles: they provide benchmark observations for training and validation, supply environmental variables required by empirical and physical retrieval models, and support the interpretation of flood inundation dynamics under extreme precipitation conditions. By integrating these multi-source datasets with spaceborne GNSS-R observations, the retrieval methods presented in this chapter can be evaluated more comprehensively over a wide range of land surface and hydrological conditions.

3.1 Data Preprocessing and Quality Control

To ensure the consistency and reliability of the multi-source datasets used in this chapter, all GNSS-R, reference, and ancillary datasets were subjected to preprocessing and quality-control procedures before retrieval and validation. Because the data originate from different satellite missions and ground-based observations, these procedures mainly include observation filtering, spatial and temporal collocation, gridding, and sample screening according to the requirements of different retrieval tasks [18–21].

For the TDS-1 and CYGNSS SM studies over Mainland China, the raw GNSS-R observations were first filtered to retain only high-quality measurements over land. For TDS-1, reflected waveforms from surfaces with elevations higher than 3000 m were excluded, while for CYGNSS, the onboard data compression limit makes valid observations mainly distributed over relatively low-elevation areas. In both cases, observations with antenna gain lower than 0 dB were removed, and only measurements with favorable observation geometry were retained. Additional quality screening was further applied using mission-specific flags, such as direct-signal contamination and low-confidence transmitter power estimates. After filtering, the TDS-1 and CYGNSS observables were collocated with SMAP SM for both daily sensitivity analysis and monthly SM retrieval [22].

For the BP-ANN monthly SM model, all input variables and reference SM data were converted to monthly averages before model training. The GNSS-R observables, vegetation variables, terrain descriptors, and precipitation data were normalized to the range of 0–1 to ensure a stable training process. The complete dataset was then divided into training, testing, and validation subsets with proportions of 60%, 20%, and 20%, respectively. Repeated training experiments were carried out to identify the optimal network structure and avoid overfitting [18].

For the CYGNSS–SMAP coupled physics-based SM retrieval algorithm, CYGNSS observations were first aggregated to the 36 km SMAP EASE-2 grid in order to match the spatial resolution of SMAP brightness temperature and SM products. High-quality SMAP retrievals were selected using the recommended quality flags, and an averaging over three consecutive days was applied to generate spatially seamless brightness temperature fields for model construction. During the calibration stage, CYGNSS surface reflectivity and SMAP SM were spatially and temporally collocated on a grid-by-grid basis, and grids with too few valid observations were excluded from the regression calibration [24].

For the PHYSER framework based on BF-1 observations, BF-1 Level-1C data were processed to derive bistatic radar cross section and surface reflectivity variables at the specular point. The BF-1 observations were collocated with SMAP and ERA5-Land products for large-scale comparison and with in situ measurements for independent validation. In the site-scale validation, only grids with sufficient temporal overlap were retained, and observations from multiple stations located within the same grid were combined to improve representativeness. In addition, land-cover-based grouping and water-body-related sample screening were used in the original study to support the calibration of correction coefficients and the assessment of retrieval performance across different environmental conditions [27–31].

For flood inundation retrieval, CYGNSS Level-1 observations were first screened using antenna gain and quality flags and then gridded to the 9 km EASE-Grid 2.0. GPM precipitation data were resampled to the same grid by simple averaging, while the VIIRS floodwater fraction product was used as the benchmark for inundation threshold determination and validation. To construct the annual threshold flood inundation index, grid cells with fewer than 30 CYGNSS observations during the calibration period were excluded. For each valid grid, the upper and lower annual reference bounds of calibrated surface reflectivity were determined from the top and bottom 5% of yearly observations, respectively. During the CYGNSS–VIIRS matchup procedure, low-quality samples were further removed using a 95% standard deviation filtering criterion to reduce the influence of outliers on inundation classification [25, 26].

Ground-based SM observations were also quality controlled before being used for validation. In the Mainland China study, monthly averaged in situ SM records from dense meteorological networks were used to match the monthly satellite products. In the PHYSER study, ISMN observations at shallow depths were averaged to daily values, and only sites with sufficient temporal sampling were retained for evaluation. These procedures help reduce inconsistencies caused by differences in temporal resolution, observation depth, and spatial representativeness between satellite retrievals and point-scale measurements.

Overall, the preprocessing and quality-control procedures adopted in this chapter are designed to improve the comparability of multi-source observations and to reduce the impact of low-quality samples, scale mismatch, and spatiotemporal inconsistency. These steps provide a consistent data foundation for the subsequent SM and flood inundation retrieval experiments.

4 GNSS-R Retrieval Methodology

The SM retrieval framework developed in this chapter consists of four main steps: surface reflectivity retrieval from GNSS-R observations, reflectivity calibration to account for vegetation and surface roughness effects, surface reflectivity formulation, and SM estimation.

4.1 Surface Reflectivity Calculation

Surface reflectivity derived from GNSS-R observations has been widely used as a proxy for SM because land surface reflectivity is closely related to the dielectric properties of soil. Since soil dielectric properties are strongly controlled by soil water content, variations in GNSS-reflected signals provide valuable information for SM estimation. In GNSS-R land studies, however, the reflected signal can be represented by several related observables, including signal-to-noise ratio (SNR), effective reflected power, bistatic radar cross section (BRCS), and surface reflectivity. These quantities are physically related, but they are derived from different observables and are formulated from different perspectives depending on the satellite mission, data product, and retrieval objective [11–13].

Among these observables, SNR is directly obtained from the delay-Doppler map (DDM) and is commonly calculated as the difference between the peak power of the reflected signal and the background noise floor. To obtain physically meaningful reflection-related variables, the raw SNR must be corrected for several factors, including antenna gain, receiver noise, and signal propagation losses. After these corrections, the reflected signal strength can be further expressed either as effective reflected power or as surface reflectivity, depending on the formulation adopted in a given study [33].

For the TDS-1 mission, the effective reflected power can be derived from the corrected SNR based on the bistatic radar equation describing the coherent component of the received signal power. Following the formulations proposed by Carreno-Luengo et al. and Chew et al., the effective reflected power is expressed as Eq. (1). In this formulation, SNR represents the difference between the peak DDM power and the noise floor, while the remaining terms account for the transmitter–receiver geometry, receiver antenna gain, and incidence angle. The resulting effective reflected power provides a reflection-related metric that can be directly compared with reference SM products in sensitivity analyses [15].

$$P_{r,eff} \propto SNR + 20\log(R_{ts} + R_{sr}) - 10\log G_r + 20\log(\cos\theta) \tag{1}$$

where SNR denotes the signal-to-noise ratio derived from the DDM, R_{ts} is the distance from the transmitter to the specular reflection point, R_{sr} is the distance from the specular reflection point to the receiver, G_r is the receiver antenna gain in the direction of the specular reflection point, and θ is the incidence angle. In this formulation, the corrected reflected signal is normalized by the bistatic observation geometry and receiver response, so that the resulting quantity can be used as a reflection-related indicator for sensitivity analysis with SM.

For CYGNSS observations, previous studies have shown that surface reflectivity can also be estimated directly from the corrected SNR while accounting for the transmitted power, antenna gains, and signal propagation geometry. In decibel form, the SR can be expressed as:

$$SR \propto SNR - 10\log P_t - 10\log G_t - 10\log G_r \\ - 20\log\lambda + 20\log(R_{ts} + R_{sr}) + 20\log(4\pi) \tag{2}$$

where P_t denotes the transmitted GNSS power, G_t and G_r represent the gains of the transmitting and receiving antennas, respectively, and λ is the wavelength of the GPS L1 signal. Assuming that land reflections are dominated mainly by the coherent component at the surface, the surface reflectivity can also be formulated from the Friis transmission equation. In decibel form, this expression can be written as:

$$SR(dB) = 10\log P_r - 10\log P_t - 10\log G_t - 10\log G_r \\ + 20\log(R_{ts} + R_{sr}) - 20\log\lambda + 20\log 4\pi \tag{3}$$

where P_r represents the peak reflected power after noise subtraction, P_t is the transmitted GNSS power, R_{ts} and R_{sr} denote the distances from the transmitter and receiver to the specular reflection point, respectively, and G_t and G_r correspond to the gains of the transmitting and receiving antennas. In this way, CYGNSS surface reflectivity is described from the perspective of received reflected power normalized by transmission geometry and antenna characteristics. To facilitate comparisons among different observations, an empirical constant, typically about 140 dB, is often subtracted from the raw SR values derived from Eq. (3), thereby constraining the reflectivity values to a physically interpretable range [34].

In addition to SNR-based and reflected-power-based formulations, GNSS-R studies frequently derive reflectivity from the bistatic radar cross section (BRCS) obtained from the DDM. In this case, BRCS describes the scattering strength of the land surface under the bistatic transmitter–receiver geometry and provides a more direct representation of the reflected signal at the specular point [35]. Based on the geometric relationship among the transmitter, receiver, and specular reflection point, the raw surface reflectivity in decibel form can be expressed as:

$$\Gamma_{raw,dB} = 10\log\sigma + 20\log(R_{st} + R_{sr}) - 20\log R_{st} \\ - 20\log R_{sr} - 10\log 4\pi \tag{4}$$

where $\Gamma_{raw,dB}$ represents the raw surface reflectivity in decibel form, BRCS denotes the bistatic radar cross section, R_{ts} is the distance from the transmitter to the specular reflection point, and R_{sr} is the distance from the receiver to the specular reflection point. For subsequent SM retrieval models, the reflectivity in decibel form can be further converted to linear scale as:

$$\Gamma_{raw} = 10^{(\Gamma_{raw,dB}/10)} \tag{5}$$

where Γ_{raw} represents the raw surface reflectivity in linear scale. In most GNSS-R observations, the reflectivity in decibel form is typically negative because the reflected signal power is much weaker than the direct transmitted GNSS signal.

From a physical perspective, the above formulations all characterize the response of the land surface to incident GNSS signals, but they differ in their mathematical definitions, normalization procedures, and application contexts. Therefore, the choice of observable depends on the available Level-1 product, the mission-specific processing chain, and the purpose of the retrieval. Because this chapter synthesizes several GNSS-R studies involving TDS-1, CYGNSS, and BF-1 observations, the reflection-related variables introduced here are not restricted to a single formulation. Instead, the specific observable adopted in each retrieval framework is stated explicitly in the corresponding subsection. Regardless of the formulation used, these variables describe the scattering strength of the land surface at or near the specular reflection point and provide the physical basis for subsequent calibration and SM retrieval. However, they are also affected by vegetation attenuation, surface roughness, and observation geometry. Therefore, additional calibration and correction procedures are required before the SM signal can be isolated more reliably, as introduced in the following section.

4.2 SM Retrieval Models

After deriving the calibrated surface reflectivity from GNSS-R observations, SM can be retrieved using different modeling approaches. In this study, three types of retrieval models are employed, including an empirical model trained using SMAP SM products, an artificial neural network (ANN) retrieval model, and the physics-based PHYSER framework. These models represent different levels of complexity and provide complementary approaches for estimating SM from GNSS-R observations.

(1) **BP-ANN-Based Monthly SM Retrieval Model**

GNSS-R reflectivity is sensitive not only to SM but also to several other geophysical parameters, such as vegetation canopy, elevation, slope, surface roughness, and precipitation. To account for these nonlinear influences, a back-propagation artificial neural network (BP-ANN) model was constructed to estimate continuous monthly SM over the study area. The structure of the proposed BP-ANN model is illustrated in Fig. 1.

The BP-ANN is a supervised learning algorithm consisting of an input layer, one or more hidden layers, and an output layer. Because it can effectively represent nonlinear relationships between input and output variables, it has been widely used

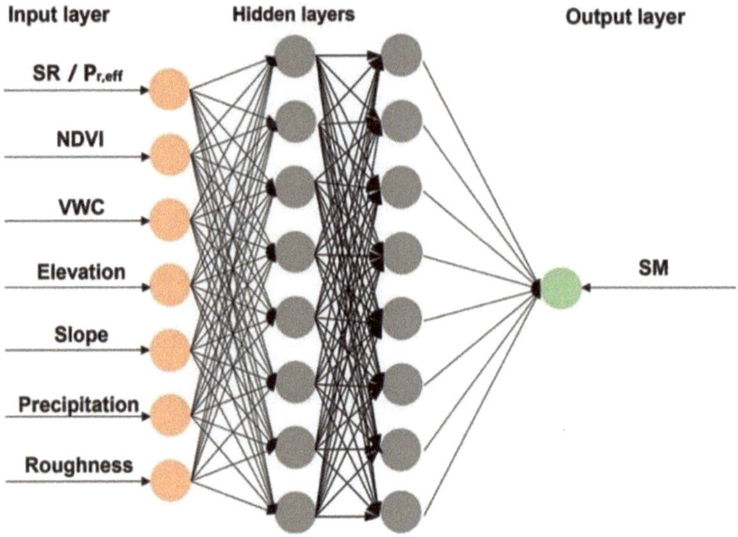

Fig. 1 The back-propagation artificial neural network (BP-ANN) model to estimate monthly SM

in regression and classification tasks, including applications in the geosciences. In this study, the BP-ANN was used as a multifactor nonlinear regression model for monthly SM retrieval [14].

During model construction, the contributions of individual variables and their combinations were assessed in order to determine the optimal inputs for SM estimation. In addition to GNSS-R reflectivity, the candidate input variables included normalized difference vegetation index (NDVI), vegetation water content (VWC), elevation, slope, surface roughness, precipitation, and the noise floor derived from the native observations. The noise floor was introduced to characterize DDM noise, while precipitation was considered because seasonal SM is strongly influenced by rainfall and may also respond to relatively small precipitation changes [15].

To identify the optimal input configuration, the contributions of individual variables and their combinations were evaluated during model construction. The results showed that the best-performing model included seven variables, namely SR, NDVI, VWC, elevation, slope, precipitation, and surface roughness. Among them, SR played the most important role in the retrieval performance, while the inclusion of noise had a negative effect and was therefore not retained in the final model [16].

The input variables were organized in the input layer of the network, while SMAP Level 3 SM data at 9 km resolution were used in the output layer during the training stage. All datasets were averaged to monthly values and normalized to the range of 0–1 before training. The samples were divided into training, testing, and validation sets with proportions of 60%, 20%, and 20%, respectively. Repeated experiments were then conducted to determine the optimal BP-ANN structure. The final configuration adopted in the original study consisted of seven input nodes, two hidden layers with eight nodes, and one output node representing monthly SM. The hyperbolic tangent function was selected as the activation function, and the final layer was implemented as a regression layer without activation [32].

(2) **Physics-Based CYGNSS–SMAP Coupled SM Retrieval Model**

Building upon the surface reflectivity derived from GNSS-R observations in Sect. 3.1, a physics-based retrieval algorithm was developed to improve SM estimation by coupling CYGNSS surface reflectivity with SMAP brightness temperature observations. As shown in the workflow, the method consists of three major steps: first, the original CYGNSS surface reflectivity is linked to a vegetation–roughness combined correction term; second, this combined parameter is derived from SMAP brightness temperature using the zeroth-order radiative transfer model; and third, the calibrated CYGNSS reflectivity is converted to SM through grid-based regression.

In most practical situations, the original CYGNSS surface reflectivity contains not only the SM signal but also the effects of vegetation attenuation and surface roughness [34]. These two effects are represented by a combined parameter, so that the original reflectivity can be written as:

$$SR = SR_{cali} \exp\left(-2\tau_p \sec\theta - h\cos^2\theta\right) \tag{6}$$

where SR_{cali} is the calibrated surface reflectivity, τ_p is the vegetation optical depth, h is the surface roughness parameter, and θ is the incidence angle. Instead of estimating vegetation and roughness separately from auxiliary land parameters, the proposed method derives the combined term $\exp\left(-2\tau_p \sec\theta - h\cos^2\theta\right)$ directly from SMAP brightness temperature observations.

To estimate this combined parameter, the method employs the zeroth-order radiative transfer model, namely the τ-ω model. In its general form, the brightness temperature at polarization p is expressed as:

$$T_{Bp} = T_s e_p \gamma_p + T_s\left(1-\omega_p\right)\left(1-\gamma_p\right) + T_v\left(1-\omega_p\right)\left(1-\gamma_p\right)R_{p\_\mathrm{rough}}\gamma_p \tag{7}$$

where T_s and T_v denote the soil and vegetation temperatures, e_p is the emissivity, ω_p is the single scattering albedo, R_{p_rough} is the rough-surface reflectivity, and $\gamma_p = \exp\left(-\tau_p \sec\theta\right)$ is the vegetation transmissivity. Under L-band conditions, ω_p is assumed to be negligible and the soil and vegetation temperatures are taken to be equal, i.e., $T_s = T_v = T$. The model can then be simplified to:

$$T_{Bp} = T\left[1 - R_{p\_\mathrm{rough}} {*} \exp\left(-2\tau_p \sec\theta\right)\right] \tag{8}$$

The rough-surface reflectivity is further related to the smooth-surface reflectivity through:

$$R_{p\_\mathrm{rough}} = R_{p\_\mathrm{smooth}} {*} \exp\left(-h\cos^2\theta\right) \tag{9}$$

By combining the above equations, the brightness temperature can be rewritten as:

$$T_{Bp} = T\left[1 - R_{p\_smooth} * \exp\left(-2\tau_p \sec\theta - h\cos^2\theta\right)\right] \quad (10)$$

For SMAP, this expression is applied to the horizontal and vertical polarizations, and the combined correction term can be solved as [36]:

$$\exp\left(-2\tau_p \sec\theta - h\cos^2\theta\right) = \frac{T_{B,V} - T_{B,H}}{T_{B,V}R_{H,smooth} - T_{B,H}R_{V,smooth}} \quad (11)$$

The smooth-surface reflectivities R_{H_smooth} and R_{V_smooth} are calculated from the Fresnel equations, and the dielectric constant required by the Fresnel formulation is obtained from SMAP SM using the Topp model:

$$SM = -0.053 + 0.0292\varepsilon - 0.00055\varepsilon^2 + 0.0000043\varepsilon^3 \quad (12)$$

Since SMAP observations are acquired at a fixed incidence angle of 40°, whereas CYGNSS incidence angles vary from about 5° to 35°, a first-order Taylor expansion is further used to transfer the correction parameter from the SMAP geometry to the CYGNSS geometry, thereby reducing the uncertainty caused by the incidence-angle mismatch.

After the combined correction parameter is obtained, the original CYGNSS reflectivity is converted into calibrated reflectivity SR_{cali}. The calibrated reflectivity is then related to SMAP SM using a grid-based linear regression. CYGNSS observations are spatially and temporally collocated with SMAP grids, and the mean values of both SR_{cali} and SMAP SM are first calculated for each grid over the calibration period. A linear relationship is then established between the anomalies of the two variables:

$$\Delta SR_{cali} = SR_{cali} - SR_{average} \quad (13)$$

$$SM_{cali} = \beta\,\Delta SR_{cali} + \alpha \quad (14)$$

where β is the grid-specific regression slope and α is the mean SMAP SM during the calibration period. In this way, the method converts the physically calibrated CYGNSS reflectivity into spatially continuous SM estimates while preserving consistency with the SMAP radiative transfer framework (Fig. 2).

(3) **Physics-Based Soil Reflectivity Retrieval Framework**

PHYSER is a physics-based framework designed to retrieve SM from spaceborne GNSS-R observations through the derivation of accurate soil reflectivity rather than through direct regression against reference SM products. As illustrated in the workflow, the framework differs fundamentally from empirical and semiempirical retrieval

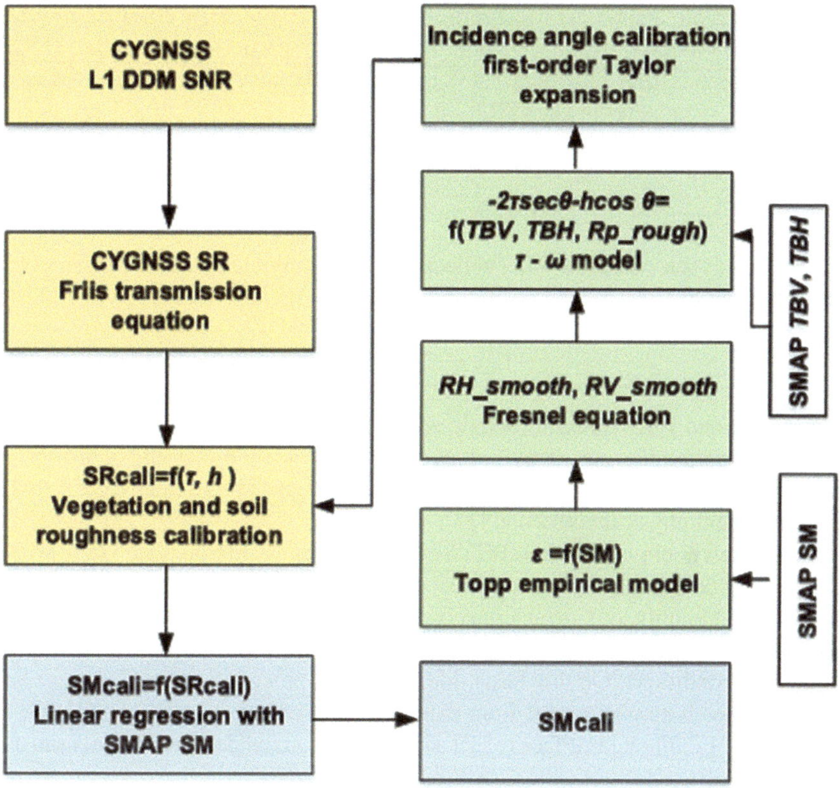

Fig. 2 Flowchart of the proposed algorithm

schemes. Instead of establishing statistical relationships between GNSS-R observables and external SM products, PHYSER retrieves SM from GNSS-R-derived accurate soil reflectivity and the resulting soil permittivity. The framework assumes that the observed GNSS-R surface reflectivity is affected by two major types of error, namely observed system errors and land surface errors, and corrects them sequentially through Surface Reflectivity CALibrating (SuR-CAL) and Soil Reflectivity CORrecting (SoR-COR) before the final SM calculation.

The first component, SuR-CAL, is used to reduce the observed system errors in the Level-1 GNSS-R measurements. Following the strategy proposed in the PHYSER study, calm and clean inland water bodies are used as calibration references, because their reflectivity is physically stable and can be theoretically estimated. The theoretical reflectivity of water can be expressed as:

$$\Gamma_{t\_w} = \left(\frac{\varepsilon_w - 1}{\varepsilon_w \sin(90^\circ - \theta) + \sqrt{\varepsilon_w - \sin^2\theta}}\right)^2 \left(\frac{\cos\theta}{\sin\theta + \sqrt{\varepsilon_w - \sin^2\theta}}\right)^2 \quad (15)$$

where Γ_{t_w} is the theoretical water reflectivity, ε_w is the dielectric constant of water, and θ is the GNSS-R incidence angle. Because each pseudo-random noise (PRN) code corresponds to an individual GNSS transmitter, the system correction coefficient is estimated separately for each PRN as:

$$CC = \frac{\Gamma_{t\_w}}{\Gamma_{o\_w}} \tag{16}$$

where Γ_{o_w} is the observed water reflectivity. The calibrated surface reflectivity is then obtained by:

$$\Gamma_{cal} = \Gamma_{raw} \times CC \tag{17}$$

where Γ_{raw} and Γ_{cal} represent the raw and calibrated surface reflectivities, respectively. This calibration step helps reduce system-dependent biases and improves the consistency of the reflectivity measurements, thereby providing a more reliable basis for subsequent physical correction [35].

After the system calibration, PHYSER introduces an intermediate land surface clustering (LSC) step before the physical correction. This step groups the global land surface into monthly lookup tables according to environmental similarity, so that the subsequent correction can better account for spatial differences in vegetation and roughness conditions. On this basis, the SoR-COR procedure is applied to isolate the soil reflectivity component from the calibrated surface reflectivity [34, 36]. In PHYSER, the calibrated surface reflectivity and the corrected soil reflectivity satisfy:

$$\Gamma_{cal} = \Gamma_{cal\_cor} * \exp\left(-2\tau \sec\theta - h\cos^2\theta\right) \tag{18}$$

where Γ_{cal_cor} is the corrected soil reflectivity, τ is the vegetation optical depth, h is the roughness parameter, and θ is the incidence angle. The combined vegetation-roughness correction term is estimated using the zeroth-order radiative transfer (τ–ω) model. For polarization p, the brightness temperature can be written as:

$$T_{Bp} = T\left[1 - \Gamma_{p\_r}\exp(-2\tau\sec\theta)\right] \tag{19}$$

and the rough-surface reflectivity is related to the smooth-surface reflectivity by

$$\Gamma_{p\_r} = \Gamma_{p\_s}\exp\left(-h\cos^2\theta\right) \tag{20}$$

Combining these two equations gives:

$$T_{Bp} = T\left[1 - \Gamma_{p\_s}\exp\left(-2\tau\sec\theta - h\cos^2\theta\right)\right] \tag{21}$$

Thus, the calibrated surface reflectivity can be corrected for vegetation attenuation and surface roughness effects, and the final corrected reflectivity Γ_{cal_cor} is regarded as the accurate soil reflectivity required for physical SM inversion.

Finally, the corrected soil reflectivity is converted to soil permittivity through the Fresnel reflection relationship, and SM is then retrieved using the Hallikainen dielectric model [37]. In the initial realization of PHYSER, this relationship is written as:

$$\begin{aligned}\varepsilon = {} & (119.006 - 0.500W_s - 0.633W_c)SM^2 \\ & + (3.803 + 0.462W_s - 0.341W_c)SM \\ & + (2.862 - 0.012W_s + 0.001W_c)\end{aligned} \quad (22)$$

where ε is the soil permittivity, SM is volumetric SM, and W_s and W_c are the sand and clay fractions, respectively. In this way, PHYSER establishes a complete physics-based retrieval pathway from GNSS-R observations to calibrated soil reflectivity, soil permittivity, and finally SM (Fig. 3).

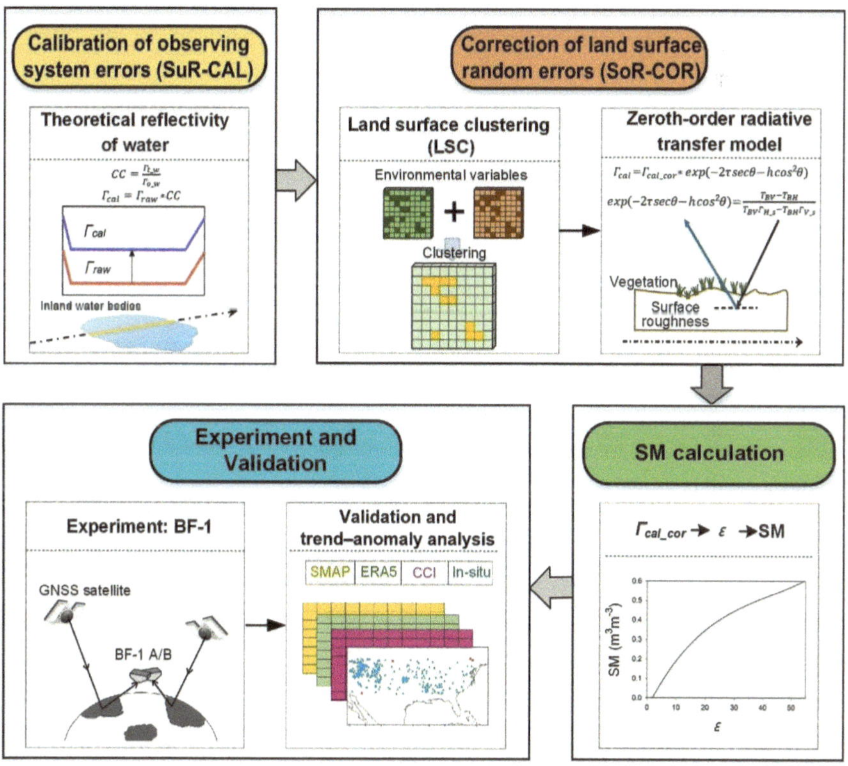

Fig. 3 The realization and validation of the proposed PHYSER SM retrieval framework. SuR-CAL, surface reflectivity calibration; CC, correction coefficient; SoR-COR, soil reflectivity correction; ERA5, fifth-generation ECMWF atmospheric reanalysis; CCI, climate change initiative

4.3 Flood Inundation Retrieval Method

To extend GNSS-R applications from SM retrieval to hydrological hazard monitoring, a flood inundation retrieval method was developed using CYGNSS observations. As illustrated in the workflow, the method mainly consists of two parts. The first part is to derive CYGNSS surface reflectivity calibrated for land surface properties, hereafter denoted as SR_{cali}. The second part is to construct an ATFII based on the temporal variation of SR_{cali}, and then classify inundation grades with the aid of VIIRS floodwater fraction data.

Because CYGNSS measurements are highly sensitive to water surfaces, increases in surface wetness and inundation generally lead to stronger reflected GNSS signals. However, the CYGNSS footprint is usually a mixture of water, soil, and vegetation, and the effects of vegetation attenuation and surface roughness may reduce the sensitivity of L-band reflectivity to floodwater. Therefore, the first step of the method is to calibrate the original CYGNSS surface reflectivity for land surface properties [21]. Following the physics-based formulation adopted in the original study, the calibrated surface reflectivity can be expressed as:

$$SR_{cali} = \frac{(4\pi)^2 (R_{ts} + R_{sr})^2 P_r}{\lambda^2 P_t G_t G_r \exp\left(-2\tau \sec\theta - h\cos^2\theta\right)} \tag{23}$$

where P_r is the peak reflected power after noise removal, R_{ts} and R_{sr} are the distances from the transmitter and receiver to the specular point, P_t is the transmitted GPS power, G_t and G_r are the transmitter and receiver antenna gains, τ is the vegetation optical depth, h is the surface roughness parameter, and θ is the incidence angle. In this study, the combined vegetation–roughness term is derived from SMAP brightness temperature data using the radiative transfer approach introduced previously, rather than being prescribed independently [36].

Based on the radiative transfer framework, the combined correction parameter of vegetation and roughness can be written as:

$$\exp\left(-2\tau \sec\theta - h\cos^2\theta\right) = \frac{TB_V - TB_H}{TB_V R_{H\_sm} - TB_H R_{V\_sm}} \tag{24}$$

where TB_V and TB_H denote the vertical- polarized and horizontal-polarized SMAP brightness temperatures, respectively, and R_{H_sm} and R_{V_sm} are the smooth-surface Fresnel reflectivities at the two polarizations. Since the incidence angle of CYGNSS differs from that of SMAP, a first-order Taylor expansion is further used in the original algorithm to reduce the uncertainty caused by the angle mismatch. Through this step, the raw CYGNSS reflectivity is converted into SR_{cali}, which is more directly related to surface wetness and inundation conditions.

After obtaining SR_{cali}, the second part of the method is to construct the annual threshold flood inundation index. This index is based on the idea that wetter and more inundated surfaces generally produce stronger reflectivity. For a CYGNSS observation at time t, the ATFII is defined as

$$ATFII = \frac{SR_{cali\_t} - SR_{cali\_\min}}{SR_{cali\_\max} - SR_{cali\_\min}} \tag{25}$$

where SR_{cali_t} is the calibrated reflectivity at time t, and $SR_{cali\_\max}$ and $SR_{cali\_\min}$ are the annual maximum and minimum reference values of SR_{cali}, respectively. In practice, CYGNSS observations are first gridded to the 9 km EASE-Grid 2.0. For each grid cell, grids with fewer than 30 observations are excluded. The upper and lower annual reference bounds are then determined using the mean values of the top 5% and bottom 5% of the yearly SR_{cali} observations, respectively. This normalization allows the ATFII to characterize inundation dynamics relative to the local annual range of reflectivity variation.

To transform the continuous ATFII into discrete inundation states, the CYGNSS-derived ATFII is collocated with the VIIRS floodwater fraction product. The VIIRS floodwater fraction represents the proportion of open water within each pixel and is used here as the benchmark for classification. Based on the matched CYGNSS–VIIRS samples, five inundation grades are defined: non-inundation, mild inundation, moderate inundation, severe inundation, and inundated grid. According to the thresholds reported in the original study, the corresponding ATFII ranges are 0–0.33, 0.33–0.47, 0.47–0.68, 0.68–0.86, and > 0.86, respectively. In this way, the proposed method not only retrieves flood inundation but also quantifies its severity using a graded classification scheme.

Overall, the flood inundation retrieval framework uses calibrated CYGNSS reflectivity as the physical basis and the ATFII as the normalized inundation indicator. Compared with directly thresholding raw reflectivity, this strategy is more suitable for regional and temporal comparisons because it accounts for local land surface conditions and annual variability. The resulting inundation grades can therefore be used to analyze both the spatial distribution and temporal evolution of flood events [38–41] (Fig. 4).

5 Results and Analysis

After developing the SM retrieval models described in the previous section, the performance of the proposed methods is evaluated using multiple datasets. This section first analyzes the sensitivity of GNSS-R observables to SM variations. The SM retrieval results derived from the proposed models are then assessed using reference datasets. Finally, the application of GNSS-R observations for flood inundation monitoring during extreme precipitation events is presented.

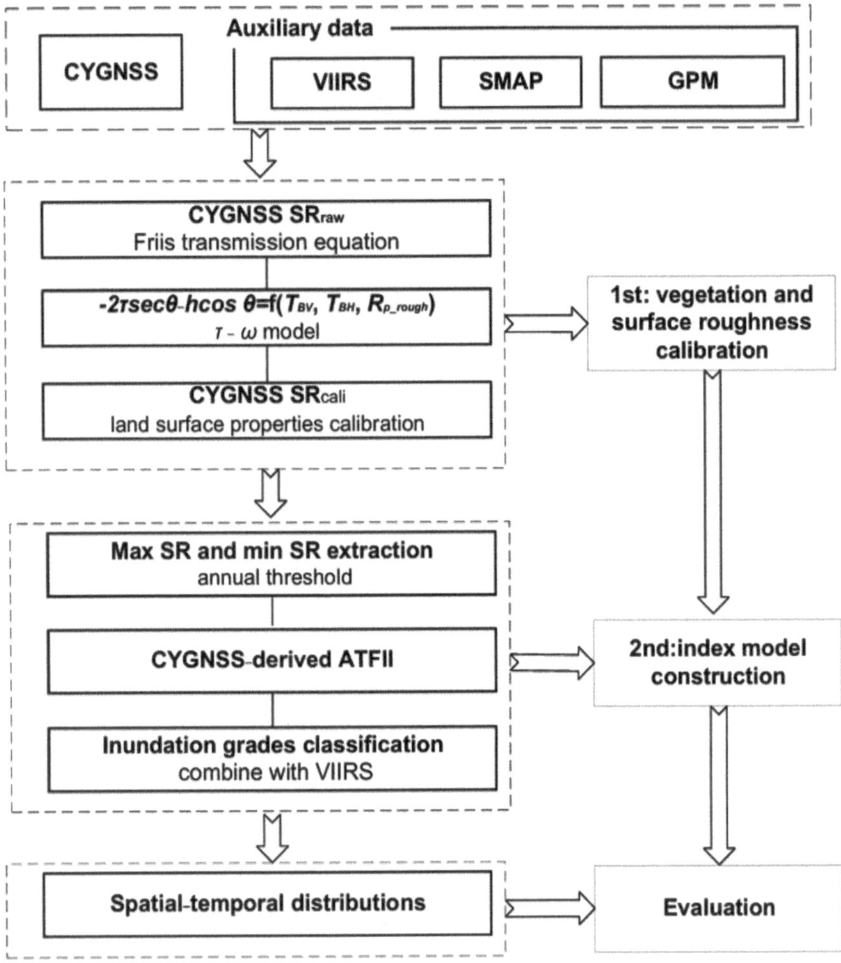

Fig. 4 Flowchart of the proposed index in this study to retrieve inundation

5.1 SM Retrieval Using CYGNSS and TDS-1 Observations

To evaluate the capability of GNSS-R observations for SM estimation, the variations of the retrieved SM derived from CYGNSS and TDS-1 are compared with the SMAP SM product. Figure 5 presents the temporal variations of the GNSS-R observables and SMAP SM for a representative grid located at (31.5°, 104.4°). This grid was selected because it contains the largest number of available CYGNSS observations during the study period. For visualization purposes, the SMAP SM data were interpolated to produce a continuous curve; this processing does not affect the statistical evaluation.

Overall, both the effective reflected power $P_{r,eff}$ derived from TDS-1 and the SR derived from CYGNSS exhibit clear consistency with the SMAP SM variations. The

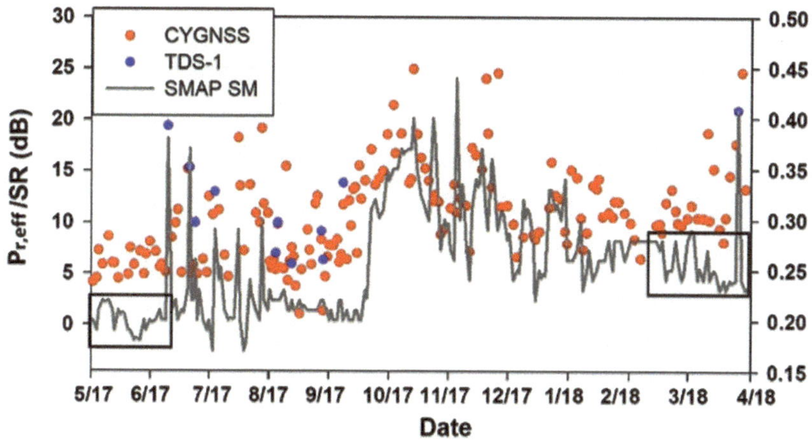

Fig. 5 Variations of CYGNSS, TDS-1, and SMAP during the observation period

correlation coefficient between the GNSS-R observations and SMAP SM reaches R = 0.682 for TDS-1 and R = 0.763 for CYGNSS. These results indicate that GNSS-R observations are capable of capturing the temporal dynamics of SM. However, it should be noted that the number of TDS-1 observations used in this analysis is relatively limited, which reduces the statistical significance of the comparison with SMAP SM. For CYGNSS observations, the relationship between SR and SMAP SM becomes less proportional when SM values are relatively low. This discrepancy may partly result from the underestimation of SM by SMAP products in regions with dense vegetation. In addition, differences in spatial resolution between SMAP pixels and CYGNSS specular points may also introduce certain biases.

The spatial distribution of SM derived from CYGNSS observations over Mainland China is illustrated in Fig. 6 for the same four-month period as the TDS-1 observations shown previously. Overall, the CYGNSS-derived SM successfully captures the seasonal dynamics of SM across the study area, showing variations consistent with the SMAP SM product. The spatial patterns indicate that SM varies considerably across different regions of China. For example, in northeastern China (highlighted in Fig. 6 a1), the estimated SM exhibits relatively small seasonal variability and remains close to 0.4 m^3 m^{-3} throughout the year, which is consistent with the SMAP observations. In contrast, the central and southern regions of Mainland China show more pronounced spatial variations in SM, reflecting the influence of regional climatic conditions.

To further evaluate the performance of the GNSS-R-based SM estimates, Fig. 7 presents the spatial distributions of three statistical metrics—correlation coefficient (R), unbiased root mean square error (ubRMSE), and mean absolute error (MAE)—between the GNSS-R retrievals and SMAP SM on a monthly basis. In general, CYGNSS observations demonstrate good performance across most regions, suggesting that CYGNSS data can be effectively used to generate SM products. The overall statistical results indicate R = 0.798, ubRMSE = 0.062 m^3 m^{-3}, and MAE =

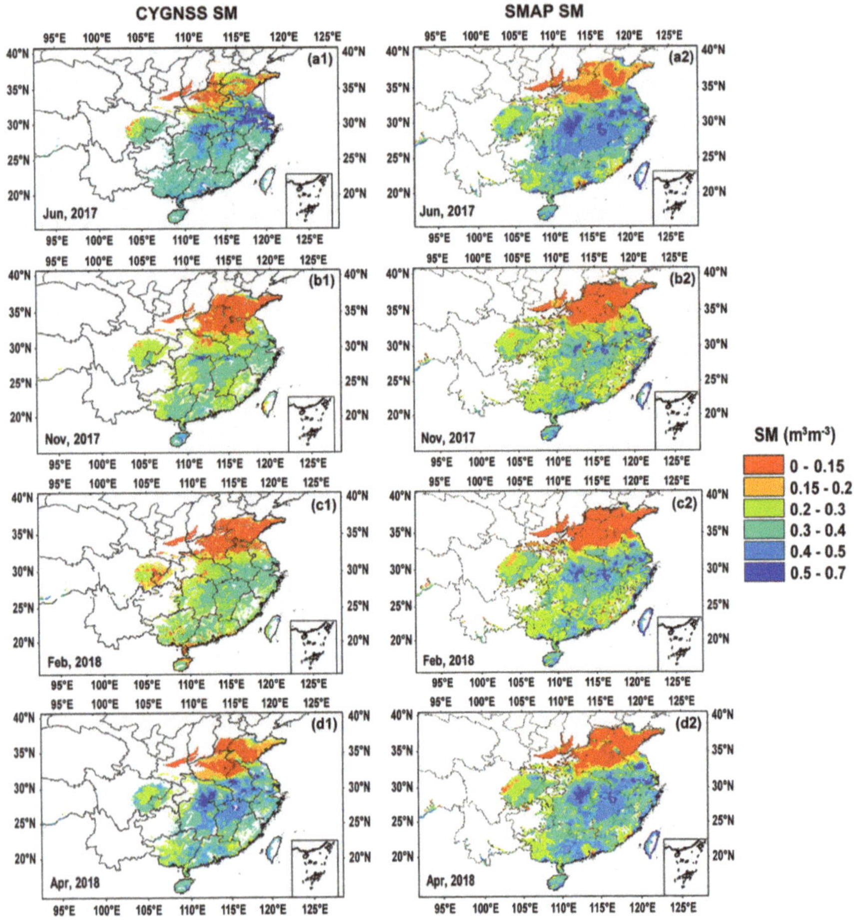

Fig. 6 Comparisons of the SM derived from CYGNSS and the SMAP SM. (a1–a2) June 2017, (b1–b2) November 2017, (c1–c2) February 2017, and (d1–d2) April 2018

0.040 $m^3\ m^{-3}$ for CYGNSS retrievals. TDS-1 also shows reasonable agreement with SMAP SM (R = 0.676, ubRMSE = 0.052 $m^3\ m^{-3}$, MAE = 0.060 $m^3\ m^{-3}$), although the spatial coverage of TDS-1 observations is relatively sparse.

The numerical distributions of the correlation coefficients for both missions are illustrated by the inner boxes in Fig. 7 a1 and a2. For TDS-1, more than 65% of the grid cells exhibit correlation coefficients greater than 0.4, while ubRMSE values remain below 0.08 $m^3\ m^{-3}$ over most regions. Larger errors are mainly observed in northeastern China, where both ubRMSE and MAE values tend to exceed 0.07 $m^3\ m^{-3}$. This pattern is generally consistent with the corresponding error distributions. Because overlapping TDS-1 observations were used for both monthly SM estimation and daily sensitivity analysis, the correlation with SMAP SM is higher than that obtained in the daily analysis.

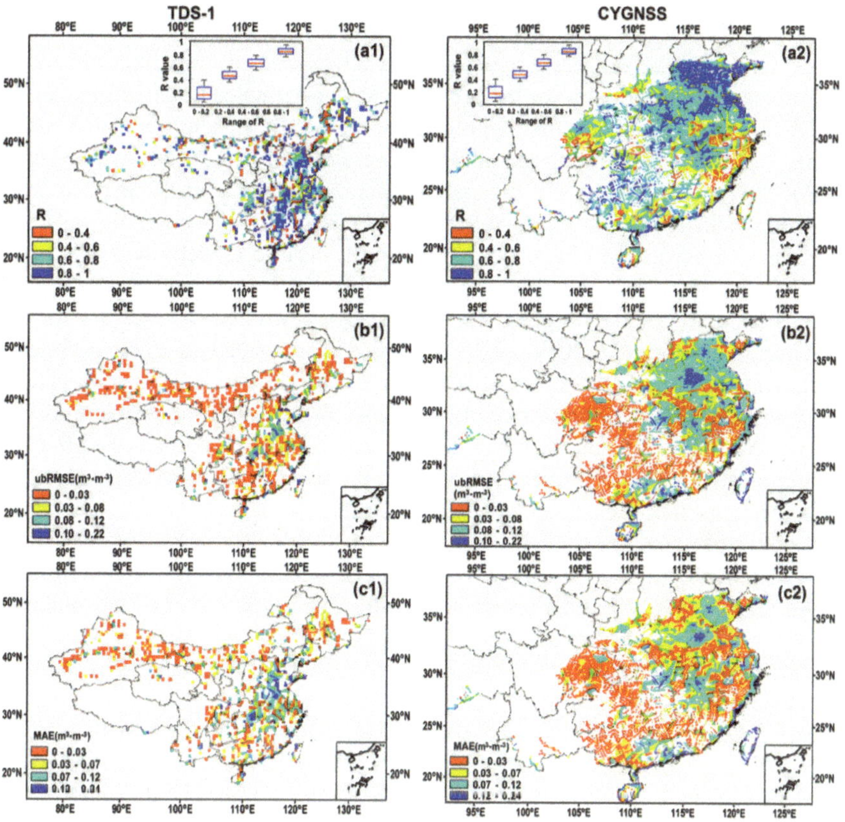

Fig. 7 Statistical indices for the SM derived from TDS-1/CYGNSS against the SMAP SM. (a1–a2) (b1–b2) ubRMSE, and (c1–c2) MAE

For CYGNSS observations, the spatial distribution of the correlation coefficients shows higher values in northern China and relatively lower values in central regions. This spatial pattern may be related to incoherent scattering effects caused by dense vegetation or increased surface roughness. In terms of error metrics, the ubRMSE values remain below 0.08 m^3 m^{-3} across more than half of Mainland China, which is comparable to the performance of TDS-1 observations. In several central regions, slightly larger ubRMSE values are observed, corresponding to areas where the correlation coefficients are also relatively lower. The MAE values for CYGNSS remain smaller than 0.03 m^3 m^{-3} over most parts of Mainland China, indicating a generally good retrieval accuracy.

Overall, these results demonstrate that GNSS-R observations from both TDS-1 and CYGNSS provide valuable information for SM retrieval. In particular, the dense spatial coverage and higher temporal resolution of CYGNSS observations enable more robust SM estimation over large regions.

5.2 SM Retrieval Using the Physics-Based Empirical Model

To evaluate the performance of the proposed physics-coupled empirical model, the CYGNSS observations are first aggregated to the spatial resolution of the SMAP EASE-2 grid (36 km) through spatial averaging, similar to the procedure used in the UCAR SM product (SM_{UCAR}). This spatial matching enables a consistent comparison between GNSS-R observations and SMAP SM products.

Figure 8 presents the relationship between the surface reflectivity derived from CYGNSS observations and SMAP SM. Specifically, Fig. 8a and b illustrate the correlations between the original SR and the SR_{cali} with SMAP SM for all grid cells during the period of 2019–2020. The statistical distributions of the correlation coefficient (R) are shown in Fig. 8c. Compared with the original SR observations, the calibrated surface reflectivity exhibits significantly improved correlations with SMAP SM, indicating that the calibration process effectively reduces the effects of vegetation attenuation and surface roughness. As a result, the calibrated reflectivity signal becomes more sensitive to SM variations.

It should be noted that the consistency between GNSS-R observations and SMAP SM varies across different climatic regions. In particular, relatively low correlation values are observed in arid regions, as highlighted in Fig. 8b. This phenomenon mainly results from the limited variability of SM in these regions during the observation period, which reduces the detectable sensitivity of the reflectivity signal to SM changes.

The SM estimates derived from the calibrated reflectivity (SM_{cali}) are illustrated in Fig. 9, together with the SM_{UCAR} product and SMAP SM. The spatial distributions on 1 January 2019 show that both GNSS-R-based products capture the general spatial patterns of global SM. The difference maps indicate that the absolute deviations between the retrieved SM and SMAP observations are relatively small in most arid regions, including North Africa and Western Australia, where SM values are typically

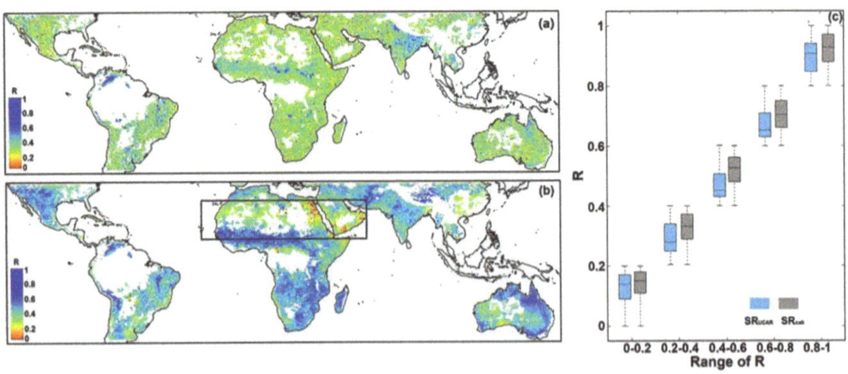

Fig. 8 Comparison of R values between SR (SR_{cali}) and SMAP. **a** R between SR and SMAP. **b** R between SR_{cali} and SMAP. **c** Boxplot of the R values in **a** and **b**

low. In these regions, the differences are generally smaller than 0.05 m^3 m^{-3}. Larger deviations occur in tropical rainforest regions, where strong vegetation scattering and insufficient open-water masking may influence the GNSS-R observations. In other regions, such as the eastern United States and southern South America, the differences range approximately from – 0.39 to 0.05 m^3 m^{-3}.

Figure 10 further compares the GNSS-R-derived SM with SMAP observations using density plots for the first day of each month in 2019. The point density is represented by color intensity. Both SM_{UCAR} and SM_{cali} show strong agreement with SMAP SM and are generally distributed close to the 1:1 line. However, the calibrated model demonstrates improved performance. The correlation coefficient varies between 0.732 and 0.909, while the root mean square error (RMSE) ranges from 0.045 to 0.057 m^3 m^{-3} for SM_{cali}, which is lower than that of the SM_{UCAR} product.

The statistical performance of the two products is summarized using three commonly used metrics, namely the correlation coefficient (R), the root mean square error (RMSE), and the mean absolute error (MAE). The monthly averaged statistics for 2019 indicate that the calibrated SM product performs better than SM_{UCAR}. The mean correlation coefficient increases from 0.834 for SM_{UCAR} to 0.875 for SM_{cali}, while the RMSE decreases from 0.060 m^3 m^{-3} to 0.050 m^3 m^{-3}. Similarly, the MAE decreases from 0.053 m^3 m^{-3} to 0.041 m^3 m^{-3}.

The spatial distributions of the three statistical metrics are shown in Fig. 11 for the period from January 2019 to November 2020. Overall, the calibrated SM estimates demonstrate better spatial performance than the SM_{UCAR} product. For SM_{cali}, more than 70% of the global grids exhibit correlation coefficients higher than 0.6, while the RMSE values remain below 0.07 m^3 m^{-3} over most regions. In addition, the MAE values are generally smaller than 0.05 m^3 m^{-3} across large areas of the globe.

The statistical distributions of these metrics are also presented in Fig. 11d. In general, the calibrated SM estimates show higher correlation coefficients and lower error metrics than those of the SM_{UCAR} product. These results demonstrate that the physics-coupled empirical model effectively improves the SM retrieval accuracy by integrating CYGNSS reflectivity observations with SMAP SM information.

5.3 SM Retrieval Using the PHYSER Framework

To evaluate the performance of the PHYSER framework, the SM retrieved using the proposed physics-based approach (denoted as BFphy) is compared with the BF-1 SM derived from an empirical method (BFemp) developed by Guo et al. [38]. The empirical approach applies a least-squares calibration algorithm to establish linear relationships between BF-1 surface reflectivity and SMAP SM. In this section, both retrieval approaches are evaluated against the SMAP SM product.

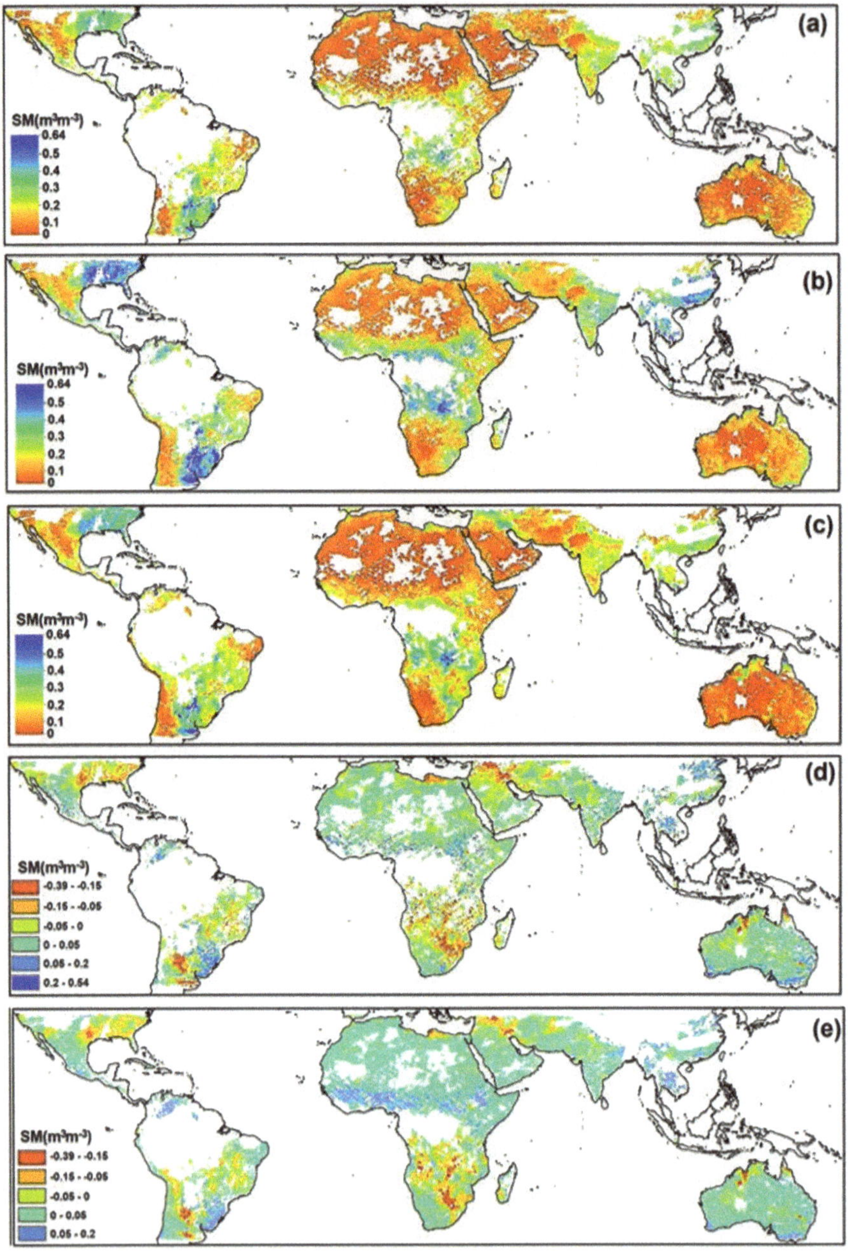

Fig. 9 Comparison of SM values on January 1, 2019. **a** SMUCAR. **b** SM_{cali}. **c** SMAP SM (an averaging over January 1–3, 2019). **d** SMUCAR minus SMAP SM. **e** SM_{cali} minus SMAP SM

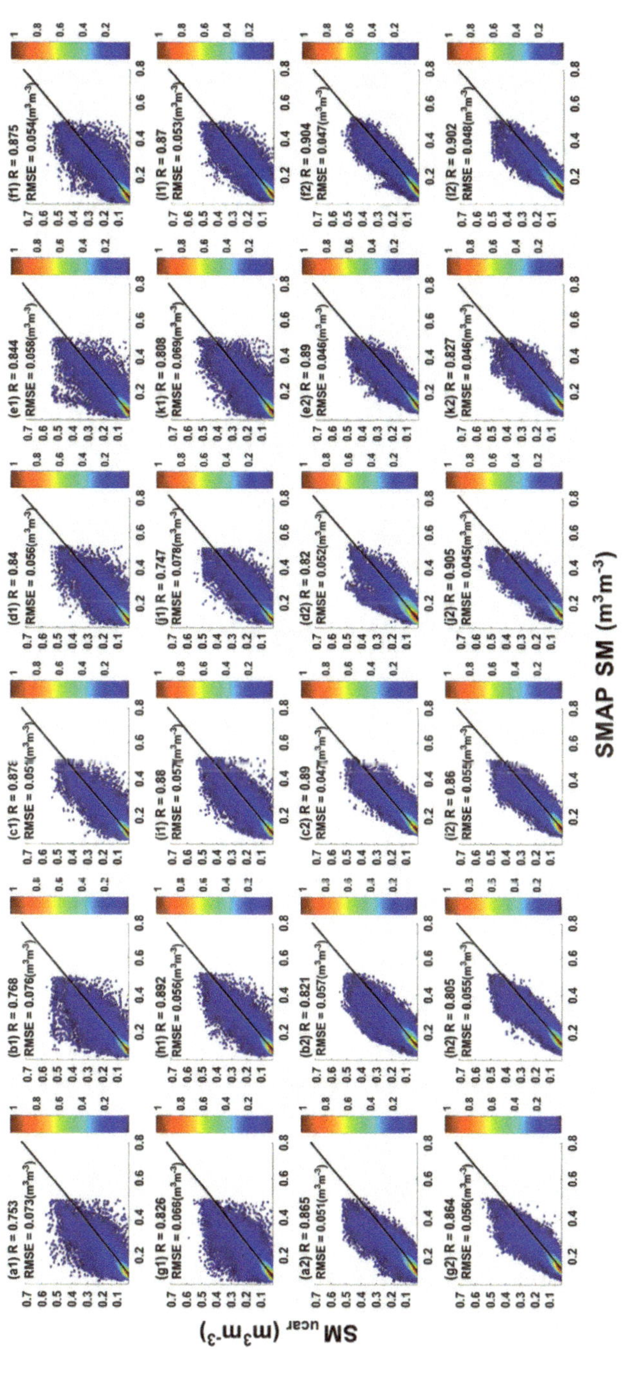

Fig. 10 Comparison of CYGNSS SM with SMAP SM on the first day of each month in 2019. (a1)–(i1) SM_{UCAR}. (a2)–(i2) SM_{cali}

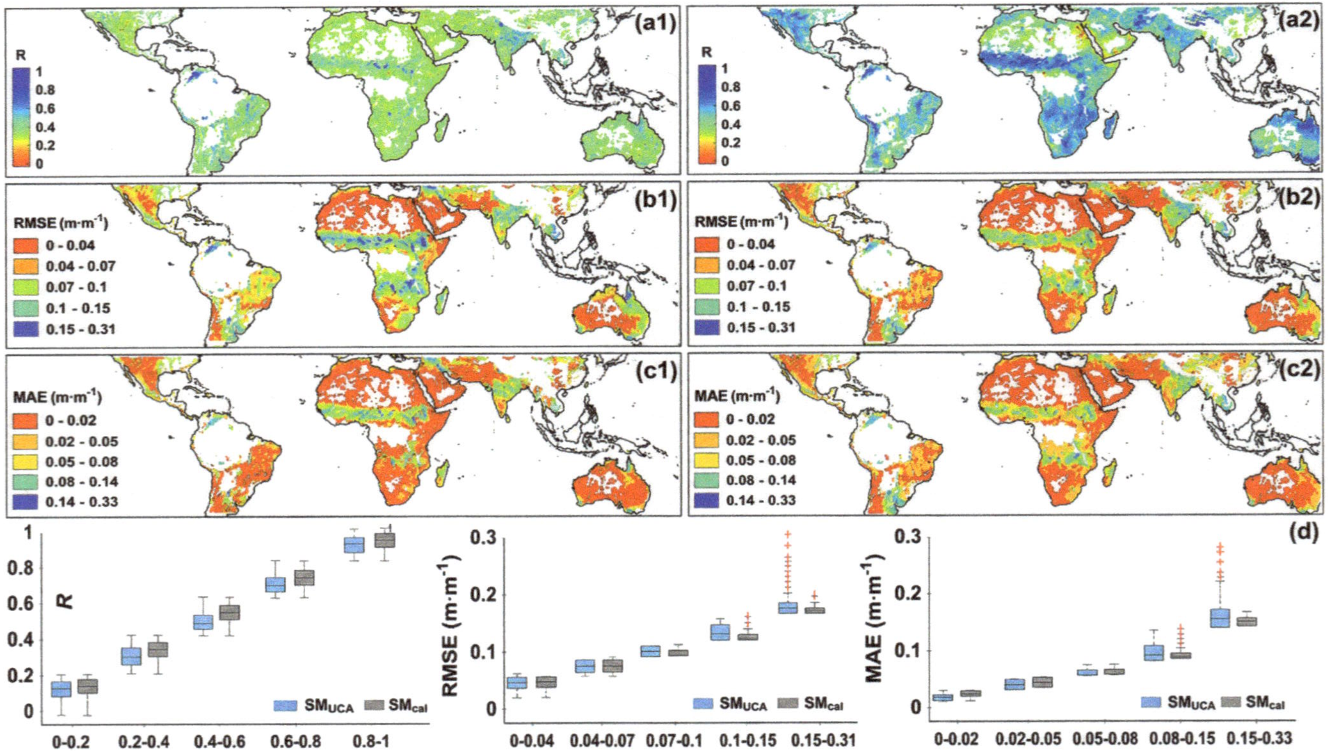

Fig. 11 Statistical indices for the SM derived from SM_{UCAR} (SM_{cali}) against the SMAP SM. (a1) and (a2) R. (b1) and (b2) RMSE. (c1) and (c2) MAE. (d) Boxplot of R, RMSE, and MAE

Three representative regions, including the eastern United States, southern Australia, and parts of South Asia, are selected to illustrate the spatial characteristics of the retrieved SM. Figure 12 shows the spatial distributions of BFphy SM, BFemp SM, and SMAP SM for October 2019.

In the eastern United States (Fig. 12 A1–A3), the spatial pattern of BFphy SM agrees well with the SMAP SM product. The correlation coefficient between BFphy SM and SMAP SM reaches R = 0.60, with ubRMSD = 0.079 $m^3\ m^{-3}$ and BIAS = 0.022 $m^3\ m^{-3}$. SM values in this region are mainly distributed between 0.05 and 0.30 $m^3\ m^{-3}$. In southern Australia (Fig. 12 B1–B3), SM values are generally lower. In this region, BFphy SM shows the lowest agreement with SMAP SM, with R = 0.47, ubRMSD = 0.058 $m^3\ m^{-3}$, and BIAS = 0.055 $m^3\ m^{-3}$. The reduced consistency is

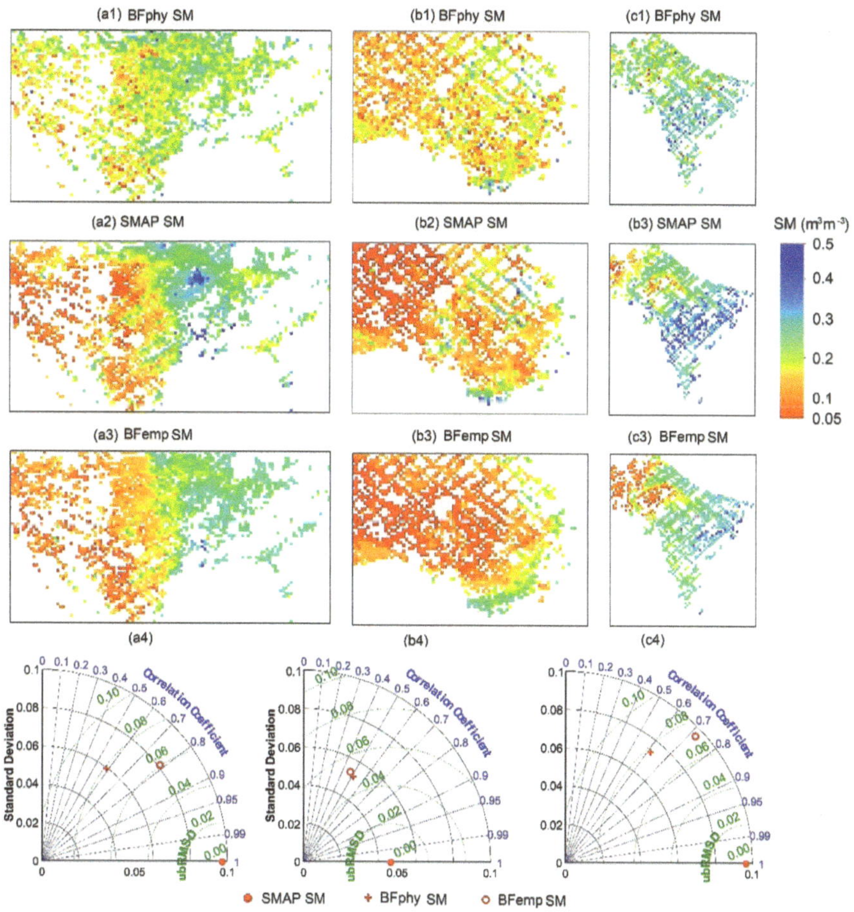

Fig. 12 Comparisons of BFphy SM and BFemp SM under 3 regions (i.e., eastern USA, southern Australia, and parts of South Asia) in the world. (A1) to (A4) is eastern USA, (B1) to (B4) is southern Australia, (C1) to (C4) is parts of South Asia

mainly attributed to dense vegetation in certain areas, where normalized difference vegetation index values range between 0.4 and 0.6, which may weaken the relationship between GNSS-R reflectivity and SM. In contrast, the eastern United States typically exhibits cooler surface temperatures and moderate vegetation cover, which contributes to the relatively higher correlation between BFphy SM and SMAP SM.

For parts of South Asia (Fig. 12 C1–C3), the correlation between BFphy SM and SMAP SM reaches R = 0.61, which is higher than that observed in southern Australia. SM values in this region are generally distributed between 0.39 and 0.50 m^3 m^{-3}. However, BFphy SM tends to underestimate high SM values, which results in larger errors in this region, with ubRMSD = 0.085 m^3 m^{-3} and BIAS = 0.020 m^3 m^{-3}. The Taylor diagrams presented in Fig. 12 A4–C4 summarize the statistical characteristics of BFphy and BFemp SM retrievals relative to SMAP observations by simultaneously considering the correlation coefficient, ubRMSD, and bias. According to the Taylor diagrams [42], the empirical approach (BFemp) shows slightly higher accuracy than the PHYSER retrieval, with correlation coefficients ranging from 0.47 to 0.78, compared with 0.51 to 0.61 for BFphy. Nevertheless, the PHYSER framework represents an important step toward developing a physically consistent GNSS-R SM retrieval method that does not directly rely on reference SM data for calibration.

To further evaluate the global performance of the PHYSER framework, the spatial distributions of three statistical metrics—correlation coefficient (R), unbiased root mean square difference (ubRMSD), and mean absolute error (MAE)—between BF-1 SM and SMAP SM are presented in Fig. 13 for the period from August 2019 to July 2020. Overall, the BF-1 observations provide reliable quasi-global daily SM estimates with reasonable spatial consistency. The global mean statistics obtained when compared with SMAP SM are R = 0.57, ubRMSD = 0.067 m^3 m^{-3}, and MAE = 0.073 m^3 m^{-3}. More than 60% of the grid cells exhibit correlation coefficients greater than 0.5. The ubRMSD values remain below 0.06 m^3 m^{-3} across most regions, although relatively larger errors occur in central Africa and southern Australia. The spatial distribution of MAE shows a similar pattern, with values generally smaller than 0.06 m^3 m^{-3} over most areas.

The PHYSER SM estimates are also compared with the ERA5-Land SM dataset. In general, the PHYSER retrieval maintains a strong correlation with ERA5-Land SM across most regions of the globe, with global mean statistics of R = 0.49, ubRMSD = 0.079 m^3 m^{-3}, and MAE = 0.088 m^3 m^{-3}. Lower correlation values are observed in highly vegetated regions such as the Amazon and Congo rainforests, where the correlation coefficient is approximately 0.4. The ubRMSD values are generally smaller than 0.07 m^3 m^{-3} across more than half of the globe, while MAE values remain below 0.08 m^3 m^{-3} over most areas. The slightly lower correlation with ERA5-Land compared with SMAP is mainly attributed to differences in spatial resolution and soil depth representation. Specifically, ERA5-Land SM represents the upper 0–7 cm soil layer, whereas GNSS-R L-band signals can penetrate deeper into the soil profile, typically ranging from 0 to 20 cm, depending on SM conditions.

Figure 14 present the comparison between the 9 km BF-1 SM product and in situ observations from several ground stations. In total, six representative sites are used for validation, while eight dense sites from the Hebi monitoring network are located

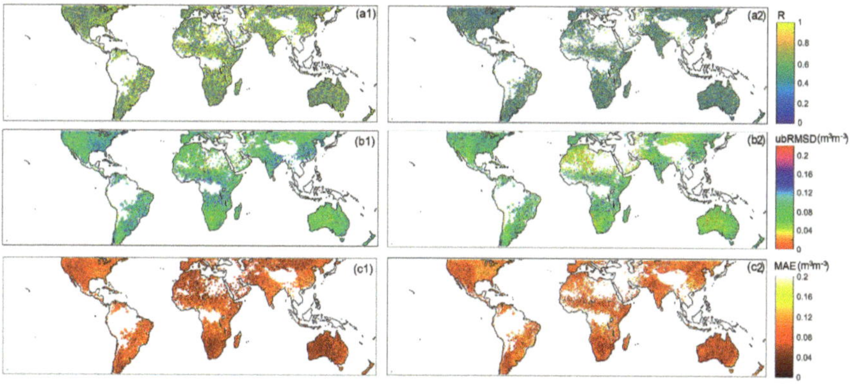

Fig. 13 Statistical indices for the BF-1 SM and reference SM datasets. (A1) R between BF-1 SM and SMAP SM. (B1) ubRMSD between BF-1 SM and SMAP SM. (C1) MAE between BF-1 SM and SMAP SM. (A2) R between BF-1 SM and ERA5-Land SM. (B2) ubRMSD between BF-1 SM and ERA5-Land SM. (C2) MAE between BF-1 SM and ERA5-Land SM

within the same 9 km grid cell. Although the BF-1 observations are spatially sparse, they capture the temporal dynamics of SM effectively, particularly during periods of significant SM change. In general, the BF-1 SM slightly overestimates SM under dry conditions and underestimates SM under wet conditions. For example, when in situ SM values are relatively low, BF-1 estimates range between 0.077 and 0.112 $m^3 m^{-3}$, whereas in situ values range from 0.045 to 0.046 $m^3 m^{-3}$. Conversely, BF-1 tends to underestimate SM at sites with relatively high in situ values.

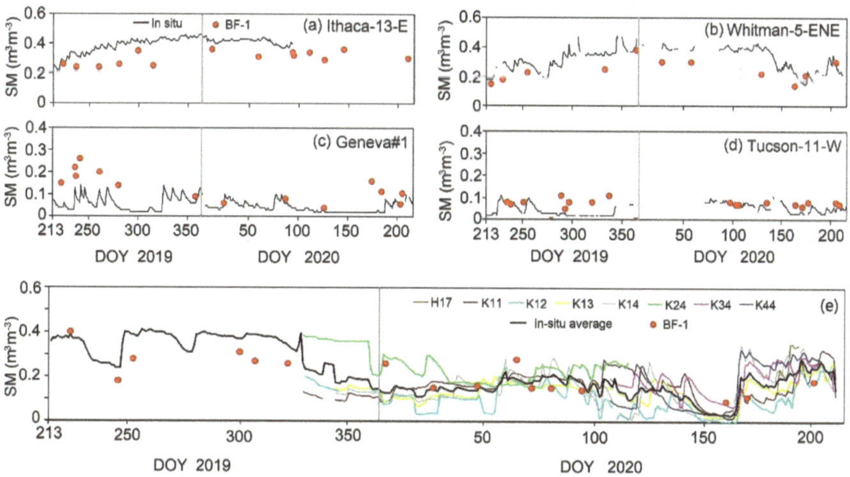

Fig. 14 Comparisons of the BF-1 SM and the in situ SM for different sites. **a** Ithaca-13-E. **b** Whitman-5-ENE. **c** Geneva#1. **d** Tucson-11-W. **e** Henan network. DOY, day of the year

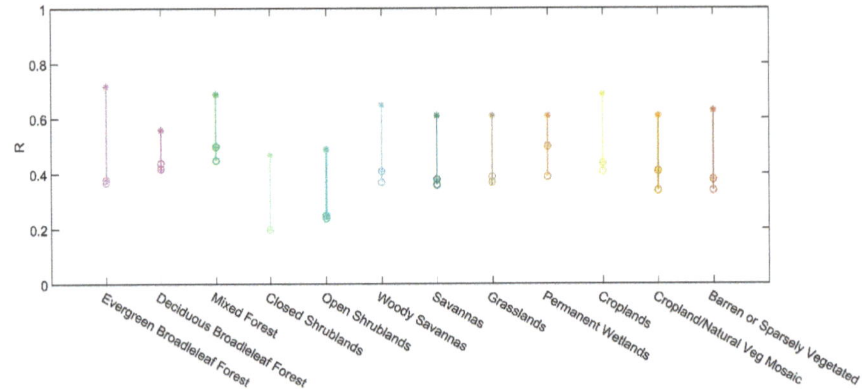

Fig. 15 The improvement of the R values between the 3 SoR versions (i.e., Γ_{raw}, Γ_{cal}, and Γ_{cal_cor}) and the SMAP SM on different IGBP land types. The hollow circle represents Γ_{raw} vs. SMAP SM, the hollow circle with a plus sign represents Γ_{cal} vs. SMAP SM, and the asterisk denotes Γ_{cal_cor} vs. SMAP SM

For the eight stations located within the same grid cell in the Hebi network (Fig. 14 E), a good agreement is observed between the two datasets, with an unbiased RMSE of 0.069 m^3 m^{-3} and a bias of − 0.023 m^3 m^{-3}. The coefficient of variation (CV) is also calculated to evaluate the variability of SM across representative sites. The CV values for in situ observations range from 13.34 to 56.32%, indicating diverse hydrological dynamics among different regions. The BF-1 SM retrieval shows a similar variability pattern, with an average CV of 31.25%, compared with 39.53% for in situ observations. This consistency demonstrates that the PHYSER retrieval effectively preserves the relative variability of SM across heterogeneous environments.

Finally, the contributions of the SuR-CAL and SoR-COR correction steps are statistically evaluated based on the correlations between SMAP SM and the reflectivity variables Γ_{raw}, Γ_{cal}, and Γ_{cal_cor} across 12 IGBP land cover types, as shown in Fig. 15 The results indicate that the correlation coefficient gradually increases after each correction step. Specifically, the average correlation improves from 0.36 for Γ_{raw} to 0.39 after the SuR-CAL correction, representing approximately 7% improvement, and further increases to 0.60 after the SoR-COR correction, corresponding to an improvement of approximately 36%. These results demonstrate that considering land surface properties significantly enhances the performance of GNSS-R SM retrieval compared with correcting only systematic errors.

5.4 *Flood Inundation Retrieval*

To evaluate the capability of GNSS-R observations for flood monitoring, the sensitivity of the calibrated CYGNSS surface reflectivity to precipitation-induced surface wetting is first examined. Figure 16 presents spatial comparisons of GPM precipitation, SMAP SM (SM), and CYGNSS SR_{cali} for two representative dates in southeast China, namely 17 July 2021 and 21 July 2021. These two dates were selected because an extreme precipitation event occurred between 19 and 22 July 2021. To provide continuous spatial patterns over the study region, both CYGNSS SR_{cali} and SMAP SM were aggregated to a 36-km grid, although this resolution is coarser than the nominal spatial resolution of CYGNSS observations.

Overall, the spatial pattern of CYGNSS SR_{cali} shows good agreement with the SMAP SM distribution. On 17 July 2021, relatively low SR_{cali} values are observed in the northern part of southeast China. After the extreme precipitation event, on 21 July 2021, SR_{cali} values in the northern region increase significantly, while the central region shows relatively small variations. These spatial variations are consistent with the corresponding changes in SMAP SM and GPM precipitation. The difference map of CYGNSS SR_{cali} (Fig. 16 c3) indicates positive anomalies in the northeastern part

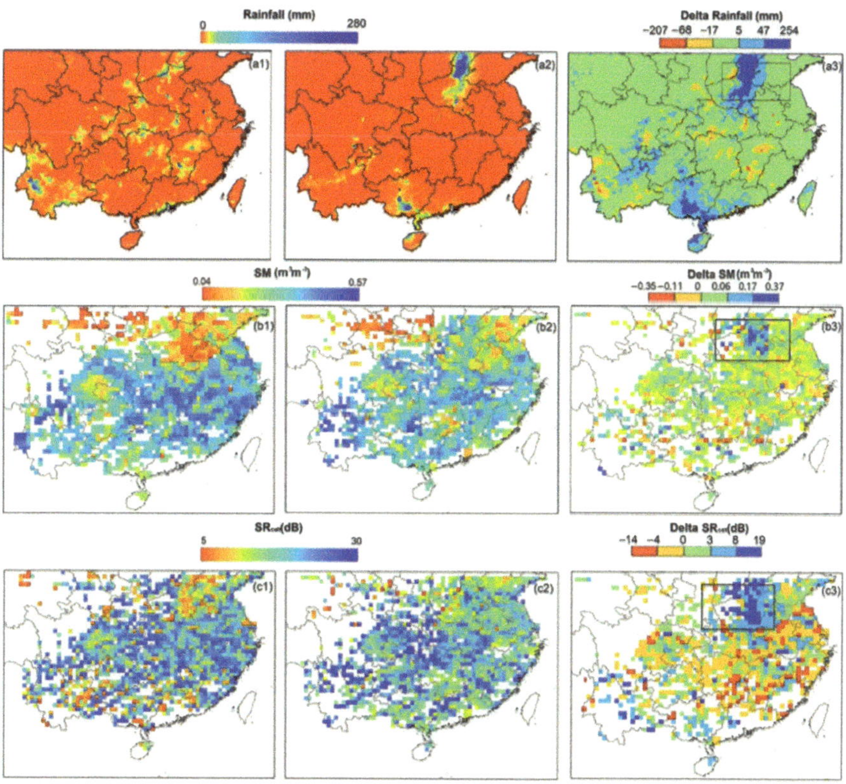

Fig. 16 Spatial patterns of the GPM precipitation, SMAP SM, and CYGNSS SR_{cali} in southeast China. (a1–a3) Precipitation on 17 July 2021 and 21 July 2021, and their differences (Delta GPM precipitation), (b1–b3) SMAP SM on 17 July 2021 and 21 July 2021, and their differences (Delta SM), and (c1–c3) CYGNSS SR_{cali} on 17 July 2021 and 21 July 2021, and their differences (Delta SR_{cali})

of the region and negative anomalies in the central area. This spatial pattern closely matches the differences observed in GPM precipitation and SMAP SM (Fig. 16 a3 and b3), demonstrating the sensitivity of GNSS-R reflectivity to precipitation-induced SM changes and potential flooding conditions.

To further evaluate the ability of CYGNSS observations to detect inundated surfaces, the Adaptive Threshold Flood Inundation Index derived from CYGNSS observations is compared with the VIIRS floodwater fraction product, as shown in Fig. 17. The color scale in the density plots represents the density of matched observations. To represent seasonal variability, data from one month in each season of 2021 were selected. For each VIIRS pixel, CYGNSS observations located within 187.5 km from the pixel center were used for comparison. The correlation coefficient (R) was used to assess the linear relationship between the CYGNSS-derived ATFII and the VIIRS floodwater fraction. All correlations are statistically significant at the 0.05 significance level ($p < 0.05$). However, it should be noted that the spatial resolutions of the two datasets differ substantially (375 m for VIIRS versus approximately 3×0.5 km for CYGNSS). This spatial mismatch may introduce dispersion in the comparison results because datasets with different spatial resolutions capture surface characteristics at different scales.

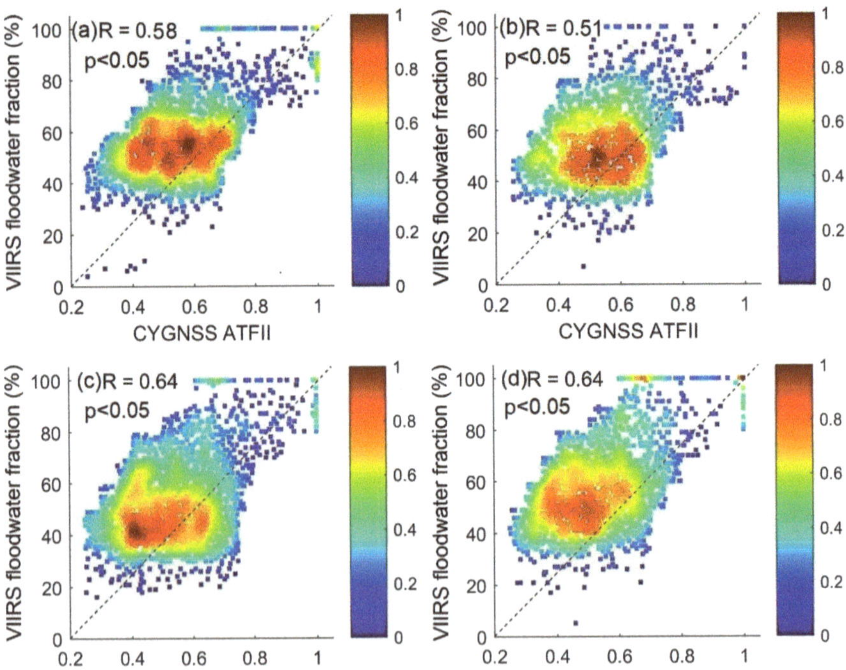

Fig. 17 Comparison of the CYGNSS ATFII with VIIRS floodwater fraction in **a** January 2021, 10
b April 2021, **c** July 2021, and **d** October 2021

To determine the thresholds for different inundation levels, one year of CYGNSS ATFII data collected in 2020 was matched with the corresponding VIIRS floodwater fraction data. After matching the datasets, low-quality observations were removed using a 95% standard deviation filtering criterion. The resulting ranges of ATFII values corresponding to different inundation grades are summarized in Table 1.

o further investigate the performance of the CYGNSS-derived inundation classification, two extreme precipitation events that occurred in southeast China during 2021 were analyzed: 17–23 July and 19–25 August. In some regions, the cumulative precipitation during these events exceeded 500 mm. Figure 18 illustrates the spatial distributions of the CYGNSS-derived inundation grades before and after the two extreme precipitation events. The results show that the GNSS-R observations successfully capture the spatial evolution of high-intensity precipitation, particularly in the central and eastern regions, which are highlighted by red circles in the figure.

The corresponding area-ratio statistics of the inundation grades are presented in Fig. 18 c1 and c2. For the July precipitation event, the proportion of severe and inundated areas increased from 14 to 22% after the event. A similar trend is observed for the August event, where the proportion of severe and inundated areas increased from 15 to 20%. These results demonstrate that the CYGNSS-derived inundation index can effectively capture large-scale flood expansion associated with extreme precipitation events.

To provide a more detailed view of the temporal evolution of the flood event, Fig. 19 shows the CYGNSS-derived inundation grades in Henan Province from 19 to 24 July 2021. GPM precipitation data are also included for comparison. Significant variations in the inundation grades are observed between 20 and 23 July, corresponding to the peak of the extreme precipitation event. As shown in Fig. 19 a1–a6, severe inundation mainly occurs in the northern and central regions, where large areas are classified as severe or fully inundated grids. The area ratios of severe and inundated grids during this period increase from 0, 19, 52, and 56%, before gradually decreasing to 25% and 9% as the flood recedes. In contrast, the southern part of the region experiences relatively lighter inundation conditions, mainly classified as mild, moderate, or non-inundated soils, which is consistent with the spatial distribution of precipitation observed in the GPM dataset.

It should be noted that in Fig. 19 a4 and a5, severe inundation appears in certain southern areas even though the corresponding GPM precipitation values (Fig. 19 b4 and b5) are relatively small. This discrepancy may be explained by surface water accumulation and delayed drainage caused by persistent precipitation, which can lead to temporary water retention even after rainfall intensity decreases.

Table 1 Ranges of ATFII values for different inundation grades

	Non-inundation	Mild-inundation	Moderate-inundation	Severe-inundation	Inundated area
ATFII value	0–0.33	0.33–0.47	0.47–0.68	0.68–0.86	> 0.86

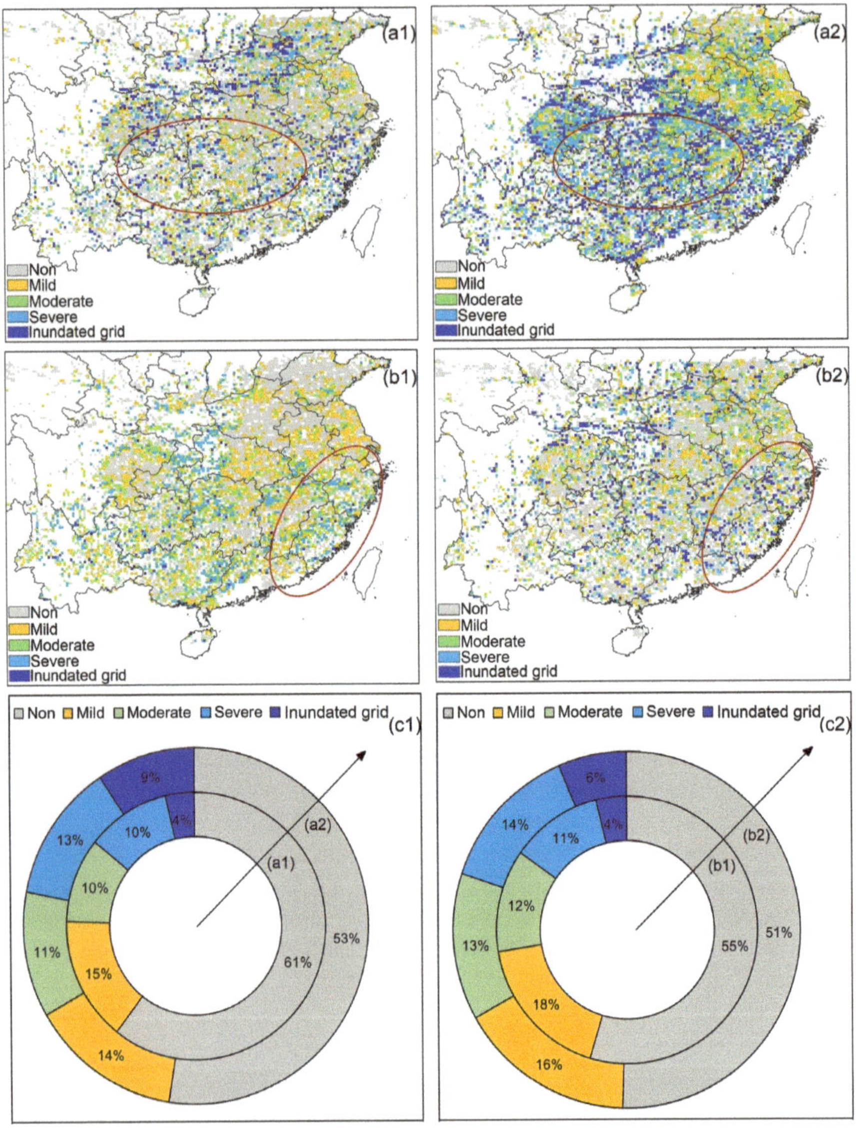

Fig. 18 The CYGNSS-derived inundation grades in southeast China before and after two extreme precipitation events. (a1, a2) Inundation grade distributions before and after the extreme precipitations which occurred from 17 to 23 July 2021, respectively. (b1, b2) Inundation grade distributions before and after the precipitation which occurred from 19 to 25 August 2021, respectively. (c1, c2) The area ratios of (a1, a2, b1, b2), respectively

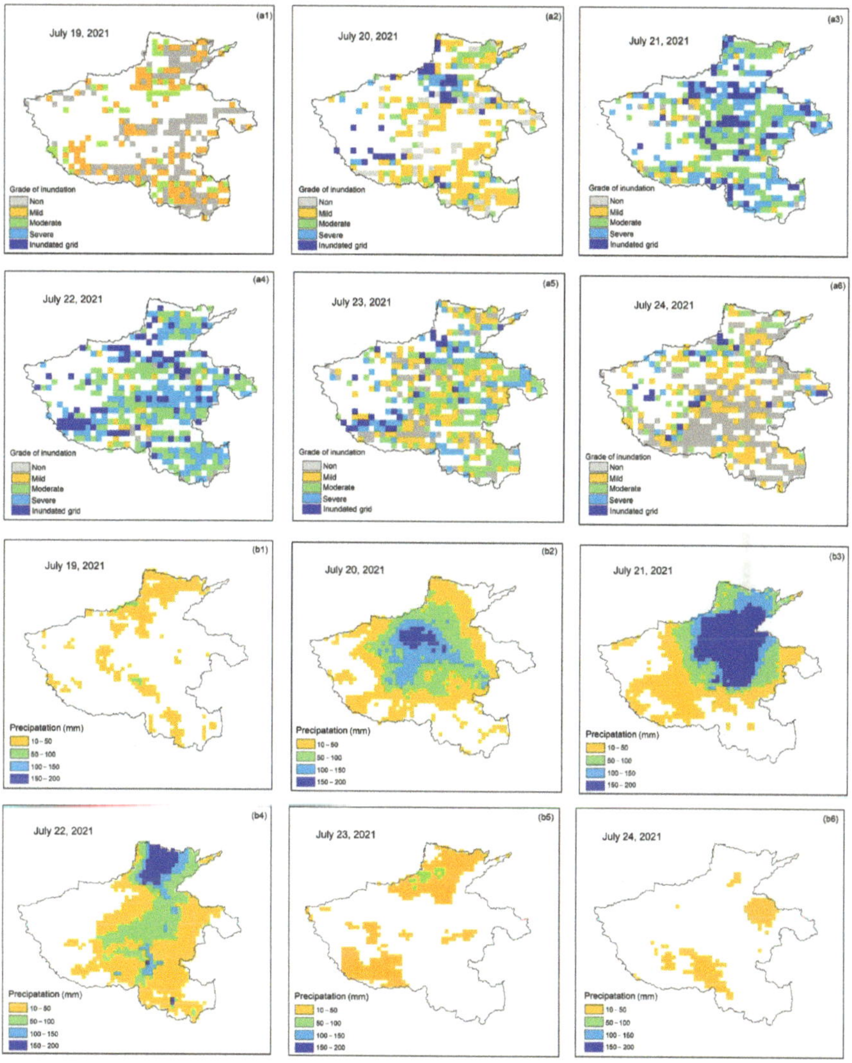

Fig. 19 The results of daily CYGNSS-derived inundation grades and GPM precipitation in Henan Province from 19 July 2021 to 24 July 2021. (a1–a6) CYGNSS-derived inundation grade distributions and (b1–b6) GPM precipitation distributions

6 Summary

This chapter has reviewed and integrated several GNSS-R-based approaches for soil moisture retrieval and flood inundation monitoring using TDS-1, CYGNSS, and BF-1 observations. The results show that spaceborne GNSS-R can provide meaningful hydrological information over large areas and can complement conventional microwave remote sensing products. The chapter structure in the revised draft also reflects this progression from data and reflectivity modeling to retrieval methods and applications. For soil moisture retrieval, both empirical and machine-learning approaches demonstrate that GNSS-R observations are sensitive to soil moisture variations. Over Mainland China, CYGNSS generally provides better performance than TDS-1 because of its denser temporal sampling and broader spatial coverage, while both missions show good consistency with SMAP and in situ measurements.

The chapter also highlights the importance of physically based correction of vegetation and surface roughness effects. The CYGNSS–SMAP coupled algorithm improves soil moisture estimation by calibrating GNSS-R reflectivity with microwave radiometric information, while the PHYSER framework further demonstrates that physics-based GNSS-R retrieval can be achieved without directly relying on reference soil moisture products during retrieval. The PHYSER results show that correcting land surface effects is particularly important for improving reflectivity accuracy and final soil moisture performance. n addition, the chapter extends GNSS-R applications from soil moisture to flood inundation monitoring. By constructing the ATFII from calibrated CYGNSS reflectivity, the study shows that GNSS-R observations can capture daily inundation dynamics and distinguish inundation grades during extreme precipitation events. The flood case study further demonstrates the value of GNSS-R for monitoring rapid hydrological changes when frequent observations are required.

Overall, this chapter demonstrates that GNSS-R is a promising technique for integrated hydrological remote sensing. The combination of empirical retrieval, ANN modeling, physics-based correction, and flood index construction provides a useful framework for future soil moisture and flood applications based on current and next-generation GNSS-R missions.

References

1. Owe M, de Jeu R, Walker J (2001) A methodology for surface soil moisture and vegetation optical depth retrieval using the microwave polarization difference index. IEEE Trans Geosci Remote Sens 39(8):1643–1654
2. Rahimzadeh-Bajgiran P, Berg AA, Champagne C, Omasa K (2013) Estimation of soil moisture using optical/thermal infrared remote sensing in the Canadian Prairies. ISPRS J Photogramm Remote Sens 83:94–103
3. Brocca L, Ciabatta L, Massari C, Camici S, Tarpanelli A (2017) Soil moisture for hydrological applications: open questions and new opportunities. Water 9(2):140
4. Draper CS, Reichle RH, De Lannoy GJM, Liu Q (2012) Assimilation of passive and active microwave soil moisture retrievals. Geophys Res Lett 39:L04401
5. Chew C, Reager JT, Small E (2018) CYGNSS data map flood inundation during the 2017 Atlantic hurricane season. Sci Rep 8:1–8
6. Wan W, Liu B, Zeng Z, Chen X, Wu G, Xu L, Chen X, Hong Y (2019) Using CYGNSS data to monitor China's flood inundation during typhoon and extreme precipitation events in 2017. Remote Sens 11:854
7. Clarizia MP, Ruf CS (2016) Wind speed retrieval algorithm for the Cyclone Global Navigation Satellite System (CYGNSS) mission. IEEE Trans Geosci Remote Sens 54:4419–4432

8. Unwin M, Jales P, Tye J, Gommenginger C, Foti G, Rosello J (2018) Spaceborne GNSS-reflectometry on TechDemoSat-1: early mission operations and exploitation. IEEE J Sel Top Appl Earth Obs Remote Sens 11(12):4525–4539
9. Jing C, Niu X, Duan C, Lu F, Di G, Yang X (2019) Sea surface wind speed retrieval from the first Chinese GNSS-R mission: technique and preliminary results. Remote Sens 11(24):3013
10. Camps A, Park H, Pablos M, Foti G, Gommenginger CP, Liu PW, Judge J (2016) Sensitivity of GNSS-R spaceborne observations to soil moisture and vegetation. IEEE J Sel Top Appl Earth Obs Remote Sens 9(10):4730–4742
11. Chew C, Shah R, Zuffada C, Hajj G, Masters D, Mannucci AJ (2016) Demonstrating soil moisture remote sensing with observations from the UK TechDemoSat-1 satellite mission. Geophys Res Lett 43(7):3317–3324
12. Chew CC, Small EE (2018) Soil moisture sensing using spaceborne GNSS reflections: comparison of CYGNSS reflectivity to SMAP soil moisture. Geophys Res Lett 45(9):4049–4057
13. Kim H, Lakshmi V (2018) Use of Cyclone Global Navigation Satellite System (CYGNSS) observations for estimation of soil moisture. Geophys Res Lett 45(16):8272–8282
14. Eroglu O, Kurum M, Boyd D, Gurbuz AC (2019) High spatio-temporal resolution CYGNSS soil moisture estimates using artificial neural networks. Remote Sens 11(19):2272
15. Clarizia MP, Pierdicca N, Costantini F, Floury N (2019) Analysis of CYGNSS data for soil moisture retrieval. IEEE J Sel Top Appl Earth Obs Remote Sens 12(7):2227–2235
16. Chew C, Small E (2020) Description of the UCAR/CU soil moisture product. Remote Sens 12(10):1558
17. Yan Q, Huang W, Jin S, Jia Y (2020) Pan-tropical soil moisture mapping based on a three-layer model from CYGNSS GNSS-R data. Remote Sens Environ 247:111944
18. Yang T, Wan W, Sun Z, Liu B, Li S, Chen X (2020) Comprehensive evaluation of using TechDemoSat-1 and CYGNSS data to estimate soil moisture over mainland China. Remote Sens 12(11):1699
19. Yang T, Wan W, Wang J, Liu B, Sun Z (2022) A physics-based algorithm to couple CYGNSS surface reflectivity and SMAP brightness temperature estimates for accurate soil moisture retrieval. IEEE Trans Geosci Remote Sens 60:4409715
20. Yang T, Guo Z, Lu F, Wan W (2026) Concept and initial realization of PHYSER—a physics-based framework for spaceborne GNSS-R soil moisture retrieval with accurate soil reflectivity. J Remote Sens 6:0939
21. Yang T, Sun Z, Jiang L (2023) A novel index for daily flood inundation retrieval from CYGNSS measurements. Remote Sens 15(2):524
22. Gleason S, Ruf CS, O'Brien AJ, McKague DS (2019) The CYGNSS Level 1 calibration algorithm and error analysis based on on-orbit measurements. IEEE J Sel Top Appl Earth Obs Remote Sens 12:37–49
23. Wan W, Liu B, Guo Z, Lu F, Niu X, Li H, Ji R, Cheng J, Li W, Chen X et al (2021) Initial evaluation of the first Chinese GNSS-R mission BuFeng-1 A/B for soil moisture estimation. IEEE Geosci Remote Sens Lett 19:8017305
24. O'Neill PE et al (2018) Soil Moisture Active Passive (SMAP) algorithm theoretical basis document: level 2 & 3 soil moisture (Passive) data products. NASA Goddard Space Flight Center, Tech. Rep. JPL D-66480
25. Li S, Sun D, Goldberg MD, Sjoberg B, Santek D, Hoffman JP, DeWeese M, Restrepo P, Lindsey S, Holloway E (2018) Automatic near real-time flood detection using Suomi-NPP/VIIRS data. Remote Sens Environ 204:672–689
26. Skofronick-Jackson G, Petersen WA, Berg W, Kidd C, Stocker EF, Kirschbaum DB, Kakar R, Braun SA, Huffman GJ, Iguchi T et al (2017) The global precipitation measurement (GPM) mission for science and society. Bull Am Meteor Soc 98:1679–1695
27. Munoz-Sabater J, Dutra E, Agustí-Panareda A, Albergel C, Arduini G, Balsamo G, Boussetta S, Choulga M, Harrigan S, Hersbach H et al (2021) ERA5-Land: a state-of-the-art global reanalysis dataset for land applications. Earth Syst Sci Data 13:4349–4383
28. Dorigo W, Wagner W, Albergel C, Albrecht F, Balsamo G, Brocca L, Chung D, Ertl M, Forkel M, Gruber A et al (2017) ESA CCI soil moisture for improved earth system understanding: state-of-the-art and future directions. Remote Sens Environ 203:185–215

29. Pekel JF, Cottam A, Gorelick N, Belward AS (2016) High-resolution mapping of global surface water and its long-term changes. Nature 540(7633):418–422
30. Wang S, Li J, Zhang B, Spyrakos E, Tyler AN, Shen Q, Zhang F, Kuster T, Lehmann MK, Wu Y et al (2018) Trophic state assessment of global inland waters using a MODIS-derived Forel-Ule index. Remote Sens Environ 217:444–460
31. Nachtergaele F, Batjes N (2012) Harmonized world soil database. FAO, Rome
32. Santi E et al (2017) Vegetation water content retrieval by means of multifrequency microwave acquisitions from AMSR2. IEEE J Sel Top Appl Earth Obs Remote Sens 10(9):3861–3873
33. Carreno-Luengo H, Amezaga A, Vidal D, Olive R, Munoz JF, Camps A (2015) First polarimetric GNSS-R measurements from a stratospheric flight over boreal forests. Remote Sens 7(10):13120–13138
34. Choudhury BJ, Schmugge TJ, Chang A, Newton RW (1979) Effect of surface roughness on the microwave emission from soils. J Geophys Res 84(C9):5699–5706
35. Loria E, O'Brien A, Zavorotny V, Downs B, Zuffada C (2020) Analysis of scattering characteristics from inland bodies of water observed by CYGNSS. Remote Sens Environ 245:111825
36. Panciera R, Walker JP, Jackson TJ, Gray DA, Tanase MA, Ryu D, Monerris A, Yardley H, Rüdiger C, Wu X et al (2013) The Soil Moisture active passive experiments (SMAPEx): toward soil moisture retrieval from the SMAP mission. IEEE Trans Geosci Remote Sens 52(1):490–507
37. Hallikainen MT, Ulaby FT, Dobson MC, El-Rayes MA, Wu LK (1985) Microwave dielectric behavior of wet soil—part 1: empirical models and experimental observations. IEEE Trans Geosci Remote Sens GE 23(1):25–34
38. Unnithan SLK, Biswal B, Rüdiger C (2020) Flood inundation mapping by combining GNSS-R signals with topographical information. Remote Sens 12:3026
39. Zhang S, Ma Z, Li Z, Zhang P, Liu Q, Nan Y, Zhang J, Hu S, Feng Y, Zhao H (2021) Using CYGNSS data to map flood inundation during the 2021 extreme precipitation in Henan Province, China. Remote Sens 13:5181
40. Yang W, Gao F, Xu T, Wang N, Tu J, Jing L, Kong Y (2021) Daily flood monitoring based on spaceborne GNSS-R data: a case study on Henan, China. Remote Sens 13:4561
41. Gerlein-Safdi C, Ruf CS (2019) A CYGNSS-based algorithm for the detection of inland waterbodies. Geophys Res Lett 46(21):12065–12072
42. Taylor KE (2001) Summarizing multiple aspects of model performance in a single diagram. J Geophys Res 106(D7):7183–7192

Chapter 9
GNSS-R Snow Monitoring

Wei Wan and Jie Zhang

Abstract This chapter provides an introduction of Global Navigation Satellite System Reflectometry (GNSS-R) techniques for snow monitoring, with emphasis on their physical principles, retrieval methods, and applications. In this context, GNSS-R is defined as a broader term that encompasses two branches: GNSS-IR, which uses a geodetic receiver with phase and frequency as its primary observables, and a more specific form of GNSS-R that explicitly uses a geodetic receiver with signal strength (or amplitude) as its main observable. We first introduce the mechanisms of GNSS signal propagation in a snow-covered layer and the primary observables used for snow monitoring. We then present methods for GNSS-IR retrieving snow depth, including SNR-based and carrier-phase-based models. Furthermore, we describe methods for retrieving snow density and liquid water content from geodetic GNSS receivers, extending the classic GNSS-IR technique into a customized reflectometry approach (i.e., GNSS-R). Overall, using geodetic GNSS sites to observe multiple snow properties has becoming an efficient and effective approach that supports both scientific research and operational applications. By integrating with other snow monitoring techniques, it is encouraging to consider the production of more comprehensive and precise products for practical use.

1 Introduction

Snow is one of the most active components of the cryosphere, and the maximum snow-covered area during winter occupies nearly 50% of the total land surface area of the Northern Hemisphere [1–3]. Variations in snow substantially influence surface energy balance and runoff generation, thereby affecting hydrological, ecological, and climatic systems [4]. Consequently, accurate estimation of snow hydrology parameters and their temporal changes is fundamental to studies on climate, hydrology, and water resources management.

W. Wan (✉) · J. Zhang
School of Earth and Space Sciences, Peking University, Beijing, China
e-mail: w.wan@pku.edu.cn

K. Yu and N. Zheng (eds.), *Advances in GNSS Reflectometry*, Navigation: Science and Technology 19, https://doi.org/10.1007/978-981-92-2325-1_9

Snow parameters primarily include snow coverage, snow depth, snow density, liquid water content (LWC), and snow water equivalent (SWE). Existing snow monitoring approaches are primarily based on remote sensing and ground measurements. Optical remote sensing sensors achieve high-accuracy snow coverage detection but are affected by cloud contamination and solar illumination conditions [5]. Passive microwave sensors, such as AMSR-E and AMSR2, provide long-term, global retrievals of snow depth but suffer from coarse spatial resolution (>25 km) [6]. Active microwave techniques can improve spatial detail but are restricted by high costs and long revisit periods (>20 days). Ground-based techniques such as ultrasonic or laser sensors provide highly accurate point-scale snow depth observations, including those from meteorological and hydrological networks [7]. Other parameters (e.g., snow density, SWE, and LWC) can be measured by the Snow Pillow or Snow Pack Analysing System (SPA) [8]. However, these methods are labor-intensive, have a limited observation range, and are relatively costly [9].

In recent years, using signals transmitted by the 100+ navigation satellites currently in operation from the four Global Navigation Satellite Systems (GNSS), such as GPS, BeiDou, GLONASS, and Galileo, has emerged as a promising alternative for snow monitoring [10]. The most mature technique is GNSS Interferometric Reflectometry (GNSS-IR), first proposed for soil moisture [11] and later extended to snow depth [12]. The ground-based GNSS antenna receives a composite signal formed by the interference between the direct and land-surface-reflected signals. Its electromagnetic metrics, such as phase, frequency (i.e., phase change rate), and amplitude derived from the Signal-to-Noise Ratio (SNR) observations, are sensitive to geophysical components at the reflecting surface. The "frequency" of the SNR or other observables (such as carrier phase) at low elevation angles has been widely used to estimate snow depth. Unlike conventional active microwave remote sensing methods, GNSS-IR uses existing ground-based GNSS stations and requires no dedicated transmitters, making it a cost-effective approach. The footprint of this pattern is ~1000 m^2 at typical sites [13] and could provide hourly products [14], making it a new tool for complementing existing observing scales between in-situ and satellite data.

Beyond snow depth, other snow parameters can also be retrieved from the GNSS signal. A new technique using the refracted GNSS signal was recently developed using GNSS sensors installed above and below the target medium. It was first investigated to estimate the liquid water content (LWC) of wet snow using signal attenuation [9], and then to estimate SWE by measuring the phase delay of GNSS signals in the snowpack [11, 12]. The specially designed instrument for this technique has been commercialized as SnowSense (https://business.esa.int/projects/snowsense-dp) to measure SWE and LWC [14], and the instrument has been tested under different snowpack conditions in alpine regions [13].

Based on the above summary of the development of GNSS refractometry and interferometric reflectometry, it is clear that these two techniques have distinct advantages for directly observing specific snow properties: the former for SWE and LWC, and the latter for snow depth. Additionally, for a traditional geodetic GNSS receiver, the "amplitude" is a proxy of the signal strength directly related to the dielectric property of the reflected surface, assuming the geodetic GNSS receiver as a customized reflectometry. Several previous studies have investigated the correlation between vegetation water content and the SNR amplitude [14, 15]. More recently, the SNR amplitude has also been explored to retrieve snow density from geodetic GNSS receivers [16].

This chapter introduces the principles, methods, and applications of GNSS-R snow monitoring. It is worth noting that GNSS-R here is a broader term that encompasses two branches: GNSS-IR, which uses a geodetic receiver and phase and frequency as its primary observables, and GNSS-R, which refers explicitly to GNSS-R that uses a geodetic receiver and signal strength (or amplitude) as its main observable. Section 2 introduces the GNSS signal propagation mechanisms in a snow-covered layer and the primary observables used for snow monitoring. Sect. 3 presents methods for GNSS-IR retrieving snow depth, including SNR-based and carrier-phase-based models. Section 4 presents methods for retrieving snow density and liquid water content from geodetic GNSS receivers, extending the classic GNSS-IR technique into customized reflectometry. Finally, Sect. 5 summaries conclusions and discusses the future perspectives.

2 GNSS Signal Propagation Mechanisms in a Snow-Covered Layer

2.1 GNSS Signal Propagation in Dry/Wet Snow Conditions

The propagation of GNSS signals through a snow layer is illustrated in Fig. 1. The reflection of incident waves in a layered medium can be divided into four components: reflection at the air–snow interface, incoherent scattering by ice grains, reflection at the snow–soil interface, and multiple reflections between the air–snow and snow–soil interfaces. For GNSS L-band signals (~1.4 GHz), the incoherent scattering by ice grains is typically negligible. The penetration depth of GNSS signals varies with snow conditions, particularly between dry and wet snow, and different layered reflection characteristics must be considered.

Dry snow consists primarily of ice crystals and air, resulting in very low microwave signal attenuation. In contrast, wet snow comprises ice crystals, air, and varying amounts of liquid water, which strongly absorb electromagnetic waves and cause significant attenuation. As illustrated in Fig. 1, under dry snow conditions, GNSS signals can generally penetrate through the snowpack and reach the underlying soil layer. The GNSS antenna positioned above the snow primarily captures a composite reflected signal from both the air–snow and snow–soil interfaces, consistent with the dual-interface reflection theory.

Under wet snow conditions, GNSS signal penetration into the snowpack depends on both LWC and snow depth. When the LWC or snow depth is relatively high,

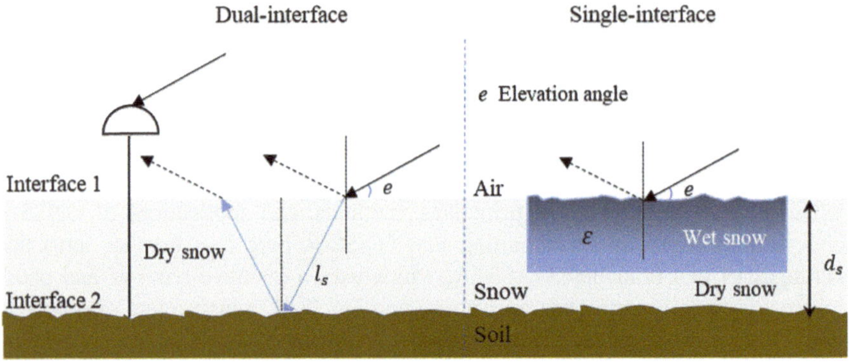

Fig. 1 The observation geometry of GNSS signal propagation through a snow layer

the signal is limited to the air–snow interface, corresponding to a single-interface reflection, as it cannot penetrate the snowpack to reach the soil layer. Conversely, when the LWC or snow depth is relatively low, the signal can penetrate the snowpack and reflect at the air–snow–soil interface, consistent with dual-interface reflection theory.

As GNSS signals propagate through snow, electromagnetic attenuation occurs. The signal power after passing through snow can be expressed as [9]:

$$p_s = p_a e^{-2\gamma l_s} \tag{1}$$

$$\gamma = \varepsilon_s'' \varepsilon_0 \pi f \sqrt{\frac{\mu_0}{\varepsilon_s' \varepsilon_0}} \tag{2}$$

$$l_s = \frac{d_s}{\cos(90° - e)} \tag{3}$$

Where p_a is the GNSS signal power in free space, p_s is the attenuated power after traveling a distance l_s through snow, γ is the attenuation coefficient, related to the real and imaginary parts of the complex dielectric constant of snow, ε_s' is the real part of snow permittivity, ε_s'' is the imaginary part of the snow permittivity, $\varepsilon_0 = 8.8542 \times 10^{-12}\ F/m$ is the vacuum permittivity, $\mu_0 = 4\pi \times 10^{-7}\ NA^{-2}$ is the vacuum permeability, f is the signal frequency, e is the satellite elevation angle, and d_s is the snow depth.

The maximum penetration depth is defined as the distance at which the signal power attenuates to 1/e of its initial value (i.e., $p_a e^{-1}$). The corresponding theoretical maximum penetration depth can be expressed as:

$$d_s = \frac{\sin(e)}{\varepsilon_s'' 2\pi f} \sqrt{\frac{\varepsilon_s'}{\mu_0 \varepsilon_0}} \tag{4}$$

This relationship quantitatively describes how signal penetration is constrained by snow dielectric properties and observation geometry, providing the theoretical basis for applying the single- or dual-interface theory under varying snow conditions.

2.2 Observables Used for Snow Monitoring

The geodetic GNSS receiver receives an interferometric signal composed of direct and reflected signals, forming an interferometric pattern in the SNR observations. The snow depth can be determined by estimating the frequency of SNR observable measurements collected by the geodetic GNSS receiver at low elevation angles. Changes in reflector height directly affect frequency, i.e., the greater the height, the higher the frequency, as shown in Fig. 2a. Unlike snow depth, snow density and LWC are directly related to snow permittivity and the resulting power reflectivity, which the SNR amplitude can generally reflect, i.e., higher snow density results in greater amplitude, as shown in Fig. 2b.

The SNR is defined as the ratio of interferometric signal power (P_i) to noise power (P_n) and is represented as $SNR = P_i/P_n$. The noise power, P_n, is related to the noise bandwidth, noise density, and unitless factor.

As described in the forward model proposed by Nievinski and Larson [17], the GNSS multipath SNR is expressed as:

$$SNR = P_i/P_n = \left(P_d + P_r + 2(P_d P_r)^{\frac{1}{2}} \cos\psi\right)/P_n \tag{5}$$

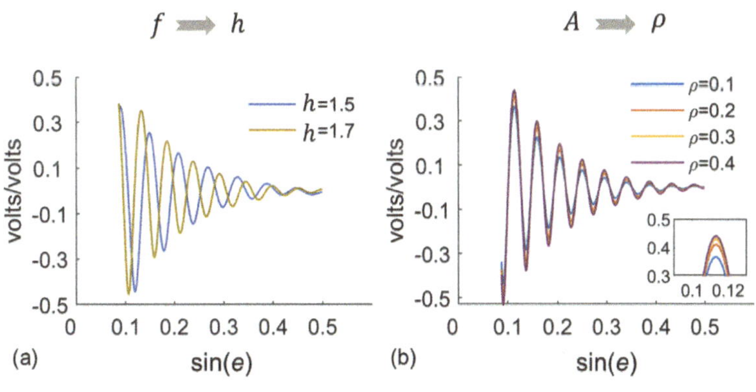

Fig. 2 The observables used for snow depth, snow density, and LWC retrieval, along with the frequency of the SNR oscillation term (f) for snow depth and the amplitude of the SNR oscillation term (A) for snow density and LWC. **a** Changes of f with the reflector height (h), the greater the height, the higher the frequency; **b** changes of A with snow density (ρ), higher snow density results in greater amplitude

Where P_d and P_r represent the power of the direct and reflected signals, respectively, and ψ is the phase of interferometric signal.

Since $P_i = A_i^2$, where A_i is the amplitude of the interferometric signal, Eq. (5) can also be expressed as [17]:

$$SNR = \frac{A_i^2}{P_n} = \frac{A_d^2}{P_n} + 2\frac{A_d A_r}{P_n}\cos\psi + \frac{A_r^2}{P_n} \tag{6}$$

Here, A_d is the amplitude of the direct signal, A_r is the amplitude of the reflected signal. The trend term (SNR_{trend}) can be expressed as follows [17]:

$$SNR_{trend} = \frac{A_d^2}{P_n} + \frac{A_r^2}{P_n} \tag{7}$$

Low-order polynomial fitting is executed to identify the trend term from the raw SNR oscillation. After removing SNR_{trend} from Eq. (6), the residual oscillation term ($SNR_{detrend}$) is expressed as follows:

$$SNR_{\mathrm{det}rend} = 2\frac{A_d A_r}{P_n}\cos\psi \tag{8}$$

Where ψ is a function of the receiver satellite geometry, phase difference between reflected and direct signals [18].

(1) From f to h_{snow}

The phase difference ψ is given by:

$$\psi = \frac{2\pi}{\lambda}2h\sin(e) \tag{9}$$

where h is the effective reflector height, e is satellite elevation angle, λ is the GNSS signal wavelength. The derivative of ψ with respect to e is:

$$\frac{d(\psi)}{d(e)} = \frac{2\pi}{\lambda}2h\cos(e) \tag{10}$$

Alternatively, its derivative with respect to sin(e) is constant:

$$\frac{d(\psi)}{d(\sin e)} = \frac{4\pi h}{\lambda} \tag{11}$$

It can be seen from (11) that the frequency of the $SNR_{detrend}$ is:

$$f = \frac{2h}{\lambda} \tag{12}$$

Thus, the effective reflector height (h) can be estimated by determining the frequency through the spectrum analysis of the SNR time series as a function of satellite elevation angle.

The effective reflector height (h) represents the vertical distance between the antenna and the reflecting surface. The snow depth (h_{snow}) is determined by tracking the change in the effective multipath reflector height (the snow surface) relative to the snow-free reference surface. The detailed retrieval methods for snow depth are discussed in Sect. 3.

(2) From A to ρ

The snow density and LWC are inferred from variations in the SNR amplitude. However, two primary challenges exist in quantitatively retrieving snow density or LWC using SNR amplitude to derive power reflectivity. It is difficult to separate the reflected amplitude from the raw SNR observation, and the amplitude also attenuates with increasing GNSS satellite elevation angle, as shown in Fig. 2b. Overcoming this issue by taking an indirect path, the amplitude ratio α ($\alpha = \frac{A_r}{A_d}$) is introduced as a proxy of the multipath reflection coefficient, as previously defined by Larson and Small [19].

The theoretical α was deduced based on signal power: $\alpha = \frac{A_r}{A_d} = (P_r/P_d)^{\frac{1}{2}}$, where P_d and P_r represent power of the direct signal and the reflected signal, respectively, which is given by Nievinski and Larson [17]:

$$\begin{cases} P_d = P_0 \mathrm{G}_\mathrm{d}^\mathrm{R} \mathrm{W}_\mathrm{d}^2 \\ P_r = P_0 |XS\mathrm{W}_\mathrm{r}|^2 \end{cases} \tag{13}$$

where P_0 represents incident power of the direct signal at the receiving antenna, G_d^R is the antenna gain for the direct signal. W_d and W_r are the Woodward ambiguity functions for the direct and reflected signals, respectively, and here these two variables are assumed to be equal. The term S accounts for the impact of surface roughness. X represents the reflection coefficient, incorporating the effects of antenna gain.

Thus, the theoretical amplitude ratio (α_t) can be expressed as:

$$\alpha_t = \frac{A_r}{A_d} = \left(\frac{P_r}{P_d}\right)^{\frac{1}{2}} = \left(\frac{|XS|^2}{\mathrm{G}_\mathrm{d}^\mathrm{R}}\right)^{\frac{1}{2}} \tag{14}$$

The term S accounts for the impact of surface roughness, which is ignored during the dry snow period. Here, the dry snow period is characterized as the period during which air temperatures generally remain below 0 °C. In contrast, periods with air temperatures rising to around or above 0 °C, where melt processes can occur, are identified as wet-snow. As the snow melts, the roughness of its surface will change, which makes the term S cannot be ignored. In geodetic GNSS receivers, the parameter X in the reflected signal is determined by the composite reflection coefficients of right- and left-hand circular polarizations, which are functions of the dielectric constants of the air–snow and snow–soil interfaces.

For dual-interface reflections, X can be expressed as:

$$X = R_{RR}^{comp} \sqrt{\mathrm{G}_\mathrm{r}^\mathrm{R}} + R_{RL}^{comp} \sqrt{\mathrm{G}_\mathrm{r}^\mathrm{L}} \tag{15}$$

where R_{RR}^{comp} and R_{RL}^{comp} are the composite reflection coefficients of right-hand circularly polarized (RHCP) and left-hand circularly polarized (LHCP) signal components, respectively, and G_r^L and G_r^R are the corresponding antenna gains.

The composite reflection coefficient (R_x^{comp}) for the dual-interface model [20, 21] is as follows:

$$R_x^{comp} = \frac{R_x^{(1)} + R_x^{(2)} e^{-2\gamma_2 d \cos\theta_2}}{1 + R_x^{(1)} R_x^{(2)} e^{-2\gamma_2 d \cos\theta_2}} \tag{16}$$

where $R_x^{(1)}$ and $R_x^{(2)}$ denote the x-polarized reflection coefficients at the air-snow and snow-soil interfaces, respectively, where x represents RHCP or LHCP. The parameter d is the snow depth and θ_2 is the incidence angle at the snow-soil interface. $\gamma_2 = \frac{j2\pi}{\lambda}\sqrt{\varepsilon_{\text{snow}}}$ is the propagation constant, and ε_{snow} is the permittivity of the snow layer. The term $e^{-\gamma_2 d \cos\theta_2}$ represents the propagation factor within the snow layer. The factor is squared because the signal traverses the snow layer twice, once during downward propagation and once during upward propagation. For dry snow, the dielectric loss is low, resulting in a negligible imaginary part of the snow permittivity and a small attenuation constant. The quantity $2\gamma_2 d \cos\theta_2$ is close to zero and therefore the exponential attenuation term can be approximated as $e^{-2\gamma_2 d \cos\theta_2} \approx 1$.

The x-polarized reflection coefficients at the air-snow and snow-soil interfaces (i.e., $R_x^{(1)}$ and $R_x^{(2)}$ in Eq. (16) can be expressed as:

$$\begin{cases} R_{RR}^{(k)} = \frac{1}{2}\left(R_v^{(k)} + R_h^{(k)}\right) \\ R_{RL}^{(k)} = \frac{1}{2}\left(R_v^{(k)} - R_h^{(k)}\right) \end{cases} \tag{17}$$

where $R_v^{(k)}$ and $R_h^{(k)}$ are the vertical and horizontal polarized reflection coefficients [21, 22], respectively. In the snow case, when at the air-snow interface, $k = 1$; when at the snow-soil interface, $k = 2$.

For single-interface reflections, X can be expressed as:

$$X = R_{RR}^{(1)}\sqrt{\mathrm{G_r^R}} + R_{RL}^{(1)}\sqrt{\mathrm{G_r^L}} \tag{18}$$

Note that the observed amplitude ratio (using α_0 to represent) often deviates from the theoretical value α_t due to antenna gain and the impact of underlying soil moisture, which makes it an inaccurate quantity for calculating snow permittivity. To address this issue, a correction strategy is required to minimize their differences. The detailed processing approach and the resulting corrected α value used for snow density and LWC retrieval are described in Sect. 4.

3 GNSS-IR Snow Depth Retrieval

As shown in Sect. 2.2, snow depth is retrieved from the difference in reflector heights between the snow surface and the snow-free surface. Both the SNR and the carrier phase (CP) from GNSS RINEX raw data can be used to calculate the frequency (f) and the resulting reflector height (h).

3.1 The SNR Model

The principle of the SNR model is to establish a linear relationship between the oscillation frequency of the SNR-observed reflected signal and the height of the reflecting surface [10]. This model was later derived into several variants: e.g., the triple-frequency SNR combination model (SNR_COM) [23], the SNR model based on raw SNR sequences [24], the SNR model based on horizontal polarization antenna [25], the SNR model considering the influence of construction facilities [26], and the SNR model considering the influence of terrain [27].

The geometry and principle of the SNR model are shown in Fig. 3. As shown in Fig. 3a, the snow depth (h_{snow}) is calculated using a simple equation:

$$h_{snow} = h_0 - h \tag{19}$$

where h_0 is the reflector height of the snow-free surface, and h is the reflector height of the snow-covered surface. The approaches to derive h_0 and h are similar, with Fig. 3 b1 ~ b3 showing the general technical process. First, the time series of GNSS SNR observation is shown as a function of sine (elevation angle), and the trend signal is removed using the polynomial fitting method. The remaining is treated as the contribution of the reflected signal from the land surface. Second, the reflected signal is converted from dB-Hz to Volts/Volts. Third, the Lomb-Scargle periodogram (LSP) analysis is applied to the reflected signal curve to determine the dominant frequency of the transformation. The Peak-to-Noise Ratio (PNR) of the LSP is usually set to exceed a certain threshold to filter out low-quality satellite tracks. The h_0 or h can be calculated by [10]:

$$h = \lambda f / 2 \tag{20}$$

where λ is the wavelength of the GNSS signal and f is the dominant frequency.

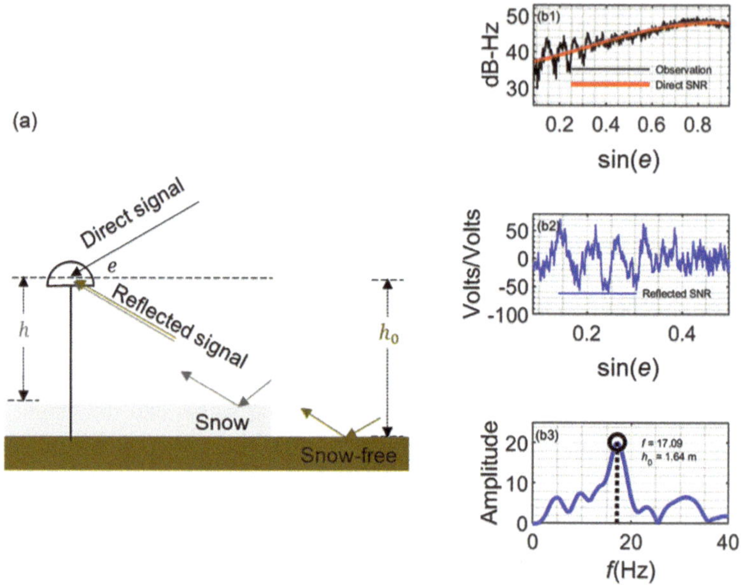

Fig. 3 Geometry and principle of the SNR model. **a** The geometry of the direct and reflected signal over the snow surface and the bare soil surface when there is no snow on the ground; (b1) example of the recorded GNSS SNR data and the removal of the direct signal with a second-order polynomial; (b2) residual of (b1) below elevation angle (*e*) of 30°, converted from dB to linear units (for simplicity, Volts); (b3) Lomb-Scargle analysis of (b2) to find out the dominant frequency of the transformation and the resulting reflector height

3.2 *The Carrier Phase Model*

The carrier-phase combination model was initially proposed to estimate snow depth when no SNR data were available in the raw GNSS observation file [28]. The initial form of this model used the geometry-free linear combinations of the phase measurements (L4), and [29, 30] extended the model to include triple-frequency carrier phase observations (F3) as well as the combination of pseudorange and carrier phase for dual-frequency signals (F2C).

The main formulas and the applicability of the five models mentioned above are listed in Table 1, and Table 2 further explains the meanings of the variables for the models in Table 1. The SNR, L4, and F2C models are suitable for all sites because the input observables are available in GNSS raw data. The SNR model has been verified to have higher accuracy than the L4 and F2C models [31]. The applicability of the SNR_COM and F3 models is limited because most of the GNSS sites do not contain three SNR or CP observables in a single raw data file.

Table 1 Snow depth models and their corresponding formulas

Model	Main formulas
SNR [10]	$SNR^2 = A_c^2 = A_d^2 + A_r^2 + 2A_d A_r \cos Q$ $A_i = A\cos\left(\frac{4\pi h}{\lambda}\sin e + \varphi\right)$ $f = \frac{2h}{\lambda}$
SNR_COM [23]	$SNR_{com,\,i} = [SNR_{1,\,i} SNR_{2,\,i} SNR_{3,\,i}]$
L4 [28]	$L_1 = \rho + I(f_1) + T + M_{L1} + \text{noise}_1$ $L_2 = \rho + I(f_2) + T + M_{L2} + \text{noise}_2$ $L_4 = L_1 - L_2 = I(f_1) - I(f_2) + M_{L1} - M_{L2} + \text{noise}_1 - \text{noise}_2$
F3 [29]	$L_3 = \rho + I(f_3) + T + M_{L3} + \text{noise}_3$ $f_3 = \lambda_3^2(L_1 - L_2) - \lambda_2^2(L_1 - L_3) + \lambda_1^2(L_2 - L_3)$
F2C [30]	$c_1 = \rho + I(f_1) + T + M_{c1}$ $f_{2c} = \frac{\lambda_1^2+\lambda_2^2}{\lambda_1^2-\lambda_2^2}(c_1 - L_1) - \frac{2\lambda_1^2}{\lambda_1^2-\lambda_2^2}(c_1 - L_2)$

Table 2 Meanings of variables for the models in Table 1

Variables	Meanings
A_d	Amplitudes of the direct signal
A_r	Amplitudes of the reflected signal
A_i	Amplitudes of the synthetic signal
$\cos Q$	Cosine value of the angle between the direct signal and the reflected signal
λ	Carrier wavelength
e	Satellite elevation angle
h	Vertical reflection distance
f	Frequency of GNSS multipath reflection signal
φ	Phase values less than an entire period
$SNR_{com,\,i}$	SNR observation values of triple-frequency
λ_i	Wavelength
ρ	The true geometric range between the satellite and receiver
T	Tropospheric delay
$I(f_i)$	Ionospheric delay for the signal
M_{Li}	Multipath error for the signal
$noise_i$	Integer ambiguities for the signal
L_4	Multipath error sequence of L4
f_3	Multipath error sequence of F3
f_{2c}	Multipath error sequence of F2C

4 Snow Density and Liquid Water Content Retrieval Using Geodetic GNSS Receivers

As discussed in Sect. 2.2, the amplitude ratio of the reflected-to-direct signals observed at the snow surface can be used to retrieve snow density and LWC. However, due to the influence of antenna gain and the underlying surface conditions, the observed initial amplitude ratio (α_0) deviates from the theoretical amplitude ratio (α_t).

The primary factors affecting α_0 differ between snow conditions. During dry snow periods, the snow surface is generally smooth, and the attenuation of the reflected signal due to surface roughness can be neglected. In this case, the main uncertainty arises from variations in soil moisture. During wet snow periods, the roughness of its surface increases as melting progresses, ranging from millimeters to meters, and it cannot be ignored.

To address this issue, we proposed a step-wise strategy to correct systematic errors from antenna gain and errors from soil moisture or roughness error from the snow surface in α_0 to achieve the finalized corrected α, which is used to retrieve the snow permittivity and the resulting snow density/LWC (as described in Sect. 4.3). The calculation of the initial amplitude ratio (α_0) from SNR observations (Sect. 4.1), together with its subsequent correction to obtain the finalized amplitude ratio α (Sect. 4.2) is collectively summarized in the novel GNSS Signal Amplitude Ratio Model (GSARM). Beyond snow applications, GSARM can also be extended to retrieve other geophysical parameters linked to permittivity.

4.1 The GSARM Model Part I: Calculation of the Initial Amplitude Ratio (α_0) from SNR Observations

The amplitude ratio α ($\alpha = \frac{A_r}{A_d}$) is introduced as a proxy of the multipath reflection coefficient, as previously defined by Larson and Small [19]. To calculate α using the waveform (Fig. 4), $R_{\frac{r}{d}}$ is defined as the ratio of the detrended oscillation term to the trend term. The following Equation can be achieved by calculating the ratio using Eqs. (7) and (8):

$$R_{\frac{r}{d}} = \frac{SNR_{detrend}}{SNR_{trend}} = \frac{2A_dA_r}{A_d^2+A_r^2}\cos\psi = \frac{2\alpha}{1+\alpha^2}\cos\psi \tag{21}$$

Next, using the Hilbert transform [15, 32], we obtain the envelope, i.e., the instantaneous amplitude:

$$\hat{R_{\frac{r}{d}}} = H\left[R_{\frac{r}{d}}\right] \tag{22}$$

The representation of the upper envelope magnitude and phase is as follows [32]:

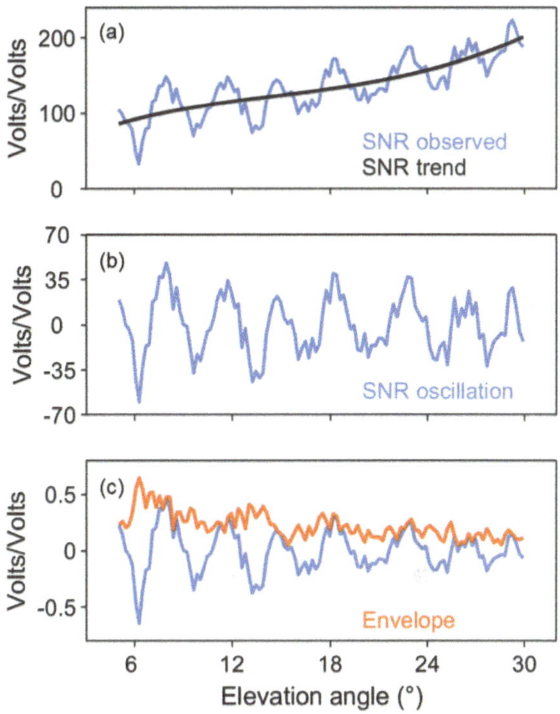

Fig. 4 The process from the SNR waveform to the envelope of the Hilbert transform; **a** the original signal and SNR trend term received by the receiver; **b** the SNR oscillation term after removing the trend term; **c** the upper envelope of normalized SNR oscillation term

$$|R_{\frac{r}{d}}| - \sqrt{R_{\frac{r}{d}}^2 + (\hat{R_{\frac{r}{d}}})^2} \tag{23}$$

$$\psi = a\tan(\hat{R_{\frac{r}{d}}} / R_{\frac{r}{d}}) \tag{24}$$

Finally, by substituting Eq. (24) into Eq. (21) and combining it with Eq. (23), the initial amplitude ratio (α_0) can be obtained, as shown in Fig. 4.

$$\frac{A_r}{A_d} = \alpha_0 = \left(1 - \sqrt{1 - |R_{\frac{r}{d}}|^2}\right) / |R_{\frac{r}{d}}| \tag{25}$$

4.2 *The GSARM Model Part II: Correction of the Initial Amplitude Ratio (α_0) to Derive the Finalized Amplitude Ratio α*

The theoretical amplitude ratio (α_t) in Eq. (14) is used as the benchmark for the systematic and land-surface error corrections of the initial amplitude ratio (α_0). It is evident that the α_t is influenced by the antenna gain pattern, the Fresnel reflection coefficient, and surface roughness. These factors originate from different sources and can be broadly categorized into two types of errors affecting α_0: systematic errors related to the GNSS receiver system and land-surface errors associated with the surface or subsurface properties. Accordingly, a two-step correction strategy is adopted to derive the finalized amplitude ratio (α).

1. Systematic errors correction

Geodetic GNSS antennas exhibit non-uniform gain patterns that vary with elevation angle and differ across antenna types. As a result, the observed amplitude ratio (α_0) contains elevation angle distortions. To mitigate this effect, the correction is applied to remove the elevation-angle bias and to adjust the absolute level of α_0.

2. Land-surface error correction

Under dry snow conditions, α_t is primarily controlled by the composite Fresnel reflection coefficient, which is influenced by the permittivity of the underlying soil. Soil permittivity is primarily influenced by soil moisture content. We therefore quantify the expected contribution of soil moisture to α_t and subtract this component from α_0 to reduce soil-moisture related bias.

Under wet snow conditions, the snow is deep enough or the LWC is high enough, the signal cannot penetrate the snow layer and reach the soil. In this case, there is no need to consider the influence of the soil on α_t, and only the influence of the snow surface roughness needs to be considered for α_t. Whether the GNSS signal penetrates the snow layer depends on the snow depth and other snow properties (see Sect. 2.1). If the signal does not penetrate the snowpack, a single-interface reflection theory is applied; if it does penetrate, a dual-interface reflection theory is used to account for the underlying soil. The correction steps for the initial amplitude ratio (α_0) mainly involve the correction of systematic errors and roughness errors.

The detailed theoretical derivations and experimental validations can be found in [16] for dry snow and [33] for wet snow.

4.3 Snow Density and Liquid Water Content Retrieval

After using the Hilbert transform to calculate the initial α_0 of the snow season from the observed SNR values, the systematic error corrections and land surface error corrections are applied to adjust α_0 to the finalized amplitude ratio (α), making α match the theoretical amplitude ratio (α_t).

Subsequently, the snow permittivity is derived from α using Eqs. (14)–(18). The snow density is then calculated based on the snow permittivity Equation in Table 3. Note that the calculation of snow permittivity differs for dry and wet snow. Also, as shown in Fig. 5, at low elevation angles, curves corresponding to different permittivity values are more easily distinguished (Table 4).

Table 3 Equations to calculate the permittivity of dry and wet snow

Category	Parameter	Equation
Dry snow	Snow permittivity	$\varepsilon_{snow} = \varepsilon_s' - i\varepsilon_s''$ $\varepsilon_s' = 1 + 1.7\rho_{ds} + 0.7{\rho_{ds}}^2$ [34] $\varepsilon_s'' =$ $1.59^*10^{6*}(0.52\rho_{ds} + 0.62{\rho_{ds}}^2) * (f^{-1} + 1.23^*10^{-14*}sqrt(f))e^{0.036T}$ [35]
Wet snow	Snow permittivity	$\varepsilon_s' = 1 + 1.92^*10^{-3}\rho_{ws} + 4.4^*10^{-7}\rho_{ws}^2$ $+1.87^*10^{-1}L_w + 4.5^*10^{-3}L_w^2$ [36] $\varepsilon_s'' = \frac{f}{10^9 Hz}(1.0^*10^{-3}L_w + 8.0^*10^{-5}L_w^2)\varepsilon_w''$ [34]
	Snow density	$\rho_{ws} = \rho_{ds0} + 0.01L_w\rho_w$ [36] $\rho_w = 1000$ kg/m^3

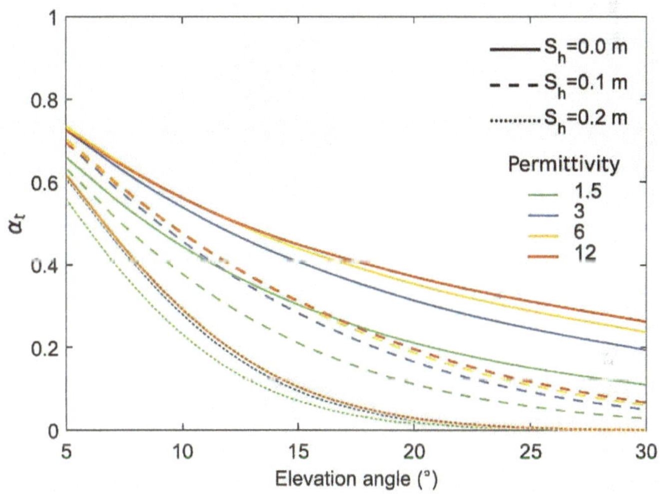

Fig. 5 Variations of the simulated theoretical amplitude ratio (α_t) along with the elevation angles under different land surface permittivity and roughness. S_h: surface height standard deviation (in meters)

5 Conclusions and Future Perspectives

Using geodetic GNSS sites to observe multiple snow properties provides an efficient and effective approach that supports both scientific research and operational applications. GNSS Interferometric Reflectometry (GNSS-IR) for snow depth monitoring has evolved from a theoretical concept into a mature operational product. This

Table 4 The meanings of the variables in Table 3

Variables	Meanings
ε_s'	The real part of snow permittivity
ε_s''	The imaginary part of snow permittivity
L_w	The liquid water content (LWC, volumetric water content in %) of wet snow
ρ_{ws}	The snow density of wet snow
ρ_{ds}	The snow density of dry snow
ρ_{ds0}	The average maximum dry snow density at the study site
ρ_w	Density of water
f	Microwave frequency
T	The temperature of snow in degrees Celsius

technique bridges the scale gap between in-situ measurements and satellite observations, facilitating the generation of long-term snow depth datasets, particularly in remote and data-sparse regions. Beyond snow depth, geodetic GNSS receivers can provide complementary measurements of snow density, liquid water content (LWC), and snow water equivalent (SWE). These applications extend beyond classic GNSS-IR, positioning geodetic GNSS receivers as a form of customized reflectometry. The technique has also demonstrated promise in monitoring snow disasters [37] and snow droughts/deluges [38].

Looking beyond snow, the functionality of a single geodetic GNSS site is becoming increasingly integrated, enabling the monitoring of multiple environmental variables such as water vapor, soil moisture, vegetation, and snow. This represents a significant expansion beyond its original purpose. The future development of ground-based GNSS-R for environmental monitoring is advancing in several directions: toward low-cost solutions [39], dedicated instruments designed specifically for remote sensing applications [9, 11], and more accessible platforms like smartphones for observing atmospheric and ocean/land components [40, 41]. Another emerging trend is the integration of GNSS-R with other microwave remote sensing techniques, such as radiometers and scatterometers, to produce more comprehensive and precise products for practical applications.

References

1. Frei A, Robinson DA (1999) Northern Hemisphere snow extent: regional variability 1972–1994. Int J Climatol 19(14):1535–1560
2. Armstrong RL, Brodzik MJ (2001) Recent Northern Hemisphere snow extent: a comparison of data derived from visible and microwave satellite sensors. Geophys Res Lett 28(19):3673–3676
3. Robinson DA, Dewey KF, Heim RR (1993) Global snow cover monitoring: an update. Bull Am Meteorol Soc 74(9):1689–1696
4. Henderson GR, Peings Y, Furtado JC et al (2018) Snow–atmosphere coupling in the Northern Hemisphere. Nat Clim Chang 8(11):954–963

5. Dai L, Che T, Ding Y et al (2017) Evaluation of snow cover and snow depth on the Qinghai–Tibetan Plateau derived from passive microwave remote sensing. Cryosphere 11(4):1933–1948
6. Lievens H, Demuzere M, Marshall H-P et al (2019) Snow depth variability in the Northern Hemisphere mountains observed from space. Nat Commun 10(1):4629
7. Kinar NJ, Pomeroy JW (2015) Measurement of the physical properties of the snowpack. Rev Geophys 53(2):481–544
8. Fiel R, Sommer W, Rentsch T (2009) SPA—snow pack analysing system. In: Paper presented at the ISSW 09 Europe: international snow science workshop, proceedings, Davos, Switzerland
9. Koch F, Prasch M, Schmid L et al (2014) Measuring snow liquid water content with low-cost GPS receivers. Sensors 14(11):20975–20999
10. Larson KM, Gutmann ED, Zavorotny V et al (2009) Can we measure snow depth with GPS receivers. Geogr Res Lett 36:L17502
11. Koch F, Henkel P, Appel F et al (2019) Retrieval of snow water equivalent, liquid water content, and snow height of dry and wet snow by combining GPS signal attenuation and time delay. Water Resour Res 55(5):4465–4487
12. Steiner L, Meindl M, Marty C et al (2020) Impact of GPS processing on the estimation of snow water equivalent using refracted GPS signals. IEEE T Geosci Remote 58(1):123–135
13. Capelli A, Koch F, Henkel P et al (2022) GNSS signal-based snow water equivalent determination for different snowpack conditions along a steep elevation gradient. Cryosphere 16(2):505–531
14. Wan W, Larson KM, Small EE et al (2015) Using geodetic GPS receivers to measure vegetation water content. GPS Solut 19(2):237–248
15. Li J, Yang D, Wang F et al (2023) A new algorithm for measuring vegetation growth using GNSS interferometric reflectometry. IEEE J-Stars 16:1033–1041
16. Zhang J, Wan W, Liang H et al (2025) Feasibility demonstration of using the signal-to-noise ratio observations from geodetic GNSS receivers to retrieve dry snow density. Water Resour Res 61(4):e2023WR036657
17. Nievinski FG, Larson KM (2014) Forward modeling of GPS multipath for near-surface reflectometry and positioning applications. GPS Solut 18(2):309–322
18. Bilich A, Larson KM (2007) Mapping the GPS multipath environment using the signal-to-noise ratio (SNR). Radio Sci 42(6)
19. Larson KM, Small EE (2014) Normalized microwave reflection index: a vegetation measurement derived from GPS networks. IEEE J-Stars 7(5):1501–1511
20. Chen Q, Won D, Akos DM (2017) Snow depth estimation accuracy using a dual-interface GPS-IR model with experimental results. GPS Solut 21(1):211–223
21. Ulaby FT, Long DG, Blackwell W et al (2015) Microwave radar and radiometric remote sensing. The University of Michigan Press
22. Chen Q (2016) Optimization of GPS interferometric reflectometry for remote sensing. Doctoral dissertation. University of Colorado, Boulder
23. Zhou W, Liu L, Huang L et al (2019) A new GPS SNR-based combination approach for land surface snow depth monitoring. Sci Rep 9(1):3814
24. Peng X, Wan W, Chen X (2016) Using GPS SNR data to estimate soil moisture variations: proposing a new interference model. In: 2016 IEEE international geoscience and remote sensing symposium (IGARSS), 10–15 July 2016, pp 4819–4822
25. Chen Q, Won D, Akos DM (2014) Snow depth sensing using the GPS L2C signal with a dipole antenna. EURASIP J Adv Signal Process 1:1–10
26. Vey S, Güntner A, Wickert J et al (2016) Monitoring snow depth by GNSS reflectometry in built-up areas: a case study for Wettzell, Germany. IEEE J Sel Top Appl Earth Obs Remote Sens 9(10):4809–4816
27. Zhang S, Wang X, Zhang Q (2017) Avoiding errors attributable to topography in GPS-IR snow depth retrievals. Adv Space Res 59(6):1663–1669
28. Ozeki M, Heki K (2012) GPS snow depth meter with geometry-free linear combinations of carrier phases. J Geod 86(3):209–219

29. Yu K, Ban W, Zhang X et al (2015) Snow depth estimation based on multipath phase combination of GPS triple-frequency signals. IEEE Trans Geosci Remote Sens 53(9):5100–5109
30. Yu K, Li Y, Chang X (2018) Snow depth estimation based on combination of Pseudorange and carrier phase of GNSS dual-frequency signals. IEEE Trans Geosci Remote Sens 57:1–12
31. Liu S, Zhang J, Wan W et al (2022) A comprehensive evaluation of utilizing BeiDou data to estimate snow depth from two ground-based stations. GPS Solut 26(4):118
32. Johansson M (1999) The hilbert transform. Mathematics Master's Thesis. Växjö University, Suecia, p 19
33. Liang C, Wan, W, Liu, S, Zhang, J, Ma, W (2026) Exploring the feasibility of using the signal-to-noise ratio observations from geodetic GNSS receivers to retrieve liquid water content of snow. (under review).
34. Sihvola A, Tiuri M (1986) Snow fork for field determination of the density and wetness profiles of a snow pack. IEEE T Geosci Remote 24(5):717–721
35. Tiuri ME, Sihvola AH, Nyfors EG et al (1984) The complex dielectric-constant of snow at microwave-frequencies. IEEE J Ocean Eng 9(5):377–382
36. Denoth A, Foglar A, Weiland P et al (1984) A comparative-study of instruments for measuring the liquid water-content of snow. J Appl Phys 56(7):2154–2160
37. Zhang J, Liu S, Liang H et al (2022) Using GNSS-IR snow depth estimation to monitor the 2022 early February snowstorm over southern China. Remote Sens-Basel 14(18):4530
38. Zhang H, Zhang Y, Chang M et al (2025) Characterizing snow droughts and deluges in the Sacramento River Basin, California using GNSS-derived snow depth (2009–2023). J Hydrol 664:134336
39. Zheng N, Chen L, Chai H et al (2022) Research on GNSS-IR height measurement performance of smartphone platform. In: China satellite navigation conference, pp 123–135. Springer
40. Karegar MA, Kusche J, Geremia-Nievinski F et al (2022) Raspberry pi reflector (RPR): a low-cost water-level monitoring system based on GNSS interferometric reflectometry. Water Resour Res 58(12):e2021WR031713
41. Smith J, Kast A, Geraschenko A et al (2024) Mapping the ionosphere with millions of phones. Nature 635(8038):365–369

Chapter 10
Snow Depth Retrieval Based on Fusion Algorithms Using Multi-GNSS Interferometric Reflectometry under Complex Terrain Conditions

Naiquan Zheng

Abstract This chapter introduces two multi-GNSS fusion approaches for snow depth retrieval based on GNSS interferometric reflectometry (GNSS-IR), which effectively address the limitations of single-system retrieval. Under conditions of gentle terrain, a robust estimation algorithm was applied to fuse snow depth retrievals derived from different signal-to-noise ratio (SNR) types and GNSS constellations. At the SG27 site, the fused results achieved a correlation coefficient (R), root mean square error (RMSE), and mean error (ME) of 0.759, 3.7 cm, and 1.4 cm, respectively, relative to the nearest in situ observations. These results indicate good agreement and confirm the reliability of the robust estimation approach. In areas with pronounced topographic variability, terrain tilt correction (TTC) and Helmert variance component estimation (HVCE) were employed to fuse snow depth retrievals from multiple GNSS systems. Validation at stations P351 and P676 showed that the retrieved snow depths exhibited R values above 0.9, RMSE below 20.0 cm, and absolute ME less than 15.0 cm compared with in situ measurements. Considering that the true snow depth at these two sites is around 1.2 m, these error magnitudes correspond to relative errors of less than $\sim 15\%$, thereby demonstrating the robustness of the proposed methods.

Keywords Snow Depth Estimation · GNSS-IR · Terrain Tilt Correction · Multi-GNSS · Helmert Variance Component Estimation · Robust Estimation

---

N. Zheng (✉)
College of Geodesy and Geomatics, Shandong University of Science and Technology, Qingdao, China

K. Yu and N. Zheng (eds.), *Advances in GNSS Reflectometry*, Navigation: Science and Technology 19, https://doi.org/10.1007/978-981-92-2325-1_10

1 Introduction

Snow depth, as one of the key indicators of global climate observations, provides important insights into climate variability. Current observation methods include manual measurements, ultrasonic snow depth sensors, laser-based instruments, satellite remote sensing, and GNSS interferometric reflectometry (GNSS-IR). Manual measurements using a ruler are intuitive, inexpensive, and durable, but they are labor-intensive, difficult to perform under harsh weather conditions. Ultrasonic sensors offer advantages such as long-term time series, high accuracy, compact design, continuous tracking, and solar-powered operation. However, their relatively high cost prevents widespread deployment. Laser snow depth instruments, which provide high accuracy, continuous observations, real-time display, and automatic recording, are gaining attention but still suffer from high cost. Both ultrasonic and laser instruments are also unstable under severe weather conditions and often require manual intervention. In recent years, satellite altimetry has emerged as a reliable approach to snow depth monitoring, with the advantage of abundant data and wide spatial coverage, thereby complementing meteorological stations and in situ field measurements. Nevertheless, its relatively long revisit cycle results in low temporal resolution. Overall, traditional snow depth measurement techniques remain costly, environmentally constrained, and insufficient for current monitoring needs. With the rapid expansion of GNSS applications, ground-based GNSS-IR has become a promising alternative for snow depth monitoring. GNSS-IR exploits the interference between direct and reflected signals to estimate the height of the reflecting surface, where the interference manifests as oscillations in the signal-to-noise ratio (SNR). By modeling the relationship between SNR frequency and reflector height, snow depth surrounding a GNSS station can be retrieved. Owing to its low site requirements, continuous operation, cost efficiency, and high accuracy, GNSS-IR has attracted increasing attention in recent years.

Although significant progress has been made in snow depth retrieval in previous studies, two major challenges remain. The first concerns how to select representative values from the numerous daily retrievals at sites with flat terrain. The second involves terrain tilt correction (TTC) and multi-GNSS weighting strategies at sites with complex topography.

Regarding the first challenge, extensive progress has been made in applying GNSS-IR to snow depth monitoring. Larson et al. [1] measured snow depth at Marshall, Colorado, and demonstrated strong consistency between in situ measurements and GPS SNR-derived estimates, providing the first evidence that GPS receivers can be used for snow depth retrieval. Larson and Nievinski [2] further detailed the SNR-based retrieval process using Plate Boundary Observatory (PBO) GPS stations, where daily snow depth was obtained by subtracting snow-free reflector heights from snow-covered values. Nievinski and Larson [3, 4] extended this approach by proposing the use of median statistics to improve robustness in estimating representative daily snow depth. Najibi et al. [5] modeled reflected GPS signals at northern Canadian stations (BAKE and KUUJ), using daily GPS L4 averages to determine snow depth. Jin et al. [6] analyzed L2P SNR data from three Alaskan stations, estimating reflector height through Lomb–Scargle periodogram (LSP) spectral analysis, and calculating daily snow depth as the average of all valid GPS tracks. Their results showed correlation coefficients (R) of 0.79, 0.88, and 0.98, with RMSE values of 0.14 m, 0.12 m, and 0.12 m for sites SG27, AB39, and AB33, respectively. Subsequent studies further improved retrieval strategies. Larson and Small [7] introduced median filtering to remove outliers before computing daily averages. Roesler and Larson [8] released open-source software for GNSS-IR snow depth studies, enabling GNSS file conversion, reflector zone mapping, Nyquist frequency calculation, and reflector height estimation, while adopting median statistics instead of mean values for daily snow depth. Wei et al. [9] proposed the Tilted Surface Strategy (TSS) using GPS-IR, which determined daily reflector height averages from all available reflections. Although mean- and median-based methods achieved promising results, multi-GNSS retrieval experiments had not yet been performed at that time. More recently, Tu et al. [10] applied multi-frequency averaging and multi-constellation fusion (GPS, GLONASS, Galileo, and BDS) using PBO station P387. Their results showed that multi-frequency averaging improved retrieval accuracy to better than 7.0 cm, while multi-GNSS fusion further enhanced accuracy to better than 3.0 cm, confirming the feasibility and precision of multi-system, multi-frequency snow depth retrieval. These studies demonstrate that mean fusion yields reliable results; however, most previous work has overlooked its lack of robustness against outliers [11].

Regarding the second challenge, GNSS-IR applications to snow depth retrieval have also yielded substantial progress. The availability of multi-GNSS observations introduces additional snow depth estimates, which can enhance retrieval stability. Wang et al. [12] employed data from three MGEX stations (PRDS, FRDN, and XJJC) and applied a robust regression-based combination approach to integrate SNR-derived snow depth estimates from four constellations, achieving approximately 50% higher accuracy compared to single-constellation results. Hu et al. [2] proposed a variational mode decomposition (VMD) method to extract SNR residuals (δSNR). Compared with least squares fitting (LSF), their approach reduced RMSE by 20–40% and achieved a correlation coefficient (R) of 0.99 [13]. Liu et al. [14] assessed snow depth models during the 2020 snow season at two stations using BDS data. Their results demonstrated that estimates obtained from SNR, triple-frequency SNR combination (SNR_COM) [15], and triple-frequency phase combination [10] models were comparable, with $R = 0.93$, thereby providing supporting evidence for the use of BDS signals in snow depth retrieval. An improved retrieval algorithm was further proposed, in which δSNR was extracted via wavelet decomposition, outliers were removed using the RANdom SAmple Consensus (RANSAC) algorithm, and terrain-related errors were carefully corrected. Validation at three stations (AB33, AB39, and SG27) showed RMSE values ranging from 4.0 to 8.0 cm and correlation coefficients exceeding 0.95, confirming the effectiveness of this approach.

In summary, GNSS-IR provides valuable technical support for snow depth monitoring at meteorological stations, along railway lines, at ski resorts, and on roads in remote communities affected by snowstorms. Moreover, it plays an important role in complementing existing polar snow depth measurement systems and in acquiring high-accuracy, high-resolution snow depth data in polar regions.

2 Fundamental Principles of Multi-GNSS Fusion Algorithms

The fundamental principles of multi-GNSS fusion algorithms can be divided into two categories: the robust estimation algorithm and the Helmert variance component estimation (HVCE) algorithm.

2.1 *Robust Estimation Algorithm*

To mitigate the influence of outliers and gross errors on GNSS-IR snow depth retrievals, a robust estimation algorithm is introduced to fuse results from multiple GNSS systems and signal-to-noise ratio (SNR) types. Robust estimation, derived from statistical theory and adjustment computations in geodesy, can effectively reduce the impact of gross errors while maintaining computational efficiency. Its fundamental principle is to fully utilize valid information, limit the use of questionable data, and exclude harmful observations. Considering the weak correlations among different GNSS systems and SNR types, the M-estimation method is adopted, using the IGGIII weighting function proposed by Yang [16] as the equivalent weight function.

First, the observation error equation is expressed as:

$$V = \widehat{h_R} - h_R \tag{1}$$

In the equation, $\widehat{h_R} = \widetilde{h_R} + \delta\hat{x}$, $\widehat{h_R}$ represents the optimal estimate to be solved for h_R, $\widetilde{h_R}$ is the approximate value, i.e., the initial value. In this study, two types of initial values are used for the experiment: the mean value $\left(\widetilde{h}_{R_{mean}}\right)$ and the median value $\left(\widetilde{h}_{R_{median}}\right)$. $\delta\hat{x}$ represents the correction term for h_R. Thus, the error equation can be derived as:

$$V = \delta\hat{x} - \left(h_R - \widetilde{h_R}\right) \tag{2}$$

Let $h_R - \widetilde{h_R} = L$, then

$$V = \delta\hat{x} - L \tag{3}$$

Here V is the residual vector of $n \times 1$, $\delta\hat{x}$ is the unknown parameter vector, and L is the observation vector of $n \times 1$. Assuming that the observations are independent, the weight matrix P is a diagonal matrix corresponding to $n \times n$. In this study, the initial equivalent weight matrix is constructed using the inverse of h_R. Then, the normal equation and the least-squares solution for the unknown parameters are given by [17]:

$$\begin{cases} A^T PA\delta\hat{x} - A^T PL = 0 \\ \delta\hat{x} = \left(A^T PA\right)^{-1} A^T PL \end{cases} \tag{4}$$

The robust M-estimation algorithm usually requires multiple iterations to obtain the optimal estimate. After $k + 1$ iterations, the resulting robust M-estimate is given by:

$$\delta\hat{x}^{k+1} = \left(A^T \overline{P}^k A\right)^{-1} A^T \overline{P}^k L \tag{5}$$

where $\delta\hat{x}^{k+1}$ is the solution obtained after $k + 1$ iterations. To achieve higher accuracy, the iteration termination criterion is set as $\left|\delta\hat{x}^{k+1} - \delta\hat{x}^k\right| < 0.01$ m. $\overline{P}^k$ is the equivalent weight matrix used in the $k + 1$ iteration, constructed according to the IGGIII scheme.

After solving for $\delta\hat{x}$ from the above equation, substituting it into Formula (3) yields the residual vector V. The approximate value of the unit weight standard deviation σ_0 is then given by [18]:

$$\sigma_0 = \sqrt{\frac{V^T PV}{n - 1}} \tag{6}$$

From the cofactor propagation law, it can be seen that:

$$Q_{vv} = Q - AN_{AA}^{-1} A^T \tag{7}$$

where Q is the inverse matrix of the equivalent weight matrix $\overline{P}$.

Robust estimation algorithms can basically be divided into three types: maximum likelihood estimation (robust M-estimation), linear combination of order statistics estimation (robust L-estimation), and rank-test-type estimation (robust R-estimation). The main difference among them lies in the equivalent weight function. In this study, the solution process for the equivalent weight function of the robust M-estimation is as follows. The equivalent weight matrix constructed by the IGGIII scheme is:

$$\overline{p}_i = \begin{cases} p_i & |\widetilde{v}_i| \leqslant k_0 \\ p_i \frac{k_0}{|\widetilde{v}_i|} \left(\frac{k_1 - |\widetilde{v}_i|}{k_1 - k_0}\right)^2 & k_0 < |\widetilde{v}_i| \leqslant k_1 \\ 0 & |\widetilde{v}_i| > k_1 \end{cases} \tag{8}$$

where $\widetilde{v}_i = \frac{v_i}{\sigma_{v_i}} = \frac{v_i}{\sigma_0 Q_{vv}(i,i)}$ is the standardized residual. k_0 and k_1 are constants, which are usually chosen as $k_0 = 1.0$ and $k_1 = 2.5$, respectively.

When the iteration termination criterion is satisfied, the robust M-estimation iterative solution of snow depth h_R can be further obtained from the result of Formula (3.5):

$$\begin{cases} \widehat{h}_{R_{mean}} = \widetilde{h}_{R_{\text{mean}}} + \delta\widehat{x}^{k+1} \\ \widehat{h}_{R_{median}} = \widetilde{h}_{R_{\text{median}}} + \delta\widehat{x}^{k+1} \end{cases} \tag{9}$$

When using satellite SNR data for snow depth inversion, an azimuth angle range and an elevation angle range need to be selected. To avoid the influence of terrain undulation on the robust M-estimation algorithm, this study selects azimuth angles in areas with negligible terrain variation. The robust M-estimation algorithm operates on datasets where each estimate has one and only one true value. By selecting a relatively flat area, it can be ensured that the true height corresponding to each azimuth track in that area is unique, allowing the robust M-estimation algorithm to compute the optimal estimate of the reflection height in that area. For the SG27 station, when the azimuth angle range is 90°–180°, the terrain within a 100 m radius centered on the station is flat, meeting the experimental requirements. To utilize as much SNR data as possible, the elevation angle range is chosen as 5°–30°. Using the formula for determining the reflection area, the distance corresponding to a 5° elevation angle is 43.4 m, and the distance corresponding to a 30° elevation angle is 6.6 m. Therefore, the snow depth inversion area in this study is a sector with the station at the center and radii of 6.6 m and 43.4 m. The specific formula is as follows:

$$\frac{h}{\tan(e_{\max})} \leq L \leq \frac{h}{\tan(e_{\min})} \tag{10}$$

where L is the horizontal distance from the scattering point to the station, $e_{\min}$ represents the minimum cutoff elevation angle, $e_{\max}$ represents the maximum cutoff elevation angle, and h is the station height; for the SG27 station, which is approximately 3.8 m.

The satellite reflection point trajectory is modeled using the elevation angle e, azimuth angle α, and the height H of the reflection point relative to the phase center of the receiver antenna. This reflects the position of a reflection point at a specific satellite epoch relative to the GNSS station. It can be expressed as:

$$\begin{cases} x = \frac{H}{\tan(e)} \sin(\alpha) \\ y = \frac{H}{\tan(e)} \cos(\alpha) \end{cases} \tag{11}$$

where x and y represent the coordinates of the satellite signal specular reflection point trajectory along the X-axis and Y-axis, respectively.

During the robust estimation process, the number of data points involved in the computation must be at least 4 in order to apply the robust estimation algorithm.

Table 1 Overview of the four experimental schemes

Experimental scheme	Symbol	Description
Average value	S_{avg}	Compute the final snow depth solution using the average of the retrieved snow depths
Median value	S_{med}	Compute the final snow depth solution using the median of the retrieved snow depths
M-estimation with average as initial value	S_{RE_avg}	Use the average of the retrieved snow depths as the initial value and compute the final solution using robust M-estimation
M-estimation with median as initial value	S_{RE_med}	Use the median of the retrieved snow depths as the initial value and compute the final solution using robust M-estimation

When the number of outliers is fixed, the more estimated values available, the more pronounced the advantages of the robust estimation algorithm, and the stronger its ability to resist outliers. To evaluate the effectiveness of the algorithm proposed in this study, Table 1 briefly describes four comparative estimation methods, including one using the robust estimation algorithm.

2.2 *Helmert Variance Component Estimation (HVCE) Algorithm*

With the continuous improvement of measurement accuracy and the development of modern measurement techniques, the objects of adjustment have evolved from a single type of homogeneous observations to multiple types of observations with different accuracies. In this study, we assume that the snow depth retrieval results from different GNSS represent five types of observations with different precisions: GPS (G), GLONASS (R), Galileo (E), BDS2 (C2), and BDS3 (C3). Before adjustment, it is difficult to accurately determine the weights of snow depth values retrieved from different GNSS. If the assigned weights contain errors or the weight ratios among GNSS are imbalanced, the accuracy and reliability of the snow depth retrieval results will be directly affected. To improve the reliability of the adjustment results and accurately determine the weight ratios among different types of observations, this study introduces the HVCE algorithm, which is primarily used to determine the weights among independent random observations. Furthermore, to avoid singular solutions, the weighting process in this study is performed only for two types of observations at a time. Specifically, GLONASS, Galileo, BDS2, and BDS3 are each weighted relative to GPS, which yields four system-specific weights relative to GPS. Below, the HVCE weighting procedure is illustrated using GPS and GLONASS as an example for one day of snow depth retrieval results.

Assume the error equation is:

$$\begin{cases} V_G = \widehat{h_{R_G}} - h_{R_G} \\ V_R = \widehat{h_{R_R}} - h_{R_R} \end{cases} \tag{12}$$

In the equation, $\widehat{h_{R_G}} = \widetilde{h_{R_G}} + \delta\widehat{x}_G$, $\widehat{h_{R_R}} = \widetilde{h_{R_R}} + \delta\widehat{x}_R$, $\widehat{h_{R_G}}$ and $\widehat{h_{R_R}}$ are unknowns representing the optimal snow depth solutions for that day derived from GPS and GLONASS. $\widetilde{h_{R_G}}$ and $\widetilde{h_{R_R}}$ represent the approximate solutions of snow depth from GPS and GLONASS, which are calculated here using the median. $\delta\widehat{x}_G$ and $\delta\widehat{x}_R$ denote the correction terms for the GPS and GLONASS snow depths, while h_{R_G} and h_{R_R} correspond to the observed snow depth values retrieved from GPS and GLONASS. Therefore, the observation equation can be rewritten as:

$$\begin{cases} V_G = \delta\widehat{x}_G - \left(h_{R_G} - \widetilde{h_{R_G}}\right) \\ V_R = \delta\widehat{x}_R - \left(h_{R_R} - \widetilde{h_{R_R}}\right) \end{cases} \tag{13}$$

Assuming $h_{R_G} - \widetilde{h_{R_G}} = L_G$, $h_{R_R} - \widetilde{h_{R_R}} = L_R$, then:

$$\begin{cases} V_G = \delta\widehat{x}_G - L_G \\ V_R = \delta\widehat{x}_R - L_R \end{cases} \tag{14}$$

where V_G is the residual vector of $n \times 1$, and V_R is the residual vector of $m \times 1$. $\delta\widehat{x}_G$ and $\delta\widehat{x}_R$ are the unknown parameter vectors. L_G and L_R are the observation vectors of $n \times 1$ and $m \times 1$, respectively. Since the snow depth inversion results from GPS and GLONASS are independent of each other, the initial weight ratio of GPS to GLONASS is set to 1:1. Therefore, the initial weight matrices P_G and P_R for GPS and GLONASS are unit diagonal matrices of $n \times n$ and $m \times m$, respectively.

Thus, the normal equations can be derived as:

$$\begin{cases} N_G\delta\widehat{x}_G - W_G = 0 \\ N_R\delta\widehat{x}_R - W_R = 0 \end{cases} \tag{15}$$

where the coefficient matrices of the normal equations are $N_G = A_G{}^T P_G A_G$ and $N_R = A_R{}^T P_R A_R$. The free term vectors of the normal equation are $W_G = A_G{}^T P_G L_G$ and $W_R = A_R{}^T P_R L_R$.

Based on the quadratic expectation formula, the rigorous HVCE can be derived as [19]:

$$\widehat{\sigma}^2 = S^{-1} W \tag{16}$$

$$\begin{cases} S = \begin{bmatrix} n - 2tr(N^{-1}N_G) + tr(N^{-1}N_G)^2 & tr(N^{-1}N_G N^{-1}N_R) \\ tr(N^{-1}N_G N^{-1}N_R) & m - 2tr(N^{-1}N_R) + tr(N^{-1}N_R)^2 \end{bmatrix} \\ N = A^T PA = \begin{bmatrix} A_G \\ A_R \end{bmatrix}^T \begin{bmatrix} P_G & \\ & P_R \end{bmatrix} \begin{bmatrix} A_G \\ A_R \end{bmatrix} = N_G + N_R \\ W = \begin{bmatrix} W_G \\ W_R \end{bmatrix} \\ \hat{\sigma}^2 = \begin{bmatrix} \hat{\sigma}^2{}_G \\ \hat{\sigma}^2{}_R \end{bmatrix} \end{cases} \tag{17}$$

where $tr(.)$ refers to the rank of the matrix to be determined, and $\hat{\sigma}^2{}_G$ and $\hat{\sigma}^2{}_R$ represent the unit weight variance estimates for GPS and GLONASS, respectively.

After solving for $\hat{\sigma}^2{}_G$ and $\hat{\sigma}^2{}_R$, the weight matrices between the two systems' observations are readjusted.

$$\begin{cases} P_G^{k+1} = \frac{P_G^k}{\hat{\sigma}^2{}_G} \\ P_R^{k+1} = \frac{P_R^k}{\hat{\sigma}^2{}_R} \end{cases} \tag{18}$$

where k represents the number of iterations.

After readjusting the weights, the adjustment and HVCE are performed again. In this study, the iteration termination condition is set as [20]:

$$\frac{1 - \min\left(\hat{\sigma}^2\right)}{\max\left(\hat{\sigma}^2\right)} < 0.01 \tag{19}$$

That is, when $\hat{\sigma}^2{}_G \approx \hat{\sigma}^2{}_R$, the HVCE process is complete. The weight ratio that satisfies $\hat{\sigma}^2{}_G \approx \hat{\sigma}^2{}_R$ represents the relative contribution of each system to the snow depth retrieval when combining GPS and GLONASS.

Similarly, the contribution ratios can be obtained for GPS–GLONASS, GPS–Galileo, GPS–BDS2, and GPS–BDS3 pairs. Taking GPS as the reference, these can be converted into the contribution ratios of GPS versus GLONASS, Galileo, BDS2, and BDS3, denoted as r_G: r_R: r_E: r_{C2}: r_{C3}. The sum of the snow depth contribution ratios for all satellite systems on a given day is:

$$r_S = r_G + r_R + r_E + r_{C2} + r_{C3} \tag{20}$$

Based on the daily median snow depth results from different satellite systems—GPS, GLONASS, Galileo, BDS2, and BDS3 $\left(h_{G_G}, h_{G_R}, h_{G_E}, h_{G_{C2}} \text{ and } h_{G_{C3}}\right)$—and their corresponding contribution ratios (r_G: r_R: r_E: r_{C2}: r_{C3}), a multi-system snow depth retrieval fusion can be performed. The daily snow depth retrieval result $h_{G_{HVCE}}$

obtained via HVCE is then given by:

$$h_{G_{HVCE}} = \frac{r_G}{r_S}h_{G_G} + \frac{r_R}{r_S}h_{G_R} + \frac{r_E}{r_S}h_{G_E} + \frac{r_{C2}}{r_S}h_{G_{C2}} + \frac{r_{C3}}{r_S}h_{G_{C3}} \quad (21)$$

3 Experimental Analyses in Areas without Terrain Variation

The SG27 station belongs to the PBO network and is located in Alaska, USA (latitude: 71.3229°, longitude: – 156.6103°). The receiver is SEPT POLARX5, and the antenna is TRM59800.80 SCIS. GNSS observation files were sampled at 15 s intervals, and the data were provided by the University NAVSTAR Consortium (UNAVCO). The snow depth reference site is PABR (latitude: 71.2833°, longitude: – 156.7667°), approximately 7.2 km from the SG27 station. Because this region is a vast, flat permafrost area spanning hundreds of kilometers, the snow depth observations at this site can serve as a reliable reference [10]. The National Operational Hydrologic Remote Sensing Center (NOHRSC) of the National Oceanic and Atmospheric Administration (NOAA) provides daily snow depth measurements for comparison. The SG27 station environment and DEM are shown in Fig. 1. The station is located in a snow-prone area suitable for snow depth retrieval. The precise ephemeris files used in this study were provided by the IGS Analysis Center. The area enclosed by black lines in Fig. 1 represents the reflection zone. It was observed that terrain variations are minimal and can be neglected. The experiment used data from DOY 176–365 in 2018. Snow-free and snow-covered periods were processed separately: Snow-free period: DOY 176–290 (115 days), mainly used to determine the station height. Snow-covered period: from DOY 291–365 in 2018 to DOY 001–050 in 2019 (120 days in total). The days DOY 301, 322, 355, 356 in 2018 and DOY 041 in 2019 lacked snow depth measurements and were excluded from the snow depth retrieval experiments.

3.1 Accuracy Assessment of Different SNR Types

To compare the snow depth retrieval accuracy of different SNR types during snow-free and snow-covered periods, this section presents a schematic comparison of the accuracies of the four experimental schemes. Table 2 shows the GNSS data received during different time periods. Table 3 lists the frequency information of different GNSS and SNR types used at the GNSS station in this study. It is noteworthy that previous studies on GPS S1W and S2W have demonstrated that these signals are not suitable for snow depth retrieval [21, 22, 23]. It is speculated that this may be related to their channels or the use of Z-tracking and similar (AS on), where the

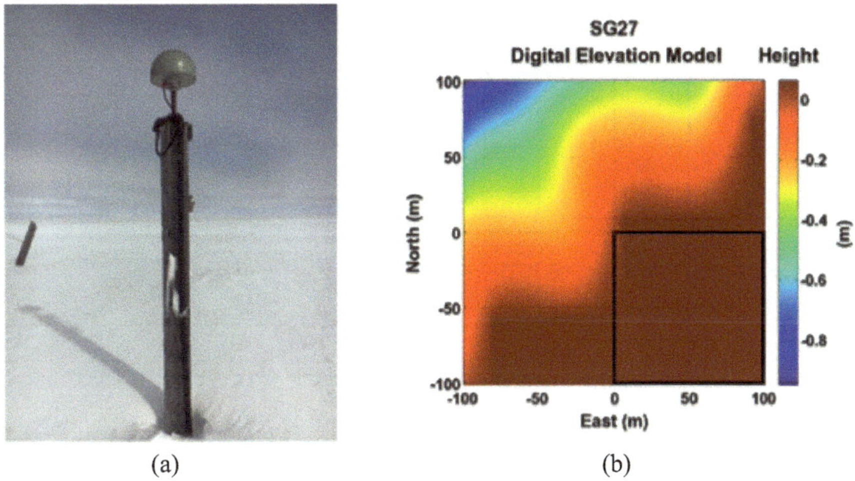

Fig. 1 Geographic location and surrounding environment of SG27 station

Anti-Spoofing (AS) policy adds the P-code modulo-2 multiplied to the W-code to form a new Y-code. This process may alter the signal frequency, making it unsuitable for snow depth retrieval [24]. Therefore, in this study, the retrieval results from GPS S1W and S2W are directly discarded, and invalid values are removed to prevent their impact on the snow depth retrieval results.

On the same day, snow depth retrieval results from different GNSS and SNR types include h_R values retrieved from multiple satellites. However, for a single SNR type on a given day, there is only one true value of h_R, i.e., only one true snow depth. To obtain the optimal estimate of h_R for a single SNR type, this study introduces a robust estimation method and establishes a quality control experiment. Four solutions are used to derive h_R from multiple satellites of the same SNR type on the same day. Subsequently, the daily evaluation results of h_R for each of the 13 SNR types are obtained. Figure 2 shows the comparison of R for different SNR types during the snow-covered period. Since the observed snow depth is zero during the snow-free period, R cannot be calculated. Therefore, no correlation analysis of R is conducted for the snow-free period. No correlation is observed between C3-MEO-S2I and C3-MEO-S6I because BDS3 data are only available from 2019, and the experiment in 2019 used only the first 50 days, which is a short period and does not show good correlation. Among the remaining SNR types, R-S2C exhibits the highest R across

Table 2 GNSS observations received at different time periods

Snow period	Year	DOY	GNSS
Snow-free period	2018	DOY 176 ~ 290	G, R, E, C2-MEO
Snow-covered period	2018	DOY 291 ~ 365	G, R, E, C2-MEO
	2019	DOY 001 ~ 050	G, R, E, C2-MEO, C3-MEO

Table 3 Channel parameters of different SNR types for four GNSS

GNSS	Band/Frequency(MHz)	Channel	SNR type
GPS	L1/1575.42	Z-tracking and similar	S1W
		C/A	S1C
	L2/1227.60	Z-tracking and similar	S2W
		L2C (L)	S2L
	L5/1176.45	Q	S5Q
GLONASS	G1/1602 + n * 0.5625	C/A	S1C
	G2/1246 + n * 0.4375	C/A	S2C
Galileo	E1/1575.42	C OS pilot	S1C
	E5a/1176.45	Q no data	S5Q
	E6/1278.75	C no data	S6C
	E5b/1207.140	Q no data	S7Q
	E5(E5a + E5b)/1191.795	Q	S8Q
BDS	B1/1561.098(BDS-2/3 signal)	I (B1I signal)	S2I
	B3/1268.52(BDS-2/3 signal)	I	S6I
	B2/1207.140(BDS-2 signal)	I (B2I signal)	S7I

the four solutions (Savg: 0.732, Smed: 0.756, SRE_avg: 0.778, SRE_med: 0.778), while C2-MEO-S2I shows the lowest RRR (Savg: 0.503, Smed: 0.511, SRE_avg: 0.503, SRE_med: 0.511).

The R only reflects the consistency of trends between retrieved and observed snow depths. To evaluate retrieval accuracy, it is necessary to further compare RMSE and ME. Figure 3 presents 3D histograms of RMSE for four solutions across different SNR types. From Fig. 3a, it can be seen that GPS-S1C, GPS-S2L, and GPS-S5Q exhibit relatively small RMSE values during the snow-free period, indicating that

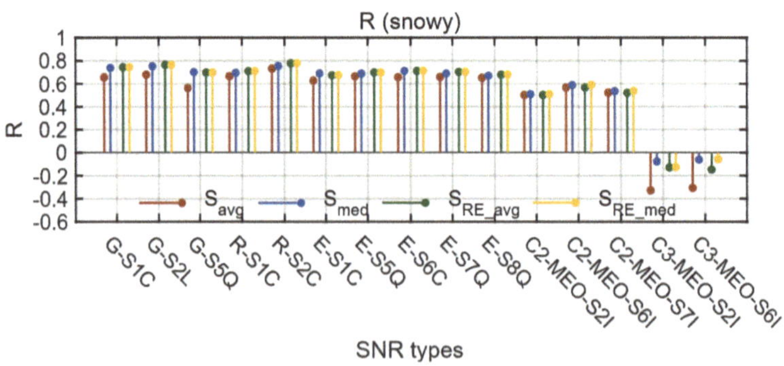

Fig. 2 Comparison of R values for four solutions under different SNR types during the snow-covered period

surface heights retrieved from these three SNR types have comparatively good accuracy. In contrast, C2-MEO-S2I, C2-MEO-S6I, and C2-MEO-S7I show larger RMSE, suggesting poorer retrieval accuracy for these SNR types. Comparing the four solutions, it is found that Savg has relatively large RMSE across all 13 SNR types, indicating certain shortcomings in snow depth retrieval. In comparison, SRE_avg exhibits lower RMSE, demonstrating the advantage of robust estimation. Both Smed and SRE_med show relatively small RMSE across the 13 SNR types, reflecting the robustness of the median against large or small datasets. From Fig. 3b, snow depths retrieved from 15 SNR types during the snow-covered period maintain relatively consistent accuracy. Compared with Fig. 3a, it can be concluded that RMSE for C2-MEO-S2I, C2-MEO-S6I, and C2-MEO-S7I remains relatively stable in both snow-free and snow-covered periods.

Figure 4 presents the comparison histograms of another evaluation metric, ME. During the snow-free period, ME varies between − 2.0 cm and 1.0 cm, while during the snow-covered period, ME ranges approximately from − 3.0 cm to 0.0 cm. It can be seen that, although RMSE and ME differ across different SNR types, RMSE remains consistently below 5.0 cm, and ME fluctuates around 0.0 cm. This demonstrates the reliability and accuracy of GNSS-IR snow depth retrieval. When there is a low-density fresh snow layer (such as a layer with low density just after snowfall) or a multi-layer snow structure, GNSS reflected signals will penetrate the surface snow layer and reach the deeper snow-ground interface or lower snow surface. At this time, the retrieval algorithm will mistakenly take the "deeper reflecting surface" as the snow surface, resulting in the calculated snow depth being smaller than the actual depth. For example, fresh snow has low density and many pores, so signals can easily transmit through it, and this penetration phenomenon is more significant especially during periods of rapid snow accumulation (such as continuous snowfall periods).

3.2 Accuracy Assessment for Different GNSS

Figure 5 presents the comparison of retrieval accuracy for the four schemes (Savg, Smed, SRE_avg, and SRE_med) across different GNSS during snow-free and snow-covered periods, with R, RMSE, and ME shown on three Y-axes.

Notably, for all four experimental schemes, the R values of C3-MEO during snow-covered periods are negative, consistent with previous observations. Apart from C3-MEO, C2-MEO has the smallest R values, which are 0.542, 0.556, 0.557, and 0.557. GLONASS achieves the largest R values, namely 0.715, 0.743, 0.753, and 0.753. Regarding RMSE, during snow-free periods, G exhibits the smallest RMSE values: 0.008 m, 0.009 m, 0.007 m, and 0.007 m, while C2-MEO shows the largest RMSE values: 0.036 m, 0.035 m, 0.035 m, and 0.035 m. During snow-covered periods, the minimum and maximum values of Savg for R and C3-MEO are 0.040 m and 0.052 m, respectively. For G, the minimum values of Smed, SRE_avg, and SRE_med are 0.037 m, 0.036 m, and 0.036 m, while the maximum values for C2-MEO are 0.049 m, 0.048 m, and 0.049 m. The results also indicate the presence

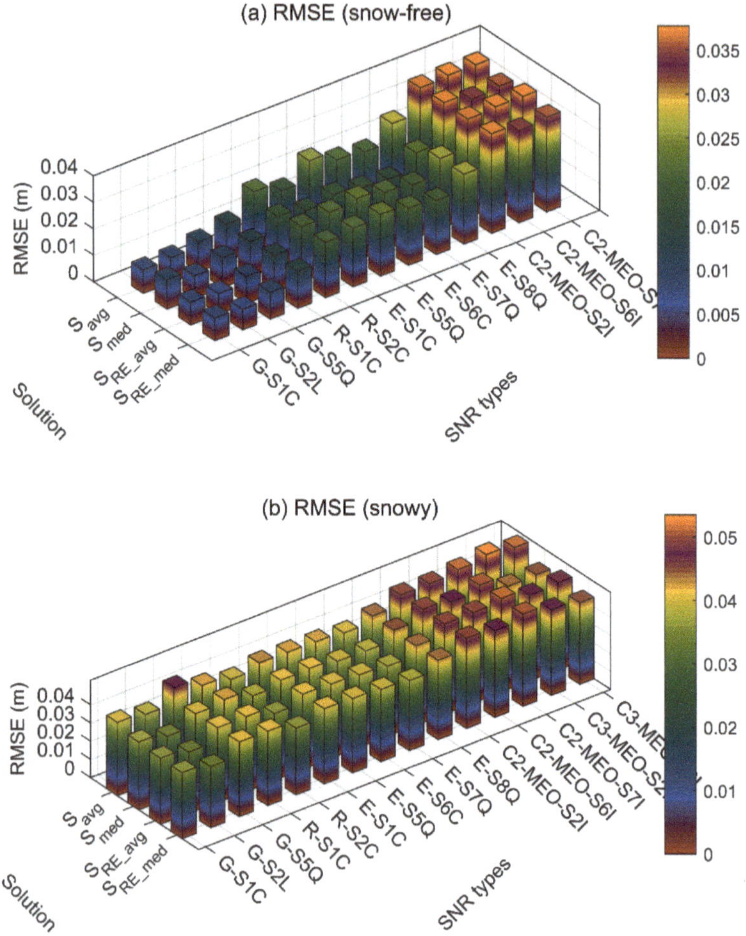

Fig. 3 3D histograms of RMSE for four solutions under different SNR types during snow-free and snow-covered periods

of negative errors across all four schemes. This is likely due to uncertainties in the station height. During snow-free periods, the minimum and maximum values of Smed for C2-MEO and R are – 0.001 m and – 0.008 m, for SRE_avg are – 0.001 m and – 0.001 m, and for SRE_med for G and R are – 0.001 m and – 0.010 m. During snow-covered periods, the minimum ME values for G are – 0.007 m, – 0.012 m, – 0.009 m, and – 0.009 m, whereas the maximum ME values for C3-MEO are – 0.025 m, – 0.026 m, – 0.030 m, and – 0.030 m. In summary, the retrieved snow depth achieves centimeter-level accuracy, demonstrating that GNSS-IR snow depth retrieval is highly reliable.

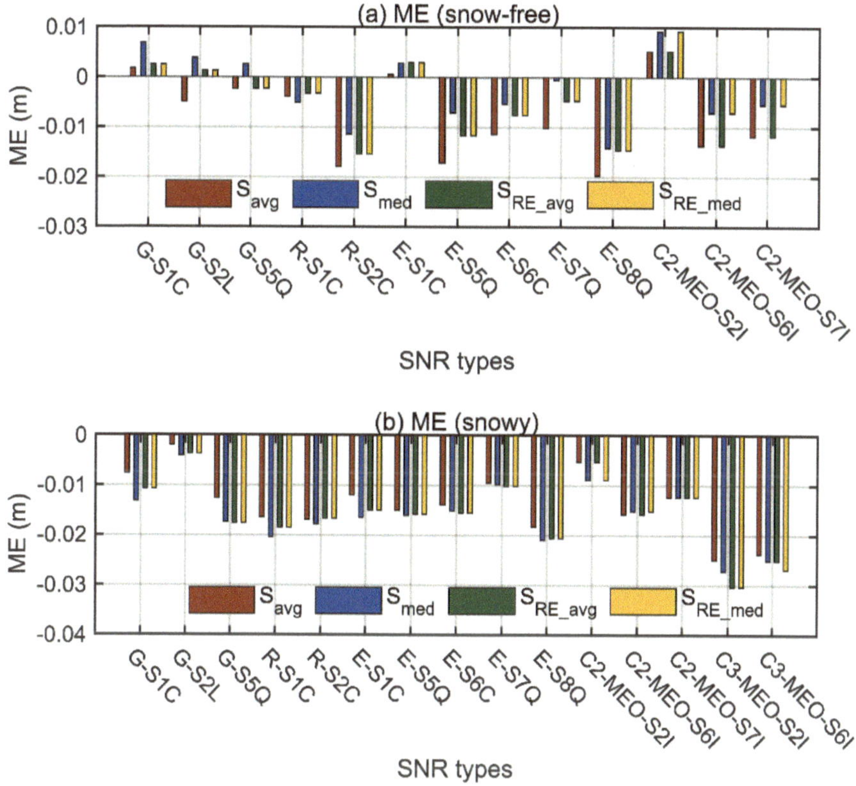

Fig. 4 Histograms of ME values for four solutions under different SNR types during snow-free and snow-covered periods

To illustrate the snow depth retrieval performance of different GNSS more intuitively, Fig. 6 shows the comparison between retrieved snow depth and measured snow depth using SRE_med as an example, along with histograms of retrieval errors. Panels (a), (b), (c), and (d) display the comparison between measured and retrieved snow depth, with orange representing snow-free periods and light green representing snow-covered periods. Panels (f), (g), (h), and (i) show the histograms of deviations during snow-free periods, while panels (k), (l), (m), (n), and (o) show the histograms of deviations during snow-covered periods. The snow depth retrieval results for the four GNSS systems (G, R, E, and C2-MEO) exhibit slight fluctuations, with C2-MEO showing larger deviations. From the deviation plots, during snow-free periods, the errors for G are mostly within − 5 to 5 cm, while C2-MEO errors range from − 15 to 10 cm. During snow-covered periods, G errors are within − 10 to 10 cm, and C2-MEO errors range from − 15 to 15 cm. This indicates that G provides significantly higher snow depth retrieval accuracy than C2-MEO, and the retrieval accuracy during snow-free periods is better than during snow-covered periods. Although the number of C3-MEO observations is small, its retrieval errors during snowfall periods are

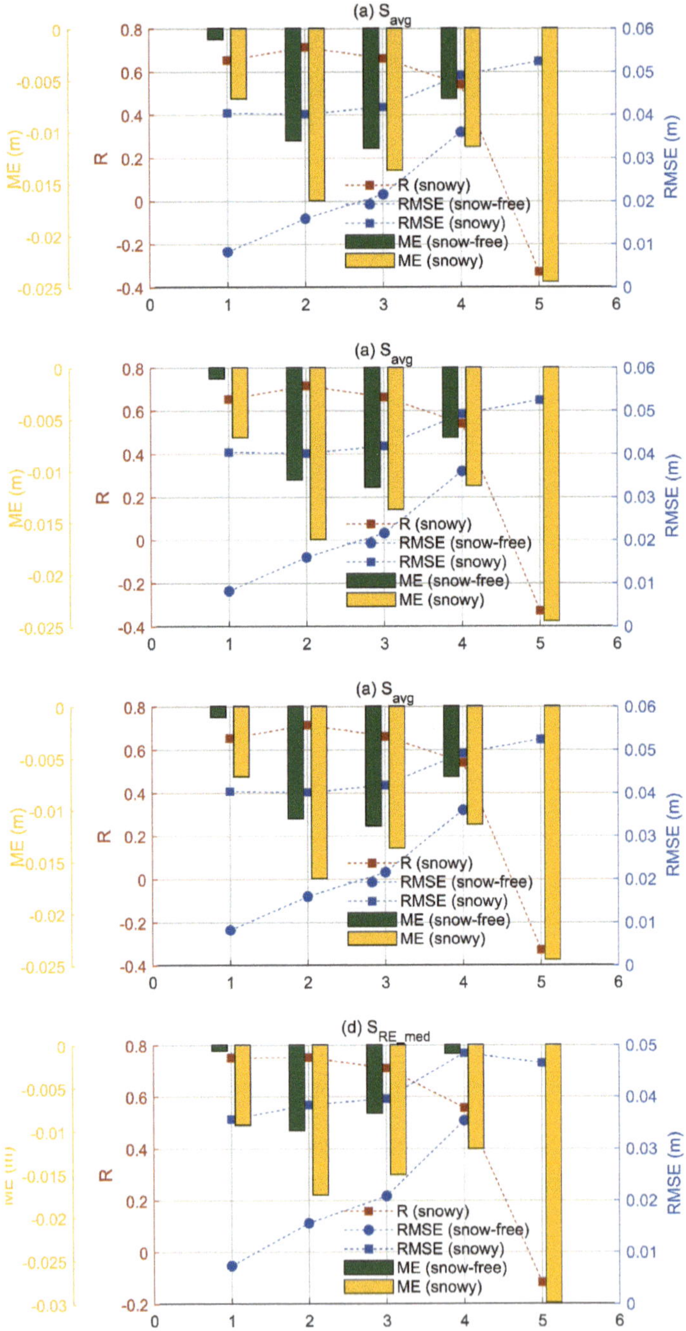

Fig. 5 Comparison of snow depth retrieval accuracy for four solutions from different GNSS during snow-free and snow-covered periods

also concentrated within − 10 to 10 cm, showing a similar error range as G. Overall, all four solutions exhibit good consistency between retrieved and measured snow depths.

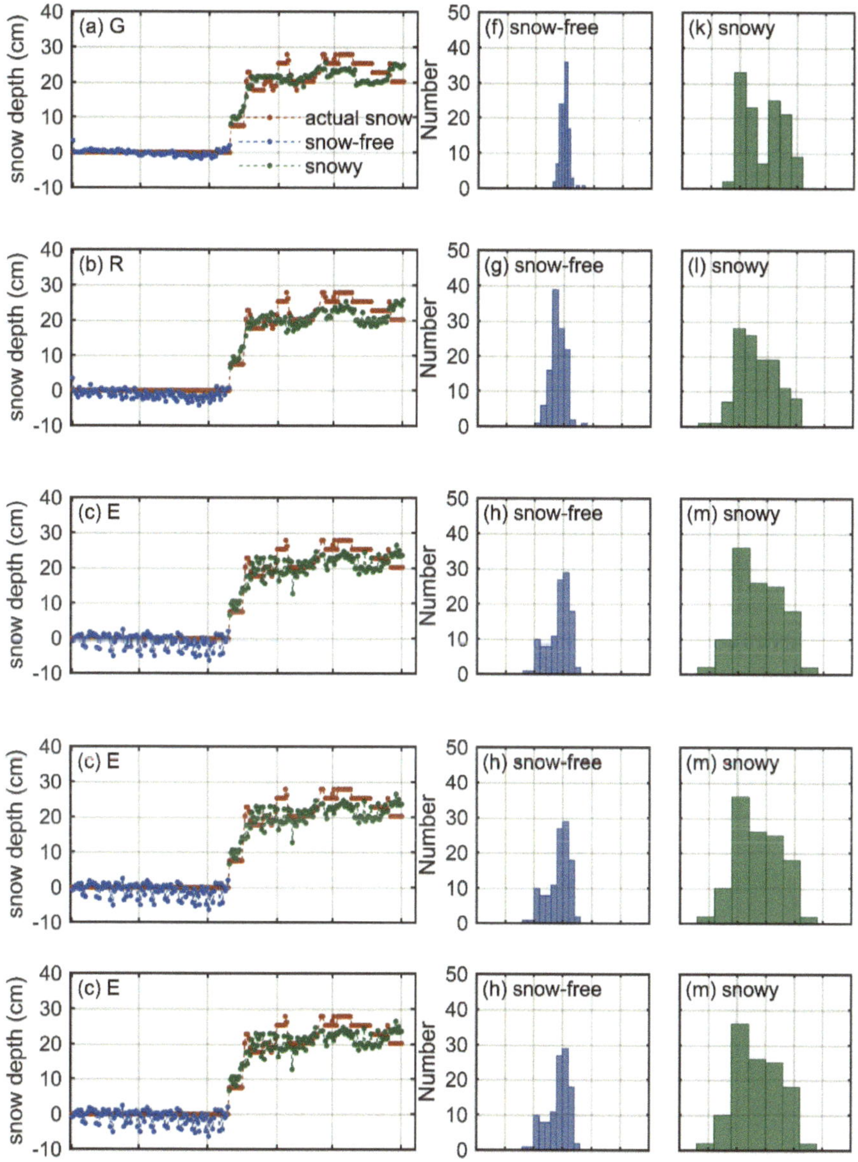

Fig. 6 GNSS-IR snow depth retrievals for different GNSS and histograms of deviations during snow-free and snow-covered periods

3.3 Accuracy Assessment of Multi-System Fusion

Zhang et al. [25] proposed an improved snow depth retrieval algorithm based on a dynamic clustering approach and conducted experiments at the SG27 station. Their results from 2015–2017 showed a mean error (ME) of 2.1 cm and an RMSE of 3.2 cm, demonstrating good accuracy [25]. Although the present study was also carried out at SG27, the selected snow depth retrieval periods differ, yet comparable retrieval accuracy was obtained. To evaluate the snow depth retrieval accuracy of the four solutions more intuitively, Fig. 7 presents the overall comparison for both snow-free and snow-covered periods, with R, RMSE, and ME shown in the 3Y-axis plots.

The Savg solution produced relatively unreliable results, with lower R and higher RMSE, indicating that averaging in the fusion process significantly reduces accuracy compared to the other three solutions. This further confirms that Savg is not suitable for snow depth estimation. In contrast, Smed, SRE_avg, and SRE_med achieved higher R values, suggesting that median-based and robust estimation approaches effectively resist the influence of gross errors, ensuring the correlation of retrieval results. During the snow-free period, the RMSE values for Savg, Smed, SRE_avg, and SRE_med were 1.3 cm, 0.9 cm, 1.2 cm, and 1.2 cm, respectively, while the corresponding ME values were 0.7 cm, 0.1 cm, 0.6 cm, and 0.6 cm. During the snow-covered period, R values for Savg, Smed, SRE_avg, and SRE_med reached 0.686, 0.744, 0.759, and 0.759, respectively; RMSE values were 4.0 cm, 3.8 cm, 3.7 cm, and 3.7 cm; and ME values were − 1.2 cm, − 1.5 cm, − 1.4 cm, and − 1.4 cm. The negative ME in both periods may be attributed to tropospheric effects, which cause measured heights to be lower than true geometric heights [26]. In most snow depth studies, Smed is commonly used for data processing. However, in this study, SRE_avg and SRE_med exhibited superior snow depth retrieval performance. Compared to Smed, R improved by 2.0%, RMSE by 2.6%, and ME decreased by 6.7%. Among the four experimental schemes, these two solutions achieved the best results in terms

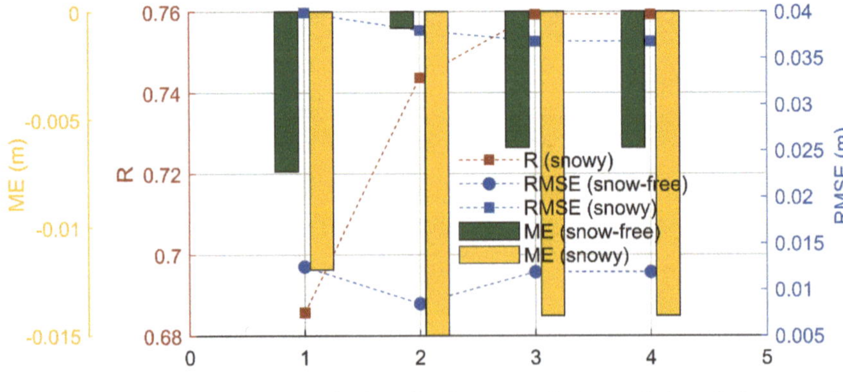

Fig. 7 Comparison of overall snow depth retrieval accuracy for four solutions during snow-free and snow-covered periods

of R, RMSE, and ME. Therefore, SRE_avg and SRE_med are considered the most effective approaches for snow depth retrieval in this study.

Figure 8 shows the comparison between the four solutions—Savg, Smed, SRE_avg, and SRE_med—and the observed snow depth, along with histograms of the error distributions. As seen in panels (a), (b), (c), and (d), all four solutions agree well with the observed values, with no significant bias. Moreover, the retrievals during the snow-free period are more consistent than those during the snow-covered period. Panels (e), (f), (g), and (h) represent the deviation histograms for the snow-free period, while panels (i), (j), (k), and (l) correspond to the snow-covered period. The deviations are concentrated within − 5 to 5 cm for the snow-free period and − 10 to 10 cm for the snow-covered period, confirming the high accuracy and stability of the four solutions in snow depth retrieval. In 2020, Zhang et al. compared snow depths derived from phase combinations with in-situ observations at SG27, showing correlation coefficients (R) of 0.98, 0.97, and 0.95 for GPS, Galileo, and BDS, respectively, with corresponding RMSE values of 2.5 cm, 2.6 cm, and 3.5 cm [10]. In 2022, Zhan et al. applied the self-learning capability of a BP neural network to adaptively weight the contributions of different P351 GPS satellites, obtaining an RMSE of 3.0 cm and ME of 2.2 cm [27]. Although the methods differ, the RMSE and ME values obtained in this study are comparable, indicating that the robust estimation algorithm achieves superior snow depth retrieval accuracy and reliability in multi-SNR and multi-system fusion.

4 Experimental Analyses in Areas with Terrain Variation

The P351 station is located in Ketchum, USA (latitude: 43.87441°, longitude: − 114.71916°, elevation: 2692.608 m). The corresponding snow depth observation site is Galena Summit, with station ID 490, providing daily snow depth measurements from the Natural Resources Conservation Service National Water and Climate Center (NRCSNWCC) of the United States Department of Agriculture (USDA). The observed snow depth varies between 0 and 160 cm. The P676 station is located in Island Park, USA (latitude: 44.65422°, longitude: − 111.33795°, elevation: 2189.778 m), with the snow depth observation site named Island Park, station ID 546. Similarly, daily snow depth measurements from USDA NRCSNWCC are used for comparison and validation, with observed snow depths ranging from 0 to 120 cm.

Both P351 and P676 stations are equipped with TRIMBLE NETR9 receivers and AOAD/M_B DOME antennas. The GNSS observation data, provided by UNAVCO in RINEX 3.0 format, have a sampling interval of 15 s. The station height is assumed to be 2.0 m. The station environments and geographical locations are shown in Fig. 9, situated in high-latitude, snow-covered areas with minimal obstruction from tall buildings, making them suitable for snow depth retrieval studies. Figure 9a depicts the surrounding environment of P351 in the northwest, northeast, southeast, and southwest directions, while Fig. 9b shows the corresponding environment of P676.

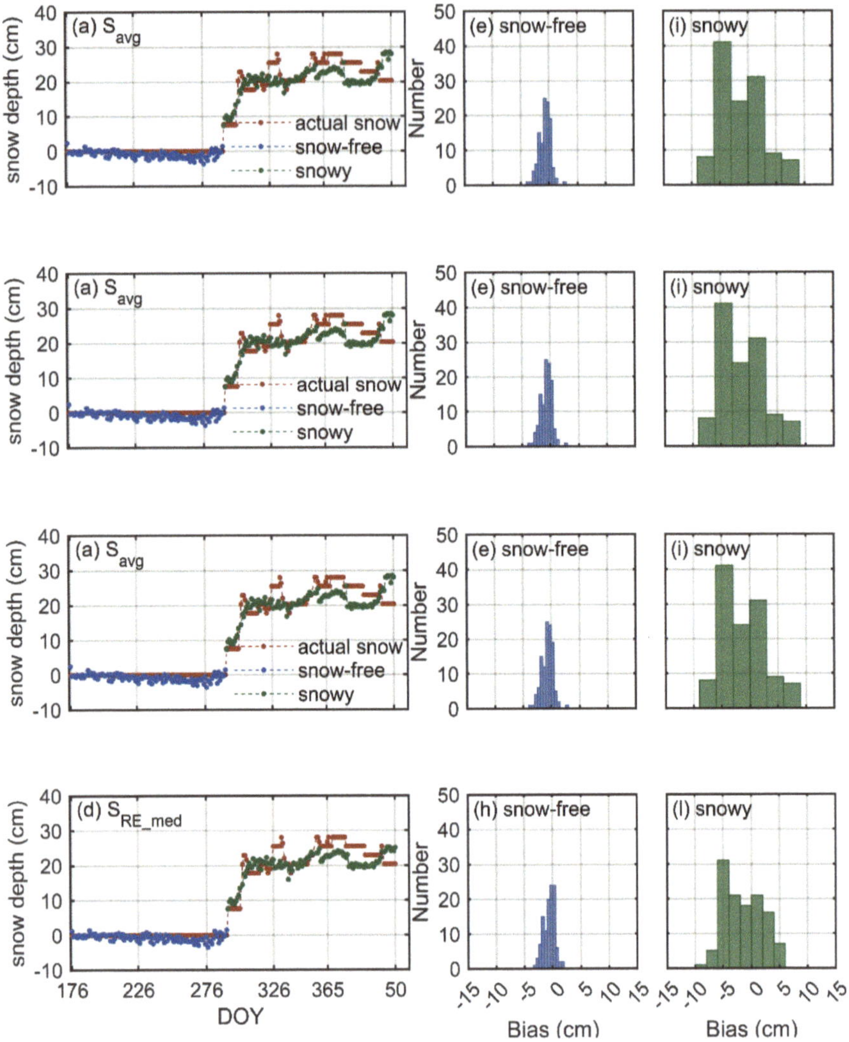

Fig. 8 Multi-system fused snow depth retrievals for four solutions and histograms of deviations during snow-free and snow-covered periods

The experimental data used in this study are summarized in Table 4, distinguishing between snow-free and snow-covered periods for separate processing. Precise GNSS ephemerides are provided by the Helmholtz-Centre Potsdam—German Research Centre for Geosciences (GFZ).

It is evident that tall trees and other objects around the stations may cause multipath effects due to reflections from trees, roads, and passing vehicles. To mitigate such influences, azimuth sectors less affected by the surrounding environment were selected for the experiments. Based on the station environments and the projections

Fig. 9 First Fresnel reflection zones of P351 and P676 stations projected on Google Earth at elevation angles of 5°, 10°, 15°, and 20°

Table 4 Snow-free and snow-covered periods of P351 and P676 stations in 2020

Station	Snow-covered periods(DOY)	Snow-free periods(DOY)
P351	001 ~ 148	191 ~ 280
P676	001 ~ 132	151 ~ 280

of the Fresnel reflection zones on Google Maps (Fig. 9), the suitable azimuth sectors are 230°–270° for P351 and 180°–240° for P676. For both stations, the elevation angle range of 5°–20° was used for the experiments.

4.1 *Implementation Procedure of TTC*

In many snow depth retrieval studies, it is often assumed that the area surrounding the GNSS station is an ideal reflection surface or that the azimuth sectors around the station have minimal terrain variation. However, for stations with significant surrounding terrain undulations, the impact on snow depth retrieval accuracy cannot be ignored, making it necessary to conduct Terrain Tilt Correction (TTC) research.

To quantify terrain errors, the ground height retrievals during the snow-free period were divided into azimuth sectors, with each sector spanning 10°, and the median of the retrievals within each sector was calculated. The difference between the median and the station height effectively reflects the terrain error, representing the concentrated influence of terrain on δSNR within the given azimuth sector and the elevation angle range of 5°–20°. To compare snow depth retrieval accuracy across different GNSS systems and reduce inter-system biases, the experiments were conducted separately for GPS, GLONASS, Galileo, BDS2, and BDS3.

Table 5 Distances from the station to reflection points at different reflection heights for P351 and P676, at 5° and 20° elevation angles, and the center distance of the elevation angle range

Reflection height (m)	20° (m)	5° (m)	Central distance (m)
0.5	1.4	5.7	3.5
1.0	2.7	11.4	7.1
1.5	4.1	17.1	10.6
2.0	5.5	22.9	14.2

Assuming no terrain variation around the station and that the station height is equal to the nominal station height, the distance from the reflection point to the station center is 22.9 m at a 5° elevation angle and 5.5 m at a 20° elevation angle. Since δSNR is derived from a series of SNR measurements within the 5°–20° elevation angle range, the terrain error within a fixed azimuth sector can be considered as representing the terrain variation at the average distance of the reflection points for 5° and 20° elevation angles (i.e., 14.2 m). Under different reflection surface heights (i.e., varying snow depths), the distances of reflection points from the station and the central distances corresponding to the elevation angle range also differ. Table 5 provides the distances for reflection surface heights of 0.5 m, 1.0 m, 1.5 m, and 2.0 m.

The terrain error at a fixed azimuth sector can be obtained using the following formula.

$$h_T = h_{azi} - h_S \tag{22}$$

In the equation, h_T represents the terrain error during the snow-free period, h_{azi} is the single ground-reflected height in the fixed azimuth sector during the snow-free period, and h_S is the receiver station height.

The lower the reflection surface height, the closer the center of the elevation angle range is to the station. Different reflection surface heights correspond to different terrain slope corrections: the lower the reflection surface, the smaller the terrain slope correction. There exists a linear relationship between the terrain slope correction and the snow-free terrain error h_T, which can be modeled using the following function.

$$h_{T_{cor}} = \frac{\frac{\frac{h_R}{\tan(e_{\min})} + \frac{h_R}{\tan(e_{\max})}}{2}}{\frac{\frac{h_S}{\tan(5^o)} + \frac{h_S}{\tan(20^o)}}{2}} h_T \tag{23}$$

In the equation, $h_{T_{cor}}$ represents the terrain slope correction value for different reflection surface heights during the snow period.

Subsequently, the snow depth h_{snow} after applying the terrain slope correction can be calculated.

$$h_{snow} = h_S - h_R + h_{T_{cor}} \tag{24}$$

4.2 Accuracy Assessment of Different GNSS and Multi-system Weighted Fusion

The above experiments on ground-reflected heights for different azimuth sectors during the snow-free period indicate that it is essential to conduct TTC research in areas with relatively steep terrain. It is hypothesized that at the P351 station, the selected azimuth sectors exhibit minimal terrain variation, with reflection area heights nearly equal to the station height; therefore, terrain correction is expected to have little impact on retrieval accuracy. In contrast, the selected azimuth sectors at the P676 station have significant terrain variation, so the TTC algorithm proposed herein is expected to substantially improve snow depth retrieval accuracy. To test this hypothesis, detailed experiments were conducted.

For a more intuitive presentation of snow depth retrieval performance, Figs. 10 and 12 show the comparison between retrieved and observed snow depths at the P351 and P676 stations, respectively, while Figs. 11 and 13 present the corresponding snow depth errors. In these figures, the blue line represents observed snow depths from the meteorological station, the orange line represents TTC-corrected snow depth retrievals, and the green line represents snow depth retrievals without TTC. Due to the limited number of BDS2 satellites, some DOYs within the study period lack retrievals; consequently, the number of valid snow depth retrieval days for BDS2 is 100 at P351 and 117 at P676.

At the P351 station, the snow depth retrievals with and without TTC showed little difference across GNSS constellations, which explains the overlapping of the orange and green lines observed in Figs. 10 and 11. This confirms the earlier assumption that the terrain height in the snow reflection area is approximately equal to the station height, resulting in negligible terrain variation; thus, the TTC algorithm does not significantly improve snow depth retrieval accuracy at P351. Comparing the TTC-based retrievals for individual GNSS, GLONASS achieved the highest accuracy with R = 0.92, RMSE = 18.8 cm, and ME = − 11.2 cm, followed by BDS3 with R = 0.92, RMSE = 18.9 cm, and ME = − 10.8 cm. BDS2 showed slightly lower accuracy with R = 0.90, RMSE = 22.8 cm, and ME = − 16.0 cm, but it was still used in the retrieval process. Error comparison plots indicate that most snow depth errors were concentrated between − 30.0 and 0.0 cm, gradually increasing to positive errors during the snowmelt period. This transition from negative to positive errors is likely caused by temporal and spatial differences between the GNSS station and the in situ meteorological measurements. Under normal circumstances, snow depth errors should fluctuate around 0.0 cm, whereas at P351, the negative bias indicates that the in situ measurements were generally larger than the retrieved snow depths.

At the P676 station, the difference between TTC and non-TTC snow depth retrievals was significant. Retrievals with TTC closely matched the observed snow depths, supporting the assumption that the snow reflection area exhibits substantial terrain variation compared to the station height. For TTC-based individual GNSS retrievals, GLONASS again achieved the highest accuracy with R = 0.97, RMSE = 7.8 cm, and ME = − 2.5 cm, followed by BDS3 with R = 0.96, RMSE = 9.1 cm, and

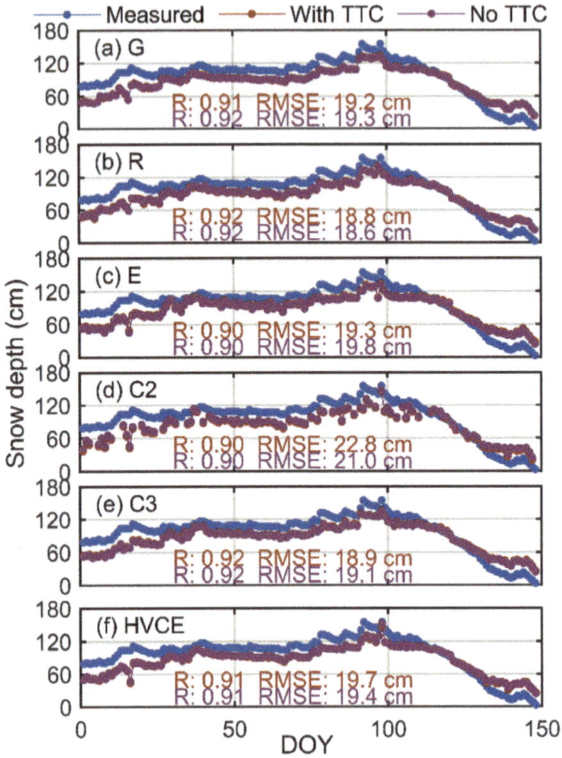

Fig. 10 Comparison of observed and retrieved snow depths at P351 station

ME = − 4.8 cm. BDS2 remained the least accurate, with R = 0.92, RMSE = 13.3 cm, and ME = − 6.2 cm. Compared to pre-TTC retrievals, the R values remained nearly constant across all systems, while RMSE and ME improved substantially. Specifically, for GLONASS, RMSE decreased by 64.9% and the absolute ME decreased by 87.9%; for BDS3, RMSE decreased by 56.5% and the absolute ME by 74.9%; for BDS2, RMSE decreased by 31.1% and the absolute ME by 65.9%. Post-TTC snow depth errors were concentrated between − 30.0 and 0.0 cm and mostly close to 0.0 cm, further demonstrating the effectiveness of the TTC algorithm and indicating minimal temporal and spatial discrepancies between in situ and P676 station measurements. During the late snowmelt stage, retrievals from different GNSS maintained similar snow depth values, with observable differences from the measured values, likely due to local anthropogenic factors, such as maintenance activities or vehicle parking within the selected azimuth sectors affecting SNR signals.

Figure 11f presents the P351 station results after HVCE weighting based on TTC. The multi-system fused snow depth retrievals achieved R = 0.91, RMSE = 19.7 cm, and ME = − 12.2 cm. Among GNSS, R ranked GLONASS = BDS3 > HVCE = GPS > Galileo = BDS2, RMSE ranked GLONASS < BDS3 < GPS < Galileo < HVCE < BDS2, and ME (negative errors) ranked Galileo > BDS3 > GLONASS > GPS > HVCE > BDS2. Figure 2f shows the P676 station results after HVCE

Fig. 11 Comparison of snow depth errors at P351 station with and without TTC

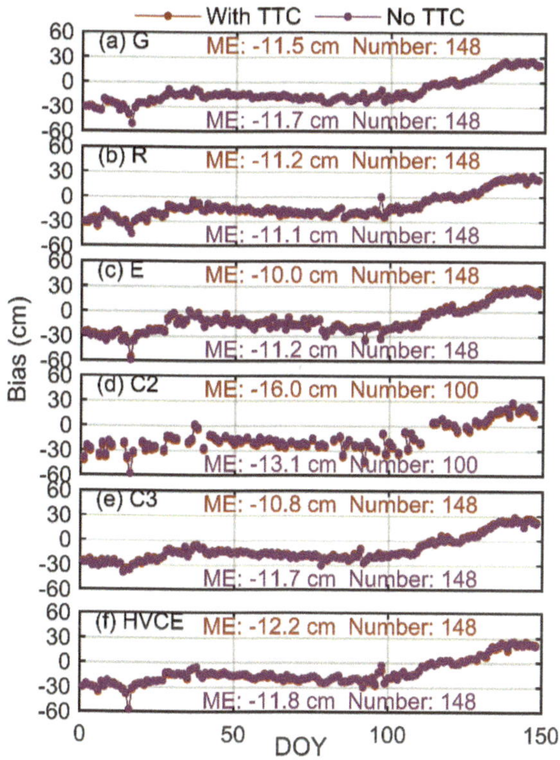

weighting with TTC, achieving R = 0.95, RMSE = 9.9 cm, and ME = − 4.9 cm. Compared to pre-TTC HVCE results, RMSE decreased by 53.3% and the absolute ME decreased by 74.2%. For R and RMSE, the GNSS ranking was GLONASS > BDS3 > GPS > HVCE > Galileo > BDS2; for ME (negative errors), the ranking was GLONASS > GPS > BDS3 > HVCE > BDS2 > Galileo. These results indicate that GLONASS and BDS3 consistently perform better in snow depth retrieval, while HVCE weighting improves the performance of less accurate systems while maintaining good correlation.

Furthermore, the advantage of HVCE weighting is evident from the daily snow depth retrieval counts for different GNSS during the snow-covered period. Figures 14 and 15 show histograms for P351 and P676 stations, with the x-axis representing the number of daily retrievals per satellite system and the y-axis representing the number of days corresponding to each count. GPS, GLONASS, Galileo, and BDS3 exhibit 10–70 retrievals per day, providing enough valid snow depth measurements. BDS2, however, has only a few retrievals, which may introduce bias when computing daily medians. This further confirms that HVCE weighting balances GNSS performance and ensures stability in snow depth retrievals. To compare the snow depth retrieval accuracy with and without TTC across GNSS, Fig. 16 presents RMSE and R for P351

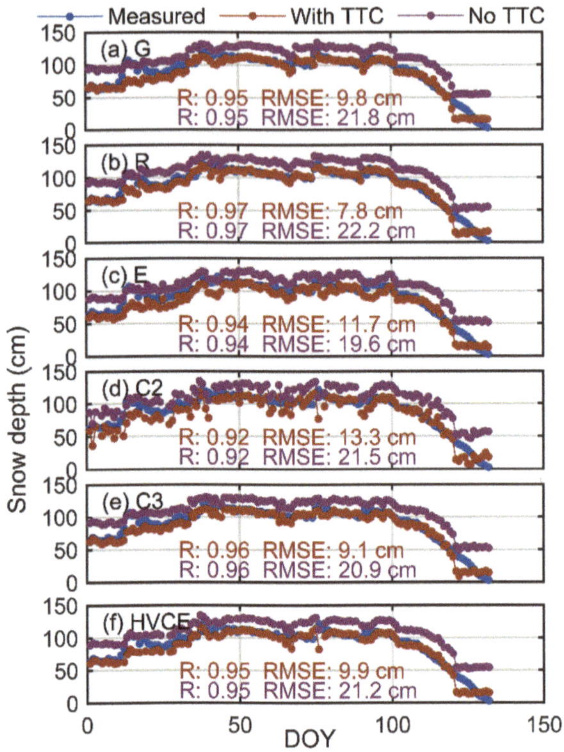

Fig. 12 Comparison of observed and retrieved snow depths at P676 station

and P676, providing a clear assessment of retrieval performance among different GNSS constellations.

In summary, for the two stations investigated in this study, the snow depth retrieval accuracy of different GNSS under TTC follows the descending order: GLONASS > BDS3 > GPS > Galileo > BDS2. GLONASS produced the lowest RMSE, approximately 18.8 cm at P351 and 7.8 cm at P676. This performance is likely attributable to two factors. First, GLONASS employs a lower Direct Sequence Spread Spectrum (DSSS) [13]. Second, the primary distinction between GLONASS and other GNSS lies in their channel access methods: GLONASS uses Frequency Division Multiple Access (FDMA) with each satellite transmitting on distinct RF subchannels, whereas GPS, Galileo, and BDS employ Code Division Multiple Access (CDMA) to differentiate satellites [22]. The FDMA method mitigates inter-satellite interference, rendering GLONASS signals less susceptible to spurious noise, which appears advantageous for GNSS-IR snow depth retrievals [28]. Both stations exhibited noticeable negative mean errors (ME), which may also be explained by two factors. First, the incident waves can penetrate the snow surface, leading to longer signal paths and higher apparent reflector heights, ultimately causing underestimation of snow depth [29]. Second, differences in temporal and spatial scales between in situ meteorological snow depth measurements and GNSS-IR retrievals contribute to discrepancies.

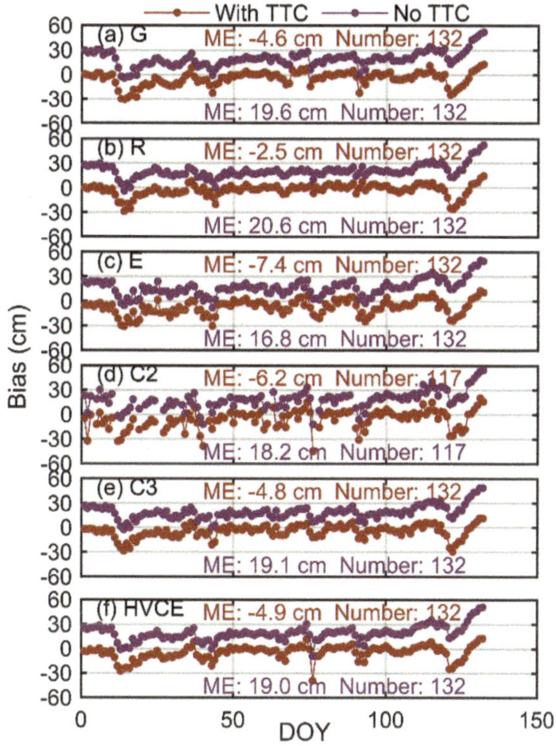

Fig. 13 Comparison of snow depth errors at P676 station with and without TTC

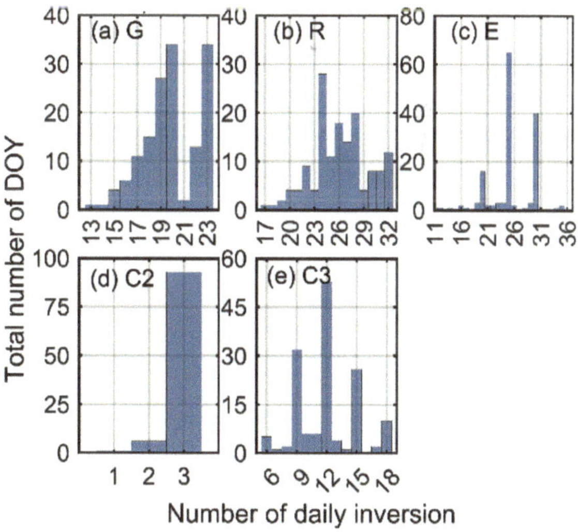

Fig. 14 Daily snow depth retrieval counts for different GNSS at P351 station during the snow-covered period

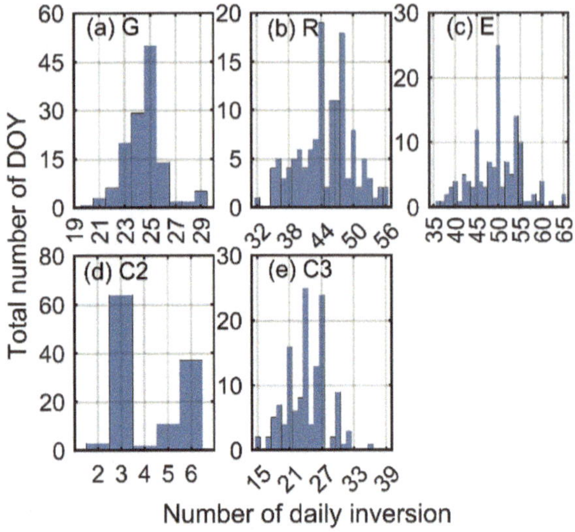

Fig. 15 Daily snow depth retrieval counts for different GNSS at P676 station during the snow-covered period

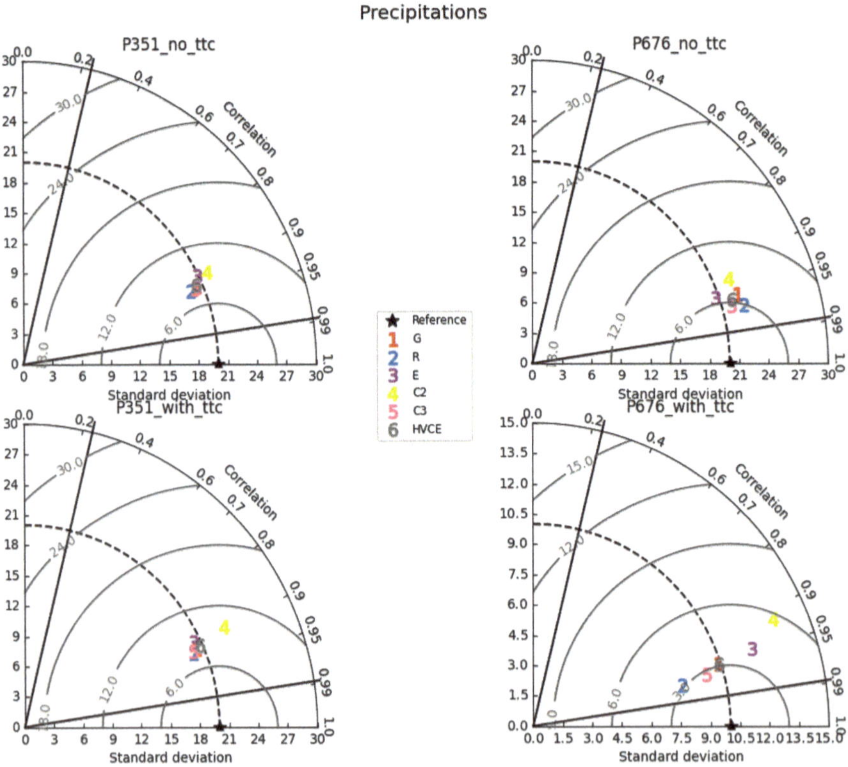

Fig. 16 Comparison of RMSE and correlation coefficient (R) at P351 and P676 stations with and without TTC

No significant systematic bias was observed among the GNSS constellations, indicating that the TTC algorithm, applied across different azimuth sectors for various GNSS, not only effectively corrects terrain effects but also reduces inter-system biases. Finally, comparison of snow depth errors during the snowmelt stage reveals a noticeable increase in retrieval errors. This phenomenon can be attributed to two main reasons. First, due to rapid snowmelt, the snow depth at 0 h and 24 h at the GNSS reflector area may differ substantially, while all retrievals within a 24 h period are treated as a single daily median, implicitly assuming constant snow depth and neglecting temporal systematic bias. Second, the retrieval times for different satellites under various GNSS are determined by the average observation times of δSNR arcs at low elevation angles. During snowmelt, the corresponding reflector surface gradually lowers, which diminishes the significance of frequencies in the Lomb–Scargle periodogram (LSP) analysis, thereby affecting the snow depth retrieval results.

5 Conclusion

This chapter investigates terrain-related issues in snow depth retrieval and presents two main approaches. For sites with gentle terrain, a robust estimation algorithm was introduced to fuse snow depth retrievals from different GNSS systems and SNR types. For sites with pronounced topographic variation, terrain tilt correction (TTC) and Helmert variance component estimation (HVCE) algorithms were applied to fuse multi-GNSS retrievals.

1. Experiments at SG27 and Evaluation of Robust Estimation

Four experimental schemes were conducted at SG27 to verify and extend a novel method for selecting optimal daily snow depth values: the robust estimation approach. Detailed analysis demonstrated improvements in snow depth monitoring accuracy. Key findings include:

(1) The S1W and S2W data at SG27 were identical and therefore invalid. Simulations showed substantial differences in SNR across carrier frequencies. Accordingly, invalid SNR types (S1W and S2W) were discarded to avoid compromising retrieval accuracy.
(2) RMSE and mean error (ME) values varied among different SNR types; however, RMSE remained below 5.0 cm and ME fluctuated around 0.0 cm, confirming the reliability of GNSS-IR snow depth retrievals. Comparisons of four retrieval solutions—Savg, Smed, SRE_avg, and SRE_med—revealed that Savg exhibited noticeably larger errors than the other three solutions, indicating that S_avg is unsuitable for snow depth estimation. This further validates the accuracy and stability of the robust estimation algorithm.
(3) Across different GNSS constellations, all four solutions achieved centimeter-level retrieval accuracy. Snow-free periods exhibited higher precision than snow-covered periods, likely due to signal absorption by the snow surface affecting

path delay. The C3-MEO constellation showed slightly poorer performance due to limited data availability, suggesting that future studies with more extensive datasets are warranted.

(4) After multi-system fusion, the RMSEs during snow-free periods were 1.3 cm, 0.9 cm, 1.2 cm, and 1.2 cm, and MEs were 0.7 cm, 0.1 cm, 0.6 cm, and 0.6 cm for the four solutions, respectively. During snow-covered periods, correlation coefficients (R) were 0.686, 0.744, 0.759, and 0.759, RMSEs were 4.0 cm, 3.8 cm, 3.7 cm, and 3.7 cm, and MEs were − 1.2 cm, − 1.5 cm, − 1.4 cm, and − 1.4 cm, respectively. SRE_avg and SRE_med achieved the best performance, indicating superior snow depth retrieval capability.

2. Experiments at P351 and P676 Using TTC and HVCE

Snow depth retrieval experiments incorporating TTC and HVCE were conducted at P351 and P676 to evaluate GNSS-specific performance. The key contributions of this chapter include:

(1) A novel TTC algorithm based on varying azimuth sectors of the reflector area was proposed. Applying TTC corrected terrain-induced effects around the station and demonstrated its effectiveness across different GNSS constellations. Compared with experiments without TTC, the algorithm significantly improved retrieval accuracy at sites with pronounced topography.
(2) Comparison of TTC-based retrievals across GNSS constellations at the two stations showed descending accuracy as follows: GLONASS > BDS3 > GPS > Galileo > BDS2. Post-TTC correlation coefficients exceeded 0.90 for all GNSS, indicating strong agreement with observed snow depth. RMSE was lowest for GLONASS (~ 18.8 cm at P351, ~ 7.8 cm at P676) and highest for BDS2 (~ 22.8 cm at P351, ~ 13.3 cm at P676).
(3) An HVCE-based weighting method was applied to combine TTC-corrected snow depth retrievals from different GNSS. Multi-system weighted retrievals yielded $R = 0.91$ and $RMSE = 19.7$ cm at P351, and $R = 0.95$ and $RMSE = 9.9$ cm at P676. Both stations exhibited strong correlation and high accuracy, confirming the feasibility of HVCE-based weighting. The multi-system fusion strategy enhances retrieval stability, providing a reference for further multi-GNSS studies and other GNSS-IR applications, as well as a practical approach for integrating diverse surface monitoring methods.

References

1. Larson KM, Gutmann ED, Zavorotny VU, Braun JJ, Williams MW, Nievinski FG (2009) Can we measure snow depth with GPS receivers?. Geophys Res Lett 36:L17502. https://doi.org/10.1029/2009GL039430
2. Larson KM, Nievinski FG (2013) GPS snow sensing: results from the EarthScope Plate Boundary Observatory. GPS Solut 17:41–52. https://doi.org/10.1029/2009GL039430

3. Nievinski FG, Larson KM (2014) Inverse modeling of GPS multipath for snow depth estimation—part I: formulation and simulations. IEEE Trans Geosci Remote Sens 52:6555–6563. https://doi.org/10.1029/2009GL039430
4. Nievinski FG, Larson KM (2014) Inverse modeling of GPS multipath for snow depth estimation—part II: application and validation. IEEE Trans Geosci Remote Sens 52:6564–6573. https://doi.org/10.1029/2009GL039430
5. Najibi N, Jin S, Wu X (2015) Validating the variability of snow accumulation and melting from GPS-reflected signals: forward modeling. IEEE Trans Antennas Propagat 63:2646–2654. https://doi.org/10.1029/2009GL039430
6. Jin S, Qian X, Kutoglu H (2016) Snow depth variations estimated from GPS-reflectometry: a case study in Alaska from L2P SNR data. Remote Sens 8:63. https://doi.org/10.3390/rs8010063
7. Larson KM, Small EE (2016) Estimation of snow depth using L1 GPS signal-to-noise ratio data. IEEE J Sel Top Appl Earth Obs Remote Sens 9:4802–4808. https://doi.org/10.1109/JSTARS.2015.2508673
8. Roesler C, Larson KM (2018) Software tools for GNSS interferometric reflectometry (GNSS-IR). GPS Solut 22:80. https://doi.org/10.1007/s10291-018-0744-8
9. Wei H, He X, Feng Y, Jin S, Shen F (2019) Snow depth estimation on slopes using GPS-interferometric reflectometry. Sensors 19:4994. https://doi.org/10.3390/s19224994
10. Zhang Z, Guo F, Zhang X (2020) Triple-frequency multi-GNSS reflectometry snow depth retrieval by using clustering and normalization algorithm to compensate terrain variation. GPS Solut 24:52. https://doi.org/10.3390/rs12182901
11. Tu J, Wei H, Zhang R, Yang L, Lv J, Li X, Nie S, Li P, Wang Y, Li N (2021) GNSS-IR snow depth retrieval from multi-GNSS and multi-frequency data. Remote Sens 13:4311. https://doi.org/10.3390/rs13214311
12. Wang X, Zhang S, Wang L, He X, Zhang Q (2020) Analysis and combination of multi-GNSS snow depth retrievals in multipath reflectometry. GPS Solut 24:77. https://doi.org/10.1007/s10291-022-01308-1
13. Hu Y, Yuan X, Liu W, Wickert J, Jiang Z (2022) GNSS-R snow depth inversion based on variational mode decomposition with multi-GNSS constellations. IEEE Trans Geosci Remote Sens 60:1–12. https://doi.org/10.1007/s10291-022-01308-1
14. Liu S, Zhang J, Wan W, Liang H, Liu B, Guo Z (2022) A comprehensive evaluation of utilizing BeiDou data to estimate snow depths from two ground-based stations. GPS Solut 26:118. https://doi.org/10.1007/s10291-022-01308-1
15. Zhou W, Liu L, Huang L, Yao Y, Chen J, Li S (2019) A new GPS SNR-based combination approach for land surface snow depth monitoring. Sci Rep 9:3814. https://doi.org/10.3390/rs12182901
16. Yang Y (1994) Robust estimation for dependent observation. Manuscra Geod 19:10–17
17. Yang Y, Cheng MK, Shum CK, Tapley BD (1999) Robust estimation of systematic errors of satellite laser range. J Geod 73:345–349. https://doi.org/10.3390/rs12182901
18. Yang Y, Song L, Xu T (2002) Robust estimator for correlated observations based on bifactor equivalent weights. J Geod 76:353–358. https://doi.org/10.3390/rs12182901
19. Xu B, Chen Y, Zhang S, Wang J (2020) Improved point–line visual–inertial odometry system using Helmert variance component estimation. Remote Sens 12:2901. https://doi.org/10.3390/rs12182901
20. Li M, Nie W, Xu T, Rovira-Garcia A, Fang Z, Xu G (2020) Helmert variance component estimation for multi-GNSS relative positioning. Sensors 20:669. https://doi.org/10.3390/s20030669
21. Meurer M, Antreich F (2017) Signals and Modulation. In: Teunissen PJG, Montenbruck O (eds) Springer handbook of global navigation satellite systems, springer international publishing, Cham pp 91–119. https://doi.org/10.1007/978-3-319-42928-1_4
22. Tabibi S, Geremia-Nievinski F, van Dam T (2017) Statistical comparison and combination of GPS, GLONASS, and multi-GNSS multipath reflectometry applied to snow depth retrieval. IEEE Trans Geosci Remote Sens 55:3773–3785. https://doi.org/10.1109/TGRS.2017.2679899

23. Wang X, He X, Xiao R, Song M, Jia D (2021) Millimeter to centimeter scale precision water-level monitoring using GNSS reflectometry: Application to the South-to-North Water Diversion Project, China. Remote Sens Environ 265:112645. https://doi.org/10.1016/j.rse.2021.112645
24. Zheng N, Chai H, Chen L, Ma Y, Tian X (2022) Snow depth retrieval by using robust estimation algorithm to perform multi-SNR and multi-system fusion in GNSS-IR. Adv Space Res S0273117722009358. https://doi.org/10.1016/j.asr.2022.10.014
25. Zhang S, Zhang C, Zhao Y, Li H, Liu Q, Pang X (2021) Snow depth estimation based on GNSS-IR cluster analysis, Meas Sci Technol 32:095801. https://doi.org/10.1088/1361-6501/abee54
26. Williams SDP, Nievinski FG (2017) Tropospheric delays in ground-based GNSS multipath reflectometry-Experimental evidence from coastal sites: Tropospheric Delays in GNSS-MR, J Geophys Res Solid Earth 122:2310–2327. https://doi.org/10.1002/2016JB013612
27. Zhan J, Zhang R, Tu J, Lv J (2022) GNSS-IR Snow depth retrieval based on the fusion of multi-satellite snr data by the bp neural network. Remote Sens 16. https://doi.org/10.3390/rs14061395
28. Dale SA, Daly P, Kitching ID (1989) Understanding signals from glonass navigation satellites. Int J Satell Commun 7:11–22. https://doi.org/10.1002/sat.4600070105
29. Zhou Z, Yu K, Bu J., Li Y, Han S (2022) Snow depth estimation based on combination of pseudorange measurements of GNSS geodetic receivers. Adv Space Res 69:1439–1450. https://doi.org/10.1016/j.asr.2021.11.004

Chapter 11
Ground-Based GNSS-MR Sea Level Inversion

Xiaolei Wang

Abstract Sea level is an important parameter for ensuring coastal safety, monitoring marine climate, and maintaining height reference benchmarks. Effectively monitoring sea-level changes and predicting its rising trend has become a hot research topic in recent years. The ground-based Global Navigation Satellite System (GNSS) reflection signal remote sensing method has been proven to be applicable to sea level monitoring. Compared with traditional tide gauge measurements, the GNSS-Multipath Reflectometry (GNSS-MR) technique based on Signal-to-Noise Ratio (SNR) observations offers advantages such as low cost, continuous tracking, and all-weather capability. However, this technique still suffers from problems such as limited inversion accuracy and low temporal resolution. To address these issues, this chapter proposes a combined sea level inversion method using multi-constellation and multi-frequency GNSS SNR data, improving both accuracy and temporal resolution of tide monitoring through data fusion and multi-frequency combination. The study develops a fusion algorithm for multi-constellation, multi-frequency GNSS-MR sea level inversion by integrating GPS, GLONASS, Galileo, and BeiDou observation data. A robust regression approach based on the IGGIII model is applied to fuse the sea level inversion results of the four systems, where both sea surface dynamics and tropospheric delay correction terms are simultaneously adjusted. Special emphasis is given to analyzing the characteristics of new BeiDou frequency signals in GNSS-MR sea level inversion.

1 Introduction

With the continuous development and improvement of GNSS technology, it has been demonstrated that GNSS reflected signals can be exploited for remote sensing studies. The GNSS-Reflectometry (GNSS-R) technique, which utilizes GNSS satellite reflection signals, has proven valuable in monitoring various environmental parameters, such as soil moisture, vegetation index, sea level variations, snow depth,

X. Wang (✉)
School of Earth Sciences and Engineering, Hohai University, NanKing, Jiangsu, China
e-mail: gnsswxl@hhu.edu.cn

K. Yu and N. Zheng (eds.), *Advances in GNSS Reflectometry*, Navigation: Science and Technology 19, https://doi.org/10.1007/978-981-92-2325-1_11

volcanic ash, and permafrost changes [1–4]. In the field of tide monitoring, a subdiscipline of ground-based GNSS-R research—single-antenna GNSS-MR (GNSS Multipath Reflectometry)—uses standard geodetic receivers to analyze oscillatory frequency variations in SNR time series caused by multipath interference, thereby enabling coastal sea level monitoring. The technique offers advantages such as low cost, continuous tracking, and all-weather capability [5, 6]. At present, the major GNSS constellations—GPS, Galileo, GLONASS, and BeiDou—as well as regional systems such as Japan's QZSS (Quasi-Zenith Satellite System) and India's IRNSS (Indian Regional Navigation Satellite System), have become increasingly mature. The four GNSS constellations have already provided global services for long periods of time. The number of available satellites and signal frequencies continues to grow, providing abundant data sources for applying GNSS-MR technology to tide monitoring. Consequently, GNSS-MR sea level monitoring is emerging as a research hotspot in GNSS-based remote sensing.

In 2000, Anderson et al. used SNR data from a single-antenna geodetic receiver to derive a time series of dominant frequencies proportional to antenna height [7]. In 2002, Lowe et al. conducted an airborne reflection altimetry experiment and achieved 5 cm accuracy with 5 km spatial resolution using P(Y)-code delay information [8]. Bilich et al. [9] studied the relationship between oscillatory characteristics in GPS SNR data and reflecting surface properties, further identifying the mapping relationship between SNR observations and GNSS site environments. In 2008, Larson systematically proposed a GNSS remote sensing technique based on SNR observations and successfully retrieved soil moisture from GPS data [10]. Subsequent research by Larson and others gradually expanded the technique to retrieve vegetation index, sea surface height, snow depth, volcanic ash, and permafrost, achieving a series of valuable results [2, 11–19].

In ground-based GNSS-MR applications, sea level inversion mainly depends on SNR frequency analysis. The SNR is the ratio between the power of the received signals and the noise; these signals are the interference signals between the direct and reflected signals. Different reflecting surfaces can lead to different interference signals and SNR arcs. The GNSS-MR technique relates the characteristics of SNR to the parameters of the reflecting surface, making geodetic GNSS receivers capable of estimating the water level [17]. Currently, most GNSS signals demonstrate the effectiveness of water-level GNSS-MR retrieval. In 2013, Larson et al. [17] used SNR data from GPS L1C/A signals to estimate the water level. Löfgren and Haas [20] used L1C/A and L2P data from GPS, and L1 and L2 data from GLONASS to estimate the water level. Strandberg et al. [21] used L1C/A, L2P, and L2C GPS data to retrieve the water level. Jin et al. [22] used L2I (B1I), L6I (B3), and L7I (B2I) signal data of BDS, and L2P and L5C signal data of GPS to estimate the water level, and evaluated their results based on tide gauge records. Wang et al. [23] used L1C/A, L2P, L2C, and L5C signals of GPS; L1C, L1P, L2C, and L2P signals of GLONASS; L1, L5, L7, and L8 signals of Galileo; and L2 (B1I), L6 (B3), and L7 (B2I) of BDS to estimate the water level, and proposed a combination method to combine multi-GNSS retrievals.

Currently, the GNSS-MR sea-level monitoring technique has developed a relatively mature framework for sea level inversion. During inversion, height deviations caused by sea surface dynamics and tropospheric refraction-induced delay must be corrected. Even after these corrections, the practical application of GNSS-MR still faces two key challenges: insufficient accuracy and low temporal resolution. Further improvements in inversion principles and the identification of unknown error sources are needed. With the continuous advancement of global and regional satellite navigation constellations, richer datasets are now available for GNSS-MR applications in tide monitoring. This chapter focuses on multi-constellation and multi-frequency GNSS data processing methods to enhance the accuracy and temporal resolution of GNSS-MR sea Level inversion.

2 Theoretical Basis of GNSS-MR Sea Level Monitoring

In typical geodetic GNSS measurements with antenna facing up, the signals received by the GNSS receiver contain not only the direct signal from the satellite but also reflected signals from surrounding surface media. These reflected signals interfere with the direct signals, introducing errors into the measurements for positioning, navigation, and timing. This phenomenon, known as the multipath effect, occurs when multiple signal paths interfere, leading to decreased GNSS positioning accuracy. Multipath interference is therefore recognized as one of the main sources of error in GNSS measurements.

Consider only a single multipath reflection exists (as illustrated in Fig. 1), and let the direct signal be S and the reflected signal be S'. When the antenna is placed on a tripod or top of a pole, i.e., the ground-based case, the reflected signal carries information about the ground surface (Fig. 1). Similarly, if the receiver is positioned near the sea, the reflected signal carries information about the sea surface, which interferes with the direct signal at the antenna. Figures 1 and 2 shows a schematic diagram of the GNSS-MR sea level inversion method. The path length difference D between the direct and reflected signals can be expressed as [17]:

$$D = D_1 - D_2 = h/\sin e - D_1 \cos 2e = 2h \sin(e) \quad (1)$$

Here h represents the vertical distance from the instantaneous sea surface to the antenna phase center, i.e., the reflector height (RH), and e is the satellite elevation angle.

Due to the existence of this path difference D, a phase delay ψ occurs between the direct and reflected signals, which can be expressed as:

$$\psi = 2\pi D/\lambda \quad (2)$$

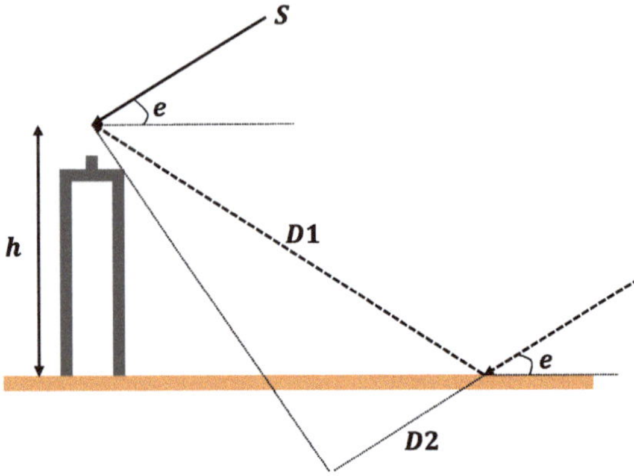

Fig. 1 Principle of single reflection

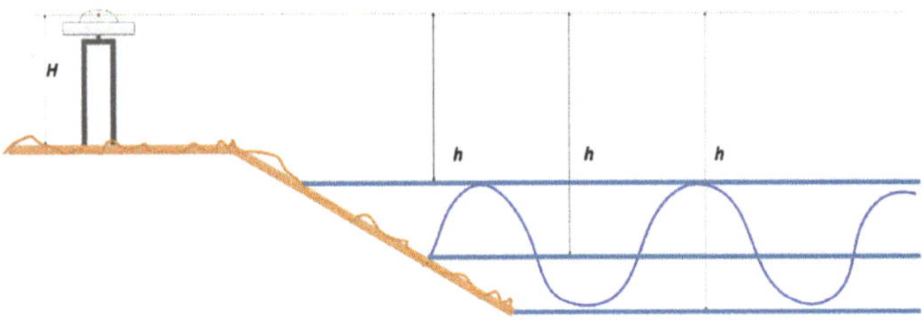

Fig. 2 Schematic diagram of GNSS-MR sea level inversion. The sea level is calculated as a reference datum minus the RH

Here λ denotes the wavelength of the GNSS signal. From this equation, the phase delay caused by multipath effects is linearly related to the sine of the satellite elevation angle. Based on this theoretical relationship, SNR data can be applied to retrieve h.

The phase difference between the direct and reflected signals can be expressed as:

$$\psi = \frac{2\pi}{\lambda}D = \frac{4\pi h}{\lambda}\sin(e) \tag{3}$$

According to the vector relationship in the figure, the power of superimposed wave signal P_R can be expressed as:

$$P_R = P_d + P_r + 2\sqrt{P_d P_r}\cos\varphi_i \tag{4}$$

Here P_d is the power of the direct signal and P_r is the power of the reflected signal. According to the definition of SNR, we can obtain:

$$SNR = \frac{1}{P_N}\left(P_d + P_r + 2\sqrt{P_d P_r}\cos\varphi_i\right) = \frac{P_d + P_r}{P_N} + \frac{2\sqrt{P_d P_r}}{P_N}\cos\varphi_i \tag{5}$$

Equation (5) divides the SNR into two parts: the first part $\overline{s}$, mainly containing the direct signal component $\left(\frac{P_d+P_r}{P_N}, P_r \ll P_d\right)$; the second part $\left(\left(\frac{2\sqrt{P_d P_r}}{P_N}\cos\varphi_i\right)\right)$ is the major interferometric component, which contains the antenna height information [47].

Figure 3a shows the variation of SNR for GPS satellite PRN 04 (L1 carrier) observed at a coastal CORS station. The SNR amplitude is mainly affected by multipath interference, antenna gain, and satellite transmission power. Figure 3a illustrates the variation in SNR as a satellite rises from a low elevation angle, is acquired by the receiver, moves toward the zenith direction, and then gradually descends. Throughout this process, the SNR generally exhibits a convex parabolic shape, with the highest SNR value occurring when the satellite is near the zenith direction. This phenomenon is primarily related to the antenna gain, which is higher in the zenith direction and lower at low elevation angles. Due to the lower gain of the direct signal at low elevation angles and the relatively stronger energy of the reflected signal at these angles, interference oscillations are observed in the low elevation angle range. This is precisely why GNSS-MR technology typically utilizes data from low elevation angle intervals.

For G04 satellite, the original SNR sequence is fitted by a low-order polynomial (the red curve in Fig. 3a), and then subtracted from the original sequence, to obtain the data affected by multipath, that is, the SNR residual sequence in Fig. 3b. The horizontal axis corresponds to the sine of the satellite elevation angle (a function of time), and the vertical axis represents the SNR amplitude (converted from dB-Hz to linear units, V/V). For low elevation angles, the SNR residual amplitude satisfies:

$$A_m = A\cos\left(\frac{4\pi \mathrm{h}}{\lambda}\sin e + \phi\right) \tag{6}$$

Here, A is the oscillation amplitude and ϕ is the phase offset.

The time-varying phase $\psi = 4\pi h \sin e/\lambda$ varies with the elevation angle, which hides a frequency information f of the time series A_m:

$$2\pi f = \frac{d\psi}{d(\sin e)} \tag{7}$$

where $\sin e$ is treated as an independent time variable. Under the assumption of a static reflecting surface (ignoring temporal variations of h), we have:

$$2\pi f = 4\pi\frac{\overline{h}}{\lambda} \tag{8}$$

Therefore, a spectral analysis is performed on the time series (6) to obtain the dominant frequency f, from which the vertical distance $\overline{h}$ between the antenna phase center and the sea surface can be derived under the static assumption. At present, in GNSS-MR sea level inversion theories based on spectral analysis, the primary method used for frequency extraction is the Lomb-Scargle Periodogram (LSP). However, two

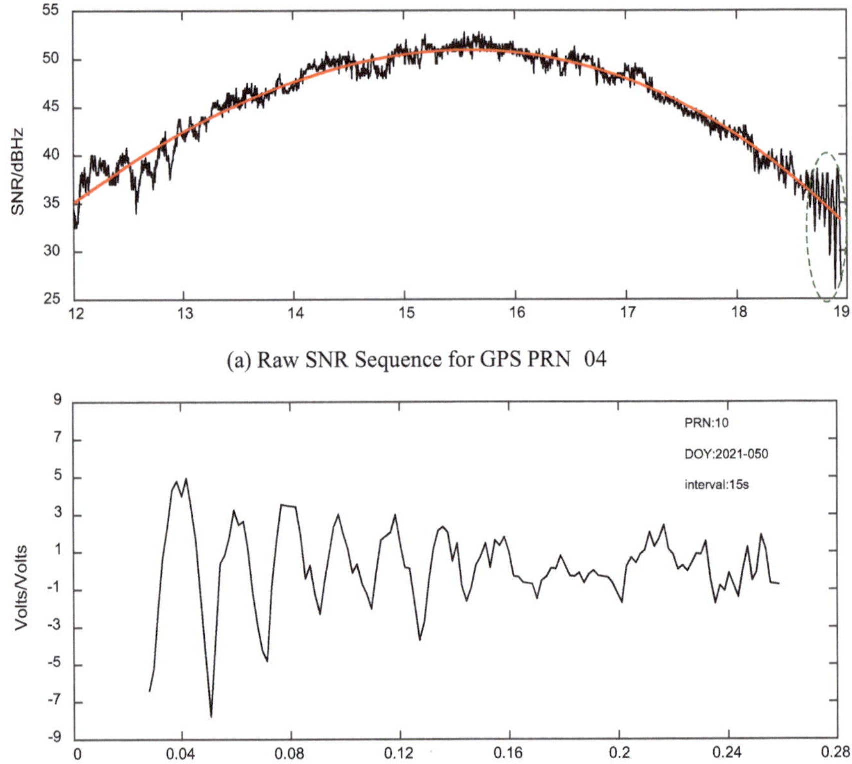

(a) Raw SNR Sequence for GPS PRN 04

(b) Detrended SNR Residual Sequence

Fig. 3 Variation of SNR sequences

systematic errors have been widely recognized in this process—namely, sea surface dynamic variation error and tropospheric delay error.

2.1 Sea Surface Dynamic Correction

The classical GNSS-MR sea level retrieval principle described in Eqs. (6–8) is derived under the assumption of a static and invariant RH. In practice, the sea surface is constantly changing, especially in regions where the tidal amplitude exhibits significant diurnal variations, making this static assumption unreasonable. Larson et al. [18] were the first to propose a correction for the height errors caused by dynamic sea surface changes. In their retrieval model, they introduced two time-varying dynamic parameters: the rate of satellite elevation change $\dot{e}\left(\frac{\mathrm{d}e}{\mathrm{d}t}\right)$ and the rate of sea surface height change $\dot{h}\left(\frac{\mathrm{d}h}{\mathrm{d}t}\right)$. Differentiating the phase term φ_i in terms of both sin e and antenna height h yields:

$$\omega_{\varphi} = 2\pi f = \frac{4\pi}{\lambda}\frac{\tan e}{\dot{e}}\dot{h} + \frac{4\pi}{\lambda}h \tag{9}$$

which can be rewritten as:

$$\frac{\lambda f}{2} = \frac{\tan e}{\dot{e}}\dot{h} + h \tag{10}$$

By substituting Eq. (8) into Eq. (10), we obtain:

$$\overline{h} = \frac{\tan e}{\dot{e}}\dot{h} + h \tag{11}$$

Here h represents the effective height under dynamic sea surface conditions. The correction of the parameter $\overline{h}$ to obtain h constitutes the dynamic sea surface correction in sea level retrieval.

In classical strategies, dynamic height correction involves first deriving the time series of $\overline{h}$, then estimating $\dot{h}$, and finally subtract $\Delta h_{\dot{h}} = \frac{\tan(e)}{\dot{e}}\dot{h}$ from the time series $\overline{h}$ to get the corrected time series h. Larson et al. [19] employed tidal harmonic analysis to model sea-level changes using the static inversion $\overline{h}$ sequence. The harmonic model is given by:

$$\overline{h} = \sum_{i=1}^{N} C_i f_i \cos(w_i t + v_i) + S_i f_i \sin(w_i t + v_i) \tag{12}$$

Here S_i andC_i represent the sine and cosine coefficients, N denotes the data length, w_i is the tidal frequency, v_i is the tidal phase, and f_i is the amplitude adjustment factor. After obtaining these parameters through the tidal harmonic analysis model, differentiating this model yields the sea surface height rate $\dot{h}$:

$$\dot{h} \approx \dot{\overline{h}} = \sum_{i=1}^{N} -w_i C_i f_i \sin(w_i t + v_i) + S_i f_i \cos(w_i t + v_i) \tag{13}$$

Subtracting the dynamic term $\frac{\tan(e)}{\dot{e}}\dot{h}$ from the uncorrected series $\overline{h}$ produces the corrected time series h:

$$h = \overline{h} - \frac{\tan e}{\dot{e}}\dot{h} = \overline{h} - \Delta h_{\dot{h}} \tag{14}$$

where $\Delta h_{\dot{h}} = \frac{\tan e}{\dot{e}}\dot{h}$.

2.2 Tropospheric Delay Correction

As GNSS signals propagate through the troposphere, refraction causes the signal path to bend, introducing delays that affect height estimations [24]. Anderson [7] first noted that atmospheric refraction could affect sea level inversion accuracy. Santamaría-Gómez et al. [25] applied an empirical astronomical refraction model based on atmospheric pressure and temperature to eliminate tropospheric delay in GPS-MR sea-level estimation. Williams et al. [24] employed the GPT2w (Global Pressure and Temperature 2 Wet) and VMF1 (Vienna Mapping Functions 1) models to mitigate tropospheric effects.

The GPT2w model improves upon GPT2 by providing grid-based meteorological parameters (pressure, temperature, water vapor pressure, lapse rate, and VMF1 coefficients) with a resolution of (1° × 1°) [26]. These parameters are estimated using trigonometric functions:

$$\begin{aligned} r(t) = A_0 + A_1 \cos\left(\frac{\text{doy}}{365.25} 2\pi\right) + B_1 \sin\left(\frac{\text{doy}}{365.25} 2\pi\right) \\ + A_2 \cos\left(\frac{\text{doy}}{365.25} 4\pi\right) + B_2 \cos\left(\frac{\text{doy}}{365.25} 4\pi\right) \end{aligned} \tag{15}$$

Here A_0, A_1, B_1, A_2, B_2 are model coefficients, and doy is the day of the year. Using the external grid file gpt2_1w.grd and the station coordinates, the model coefficients can be calculated by applying the bilinear interpolation method.

The meteorological parameters obtained from gpt2_1w.grd correspond to the four grid points closest to the station. Data measured at different heights can be obtained using the following equation:

$$\begin{cases} T = T_0 + dT \cdot dh \\ P = P_0 \cdot \exp\left\{-g_m \cdot \frac{dM_{tr}}{R_g \cdot T_0 \cdot (1+0.6077 \cdot Q)} \cdot dh\right\} / 100 \\ e_0 = Q \cdot \frac{P}{0.622+0.378 \cdot Q} \\ e = e_0 \cdot \left(100 \frac{P}{P_0}\right)^{\lambda+1} \end{cases} \tag{16}$$

Here T_0, P_0 and e_0 denote temperature (unit°C), air pressure (unit hPa) and water vapor pressure (unit hPa) of the grid network respectively; T, P and e are the temperature, pressure and water vapor pressure normalized to the station height; dT is the temperature drop rate (unit K/km); h is the height (unit m); g_m is gravitational acceleration, $g_m = 9.80665$ m/s^2; dM_{tr} is molar mass of dry air, $dM_{tr} = 28.965 \times 10^{-3}$ kg/mol; R_g is the universal gas constant, $R_g = 8.3143$ J/(K · mol); Q is the specific humidity (unit g/kg); and λ is the water vapor descent factor. After normalizing the grid point data to the station height, the meteorological parameters at the station can be obtained by bilinearly interpolating the parameters from the surrounding grid points. Finally, the meteorological data at the station computed by the GPT2w model can be substituted into the Saastamoinen and Askne-Nordius models to calculate the zenith hydrostatic delay and zenith wet delay components at the station [27].

The VMF1 model is an improved version developed by Boehm et al. [28] based on the VMF model, utilizing observational meteorological data from the European Centre for Meteorological Warning (ECMWF). This enhanced model re-fits the coefficients of the hydrostatic mapping function's b_h and c_h, yielding a $b_h = 0.0029$. The expression for c_h is:

$$c_h = c_0 + \left[\left(\cos\left(\frac{\text{doy} - 28}{365} \cdot 2\pi + \Psi\right) + 1\right) \cdot \frac{c_{11}}{2} + c_{10}\right] \cdot (1 - \cos\varphi) \quad (17)$$

where $c_0 = 0.062$, constant c_{11}, c_{10}, and Ψ values differ between the Northern and Southern Hemispheres. In the Northern Hemisphere, $c_{11} = 0.005$, $c_{10} = 0.001$, $\Psi = 0$; in the Southern Hemisphere, $c_{11} = 0.007$, $c_{10} = 0.002$, $\Psi = \pi$.

Combining GPT2w and VMF1, the tropospheric delay correction is given by:

$$\tau_T = 2\Delta\tau_h^z \cdot m_h(e) + 2\Delta\tau_w^z \cdot m_w(e) \quad (18)$$

where $\Delta\tau^z = \tau^z(-H) - \tau^z(0)$ represents the zenith delay difference between the antenna and ground positions. m_h and m_w denote the dry delay mapping function and wet delay mapping function, respectively. These parameters can be calculated using the GPT2w model and the VMF1 model, respectively. Consequently, the inversion height deviation Δh_T caused by tropospheric delays τ_T is expressed as:

$$\Delta h_T = \frac{1}{2}\frac{\partial \tau_T}{\partial \sin e} \quad (19)$$

where Δh_T always negative, under the influence of tropospheric delay, the vertical inversion distance $\overline{h}$ and the vertical distance under the static assumption have the following relationship:

$$\overline{h} = \frac{\tan(e)}{\dot{e}}\dot{h} + h + \Delta h_T \quad (20)$$

After applying both sea surface dynamic and tropospheric delay corrections, the actual vertical distance h from the GNSS antenna to the sea surface can be obtained. Subtracting this value from the receiver's ellipsoidal height yields the true sea level.

3 Multi-mode and Multi-frequency Combination Method

For each frequency band within a GNSS system, SNR data can be used to monitor sea levels following the classical GNSS-MR inversion theory and correction procedures described in the previous section. During the inversion process, it is first necessary to define the elevation and azimuth angle ranges to ensure that only reflected signals from the sea surface are included. The detrended SNR residual sequences are then analyzed using the LSP method to derive effective reflector heights. After quality control, vertical height corrections are performed from two aspects: sea surface dynamics and tropospheric delay. Using multi-constellation and multi-frequency data for sea level inversion can effectively improve the temporal resolution of GNSS-MR-based sea level estimation. However, the final accuracy still depends on the quality of SNR data from each frequency band and cannot be substantially improved through multi-frequency observations alone. To overcome this limitation, a multi-mode, multi-frequency GNSS-MR sea level inversion fusion algorithm is proposed. This algorithm establishes a state equation linking the sea surface dynamic correction term, the tropospheric delay correction term, and the effective reflector height, and applies robust estimation to achieve data fusion. The optimal solution of the state equation is obtained through iterative adjustment. The basic principle is similar to the least-squares adjustment method. First, a sliding window is defined to contain sufficient sea level inversion points for equation establishment and parameter estimation. Suppose the window length is T, and the sliding window has a step size of one-sixth of an hour; The time for the ith window is $t_i \in \left\{0, \frac{1}{6}, \frac{1}{3}, \frac{1}{2}, \frac{2}{3}, \frac{5}{6}, 1, \cdots\right\}$ hours. According to Eq. (20), a state equation can be established within the ith window:

$$\overline{h}_{i,j,l}\left(t_{i,j,l}\right) - \Delta h_{Ti,j,l}\left(t_{i,j,l}\right) = \frac{\tan\left(e_{i,j,l}\right)}{\dot{e}_{i,j,l}}\dot{h}_i(t_i) + \dot{h}_i(t_i) \times \left(t_{i,j,l} - t_i\right) + h_i(t_i) \quad (21)$$

Here j denotes the index of signal type, l represents the serial number of the inversion value within the window, and $|t_l - t_i| < T/2$.The inversion results from all systems within the ith window can be written as:

$$\begin{cases} \cdots \\ \overline{h}_{i,j,l-1}\left(t_{i,j,l-1}\right) - \Delta h_{Ti,j,l-1}\left(t_{i,j,l-1}\right) = \left[\frac{\tan\left(e_{i,j,l-1}\right)}{\dot{e}_{i,j,l-1}} + \left(t_{i,j,l-1} - t_i\right)\right]\dot{h}_i(t_i) + h_i(t_i) \\ \overline{h}_{i,j,l}\left(t_{i,j,l}\right) - \Delta h_{Ti,j,l}\left(t_{i,j,l}\right) = \left[\frac{\tan\left(e_{i,j,l}\right)}{\dot{e}_{i,j,l}} + \left(t_{i,j,l} - t_i\right)\right]\dot{h}_i(t_i) + h_i(t_i) \\ \overline{h}_{i,j,l+1}\left(t_{i,j,l+1}\right) - \Delta h_{Ti,j,l+1}\left(t_{i,j,l+1}\right) = \left[\frac{\tan\left(e_{i,j,l+1}\right)}{\dot{e}_{i,j,l+1}} + \left(t_{i,j,l+1} - t_i\right)\right]\dot{h}_i(t_i) + h_i(t_i) \\ \cdots \end{cases} \quad (22)$$

which can be simplified into the following matrix equation:

$$\overline{H}_i = M_i\dot{h}_i + h_i = A_iX_i \quad (23)$$

Here $\overline{H}_i = \begin{pmatrix} \cdots \\ \overline{h}_{i,j,l}\left(t_{i,j,l}\right) - \Delta h_{Ti,j,l}\left(t_{i,j,l}\right) \\ \cdots \end{pmatrix}$, $M_i = \begin{pmatrix} \cdots \\ \left(\frac{e_{i,j,l}}{\dot{e}_{i,j,l}} + \left(t_{i,j,l} - t_i\right)\right) \\ \cdots \end{pmatrix}$, $A_i = (M_i \; 1)$, $X_i = \begin{pmatrix} \dot{h}_i \\ h_i \end{pmatrix}$.

If the inverse values $\overline{h}$ of different SNR sequences have the same precision, the least-squares solution of Eq. (23) is:

$$X_i = \left(A_i^T A_i\right)^{-1} A_i^T \overline{H}_i \tag{24}$$

However, the SNR quality varies across systems and frequencies, causing differences in dominant frequencies and inversion accuracy even within the same constellation. Robust estimation is essentially an anti-difference estimation. In the context of data fusion, its principle is as follows: Firstly, an initial least-squares adjustment is performed to obtain preliminary residuals for each inversion value; then based on these residuals, new weight factors are computed for the next iteration; finally, the process is repeated iteratively until the change between consecutive estimates falls below a predefined threshold.

At present, most multi-mode and multi-frequency GNSS-MR combination algorithms employ the "bisquare" weighting method. Considering that the IGGIII weighting model allows for a more refined and reasonable treatment of deviations of different magnitudes, this chapter adopts a robust estimation method based on the IGGIII model for multi-mode and multi-frequency GNSS-MR data combination.

During the robust estimation process, the first iteration is carried out using:

$$\hat{X}_i^{(1)} = \left(A_i^T A_i\right)^{-1} A_i^T \overline{H}_i \tag{25}$$

The estimation residuals of $\hat{X}$ for the first iteration are then computed as:

$$v_i^{(1)} = A_i \hat{X}_i^{(1)} - \overline{H}_i \tag{26}$$

In subsequent iterations, the inversion values with lower precision are reweighted, gradually reducing their influence on the estimation of the vertical distance h_i and thereby improving the overall accuracy of sea level inversion. Most inversion values are concentrated, while outliers that deviate from the main cluster constitute only a small fraction. In robust estimation, the reliable main data should be fully utilized; therefore, the IGGIII model is adopted as the resistant weighting model to automatically assign appropriate weights. The IGGIII weighting model is iteratively implemented as follows:

$$\overline{p}_{ii}^{(k+1)} = \begin{cases} \overline{p}_{ii}^{(k)} & \left|\tilde{v}_i^{(k)}\right| \le k_0 \\ \frac{k_0}{\left|\tilde{v}_i^{(k)}\right|}\left(\frac{k_1-\left|\tilde{v}_i^{(k)}\right|}{k_1-k_0}\right)^2 \overline{p}_{ii}^{(k)} & k_0 < \left|\tilde{v}_i^{(k)}\right| \le k_1 \\ 0 & \left|\tilde{v}_i^{(k)}\right| > k_1 \end{cases} \tag{27}$$

Here $\tilde{v}_i = \frac{v_i}{\hat{\sigma}_0\sqrt{(Q_{vv})_{ii}}}$, σ_0 represents the error in the correction term, $\overline{p}_{ii}$ is an element of the weighting matrix $\overline{P}$, $Q_{vv} = \overline{P}^{-1} - A\left(A^T\overline{P}A\right)^{-1}A^T$, k_0 is the quantile, and k_1 is the rejection threshold. Typically, $k_0 \in [2.0, 3.0]$, $k_1 \in [4.5, 8.5]$.

At iteration $k + 1$:

$$\hat{X}_i^{(k+1)} = \left(A_i^T\overline{P}_i^{(k+1)}A_i\right)^{-1}\left(A_i^T\overline{P}_i^{(k+1)}\right)\overline{H}_i \tag{28}$$

The iteration terminates when the difference between two consecutive estimates satisfies $|\ \hat{X}_i^{(k+1)} - \hat{X}_i^{(k)}\ | < \varepsilon$. The vertical height h_i and vertical height variation rate $\dot{h}_i$ within the ith time window are obtained after applying the sea surface dynamic correction and tropospheric delay correction. The sea level value can then be determined by subtracting the corrected vertical height h_i from the ellipsoidal height of the receiver.

4 Experimental Analyses

4.1 *Station Overview*

To verify the effectiveness of the proposed multi-mode and multi-frequency GNSS-MR technique for sea level monitoring, two experimental sites, BRST and HKQT, were selected. Both stations can receive satellite observations from the four major GNSS constellations. The BRST station (48.4° N, 4.5° W) is located on the western coast of France near the Brest harbor, where the daily tidal range is approximately 7 m [25]. Figure . 4a shows the observational environment of the BRST station. The station is equipped with a Trimble NETR9 geodetic receiver and a Trimble TRM57971.00 antenna, capable of collecting GPS, GLONASS, Galileo, BeiDou, and SBAS observations with a 30 s sampling interval. To assess the accuracy of the GNSS-MR-based sea level retrieval, the results were compared with the Brest tide gauge, located 292 m from the BRST station.

Since GNSS-MR requires only sea-surface-reflected signals, the first Fresnel reflection zone was mapped using Larson's googleEarthFresnel.m toolbox [36] to identify the sea-surface reflection area for the BRST station. Figure . 4b shows the Fresnel zones for elevation angles of 5°, 10°, 20°, 30°. Based on this analysis, two azimuth-elevation angle ranges were defined to exclude land reflections: Azimuth: 130°-165°, Elevation: 5°-20°; Azimuth: 165°-330°, Elevation: 12°-25°.

The HKQT station (22.3° N, 114.1° E) is located on the northern coast of Hong Kong and is equipped with a Trimble NETR5 geodetic receiver. It provides GPS, GLONASS, Galileo, BeiDou, QZSS, and SBAS observations with sampling intervals of 1 s, 5 s, and 30 s. Figure 5 shows the observation environment and Fresnel reflection zones of the HKQT station. The azimuth range for valid sea-surface reflections is

(a) Observation environment of the BRST station

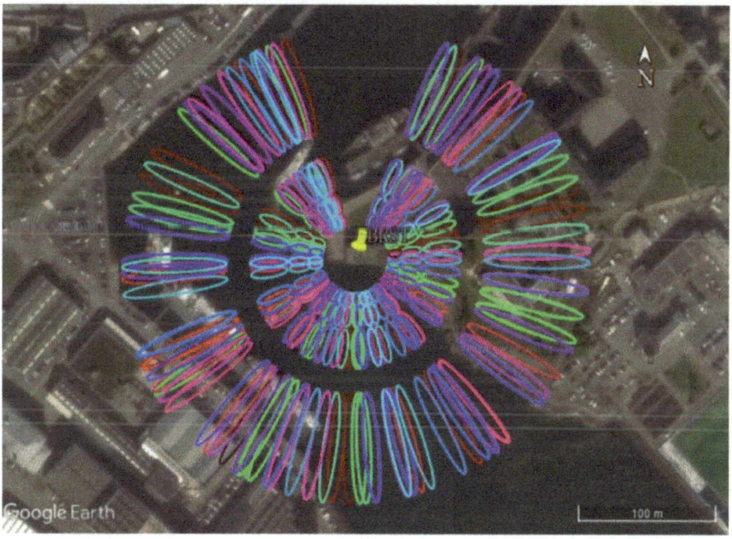

(b) The first Fresnel reflection zones of the BRST station

Fig. 4 Observation environment and first Fresnel reflection zone of BRST station

– 60°–105°, and the effective elevation angle range is 4°–9°. Tide gauge data from Quarry Bay, located 2 m from the HKQT station, were used for accuracy validation.

For the BRST station, SNR data collected during Day 197–253 of 2020 were used. The available signals for each GNSS constellation were: GPS: S1C, S2W, S2X, S5X; GLONASS: S1C, S1P, S2C, S2P; Galileo: S1X, S5X, S7X, S8X, S6X; BeiDou: S1X,

(a) Observation environment of the HKQT station

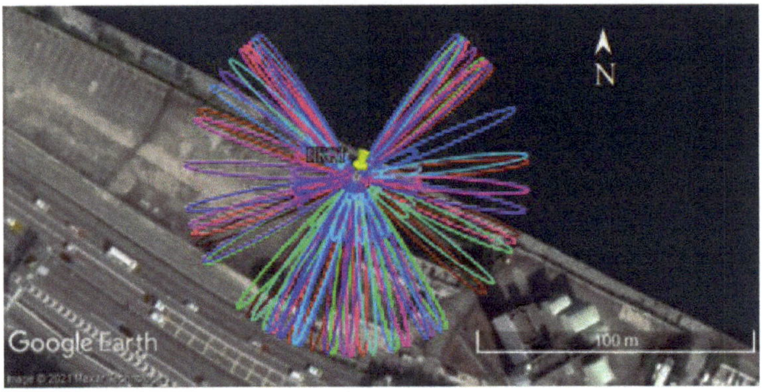

(b) The first Fresnel reflection zones of the HKQT station

Fig. 5 Observation environment and first Fresnel reflection zone of HKQT station

S5X, S2I, S7I, S6I. A summary of multi-band signal information for the four GNSS constellations is provided in Table 1.

The number of available SNR arcs for each constellation per day is shown in Figs. 6 and 7 displays examples of SNR sequences and corresponding LSP results for all four systems.

Table 1 Satellite signal information for GPS, BeiDou, GLONASS, and Galileo systems

Satellite system	Frequency band	Frequency	Signal	Pseudorange code	Phase code	SNR code
GPS	L1	1575.42	C/A	C1C	L1C	S1C
	L2	1227.60	Z-tracking and similar (AS on)	C2W	L2W	S2W
			L2C(M + L)	C2X	L2X	S2X
			I + Q	C5X	L5X	S5X
GLONASS	G1	1602 + k × 9/16 k = −7 … + 12	C/A	C1C	L1C	S1C
			P	C1P	L1P	S1P
	G2	1246 + k × 7/16	C/A	C2C	L2C	S2C
			P	C2P	L2P	S2P
Galileo	E1	1575.42	B + C	C1X	L1X	S1X
	E5a	1176.45	I + Q	C5X	L5X	S5X
	E5b	1207.140	I + Q	C7X	L7X	S7X
	E5 (E5a + E5b)	1191.795	I + Q	C8X	L8X	S8X
	E6	1278.75	B + C	C6X	L6X	S6X
BeiDou	B1–2	1561.098	I	C2I	L2I	S2I
	B1(BDS-3 signals)	1575.42	Data + pilot	C1X	L1X	S1X
	B2a (BDS-3 signals)	1176.45	Data + pilot	C5X	L5X	S5X
	B2b	1207.14	I	C7I	L7I	S7I
	B3	1268.52	I	C6I	L6I	S6I

4.2 *Multi-constellation and Multi-frequency GNSS-MR Sea Level Inversion Time Series*

By performing LSP analysis on the detrended SNR sequences, the time series of vertical reflector height $\overline{h}$ for each signal can be obtained. To further evaluate the inversion performance of SNR sequences at different frequency bands, and to correct for both sea surface dynamic variation error $\Delta h_{\dot{h}}$ and tropospheric delay error Δh_T, the root mean square error (RMSE) statistics of the BRST station's inversion sequences $\overline{h}, \overline{h}-\Delta h_{\dot{h}}$, and $\overline{h}-\Delta h_{\dot{h}}-\Delta h_T$ are summarized in Table 2. The accuracy improvement percentages after the two correction steps are illustrated in Fig. 8.

According to Table 2 and Fig. 8, applying sea surface dynamic correction and tropospheric delay correction can effectively improve the accuracy of GNSS-MR

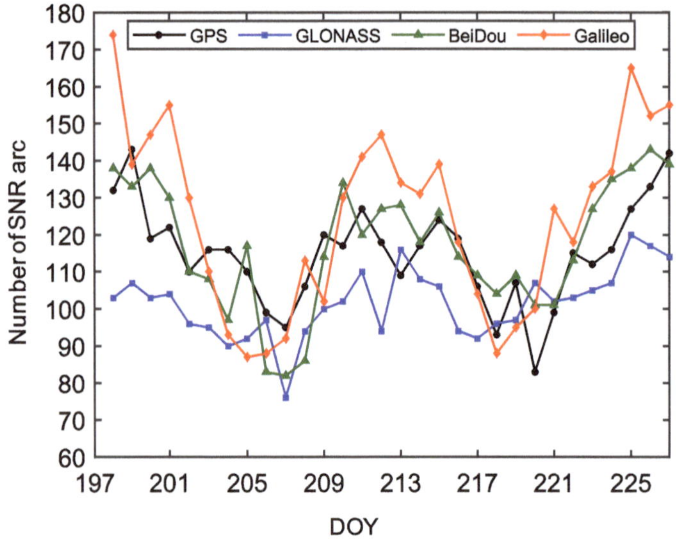

Fig. 6 Number of daily SNR arcs for GPS, GLONASS, Galileo, and BeiDou

sea level inversion. The inversion accuracy of all four GNSS constellations improved by approximately 50%. Among these, sea surface dynamics have a major impact on GNSS-MR sea level inversion and represent an essential correction term, while the tropospheric delay exerts a relatively smaller influence. Overall, the average RMSE across the four systems is approximately 24 cm (Fig. 9).

Similarly, the RMSEs of the signal retrieval sequences $\overline{h}$, $\overline{h} - \Delta h_{\dot{h}}$, and $\overline{h} - \Delta h_{\dot{h}} - \Delta h_T$ at the HKQT station are summarized in Table 3. After applying both dynamic sea surface and tropospheric delay corrections, the overall sea level retrieval accuracy of the four systems is approximately 25.42 cm.

4.3 Experimental Study of the Multi-mode and Multi-frequency GNSS-MR Combination Algorithm

As shown above, single-system, single-frequency inversions generally yield limited accuracy, and even multi-frequency inversions within a single system may result in uneven temporal coverage or overlapping results. To address these issues, the proposed IGGIII-model-based combination algorithm was applied using a sliding time window to build the state equation and perform robust regression for fusing multi-system inversion results. The selection of window length is crucial. Theoretically, shorter windows better approximate linear tide variations but contain fewer data points, reducing estimation reliability. Longer windows include more data but increase nonlinearity errors. Therefore, both accuracy and data loss rate (i.e., missing

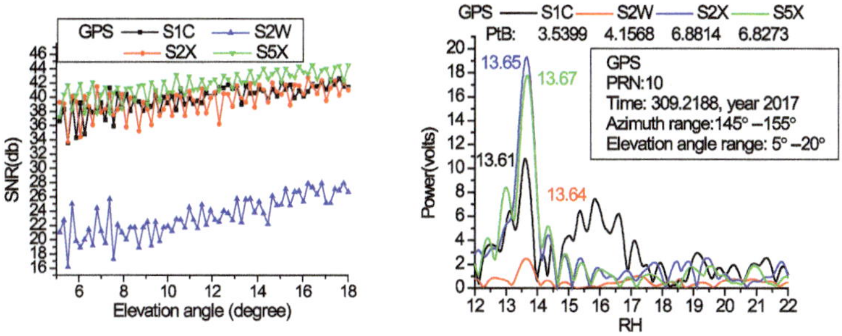

(a) The GPS SNR sequence and LSP analysis results

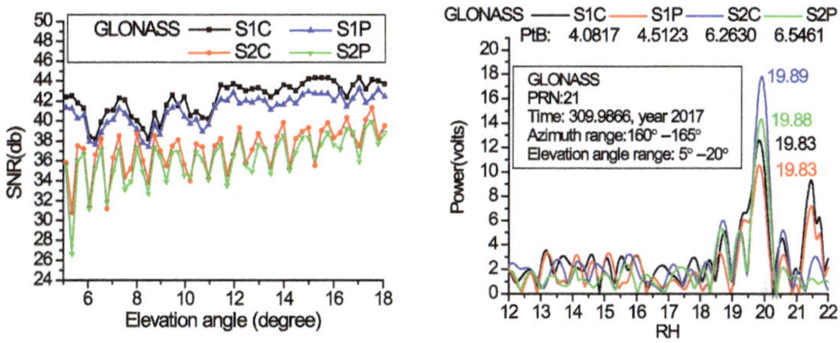

(b) The GLONASS SNR sequence and LSP analysis results

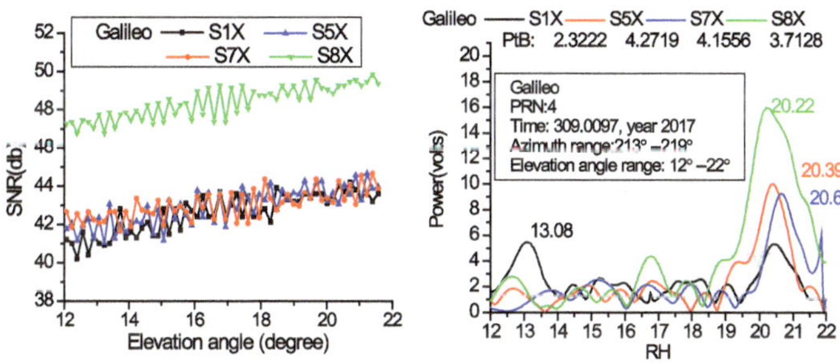

(c) The Galileo SNR sequence and LSP analysis results

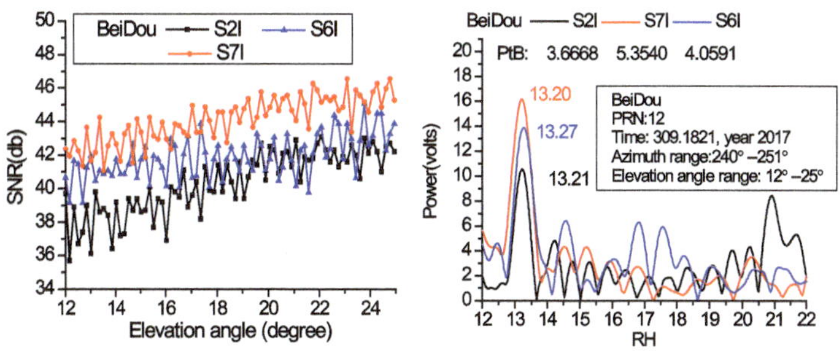

(d) The BeiDou SNR sequence and LSP analysis result

Fig. 7 SNR sequences (left) and corresponding LSP analysis (right) for the four systems for DOY 310

Table 2 Accuracy comparison of BRST station inversion results

System	Type	Number (PCs/day)	RMSE (m) of retrievals from		
			$\overline{h}$	$\overline{h} - \Delta h_{\dot{h}}$	$\overline{h} - \Delta h_{\dot{h}} - \Delta h_T$
GPS	S1C	38.5333	0.4640	0.3006	0.2758
	S2W	28.5333	0.4669	0.2955	0.2652
	S2X	25.9667	0.4730	0.3045	0.2716
	S5X	19.4667	0.4517	0.2867	0.2502
GLONASS	S1C	25.0333	0.4636	0.2382	0.2045
	S1P	25.2000	0.4748	0.2491	0.2068
	S2C	24.9333	0.4460	0.2308	0.2040
	S2P	25.9333	0.4487	0.2272	0.2008
Galileo	S1X	20.5333	0.4633	0.2617	0.2400
	S5X	25.8667	0.5070	0.2675	0.2106
	S7X	25.1000	0.4974	0.2733	0.2272
	S8X	24.0333	0.5322	0.2670	0.2254
	S6X	25.1333	0.4900	0.2593	0.2316
BeiDou	S1X	23.9667	0.4752	0.2731	0.2488
	S5X	28.0333	0.4994	0.2686	0.2360
	S2I	26.4667	0.4681	0.3009	0.2749
	S7I	3.8333	0.4746	0.3523	0.3202
	S6I	31.2333	0.4900	0.2590	0.2268

data ratio per window) were analyzed for various window lengths (0.5, 1, 1.5, 2, 2.5, 3 h). The comparison for the BRST and HKQT stations is shown in Tables 4 and 5, respectively. The optimal window length was determined to be 2 h in terms of RMSE.

Figure 10 shows the multi-constellation combination inversion results for the BRST station, where the black solid line represents the tide gauge observations and the blue dots represent the fused GNSS-MR results. The combination results match the tide gauge data well, resolving the overlapping issue of previous results and achieving uniform sampling at 10 min intervals, meeting the requirements of routine tide monitoring. The inversion accuracy reached 12.43 cm, improving by 40–60% compared with single-signal results.

Similarly, Fig. 11 shows the multi-constellation combination inversion results for the HKQT station. The black line represents tide gauge observations, and the red dots represent the combination results. The results accurately reflect tidal fluctuations, with an inversion accuracy of 7.09 cm, representing an improvement of approximately 72% compared to individual systems.

Experiments at the two stations indicate that, when considering dynamic sea surface errors and tropospheric delay in the robust regression method based on the IGGIII model, sea level retrieval using multi-mode and multi-frequency GNSS-MR

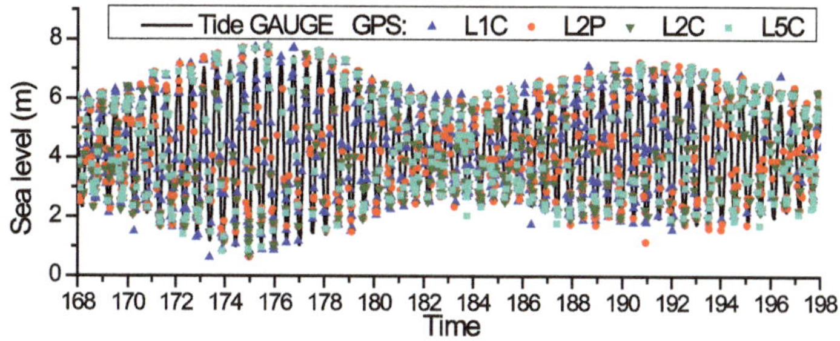

(a) GPS inversion results

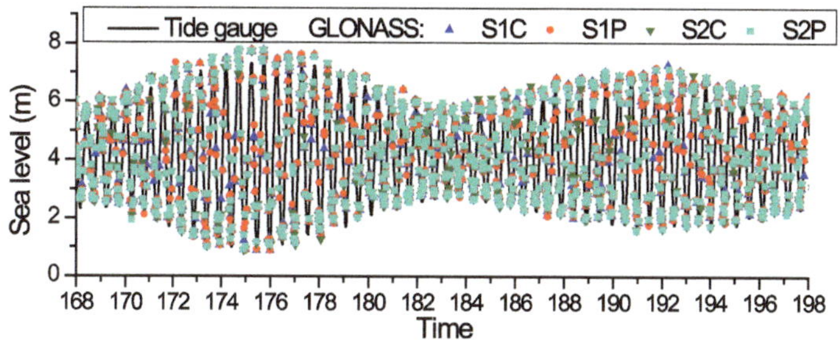

(b) GLONASS inversion results

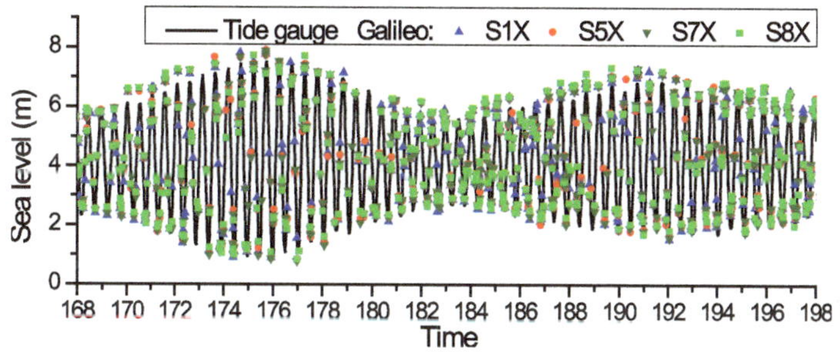

(c) Galileo inversion results

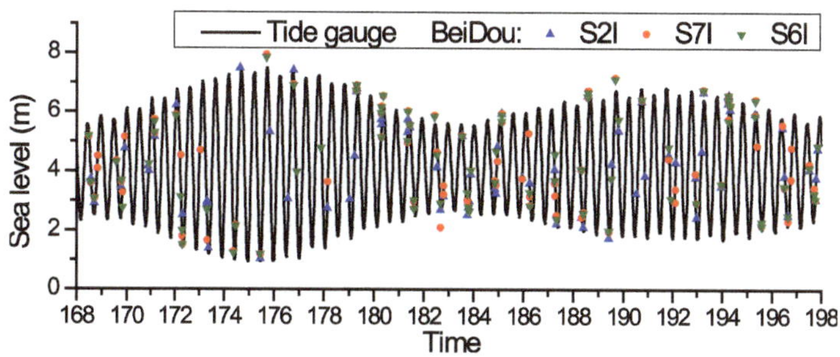

(d) BeiDou inversion results

Fig. 8 GNSS-MR sea level inversion results at BRST station

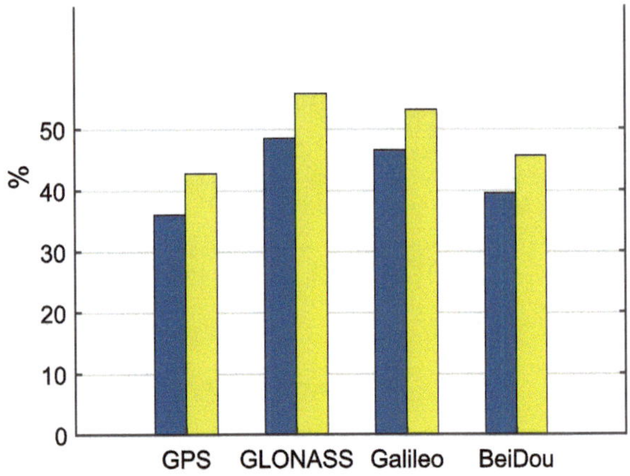

Fig. 9 Accuracy improvement after corrections for the four GNSS constellations

Table 3 Accuracy comparison of HKQT station inversion results

System	Type	Number (PCs/day)	RMSE (m) of retrievals from		
			$\overline{h}$	$\overline{h} - \Delta h_{\dot{h}}$	$\overline{h} - \Delta h_{\dot{h}} - \Delta h_T$
GPS	S1C	28.53	0.3758	0.3045	0.2822
	S2W	5.30	0.3608	0.3167	0.2949
	S2X	18.30	0.3248	0.2510	0.2322
	S5X	16.03	0.3164	0.2537	0.2304
GLONASS	S1C	21.00	0.3752	0.2893	0.2736
	S1P	20.07	0.3819	0.2682	0.2575
	S2C	16.67	0.3481	0.2508	0.2394
	S2P	17.77	0.3602	0.2642	0.2321
Galileo	S1X	12.63	0.3786	0.2654	0.2464
	S5X	12.00	0.3844	0.2775	0.2630
	S7X	11.70	0.4138	0.2873	0.2507
	S8X	12.40	0.4010	0.2776	0.2563
BeiDou	S2I	6.93	0.3643	0.2517	0.2454
	S7I	0.80	0.3804	0.2703	0.2546

Table 4 Comparison of inversion accuracy and data loss rate for BRST station

Window length (h)	RMSE (m)	Data loss (%)
0.5	0.2980	74.91
1	0.1983	19.37
1.5	0.1387	4.21
2	0.1243	0.20
2.5	0.1278	0.05
3	0.1377	0

Table 5 Comparison of inversion accuracy and data loss rate for HKQT station

Window length (h)	RMSE (m)	Data loss (%)
0.5	0.1012	46.36
1	0.0698	18.37
1.5	0.0690	12.69
2	0.0709	6.87
2.5	0.0772	3.31
3	0.0856	1.37

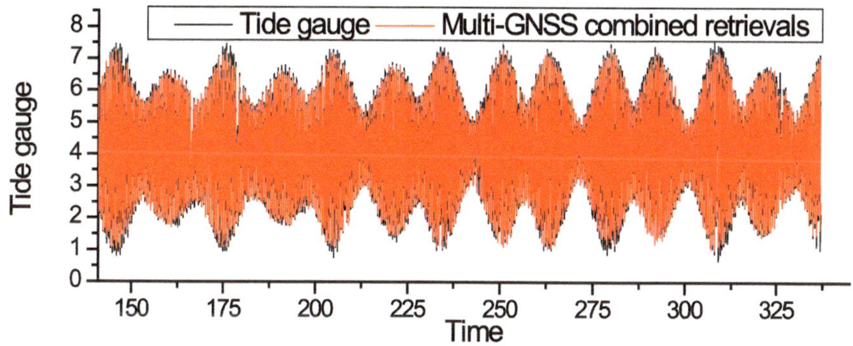

Fig. 10 Multi-mode and multi-frequency GNSS sea level retrieval at the BRST station

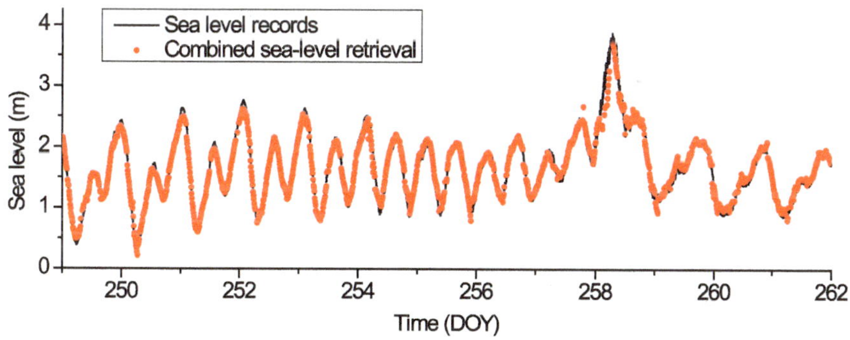

Fig. 11 Multi-mode and multi-frequency GNSS sea level retrieval at the HKQT station

technology yields RMSEs of 12.43 and 7.09 cm at the BRST and HKQT stations, respectively, representing an improvement of approximately 40%–70% compared with traditional methods. In conventional approaches, the sea level values cannot achieve uniform sampling due to the irregular SNR data acquisition times. Using a data combination algorithm based on the IGGIII model, equations can be established within a specific time window to retrieve sea levels, enabling both stations to achieve uniform 10 min sampling. This significantly improves the temporal resolution of

GNSS-MR technology for sea-level monitoring and meets the requirements of sea level observation.

5 Conclusion

To address the two major issues of low accuracy and limited temporal resolution in the practical application of GNSS-MR technology for sea-level monitoring, this chapter proposes a robust regression method based on the IGGIII model. By comprehensively considering dynamic sea surface variations and tropospheric delay, the method fuses multi-mode and multi-frequency GNSS-MR retrievals, effectively improving both the accuracy and temporal resolution of sea level retrievals. The retrieval accuracies at the BRST and HKQT stations are 12.43 cm and 7.09 cm, respectively, representing an improvement of approximately 40%–70% compared with traditional methods, while also achieving uniformly spaced sampling.

References

1. Jin SG, Zhang QY, Qian XD (2017) New progress and application prospects of global navigation satellite system reflectometry (GNSS+R). Acta Geodaetica et Cartographica Sinica 46(10):1389–1398
2. Larson KM, Small EE (2014) Normalized microwave reflection index: a vegetation measurement derived from GPS networks. IEEE J Sel Top Appl Earth Obs Remote Sens 7(5):1501–1511
3. Larson KM, Ray RD, Williams SDP (2017) A ten-year comparison of water levels measured with a geodetic GPS receiver versus a conventional tide gauge. J Atmos Ocean Technol 34(2):295–307
4. Martin-Neira M (1993) A passive reflectometry and interferometry system (PARIS): application to ocean altimetry. ESA J 17(4):331–355
5. Wang X, He X, Xiao R et al (2021) Millimeter to centimeter scale precision water-level monitoring using GNSS reflectometry: application to the south-to-north water diversion project, China. Remote Sens Environ 265:112645
6. Wang X, Song M, He X et al (2025) Enhancing Sea level inversion accuracy with a novel phase-based error correction method and multi-GNSS combination approach. GPS Solutions 29:34
7. Anderson KD (2000) Determination of water level and tides using interferometric observations of GPS signals. J Atmos Ocean Technol 17(8):1118–1127
8. Lowe ST, Zuffada C, Chao Y et al (2002) 5-cm-precision aircraft ocean altimetry using GPS reflections. Geophys Res Lett 29(10):13–1-13-4
9. Bilich A, Larson KM, Axelrad P (2004) Observations of signal-to-noise ratios (SNR) at geodetic GPS site CASA: implications for phase multipath. In: Proceedings of the Centre for European Geodynamics and Seismology, vol 23, pp 77–83
10. Larson KM, Small EE, Gutmann ED et al (2008) Using GPS multipath to measure soil moisture fluctuations: initial results. GPS Solutions 12(3):173–177
11. Larson KM, Small EE, Gutmann ED et al (2008) Use of GPS receivers as a soil moisture network for water cycle studies. Geophys Res Lett 35(24):L24405

12. Larson KM (2016) GPS interferometric reflectometry: applications to surface soil moisture, snow depth, and vegetation water content in the western United States. Wiley Interdiscip Rev Water 3(6):775–787
13. Larson KM, Gutmann ED, Zavorotny VU et al (2009) Can we measure snow depth with GPS receivers? Geophys Res Lett 36(17):L17502
14. Larson KM, Poland M, Miklius A (2010) Volcano monitoring using GPS: developing data analysis strategies based on the June 2007 Kīlauea volcano intrusion and eruption. J Geophys Res Solid Earth 115(B7):B07407
15. Larson KM (2013) A new way to detect volcanic plumes. Geophys Res Lett 40(11):2657–2660
16. Larson KM (2013) A methodology to eliminate snow-and ice-contaminated solutions from GPS coordinate time series. J Geophys Res Solid Earth 118(8):4503–4510
17. Larson KM, Löfgren JS, Haas R (2013) Coastal Sea level measurements using a single geodetic GPS receiver. Adv Space Res 51(8):1301–1310
18. Larson KM, Ray RD, Nievinski FG et al (2013) The accidental tide gauge: a GPS reflection case study from Kachemak Bay, Alaska. IEEE Geosci Remote Sens Lett 10(5):1200–1204
19. Larson KM, Ray RD, Williams SDP (2017) A 10-year comparison of water levels measured with a geodetic GPS receiver versus a conventional tide gauge. J Atmos Ocean Technol 34(2):295–307
20. Löfgren JS, Haas R (2015) sea level measurements using multi-frequency GPS and GLONASS observations. EURASIP J Adv Signal Process 2015(1):50
21. Strandberg J, Hobiger T, Haas R (2016) Improving GNSS-R Sea level determination through inverse modeling of SNR data. Radio Sci 51(9):1286–1296
22. Jin SG, Qian XD, Wu X (2017) Sea level change from BeiDou navigation satellite system-reflectometry (BDS-R): first results and evaluation. Glob Planet Change 149:20–25
23. Wang X, He X, Zhang Q (2019) Evaluation and combination of quad-constellation multi-GNSS multipath reflectometry applied to sea level retrieval. Remote Sens Environ 231:111229
24. Williams SDP, Nievinski FG (2017) Tropospheric delays in ground-based GNSS multipath reflectometry—experimental evidence from coastal sites. J Geophys Res Solid Earth 122(3):2310–2327
25. Santamaría-Gómez A, Watson C, Gravelle M et al (2015) Levelling co-located GNSS and tide gauge stations using GNSS reflectometry. J Geod 89(3):241–258
26. Böhm J, Möller G, Schindelegger M et al (2015) Development of an improved empirical model for slant delays in the troposphere (GPT2w). GPS Solutions 19(3):433–441
27. Lagler K, Schindelegger M, Böhm J et al (2013) GPT2: Empirical slant delay model for radio space geodetic techniques. Geophys Res Lett 40(6):1069–1073
28. Boehm J, Werl B, Schuh H (2006) Troposphere mapping functions for GPS and very long baseline interferometry from European centre for medium-range weather forecasts operational analysis data. J Geophys Res Solid Earth 111(B2):B02406

Chapter 12
CYGNSS-Based Mapping of Surface Water Fraction in the Pan-Tropical Belt

Qingyun Yan and Huan Feng

Abstract A novel method, which integrates multiple variables including Soil Moisture (SM) derived from Soil Moisture Active Passive (SMAP) and vegetation optical depth, water seasonality, geolocation, digital elevation model, slope, and biomass as inputs, and employs the technique of Bootstrap Aggregation of Regression Trees (BARTs), has been proposed for retrieving monthly surface water fraction (SWF) at a spatial resolution of 0.025° from Cyclone Global Navigation Satellite System (CYGNSS) data. The model is trained with Surface Water Microwave Product Series (SWAMPS) data at a coarser resolution of 25 km and then applied to CYGNSS data with an enhanced resolution of 0.025° to produce high-resolution water maps. The resulting CYGNSS SWF (CSWF) maps are evaluated by comparison with other water data sources, namely SWAMPS, Global Surface Water (GSW), and Global surface water dynamics (GLAD), as well as ground measurements. A quadruple collocation analysis indicates that the CSWF results have the lowest error variance among the four SWF datasets. Furthermore, additional testing with water level measurements demonstrates a strong correlation with station data and clear seasonal patterns. Notably, the CSWF estimates achieve significantly better spatial coverage than both the high-resolution optical data (such as GSW and GLAD) and the coarser-resolution SWAMPS data. This study highlights the effectiveness and efficiency of CSWF estimates, emphasizing their potential as a valuable addition to existing microwave- and optical-based surface water products.

Keywords GNSS-Reflectometry · CYGNSS · Surface water · GSW · GLAD

Q. Yan (✉) · H. Feng
School of Remote Sensing and Geomatics Engineering, Nanjing University of Information Science and Technology, Nanjing, China
e-mail: qy2543@mun.ca

H. Feng
e-mail: 202412482419@nuist.edu.cn

K. Yu and N. Zheng (eds.), *Advances in GNSS Reflectometry*, Navigation: Science and Technology 19, https://doi.org/10.1007/978-981-92-2325-1_12

1 Introduction

Surface water plays a pivotal role in conjunction with greenhouse gas emissions [1], biodiversity [2], and diverse life forms [3]. Despite its importance, it faces threats from climate changes and human activities [4]. Moreover, the variability of surface water in terms of presence, extent, and quantity is considerable across different spaces and times. Consequently, accurately, effectively, and promptly monitoring surface water and its spatiotemporal changes is essential yet challenging.

Remote sensing (RS) technology offers a reliable method for monitoring surface water across various scales, from regional to global. It employs two main types of sensors: optical and microwave, which are essential for mapping surface water bodies. Optical sensors, such as those utilized in the GSW [5] and GLAD [6] datasets derived from Landsat imagery, offer high spatial resolutions of 30 m. Despite their high resolution, optical RS faces challenges due to cloud cover and vegetation, which often reduce the availability of cloud-free satellite images, particularly during wet seasons compared to dry ones [5]. This issue leads to data scarcity and limits the application of optical RS in tropical regions during their wet seasons. As illustrated in Fig. 1, the temporal coverage rate for these two datasets is relatively low in tropical and South Asian regions, with infrequent revisits. The spectral characteristics of water surfaces also add complexity, as they can vary with depth, dissolved substances, and chlorophyll levels [5]. In contrast, microwave RS provides surveillance capabilities that are less impacted by these factors, allowing for continuous observation regardless of weather conditions. While passive microwave sensors generally have lower spatial resolutions, ranging from 25 to 50 km, active microwave platforms like SAR and radar altimeters offer higher resolutions. However, these platforms often have longer revisit periods. For instance, Sentinel-1 has a spatial resolution that varies from several to tens of meters and a revisit time of 6 days at the equator, which increases with latitude [7]. The SWAMPS dataset [8] stands out as one of the most extensive records of global water coverage, spanning over 20 years, and is obtained by combining passive and active microwave observations. It provides data with a revisit time of about 3 days and a spatial resolution of 25 km. Although SWAMPS has been successfully applied in various contexts, its relatively low detail level makes it less suitable for detailed or regional water mapping.

In recent years, the Global Navigation Satellite System-Reflectometry (GNSS-R) has emerged as an essential tool in the field of RS. Operating in a multistatic configuration, satellite-based GNSS-R captures signals that are scattered from the Earth's surface around the specular point (SP). The trajectory of these SPs appears quasi-random due to the satellite's orbit. Unlike conventional RS methods that gather data in broad swaths, GNSS-R offers broader coverage, albeit with occasional gaps in certain areas. The CYGNSS mission, the first of its kind, exemplifies this by delivering comprehensive coverage of tropical regions with a spatial resolution of 0.025° C and on a monthly basis (refer to Fig. 1 for a comparison and Sect. 4.3 for further details). GNSS-R has proven effective in various global applications, including ocean wind speed measurement [9, 10], altimeter data collection [11, 12], soil moisture assessment [13, 14], and ice monitoring [15, 16]. Lately, GNSS-R has shown remarkable

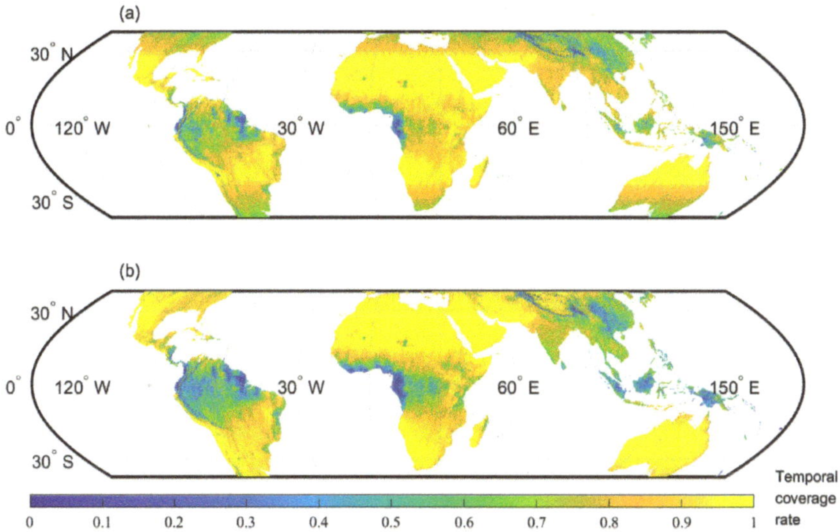

Fig. 1 Temporal coverage rates (i.e. monthly revisit rates) of **a** GSW [5] and **b** GLAD [6] data for the pan-tropical regions over the period from August 2018 to December 2021

potential in mapping surface water. It utilizes L-band GNSS signals, which have superior penetration capabilities through vegetation compared to higher frequency signals. This is demonstrated by CYGNSS's ability to detect small Amazonian tributaries beneath dense foliage [17]. Subsequently, numerous hydrological applications have been conducted, such as flood detection [18–20], wetland monitoring [21–23], and mapping of inundated areas and inland water bodies [24–31]. It is crucial to compare surface water data from CYGNSS with that from other instruments to highlight the unique benefits of CYGNSS-derived products. Chapman et al. [32] analyzed SAR and CYGNSS entropy and found that GNSS signals are highly sensitive to inundation, even in vegetated wetlands. Downs et al. [33] evaluated the sensitivity of CYGNSS's normalized signal-to-noise ratio to inundation levels, comparing it with data from VIIRS, SAR, and optical sensors. They concluded that CYGNSS is overly sensitive to small water bodies, suggesting that a simple threshold method is not reliable for precise inland water mapping.

Despite some successful applications yielding positive outcomes, the majority of current research has concentrated on classification tasks, which involve assigning pixels to specific classes (such as water or land). However, these studies encounter two significant limitations. On one hand, the precision of classification can be compromised by inaccuracies in label assignment [6]. On the other hand, classification-based techniques often fail to integrate additional pixel information beyond its class label. Consequently, there is a need for a more inclusive strategy that quantifies the nature of surface features and also considers their related attributes, particularly those concerning surface water characteristics. This broader perspective will enhance our understanding of the area under study. Specifically, Loria et al. [34] analyzed the reflection properties of CYGNSS signals from inland water bodies and calculated the SWF for a section of the Amazon River Basin as a pilot study. However, the applicability of their approach for global analysis remains untested, and its precision could be further evaluated and refined. Additionally, Chew et al. [35] created a dielectric model to determine SWF by accounting for soil and water conditions. They presented flood maps for four distinct locations (Amazon, Mozambique, Mali, and Australia) and compared these with other datasets (e.g., SWAMPS, Dartmouth Flood Observatory, and GSW). Nonetheless, there are uncertainties in their findings that primarily stem from the assumptions made about soil and water surface roughness, leading to a potential underestimation of SWF, particularly in regions with high SWF values.

The objective of this research is to estimate the SWF across the tropics with a spatial precision of 0.025° C and on a monthly basis, utilizing data from the CYGNSS. To accomplish this, a novel retrieval model leveraging the bootstrap aggregation of regression trees (BART) approach has been constructed. Initially, the model undergoes training with the SWAMPS dataset, which features a resolution of 25 km, before being adapted to handle inputs scaled to 0.025° C to generate higher-resolution SWF maps. The outcomes are then subjected to a thorough assessment and benchmarked against the microwave-based SWAMPS dataset [8], along with two optical datasets [5, 6], as well as in situ observations. The structure of this chapter is as follows: Sect. 2 details the CYGNSS, reference, and supplementary datasets employed. Section 3 outlines the BART algorithm-based SWF retrieval model that has been developed. Section 4 presents the experimental validation and discussions, while the key findings are encapsulated in Sect. 5.

2 Datasets

This part provides a detailed description of the CYGNSS remote sensing (RS) data, the comparative surface water datasets, and additional supportive data that are utilized in the study.

2.1 CYGNSS Remote Sensing Data

CYGNSS covers the latitude range from 38° S to 38° N. The study utilizes the CYGNSS Level 1 (L1) Version 3.1 datasets spanning 41 months, from August 2018 to December 2021, which can be accessed at https://cmr.earthdata.nasa.gov/virtual-directory/collections/C2146321631-POCLOUD. The surface reflectivity Γ, which is crucial for our analysis, is derived under the assumption that coherent reflections are predominant over land surfaces, as detailed in previous works [13, 14, 36–38]:

$$\Gamma = \frac{\sigma (R_t + R_r)^2}{4\pi (R_t R_r)^2}. \tag{1}$$

Here, σ represents the CYGNSS bistatic radar cross section, and R_t and R_r denote the distances from the SP to the transmitter and receiver, respectively. The CYGNSS L1 data include these measurements along with supplementary details about the observations and their geometrical setup. Additionally, a delay-Doppler map observable, known as the pixel number observable (PN), which characterizes surface roughness [15], is incorporated into the analysis. The data quality control adheres to the methods outlined in [14], and observations flagged with "SP in the sidelobe" are omitted due to their unreliable antenna gain. The CYGNSS data are compiled into monthly aggregates with a spatial resolution of 0.025° ×0.025° over the region between 37.5° S and 37.5° N, based on the geolocation of the specular points.

2.2 *Reference Data*

SWAMPS Data

The SWAMPS dataset [8] serves as the benchmark for our training process. This dataset, which consists of a gridded representation of inundation fractions, is compiled from microwave sensor data obtained from various sources. It is recognized for being among the most extensive publicly accessible collections of microwave remote sensing data, offering daily records from 1992 through 2020. The dataset features a revisit interval of about three days, with a spatial resolution of 25 km. Utilized for training purposes, it also functions as a comparative standard for low-resolution assessments. Furthermore, the SWAMPS dataset has been condensed from daily to monthly mean values for this analysis.

Optical-Based Data

In this study, for high-resolution evaluation, we utilize two distinct surface water datasets derived from Landsat optical images: the GSW [5] and the GLAD [6]. These datasets are renowned for their extensive collection of surface water information over extended time frames, with GSW covering from 1984 to 2021 and GLAD from 1999 to 2021. They offer spatial resolutions of 30 m and temporal resolutions of one month. Despite their comprehensive nature, they can encounter issues like inadequate coverage due to cloud cover or dense foliage. The datasets from August 2018 to December 2021, totaling 41 months, are analyzed. The temporal coverage is determined by calculating the ratio of valid observations (excluding NaN values) to the total number of months, as depicted in Fig. 1. Discrepancies in the temporal coverage of the two datasets can be observed in Fig. 1. It has been noted that the GLAD dataset tends to detect more water bodies than the GSW dataset [6]. Given the lack of agreement on which dataset is superior, both GSW and GLAD are employed as comparative references in our analysis.

Hydro Station Data

For the purpose of local-scale assessment, this study employs two distinct sets of water level (WL) data. The initial dataset comprises daily WL readings from a network of 1491 monitoring stations situated throughout Australia. These data are sourced from the Bureau of Meteorology's Water Data Online platform (http://www.bom.gov.au/waterdata/). While these measurements are taken at a high frequency of once per day, they are confined to discrete locations.

The second dataset includes WL measurements from 2173 virtual stations, which are generated from satellite altimetry and are specific to the Amazon River basin within the geographical bounds of [(15° S–5° N), (80° W–30° W)]. These data are accessible through the Copernicus Global Land Service (https://land.copernicus.eu/global/products/wl). It is important to note that satellite altimeters are capable of measuring WL only for water bodies that are situated directly beneath the satellite's path (nadir position). Furthermore, the accuracy of these measurements is influenced by both the dimensions of the water body and the characteristics of the surrounding terrain. The frequency at which virtual stations are revisited is contingent upon the satellite's orbital period. In the Amazon basin, there is a median correlation coefficient of 0.95 between the time series data from virtual and in situ stations, with an average temporal sampling interval of 27.9-days. Although the virtual station data exhibit a strong correlation with in situ measurements, the relatively infrequent revisits represent a drawback of this method.

2.3 Ancillary Data

To account for the impacts of geolocation, topography, canopy attenuation, and surface soil moisture (SM), latitude/longitude, digital elevation model (DEM)/slope, biomass/vegetation water content (VWC) and SM data are respectively adopted as auxiliary data and they are described as follows.

DEM

The Global Land One-kilometer Base Elevation (GLOBE) DEM dataset, accessible at https://www.ngdc.noaa.gov/mgg/fliers/globedem.html, is incorporated into the study. Alongside the DEM data, we also make use of surface slope data derived from these models. Each of these datasets is characterized by a resolution of 1 km.

Water Seasonality

The GSW Seasonality map, accessible at https://jeodpp.jrc.ec.europa.eu/ftp/jrc-opendata/GSWE/Aggregated/LATEST/seasonality, features a spatial resolution of 30 m. This map indicates the number of months that water is observed within a given year and is based on GSW data [5]. While it might not accurately capture the full extent of semi-permanent water bodies hidden under dense vegetation, the map provides essential a priori insights into the typical locations of surface water occurrence.

Biomass

The dataset for above-ground biomass from the European Space Agency's (ESA) Climate Change Initiative (CCI) for the year 2018, found at https://data.ceda.ac.uk/neodc/esacci/biomass/data/agb/maps/v3.0/, is employed in this analysis. It is important to note that these supplementary datasets are static and do not account for temporal changes. While the biomass data cover the globe, they might exhibit regional biases. Furthermore, the primary data source, which relies on radar backscatter, may not precisely represent the intricate biomass structure of forests, potentially causing discrepancies in regions with thick undergrowth or substantial deadwood.

VWC and SM

The Soil Moisture Active Passive (SMAP) Enhanced Level 3 Radiometer Global and Polar Grid Daily 9 km EASE-Grid Soil Moisture dataset, Version 5 [39], is incorporated into our study. The dataset covers the period from August 2018 to December 2021. We utilize Google Earth Engine (GEE), a powerful cloud-based computing platform [40], to access and manipulate the SMAP data, which includes daily morning and evening SM and VWC measurements from August 1, 2018, to December 31, 2021. The daily morning and evening data for these parameters are combined to form a daily composite, which is subsequently summarized to a monthly average, maintaining a spatial resolution of 9 km. The GEE Python API [41] is leveraged to facilitate the data retrieval and processing tasks.

2.4 Data Collocation

Training Data Preparation

In the training phase, all relevant parameters are mapped onto the geographic framework supplied by the SWAMPS dataset, which has a spatial resolution of 25 km. The data are compiled and calculated on a monthly average basis.

Table 1 Resolutions of ancillary input data sources and references

Data source	Native resolution	Resolution during training	Resolution during retrieval
SWAMPS	25 km	25 km	0.025°
DEM	1 km		
Water seasonality	30 m		
Biomass	100 m		
VWC	9 km		
SM	9 km		
GSW	30 m		
GLAD	30 m		

Finer-Resolution Retrieval

The objective of this study is to achieve SWF estimation with a spatial resolution of 0.025° and a monthly temporal resolution. Consequently, the referenced datasets are aligned to these specifications. High-resolution DEM, slope, and biomass datasets, along with global surface water reference datasets, are spatially aggregated into 0.025° grids. The lower-resolution SM and VWC datasets are subjected to linear interpolation and re-projected to match the chosen spatial grids. The monthly mean water percentage from the GLAD dataset [6] is utilized as is. The water mask data derived from the GSW dataset [5] are transformed into a comparable metric through pixel counting. These datasets serve as comparative benchmarks. A summary of the native resolutions for all input and reference optical SWF products is presented in Table 1.

3 SWF Estimation Method

Regression trees (RTs), as introduced by Breiman et al. [42], are capable of recursively segmenting the feature space. In our study, this includes variables such as CYGNSS Γ and PN, SMAP SM and VWC, water seasonality, along with geolocation, DEM, slope, and biomass data, to predict the SWF. Individual RTs tend to overfit, which can lead to poor generalization. To address this, the Bagging method [43] is applied to enhance the predictive performance of RTs. This approach decreases variance and mitigates overfitting, thus improving model stability and accuracy. Moreover, it is effective in managing high-dimensional datasets and capturing intricate feature interactions, which is pertinent given the 10 different input elements in our analysis.

In this context, Bootstrap Aggregation (Bagging) of Regression Trees (BARTs) is the chosen method for establishing the quantitative link between SWF and the array of inputs, which include CYGNSS data and other influential factors. It should be highlighted that BART has demonstrated its efficacy across various remote sensing tasks, such as enhancing the resolution of SMAP SM products [44] and retrieving soil moisture [45]. For the purpose of SWF estimation, we consider a training dataset denoted as $T = \{(x_1, y_1), (x_2, y_2), ..., (x_N, y_N)\}$, where x_i represents the ith input vector with 10 components, y_i is the corresponding SWF value, and N signifies the total count of training instances.The approach to implementing BARTs in this study is outlined below. The primary objective is to construct a predictive model $f(x)$ that minimizes the loss function, as

$$\text{object} = \min\left(\text{Loss}(f(x_i), y_i)\right). \tag{2}$$

Generally, the loss function is defined as the mean square error (MSE) between the estimated and targeted outputs

$$\text{MSE} = \frac{1}{n}\sum_{i=1}^{n}(f(x_i) - y_i)^2. \tag{3}$$

Assuming the designed RTs are allocated with M leaves, consequently there will be at most M varying estimates (or equivalent M units R_1,...,R_m), each leaf node R_m contains a set of sample points x_i , and these sample points are assigned the same predicted value c_m within the model, minimizing the MSE in Equation (3) is equivalent to minimizing the function

$$\frac{1}{n}\sum_{m=1}^{M}\sum_{x_i \in R_m}(c_m - y_i)^2, \tag{4}$$

where c_m is the prediction from the mth leaf. As such, the splitting parameter and point for each leaf that minimize function (4) should be recorded. This is achieved through traversing a parameter p, scanning the splitting point q, and determining the (p, q) pair that meets

$$\min_{p,q}\left[\min_{c_1}\sum_{x_i \in R_1\{p,q\}}(y_i - c_1)^2 + \min_{c_2}\sum_{x_i \in R_2\{p,q\}}(y_i - c_2)^2\right]. \tag{5}$$

Based on the determined (p, q) pair, one can further calculate the outputs:

$$\hat{c_m} = \frac{1}{N_m}\sum_{x_i \in R_m(j,s)} y_i, x \in R_m, m = 1, 2, \tag{6}$$

with

$$R_1\{j, s\} = \{x \mid x^{(j)} \leq s\},$$
$$R_2\{j, s\} = \{x \mid x^{(j)} > s\}, \tag{7}$$

and N_m is the number of elements for R_m. When the total M leaves are determined, the overall RT is thus in the following form:

$$f(x) = \begin{cases} \hat{c_1}, x_i \in R_1(j, s) \\ \hat{c_2}, x_i \in R_2(j, s) \\ \hat{c_3}, x_i \in R_3(j, s) \\ \dots \\ \hat{c_M}, x_i \in R_M(j, s). \end{cases} \tag{8}$$

For details of the implementation of Bagging, one can refer to [43], the formula can be given as

$$\hat{h}(x) = \frac{1}{T} \sum_{t=1}^{T} \hat{f}_t(x), \tag{9}$$

where T is the number of new sample sets and $\hat{f}_t$ is the predication from the tth tree.

To conclude, the BART model leverages a diverse set of inputs including CYGNSS Γ and PN, SMAP SM and VWC, water seasonality indices, geographic coordinates, DEM data, slope measurements, and biomass estimates to predict SWF. It is important to note that different variable types, such as the inundation fraction from SWAMPS and CYGNSS Γ, can serve as interchangeable inputs for the BART model throughout both the training and application phases. The development and analysis of the model in this study are executed utilizing the regression learner toolbox available in MATLAB R2021b.

4 Experiments and Evaluation

Here, the proposed scheme for retrieving CSWF is implemented and evaluated using multi-source data, including SWAMPS [8], rescaled GSW [5], and GLAD [6] global products, as well as field measurements.

4.1 *Model Training*

While the SWAMPS dataset features a relatively low resolution of 25 km, it offers the advantage of cloud penetration, capturing observations that are not available in the GSW/GLAD datasets. For this reason, SWAMPS was chosen as the reference for training. However, it is important to acknowledge that SWAMPS could potentially

Table 2 Accuracy of SWF estimates at 25 km-resolution (* indicates p-value is < 0.05)

	Measure	R^2	RMSE
CSWF	Training	0.976*	0.0147
Versus	Test	0.964*	0.0179
SWAMPS	Overall	0.974*	0.0153
GSW versus SWAMPS		0.532*	0.0618
GLAD versus SWAMPS		0.546*	0.0552
CIF versus SWAMPS		0.583*	0.0612
CSWF versus GLAD		0.523*	0.0485
GSW versus GLAD		0.964*	0.0099
CIF versus GLAD		0.842*	0.0196

underestimate SWF in regions with thick canopies and may not fully capture seasonal variations [8, 23, 46], which could impact the accuracy of the retrievals in this study. As of the study's commencement, the SWAMPS dataset was current up to December 2020, overlapping with CYGNSS data for 29 months. To bolster the model's ability to represent diverse seasonal patterns, the initial 24 months of data were designated for training, with the subsequent months used for testing the BART model (further details are provided in Sect. 3). To mitigate computational load, the pan-tropical regions of interest (ranging from ±37.5° latitude) were segmented according to level 2 sub-basin divisions (refer to https://www.hydrosheds.org/products/hydrobasins) and the training was conducted independently for each sub-basin. The model's training and testing performance, measured by root mean square error (RMSE) and the coefficient of determination (R^2), are detailed in Table 2. In both cases, R^2 values exceed 0.96, and RMSE values are below 0.018. The minor discrepancies between training and testing sets suggest the model's capability to generalize. Figure 2 illustrates the overall strong agreement as a heat map.

4.2 Coarse-Resolution Test

Initially, the retrieval outcomes are evaluated within the SWAMPS geographic framework, which operates at a 25-km resolution. The GSW and GLAD datasets, which offer higher resolutions, are adjusted to match this framework. A comparative visualization of these datasets for December 2020 is presented in Fig. 3. Generally, the spatial patterns among the four datasets show a good degree of consistency, and detailed accuracy metrics are compiled in Table 2. To quantitatively assess the discrepancies among these SWF products, the quadruple collocation (QC) method [47] is employed. The spatial distribution of errors (as shown in Fig. 4) reveals that the CYGNSS-derived dataset exhibits the least variance in errors on a global scale.

Furthermore, to assess the significance of the dynamic parameters related to CYGNSS and SMAP satellites over time, feature selection was conducted on input

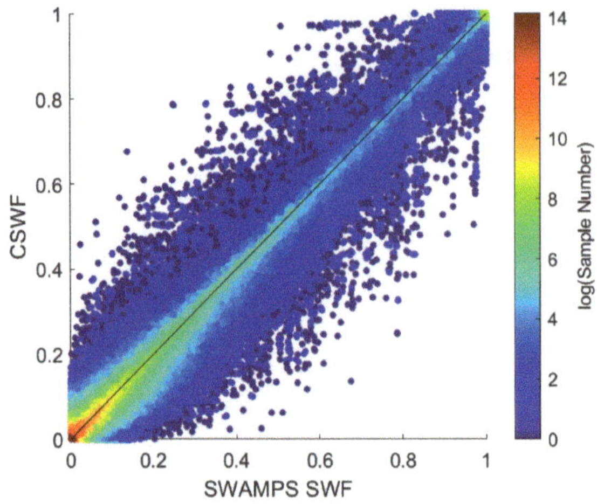

Fig. 2 Scatter plot of SWF results: CSWF versus SWAMPS

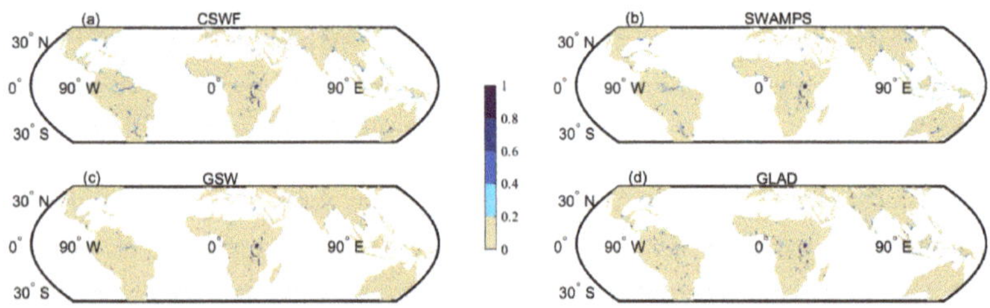

Fig. 3 25 km-resolution SWF for December 2020: **a** CSWF, **b** SWAMPS, **c** GSW, and **d** GLAD

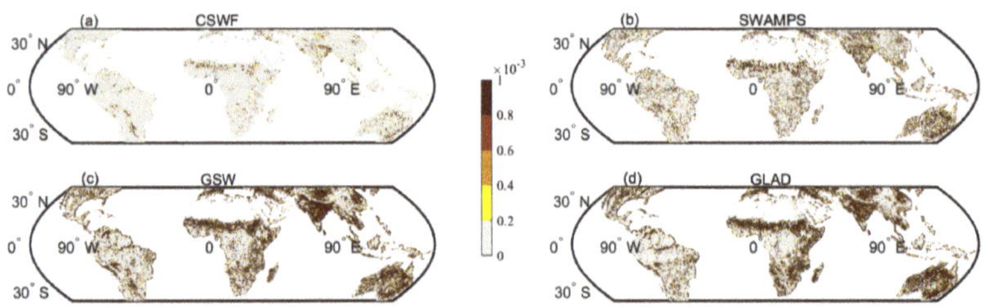

Fig. 4 Error variance of SWF based on: **a** CSWF, **b** SWAMPS, **c** GSW, and **d** GLAD data from August 2018 to December 2020

sets that omitted CYGNSS Γ and PN, as well as those that lacked SMAP SM and VWC. Both configurations showed a notable decline in performance (initial RMSE of 0.0153 and R^2 of 0.974), with the exclusion of CYGNSS data leading to a more pronounced decrease (RMSE rising to 0.0532 and R^2 to 0.778) compared to the exclusion of SMAP data (RMSE of 0.0470 and R^2 of 0.792), highlighting the greater influence of CYGNSS data. To determine the impact of static data, a variable importance

Table 3 Decrease of performance by removing variables

Remove	CYGNSS	SMAP	Seasonality	Biomass	Lat/lon	Dem/slope
ΔRMSE	0.0379	0.0317	0.0490	0.0114	0.0267	0.0337
ΔR^2	0.196	0.182	0.327	0.042	0.153	0.201

analysis was carried out by sequentially removing DEM/slope, latitude/longitude, biomass, and water seasonality data from the inputs and monitoring the subsequent reduction in retrieval accuracy for each model. The overall contribution of each input can be gauged from the corresponding performance degradation detailed in Table 3, indicating that all variables contributed positively to the model's performance, with water seasonality, CYGNSS, and topographic data exerting the most significant effects.

4.3 *Fine-Resolution Examination*

By utilizing the custom BARTs on the 0.025° gridded datasets, more detailed SWF maps were produced. For comparative analysis, SWF products derived from both the GSW and GLAD datasets were used. The level of concordance between these datasets, as measured by R^2 and RMSE, is summarized in Table 4. The CYGNSS-based SWF maps showed a good level of agreement with the optically-derived SWF maps, despite the disparity in the data sources. Moreover, the RMSE and R^2 values obtained when comparing CYGNSS and optical SWFs at two distinct spatial scales were similar, suggesting that the CYGNSS results are reliable when scaled down from 25 km to 0.025°.

The discrepancies between optically-derived and CYGNSS-based SWF estimates, in relation to water seasonality as discussed in Sect. 2.3, are depicted in Fig. 5. When contrasted with the optical data, CYGNSS appears to consistently underestimate the SWF of permanently waterlogged areas that exhibit a seasonality of 12 months. This underestimation may stem from unresolved surface roughness challenges in these

Table 4 Consistency between 0.025°-gridded SWF maps (* indicates p-value is < 0.05)

Measure	R^2	RMSE
CYGNSS (bef. int.) versus GLAD	0.494*	0.0607
CYGNSS (aft. int.) versus GLAD	0.495*	0.0608
GSW versus GLAD	0.829*	0.0193
CIF versus GLAD	0.695*	0.0442

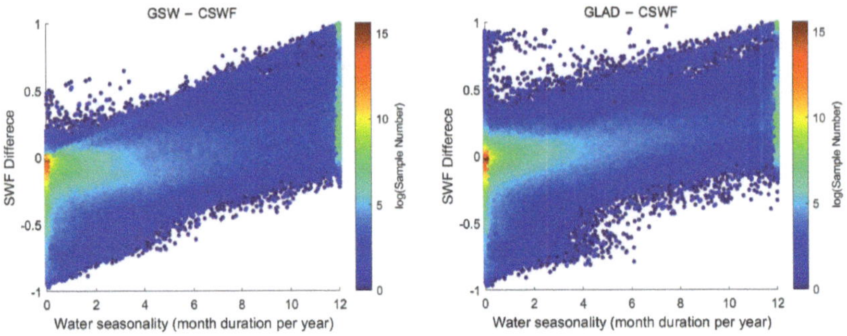

Fig. 5 Heat maps of SWF differences in terms of water seasonality: **a** GSW and **b** GLAD versus CSWF

regions, as noted in [35]. It should be highlighted that areas with zero seasonality do not prevent CYGNSS from generating nonzero SWF values, indicating its potential to identify semi-permanent water bodies that optical sensors might miss in areas with substantial biomass. To substantiate this, the maximum SWF values were determined, and density plots concerning biomass are displayed in Fig. 6, revealing higher SWF estimates from CYGNSS-based data. The mean SWF values in

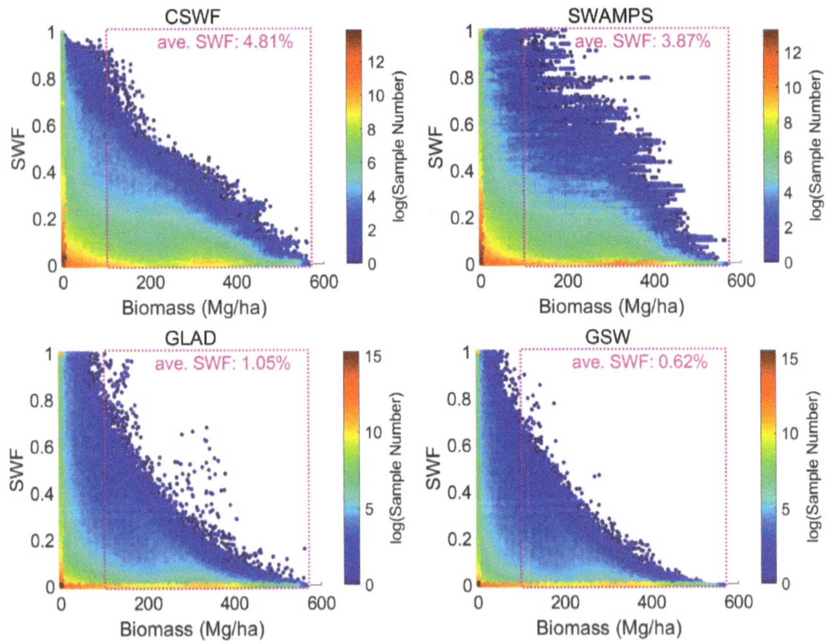

Fig. 6 Heat maps of SWF with respect to biomass: **a** CSWF, **b** SWAMPS, **c** GLAD and **d** GSW. The mean SWFs for regions with high biomass (over 100 Mg/Ha) are 4.81%, 3.87%, 1.05% and 0.62%, respectively

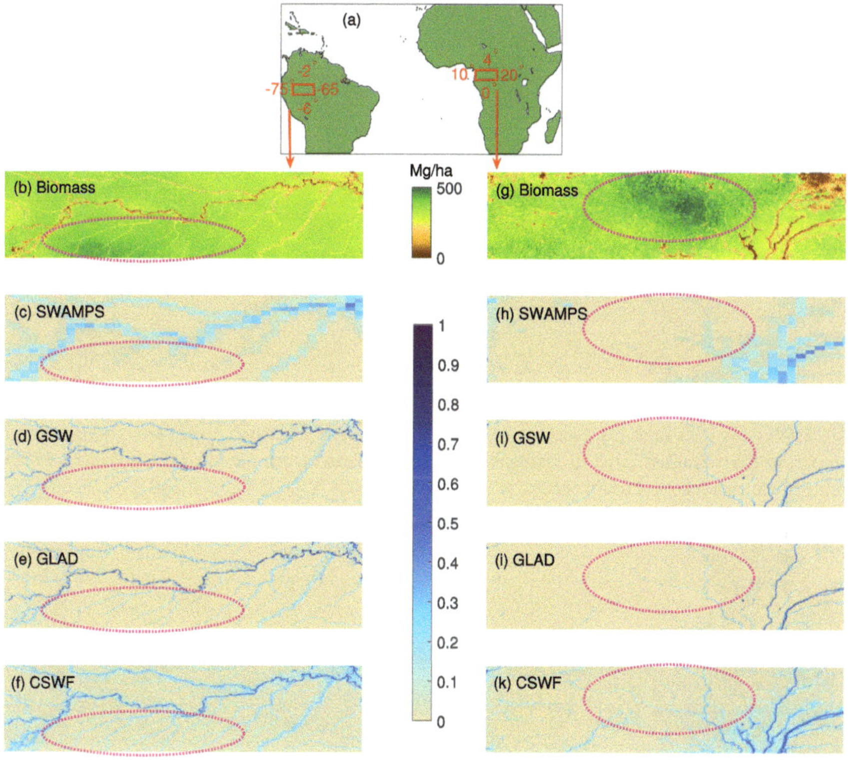

Fig. 7 Examples of maximum SWF and biomass over two regions: **a** the geolocation, **b** biomass, **c** SWAMPS, **d** GSW, **e** GLAD, and **f** CSWF in Amazon and **g** biomass, **h** SWAMPS, **i** GSW, **j** GLAD, and **k** CSWF in Congo

regions where biomass exceeds 100 Mg/ha are 4.81%, 3.87%, 1.05% and 0.62% for CSWF, SWAMPS, GLAD, and GSW, respectively. Two case studies of SWF estimations in high-biomass areas (as shown in Fig. 7) further illustrate the effectiveness of CYGNSS-based SWF in mapping water under dense canopies.

It bears emphasis that datasets such as GSW and GLAD, which rely on optical imagery, may exhibit significant data gaps owing to cloud cover, as depicted by the temporal coverage in Fig. 1. Moreover, these datasets tend to show pronounced data gaps during flood seasons. For instance, in the case of GLAD, the temporal missing data rates across the selected regions—East Asia, South Asia, Nigeria, and the Amazon—are graphed in Fig. 8.

For the sake of illustration, the Nigerian region, labeled as Area C in Fig. 8, serves as the focal point for this case study. Figure 9e–h exhibit the SWF maps for August 2018, a period marked by flooding. It is apparent that the datasets relying on optical imagery suffer from gaps in data, while microwave-based datasets like CYGNSS and SWAMPS are unaffected by these interruptions. Furthermore, when compared to the SWAMPS data, which has a resolution of 25 km, the CYGNSS retrievals at

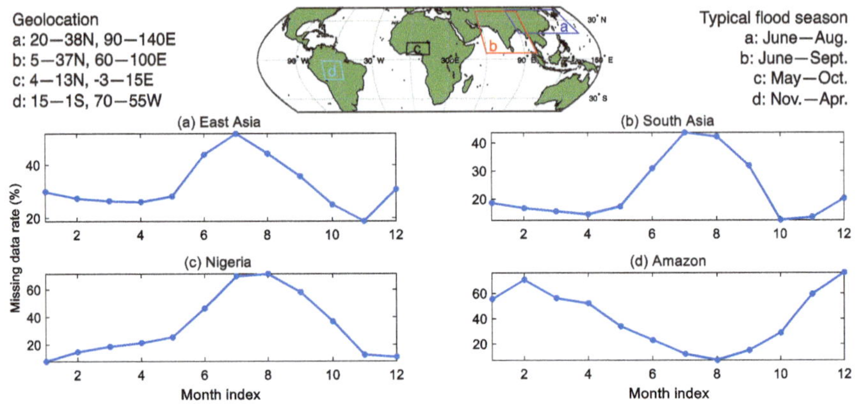

Fig. 8 Missing data rates (i.e., the average spatial fraction of areas that do not contain data during the period from August 2018 to December 2021) calculated from the GLAD [6] dataset for the pan-tropical regions over four regions: **a** East Asia, **b** South Asia, **c** Nigeria, and **d** Amazon

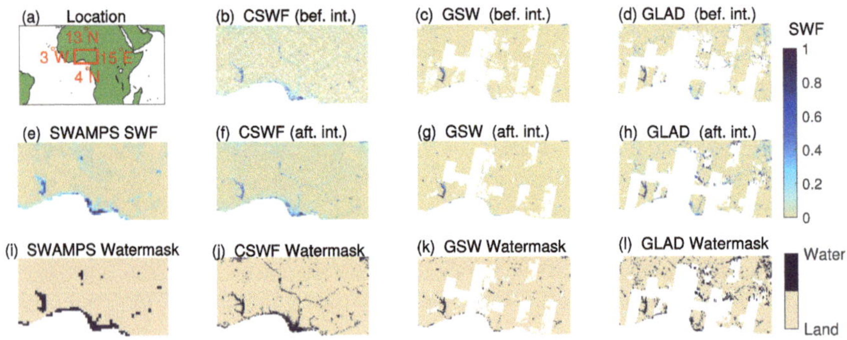

Fig. 9 SWF maps: **a** Study area, **b** CSWF (0.025°) before interpolation (BI), **c** GSW (0.025°) BI, and **d** GLAD (0.025°) BI, **e** SWAMPS (25 km), **f** CSWF (0.025°) after interpolation (AI), **c** GSW (0.025°) AI, and **d** GLAD (0.025°) AI and corresponding water masks: **i** SWAMPS, **j** CSWF, **k** GSW, and **l** GLAD for August 2018

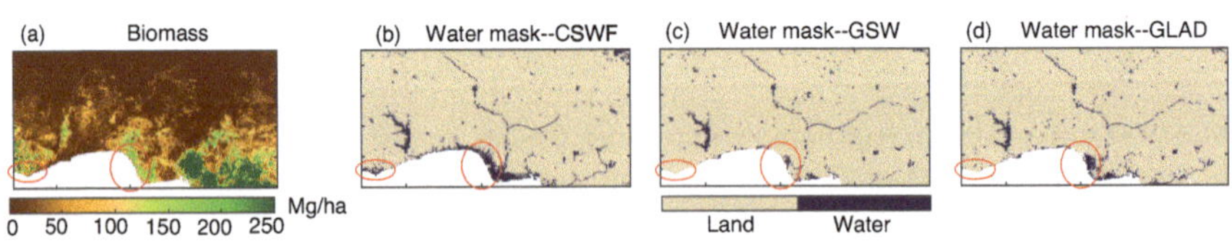

Fig. 10 Biomass **a** and water masks from: **b** CSWF, **c** GSW, and **d** GLAD for January 2019

a 0.025° grid provide a more nuanced and textured depiction of the surface water. Figure 9i–l show the binary water masks generated by applying a threshold of 0.1 to the SWF data. Figure 10b–d illustrate the water masks for January 2019—a non-flood period—where the optical datasets achieve broader spatial coverage but detect

Table 5 Coverage rate for 0.025° SWF products

Data	Original coverage	Coverage after gap filling
GSW	90.30%	91.97%
GLAD	93.54%	95.87%
CYGNSS (before 201907)	73.27%	99.41%
CYGNSS (after 201907)	91.35%	99.94%

less surface water than their microwave counterparts, as indicated by the red circles (the SWAMPS water mask closely resembles that of Fig. 9b and is thus not shown). This discrepancy may stem from the high biomass levels that could cause optical datasets to underestimate SWF.

It is important to mention that while the initial spatial coverage of the four datasets was approximately 90%, applying a straightforward yet efficient 10-by-10 moving window technique for filling gaps led to nearly complete coverage for CSWF. In comparison, the enhancement for the optical-based GSW and GLAD datasets was less pronounced. This difference is likely due to the significant gaps between the swaths and cloud cover in Landsat imagery, whereas CYGNSS data cover the Earth's surface more randomly with broader swaths and smaller gaps. Figure 9b–d and f–h provide a visual comparison of the results with and without gap filling, and Table 5 summarizes the overall coverage. Additionally, the gap-filled CYGNSS data performed comparably to the unaltered data, as detailed in Table 4.

4.4 Case Studies in Amazon and Australia: Seasonality and Validation with WL Measurement

To demonstrate, the seasonal fluctuations in the retrieved CSWF for the Amazon Basin and Australia are shown in Figs. 11 and 12 for the months of January, April, July, and October in 2019. The data are presented as deviations from the multi-year mean. It is evident that there are pronounced seasonal cycles, with each region showing unique patterns. For instance, the Orinoco River's behavior contrasts with that of the Madeira River, and the Amazon River exhibits a delayed response (as seen in Fig. 11), likely due to the movement of the Inter Tropical Convergence Zone. The variation in SWF across Australia (Fig. 12) highlights the significant spatial and temporal changes in SWF, particularly in areas like the Great Artesian Basin (GAB). This variability can be attributed to the ephemeral nature of water bodies in the GAB, influenced by a range of climatic conditions and the fluctuating connection between surface and groundwater. For example, the tropical climate in the northern GAB (e.g., the Cape York Peninsula) results in most precipitation falling in the summer, leading to an increased SWF (Fig. 12b). In contrast, the arid southern GAB (e.g., Lake Eyre and the Cooper River) relies on groundwater recharge to sustain surface

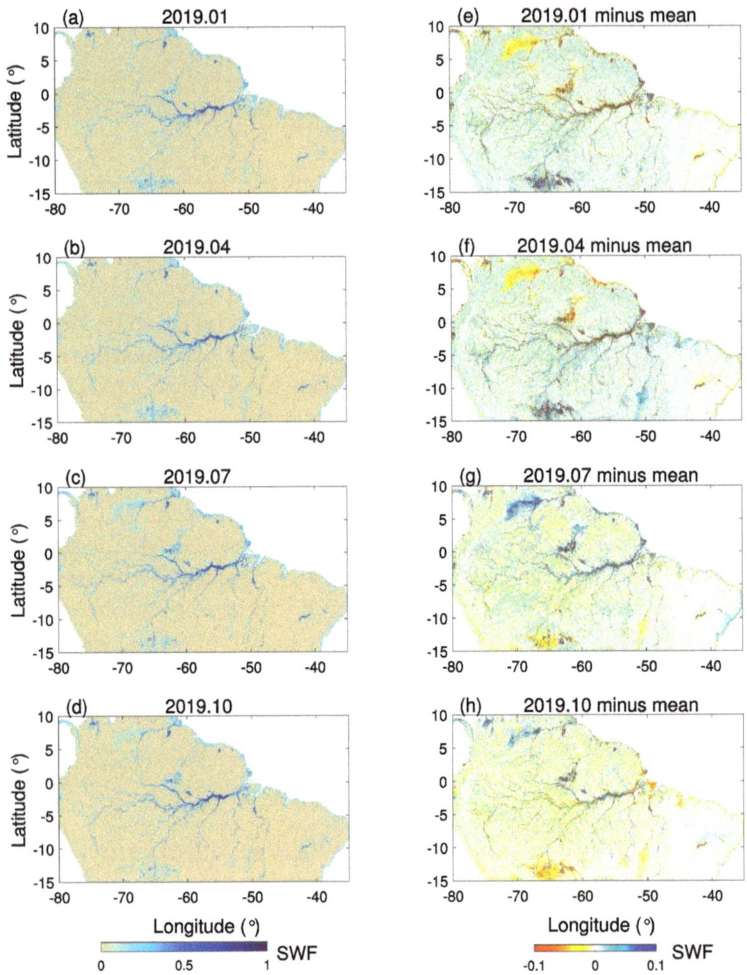

Fig. 11 Retrieved monthly CSWF maps in Amazon Basin for: **a** 2019.01, **b** 2019.04, **c** 2019.07, and **d** 2019.10 along with their corresponding values being subtracted by the yearly mean: (**e**)–(**h**)

water even in dry seasons. However, it should be noted that the seasonal variations captured by the CSWF might be somewhat underestimated due to the limitations of the SWAMPS dataset in capturing seasonal dynamics [23].

To assess the seasonal characteristics of SWF data and its utility at regional scales, WL measurements from monitoring stations were utilized as an additional reference. While WL does not directly measure SWF, there is a strong correlation between the two [48]. Data from two distinct regions, the Amazon River and Australia, were analyzed; the former was sourced from radar altimetry (which tracks water surface elevation relative to a reference point), and the latter was gathered from ongoing station monitoring. The spatial distribution of these stations (indicated by dots) and selected examples (noted by circles) of WL and SWF measurements over 41 months

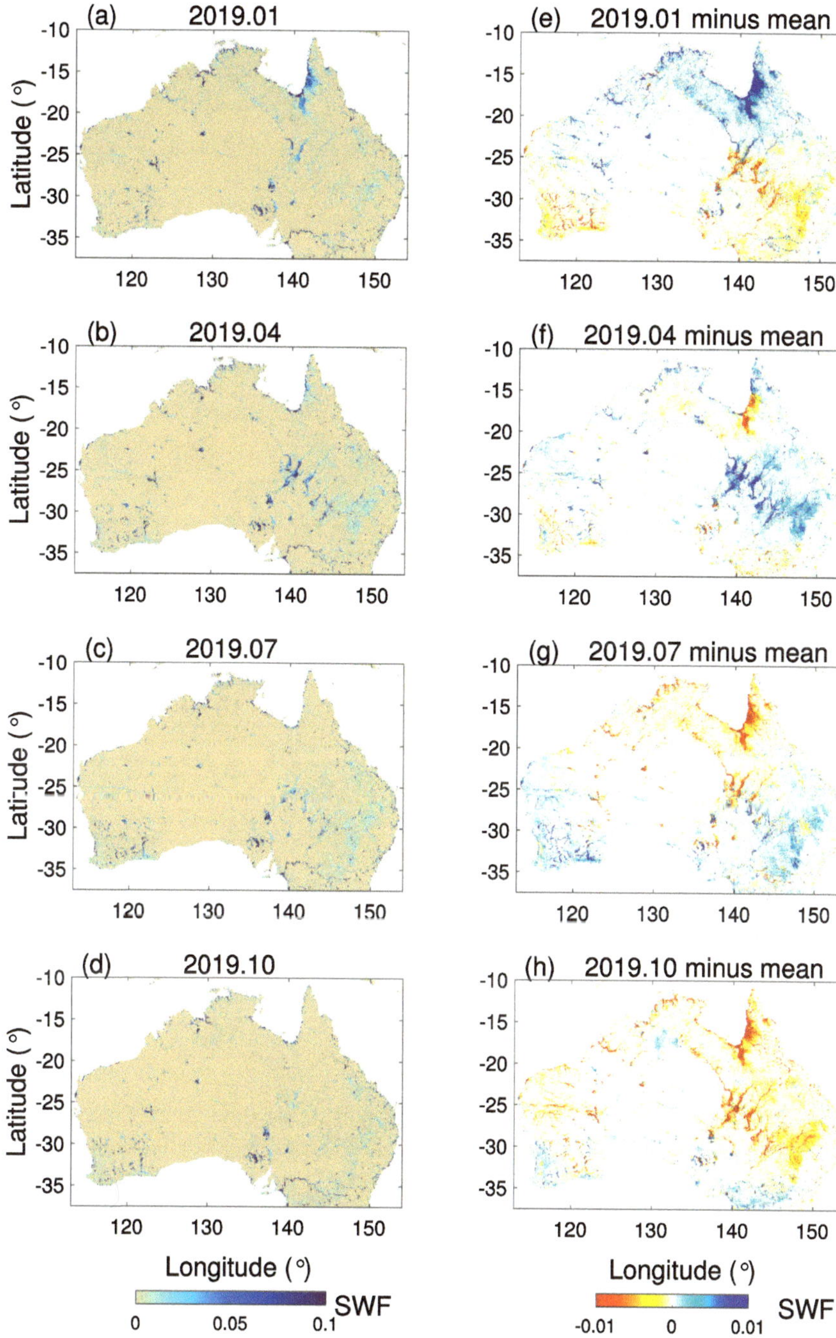

Fig. 12 Monthly CSWF retrievals in Australia for: **a** 2019.01, **b** 2019.04, **c** 2019.07, and **d** 2019.10 along with their values being subtracted by the yearly mean: (**e**)–(**h**)

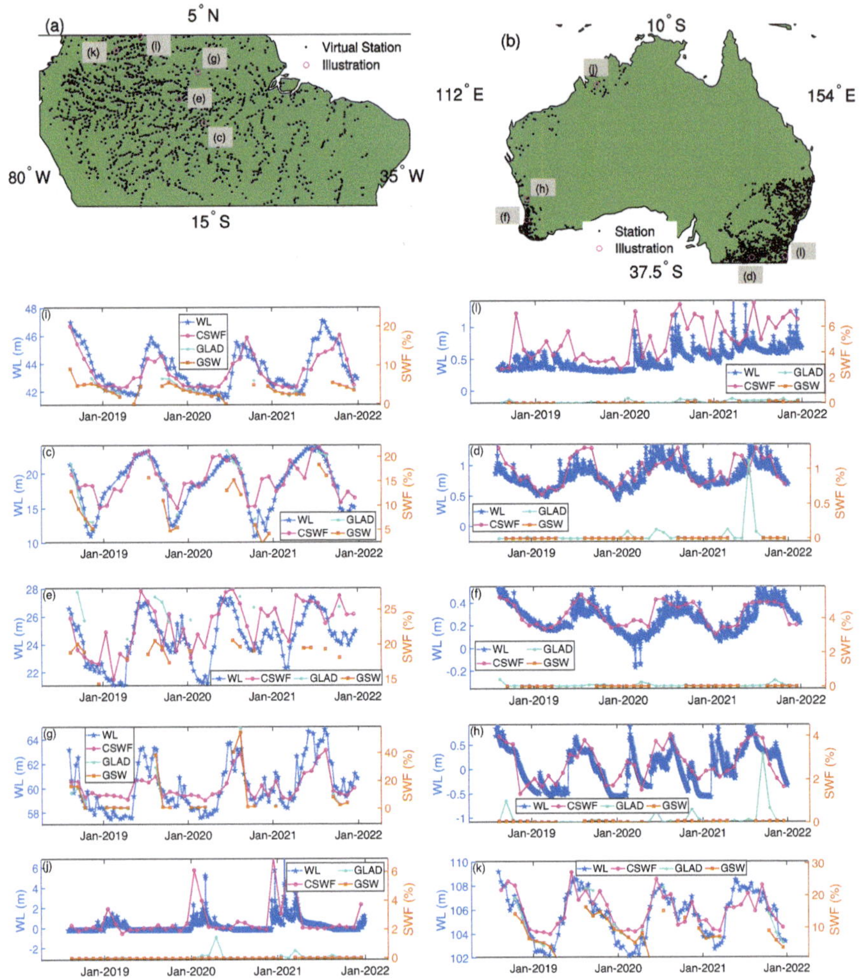

Fig. 13 Distribution of WL (virtual) stations in: **a** the Amazon River and **b** Australia and examples of WL/SWF data in both regions

in both areas are illustrated in Fig. 13. Each point in the time-series graphs corresponds to a WL sample. The WL stations in Australia record data daily, whereas those in the Amazon do so every few days, resulting in a sparser dataset for the latter (as shown in Fig. 13c–l). The Amazon shows seasonal WL changes, with an expansion (rise in WL) during the wet season and a contraction (drop in WL) during the dry season. Australia also experiences seasonal WL fluctuations, but with a notable lack of water in the dry season, especially in arid inland areas. In terms of monthly SWF time series, SWAMPS data are not included here due to its lower spatial resolution (25 km) and a brief overlap period (29 months) with CSWF. While CSWF offers continuous observations, both GLAD and GSW show data gaps in these regions.

In the Amazon, CSWF shows distinct seasonal changes that mirror those of WL. In some Australian cases, WL seasonal patterns may be less distinct, yet CSWF maintains a strong correlation. It is important to recognize that relying on data from only two annual cycles to capture the full range of SWF seasonality can be difficult. This difficulty arises because, while some areas may have consistent seasonal patterns year after year, others might experience considerable year-to-year variability due to various factors including climate anomalies, shifts in land use, and severe weather events. Moreover, by calculating monthly averages, the SWF products may not capture all the short-term variability evident in the WL data.

To evaluate the concordance between SWF and WL data more precisely, WL measurements were adjusted to monthly intervals with a spatial resolution of 0.025° for both the Amazon and Australia. Given that the SWF products also possess a 0.025° resolution on a monthly basis, the gauge-based point measurements were compiled into monthly series and assigned to the corresponding 0.025° grid cells as representative values. When multiple stations were present within a single grid cell,

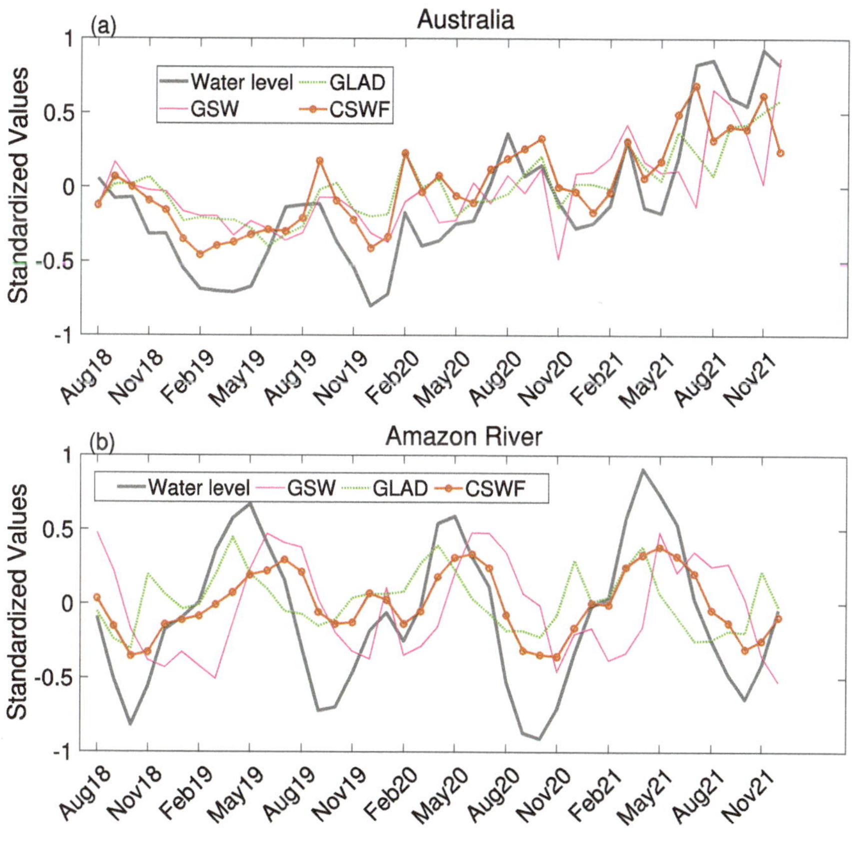

Fig. 14 Standardized WL measurements and SWF estimates: **a** Australia and **b** Amazon River

Table 6 Determination coefficient R^2 between SWF and WL (* indicates p-value is < 0.05)

Region	Dataset	R^2
Australia	GSW	0.739*
	GLAD	0.592*
	CYGNSS	0.810*
Amazon River	GSW	0.369*
	GLAD	0.795*
	CYGNSS	0.845*

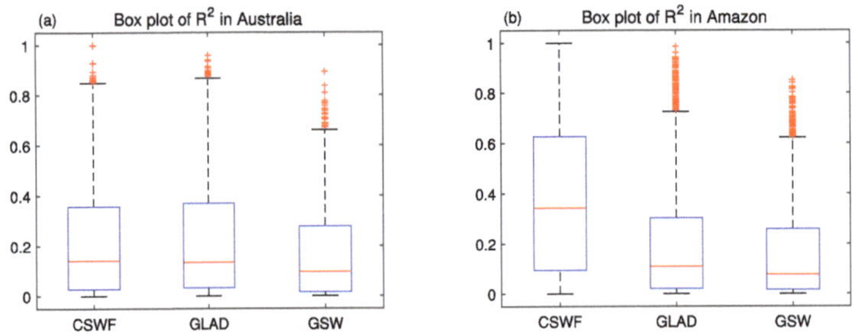

Fig. 15 Box plot of R^2 between SWF estimates and WL measurements in: **a** Australia and **b** Amazon River

the data from these stations were averaged. Furthermore, both SWF and WL datasets were standardized by subtracting the mean and dividing by the standard deviation for each. The spatially averaged time series of the normalized SWF and WL over the 41-month period for both regions with WL station data are presented in Fig. 14, with the R^2 values detailed in Table 6. The CYGNSS-based SWF demonstrates the strongest correlation with WL. As observed in Fig. 14, from August 2018 to mid-2020, both WL and SWF recorded negative values, signifying drought conditions in Australia, consistent with [49]. Conversely, an upward trend in WL and SWF has been noted since the start of 2021, supported by 2021 being the wettest year since 2016, with national rainfall 9% above the 1961 to 1990 average [50]. A pixel-level comparison was also performed for each 0.025° grid by calculating the R^2 between the corresponding WL and SWF data. The box plots of R^2 values for each individual 0.025° pixel in both regions are depicted in Fig. 15. It is evident that the performance in Australia is satisfactory across all three SWF datasets, whereas in the Amazon River region, where vegetation and cloud cover are substantial, the CYGNSS-based (microwave) SWF shows a better agreement compared to the Landsat-based (optical) GLAD and GSW datasets.

5 Conclusions

This study introduces a novel approach for determining SWF utilizing CYGNSS L1 data with the aid of BARTs. Our model integrates a variety of factors including DEM, slope, geographic coordinates, biomass, water seasonality, SMAP SM and VWC, and CYGNSS Γ. We train the model using SWAMPS data, characterized by a coarse resolution of 25 km, and then refine it for application on higher-resolution CYGNSS data at 0.025° to create more detailed SWF maps. Through extensive assessment, we have verified the method's reliability and efficacy. The correlation between global benchmarks and the derived maps was found to be quite satisfactory, with an overall r reaching up to 0.987 and RMSE of 1.53% when contrasted with the SWAMPS 25-km dataset. The comparison between the SWF results from GSW, GLAD, and CYGNSS at both 25-km and 0.025° resolutions indicates a strong consistency, affirming the accuracy of the high-resolution retrievals. Subsequent analysis using hydrological station data confirms that the CYGNSS-derived dataset has the strongest relationship with water level measurements, evidenced by Australian hydrological station records and altimeter readings over the Amazon River. To conclude, the benefits of CYGNSS-derived SWF, such as improved spatial coverage over optical data and higher resolution compared to SWAMPS, are clearly demonstrated.

The precision of the dataset is contingent upon the quality and precision of the training reference data, specifically SWAMPS, which can sometimes lead to an underestimation of SWF in regions with thick canopies and may not fully capture the range of seasonal changes. Achieving dependable and precise validation in these areas poses a considerable challenge. Looking ahead, it may be beneficial to explore more suitable references and conduct assessments with an expanded set of ground truth data, independent validation resources, and alternative methods like the radiometric model approach [35]. As part of future work, this approach aims to enhance the method for achieving finer spatiotemporal resolutions to facilitate more accurate SWF estimations in hydrological modeling and similar applications. This enhancement can be realized by incorporating data from additional GNSS-R missions, as incorporating a broader range of data can lead to improved spatiotemporal coverage. Moreover, while this study utilized and interpolated 9-km resolution SMAP data as supplementary information, which could introduce uncertainties, employing SM/VWC data with higher resolution could be advantageous. Although the current research is limited to static data such as biomass due to data availability constraints, future studies may take advantage of higher temporal resolution data once it becomes available.

References

1. Holgerson MA, Raymond PA (2016) Large contribution to inland water CO_2 and CH4 emissions from very small ponds. Nat Geosci 9. https://doi.org/10.1007/10.1038/ngeo2654
2. Haig SM, Murphy SP, Matthews JH, Arismendi I, Safeeq M (2019) Climate-altered wetlands challenge waterbird use and migratory connectivity in arid landscapes. Sci Rep 9. https://doi.

org/10.1007/10.1038/s41598-019-41135-y
3. Karpatne A, Khandelwal A, Chen X, Mithal V, Faghmous J, Kumar V, Global monitoring of inland water dynamics: state-of-the-art, challenges, and opportunities. https://doi.org/10.1007/10.1007/978-3-319-31858-5_7
4. Vörösmarty CJ, Green P, Salisbury J, Lammers RB (2000) Global water resources: vulnerability from climate change and population growth. Science 289. https://doi.org/10.1007/10.1126/science.289.5477.284
5. Pekel JF, Cottam A, Gorelick N, Belward AS (2016) High-resolution mapping of global surface water and its long-term changes. Nature 540:418–422. https://doi.org/10.1007/10.1038/nature20584
6. Pickens AH, Hansen MC, Hancher M, Stehman SV, Tyukavina A, Potapov P, Marroquin B, Sherani Z (2020) Mapping and sampling to characterize global inland water dynamics from 1999 to 2018 with full Landsat time-series. Remote Sens Environ 243. https://doi.org/10.1007/10.1016/j.rse.2020.111792
7. User Guides—Sentinel-1 SAR—Revisit and Coverage—Sentinel Online. https://sentinel.esa.int/web/sentinel/user-guides/sentinel-1-sar/revisit-and-coverage Accessed 2023-11-08
8. Jensen K, Mcdonald K (2019) Surface water microwave product series version 3: a near-real time and 25-year historical global inundated area fraction time series from active and passive microwave remote sensing. IEEE Geosci Remote Sens Lett 16(9):1402–1406. https://doi.org/10.1007/10.1109/LGRS.2019.2898779
9. Clarizia MP, Ruf CS, Jales P, Gommenginger C (2014) Spaceborne GNSS-R minimum variance wind speed estimator. IEEE Trans Geosci Remote Sens 52(11):6829–6843. https://doi.org/10.1007/10.1109/TGRS.2014.2303831
10. Foti G, Gommenginger C, Jales P, Unwin M, Shaw A, Robertson C, Roselló J (2015) Spaceborne GNSS reflectometry for ocean winds: first results from the UK TechDemoSat-1 mission. Geophys Res Lett 42(13):5435–5441. https://doi.org/10.1007/10.1002/2015GL064204
11. Cardellach E, Rius A, Martin-Neira M, Fabra F, Nogues-Correig O, Ribo S, Kainulainen J, Camps A, D'Addio S (2014) Consolidating the precision of interferometric GNSS-R ocean altimetry using airborne experimental data. IEEE Trans Geosci Remote Sens 52(8):4992–5004. https://doi.org/10.1007/10.1109/TGRS.2013.2286257
12. Li W, Cardellach E, Fabra F, Ribo S, Rius A (2020) Assessment of spaceborne GNSS-R ocean altimetry performance using CYGNSS mission raw data. IEEE Trans Geosci Remote Sens 58(1):238–250. https://doi.org/10.1007/10.1109/TGRS.2019.2936108
13. Chew C, Small EE (2018) Soil moisture sensing using spaceborne GNSS reflections: comparison of CYGNSS reflectivity to SMAP soil moisture. Geophys Res Lett 45(9):4049–4057. https://doi.org/10.1007/10.1029/2018GL077905
14. Yan Q, Huang W, Jin S, Jia Y (2020) Pan-tropical soil moisture mapping based on a three-layer model from CYGNSS GNSS-R data. Remote Sens Environ 247:111947
15. Yan Q, Huang W (2016) Spaceborne GNSS-R sea ice detection using delay-Doppler maps: first results from the U.K. TechDemoSat-1 mission. IEEE J Sel Top Appl Earth Obs Remote Sens 9(10):4795–4801. https://doi.org/10.1007/10.1109/JSTARS.2016.2582690
16. Yan Q, Huang W (2020) Sea ice thickness measurement using spaceborne GNSS-R: first results with TechDemoSat-1 data. IEEE J Sel Top Appl Earth Obs Remote Sens 13:577–587. https://doi.org/10.1007/10.1109/JSTARS.2020.2966880
17. Ruf CS, Chew C, Lang T, Morris MG, Nave K, Ridley A, Balasubramaniam R (2018) A new paradigm in earth environmental monitoring with the CYGNSS small satellite constellation. Sci Rep 8:1–13. https://doi.org/10.1007/10.1038/s41598-018-27127-4
18. Chew C, Reager JT, Small E (2018) CYGNSS data map flood inundation during the 2017 Atlantic hurricane season. Sci Rep 8(1). https://doi.org/10.1038/s41598-018-27673-x
19. Ghasemigoudarzi P, Huang W, Silva OD, Yan Q, Power DT (2020) Flash flood detection from CYGNSS data using the RUSBoost algorithm. IEEE Access 8:171864–171881. https://doi.org/10.1007/10.1109/ACCESS.2020.3025302
20. Zhang S, Ma Z, Liu Q, Hu S, Feng Y, Zhao H, Guo Q (2022) POBI interpolation algorithm for CYGNSS near real time flood detection research: a case study of extreme precipitation events in Henan, China in 2021. Adv Space Res. https://doi.org/10.1007/10.1016/j.asr.2022.11.016

21. Rodriguez-Alvarez N, Podest E, Jensen K, McDonald KC (2019) Classifying inundation in a tropical wetlands complex with GNSS-R. Remote Sens 11(9):1053. https://doi.org/10.1007/10.3390/rs11091053
22. Morris M, Chew C, Reager JT, Shah R, Zuffada C (2019) A novel approach to monitoring wetland dynamics using CYGNSS: Everglades case study. Remote Sens Environ 233. https://doi.org/10.1007/10.1016/j.rse.2019.111417
23. Gerlein-Safdi C, Bloom AA, Plant G, Kort EA, Ruf CS (2021) Improving representation of tropical wetland methane emissions with CYGNSS inundation maps. Global Biogeochem Cycles 35. https://doi.org/10.1007/10.1029/2020GB006890
24. Gerlein-Safdi C, Ruf CS (2019) A CYGNSS-based algorithm for the detection of inland waterbodies. Geophys Res Lett 46(21):12065–12072. https://doi.org/10.1007/10.1029/2019gl085134
25. Ghasemigoudarzi P, Huang W, Silva OD, Yan Q, Power D (2022) A machine learning method for inland water detection using CYGNSS data. IEEE Geosci Remote Sens Lett 19:1–5. https://doi.org/10.1007/10.1109/lgrs.2020.3020223
26. Al-Khaldi MM, Johnson JT, Gleason S, Chew CC, Gerlein-Safdi C, Shah R, Zuffada C (2021) Inland water body mapping using CYGNSS coherence detection. IEEE Trans Geosci Remote Sens. https://doi.org/10.1007/10.1109/TGRS.2020.3047075
27. Li W, Cardellach E, Ribó S, Rius A, Zhou B (2021) First spaceborne demonstration of BeiDou-3 signals for GNSS reflectometry from CYGNSS constellation. Chin J Environ 34:1–16
28. Zeiger P, Frappart F, Darrozes J, Prigent C, Jiménez C (2022) Analysis of CYGNSS coherent reflectivity over land for the characterization of pan-tropical inundation dynamics. Remote Sens Environ 282:113278. https://doi.org/10.1007/10.1016/j.rse.2022.113278
29. Liu B, Wan W, Tang G, Li H, Guo Z, Chen X, Hong Y (2022) Statistical analysis of CYGNSS speckle and its applications to surface water mapping. IEEE Trans Geosci Remote Sens 60:1–15. https://doi.org/10.1007/10.1109/TGRS.2022.3196012
30. Yan Q, Chen Y, Jin S, Liu S, Jia Y, Zhen Y, Chen T, Huang W (2023) Inland water mapping based on GA-LinkNet from CyGNSS data. IEEE Geosci Remote Sens Lett 20. https://doi.org/10.1007/10.1109/LGRS.2022.3227596
31. Loria E, Russo IM, Wang Y, Giangregorio G, Galdi C, Bisceglie MD, Wilson-Downs B, Lavalle M, O'Brien AJ, Morton YJ, Zuffada C (2023) Comparison of GNSS-R coherent reflection detection algorithms using simulated and measured CYGNSS data. IEEE Trans Geosci Remote Sens 61. https://doi.org/10.1007/10.1109/TGRS.2023.3277411
32. Chapman BD, Russo IM, Galdi C, Morris M, Bisceglie MD, Zuffada C, Downs B, Lavalle M, Loria E, Oabrien AJ (2022) Comparison of SAR and CYGNSS surface water extent metrics. IEEE J Sel Top Appl Earth Obs Remote Sens 15. https://doi.org/10.1007/10.1109/JSTARS.2022.3162764
33. Downs B, Kettner AJ, Chapman BD, Brakenridge GR, O'Brien AJ, Zuffada C (2023) Assessing the relative performance of GNSS-R flood extent observations: case study in south Sudan. IEEE Trans Geosci Remote Sens 61. https://doi.org/10.1007/10.1109/TGRS.2023.3237461
34. Loria E, O'Brien A, Zavorotny V, Downs B, Zuffada C (2020) Analysis of scattering characteristics from inland bodies of water observed by CYGNSS. Remote Sens Environ 245:1–16
35. Chew C, Small E, Huelsing H (2023) Flooding and inundation maps using interpolated CYGNSS reflectivity observations. Remote Sens Environ 293. https://doi.org/10.1007/10.1016/j.rse.2023.113598
36. Carreno-Luengo H, Camps A, Querol J, Forte G (2016) First results of a GNSS-R experiment from a stratospheric balloon over boreal forests. IEEE Trans Geosci Remote Sens 54(5):2652–2663. https://doi.org/10.1007/10.1109/TGRS.2015.2504242
37. Clarizia MP, Pierdicca N, Costantini F, Floury N (2019) Analysis of CYGNSS data for soil moisture retrieval. IEEE J Sel Top Appl Earth Obs Remote Sens 12(7):2227–2235. https://doi.org/10.1007/10.1109/JSTARS.2019.2895510
38. Eroglu O, Kurum M, Boyd D, Gurbuz AC (2019) High spatio-temporal resolution CYGNSS soil moisture estimates using artificial neural networks. Remote Sens 11(19):2272. https://doi.org/10.1007/10.3390/rs11192272

39. O'Neill PE, Chan S, Njoku EG, Jackson T, Bindlish R, Chaubell J, Colliander A (2021) SMAP enhanced L3 radiometer global and polar grid daily 9 km EASE-Grid soil moisture, Version 5. NASA National Snow and Ice Data Center Distributed Active Archive Center https://doi.org/10.1007/10.5067/4DQ54OUIJ9DL. https://nsidc.org/data/SPL3SMP_E/versions/5
40. Gorelick N, Hancher M, Dixon M, Ilyushchenko S, Thau D, Moore R (2017) Google Earth Engine: planetary-scale geospatial analysis for everyone. Remote Sens Environ 202:18–27. https://doi.org/10.1007/10.1016/j.rse.2017.06.031
41. Wu Q (2020) geemap: a Python package for interactive mapping with Google Earth Engine. J Open Sour Softw 5(51):2305
42. Breiman L, Friedman JH, Olshen RA, Stone CJ (1984) Classification and regression trees. Routledge, New York. https://doi.org/10.1007/10.1201/9781315139470
43. Breiman L (1996) Bagging predictors. Mach Learn 24(2):123–140. https://doi.org/10.1007/10.1007/bf00058655
44. Chakrabarti S, Judge J, Bongiovanni T, Rangarajan A, Ranka S (2018) Spatial scaling using temporal correlations and ensemble learning to obtain high-resolution soil moisture. IEEE Trans Geosci Remote Sens 56(3):1238–1250. https://doi.org/10.1007/10.1109/TGRS.2017.2722236
45. Yan Q, Gong S, Jin S, Huang W, Zhang C (2022) Near real-time soil moisture in China retrieved from CyGNSS reflectivity. IEEE Geosci Remote Sens Lett 19:1–5. https://doi.org/10.1007/10.1109/LGRS.2020.3039519
46. Zhang Z, Fluet-Chouinard E, Jensen K, McDonald K, Hugelius G, Gumbricht T, Carroll M, Prigent C, Bartsch A, Poulter B (2021) Development of the global dataset of wetland area and dynamics for methane modeling (WAD2M). Earth Syst Sci Data 13(5):2001–2023
47. Pierdicca N, Fascetti F, Pulvirenti L, Crapolicchio R, Muñoz-Sabater J (2015) Quadruple collocation analysis for soil moisture product assessment. IEEE Geosci Remote Sens Lett 12. https://doi.org/10.1007/10.1109/LGRS.2015.2414654
48. Prigent C, Jimenez C, Bousquet P (2020) Satellite-derived global surface water extent and dynamics over the last 25 years (GIEMS-2). J Geophys Res Atmos 125. https://doi.org/10.1007/10.1029/2019JD030711
49. Fang B, Kansara P, Dandridge C, Lakshmi V (2021) Drought monitoring using high spatial resolution soil moisture data over Australia in 2015–2019. J Hydrol 594. https://doi.org/10.1007/10.1016/j.jhydrol.2021.125960
50. Annual Climate Statement 2021: wetter and warmer than average. Social Media Blog. Bureau of Meteorology. https://media.bom.gov.au/social/blog/2669/annual-climate-statement-2021-wetter-and-warmer-than-average/

Chapter 13
Flood Mapping and Water Level Estimation Using Spaceborne GNSS Reflectometry

Zhongmin Ma and Shuangcheng Zhang

Abstract Flood disasters demand rapid and reliable monitoring methods with high spatial and temporal resolution. This study investigates the potential of CYGNSS GNSS-R observations for flood detection and water-level retrieval using multi-source remote sensing datasets, including SMAP, SMOS, GFM, ICESat-2, and GEDI. CYGNSS surface reflectivity (SR) was used to identify inundation areas via a threshold-based approach. Results show that SR is sensitive to surface moisture and flooding, with − 13 dB as the optimal threshold for distinguishing flooded regions. The temporal variation of SR closely matched soil moisture changes (r = 0.7), capturing the evolution of flood events effectively. A spatial interpolation method based on previous observations (POBI) improved CYGNSS spatial coverage and consistency. The derived surface water fraction (Wf) agreed well with SMAP/SMOS brightness temperature (Tb) data and high-resolution GFM flood maps, achieving an overall accuracy of 64.44%. A comparison with LiDAR altimetry data (ICESat-2 and GEDI) further validates the effectiveness of the CYGNSS based water level retrieval method that incorporates topographic information, with correlation coefficients exceeding 0.97 and root-mean-square errors below 1.5 m. These findings demonstrate that CYGNSS provides a robust, all-weather approach for large-scale flood monitoring and complements traditional microwave and radar observations.

1 Introduction

Flood disasters are typically triggered by extreme rainfall, river breaches, flash floods, or failures of urban drainage systems, and fundamentally involve abnormal accumulation and overflow of water [1]. Depending on their formation mechanisms, floods can be classified into heavy-rain floods, river floods, flash floods, and urban inundations

Z. Ma · S. Zhang (✉)
College of Geological Engineering and Geomatics, Chang'an University, Xi'an, China
e-mail: shuangcheng369@chd.edu.cn

Z. Ma
e-mail: zhongminma@chd.edu.cn

K. Yu and N. Zheng (eds.), *Advances in GNSS Reflectometry*, Navigation: Science and Technology 19, https://doi.org/10.1007/978-981-92-2325-1_13

[2]. Such events are often accompanied by secondary hazards, including landslides, debris flows, infrastructure damage, and public health risks, resulting in substantial casualties and economic losses, while exerting profound impacts on regional socioeconomic development and ecosystem stability [3]. With ongoing global warming, the frequency and intensity of extreme rainfall events have increased markedly, leading to wider and more destructive floods, and making flood risk a significant global environmental challenge [4].

Flood monitoring primarily relies on ground-based observations, airborne remote sensing, and satellite-based techniques [5]. Traditional ground hydrometeorological measurements, including rain gauges, water level sensors, and streamflow stations, provide high-precision and continuous point data, but their sparse distribution and limited coverage hinder comprehensive characterization of the spatiotemporal dynamics of floods [6]. Airborne remote sensing offers high spatial resolution and operational flexibility, making it suitable for localized surveys and emergency monitoring; however, it is costly and constrained by weather and terrain conditions, limiting its applicability for large-scale and continuous observations [7]. In contrast, satellite remote sensing, with its extensive coverage, long-term time series, and multi-parameter observation capabilities, has become a core tool for flood monitoring and assessment [8].

Within the satellite observation framework, different sensor types present distinct advantages and limitations in flood detection [9]. Optical sensors offer high resolution, intuitive imagery, and ease of identifying inundation, but are vulnerable to cloud cover and rainfall, limiting their use under severe convective weather. Microwave radiometers can penetrate clouds and provide all-weather observations, showing high sensitivity to large-scale water bodies and soil moisture variations; however, their coarse spatial resolution restricts fine-scale monitoring. Synthetic Aperture Radar (SAR) is weather-independent and capable of day–night imaging, enabling precise delineation of inundated areas and water boundaries, and is widely applied in disaster response. Nevertheless, SAR data processing is complex, revisit intervals are relatively long, and interpretation accuracy depends on surface scattering characteristics. With single-polarization or single-channel data, its ability to resolve complex flood scenarios remains limited. Overall, current monitoring methods still face challenges in spatial coverage, temporal continuity, and multi-parameter observation, which constrain their effectiveness in dynamic flood monitoring and risk assessment [10].

Over the past two decades, Global Navigation Satellite System Reflectometry (GNSS-R) has emerged as a promising passive remote sensing technique [11]. Its fundamental principle involves exploiting continuously transmitted GNSS signals, simultaneously receiving both the direct signals and those scattered from surfaces such as land or sea, and retrieving physical properties of the target surface through coherent or non-coherent comparisons [12]. Compared with traditional optical and active microwave remote sensing, GNSS-R offers notable advantages, including stable signal sources, wide coverage, low cost, and flexible platform compatibility [13]. Leveraging these features, GNSS-R has been widely applied to monitor and analyze environmental parameters over oceans [14], land [15], and ice–snow regions [16].

Based on the observation platform, GNSS-R systems can be classified into ground-based, airborne, and spaceborne configurations [17]. Ground-based GNSS-R provides flexible deployment and cost-effective operation, suitable for methodological validation and local-scale monitoring, but its spatial coverage is limited. Airborne GNSS-R, mounted on aircraft, enables larger-scale observations and is often used for algorithm testing and regional applications. Spaceborne GNSS-R, relying on satellite platforms, achieves continuous, large-scale global observations and represents the core direction of current GNSS-R development. Compared with ground- and airborne systems, spaceborne GNSS-R can employ constellations of small satellites to simultaneously receive reflections from multiple GNSS satellites, offering extensive coverage, short revisit intervals, and high observation timeliness. These characteristics make it particularly suitable for monitoring dynamic, large-scale natural hazards. Owing to these technical advantages, spaceborne GNSS-R shows unique potential for flood monitoring: it can reliably capture inundation extent, water boundaries, and surface soil moisture even under severe convective weather, and, through time-series observations, reveal the onset, evolution, and recession of floods, providing critical support for rapid disaster assessment, emergency response, and risk management [18].

This chapter primarily presents methods and results for retrieving flood extent and water levels using spaceborne GNSS-R Level 1 observations. The second section introduces the CYGNSS data and auxiliary datasets used; the third section details the proposed retrieval methods; sections four to six present the results and evaluation; and the final section concludes the chapter.

2 Data Overview

2.1 CYGNSS

At the end of 2016, NASA successfully launched CYGNSS (Cyclone Global Navigation Satellite System), the world's first small-satellite constellation dedicated to spaceborne GNSS-R. Since March 2017, CYGNSS scientific data have been publicly available, freely accessible for global users. This study utilizes CYGNSS Level-1 Version 3.0 data. Compared with the previous Version 2.1, Version 3.0 features a recalibrated GPS Effective Isotropic Radiated Power (ERIP) and significantly reduces errors caused by antenna pattern uncertainties, thereby enhancing overall data accuracy and consistency [19].

The core observable of CYGNSS is the Delay-Doppler Map (DDM), which is generated by correlating the received reflected GNSS signals with a locally generated replica of a specific GNSS signal. DDM shows the power of reflected signal versus the time delay and Doppler frequency, which are generated by the relative motion among GNSS satellites, the receiver, and the Earth's surface, as well as the scattering characteristics of different locations around specular reflection points [20]. The CYGNSS constellation consists of eight microsatellites, each equipped with a four-channel reflected signal receiver, allowing simultaneous multi-directional observations and enabling up to 32 independent reflection measurements at any given time. In oceanic conditions, CYGNSS achieves a typical spatial resolution of approximately 25 km × 25 km, with a revisit period of several hours [21]. For land observations, however, effective spatial resolution remains uncertain due to complex surface scattering properties and requires further quantification. Theoretical studies suggest that for relatively smooth surfaces dominated by coherent reflection, CYGNSS can achieve a minimum spatial resolution of about 3.5 km × 0.5 km [22]. Revisit periods also vary with latitude; for example, for 3 km × 3 km grid cells, the revisit period is roughly 8 days at the edge of CYGNSS's latitudinal coverage (±38°) and approximately 14 days near the equator [23].

CYGNSS Level-1 data include a set of user-selectable quality control (QC) flags. The QC conditions applied in this study are: s_band_powered_up, large_sc_attitude_err, black_body_ddm, ddmi_reconfigured, spacewire_crc_invalid, ddm_is_test_pattern, channel_idle, direct_signal_in_ddm, low_confidence_gps_eirp_estimate, rfi_detected, gps_pvt_sp3_error, sp_non_existent_error, and bb_framing_error. To ensure data reliability, additional filtering criteria were applied [24]:

1. DDM signal-to-noise ratio (SNR) greater than 1.5 dB;
2. Receiver antenna gain greater than 0 dB;
3. Elevation angle of the specular reflection point greater than 30°.

This preprocessing effectively excludes low-quality observations, providing a solid foundation for subsequent surface reflectivity analysis and flood retrievals.

2.2 SMAP

Considering that optical remote sensing data are easily affected by cloud cover and precipitation, this study adopts flood retrieval methods based on the passive microwave satellite SMAP (Soil Moisture Active Passive) to identify surface inundation, following previous research [25]. Launched by NASA in January 2015, SMAP provides a typical spatial resolution of approximately 40 km and a revisit period of 2–3 days. Since both CYGNSS and SMAP operate in the L-band, they exhibit similar sensitivity to surface dielectric properties and soil moisture variations. Prior studies have demonstrated that SMAP data are reliable and effective for surface inundation retrieval [26].

For non-flooded terrestrial surfaces, the upper limit of volumetric soil moisture is typically around 0.4 cm^3/cm^3, whereas in flooded areas, it generally ranges from 0.4 to 0.45 cm^3/cm^3. This study utilizes SMAP Level-3 global soil moisture products with a spatial resolution of 9 km × 9 km, which are derived from the original 36 km × 36 km observations via the Backus–Gilbert interpolation algorithm [27]. SMAP data are freely accessible through NASA's National Snow and Ice Data Center (NSIDC) and are widely used for large-scale analyses of surface moisture distribution and flood dynamics.

It should be noted that SMAP experienced a system failure between August 7 and September 22, 2022, resulting in missing soil moisture (SM) and brightness temperature (Tb) data during this period. To maintain continuity in the time series, SMOS (Soil Moisture and Ocean Salinity) satellite data were used as a complementary source. Launched by ESA at the end of 2009 and also operating in the L-band, SMOS primarily monitors global soil moisture and ocean salinity. This study uses SMOS Level-3 daily soil moisture products with a spatial resolution of approximately 25 km × 25 km.

During preprocessing, SMAP and SMOS data were resampled to match the CYGNSS grid resolution, ensuring spatial consistency across datasets. This harmonization enables joint analysis and comparison, providing a unified data foundation for subsequent flood detection and surface water retrieval.

2.3 GFM

The Global Flood Monitoring (GFM) products are generated by the Copernicus Emergency Management Service (CEMS) based on Sentinel-1 satellite observations [28]. These products provide near-real-time flood monitoring results at a global scale and are freely accessible to the public.

GFM employs three independent algorithms to generate separate flood inundation maps, which are then integrated using a consistency-based fusion strategy: a given pixel is classified as flooded in the final product only if at least two algorithms simultaneously identify it as inundated [29]. This approach balances detection sensitivity and reliability while effectively reducing uncertainties caused by potential misclassifications from individual algorithms.

Currently, GFM offers 11 types of flood-related products, covering flood occurrence probability, inundation extent, and temporal dynamics. In this study, the "Observed Water Extent" (OWE) product (version 2.1.0) is employed. The OWE product provides spatial information on flood inundation derived from multi-temporal radar imagery and serves as an important reference dataset for evaluating and validating flood monitoring accuracy.

2.4 Satellite-Based LiDAR Data

To assess the reliability and accuracy of flood water level retrievals, satellite LiDAR altimetry data were employed as reference, including ICESat-2 and the Global Ecosystem Dynamics Investigation (GEDI) datasets. ICESat-2, a core mission of NASA's Earth Observing System (EOS), provides multi-source observations of ice sheet elevation, sea ice thickness, terrestrial topography, vegetation structure, and cloud properties. Its Advanced Terrain Laser Altimeter System (ATLAS) features extremely high spatial sampling density, with laser pulse emission rates reaching approximately 10,000 pulses per second. This study utilized the L3A Land and Vegetation Height product (V006), which processes data in fixed 100 m segments, each typically containing over 100 signal photons, allowing simultaneous retrieval of ground and canopy information. Ground elevation along satellite tracks was extracted from this dataset for comparison with flood water level retrievals.

GEDI, a joint mission by NASA and the University of Maryland, is mounted on the International Space Station (ISS) and primarily aims to quantify the impacts of forest disturbance on global carbon cycling and atmospheric CO_2 concentrations. Equipped with full-waveform LiDAR, GEDI provides high-precision, georeferenced terrain and vegetation height information. This study used the L2A Elevation and Height Metrics Data, Global Footprint Level (V002), selecting the ground elevation relative to the geoid for analysis.

To ensure spatial and temporal consistency across datasets and facilitate subsequent comparative analysis, ICESat-2 and GEDI data were resampled onto a 3 km × 3 km grid. Each grid cell value was computed as the mean of all observations within that cell. Subsequently, grid cells containing both CYGNSS-derived water level estimates and LiDAR observations within the same study period were directly compared to evaluate the accuracy and consistency of the retrieval results.

3 Methodology Overview

3.1 Estimation of Surface Reflectivity from Spaceborne GNSS-R

The fundamental principle of GNSS-R is analogous to that of a bistatic radar, in which L-band signals reflected from the Earth's surface are received and analyzed to retrieve geophysical properties. Unlike conventional bistatic radar, GNSS-R primarily utilizes the forward-scattered signals from navigation satellites, rendering it relatively insensitive to variations in surface roughness [30].

In this study, the key observation parameter is surface reflectivity (SR), which is derived from Delay-Doppler Maps (DDMs) provided in CYGNSS Level-1 data. Extensive research has shown that SR is predominantly influenced by the dielectric properties and physical state of the surface, including soil moisture, water body coverage, surface roughness, and vegetation density. Generally, for L-band GNSS signals used by CYGNSS, higher surface moisture corresponds to increased SR. This relationship forms the theoretical basis for flood extent detection and water level retrieval using GNSS-R data.

For low-Earth orbit bistatic satellite systems, the total echo power received by the GNSS-R receiver consists of coherent and incoherent scattering components, whose relative contributions depend on the surface dielectric constant and roughness [31].

Accordingly, the total bistatic radar scattering power P_{rl} can be expressed as:

$$P_{rl} = P_{rl}^{c} + P_{rl}^{i} \tag{1}$$

where: P_{rl}^{c} and P_{rl}^{i} denote the coherent and incoherent scattering power, respectively, given by:

$$P_{rl}^{c} = \frac{P_r^t G^t}{4\pi (R_{ts} + R_{sr})^2} \frac{G^r \lambda^2}{4\pi} \tau_{rl} \tag{2}$$

$$P_{rl}^{i} = \frac{P_r^t G^t G^r \lambda^2}{(4\pi)^3 R_{ts}^2 R_{sr}^2} \sigma_{rl} \tag{3}$$

Here, P_r^t is the transmit power of the GPS satellite antenna; G^t and G^r represent the antenna gains of the GPS satellite and the CYGNSS receiver, respectively; R_{ts} and R_{sr} are the distances from the GPS satellite to the specular reflection point and from the specular point to the CYGNSS receiver; τ_{rl} denotes the surface reflectivity (SR); and σ_{rl} is the bistatic radar cross-section (in m^2).

In inland water detection and mapping using CYGNSS, it is typically assumed that the reflected signals are dominated by the coherent component [32–34]. Under this assumption, Eq. (2) can be expressed in decibel (dB) units as:

$$\begin{aligned} 10 \log P_{rl}^{c} = {} & 10 \log P_r^t + 10 \log G^t + 10 \log G^r + 20 \log \lambda \\ & + 10 \log \tau_{rl} - 20 \log(R_{ts} + R_{sr}) - 20 \log(4\pi) \end{aligned} \tag{4}$$

From (4), the SR in dB can be written as:

$$\begin{aligned} SR = 10 \log \tau_{rl} = {} & 10 \log P_{rl}^{c} - 10 \log P_r^t - 10 \log G^t - 10 \log G^r \\ & - 20 \log \lambda + 20 \log(R_{ts} + R_{sr}) + 20 \log(4\pi) \end{aligned} \tag{5}$$

Taking into account compensation for the incidence angle effect, the SR is finally determined as:

$$SR_{corrected} = SR - 10 \log \cos^n \theta \tag{6}$$

Here, θ denotes the incidence angle, and n represents the calibration coefficient, typically taken as 1.

3.2 Flood Extent Detection Based on Surface Reflectivity Threshold

To quantitatively assess the spatial expansion of rainfall-induced inundation, this study adopts the commonly used threshold detection approach from single-receiver radar-based flood extraction studies [35]. The core idea is that a grid cell is considered flooded if its CYGNSS surface reflectivity exceeds a predetermined threshold; otherwise, it is regarded as non-flooded. This method has been widely applied in CYGNSS-based surface water and flood mapping studies, demonstrating reliable performance in multiple case studies.

However, previous research indicates that the SR threshold is not fixed and is influenced by various factors, including local topography, surface roughness, and vegetation coverage [25, 34]. For instance, under conditions of moderate vegetation density and typical surface roughness, a threshold of 12 dB was adopted [34], whereas in highly vegetated, high-soil-moisture farmland with relatively homogeneous terrain, the threshold was raised to 17 dB [25]. Therefore, to ensure accurate and reliable flood detection, the SR threshold should be dynamically adjusted according to the specific surface characteristics of the area.

Figure 1 illustrates the land use and land cover types in Henan Province and the spatial distribution of CYGNSS SR from June 1 to June 5, 2021 (scatter plot format). The figure highlights the diverse surface types, including dryland, paddy fields, forest, and grassland. Notably, SR values in the southern paddy fields are significantly higher than in other land cover types, indicating substantial reflectivity differences among different surfaces. Based on this observation, the influence of land cover type on CYGNSS SR is first analyzed, providing a basis for threshold selection and subsequent flood mapping.

All CYGNSS L1 data over Henan Province from April 1 to June 30, 2021, totaling 188,815 Delay-Doppler Maps (DDMs), were processed. Based on land use and land cover characteristics, the region was categorized into six surface types: dryland, paddy fields, forest, grassland, water bodies, and urban/industrial areas.

Spatial overlay analysis was then conducted to count the number of CYGNSS specular reflection points in each land cover class, and the corresponding surface reflectivity and signal-to-noise ratio (SNR) were analyzed. The statistical results of reflectivity characteristics across different surface types are summarized in Table 1. Significant differences in SR and SNR among land cover classes are evident. Although the current analysis does not correct for topographic variation or surface roughness, the results demonstrate that CYGNSS data are highly sensitive to the variation in surface types and thus can be used to distinguish between different surface types. For example, water bodies and paddy fields exhibit the highest SR values, averaging approximately − 13 dB and − 15 dB, respectively, whereas forests and grasslands show the lowest SR, around − 22 dB. Dryland and urban/industrial areas have intermediate SR values, approximately − 19 dB. The SNR distribution follows a similar trend: water bodies and paddy fields reach the highest SNRs of roughly

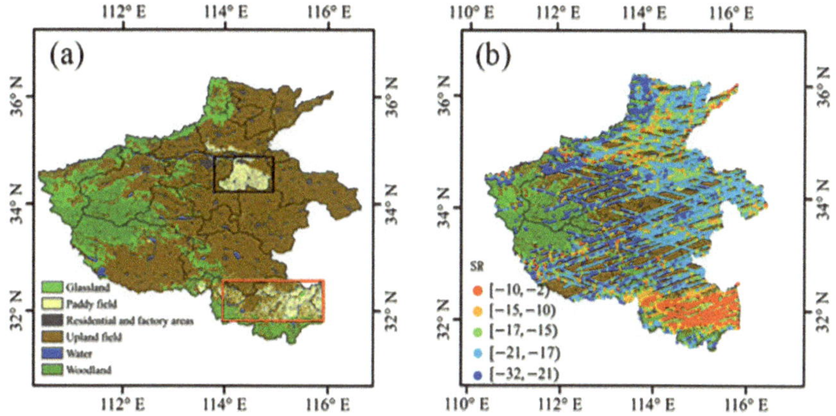

Fig. 1 **a** Land use and land cover of Henan Province. Two major rice cultivation areas are located in the south (red box) and the north (black box). The southern rice-growing season extends from early May to early September, while the northern season lasts from mid-July to mid-October. During the period, the northern rice fields were planted with upland crops and thus classified as upland areas. **b** CYGNSS-derived surface reflectivity (SR) observations (scattered form) from 1 to 5 June 2021

9 dB and 8 dB, dryland and urban/industrial areas around 5 dB, and forest/grassland the lowest at about 4 dB.

To further improve flood detection accuracy, future studies should incorporate higher-resolution land use and land cover datasets and refine surface classifications—for instance, distinguishing between tall-tree forests and low shrublands. Additionally, the influence of surrounding surface roughness on specular reflection intensity should be considered. Based on the statistical analysis in Table 1, a surface reflectivity threshold of − 13 dB was ultimately adopted for identifying flooded areas using CYGNSS data.

Table 1 Statistics of surface reflectivity (SR) and signal-to-noise ratio (SNR) across land cover types in Henan Province (1 April–30 June 2021)

Land use and cover	SR Number	Proportion (%)	SNR (dB)	SR (dB)
Upland field	118990	63.02	5.22	− 19.13
Paddy field	13586	7.20	8.33	− 14.58
Woodland	13589	7.20	4.53	− 14.58
Grassland	7043	3.73	3.98	− 22.49
Water	5388	2.85	9.16	− 13.21
Residential and factory areas	30219	16.00	5.56	− 18.49

3.3 Retrieval of Surface Water Fraction

In addition to the binary flood detection results (i.e., 0 for non-inundated and 1 for inundated areas), this study further models the CYGNSS SR to estimate the fraction of surface water (Wf) at the grid (pixel) level. Inspired by the approach proposed by Chew and Small [22], it is assumed that the observed CYGNSS SR represents a weighted sum of the reflectivity from inundated and non-inundated portions of the surface within each grid cell, expressed as:

$$SR_{surface} = Wf \times SR_{inundated} + (1 - Wf) \times SR_{not\_inundated} \quad (7)$$

where $SR_{surface}$ is the total surface reflectivity observed by CYGNSS, $SR_{inundated}$ and $SR_{not\_inundated}$ denote the reflectivities of inundated and non-inundated areas, respectively, and Wf represents the fractional water coverage within the grid cell.

Since CYGNSS receives left-hand circularly polarized (LHCP) signals, the horizontally (H) and vertically (V) polarized reflection coefficients are converted into LHCP reflection coefficients according to:

$$r_{lhcp} = 0.5\left(r_{V_{SM}} - r_{H_{SM}}\right) \quad (8)$$

where r_{H_SM} and r_{V_SM} denote the reflection coefficients for H and V polarizations, respectively, calculated as:

$$r_{H\_SM} = \frac{\sqrt{\varepsilon_{air} - \sin^2\theta} - \sqrt{\varepsilon_{surface} - \sin^2\theta}}{\sqrt{\varepsilon_{air} - \sin^2\theta} + \sqrt{\varepsilon_{surface} - \sin^2\theta}} \quad (9)$$

$$r_{V\_SM} = \frac{\varepsilon_{surface}\sqrt{\varepsilon_{air} - \sin^2\theta} - \sqrt{\varepsilon_{surface} - \sin^2\theta}}{\varepsilon_{surface}\sqrt{\varepsilon_{air} - \sin^2\theta} + \sqrt{\varepsilon_{surface} - \sin^2\theta}} \quad (10)$$

Here, θ is the incidence angle; $\varepsilon_{air} = 1.0$ and $\varepsilon_{surface}$ denotes the dielectric constant of the surface at the L-band. For soil surfaces, $\varepsilon_{surface}$ depends on soil texture and volumetric water content, which can be modeled using established dielectric mixing models such as those proposed by Dobson [36], Topp [37], and Mironov [38]. In this study, the Dobson model is applied:

$$\varepsilon_{surface} = \left[1 + \frac{\rho_b}{\rho_s}\left(\varepsilon_s^{\alpha} - 1\right) + m_v^{\beta'}\left(\varepsilon_{fw}'\right)^{\alpha} - m_v\right]^{1/\alpha} \quad (11)$$

where $\varepsilon_{surface}$ is the real part of the soil dielectric constant, m_v is the volumetric soil moisture, ρ_b and ρ_s are the bulk and particle densities of soil ($\rho_s = 2.66$ g/cm^3), ε_s is the dielectric constant of soil minerals and is a real-valued quantity, α is an empirical constant (typically 0.65), β is a soil-dependent parameter, and ε'_{fw} is the real part of the free-water dielectric constant of the soil derived from the Debye equation.

Based on these parameters, the surface reflectivity can be determined, and the LHCP reflection coefficient (r_{lhcp}) can be attributed to either inundated or non-inundated surfaces as follows:

$$SR_{cali\_inundated} = \left|r_{lhcp\_inundated}\right|^2 \tag{12}$$

$$SR_{cali\_not\_inundated} = \left|r_{lhcp\_not\_inundated}\right|^2 \tag{13}$$

By combining Eqs. (7)–(13), the reflectivity for both inundated ($SR_{inundated}$) and non-inundated ($SR_{not\_inundated}$) surfaces are derived. When computing $SR_{inundated}$,, the dielectric constant of water ($\varepsilon_{surface} = 78.9 + 4.3i$) is used [39]; for $SR_{not\_inundated}$,, soil moisture (SM) values from SMAP/SMOS are incorporated into Eq. (11) to estimate the dielectric constant of the dry surface. Finally, by substituting $SR_{inundated}$ and $SR_{not\_inundated}$ into Eq. (7), the water fraction Wf for each grid cell is calculated. A more detailed description of Eq. (7) can be found in Chew and Small [22].

3.4 Estimation of Surface Water Level Using DTM Data

After deriving the inundation extent, a method integrating digital terrain model (DTM) data is proposed to estimate surface water level and water depth. The overall workflow is illustrated in Fig. 2. By incorporating DTM data, the two dimensional flood inundation map can be transformed into a three-dimensional representation containing water level information, as shown in Fig. 2b.

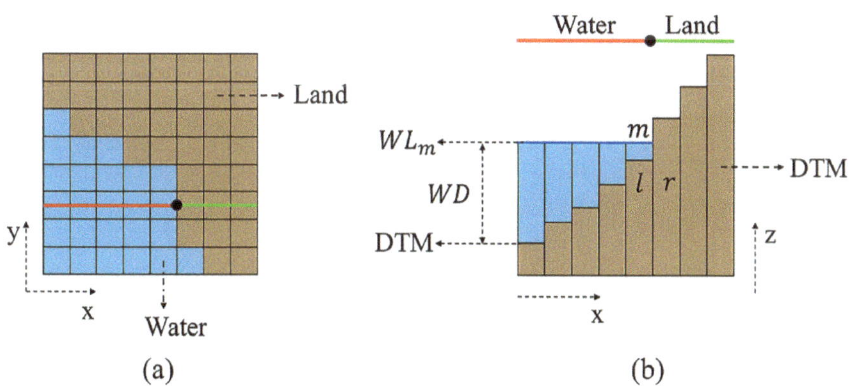

Fig. 2 Schematic of the method for estimating surface water level and depth by integrating CYGNSS observations with digital terrain model (DTM) data

Once the water boundary is determined (e.g., point m in Fig. 2b), the water level at this boundary point should lie between the DTM elevations of its adjacent terrain points (l and r). This relationship can be expressed as:

$$WL_m \in [DTM_l, DTM_r] \approx \text{mean}(DTM_l, DTM_r) \tag{14}$$

where WL_m denotes the water level at point m, and DTM_l and DTM_r represent the DTM elevations at points l and r, respectively.

Furthermore, combining the estimated water level with the corresponding DTM elevation allows for the calculation of water depth, defined as:

$$Water_{depth} = Water_{level} - DTM \tag{15}$$

Specifically, for point m:

$$WD_m = WL_m - DTM_l \tag{16}$$

where WD_m represents the water depth, WL_m is the estimated water level, and DTM_l is the ground elevation at point l.

After obtaining the boundary water levels, the water levels within the inundated area are determined using linear interpolation between boundary points. The complete procedure for estimating surface water level and depth based on this method is summarized in Fig. 3.

4 Monitoring the Flood Inundation of the 2021 Extreme Precipitation in Henan Province, China

4.1 SMAP-Based Flood Monitoring Results

Figure 4 presents the variations in SMAP-derived soil moisture before and after the flood event, as well as the inundation extent retrieved using the threshold-based method. Prior to the flood (Fig. 4a, b), soil moisture across most parts of Henan Province—except for the southern paddy field areas—remained between 0.05 and 0.30 cm^3/cm^3.

Following the intense rainfall, soil moisture in the central and northern regions of Henan increased markedly (Fig. 4c, d), with localized values exceeding 0.45 cm^3/cm^3. As expected, these results clearly demonstrate that the heavy precipitation significantly altered the regional surface moisture conditions.

Figure 4f illustrates the inundation distribution during the second phase of the rainfall event, indicating a northward expansion of the flooded area. Nevertheless, SMAP failed to capture the severe urban flooding observed in Zhengzhou, resulting in a mismatch with the actual conditions. It is also worth noting that the high-moisture

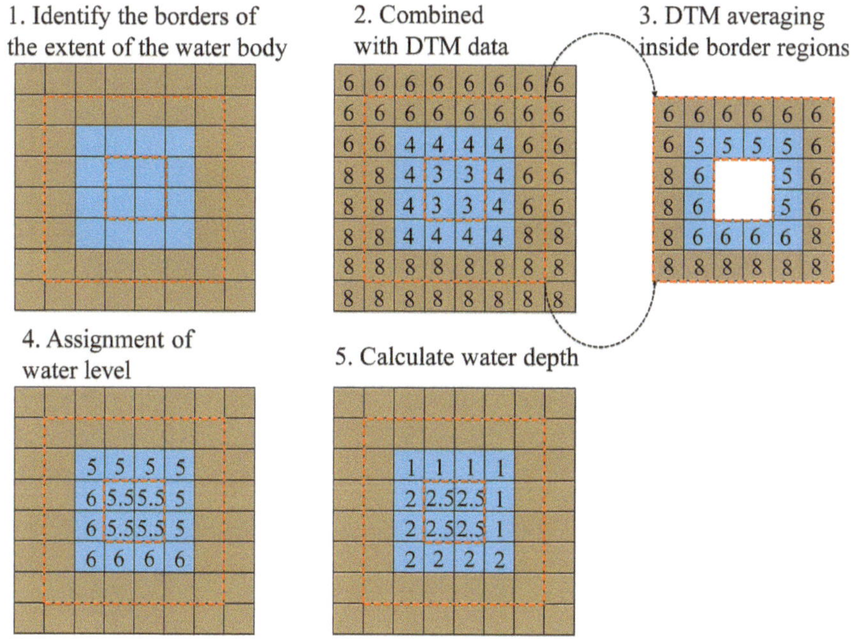

Fig. 3 Workflow illustrating the steps for estimating surface water level and water depth based on DTM data, where brown denotes land, blue denotes surface water, and red boxes indicate water boundaries. The methods for estimating water levels at the water boundaries and within the interior of water bodies are detailed in Fig. 2 and Eqs. (14)–(16)

signature of the southern paddy field zones and the location of the Danjiangkou Reservoir in the southwest are clearly distinguishable in the results.

4.2 *Flood Monitoring Results from Spaceborne GNSS-R*

The city most severely affected by the flood event included Zhengzhou, Luoyang, Xinxiang, and Hebi. Based on CYGNSS observations, this study analyzed the temporal evolution of SR and its corresponding inundation patterns before and after the heavy rainfall.

As illustrated in Fig. 5a–c, gridded CYGNSS SR maps were generated for three distinct stages, using spatial resolutions of 6 × 6 km and 3 × 3 km depending on the duration of each stage. The results indicate that SR values increased significantly in the southern and central–northern parts of Henan after the rainfall (Fig. 5b, c), suggesting a pronounced expansion of surface water coverage. Compared with SMAP data (Fig. 4), CYGNSS demonstrates a considerably higher spatial resolution, enabling more detailed discrimination between inundated and non-inundated areas—particularly during the peak flood stage in central Henan.

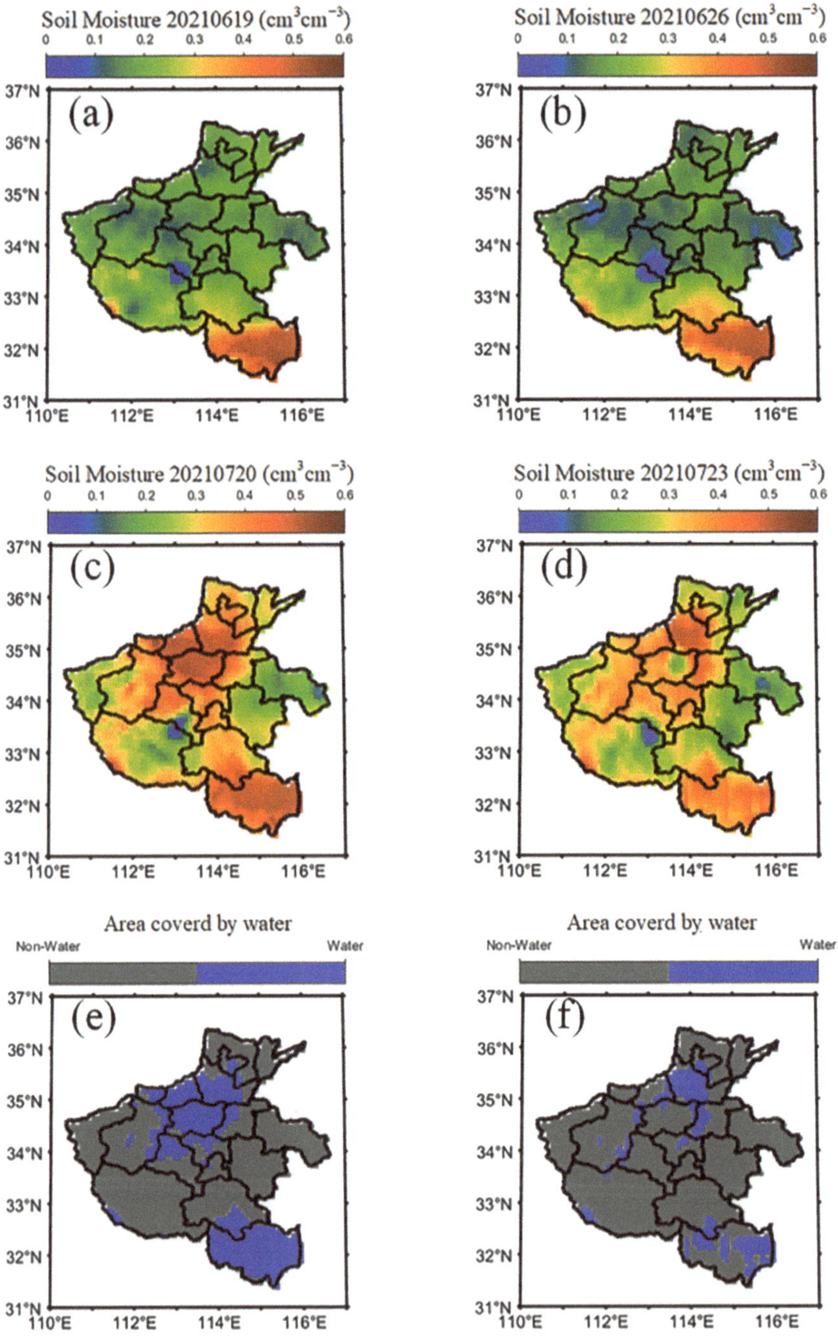

Fig. 4 Changes in Soil Moisture Active Passive (SMAP) soil moisture before and after the Henan flood and the inverted flood extent. **a–d** SMAP soil moisture on 19 June, 26 June, 20 July, and 23 July, respectively; **e** Flood extent inverted from SMAP data during the first rainfall stage (16–20 July); **f** Flood extent inverted from SMAP data during the second rainfall stage (21–25 July)

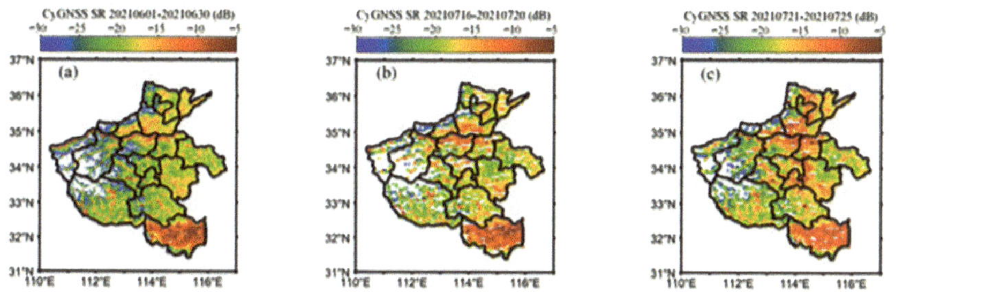

Fig. 5 Gridded Cyclone Global Navigation Satellite System (CYGNSS) surface reflectivity (SR) before and after the extreme precipitation event in Henan Province. **a** SR before the rainfall (3 × 3 km); **b** SR during the first rainfall stage (6 × 6 km); **c** SR during the second rainfall stage (6 × 6 km)

Following the analysis described in Sect. 3.2, a threshold of − 13 dB was adopted; grid cells with SR > − 13 dB was identified as inundated areas. After data preprocessing, the maximum non-anomalous SR value was approximately − 5 dB. The resulting spatial distributions of surface water coverage under this threshold are presented in Fig. 6a–c. As shown in Fig. 6a, prior to the rainfall event, the southern paddy field region of Henan maintained persistently high surface moisture, and CYGNSS effectively detected inland water bodies. Two key observations are noteworthy: (1) CYGNSS successfully captured the Yellow River in the North of the Henan (approximately 500–600 m wide), clearly delineating its spatial morphology; (2) The Danjiangkou Reservoir in the southwest was also accurately identified, further validating the capability of CYGNSS to detect small- and medium-scale inland water bodies.

Figure 6b shows the inundation distribution during the first rainfall stage. Most of Zhengzhou was flooded, parts of Xinxiang were affected, while Hebi and Luoyang

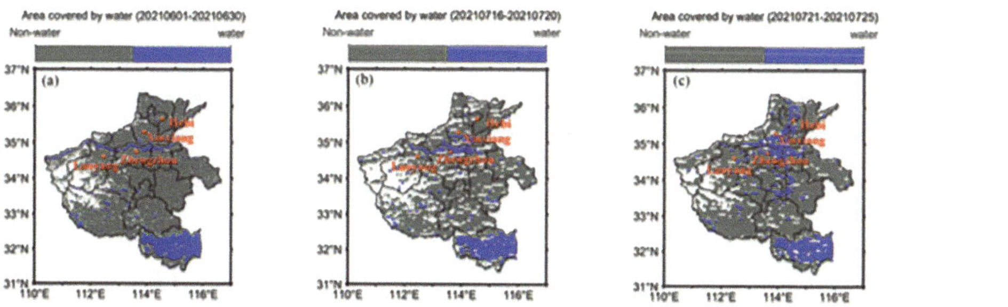

Fig. 6 Spatial distribution of surface inundation in Henan Province before and after the extreme precipitation event derived from Cyclone Global Navigation Satellite System (CYGNSS) surface reflectivity (SR). **a** Inundation extent before the event (1–30 June 2021); **b** Inundation extent during the first rainfall stage (16–20 July 2021); **c** Inundation extent during the second rainfall stage (21–25 July 2021). The red pentagrams mark the locations of Zhengzhou, Xinxiang, Hebi, and Luoyang

remained largely unaffected. This spatial pattern agrees well with ground observations, indicating that the initial rainfall was concentrated over central Henan. In the second stage (Fig. 6c), after July 20, the moisture transport shifted northward—Xinxiang and Hebi became the primary affected areas, while partial inundation persisted in Zhengzhou and Luoyang remained dry.

Overall, CYGNSS-derived results accurately capture the northward propagation of flooding from central Henan and reveal a distinct city-to-city shift of inundation within a short time span. The western city of Luoyang, located at higher elevation, was shielded by topographic uplift that impeded westward moisture transport, leading to minimal inundation during both rainfall episodes. The strong consistency between CYGNSS-derived flood extent and actual disaster reports further confirms the effectiveness and reliability of GNSS-R in flood monitoring applications.

4.3 *Time-Series Analysis*

While the previous section focused on the spatial evolution of flooding, this section further examines the temporal relationships among precipitation, soil moisture (SM), and surface reflectivity (SR). Figure 7 presents the daily variations in rainfall, SM, and SR from June 1 to August 31, 2021, for a representative area located near 34° N, 114° E (southwestern Zhengzhou, approximately 100 m above sea level). This region is characterized by flat terrain and extensive croplands, making it highly susceptible to water accumulation under heavy rainfall.

As shown in Fig. 7, prior to the July 20 rainstorm, soil moisture remained below 0.3 cm^3/cm^3, and SR fluctuated slightly with precipitation but generally stayed below − 17 dB, indicating the absence of significant surface inundation. Following the extreme rainfall on July 20, soil moisture rose sharply to approximately 0.5 cm^3/cm^3, accompanied by a synchronous and pronounced increase in SR, implying the occurrence of flooding in the area. Subsequently, as floodwaters receded, both SM and SR gradually decreased, returning to below 0.4 cm^3/cm^3 and − 16 dB, respectively.

In August, multiple fluctuations were observed in both parameters, likely associated with intermittent rainfall events. Overall, SR and SM exhibited highly consistent temporal patterns, with a correlation coefficient of $r = 0.7$, demonstrating that CYGNSS-derived surface reflectivity effectively captures the formation and recession of surface water. These results confirm the capability of GNSS-R observations to characterize the temporal dynamics of flood evolution.

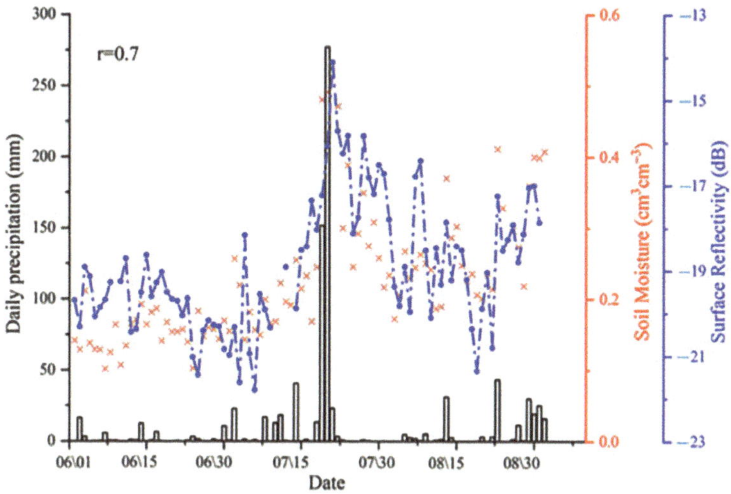

Fig. 7 Temporal variations in precipitation, soil moisture, and surface reflectivity (SR) near (34° N, 114° E) from 1 June to 31 August 2021. Rainfall, soil moisture, and SR show consistent variation patterns with a correlation coefficient of r = 0.7

5 Monitoring the Extent of the 2022 Pakistan Floods

5.1 Flood Monitoring Using Brightness Temperature Data

Figure 8 illustrates the spatiotemporal variations of brightness temperature (Tb) over Pakistan from June to November 2022. The results reveal significant changes in Tb driven by persistent heavy rainfall. After mid-July, Tb dropped below 210 K in many areas, indicating a substantial increase in surface moisture content. As shown in the black-boxed region, Tb reached its lowest values during August–September, corresponding to the peak flooding period. Thereafter, Tb gradually increased; however, by mid-November, parts of southern Pakistan had not yet recovered to pre-flood conditions (see Fig. 8l).

Based on SMAP and SMOS Tb data, the surface water fraction (*Wf*) was further retrieved, as shown in Fig. 9. Before mid-July, no evident inundation areas were observed (Fig. 9a–c). From mid-July onward, continuous rainfall caused a marked increase in *Wf*, indicating large-scale surface flooding. Within the highlighted region, *Wf* peaked during August–September, reflecting the maximum flood extent and intensity. Subsequently, *Wf* gradually decreased as floodwaters receded, yet remained higher than pre-flood levels in southern Pakistan through mid-November (Fig. 9l).

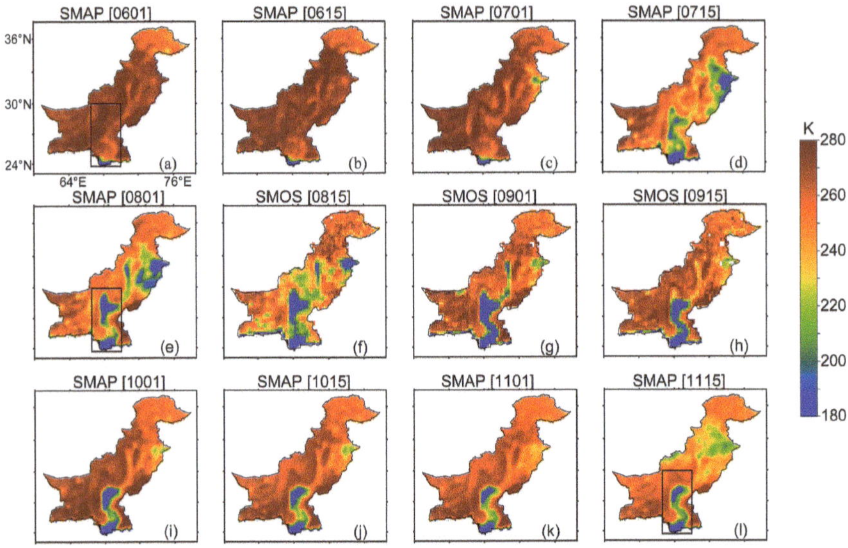

Fig. 8 Spatial and temporal variations of brightness temperature (Tb) in Pakistan from 1 June to 15 November 2022. Panels **a–e** are derived from SMAP, **f–h** from SMOS, and **i–l** again from SMAP. Note that SMAP was temporarily unavailable between 7 August and 22 September 2022, resulting in missing Tb data during this period

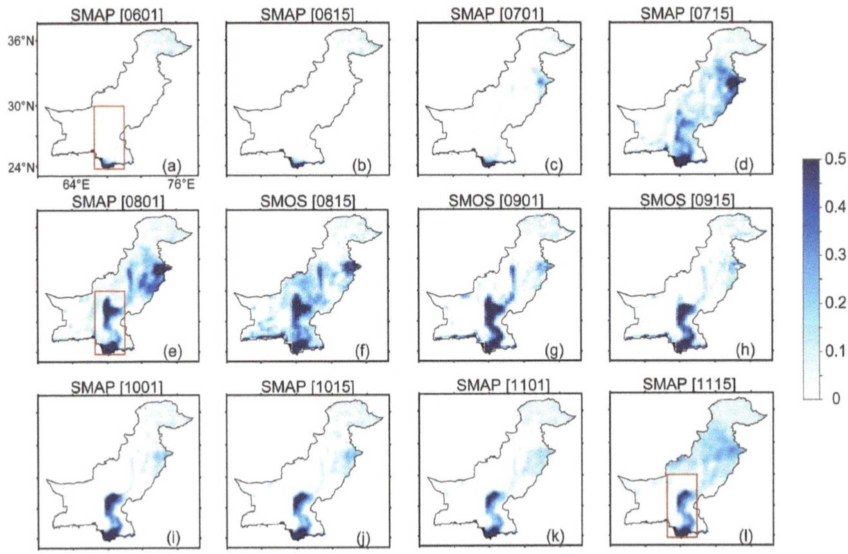

Fig. 9 Temporal variations of surface water fraction (*Wf*) in Pakistan from 1 June to 15 November 2022. Panels **a–l** are derived from SMAP and SMOS brightness temperature (Tb) data

Overall, the results demonstrate that SMAP and SMOS brightness temperature data effectively capture the spatiotemporal evolution of the 2022 Pakistan mega-flood. However, due to their coarse spatial resolution (~ 40 km), these datasets exhibit inherent limitations for high-precision flood delineation at local scales.

5.2 *Flood Monitoring Using Spaceborne GNSS-R* Wf *Retrievals*

Because spaceborne GNSS-R operates in a bistatic configuration, its observation footprints are quasi-randomly distributed on the Earth's surface. Although CYGNSS offers high temporal revisit frequency, establishing a consistent spatiotemporal reference frame requires joint spatial–temporal processing. Commonly, CYGNSS observations are gridded; however, finer spatial resolution often comes at the cost of reduced temporal coverage, constraining its potential for dynamic flood monitoring.

To address this, the POBI interpolation algorithm was applied to process CYGNSS surface reflectivity (SR) data. The results are shown in Fig. 11. Compared with the raw observations (Fig. 10), the POBI-interpolated SR achieved continuous spatial coverage while effectively preserving surface heterogeneity and discontinuities. Within the highlighted region, SR increased markedly under persistent heavy rainfall and gradually declined thereafter. Nevertheless, by mid-November, SR in some southern areas remained higher than pre-flood levels, indicating the presence of residual water bodies. Overall, the interpolated SR results exhibit high consistency with SMAP and SMOS brightness temperature (Tb) observations, confirming the feasibility and reliability of this approach for flood detection.

Figure 12 presents the temporal evolution of surface water fraction (*Wf*) derived from CYGNSS SR. Before mid-July, no significant inundation was detected, with elevated *Wf* values limited to major river channels (Fig. 12a–c). Following mid-July, continuous rainfall caused a sharp increase in *Wf*, particularly across southern and eastern Pakistan. Within the boxed region, *Wf* peaked during August–September, with the Sindh Province, northern Punjab, and eastern Balochistan experiencing the most severe inundation. Subsequently, both flood extent and *Wf* gradually decreased, though elevated *Wf* values persisted in some southern areas through mid-November.

In summary, the POBI-interpolated CYGNSS SR accurately captured the spatiotemporal dynamics of the 2022 Pakistan mega-flood, showing strong agreement with SMAP and SMOS observations while offering significantly enhanced spatial and temporal resolution. These findings demonstrate that CYGNSS SR provides a robust means for near-real-time, high-resolution flood monitoring, enabling continuous observation of rapid hydrological changes.

To further investigate the spatial–temporal characteristics of surface water dynamics during the 2022 Pakistan mega-flood, the retrieved surface water fraction (*Wf*) was classified into four categories representing different degrees of inundation: unaffected, slightly affected, moderately affected, and severely affected areas. Since

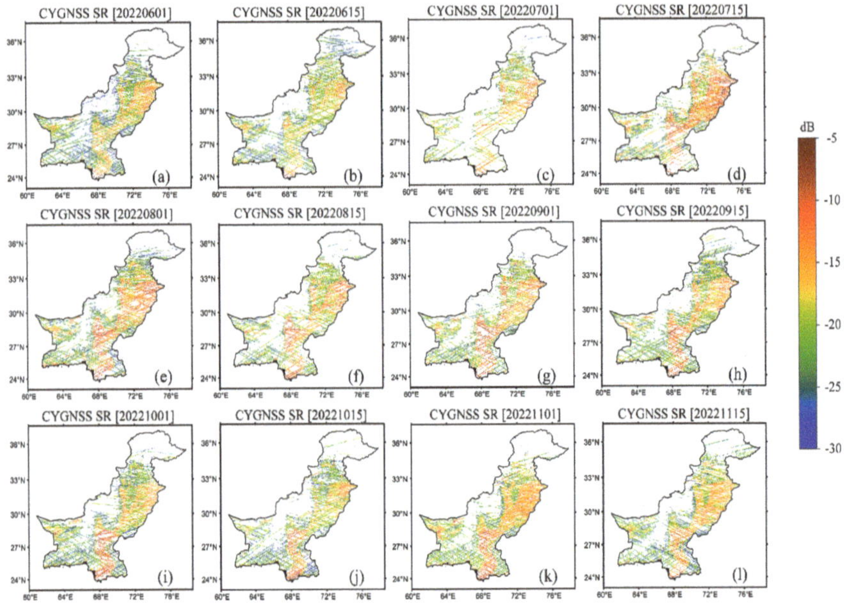

Fig. 10 Gridded Cyclone GNSS surface reflectivity (CYGNSS SR) from 1 June to 15 November 2022 (grid resolution: 3 × 3 km)

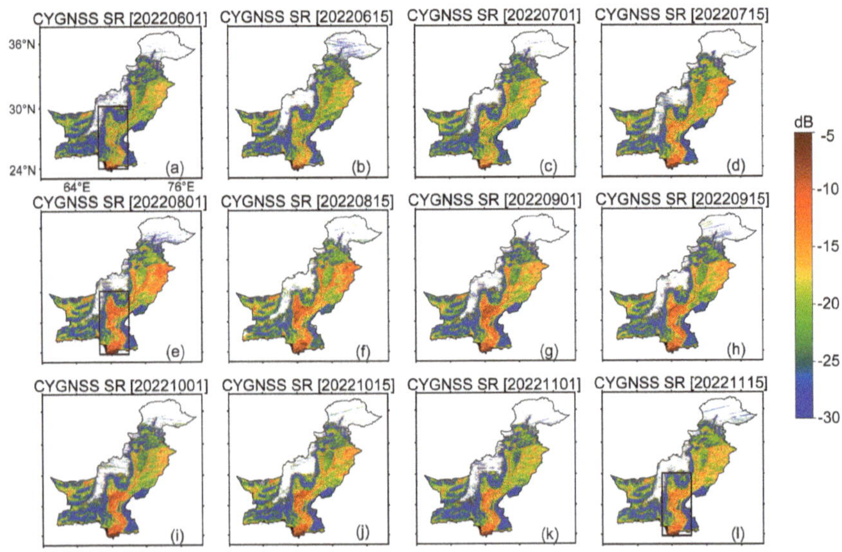

Fig. 11 Gridded Cyclone GNSS surface reflectivity (CYGNSS SR) after POBI interpolation from 1 June to 15 November 2022 (panels **a**–**l**)

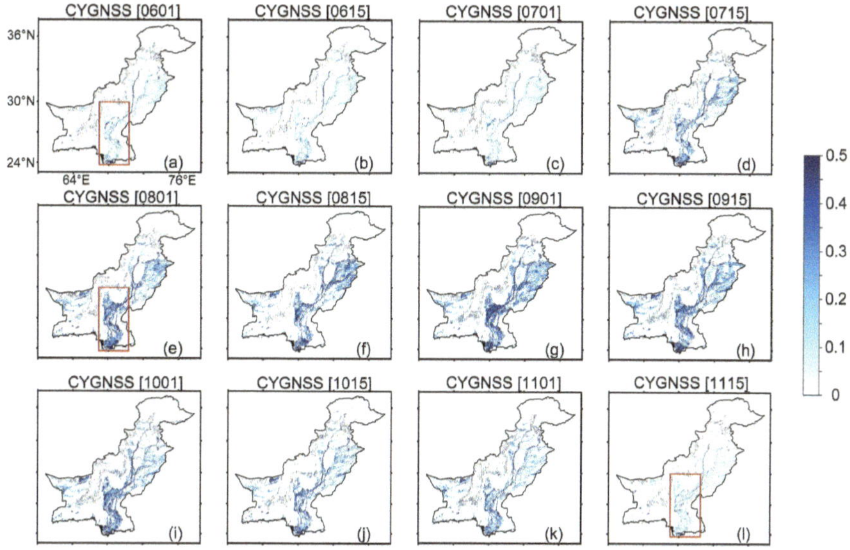

Fig. 12 Temporal variations of surface water fraction (*Wf*) in Pakistan derived from CYGNSS SR from 1 June to 15 November 2022 (panels **a**–**l**)

a universally accepted classification framework based on *Wf* is not yet available, the thresholds adopted in this study were determined empirically from preliminary statistical analyses. Future work should refine these criteria through calibration with field observations and broader regional datasets. The *Wf* ranges corresponding to each inundation level are listed in Table 2.

Figure 13 illustrates the temporal evolution of inundation severity across Pakistan from June to November 2022. In June, most regions remained unaffected by flooding. As heavy rainfall persisted, inundated areas expanded rapidly, with slightly and moderately affected regions increasing most significantly. The flood reached its peak in August and September, when approximately 72% and 61% of the national territory, respectively, experienced varying levels of inundation. Thereafter, floodwaters gradually receded, and the degree of inundation weakened progressively; however, by mid-November, some southern regions still exhibited elevated *Wf* values relative to pre-flood conditions, indicating the persistence of residual water bodies.

Overall, results from Figs. 11, 12 and 13 demonstrate that CYGNSS-derived *Wf* effectively characterizes the spatiotemporal evolution of flooding, capturing both the expansion and recession phases of inundation. Nevertheless, due to differences

Table 2 Wf ranges corresponding to different surface inundation levels

	Unaffected by flooding	Mildly affected by flooding	Moderately affected by flooding	Severely affected by flooding
Wf value	0–0.15	0.15–0.3	0.3–0.6	> 0.6

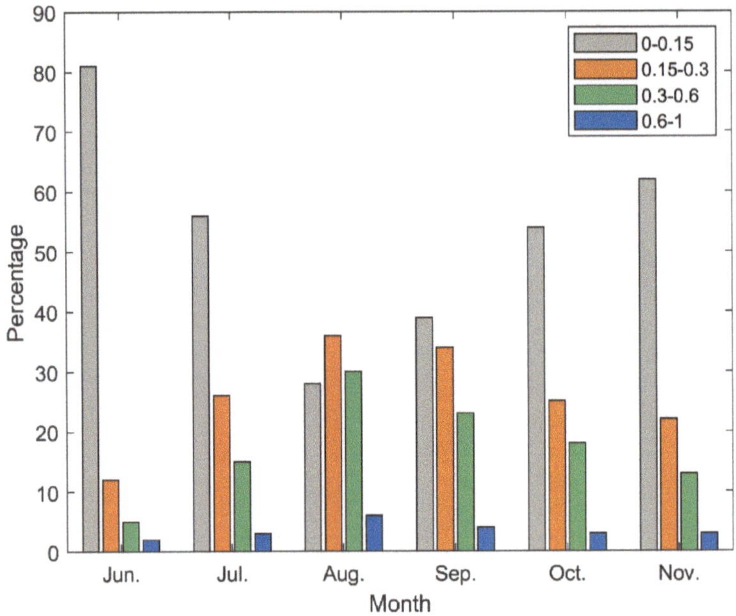

Fig. 13 Surface inundation levels in Pakistan from June to November 2022 derived from CYGNSS, presented as a bar chart

in spatial resolution between SMAP/SMOS and CYGNSS observations, a direct quantitative comparison remains challenging. To further validate the accuracy of the CYGNSS retrievals, the next section presents a pixel-level comparison with the Global Flood Monitoring (GFM) product, which provides higher-resolution flood extent observations.

5.3 *Validation Against GFM Observations*

To assess the accuracy of the CYGNSS-derived surface water fraction (*Wf*), this study conducted a comparative analysis using the Global Flood Monitoring (GFM) Open Water Extent (OWE) product. The OWE data were resampled to a 3 × 3 km grid to match the spatial resolution of the CYGNSS retrievals, and the corresponding *Wf* values were computed for each grid cell. Considering the strip-based acquisition mode and relatively low temporal resolution of Sentinel-1, two regions most severely affected by flooding were selected for analysis: Region A (northwestern Sindh and southeastern Balochistan) and Region B (western Punjab).

Figure 14 presents the spatial comparison of *Wf* derived from CYGNSS and GFM before and after the flood event. For better visualization, Fig. 15 shows enlarged subregions, while Fig. 16 illustrates the corresponding density scatterplots and probability density functions (PDFs) of the two datasets. Despite Sentinel-1's superior spatial resolution, its low revisit frequency limits temporal continuity. In contrast, CYGNSS captures flood dynamics more effectively, providing continuous temporal coverage with adequate spatial detail. The spatial comparison in Fig. 17 demonstrates good agreement between CYGNSS and GFM, with both datasets consistently delineating the post-flood inundation extent and the corresponding increase in *Wf*.

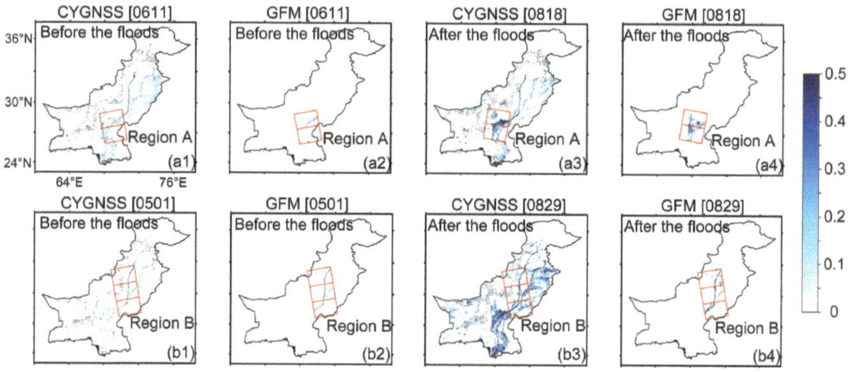

Fig. 14 Comparison of surface water fraction (*Wf*) derived from CYGNSS and GFM before and after flooding. (a1–a2) Pre-flood conditions in Region A (11 June); (a3–a4) Post-flood conditions in Region A (18 August). (b1–b2) Pre-flood conditions in Region B (1 May); (b3–b4) Post-flood conditions in Region B (29 August). Sentinel-1 footprint used by GFM is indicated by red boxes

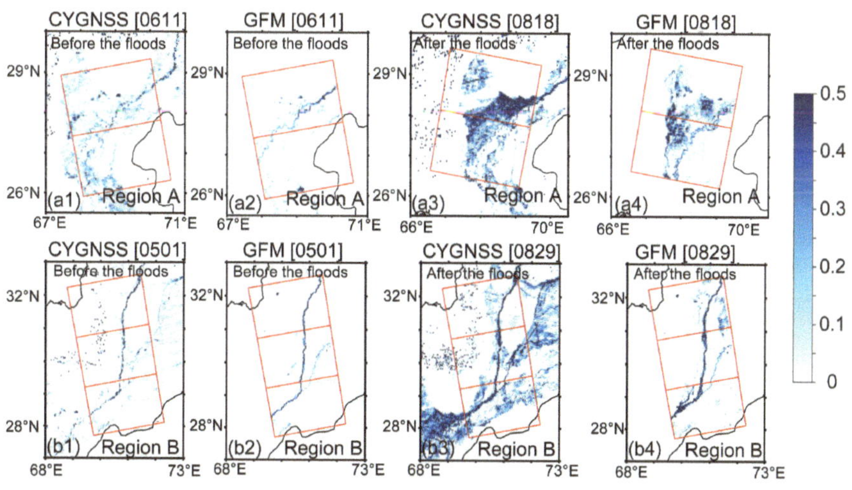

Fig. 15 Zoomed-in comparison of surface water fraction (*Wf*) derived from CYGNSS and GFM before and after flooding. (a1–a2) Pre-flood in Area A (11 June); (a3–a4) Post-flood in Area A (18 August). (b1–b2) Pre-flood in Area B (1 May); (b3–b4) Post-flood in Area B (29 August). Sentinel-1 footprint used by GFM is indicated by red boxes

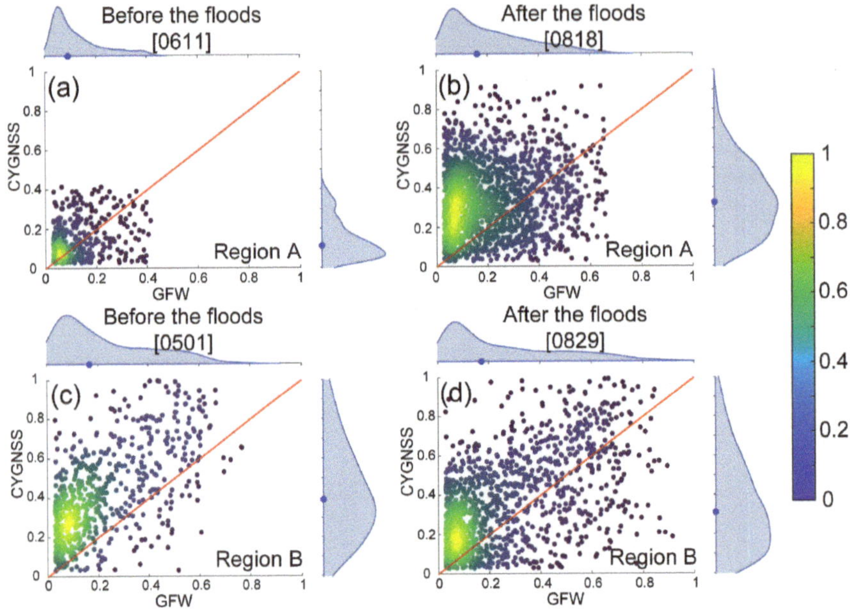

Fig. 16 Density scatter plots and probability density functions (PDFs) of CYGNSS *Wf* versus GFM *Wf* before and after flooding. **a** Pre-flood in Region A; **b** Post-flood in Region A; **c** Pre-flood in Region B; **d** Post-flood in Region B. Blue circles indicate the 50th percentile of the data

As shown in Fig. 15, the overall distribution patterns of *Wf* are similar, though CYGNSS exhibits a slight overestimation—most notably in Region A. This bias is likely attributed to vegetation effects: dense vegetation cover in Region A tends to obscure surface water in the GFM product, leading to underestimation, while CYGNSS, being more sensitive to moist and rough surfaces, produces relatively higher *Wf* values.

To quantitatively evaluate the retrieval accuracy, a confusion matrix was computed based on the results shown in Fig. 17 (Table 3). Flood severity was categorized into two classes: unaffected or slightly affected ($Wf \leq 0.3$) and moderately to severely affected ($Wf > 0.3$). The spatial comparison of inundation extent is presented in Fig. 17. Overall, CYGNSS and GFM exhibit strong spatial and temporal consistency, successfully identifying major flooded areas. However, in moderately to severely inundated regions, CYGNSS tends to overestimate *Wf*.

According to the confusion matrix results, the average overall accuracy of CYGNSS-derived *Wf* is approximately 64.44%, with a total misclassification rate of around 17.78%. The observed overestimation primarily stems from the GFM detection mechanism, which identifies only open water surfaces and is prone to omission errors under cloud cover or dense vegetation during flood events. Nevertheless, CYGNSS effectively captures surface water dynamics before and after flooding, confirming its feasibility and reliability for large-scale flood monitoring applications.

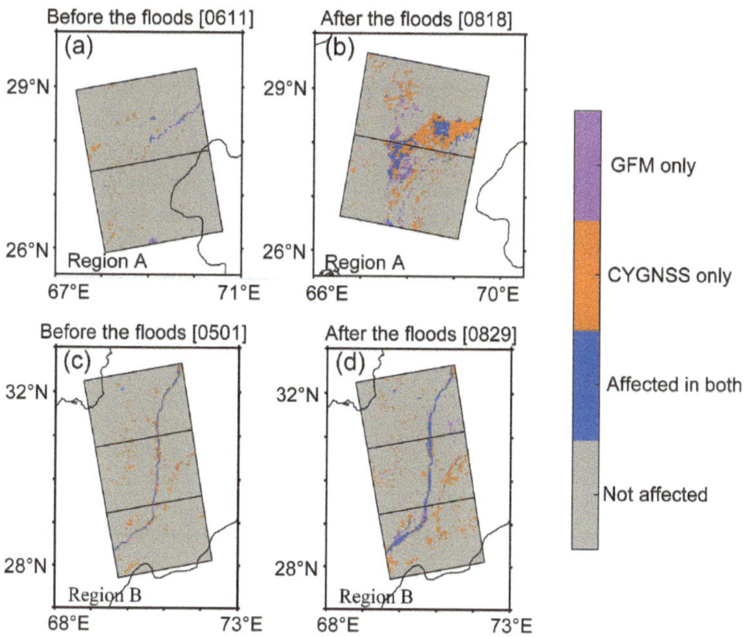

Fig. 17 Composite maps of CYGNSS *Wf* and GFM *Wf* before and after flooding. **a, b** Region A pre- and post-flood, respectively; **c, d** Region B pre- and post-flood, respectively

Table 3 Confusion matrix and overall accuracy of CYGNSS-derived *Wf*

Study Area	CYGNSS *Wf*	GFM *Wf*		Overall Accuracy
		0–0.3	> 0.3	
Region A (0611)	0–0.3	79.48	6.13	84.23
	> 0.3	9.64	4.75	
Region A (0818)	0–0.3	35.23	9.29	50.71
	> 0.3	40.00	15.48	
Region B (0501)	0–0.3	30.08	3.09	55.93
	> 0.3	40.98	25.85	
Region B (0829)	0–0.3	40.61	7.25	66.88
	> 0.3	25.87	26.27	

6 Surface Water Level Monitoring of the 2022 Pakistan Extreme Flood

Given the short duration of the flood event, the reliability of CYGNSS-derived water level retrievals was evaluated by selecting dates overlapping with ICESat-2 or GEDI observations. The average water levels estimated from CYGNSS were compared with

LiDAR measurements to obtain the water level difference, defined as the observed value minus the CYGNSS-derived mean.

Figure 18 presents the comparison results for the 2022 Pakistan flood. The analysis indicates a strong correlation between CYGNSS water level retrievals and satellite LiDAR measurements, with an average correlation coefficient (R) of 0.98. The corresponding mean bias is 0.10 m, and the root mean square error (RMSE) is 1.35 m.

Overall, the CYGNSS-derived water levels exhibit high consistency with ICESat-2 and GEDI observations, accurately reflecting temporal water level variations during

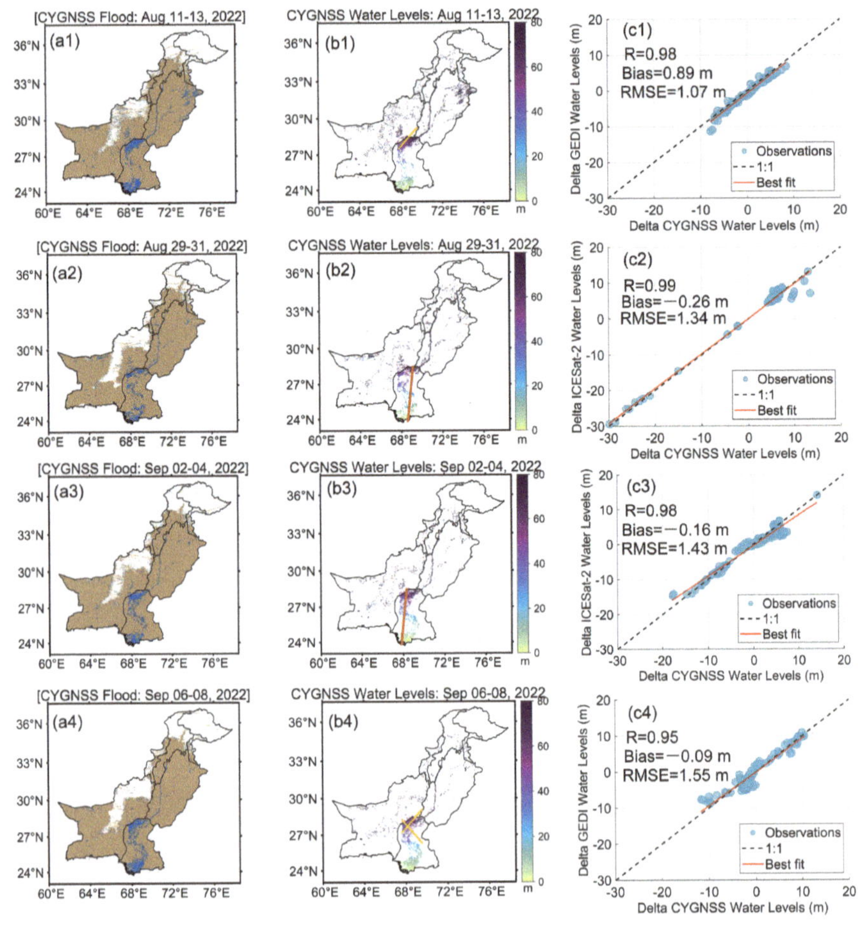

Fig. 18 Pakistan flood case results. (a1–a4) Surface inundation maps derived from CYGNSS (blue: inundated; brown: non-inundated) for observation periods 11–13 August, 29–31 August, 2–4 September, and 6–8 September 2022, respectively. (b1–b4) Corresponding water levels estimated by combining CYGNSS inundation maps with DTM data; red and orange lines indicate ICESat-2 and GEDI observation tracks, respectively. (c1–c4) Comparison of CYGNSS-derived water levels with ICESat-2 and GEDI altimetry data

the flood event. These results confirm the feasibility and accuracy of the proposed method for flood water level retrieval and provide important support for GNSS-R–based monitoring of surface hydrological dynamics.

7 Conclusion

In this chapter, the potential of CYGNSS for flood monitoring and water level retrieval was comprehensively analyzed using multi-source remote sensing data, and its performance and accuracy across different spatiotemporal scales were systematically evaluated.

First, CYGNSS surface reflectivity (SR) data were employed to identify rainfall-induced surface inundation using a threshold-based approach. The results indicate that SR exhibits a pronounced response to changes in surface water distribution, effectively distinguishing between inundated and non-inundated areas before and after flood events. Comparisons with SMAP soil moisture and precipitation time series show a high consistency between SR and soil moisture variations ($r = 0.7$), demonstrating that CYGNSS SR can accurately capture the occurrence, development, and recession of floods.

Second, by incorporating brightness temperature (Tb) data from SMAP and SMOS, the spatiotemporal evolution of surface water during the 2022 extreme flood in Pakistan was analyzed. Tb and the retrieved surface water fraction (Wf) exhibited consistent temporal and spatial patterns, successfully characterizing flood dynamics. To improve the spatial continuity of CYGNSS observations, the POBI interpolation method was applied to fuse SR data in space and time. The interpolated results preserved surface heterogeneity while substantially enhancing spatial coverage. The CYGNSS-derived Wf obtained from this method showed high agreement with SMAP and SMOS products, with clear advantages in spatiotemporal resolution.

Furthermore, comparisons between CYGNSS-derived Wf and the Copernicus Global Flood Monitoring (GFM) product demonstrated strong spatial consistency, accurately delineating the extent and intensity of surface inundation. Although CYGNSS slightly overestimated Wf in some regions, particularly in areas with dense vegetation, the overall retrieval accuracy reached 64.44%, confirming the method's reliability and robustness.

Finally, CYGNSS water level retrievals were validated against ICESat-2 and GEDI LiDAR measurements. For 2022 Pakistan flood event, the correlation coefficients were 0.98, with RMSE values below 1.5 m, indicating that CYGNSS can reliably reflect water level variations during floods.

In summary, the results of this chapter fully demonstrate the feasibility of GNSS-R–based CYGNSS observations for flood monitoring and water level retrieval. With high temporal resolution, strong penetration capability, and stable global coverage, CYGNSS provides a valuable complement to existing passive microwave and radar-based monitoring approaches, offering a novel technical pathway and robust data support for global flood dynamics observation and disaster assessment.

References

1. Ranjan R (2016) Flood disaster management. In: River system analysis and management. Springer, Singapore, pp 371–417
2. Schumacher RS (2017) Heavy rainfall and flash flooding. In: Oxford research encyclopedia of natural hazard science
3. Viglione A, Rogger M (2015) Flood processes and hazards. In: Hydro-meteorological hazards, risks and disasters. Elsevier, pp 3–33
4. Whitfield PH (2012) Floods in future climates: a review. J Flood Risk Manag 5(4):336–365
5. Rahman MS, Di L (2017) The state of the art of spaceborne remote sensing in flood management. Nat Hazards 85(2):1223–1248
6. Jafarzadegan K, Moradkhani H, Pappenberger F et al (2023) Recent advances and new frontiers in riverine and coastal flood modeling. Rev Geophys 61(2):e2022RG000788
7. Yao H, Qin R, Chen X (2019) Unmanned aerial vehicle for remote sensing applications—a review. Remote Sens 11(12):1443
8. Munawar HS, Hammad AWA, Waller ST (2022) Remote sensing methods for flood prediction: a review. Sensors 22(3):960
9. Notti D, Giordan D, Caló F et al (2018) Potential and limitations of open satellite data for flood mapping. Remote Sens 10(11):1673
10. Lopez T, Al Bitar A, Biancamaria S et al (2020) On the use of satellite remote sensing to detect floods and droughts at large scales. Surv Geophys 41(6):1461–1487
11. Rodriguez-Alvarez N, Munoz-Martin JF, Morris M (2023) Latest advances in the global navigation satellite system—reflectometry (GNSS-R) field. Remote Sens 15(8):2157
12. Camps A, Park H, Pablos M et al (2016) Sensitivity of GNSS-R spaceborne observations to soil moisture and vegetation. IEEE J Sel Top Appl Earth Obs Remote Sens 9(10):4730–4742
13. Jin S, Camps A, Jia Y et al (2024) Remote sensing and its applications using GNSS reflected signals: advances and prospects. Satellite Navig 5(1):19
14. Bu J, Liu X, Wang Q et al (2024) Ocean remote sensing using spaceborne GNSS-reflectometry: a review. IEEE J Sel Top Appl Earth Obs Remote Sens 17:13047–13076
15. Bu J, Wang Q, Wang Z et al (2024) Land remote sensing applications using spaceborne GNSS reflectometry: a comprehensive overview. IEEE J Sel Top Appl Earth Obs Remote Sens 17:12811–12841
16. Yan Q, Huang W (2019) Sea ice remote sensing using GNSS-R: a review. Remote Sens 11(21):2565
17. Yu K, Han S, Bu J et al (2022) Spaceborne GNSS reflectometry. Remote Sens 14(7):1605
18. Yang W, Gao F, Xu T et al (2021) Daily flood monitoring based on spaceborne GNSS-R data: a case study on Henan. China. Remote Sens 13(22):4561
19. Ruf C, Asharaf S, Balasubramaniam R, Gleason S, Lang TJ, McKague DS, Twigg D, Waliser DE (2019) In-orbit performance of the constellation of CYGNSS hurricane satellites. Bull Am Meteorol Soc 100(10):2009–2023
20. Huang F, Garrison JL, Rodriguez-Alvarez N et al (2019) Sequential processing of GNSS-R delay–doppler maps to estimate the ocean surface wind field. IEEE Trans Geosci Remote Sens 57(12):10202–10217
21. Chew CC, Small EE (2018) Soil moisture sensing using spaceborne GNSS reflections: comparison of CYGNSS reflectivity to SMAP soil moisture. Geophys Res Lett 45(9):4049–4057
22. Chew C, Small E (2020) Estimating inundation extent using CYGNSS data: a conceptual modeling study. Remote Sens Environ 246:111869
23. Chew C (2021) Spatial interpolation based on previously observed behavior: a framework for interpolating spaceborne GNSS-R data from CYGNSS. J Spatial Sci 1–14
24. Zhang S, Ma Z, Li Z, Zhang P, Liu Q, Nan Y, Zhang J, Hu S, Feng Y, Zhao H (2021) Using CYGNSS data to map flood inundation during the 2021 extreme precipitation in Henan Province, China. Remote Sens 13(24):5181

25. Wan W, Liu B, Zeng Z, Chen X, Wu G, Xu L, Chen X, Hong Y (2019) Using CYGNSS data to monitor China's flood inundation during typhoon and extreme precipitation events in 2017. Remote Sens 11(7):854
26. Rahman MS, Di L, Shrestha RM, Yu EG, Lin L, Zhang C, Hu L, Tang J, Yang Z (2017) Agriculture flood mapping with soil moisture active passive (SMAP) data: a case of 2016 Louisiana flood. In: Proceedings of the 6th international conference on agro-geoinformatics, pp 1–6
27. Chaubell J, Yueh SH, Entekhabi D, Peng J (2016) Resolution enhancement of SMAP radiometer data using the Backus–Gilbert optimum interpolation technique. In: IEEE IEEE International Geoscience and Remote Sensing Symposium (IGARSS), pp 284–287
28. Salamon P et al (2021) The new, systematic global flood monitoring product of the Copernicus emergency management service. In: Proceedings of the IEEE International Geoscience and Remote Sensing Symposium (IGARSS), pp 1053–1056
29. Downs B, Kettner AJ, Chapman BD, Brakenridge GR, O'Brien AJ, Zuffada C (2023) Assessing the relative performance of GNSS-R flood extent observations: case study in South Sudan. IEEE Trans Geosci Remote Sens 61:4201213
30. Ruf C, Gleason S (2025) Spaceborne GNSS-R bistatic radar remote sensing, CYGNSS, and future missions. Proc IEEE
31. Kim H, Lakshmi V (2018) Use of cyclone global navigation satellite system (CYGNSS) observations for estimation of soil moisture. Geophys Res Lett 45:8272–8282
32. Morris M, Chew C, Reager JT, Shah R, Zuffada C (2019) A novel approach to monitoring wetland dynamics using CYGNSS: everglades case study. Remote Sens Environ 233:111417
33. Gerlein-Safdi C, Ruf CS (2019) A CYGNSS-based algorithm for the detection of inland waterbodies. Geophys Res Lett 46:12065–12072
34. Chew C, Reager JT, Small E (2018) CYGNSS data map flood inundation during the 2017 Atlantic hurricane season. Sci Rep 8:9336
35. Kiage LM, Walker ND, Balasubramanian S, Babin A, Barras J (2005) Applications of RADARSAT-1 synthetic aperture radar imagery to assess hurricane-related flooding of coastal Louisiana. Int J Remote Sens 26:5359–5380
36. Dobson M, Ulaby F, Hallikainen M, El-Rayes M (1985) Microwave dielectric behavior of wet soil—part II: dielectric mixing models. IEEE Trans Geosci Remote Sens 23(1):35–46
37. Topp GC, Davis JL, Annan AP (1980) Electromagnetic determination of soil water content: measurements in coaxial transmission lines. Water Resour Res 16(3):574–582
38. Mironov VL, Kosolapova LG, Fomin SV (2009) Physically and mineralogically based spectroscopic dielectric model for moist soils. IEEE Trans Geosci Remote Sens 47(7):2059–2070
39. Kaatze U (1989) Complex permittivity of water as a function of frequency and temperature. J Chem Eng Data 34(4):371–374

Chapter 14
Maritime Target Detection and Location

Yan Li and Songhua Yan

Abstract GNSS signal used as signal of opportunity has become a prominent research focus in the field of target sensing and localization. This chapter begins by introducing a novel approach for detecting and positioning static maritime targets via ground-based GNSS-Reflectometry (GNSS-R), which primarily enhances the SNR of the reflected echo through extended coherent integration time (CIT). Subsequently, to detection maneuvering targets, this chapter presents two distinct localization techniques utilizing Doppler-range compensation and the Fractional Fourier Transform (FrFT). Furthermore, to capitalize on the global coverage capabilities of GNSS-R, a framework for the detection and localization of offshore oil rigs is detailed. This framework relies on the fusion of spaceborne delay-Doppler maps (DDMs) and incorporates advanced procedures for sea clutter suppression and the resolution of positioning ambiguities. Field experiments confirm that both groundborne and satellite-borne GNSS-R configurations hold significant promise for advanced maritime surveillance applications.

1 Introduction

As a revolutionary passive bistatic radar technology, GNSS-reflectometry (GNSS-R) has drawn considerable interest in earth remote sensing and target detection research, thanks to its broad coverage, low deployment cost, and multi-frequency characteristics [1–3]. Its core principle lies in extracting target features from Delay-Doppler Maps (DDMs) by analyzing the time delay difference, Doppler frequency shift, and power variation between direct signals and surface-reflected signals. GNSS-R is widely applied in two major scenarios: ground-based and spaceborne systems [4]. Ground-based systems leverage the stable illumination characteristics of signals

Y. Li
Hubei University of Technology, Wuhan, China
e-mail: liyan18@whu.edu.cn

S. Yan (✉)
Wuhan University, Wuhan, Hubei Province, China
e-mail: ysh@whu.edu.cn

K. Yu and N. Zheng (eds.), *Advances in GNSS Reflectometry*, Navigation: Science and Technology 19, https://doi.org/10.1007/978-981-92-2325-1_14

of navigation satellites such as Beidou Geostationary Orbit (GEO) satellites to achieve long-term monitoring of fixed areas—for instance, Beidou radar has achieved centimeter to millimeter-level deformation monitoring accuracy. Spaceborne systems (e.g., TechDemoSat-1, CYGNSS), on the other hand, focus on ocean target detection, enabling the identification of oil wells [5], oil slicks [6], sea ice [7], and red tides [8].

However, the practical application of GNSS-R technology still faces multiple challenges, which need to be addressed separately for ground-based and spaceborne scenarios. The bottlenecks of ground-based systems lie in low SNR and the compensation of target motion [9]. The power of ground reception GNSS signals is extremely low (e.g., the Beidou B3I signal is only -163 dBw); Together with the interference caused by direct and reflected signals and phase instability caused by satellite motion, which severely limits the coherent integration time [10]. For maneuvering targets, the coupling effect of Doppler frequency migration (DFM) and range migration (RM) induced by radial acceleration plunges traditional methods into difficulties [11]. The Keystone transform is only suitable for high radial velocity signals [12]; the Radon-Fourier Transform (RFT) has integration losses [13]; and the Target Motion Compensation (TMC) algorithm relies on high-precision prior information [14].

In spaceborne applications, by contrast, core difficulties concentrate on sea clutter estimation and positioning ambiguity [15]. Under complex sea conditions, target signals are easily submerged by noise: existing methods such as simulated sea clutter subtraction suffer from calculation errors in specular points (SP); adaptive filtering improves accuracy but leads to a surge in computational load due to differences in filter kernel length [16]; 3D matched filtering considers spatiotemporal dimensions but is limited by the complexity of Discrete Fourier Transform (DFT) [17]. Meanwhile, the non-unique mapping between Delay-Doppler (DD) domain coordinates and surface coordinates causes positioning ambiguity—Jacobian transformation lacks a unique solution [18], the visual inspection method relies on manual intervention [19], and the antenna tilting method lacks verification with measured data [20].

This chapter describe different solutions for ground-based and spaceborne scenarios. The ground-based field leverages the fixed coverage advantage of Beidou GEO satellites, reduces signal interference through quasi-monostatic radar configuration, and explores combined algorithms of Keystone and Fractional Fourier Transform (FrFT) to optimize energy accumulation. The spaceborne field tends to adopt multi-source data fusion (e.g., using multi-orbit DDM data from CYGNSS to resolve positioning ambiguity).

2 Static Target Dection and Location Using Ground-Based GNSS-R

2.1 Range Compression of GNSS iF Signals

A typical bistatic radar geometry is illustrated in Fig. 1 in which R_{sr}, R_{st}, and R_{tr} represent the spatial distances among the three objects: the satellite transmitter, the receiving antenna, and the target, respectively. At any instant t, the bistatic range (R_{Δ}) associated with the target can be formulated as:

$$R_{\Delta}(t) = R_{st}(t) + R_{tr}(t) - R_{sr}(t) \tag{1}$$

The signal processing procedure at the receiver is illustrated as follows. The direct signal is captured by a right-hand circularly polarized (RHCP) antenna, whereas the reflected signal is collected by a left-hand circularly polarized (LHCP) antenna. After down-conversion by the radio frequency (RF) front end, both signals are converted into intermediate-frequency (IF) signals and subsequently subjected to range compression (RC) in software. Figure 2 presents the software-based signal acquisition and range compression procedure. In this system, the direct signal is processed via closed-loop tracking to extract sufficiently strong Global Navigation Satellite System (GNSS) satellite signals, based on which the local spreading code and pseudorandom noise (PRN) code are reconstructed. These reconstructed codes are then applied to the reflected channel, where open-loop tracking is performed on the reflected signal, thereby enabling range compression.

In theory, the initial phase can generally be assumed to be zero. Accordingly, the down-converted direct intermediate-frequency signal can be expressed as:

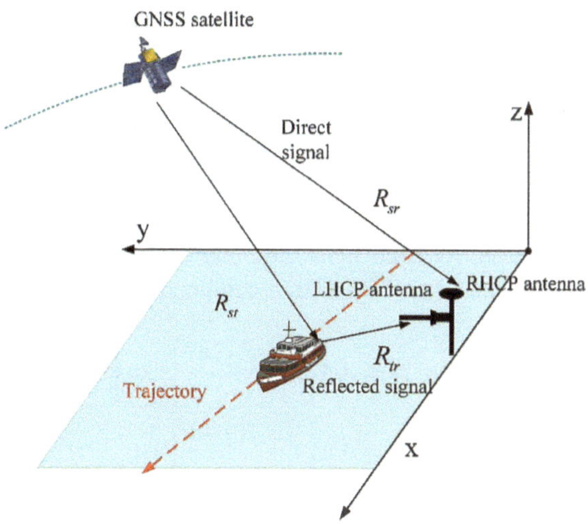

Fig. 1 Schematic diagram of the geometric relationship among GNSS satellite, receiver, and target

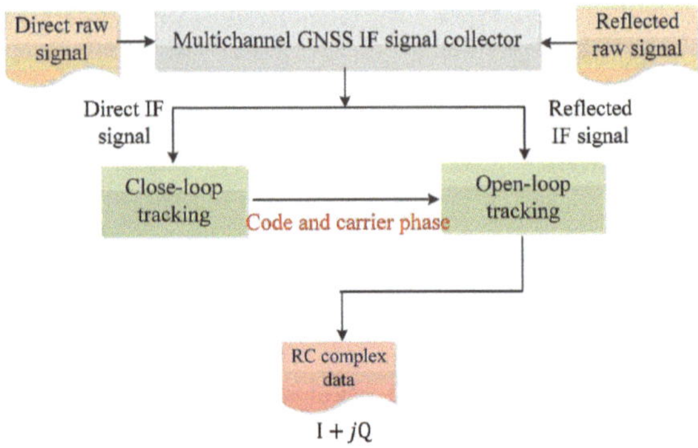

Fig. 2 The flowchart of closed-loop and open-loop processing of raw IF signals acquired by the multichannel receiver, corresponding to the range compression procedure

$$S_{IF,d}(t) = AC(t)D(t)\cos(2\pi f_{IF}t) \tag{2}$$

where direct signal is taken as a reference signal with amplitude A, where C(t) represents the pseudorandom noise (PRN) code envelope and D(t) denotes the navigation data bits that vary with time. After experiencing the delay introduced by the bistatic range difference, the signal becomes the target-reflected echo, which can be modeled as follows:

$$S_{IF,r}(t) = AC\left[t - \frac{R_\Delta}{c}\right]D\left[t - \frac{R_\Delta}{c}\right]\cos\left\{2\pi(f_{IF} + f_d)\left[t - \frac{R_\Delta}{c}\right]\right\} \tag{3}$$

In the expression, the speed of light c and the Doppler frequency shift f_d fd are commonly used parameters, which can be estimated from the direct intermediate-frequency (IF) signal after closed-loop processing, together with the pseudorandom noise (PRN) code. For the navigation data bits, a phase inversion issue may arise, which can be corrected by performing conjugate multiplication after bit extraction. Subsequently, the reflected signal is demodulated using the in-phase and quadrature components of the local carrier replica, namely [$\cos(2\pi f_{IF}t)$ and $\sin(2\pi f_{IF}t)$], so as to remove the carrier and obtain the code signal. The obtained signal is then correlated and aligned with the PRN code of the reflected signal, ultimately yielding the correctly delayed code signal.Given the robust autocorrelation properties of the PRN code, the specific code delay of PRN that produces the maximum peak of correlation value directly indicates the reflector's bistatic range. Through this mechanism, RC is achieved for each 1-ms sampling dataset. The resulting RC data, structured as a two-dimensional (2D) matrix, can be given by:

$$\begin{aligned} S(n,t) &= A\Lambda\left[t - \frac{R_\Delta(n)}{c}\right]\exp\left\{j\left[2\pi f_d(n)t + 2\pi\frac{R_\Delta(n)}{\lambda}\right]\right\} \\ &= A\Lambda\left[t - \frac{R_\Delta(n)}{c}\right]\exp\left[j2\pi f_d(n)t + j2\pi N(n) + j\phi\right] \end{aligned} \tag{4}$$

where n denotes the PRN-code period index (slow-time index), t represents the sampling instant within one PRN-code period (fast-time index). The equation shows that the amplitude Λ is obtained through the cross-correlation operation of the pseudorandom noise (PRN) code. Since the signal source is a BeiDou geostationary Earth orbit (GEO) satellite, which remains stationary with respect to the Earth's surface, no Doppler shift is introduced by the satellite itself. Consequently, the Doppler frequency shift f_d arises solely from target motion. Here, N denotes an integer number of wavelength cycles, and ϕ represents the signal phase.

2.2 SNR Enhancement Through CIT Extension

Enhancing the signal-to-noise ratio (SNR) is the first necessary step to overcome the challenge posed by weak navigation signals. For static targets, a fundamental SNR improvement technique is the coherent accumulation process (CAP), which essentially involves extending the coherent integration time (CIT). In standard navigation receivers, the maximum CIT is strictly constrained by the length of the navigation bit, for example, in GPS L1 band CIT is limited by 20 ms, as any sign reversal in the navigation bit during integration would severely degrade the correlation peak. However, within our customized software receiver architecture, the successful correction of phase inversions allows us to perform prolonged CAP well beyond the traditional 20-ms limit, thereby significantly boosting the processing gain.

By converting the target-induced time delay into the corresponding bistatic range cell, the time variable t is transformed into a cell index. Accordingly, the data $S(n, t)$ can be rearranged into a range-ordered matrix, denoted as $S(n, m)$ where m serves as the range cell index. If we define a sequence of consecutive target echoes over a duration T as a single frame, and assume that the echo experiences zero Doppler shift in the whole frame, the total CAP can be mathematically formulated as:

$$S(p, m) = \sum_{n=(\mathrm{p}-1)\times \mathrm{T}+1}^{p\times T} S(n, m) \tag{5}$$

where $p \in 1, 2, \ldots, n_{\max}$ is the frame index, T is the CIT time.

(1) **Extending CIT for moving Target**. As an example of a stationary target using a metallic reflector, the CAP results are illustrated in following figure. Figure 3a shows the isosceles triangle metal reflector with a side length of 1.5 m and three antennas mounted on a T-shaped support—one white RHCP patch antenna for direct signal and two high-gain LHCP helical antennas for reflected signal. The included angle between the metal reflector and the ground is about 60°, and the azimuth of the metal reflector is approximately 180°. The sampling frequency of the receiver is 32.738 MHz.

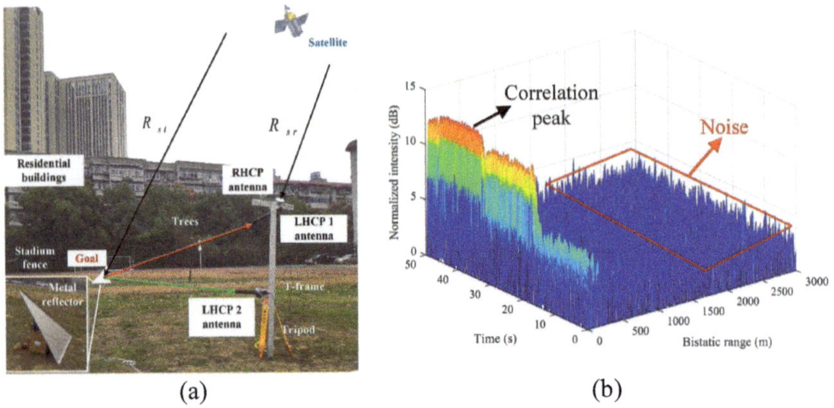

Fig. 3 Schematic diagram of stationary target detection. **a** A metallic reflector is used as the target, with a right-hand circularly polarized (RHCP) antenna employed for direct signal reception and two left-hand circularly polarized (LHCP) antennas (LHCP 1 and LHCP 2) used to receive the reflected wave; **b** three-dimensional representation of the target echo matrix

Data acquisition spans a continuous 50-s window. Within this timeframe, the metallic reflector is manually displaced toward the LHCP antennas on two occasions to emulate a realistic structural deformation scenario. It remained stationary at three distinct positions for durations of 18 s, 18 s, and 14 s, respectively. This stepwise movement ensures signal coherence within each static interval. By increasing the coherent integration time (CIT) to 200 ms—ten times the conventional integration interval of a navigation receiver—and using the BeiDou C01 satellite signal as the illumination source, the three-dimensional correlation result of the left-hand circularly polarized (LHCP 1) channel is obtained, as shown in Fig. 3b. Over the entire observation period, three distinct staircase-like continuous correlation peaks are observed at the 6th range cell, corresponding to a bistatic range of 54.96 m, accurately reflecting the three physical placements of the target. Comparable results are also observed from the LHCP 2 antenna.

(2) **Implementation and Constraints of Increasing CIT for Maritime Target**. The CAP technique is subsequently evaluated on riverine targets situated on the Changjiang River. As shown in Fig. 4a, the test scenario involves three distinct objects: two stationary pontoons (labeled A and B) and one maneuvering cargo vessel (labeled C). Signals emitted by the Beidou C01/C04 are served as the illuminators of opportunity. In terms of spatial distribution, Pontoon A, B, and the moving cargo ship C are located at approximately 55, 80 m, and beyond 700 m from the receiving station, respectively,while ship C is traveling from northeast to southwest.

After range compression (RC), as described in Sect. 2.1, the new RC data $S(n, m)$, The new dataset consists of the correlation values of m range cells accumulated over

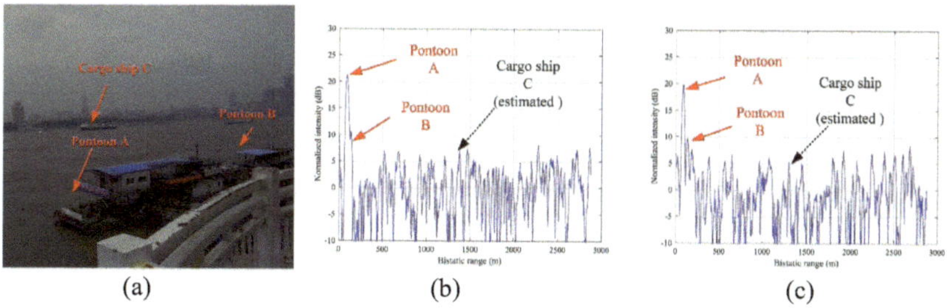

Fig. 4 **a** Overview of the riverine experimental scene; **b** correlation outputs derived from the C01 reflected signals; **c** Correlation outputs from the C04 reflected signals. The detection methodology for the maneuvering cargo ship is detailed in subsequent sections

n PRN periods. To reduce the total data volume, only 600 range cells are processed in each millisecond, i.e., m = 1, 2, …, 600. The range resolution is 4.8 m.

The optimal CIT is highly dependent on the target's radial velocity. Since pontoons A and B are zero velocity, they provided a stable baseline. In contrast, the vessel's velocity is non-zero but unknown. To address this, a parameter sweep approach is employed to determine an appropriate baseline CIT, testing values of 10, 50, and 100 ms. Empirical analysis reveales that a CIT of 50 ms yielded robust performance for the stationary pontoons, producing clearly discernible correlation peaks from their reflected signals. Consequently, the 50-ms CIT is selected, dividing the dataset into 2,000 distinct frames. Figure 4b presents cross-sectional views of the correlation peaks (spanning 0–50 ms) for the C01 satellite.The prominent peak corresponding to Pontoon A (with an SNR of approximately 19.9 dB) appears at the 19th range cell, which corresponds to a bistatic range of about 91 m. Similarly, the signal feature of Pontoon B (with an SNR of 9.4 dB) appears at the 30th range cell, corresponding to a bistatic range of approximately 144 m. In Fig. 4c, which shows the C04 satellite data, the main peak of Pontoon A with an SNR of 21.3 dB appears at the 18th range cell, corresponding to a bistatic range of approximately 86 m, whereas Pontoon B with an SNR of 10.7 dB appears at the 26th range cell, corresponding to a bistatic range of 125 m.

However, the reflected signal correlation peak or signal feature of cargo ship C cannot be found because it is buried in noise (the estimated range cell is marked by a dashed arrow). The CAP fails to enhance the SNR for ship C because its non-zero velocity causes the echo phase to vary with bistatic range.

3 Maneuvering Target Dection and Location Using Ground-Based GNSS-R

Due to the velocity of the maneuvering target, the coherence of the GNSS reflected signal is weak, and the CAP cannot complete the detection and location. Therefore, this section proposes two methods for detection of maneuvering targets.

3.1 *Method 1: Doppler-Range Compensation*

The first proposed technique mitigates the effects of target velocity by actively compensating for both Doppler and range migrations. This procedure consists of two primary stages. Initially, the system segments the continuous data stream into discrete frames and applies a Fast Fourier Transform (FFT) to calculate and estimate the corresponding Doppler shift frequencies. Subsequently, it corrects the Doppler frequency shifts and range deviations within each individual frame. Figure 5 illustrates the workflow of this compensation strategy.

(1) **Multiple Frames Doppler Processing**. The foundational assumption here is that a target does not traverse beyond a single 4.8-m range resolution cell within a brief temporal window (for instance, 2.5 s). Based on this premise, the algorithm partitions the range-compressed (RC) data along the slow-time axis into M distinct frames, where each frame spans exactly 2500 ms. Consequently, the RC data matrix corresponding to the i-th frame takes the following mathematical form:

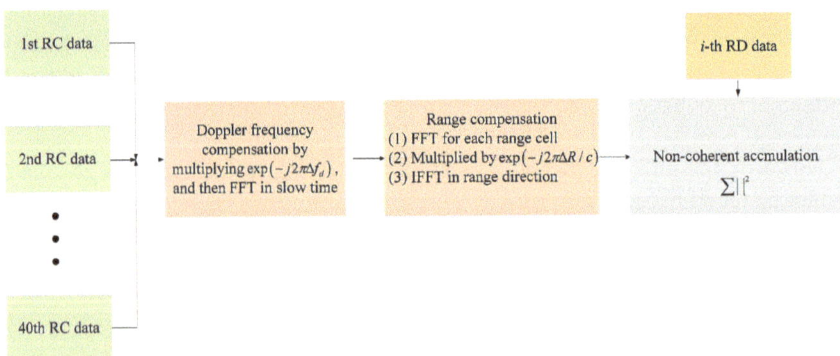

Fig. 5 Workflow of Doppler-range compensation

$$RC = \begin{bmatrix} S(T(i-1)+1,1) & S(T(i-1)+1,2) & \cdots & S(T(i-1)+1,1) \\ S(T(i-1)+2,1) & S(T(i-1)+2,2) & \cdots & S(T(i-1)+2,\mathrm{m}) \\ \vdots & \vdots & \ddots & \vdots \\ S(T(i-1)+\mathrm{T},1) & S(T(i-1)+\mathrm{T},2) & \cdots & S(T(i-1)+\mathrm{T},\mathrm{m}) \end{bmatrix} \quad (6)$$

where $i = 1, 2, \ldots, m_{\max}$; T is 2500 ms; and m is the index of range cells. Following this matrix construction, the system executes an FFT operation along the slow-time dimension for every column. This transformation generates a two-dimensional Range-Doppler (RD) map, mathematically represented as $\boldsymbol{RD}_i^j = \mathrm{FFT}\left(\boldsymbol{RC}_i^j\right)$, where j is the index of Beidou satellite.

(2) **SNR Caculate and Enhancement**. To examine the variation of Doppler frequency with time, the Doppler frequencies over multiple frames are first collected, as shown in Fig. 6 for the maneuvering target. The x-axis indicates a duration of 24 frames (60 s), while the y-axis denotes Doppler frequency. A nearly linear trend is observed, so linear regression is performed. The Doppler rate is − 0.073 Hz/s for the C01 data and − 0.061 Hz/s for the C04 data. The corresponding R^2 values are 0.96 and 0.95, respectively, confirming strong linearity.

The SNR enhancement procedure works as follows: for each RD map, the other $M - 1$ maps are first aligned by compensating for range and Doppler shifts, and then incoherent accumulation is carried out. For example, $\boldsymbol{RD}_M$ is the M-th raw map, and $factor_M^l$ is the compensation operator using the l-th frame data. The compensated M-th frame is expressed as:

$$\boldsymbol{RD}_M^{Comped} = \frac{1}{M}\left\{|\boldsymbol{RD}_M| + \sum_{l=1}^{M-1}\left|\boldsymbol{RD}_l * facter_M^l\right|\right\} \quad (7)$$

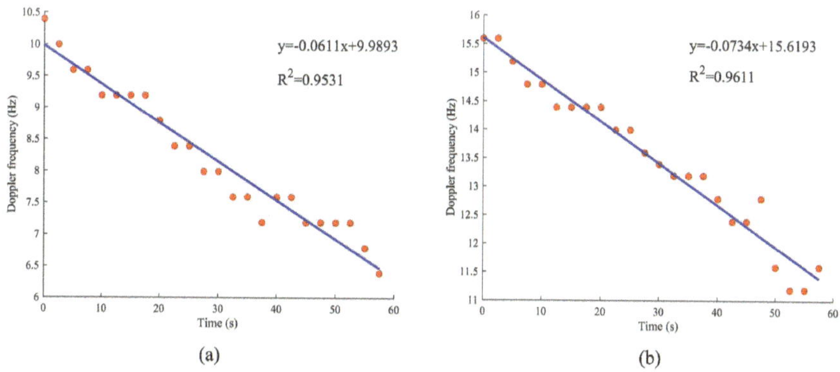

Fig. 6 Frame-wise Doppler rate over 0–60 s. **a** Doppler rate of the reflected signal from C01; **b** Doppler rate of the reflected signal from C04

where * is the compensation operation. Then, in the same way, the compensated i-th frame can be written as:

$$\boldsymbol{RD}_i^{Comped} = \frac{1}{M}\left\{|\boldsymbol{RD}_i| + \sum_{l \neq i}^{M} \left|\boldsymbol{RD}_l * facter_i^l\right|\right\} \tag{8}$$

The compensation procedure consists of two steps: Doppler-frequency correction and range correction. Let $\boldsymbol{RD}_i$ represent the data to be corrected, and let $\boldsymbol{RD}_l$ be the reference map for compensation. Because the Doppler frequency varies linearly with time, the following expression can be written as $f_d(t) = k_d \times t + f_{d0}$, where k_d is the Doppler rate, and f_{d0} is the initial Doppler frequency. Accordingly, the Doppler compensation factor is $\exp(-j2\pi\Delta f_d)$, where:

$$\Delta f_d = f_d(l) - f_d(i) = k_d \times (l - i) \tag{9}$$

The reflected-signal phases associated with the l-th and i-th frame are denoted by $\Phi_d(l)$ and $\Phi_d(i)$, respectivrely. The phase difference from l to i is written as

$$\Delta\Phi_d = \Phi_d(l) - \Phi_d(i) = 2\pi\left\{\frac{1}{2}k_d \cdot \Delta p^2 + \left[f_d(l) - k_d \cdot \Delta p\right] \cdot \Delta p\right\} \tag{10}$$

where $\Delta p = l - i$. Consequently, the range compensation factor will be $\exp(-j2\pi\Delta R/c)$, and ΔR is

$$\Delta R = \frac{\lambda\Delta\Phi_d}{2\pi} = -\lambda\left[-\frac{1}{2} \times k_d \times \Delta p^2 + f_d(l) \times \Delta p\right] \tag{11}$$

The compensation procedure for the i-th RC frame consists of two steps: (1) Doppler frequency compensation. Each column of the RC data is multiplied by factor $\exp(-j2\pi\Delta f_d)$.. A column-wise FFT is then applied to transform the RC data and get RD data. (2) Range correction or compensation. This step exploits the Fourier shift property $F\left[f(t - t_0)\right] = F(j\omega)e^{-j\omega t_0}$. Then an FFT is used in each row of the RD matrix, which convert the data into the range-frequency domain. The FFT result is multiplied by the range compensation factor along each row, after that, an inverse FFT is followed to return to the RD domain, completing the range compensation.

Figure 7 compares the SNR before and after enhancement. In Fig. 7a, the reflected signal of cargo ship C has an SNR of 6.1 dB, making it easily masked by surrounding interference. Consequently, the correlation peak's bistatic range and Doppler frequency cannot be directly identified when finding in the RD map. However, after enhancement, as shown in Fig. 7b, a distinct peak can be easily distinguished at a bistatic range of 1290 m (Doppler frequency of 6.4 Hz) with an SNR of 10.5 dB, representing an increase of 4.4 dB..

(3) **Results**. To clearly illustrate the performance of the proposed algorithm in this chapter, 40 frames of data from the same target are integrated into a single RD

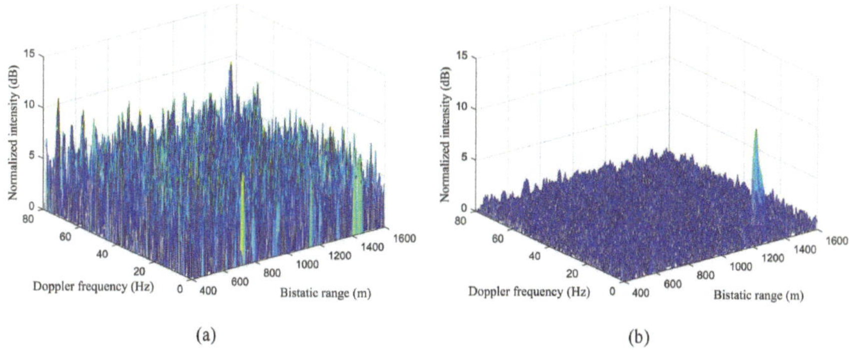

Fig. 7 SNR results of one frame. **a** The raw RD map before enhancement; **b** the RD map after enhancement

map before and after SNR enhancement. Figure 8 presents the RD maps for the C01 data. In Fig. 8a, the correlation peak corresponding to cargo ship C is not discernible (SNR < 10 dB) within the area enclosed by the white rectangle (bistatic range: 1118–1320 m). After SNR enhancement, as shown in Fig. 8b, a continuous peak emerges inside the white rectangle (SNR ≥ 10 dB), and the trajectory formed by this peak can be clearly observed in the indicated region. In addition, the Doppler frequency exhibits an approximately linear dependence on bistatic range.

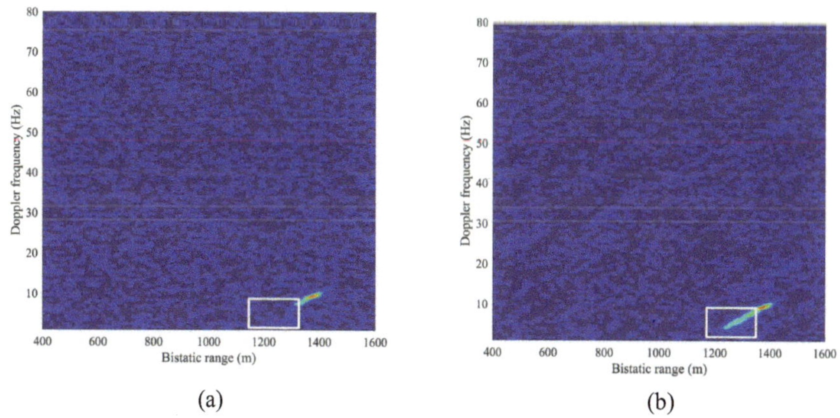

Fig. 8 SNR results of one frame. **a** Original 3D RD map before SNR enhancement; **b** 3D RD map after SNR enhancement

3.2 Method 2: Long IntegrationTime Algorithm Based on FrFT

Given the detection range, GNSS ground-based target detection is most suitable for coastal ports or inland rivers. In these areas, ships move at low speeds and may perform maneuvering operations when docking or leaving the shore. Therefore, a simple uniform motion model is insufficient. For a maneuvering target, the bistatic range (R_{Δ}) varies with time and can be expressed as:

$$R_{\Delta}(\tau) = \frac{1}{2}a_t\tau^2 + v\tau + R_{\Delta,0} \tag{12}$$

where a_t denotes radial acceleration, v denotes radial velocity, and $R_{\Delta,0}$ is the initial bistatic range. In practice applications, estimating the radial velocity and acceleration is the primary objective of target detection, because these two parameters of a non-cooperative target are unknown. Since radial velocity and acceleration correspond to frequency and chirp rate, respectively, the Doppler spectrum expansion can be expressed as:

$$f_d = \frac{\mathrm{d}R_{\Delta}(\tau)}{\lambda \mathrm{d}\tau} = \frac{d\left({a_t\tau^2}/{2} + v\tau + R_{\Delta,0}\right)}{\lambda d\tau} = f_0 + \mu\tau \tag{13}$$

where λ is the wavelength of the Beidou B3I signals (24 cm), f_0 is the Doppler central frequency, and μ is the chirp rate induced by radial acceleration. Based on the above motion model, an FrFT-based SNR enhancement algorithm is proposed, as outlined below.

Compared to the FFT, the FrFT is better suited for processing nonstationary signals because it has one additional degree of freedom, namely the transform angle. Its kernel function contains a quadratic chirp term, which allows GNSS signals to be integrated into an impulse in the corresponding FrFT domain. Let $f(t, r)$ be a two-dimensional complex function defined on the plane; the algorithm is then defined as:

$$G_r(\alpha, u) = \mathcal{F}^{\alpha}\left[f(t, r)\right] = \int_{-\infty}^{\infty} K_{\alpha}(t, u) f(t, r)\mathrm{d}t \tag{14}$$

where $\mathcal{F}^{\alpha}$ denotes the FrFT operator, $\alpha = \frac{p\pi}{2}$ is the transform angle, p is transform order, u is FrFT domain, and $K_{\alpha}(t, u)$ is the transform kernel. Specifically, the echo's FrFT can be written as

$$S_{FrFT} = \int_{-\infty}^{\infty} K_{\alpha}(\tau, u) S_{rc}\left[\left(\frac{1}{2}a_t\tau^2 + v\tau + R_{\Delta,0}\right)/c, \tau\right] d\tau \tag{15}$$

The parameters v and a_t carry explicit physical meanings, and S_{rc} represents the range-compressed data obtained from the reflected GNSS signals. To illustrate the

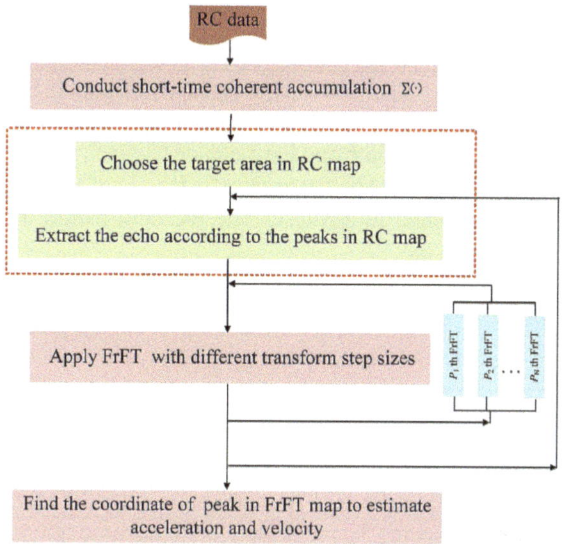

Fig. 9 Block diagram of the proposed SNR improvement algorithm

SNR enhancement process, Fig. 9 presents a flowchart for algorithm based on the FrFT using long-time (tens of seconds) integration, which consists of four steps. The details of Each processing step will be described below.

(1) **Step 1: Conduct Short-time Coherent Integration**. In a GNSS receiver, signals are sampled at two different rates: the high sampling frequency f_s corresponding to the real receiver's sampling rate and the low frequency f_r matches to the frame rate. The input RC data can be written as a 2-D matrix $\mathbf{RC} = S_{rc}(\tau, r)$ for processing, where $r \in [0, f_s]$ is range cell index, corresponding to the high sampling frequency; $\tau \in \left[-\frac{T}{2}, \frac{T}{2}\right]$ is the time range related to the low sampling frequency; and T is the dwell period of target. To improve the SNR of echo for extracting the target track, the short-time coherent integration is conducted. Note that the short time (tens of milliseconds) is significantly lower than the long time (tens of seconds). For simply we neglect the amplitude of the cross-correlation function, the Range Compressed data can be expressed as:

$$S_{rc}(\tau) = exp\left\{j\left[2\pi \frac{R_\Delta(\tau)}{\lambda}\right]\right\} + w(\tau) \tag{16}$$

where $w(\tau)$ denotes the noise. After integrated the signal is directly written as

$$z = \sum_{\tau=0}^{CIT-1} \exp\left\{j\left[2\pi \frac{R_\Delta(\tau)}{\lambda}\right]\right\} + \sum_{\tau=0}^{CIT-1} w(\tau) \approx CIT \cdot \exp\left\{j\left[2\pi \frac{R_\Delta(\tau)}{\lambda}\right]\right\} + \sum_{\tau=0}^{CIT-1} w(\tau) \tag{17}$$

where CIT denotes the coherent integration time. After basic SNR enhancement, the target echo is highlighted against the background ambient noise, enabling the

RC map to reveal the bistatic range change or history of the target and thus provide appropriate initial search parameters.

(2) **Step 2: Select the Target Area and Extract Echoes from the RC Map**. First, based on prior information from the experimental setup—such as the approximate river width and the distance between the river and the receiver—the search interval of the bistatic range $\left[-R_{\Delta,\max}, R_{\Delta,\max}\right]$ and the range resolution Δr are determined in advance. The RC data within this interval after Step 1 can be written as:

$$\mathbf{RC} = \sum_{\tau=(n-1)\times CIT+1}^{n\times CIT} S_{rc}(\tau, -r_{\max} + r) \tag{18}$$

where $n = 1, 2, \ldots, \frac{T}{CIT}, -r_{max} = \frac{-R_{\Delta,\max}}{\Delta r}$ is the starting range cell; and r is the index of range cell, $r = 1, 2, \ldots, {2R_{\Delta,\max}}/{\Delta r}$.. Then, the correlation peak data along the range direction (i.e., the row direction of the RC data matrix) are extracted to form a new data vector $\mathbf{X}_n$ for subsequent long-time integration processing, expressed as:

$$\mathrm{X}_n = \max(\mathbf{RC}(:, r)) = \left[S_{rc}(1, r_1)\ S_{rc}(2, r_2)\ \cdots\ S_{rc}(n, r_n)\right]^{\mathrm{T}} \tag{19}$$

where $\max(\cdot)$ is the operation of extracting maximum point of an array, and r is determined by the range direction of peaks.

(3) **Step 3: Apply FrFT**. The discrete FrFT with different transform angles can be implemented using the decomposition algorithm:

$$\mathcal{F}^{\alpha}(u) = \frac{\sqrt{\frac{1-j\cdot\cot(\alpha)}{2\pi}}}{2F} \sum_{\tau=-n/2}^{n/2-1} \exp\left(j\pi u^2 \cdot \cot(\alpha)\right) \cdot \exp\left(\frac{j\pi\tau^2 \cdot \cot(\alpha)}{4F^2} - \frac{j\pi u\tau \cdot \csc(\alpha)}{F}\right) S_{rc}\left(\frac{\tau}{2F}\right) \tag{20}$$

where n is the signal length, and F denotes the highest frequency component of $S_{rc}(\tau)$ in slow time.

Two loop iterations are detailed and illustrated as follows:

(a) Transform The transform order is varied in each iteration, with a one-dimensional search over p. For a given order, the FrFT produces a one-dimensional result in the u domain. By sweeping through orders $p_{\mathrm{N}} = p+N\Delta p$ (where N is determined by the step size Δp, FrFT yields a two-dimensional output. The step size Δp follows a coarse-to-fine progression, decreasing from large to small values. To reduce computational complexity, a coarse search is first performed to detect the presence of a target peak in the FrFT domain. If a peak is found, a fine search is then carried out to increase the integration gain.

The specific Δp is determined based on the acceleration, ensuring that the step size does not exceed the target's acceleration.

(b) The search area is updated for the next iteration. Typically, after the FrFT, the echo energy is concentrated into a distinct peak representing the target. However, if the echo consists solely of noise rather than target-reflected signals, no such peak appears because noise energy cannot be coherently integrated. In this case, a new search area must be selected for another iteration of transform order adjustment, since the previous area may not contain the target. This process continues until a target peak is observed.

(4) **Step 4:Extract the Coordinate of Peak and Estimate the parameters**. To achieve the optimal SNR enhancement, the transform step size is further refined compared to the previous iteration, allowing additional energy accumulation until the peak SNR shows no further or only marginal improvement. If the peak remains undetectable in the FrFT generated map after this refinement, the search area in Step 2 is reselected. The coordinates of the resulting peak in the two-dimensional (p, u) map can then be expressed as:

$$\left(p_{max}, u_{max}\right) = \text{argmax}_{p,u}|G_r(p, u)| \tag{21}$$

where function argmax$(\cdot)$ output the index of the maximum point, and function $G_r(\cdot)$ output FrFT map. Since $p = \frac{2\alpha}{\pi}$, we search p instead of searching of α.

The motion parameters at this coordinates are also calculated because the coordinates $\left(p_{\max,}u_{\max}\right)$ can match with the motion parameters of the target (including acceleration and radial velocity). The calculation is carried out based on the scale factor S, written as

$$S = \sqrt{\tau/f_r} \tag{22}$$

where τ is the number of slow time, and f_r is the sampling number in slow time after step 1 $(1000/CIT)$. The normalized chirp rate $\hat{f}_d$ can be written as

$$\hat{f}_d = -\cot(p\pi/2) \tag{23}$$

and the truth chirp rate μ can be written as

$$\mu = \hat{f}_d \cdot f_r/S^2 = \hat{f}_d \cdot f_r/\tau \tag{24}$$

The radial acceleration can be written as

$$a_t = \lambda \cdot \mu \tag{25}$$

After the acceleration, we also calculate the target velocity then the central frequency f_0 using the FrFT domain u.

$$f_0 = u \cdot \csc(p\pi/2) \tag{26}$$

and the radial velocity v can be written as

$$v = f_0 \cdot \lambda \tag{27}$$

(5) **Results.** After the range compression, the raw RC data are stored in a matrix $\mathbf{RC} = S_{rc}(\tau, r)$, where τ is the index of the PRN period ($\tau = 1, 2, \ldots, 20{,}000$), and r is the index of the range cells. Here, the CIT is set to 20 ms, depending on the target's velocity. Figure 10a shows the target area in the RC map before and after short-time coherent integration (Step 1). Only the bistatic range of interest (1000–1800 m) is displayed, as stationary targets are excluded. Two regions are visible: one containing the target and one consisting of noise. In the target region, the echo exhibits a wide bistatic range history (two parallel straight white lines) due to the PRN code point spread function in range compression. The y-axis of Fig. 10a is converted from time to frames (each frame integrates 20 ms of data), resulting in 1000 frames (20,000 ms / 20 ms). The two regions in Fig. 10a serve as distinct search areas.

The new searching area is then re-chosen as [1200, 1500 m] in the iteration for new improvement. By extracting the echo and then conducting FrFT, the result is shown in Fig. 10b. The FrFT is processed from $p = 0$ to $p = 2$ with step $\Delta p = 0.002$. In the figure, one prominent peak with SNR = 18.08 dB with the corresponding coordinates $(p,u) = (0.984, 14.85)$ can be found. This means that the target peak accumulates the gain from these lines. According to the principle of the FrFT, for a given order p, the FrFT produces a corresponding u domain. If p does not match the chirp rate of the echo, the echo energy is spread across the u domain. When p values close to the actual chirp rate are used, bright lines emerge across multiple u domains.

The parameters can be calcuted based on Step 4 for location, since the coordinates of the target peak (p, u) are matched by the motion parameters of the target (radial acceleration and radial velocity). The observation time is 20,000 ms, and the

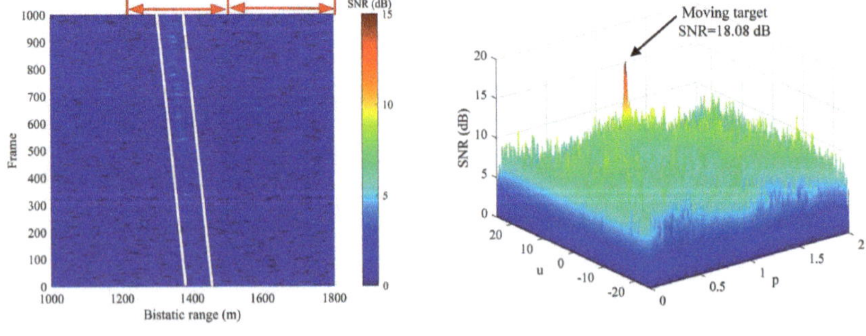

(a) (b)

Fig. 10 **a** The target area in the RC map of signals after the integration. The bistatic range history is marked by parallel straight white lines; **b** the FrFT 3D map of signals from C01 with $\Delta p = 0.002$

optimum p of target peak is 0.984. The correponding acceleration is 0.015m/s^2, and the velocity is 3.567m/s. The real target speed found in AIS website is nine knots (about 4.6 m/s). The comparison results demonstrate that the estimated motion parameters agree well with the geometric projection relationship.

4 Static Target Dection and Location Using Spaceborne GNSS-R

The target detection range of ground-based GNSS-R depends on the location where the equipment is installed, while that of spaceborne GNSS-R covers the globe. Spaceborne GNSS-R generally outputs DDM data. There are two difficulties in achieving the detection and location of maritime targets by using DDM; (1) The echo signal-to-noise ratio of maritime targets is weak and they are easily submerged by sea clutter. (2) Locaiton ambiguity exists when searching target location from the theoretical DD map to the real spatial domain. Therefore, this section describes two aspects: the first is the cancelation of sea clutter in DDM for target detection, and the second is to remove the ambiguity for target location (one DD cell coorespongds to two x–y coordinates).

4.1 DDM Subspace Projection for Cancelation of Sea Clutter

Estimating the sea clutter consists of three steps: rearranging the DDM vectors, calculating the signal domain subspace, and canceling the clutter signals in the DDMs.

(1) **Power Normalization and Rearrangement of DDM Vectors**

Due to factors such as system noise temperature fluctuations, it is necessary to consider the power-normalized DDM $\left|\overline{Y}(f_d, \tau)\right|^2$ (ranging from 0 to 1), which also facilitates subsequent data processing. Assuming that the conditions (weather, seawater dielectric constant, etc.) in the area of interest remain roughly constant, the sea clutter along a specific track can be considered approximately the same. To analyze DDM variations in both time and space, a new matrix comprising all DDMs of the area of interest is constructed. This process consists of two main steps.

(1) The 2D DDM matrix is rearranged into a one-dimensional vector $\xi_{f_d,\tau}$

$$\xi_{f_d,\tau}[n] = \left|\overline{Y}(f_d, \tau)\right|^2 \tag{28}$$

where $n = 1, 2, \ldots, 187$ is the index of cell in one DDM. The DD cells are rearranged in a Z-shaped order (first left to right, then top to bottom). The first row 11 bins correspond to a delay of − 2 chips (the first row of the DDM), occupying indices

1 to 11. The next 11 bins correspond to a delay of − 1.75 chips in the second row, occupying indices 12–22, and so on. This Z-shaped rearrangement is intended to emphasize the periodicity of the DDM, thereby facilitating the subsequent computation of the signal subspace. We refer to $\xi_{f_d,\tau}[n]$ simply as $\xi[n]$. Then we arrange the whole DDM vectors as one new matrix in chronological order.

(2) The vectors of DDMs are arranged as a new 2D matrix

$$\boldsymbol{\xi}_{T\times 187} = \begin{bmatrix} \xi_1[1] & \xi_1[2] & \cdots & \xi_1[187] \\ \xi_2[1] & \xi_2[2] & & \xi_2[187] \\ \vdots & & \ddots & \vdots \\ \xi_T[1] & \xi_T[2] & \cdots & \xi_T[187] \end{bmatrix} \tag{29}$$

where T is the number of epochs. In the newly constructed matrix, each column represents the time series of a given DD cell, while each row corresponds to an individual DDM. When estimating sea clutter, values originating from thermal noise or an oil rig (target) are considered outliers and should be removed, retaining only those due to clutter. Thus, outlier suppression is necessary for accurate sea clutter estimation.

(2) **Calculating the Signal Subspace of the 2D Matrix**

Once the 2D DDM matrix is constructed, its rows represent observations and its columns represent variables—specifically, T observations and 187 variables. Due to observed correlations among the DDMs in the area of interest, the data matrix $\boldsymbol{\xi}_{T\times 187}$ is used to form a covariance matrix. This covariance matrix is then decomposed via eigenvalue decomposition into a signal subspace and a noise subspace, which primarily contain information about sea clutter and the oil rig (target), respectively.By projecting $\boldsymbol{\xi}_{T\times 187}$ onto the signal subspace, the oil-rig contribution can be suppressed, which leads to a cleaner estimate of the sea-clutter component.

The first step is to compute the covariance matrix of $\boldsymbol{\xi}$

$$\mathbf{C} = \frac{1}{T}\hat{\boldsymbol{\xi}}\hat{\boldsymbol{\xi}}^{T} \tag{30}$$

$$\textbf{Where } \hat{\xi} = \begin{bmatrix} \xi_1[1] - \overline{\boldsymbol{\xi}[1]} & \xi_1[2] - \overline{\boldsymbol{\xi}[2]} & \cdots & \xi_1[187] - \overline{\boldsymbol{\xi}[187]} \\ \xi_2[1] - \overline{\boldsymbol{\xi}[1]} & \xi_2[2] - \overline{\boldsymbol{\xi}[2]} & & \xi_2[187] - \overline{\boldsymbol{\xi}[187]} \\ \vdots & & \ddots & \vdots \\ \xi_T[1] - \overline{\boldsymbol{\xi}[1]} & \xi_T[2] - \overline{\boldsymbol{\xi}[2]} & \cdots & \xi_T[187] - \overline{\boldsymbol{\xi}[187]} \end{bmatrix} \tag{31}$$

And $\overline{\boldsymbol{\xi}[n]}$ denotes the mean value of the n-th column. Next, an eigenvalue decomposition is applied to $\mathbf{C}$,

$$\mathbf{C} = \mathbf{Q}\sum\mathbf{Q}^{-1} \tag{32}$$

where $\sum = \text{diag}\left[\lambda_1\ \lambda_2\ \cdots\ \lambda_T\right]$ is the diagonal matrix of eigenvalues. After that, the eigenvalues are ordered according to their information contribution. This subspace projection procedure diagonalizes the covariance matrix of the DDM vectors and transforms the original measurements into a set of orthogonal vectors that are also ranked by information content. The first k eigenvectors are selected to form the signal subspace $\mathbf{Q}_k$. In this study, k is determined by the variance ratio explained by the signal subspace; once this ratio exceeds 90%, the selected k retains most of the sea-clutter information while filtering out interference from the oil rig. Applying the projection gives the orthogonal measurement space

$$\mathbf{Z} = \mathbf{Q}_k\hat{\boldsymbol{\xi}} \tag{33}$$

After subspace projection, an approximate DDM matrix ξ' representing sea clutter is obtained by projecting the raw data onto the signal subspace.

(3) **Cancelling the Clutter and Detecting the Oil rig in DDMs**

Once the two preprocessing steps are completed, a representative sea-clutter DDM vector for the area of interest can be estimated. The approximate sea-clutter vector is written as

$$\xi' = \frac{1}{T}\left(\sum_{t=1}^{T}\xi_t'[1] + \sum_{t=1}^{T}\xi_t'[2] + \cdots + \sum_{t=1}^{T}\xi_t'[187]\right) \tag{34}$$

where $t = 1, 2, \ldots, T$ is the epoch index. The DDM vector with sea clutter removed, denoted by $\boldsymbol{\xi}_0$, is then obtained as

$$\boldsymbol{\xi}_0 = \boldsymbol{\xi} - \xi' \tag{35}$$

The target DDM, which contains the delay-chip and Doppler-frequency information of the oil rig, can be recovered by reshaping $\boldsymbol{\xi}_0$ from a vector into a two-dimensional matrix.

After sea-clutter suppression, the resulting DDMs may exhibit peaks associated with the oil rig. A peak in $\boldsymbol{\xi}_0$ is classified as a target only when two conditions are satisfied. First, its value must be larger than the threshold defined as

$$th = \frac{\sum_{i=1}^{N}\overline{value_i}}{N} \tag{36}$$

where N is the number of DDMs containing the target; $\overline{value}$ is the mean power of each $\boldsymbol{\xi}_0$ over the delay range from 0 to 2 chips. Second, the peak must be the maximum value within $\boldsymbol{\xi}_0$. Once a peak is identified as a target, its coordinates (delay chips and Doppler frequency) are recorded..

Cyg02 observations on 11 July 2020 are selected to demonstrate the effect of subspace projection on sea-clutter estimation. Given the approximately 50 km ×

50 km footprint of CYGNSS, multiple DDMs near the target area are checked for oil-rig evidence. As illustrated in Fig. 11, nine clutter-suppressed DDMs reveal possible oil-rig peaks. Owing to the typical horseshoe-shaped forward-scattering pattern, only the glint region below the white boundary is retained for analysis, while the upper part is discarded using a threshold of 0.2. The excluded area mainly consists of low-amplitude instrument noise and is therefore easy to separate from the useful signal. The boundary is symmetric and follows the horseshoe contour. Subfigures represent epochs 121260–121268, and the red boxes mark the likely oil-rig locations. In most DDMs, the box center corresponds to a stationary ground point associated with the suspected rig. The location becomes less apparent several subfigures because the strongest peak falls outside the horseshoe region. The same processing applied to DDMs from other satellites also yields observable oil-rig peaks.

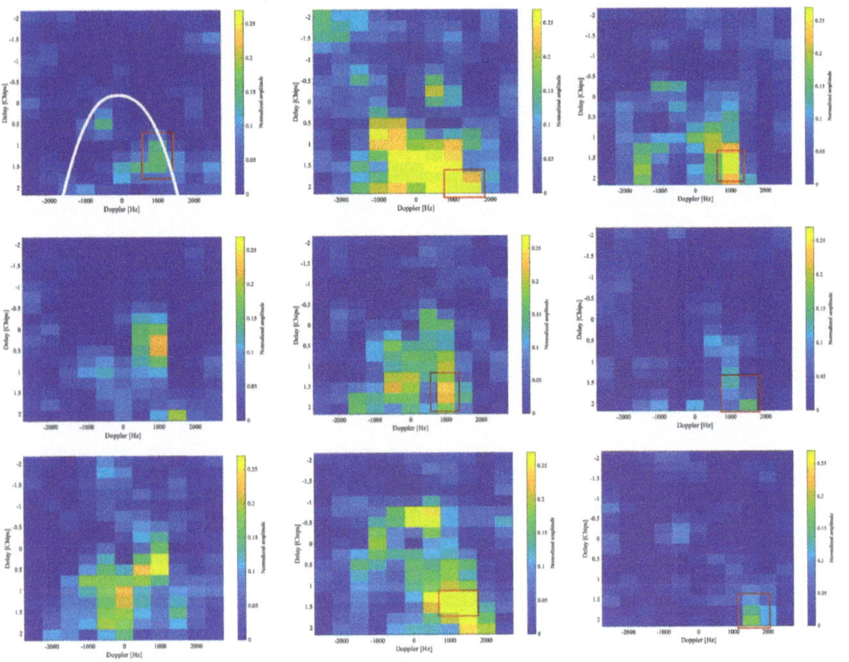

Fig. 11 Detection results from the Cyg02 DDMs after removing sea clutter. The nine subplots correspond to epochs 121260 to 121268. The red boxes enclose regions with elevated power that may be caused by the oil rig

4.2 Multiple Satellites Fusion for Rmoving Location Ambiguity

After estimating and subtracting the sea-clutter DDM of the region of interest from the raw DDMs, the delay-Doppler coordinates of a candidate oil-rig peak in \(\xi_0\) can be obtained. The remaining task is to remove the location ambiguity and identify the actual oil-rig position. To address this problem, we consider two steps to resolve the ambiguity.

(1) **Transforming Satellite Parameters from the Earth-centered Earth-fixed (ECEF) System to the local reference frame (LRF) System**

During localization, the first step is coordinate transformation, since the downloaded CYGNSS data are given in the ECEF coordinate system. Before the oil-rig coordinates can be determined in the LRF system, the satellite and receiver parameters, including positions and velocities, must first be mapped from ECEF to LRF. For convenience, the East-North-Up (ENU) frame is introduced as an intermediate coordinate system. Figure 12a shows the relationship between ECEF and ENU. In this figure, X'-Y'-Z' denotes the ECEF system, and X_{ENU}-Y_{ENU}-Z_{ENU} is the ENU system. η is the longitude of SP, and ϕ is the latitude of SP. Taking the SP location transform from ECEF to ENU as an example, the transform matrix can be written as

$$\begin{bmatrix} S_{x,ENU} \\ S_{y,ENU} \\ S_{y,ENU} \end{bmatrix} = \begin{bmatrix} -\sin\eta & \cos\eta\ 0 & \\ -\sin\phi\cos\eta & -\sin\phi\sin\eta & \cos\phi \\ \cos\phi\cos\eta & \cos\phi\sin\eta & \sin\phi \end{bmatrix} \begin{bmatrix} S_{x,ECEF} \\ S_{y,ECEF} \\ S_{z,ECEF} \end{bmatrix} \tag{37}$$

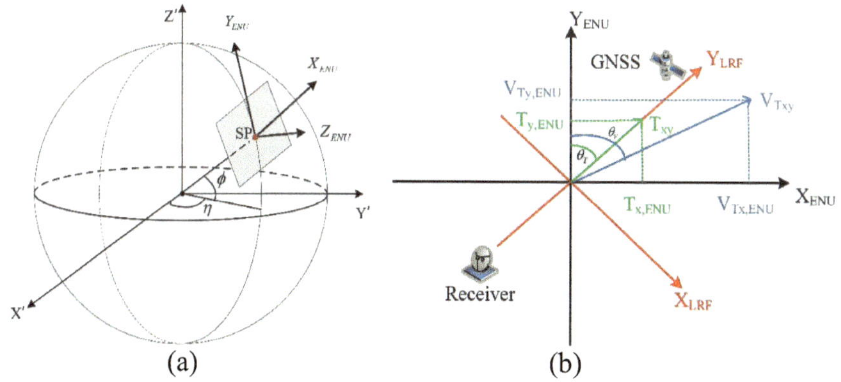

Fig. 12 **a** Geometric relationship between ECEF and ENU. X'-Y'-Z' is the ECEF system; and X_{ENU}-Y_{ENU}-Z_{ENU} represents the ENU system. η is the longitude of SP; and ϕ denotes the latitude of SP. **b** The geometry of ENU and LRF systems in X–Y plane. The included angle between the y-axis of ENU system and that of LRF-system is θ_T

where $\vec{S}_{ecef} = \left[S_{x,ECEF} \; S_{y,ECEF} \; S_{z,ECEF} \right]$ is the location of SP in the ECEF system, and $\vec{S}_{enu} = \left[S_{x,ENU} \; S_{y,ENU} \; S_{y,ENU} \right]$ denotes the location of SP in the ENU system. In accordance with the transform matrix, the locations of GNSS satellite ($\vec{T}_{ENU}$) and receiver ($\vec{R}_{ENU}$) and their velocities ($\vec{V}_{T,ENU}$ and $\vec{V}_{R,ENU}$) in the ENU system can be calculated. After the positions and velocities are transformed into the ENU system, the parameters are further converted to the LRF system. The main difference between the ENU and LRF systems lies in the orientation of the X-Y axes. Taking one GNSS satellite as an example, Fig. 12b illustrates the geometries of the ENU and LRF systems in the X-Y plane. In the LRF system, the Y_{LRF} axis is defined as the projection of the line connecting the GNSS satellite and the receiver. Therefore, the coordinates of the GNSS satellite and receiver in the LRF system can be expressed as:

$$\vec{T} = \left[0 \; \left|\vec{T}_{xy}\right| \; T_{z,ENU} \right] \tag{38}$$

$$\vec{R} = \left[0 \; \left|\vec{R}_{xy}\right| \; R_{z,ENU} \right] \tag{39}$$

where $\vec{T}_{xy} = \left[T_{x,ENU}, T_{y,ENU}\right]$ and $\vec{R}_{xy} = \left[R_{x,ENU}, R_{y,ENU}\right]$. The velocity of GNSS in the LRF system can be written as

$$\overrightarrow{V_T} = \left[\left|\vec{V}_{Txy}\right| \sin(\Delta\theta_T) \; \left|\vec{V}_{Txy}\right| \cos(\Delta\theta_T) \; V_{Tz,ENU} \right] \tag{40}$$

where $\vec{V}_{Txy} = \left[V_{Tx,ENU} \; V_{Ty,ENU} \right]$, $\Delta\theta_T = \theta_V - \theta_T$, $\theta_V = \tan^{-1}\left(\frac{V_{Tx,ENU}}{V_{Ty,ENU}}\right)$, and $\theta_T = \tan^{-1}\left(\frac{T_{x,ENU}}{T_{y,ENU}}\right)$. In a similar manner, the receiver velocity in the LRF coordinate system can be expressed as

$$\overrightarrow{V_R} = \left[\left|\vec{V}_{Rxy}\right| \sin(\Delta\theta_R) \; \left|\vec{V}_{Rxy}\right| \cos(\Delta\theta_R) \; V_{Rz,ENU} \right] \tag{41}$$

where $\vec{V}_{Rxy} = \left[V_{Rx,ENU} \; V_{Ry,ENU} \right]$, $\Delta\theta_T = \theta_{VR} - \theta_T$, and $\theta_{VR} = \tan^{-1}\left(\frac{V_{Rx,ENU}}{V_{Ry,ENU}}\right)$. After the coordinate conversion, the oil rig coordinates can then be determined in the LRF frame.

(2) **Resolving Location Ambiguity by Selecting the Minimum-Distance Cluster in Projected Coordinates**

The second stage aims to eliminate positional ambiguity through multi-satellite fusion within a unified coordinate frame. Although the first stage provides oil rig coordinates in the LRF system, each LRF map still contains one correct position and one incorrect position. In addition, the SP serves as the origin of each LRF map, while the actual locations of these SPs differ on the ground. As a result, the oil rig positions derived from different LRF maps must be projected into the same coordinate system

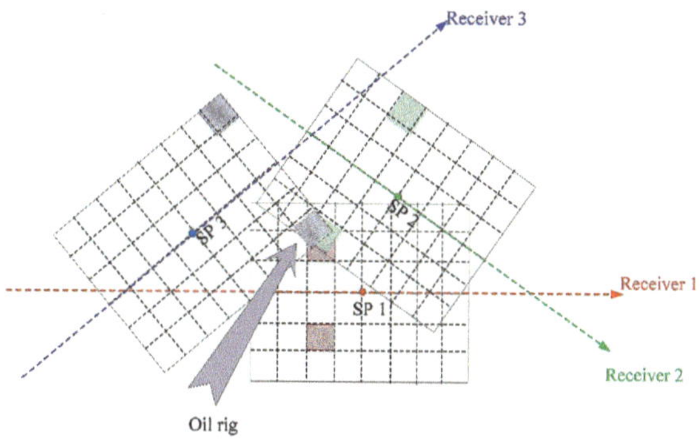

Fig. 13 Multi-satellite fusion-based ambiguity elimination for locations in LRF maps

so that the ambiguity can be removed and the final oil rig location on the Earth can be estimated.

To explain the fusion principle, Fig. 13 presents an example using LRF maps from three receivers. When the receivers' coverage areas overlap around the oil rig, each map produces one valid location and one spurious location. For convenience, only one LRF map from each receiver is included in the fusion process. As illustrated, the projected locations from the different maps gather near the oil rig and form a compact cluster, whose center corresponds to the actual oil rig position.

To integrate the localization results from multiple satellites, all LRF-derived coordinates are first converted into a common reference frame. The conversion process is carried out in four steps. First, the LRF coordinates are mapped to the ENU system. This step is convenient because both the ENU frame and the unified coordinate frame use the north direction as the y-axis. Second, one SP is selected as the origin of the unified system. Third, the positions of the oil rig estimated from the LRF maps are rotated and translated according to the relative geometry between the origin and the remaining SPs. During this procedure, the range and azimuth between the origin and each SP are calculated and then used to derive the corresponding offsets in the x and y directions. Finally, the remaining position ambiguity is removed by identifying the coordinate set with the smallest total distance.

After these operations, the projected coordinates in the unified system can be expressed as:

$$Co_i = \begin{cases} (x_{i,1}, y_{i,1}) \\ (x_{i,2}, y_{i,2}) \end{cases} \tag{42}$$

where $(x_{i,1}, y_{i,1})$ and $(x_{i,2}, y_{i,2})$ denote the two projected coordinates obtained from the i-th LRF map, and i = 1, 2, …, m_i = 1, 2, …, m, with m representing the number of LRF maps used for projection. To locate the oil rig, the total pairwise range within each coordinate set is evaluated, and the set with the minimum cumulative distance is selected as the cluster of true positions. This cluster is given by

$$Clu = min\left(\begin{array}{r}\sum_{i=1}^{m}\left|\left(x_{i,N_0}, y_{i,N_0}\right) - \left(x_{1,N_1}, y_{1,N_1}\right)\right| \\ + \left|\left(x_{i,N_0}, y_{i,N_0}\right) - \left(x_{2,N_2}, y_{2,N_2}\right)\right| \\ \cdots + \left|\left(x_{i,N_0}, y_{i,N_0}\right) - \left(x_{m,N_m}, y_{m,N_m}\right)\right|\end{array}\right) \tag{43}$$

where $N_0, N_1, \ldots, N_m = 1, 2$; and $\left|\left(x_{i,N_0}, y_{i,N_0}\right) - \left(x_{1,N_1}, y_{1,N_1}\right)\right|$ denotes the Euclidean distance between two projected points. The final oil-rig location is then determined from the coordinates contained in this cluster.

$$\overline{x}_{real} = \frac{1}{m}\sum_{i=1}^{m} x_{real,i} \tag{44}$$

$$\overline{y}_{real} = \frac{1}{m}\sum_{i=1}^{m} y_{real,i} \tag{45}$$

where $\left(x_{real,i}, y_{real,i}\right)$ are the coordinates in the minimum cluster.

This section reports the location results in a unified coordinate framework using data collected in July 2020. To determine the oil rig's position on the Earth, the LRF coordinates are first transformed into another coordinate system. The epochs selected for constructing the LRF maps, the angle between the y-axis of the ENU system and that of the LRF system, and the peak coordinates in the DD main domain after sea clutter suppression are summarized in Table 1. For localization, only several epochs from the same track are retained so that the computational burden is reduced and the visualization becomes clearer. Using the parameters listed in Table 1, the LRF positions are then mapped into the unified system. Figure 14 presents the projected locations in a coordinate system whose x-axis points north and y-axis points east. Blue dots denote the projected true LRF positions, and their clustering distance is 20.08, which is the smallest value. Red dots indicate the projected false positions. The yellow dot marks the ground-truth oil rig location, while the purple dot at the origin serves as the SP for epoch 121262. Based on the blue-dot cluster, the estimated oil rig position is (24.86 km – 5.87 km), compared with the true coordinates of (28.7 – 4.75 km). The corresponding errors are 3.64 km in the x direction and 1.4 km in the y direction.

It is reasonable to assume that the oil rig does not move noticeably between July 11 and July 13, 2020, because the operational extent of the rig is smaller than the DDM resolution. Under this assumption, the fusion results from multiple satellites are presented in Fig. 15. The Beibu Gulf is shown in the overview on the left, whereas the right panel focuses on the target area. In the enlarged view, the green arrows specify the directions associated with the white reference dots. Possible rig locations derived from the LRF maps are plotted as yellow rectangles, and the final rig position is marked by the yellow pin at the center of the green circle(left). The region outlined by the red circle contains a compact group of points close to the rig coordinates of 20°48′13.54″ N, 108°52′20.9″ E. Once the LRF maps are fused,

Table 1 Peak coordinates in the DD domain and the included angle between the ENU and LRF coordinate systems

Satellite index	Epoch	Included angle $\theta_T/^o$	Delay/chips	Doppler/Hz
Cyg02	121260	– 56	1.531	1000
	121262	– 56.1	1.276	1000
	121264	– 56.1	1.531	1000
Cyg04	14189	40.5	1.531	– 1000
	14190	40.6	1.531	– 1000
Cyg07	169971	87.9	1.786	1000
	169972	87.8	1.786	1000

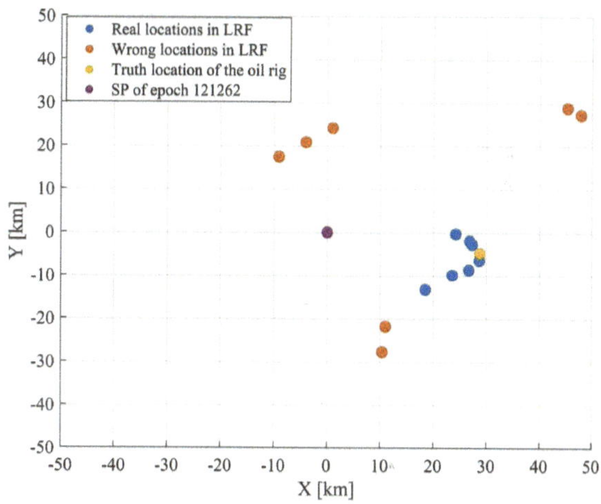

Fig. 14 Projected locations in a common coordinate system, where the y-axis points north and the origin is the reference SP of epoch 121262. The yellow dot indicates the true target position. This localization experiment uses seven DDMs from three satellites (Cyg02, Cyg04, and Cyg07) acquired in July 2020

seven valid locations emerge. In contrast, the points enclosed by the white circles are distributed more widely. This indicates that the maps consistently produce one correct location each, while the erroneous ones are not mutually aligned.

5 Summary

This chapter clarifies the metnods for GNSS-R target detection and location based on different platforms. The text begins by introducing a method for detecting and locating static maritime targets based on ground-based GNSS-R. The approach primarily employs coherent integration time (CIT) to enhance the signal-to-noise ratio (SNR) of reflected signals. Subsequently, to address maneuvering targets, two techniques are proposed: one based on Doppler-range compensation and another utilizing the Fractional Fourier Transform (FrFT). Finally, leveraging the global coverage advantage of GNSS-R, a spaceborne GNSS-R-based method is presented

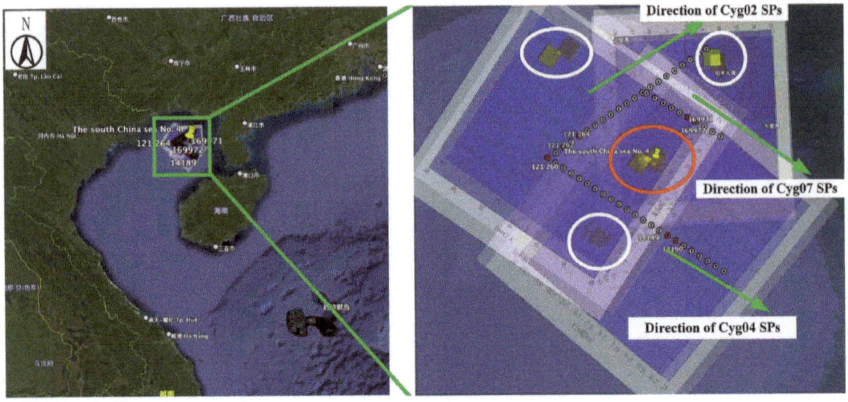

Fig. 15 Oil rig position estimated on Google Earth using fused information from Cyg02, Cyg04, and Cyg07. In the left map the yellow pin at the center of the green circle represents the true oil rig position. In the right map Green arrows denote the directions associated with the SPs. Yellow rectangles mark candidate oil rig positions in the LRF maps. The red circle indicates the cluster of correct LRF positions, while the white circles show mismatched locations

for detecting and locating offshore oil platforms. This method incorporates sea clutter suppression and resolution of location ambiguities. Field experiments have validated that both ground-based and spaceborne GNSS-R demonstrate considerable potential for maritime target detection and positioning.

References

1. Jin S, Camps A, Jia Y, Wang F, Martin-Neira M, Huang F et al (2024) Remote sensing and its applications using GNSS reflected signals: advances and prospects. Satellite Navigation 5(1):19. https://doi.org/10.1186/s43020-024-00139-4
2. Wang F, Yang D, Niu M, Yang L, Zhang B (2022) Sea Ice Detection and Measurement Using Coastal GNSS Reflectometry: Analysis and Demonstration. IEEE J Sel Top Appl Earth Observ Remote Sens 15:136–149. https://doi.org/10.1109/JSTARS.2021.3133431
3. Wang J, Li S (2023) Maritime radar target detection in sea clutter based on CNN with dual-perspective attention. IEEE Geosci Remote Sens Lett 20:1–5. https://doi.org/10.1109/LGRS.2022.3230443
4. Li Y, Yan S, Gong J, Zeng F (2021) SNR enhancement of back scattering signals for bistatic radar based on BeiDou GEO satellites. Remote Sens 13(7). https://doi.org/10.3390/rs13071254
5. Li Y, Yan S, Gong J (2023) Target detection and location by fusing delay-Doppler maps. IEEE Trans Geosci Remote Sens 61:1–14. https://doi.org/10.1109/TGRS.2022.3232164
6. Li C, Huang W, Gleason S (2015) Dual antenna space-based GNSS-R ocean surface mapping: oil slick and tropical cyclone sensing. IEEE J Sel Top Appl Earth Observ Remote Sens 8(1):425–435. https://doi.org/10.1109/JSTARS.2014.2341581
7. AlonsoArroyo A, Zavorotny VU, Camps A (2016) Sea ice detection using GNSS-R data from UK TDS-1. In: 2016 IEEE international geoscience and remote sensing symposium (IGARSS), 10–15 July 2016 2016, pp 2001–2004. https://doi.org/10.1109/IGARSS.2016.7729516

8. Ban W, Zhang K, Yu K, Zheng N, Chen S (2022) Detection of red tide over sea surface using GNSS-R spaceborne observations. IEEE Trans Geosci Remote Sens 60:1–11. https://doi.org/10.1109/TGRS.2022.3144289
9. Li Y, Yan S, Gong J, Zeng F (2021) SNR enhancement of back scattering signals for bistatic radar based on BeiDou GEO satellites. Remote Sens 13(7), https://doi.org/10.3390/rs13071254
10. Li Y, Yan S, Gong J (2022) SNR improvement for maneuvering ship using weak echo under the condition of Beidou GEO satellites. IEEE J Sel Top Appl Earth Observ Remote Sens 15:1256–1271. https://doi.org/10.1109/JSTARS.2022.3142768
11. Xu Z, Baker CJ, Pooni S (2019) Range and Doppler cell migration in wideband automotive radar. IEEE Trans Veh Technol 68(6):5527–5536. https://doi.org/10.1109/TVT.2019.2912852
12. Huang P, Liao G, Yang Z, Xia XG, Ma JT, Ma J (2016) Long-time coherent integration for weak maneuvering target detection and high-order motion parameter estimation based on keystone transform. IEEE Trans Signal Process 64(15):4013–4026. https://doi.org/10.1109/TSP.2016.2558161
13. Song Y, Jin T, Song Y, Zhang Y (2019) An improved radon-fourier transform coherent integration method. IEEE Access 7:138593–138602. https://doi.org/10.1109/ACCESS.2019.2942168
14. Pastina D, Santi F, Pieralice F, Antoniou M, Cherniakov M (2021) Passive radar imaging of ship targets with gnss signals of opportunity. IEEE Trans Geosci Remote Sens 59(3):2627–2642. https://doi.org/10.1109/TGRS.2020.3005306
15. Chen C, Yan S, Mei J (2024) A maritime target location method using geometric semantic constraints based on spaceborne GNSS-R. IEEE Geosci Remote Sens Lett 21:1–5. https://doi.org/10.1109/LGRS.2023.3345732
16. Cheong JW, Southwell BJ, Dempster AG (2019) Blind sea clutter suppression for spaceborne GNSS-R target detection. IEEE J Sel Top Appl Earth Observ Remote Sens 12(12):5373–5378. https://doi.org/10.1109/JSTARS.2019.2956183
17. Southwell BJ, Cheong JW, Dempster AG (2020) A matched filter for spaceborne GNSS-R based sea-target detection. IEEE Trans Geosci Remote Sens 58(8):5922–5931. https://doi.org/10.1109/TGRS.2020.2973142
18. Marchan-Hernandez JF, Camps A, Rodriguez-Alvarez N, Valencia E, Bosch-Lluis X, Ramos-Perez I (2009) An efficient algorithm to the simulation of delay-Doppler maps of reflected global navigation satellite system signals. IEEE Trans Geosci Remote Sens 47(8):2733–2740. https://doi.org/10.1109/TGRS.2009.2014465
19. Schiavulli D, Frappart F, Ramillien G, Darrozes J, Nunziata F, Migliaccio M (2017) Observing sea/ice transition using radar images generated from TechDemoSat-1 delay doppler maps. IEEE Geosci Remote Sens Lett 14(5):734–738. https://doi.org/10.1109/LGRS.2017.2676823
20. Valencia E, Camps A, Marchan-Hernandez JF, Park H, Bosch-Lluis X, Rodriguez-Alvarez N et al (2011) Ocean surface's scattering coefficient retrieval by delay-Doppler map inversion. IEEE Geosci Remote Sens Lett 8(4):750–754. https://doi.org/10.1109/LGRS.2011.2107500

Chapter 15
Maritime Moving Target Detection Algorithms Using GNSS-Based Passive Radar

Zhenyu He

Abstract This chapter presents maritime moving target detection and localization algorithms based on GNSS-based passive radar. To overcome the fundamental limitation of extremely weak target echo energy, a long-time hybrid integration algorithm combining keystone transform (KT) and matched filtering (MF) is proposed. This algorithm effectively compensates for the range cell migration (RCM) and Doppler frequency migration (DFM) induced by the target's bistatic motion, enabling coherent and noncoherent integration over extended observation intervals to accumulate weak target energy for reliable detection. Furthermore, the detection framework is extended to a multistatic configuration, leveraging multiple GNSS satellites to develop a novel joint detection and localization method. By exploiting the similarity of the target's Doppler frequency rate (DFR) across different satellite-receiver pairs, the method projects range-Doppler-DFR data cubes from multiple bistatic channels onto a common Cartesian plane. Subsequent multistatic noncoherent integration in the spatial domain not only enhances the signal-to-noise ratio for improved detection but also simultaneously estimates the target's position and velocity. Experimental analyses using field-measurement maritime datasets validate the effectiveness of the proposed methods. The proposed KT + MF algorithm demonstrates superior detection performance compared to existing techniques, while the joint detection and localization approach achieves accurate target state estimation with significantly reduced computational time, highlighting its practicality for real-world maritime surveillance applications.

Keywords GNSS-based passive radar, Maritime moving target detection, Long-time hybrid integration algorithm, Joint detection and localization method

Z. He (✉)
College of Information Science and Engineering, Hohai University, Changzhou, China
e-mail: zhenyuhe@hhu.edu.cn

K. Yu and N. Zheng (eds.), *Advances in GNSS Reflectometry*, Navigation: Science and Technology 19, https://doi.org/10.1007/978-981-92-2325-1_15

1 Introduction

Global Navigation Satellite Systems (GNSS)-based passive radar exploits the existing satellite constellations as non-cooperative transmitters, offering some advantages, including global coverage, continuous illumination, low cost, small size, covertness, and no electromagnetic pollution. These characteristics make this passive radar a promising technology for maritime surveillance [1, 2], ground vehicle detection [3, 4], and aircraft landing monitoring [5, 6]. However, the fundamental drawback of this passive radar is the extremely weak target echo energy, which significantly limits the radar detection performance. Existing research has shown that extending the observation time can improve the signal-to-noise ratio (SNR) of the target echo for reliable detection, along with a target motion compensation procedure to address the range cell migration (RCM) and Doppler frequency migration (DFM) effects induced by the target's bistatic motion [7–9]. Once the target detection has been declared, the measurements of the bistatic range and Doppler frequency belonging to the moving target can be output by the detection procedure. Then, the detection outputs provided by multiple bistatic channels can be fused to achieve the moving target's position and velocity through some localization and velocity estimation methods [10–13].

To achieve a better detection performance and a more accurate localization and velocity estimation, this chapter focuses on maritime moving target detection and localization algorithms. A long-time hybrid integration method combining Keystone transform (KT) and matched filtering (MF) is proposed to compensate for the RCM and DFM effects and accumulate weak target energy (see Sect. 3). Furthermore, the approach is extended to a multistatic configuration, enabling joint detection and localization of maritime targets by exploiting multiple GNSS satellites (see Sect. 4). To validate the proposed detection and localization algorithms, Experimental analyses are provided (see Sect. 5). Before proceeding further on the details of the proposed algorithms, we first briefly overview the fundamental GNSS-based passive radar system in the following section.

2 GNSS-Based Passive Radar System

This section provides a brief overview of the fundamental GNSS-based passive radar system, including bistatic acquisition geometry, signal synchronization, and signal model.

2.1 Bistatic Acquisition Geometry

Figure 1 illustrates the typical bistatic acquisition geometry of GNSS-based passive radar. The transmitter is a GNSS satellite that sends direct signals toward the area under surveillance. Since the illuminator of opportunity is non-cooperative, the waveforms of the target echoes are unknown and cannot be controlled by the receiver. In this case, the receiver, which is fixed on the sea surface, usually equips a pair of channels: a reference channel and a surveillance channel. The reference channel uses a right-hand circular polarization antenna to capture the direct signals as reference signals, while the surveillance channel adopts a high-gain left-hand circular polarization antenna to record the signals reflected from the moving target on the sea surface. The instantaneous ranges involved are: R_t, the transmitter-to-target range; R_r, the target-to-receiver range; and R_b, the baseline distance between the transmitter and the receiver. Because range compression in the passive radar system is achieved by matched filtering using the reference signal to compensate for the instantaneous delay between transmitter and receiver, the bistatic range history of the moving target over the observation time is defined as:

$$R_{bi}(u) = R_t(u) + R_r(u) - R_b(u) \tag{1}$$

Here, $u \in [-T_o/2, T_o/2]$, and T_o is the entire observation time. Correspondingly, the Doppler frequency history of the moving target is derived as:

$$f_d(u) = -\frac{1}{\lambda} \times \frac{dR_{bi}(u)}{du} \tag{2}$$

Here, λ is the central carrier wavelength. This chapter assumes a maritime target traveling at nearly constant speed, i.e., a non-maneuvering target. This assumption is reasonable for the ships sailing at cruising speeds [9]. Under this condition, the bistatic range history in Eq. (1) can be expanded using the Taylor series around $u = 0$, keeping only the first three terms. Then, the bistatic range history and Doppler frequency history can be, respectively, approximated as [5, 7–9]:

$$R_{bi}(u) = \frac{\lambda}{2} \times f_{dr} \times u^2 + f_{d,0} \times \lambda \times u + R_{bi,0} \tag{3}$$

and

$$f_d(u) = f_{dr} \times u + f_{d,0} \tag{4}$$

Here, f_{dr} is the Doppler frequency rate (DFR), $f_{d,0}$ and $R_{bi,0}$ are the Doppler frequency and bistatic range at the reference time instant, respectively.

2.2 *Signal Synchronization*

In real-world conditions, the direct signal received by the reference channel is obscured by thermal noise, with a SNR that can be as low as −30 dB [14, 15].

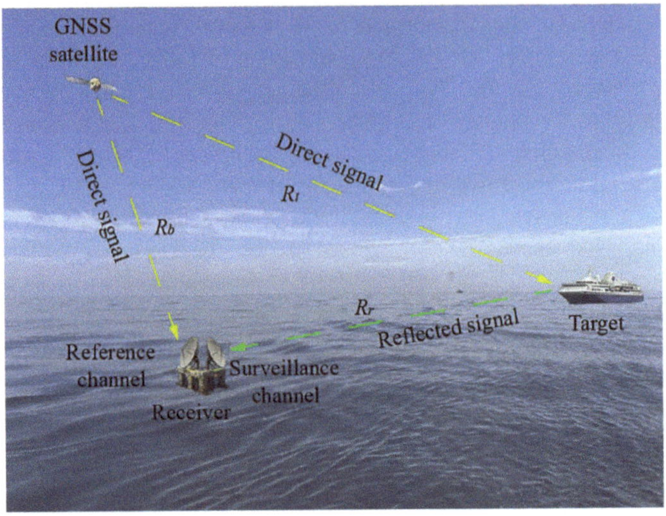

Fig. 1 Bistatic acquisition geometry

Hence, the noisy direct signal is not allowed as the reference signal for subsequent range compression processing. To address this, signal synchronization—a critical step in GNSS-based passive radar—is performed to capture the direct signal and track its observation parameters, comprising time delay, carrier Doppler frequency, carrier phase, navigation data. These observation parameters enable the creation of a noise-free replica of the direct signal. To achieve signal synchronization, the mature software receiver technique [16] emerges as a viable solution, which consists of two primary functions: signal acquisition and signal tracking.

(1) Signal acquisition

The fundamental concept of signal acquisition involves cross-correlating the incoming signal with the locally generated pseudo random noise (PRN) codes across various carrier frequencies, allowing the user to identify direct signals from all visible satellites. To achieve signal acquisition, a block diagram of a parallel code phase search algorithm with high computational efficiency is presented in Fig. 2. The acquisition process must be repeated until all frequency cases generated by the local oscillator have been tested. In this setup, 14 kHz of search band and 150 Hz step size are configured for frequency searching, which is adequate for a moving receiver or a static receiver. After search processing, a threshold is applied to confirm the presence of the direct signal. Figure 3 shows examples of signal acquisition results for PRN1 and PRN2 codes from field measurement data. In Fig. 3a, a prominent correlation peak stands out clearly from the background noise, indicating that the direct signal with the PRN1 code has been successfully acquired. Additionally, the coarse PRN code phase and carrier frequency of this direct signal are determined, serving as inputs for subsequent signal tracking processing. On the contrary, in Fig. 3b, the peaks have comparable magnitudes, indicating that the reference channel of the receiver does not capture the direct signal with the PRN2 code, or that the signal energy is extremely weak to be detected.

(2) Signal tracking

To generate a noise-free replica of the direct signal, signal tracking is implemented, which can provide the accurate observation parameters over the dwell time. The block diagram of the signal tracking algorithm is shown in Fig. 4, which mainly includes two coupled carrier tracking loop and code tracking loop. The carrier tracking loop is responsible for tracking the carrier phase of the direct signal and estimating the Doppler frequency to produce a perfectly aligned local carrier wave replica through a generated cosine wave (I-arm), sine wave (Q-arm), and a carrier loop discriminator.

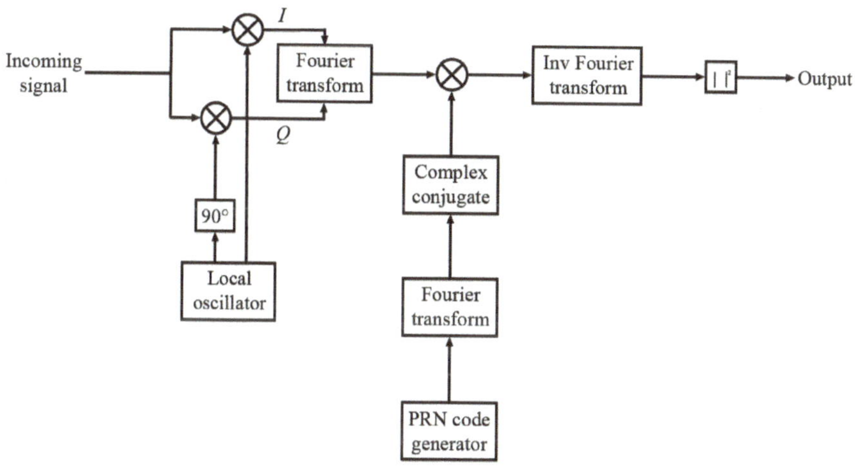

Fig. 2 Block diagram of one parallel code phase search method [11]

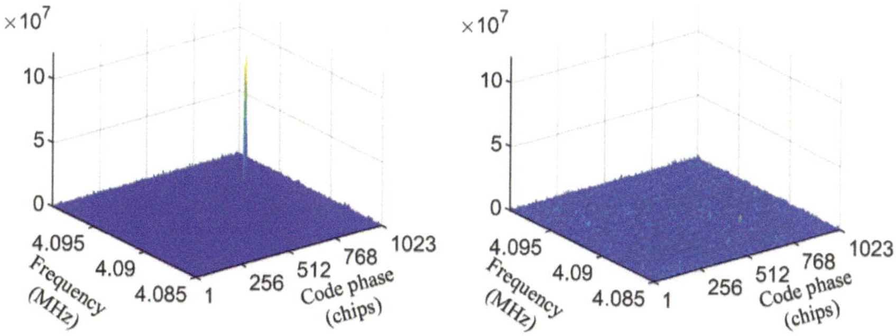

(a) PRN1 (b) PRN2

Fig. 3 Acquisition results from field measurement data

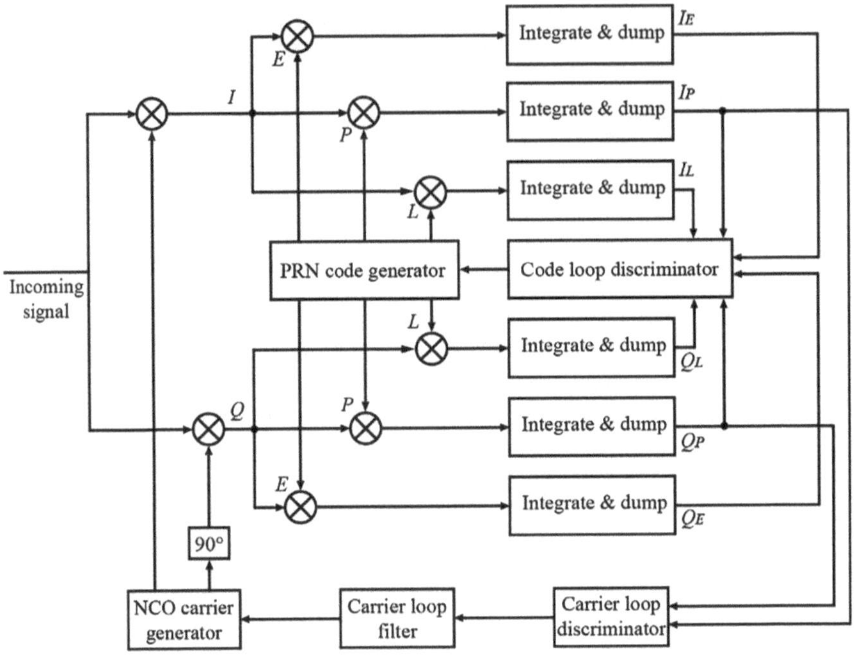

Fig. 4 Block diagram of a signal tracking algorithm [11]

Meanwhile, the code tracking loop is responsible for tracking the code phase of the direct signal to generate a perfectly aligned local code replica through three code replicas spaced $\pm$ 0.5 chip apart, produced by the PRN code generator and code loop discriminator. Once the aligned carrier wave replica and code replica are obtained, the navigation message can be demodulated. Figure 5 presents the tracking results of the direct signal from the same field-measurement data used in Fig. 3. Evidently, after an initial tracking period of about 1 s, both the carrier and code phase errors converge close to zero, as shown in Fig. 5a, b, respectively. Figure 5c displays a portion of the navigation message results during the observation time. The corresponding navigation bit data can be decoded based on the signal magnitude of the navigation message. Thus, the noise-free replica of the direct signal can be easily generated by combining the aligned carrier wave replica, code replica, and the navigation bits.

2.3 Signal Model

Because the GNSS signal is a continuous wave and cannot be directly applied to radar applications, the one-dimensional GNSS signal should be transformed into a two-dimensional radar format comprising fast-time and slow-time domains. This transformation relies on the equivalent pulse repetition interval (PRI) that matches

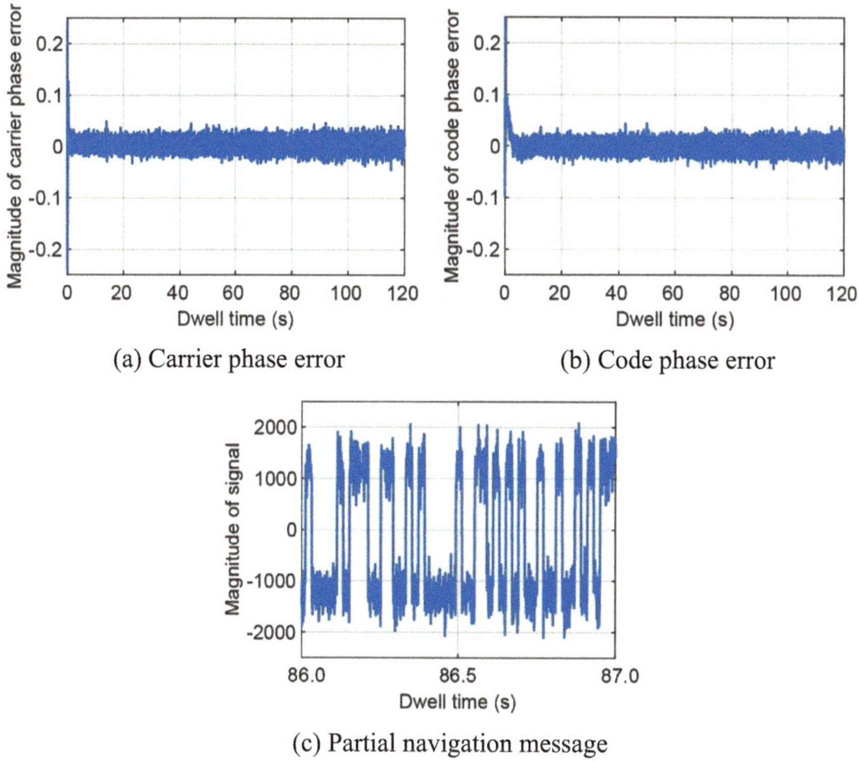

Fig. 5 Tracking results of the direct signal with PRN1 code

the ranging code length of 1 ms [7, 9]. After quadrature demodulation and ignoring constant phase and amplitude terms, the simplified form of the noise-free replica of the direct signal with a two-dimensional radar format can be written as:

$$\begin{aligned} S_{\mathrm{ref}}(t,u) &= C[t-\tau_{\mathrm{ref}}(u)-\tau_{\mathrm{e}}(u)] \times D[t-\tau_{\mathrm{ref}}(u)-\tau_{\mathrm{e}}(u)] \\ &\times \exp\{\mathrm{j}[2\pi f_{\mathrm{ref}}(u)t+\varphi_{\mathrm{ref}}(u)+\varphi_{\mathrm{e}}(u)]\} \end{aligned} \tag{5}$$

Here, $t \in [0, \mathrm{PRI}]$ is the fast-time, $u \in [-T_o/2, T_o/2]$ is the slow-time, $C[\cdot]$ is the ranging code sequence, $D[\cdot]$ is the navigation data, $\tau_{\mathrm{ref}}(u) = R_b(u)/c$, $f_{\mathrm{ref}}(u)$, and $\varphi_{\mathrm{ref}}(u) = -2\pi R_b(u)/\lambda$ are code phase delay, carrier frequency, and carrier phase term of the direct signal, respectively, c is the speed of light, τ_e and φ_e are the code phase and carrier phase errors, respectively, induced by the atmospheric delays and the receiver errors.

The signal data gathered by the surveillance channel of the receiver are complex in the real scenario, including the intruded direct signal from the sidelobes of the antenna, sea clutter, background noise, and the target echo. Because of the restricted GNSS signal power and the long dwell time, which functions as a whitening filter with respect to the background distribution, this radar system is primarily noise-limited rather than clutter-limited [9]. In this chapter, only the target echo and background noise are taken into account. Assuming a single point-like target, the target echo with a two-dimensional radar format can be expressed as:

$$\begin{aligned} S_{\text{tar}}(t,u) = C[t - \tau_{\text{tar}}(u) - \tau_{\text{e}}(u)] \times D[t - \tau_{\text{tar}}(u) - \tau_{\text{e}}(u)] \\ \times \exp\{\mathrm{j}[2\pi f_{\text{tar}}(u)t + \varphi_{\text{tar}}(u) + \varphi_{\text{e}}(u)]\} \end{aligned} \tag{6}$$

Here, $\tau_{\text{tar}}(u) = [R_{\text{t}}(u) + R_{\text{r}}(u)]/c, f_{\text{tar}}(u)$, and $\varphi_{\text{tar}}(u) = -2\pi[R_{\text{t}}(u) + R_{\text{r}}(u)]/\lambda$ are code phase delay, carrier frequency, and carrier phase term of the target echo, respectively. From Eqs. (5) and (6), it can be observed that the code phase and carrier phase errors for both the direct signal and the target echo are identical. This is because the target is typically located within a few kilometers of the receiver, causing the atmospheric delay errors for both signals to be nearly the same. Additionally, since the receiver's reference and surveillance channels share a local oscillator, the receiver's clock bias is consistent across both channels. Consequently, by cross-correlating the target echo from Eq. (6) with the direct signal from Eq. (5) in the fast-time domain for each slow-time sample, we cannot only measure the range difference between the two signals—known as the bistatic range—but also eliminate the aforementioned error terms. This process is referred to as range compression. The mathematical expression for the range-compressed target echo is given by:

$$\begin{aligned} S_{\text{rc}}(r,u) = \int_{t=0}^{t=\text{PRI}} S_{\text{tar}}(t,u) \times S^*_{\text{ref}}(t-\tau,u)dt \\ = \int_{t=0}^{t=\text{PRI}} C[t - \tau_{\text{tar}}(u) - \tau_{\text{e}}(u)] \times C[t - \tau_{\text{ref}}(u) - \tau_{\text{e}}(u) - \tau] \\ \times D[t - \tau_{\text{tar}}(u) - \tau_{\text{e}}(u)] \times D[t - \tau_{\text{ref}}(u) - \tau_{\text{e}}(u)] \\ \times \exp\left\{\mathrm{j}2\pi\left[\Delta f(u)t - \frac{R_{\text{bi}}(u)}{\lambda}\right]\right\}dt \end{aligned} \tag{7}$$

Here, $r \in [0, \tau \times c]$ is the range cell, and τ is the code phase offset. Here are some clarifications regarding Eq. (7):

(1) The product of the navigation codes from the two signals in Eq. (7) equals one because the navigation message interval is 20 ms. This corresponds to a minimum distance difference of 6000 km needed to cause a phase shift [17], which is far greater than the actual distance between the target and receiver.
(2) The term $\Delta f(u)$ in Eq. (7) represents the carrier frequency difference between the two signals. This difference can be neglected since the Doppler frequency caused by a slow-moving target relative to a fast-moving satellite during the PRI is very small [9].
(3) The computation of Eq. (7) can be implemented in the frequency domain using the convolution theorem, which allows for greater computational efficiency.

Finally, Eq. (7) can be derived as:

$$S_{\text{rc}}(r,u) = \Lambda[r - R_{\text{bi}}(u)] \times \exp\left[-\mathrm{j}2\pi\frac{R_{\text{bi}}(u)}{\lambda}\right] \tag{8}$$

Here, $\Lambda[\cdot]$ is the envelope of the cross-correlation function.

3 Moving Target Detection

This section begins with a theoretical analysis of the detection performance of GNSS-based passive radar. Following that, to tackle the problem of weak GNSS echoes, a long-time hybrid integration algorithm is proposed for maritime moving target detection.

3.1 Detection Performance Analysis

According to the bistatic radar equation [18], the SNR of the target echo output from the surveillance channel of the receiver can be derived as:

$$P_{\mathrm{SNR}} = \rho_{\mathrm{pfd}} \frac{\sigma_{\mathrm{rcs}} \lambda^2 G_{\mathrm{s}}}{(4\pi R_{\mathrm{r}})^2 K_{\mathrm{b}} T_{\mathrm{n}} B_{\mathrm{n}} F_{\mathrm{n}} L_{\mathrm{s}}} \tag{9}$$

Here, ρ_{pfd} is the GNSS signal power flux density on the earth's surface, σ_{rcs} is the bistatic radar cross section (RCS) of the moving target, G_{s} is the gain of the surveillance channel antenna, K_{b} is the Boltzmann constant, T_{n} is the receiver noise temperature, B_{n} is the receiver noise bandwidth, F_{n} is the noise factor, and L_{s} is the system loss. Equation (9) shows that the SNR of the target echo depends on the parameters of the GNSS signal, the receiving system, as well as the numerical values of the bistatic radar cross section (RCS) and the distance between the target and receiver. According to electromagnetic theory, the bistatic RCS is influenced by the configuration of the bistatic acquisition geometry. As the bistatic angle decreases, the bistatic RCS tends to increase. The bistatic RCS reaches its maximum value when the GNSS satellite, target, and receiver are aligned in a straight line, i.e., the bistatic angle is 0°. Figure 6 plots the SNR curves of the target echoes with various bistatic RCS values and target-to-receiver ranges.

In Fig. 6, the SNR of the GNSS signal reflected from the target is extremely low. For instance, for a target situated 5 km from the receiver with a bistatic RCS of 1000 m^2, the SNR is only −62.99 dB, indicating that the target echo is submerged in background noise. To achieve a detection probability of 90% and a false alarm rate of 10^{-3}, the SNR must be increased to 10.7 dB, which requires a processing gain of at least 73.69 dB. One way to enhance the strength of the reflected signal reception is by increasing the gain of the surveillance channel antenna. However, this increase narrows the antenna's beamwidth, thereby limiting the detection area. Alternatively, extending the observation time can improve the energy of weak target echoes through signal processing techniques, effectively "trading time for energy". A commonly used method is hybrid coherent integration and noncoherent integration,

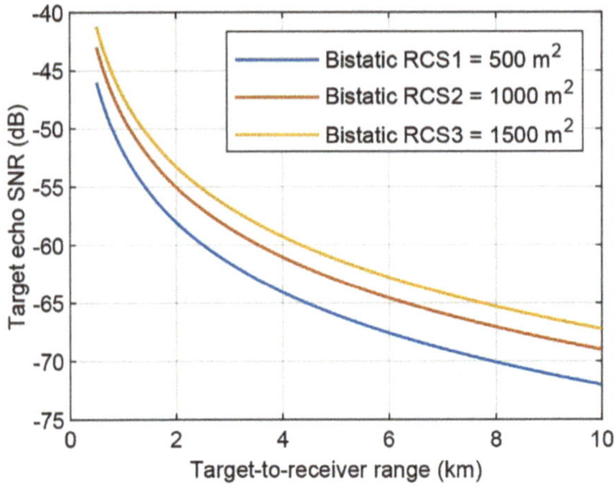

Fig. 6 The SNR curves of the target echoes

which divides the total observation time into several shorter intervals (named frames). Within each frame, the target's scattering characteristics remain relatively stable and can be treated as coherent [7, 8, 19]. Coherent integration is applied inside each frame, while noncoherent integration is performed across frames. The coherent integration gain within a frame is calculated as:

$$G_{\text{CI}} = T_{\text{f}} \times B_{\text{r}} \tag{10}$$

Here, T_{f} is the length of the frame, and B_{r} is the exploited signal bandwidth. For the square-law detector, the noncoherent integration gain can be estimated using an empirical formula [20], presented in decibel (dB) form:

$$\begin{aligned} G_{\text{NCI}} = 6.79 \times (1 + 0.253 \times p_{\text{d}}) \times \left[1 + \lg\left(\frac{1}{p_{\text{fa}}}\right)/46.6\right] \\ \times \lg(N_{\text{f}}) \times \left[1 - 0.14 \times \lg(N_{\text{f}}) + 0.0183 \times \lg^2(N_{\text{f}})\right] \end{aligned} \tag{11}$$

Here, p_{d} is the detection probability, p_{fa} is the false alarm rate, and $N_{\text{f}} = \lfloor T_{\text{o}}/T_{\text{f}} \rfloor$ is the number of frames. Following both coherent and noncoherent integrations, the SNR of the target echo can be enhanced to a level that allows for reliable detection, represented in dB form:

$$P'_{\text{SNR}} = 10\lg(P_{\text{SNR}}) + 10\lg(G_{\text{CI}}) + G_{\text{NCI}} \geq 10.7 \text{ dB} \tag{12}$$

Assuming the target is located within the 3 dB beamwidth of the surveillance channel antenna during the observation time, by inserting Eq. (9) into Eq. (12) and rearranging the terms, the maximum detection range can be derived and is given by:

$$R_{\text{det}} = \sqrt{\rho_{\text{pfd}} \frac{\sigma_{\text{rcs}} \lambda^2 G_{\text{s}} G_{\text{CI}} \times 10^{\frac{G_{\text{NCI}}}{10}}}{(4\pi)^2 K_{\text{b}} T_{\text{n}} B_{\text{n}} F_{\text{n}} L_{\text{s}} \times 10^{\frac{P'_{\text{SNR}}}{10}}}} \tag{13}$$

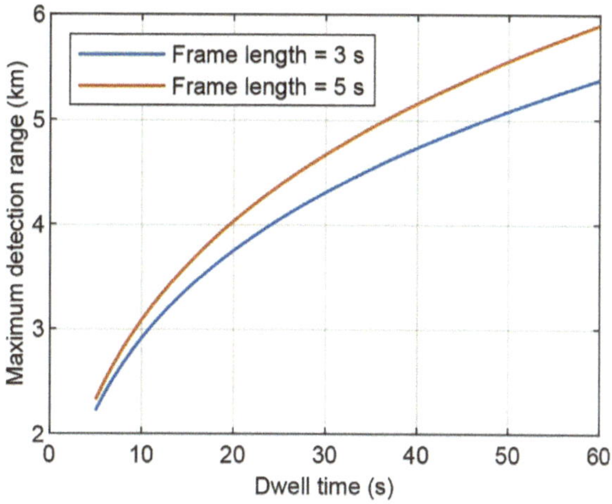

Fig. 7 Maximum detection distance curves

Figure 7 plots the maximum detection distance with various frame lengths and dwell intervals. It can be seen that when the total dwell time reaches 37 s and the frame length is 3 s, the maximum detection distance exceeds 4.6 km. By keeping the total dwell time constant and increasing the frame length to 5 s, the maximum detection distance can be extended to 5 km. Further increasing the total dwell time can lead to additional improvements in the maximum detection distance. Therefore, extending the dwell time is a viable approach to enhance the maximum detection range of GNSS passive radar using a combination of coherent and noncoherent integration processing.

3.2 *Long-Time Hybrid Integration Algorithm*

Although prolonging the observation time can theoretically increase the maximum detection distance, the effects of the RCM and DFM induced by the target's bistatic motion must be considered, as they cause the target echo energy to spread and thus reduce the detection range. Furthermore, the target's scattering characteristics cannot be assumed to remain coherent throughout the entire observation time. As a result, existing fully coherent integration approaches published in literatures [21–23] are not applicable. To tackle RCM and DFM issues and to accumulate weak target echo energy, this subsection presents a long-time hybrid integration algorithm that combines the KT and the MF [24].

(1) RCM compensation

As a non-maneuvering target is considered in this chapter, substituting Eq. (3) into Eq. (8), the range-compressed target echo can be recast as:

$$S_{\rm rc}(r,u) = \Lambda\left[r - \left(\tfrac{1}{2}\lambda f_{\rm dr}u^2 + f_{\rm d,0}\lambda u + R_{\rm bi,0}\right)\right] \times \exp\left[-{\rm j}2\pi\left(\tfrac{1}{2}f_{\rm dr}u^2 + f_{\rm d,0}u + \tfrac{R_{\rm bi,0}}{\lambda}\right)\right] \tag{14}$$

In Eq. (14), the RCM effect arises when the change in the envelope of the cross-correlation function exceeds half of the bistatic range resolution unit of a GNSS-based passive radar during the dwell time. The criteria for the occurrence of first-order and second-order RCM can be expressed as:

$$\begin{cases} \Delta R_{\rm FRCM} = \left|f_{\rm d,0}\lambda u\right| > \frac{1}{2}\Delta r \\ \Delta R_{\rm SRCM} = \left|\frac{1}{2}\lambda f_{\rm dr}u^2\right| > \frac{1}{2}\Delta r \end{cases} \tag{15}$$

Here, Δr is the range resolution unit.

First, Eq. (14) is converted into the range-frequency and slow-time domain by applying the Fourier transform (FT) in the range domain for each slow-time sample, which can be represented as:

$$S_{\rm rc}(f_{\rm r},u) = \Omega(f_{\rm r}) \times \exp\left(-{\rm j}\pi\tfrac{f_{\rm r}+f_{\rm c}}{f_{\rm c}}f_{\rm dr}u^2\right) \times \exp\left(-{\rm j}2\pi\tfrac{f_{\rm r}+f_{\rm c}}{f_{\rm c}}f_{\rm d,0}u\right) \times \exp\left(-{\rm j}2\pi\tfrac{f_{\rm r}+f_{\rm c}}{c}R_{\rm bi,0}\right) \tag{16}$$

Here, $f_{\rm r}$ is the range-frequency, $\Omega(f_{\rm r})$ is the FT of $\Lambda(f_{\rm r})$, and $f_{\rm c} = c/\lambda$ is the central carrier frequency. To eliminate the first-order and second-order RCMs, $f_{\rm r}$ must be decoupled with u-term and u^2-term in Eq. (16). The KT can remove the first-order RCM through a scaling operation without requiring prior knowledge of the target velocity. The scaling operation is written as:

$$u = \frac{f_{\rm c}}{f_{\rm r} + f_{\rm c}}u_{\rm s} \tag{17}$$

Here, $u_{\rm s}$ is the scaled slow time. The scaling operation can be implemented by the chirp-z transform [25]. Then, because the narrow bandwidth of GNSS ranging code leads to $f_{\rm c} \gg f_{\rm r}$, the approximation is given by:

$$\frac{f_{\rm c}}{f_{\rm r} + f_{\rm c}} \approx 1 - \frac{f_{\rm r}}{f_{\rm c}} \tag{18}$$

Substituting Eq. (17) and Eq. (18) into Eq. (16), Eq. (16) can be recast as:

$$S_{\rm rc}(f_{\rm r},u_{\rm s}) = \Omega(f_{\rm r}) \times \exp\left[-{\rm j}2\pi\left(\tfrac{1}{2}f_{\rm dr}u_{\rm s}^2 + f_{\rm d,0}u_{\rm s} + \tfrac{R_{\rm bi,0}}{\lambda}\right)\right] \times \exp\left(-{\rm j}2\pi\tfrac{f_{\rm r}}{c}R_{\rm bi,0}\right) \times \exp\left({\rm j}\pi\tfrac{f_{\rm r}}{f_{\rm c}}f_{\rm dr}u_{\rm s}^2\right) \tag{19}$$

As shown in Eq. (19), the coupling between $f_{\rm r}$ and $u_{\rm s}$-term has been eliminated. To remove the coupling between $f_{\rm r}$ and $u_{\rm s}^2$-term in the third exponential phase, a matched filter is generated:

$$H_{\mathrm{SRCM}}\left(f_{\mathrm{r}}, u_{\mathrm{s}}; f_{\mathrm{dr}}^{*}\right) = \exp\left(-\mathrm{j}\pi \frac{f_{\mathrm{r}}}{f_{\mathrm{c}}} f_{\mathrm{dr}}^{*} u_{\mathrm{s}}^{2}\right) \tag{20}$$

Here, f_{dr}^{*} is the DFR value under test. The limits of the spanned search interval $\left[-f_{\mathrm{dr,max}}^{*}, f_{\mathrm{dr,max}}^{*}\right]$ can be set based on the maximum DFR considered, which corresponds to a target at the closest distance moving at the highest velocity [9]. When f_{dr}^{*} exactly matches the actual unknown f_{dr}, the second-order RCM can be eliminated by multiplying Eq. (19) by Eq. (20). Following the inverse FT with respect to f_{r}, the range-compressed target echo with the RCM compensation can be expressed as:

$$S_{\mathrm{rc}}\left(r, u_{\mathrm{s}}; f_{\mathrm{dr}}^{*}\right) = \Lambda\left(r - R_{\mathrm{bi},0}; f_{\mathrm{dr}}^{*}\right) \times \exp\left[-\mathrm{j}2\pi\left(\frac{1}{2} f_{\mathrm{dr}} u_{\mathrm{s}}^{2} + f_{\mathrm{d},0} u_{\mathrm{s}} + \frac{R_{\mathrm{bi},0}}{\lambda}\right)\right] \tag{21}$$

(2) Multi-frame DFM compensation

The range-compressed target echo in Eq. (21) is segmented into consecutive N_{f} frames with respect to u_{s}. By restricting each frame to a duration of a few seconds, typically 2 to 3 s [7, 8, 19, 26], target's complex reflectivity can be assumed to remain constant inside the frame. The mathematical expression of the range-compressed target echo for the m-th frame ($m = -N_{\mathrm{f}}/2, \ldots, 0, \ldots, N_{\mathrm{f}}/2 - 1$) is given by:

$$\begin{aligned} S_{\mathrm{rc},m}\left(r, u_{\mathrm{s}}; f_{\mathrm{dr}}^{*}\right) &= \mathrm{rect}\left(\tfrac{u_{\mathrm{s}} - mT_{\mathrm{f}}}{T_{\mathrm{f}}}\right) \times \Lambda\left(r - R_{\mathrm{bi},0}; f_{\mathrm{dr}}^{*}\right) \\ &\times \exp\left[-\mathrm{j}2\pi\left(\tfrac{1}{2} f_{\mathrm{dr}} u_{\mathrm{s}}^{2} + f_{\mathrm{d},0} u_{\mathrm{s}} + \tfrac{R_{\mathrm{bi},0}}{\lambda}\right)\right] \end{aligned} \tag{22}$$

Here, $\mathrm{rect}(\cdot)$ is a window function that is zero outside the interval $[-1/2, 1/2]$ and unity inside. In Eq. (22), the quadratic exponential phase term gives rise to the Doppler shifts both inside each frame (intraframe) and among different frames (interframe). When the Doppler shift exceeds half of the Doppler resolution, the DFM effect arises. According to Eq. (4), the conditions resulting from the intraframe DFM and the interframe DFM can be written as:

$$\begin{cases} \delta D_{m} = \left|f_{\mathrm{dr}}(u_{s} - u_{m})\right| > \frac{1}{2}\Delta d \\ \Delta D_{m} = \left|f_{\mathrm{dr}}(m - l)T_{\mathrm{f}}\right| > \frac{1}{2}\Delta d \end{cases} \tag{23}$$

Here, u_{m} is the reference time instant of the m-th frame, l is the index of the reference frame, and Δd is the Doppler resolution unit.

To eliminate the intraframe DFM, the intraframe phase compensation term for the m-th frame is defined as:

$$\delta\varphi_{m}\left(u_{\mathrm{s}}; f_{\mathrm{dr}}^{*}\right) = \mathrm{rect}\left(\frac{u_{\mathrm{s}} - mT_{\mathrm{f}}}{T_{\mathrm{f}}}\right) \times \frac{1}{2} f_{\mathrm{dr}}^{*} (u_{\mathrm{s}} - u_{m})^{2} \tag{24}$$

Likewise, to remove the interframe DFM between the m-th frame and the l-th frame, the interframe compensation term for the m-th frame is defined as:

$$\Delta\varphi_m\left(u_s; f_{dr}^*\right) = \text{rect}\left(\frac{u_s - mT_f}{T_f}\right) \times f_{dr}^*(u_s - u_m)(m - l)T_f \tag{25}$$

For convenience, u_m and l can be set to the starting time instant of the m-th frame and the zeroth frame, respectively. Combing Eqs. (24) and (25), the matched filer to address the intraframe and interframe DFMs for the m-th frame can be constructed:

$$\begin{gathered} H_{\text{DFM},m}\left(r, u_s; f_{dr}^*\right) = \text{rect}\left(\frac{u_s - mT_f}{T_f}\right) \\ \times \exp\left\{-j2\pi f_{dr}^*\left(\tfrac{1}{2}(u_s - u_m)^2 + (u_s - u_m)(m - l)T_f\right)\right\} \end{gathered} \tag{26}$$

when f_{dr}^* exactly matches the unknown f_{dr}, the intraframe and interframe DFM compensations can be completed by multiplying Eq. (22) by Eq. (26) for each range sampling.

(3) Hybrid coherent and noncoherent integration

After compensating for the first-order and second-order RCMs as well as the intraframe and interframe DFMs for each frame, intraframe coherent integration is performed using the FT in the slow-time domain for each range sample. It is important to mention that the FT and inverse FT operations described above can be substituted with the fast FT (FFT) and inverse FFT to enhance computational efficiency. The range-Doppler (RD) map for the m-th frame is then derived as:

$$S_{\text{RD},m}\left(r, f_{u_s}; f_{dr}^*\right) = \Lambda\left(r - R_{\text{bi},0}; f_{dr}^*\right) \times \sin\text{c}\left[\pi\frac{f_{u_s} - f_{u_m,l}}{T_f}; f_{dr}^*\right] \tag{27}$$

Here, f_{u_s} is the Doppler frequency, and $f_{u_m,l}$ is the aligned Doppler frequency at the reference time instant of the reference frame. Next, interframe noncoherent integration is performed to cope with the variations in the target's complex reflectivity [7, 8] and to further improve the integration gain. The resulting noncoherent-integrated RD map is represented as:

$$S_{\text{NRD}}\left(r, f_{u_s}; f_{dr}^*\right) = \frac{1}{N_f}\sum_{m=1}^{m=N_f}\left|S_{\text{RD},m}\left(r, f_{u_s}; f_{dr}^*\right)\right|^2 \tag{28}$$

Since the actual target's DFR is unknown prior to detection, Eq. (28) corresponds to a three-dimensional range-Doppler-DFR (RDD) data cube. This cube consists of a series of noncoherent-integrated RD maps, each corresponding to a different tested DFR value. Evidently, when one of these tested DFR values matches the true DFR, the effects of RCM and DFM are compensated, allowing the associated RD map to provide the maximum integration gain, thereby enhancing the target echo's SNR. Consequently, the RD map from the data cube with the highest peak value can be selected as input for a detection threshold method, such as the constant false alarm rate (CFAR) technique [27], to decide whether the target is present.

4 Moving Target Joint Detection and Localization

This section first analyzes the DFR of a non-maneuvering target in the multistatic geometric configuration. Building on this analysis, the long-time hybrid integration algorithm proposed in Sect. 3 is extended from a bistatic geometric configuration to a multistatic one, allowing for the joint detection and localization of moving maritime targets.

4.1 DFR Analysis in a Multistatic Geometric Configuration

Figure 8 illustrates the general multistatic acquisition geometry of GNSS-based passive radar, where K GNSS satellites serve as the illuminators of opportunity, and a single receiver collects K direct signals and K echoes reflected from a moving target. In Fig. 8, a three-dimensional Cartesian coordinate system is established with the receiver's position as the reference point, and the X-Y plane overlaps with the sea surface. The position vectors of the k-th satellite ($k = 1, 2, \ldots, K$), the moving target, and the receiver are denoted as $\mathbf{P}_{\mathrm{s}}^{k}(u) = \left[x_{\mathrm{s}}^{k}(u), y_{\mathrm{s}}^{k}(u), z_{\mathrm{s}}^{k}(u)\right]^{\mathrm{T}}$, $\mathbf{P}_{\mathrm{t}}(u) = [x_{\mathrm{t}}(u), y_{\mathrm{t}}(u), 0]^{\mathrm{T}}$, and $\mathbf{P}_{\mathrm{r}}(u) = [x_{\mathrm{r}}(u), y_{\mathrm{r}}(u), z_{\mathrm{r}}(u)]^{\mathrm{T}}$ at time instant u within the interval $[-T_{\mathrm{o}}/2, T_{\mathrm{o}}/2]$, respectively. Since the receiver is fixed on the sea surface, the velocity vectors of the k-th satellite and the moving target are denoted as $\mathbf{V}_{\mathrm{s}}^{k}(u) = \left[v_{\mathrm{x,s}}^{k}(u), v_{\mathrm{y,s}}^{k}(u), v_{\mathrm{z,s}}^{k}(u)\right]^{\mathrm{T}}$ and $\mathbf{V}_{\mathrm{t}}(u) = [v_{\mathrm{x,\,t}}(u), v_{\mathrm{y,\,t}}(u), 0]^{\mathrm{T}}$ at time instant u within the interval $[-T_{\mathrm{o}}/2, T_{\mathrm{o}}/2]$, respectively.

According to Eq. (1) and Eq. (2), the target's bistatic range and Doppler frequency in the k-th bistatic acquisition geometry at time instant u can be, respectively, written

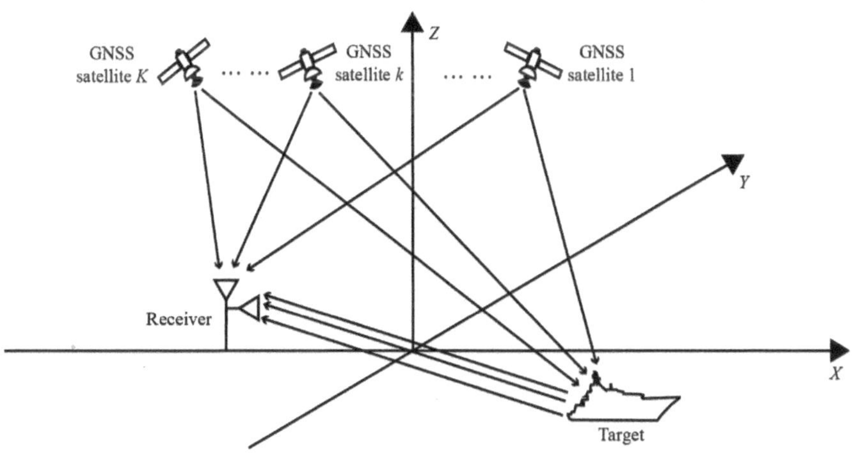

Fig. 8 Multistatic acquisition geometry

as:

$$R_{\mathrm{bi}}^{k} = \left\|\mathbf{P}_{\mathrm{s}}^{k} - \mathbf{P}_{\mathrm{t}}\right\| + \|\mathbf{P}_{\mathrm{t}} - \mathbf{P}_{\mathrm{r}}\| - \left\|\mathbf{P}_{\mathrm{s}}^{k} - \mathbf{P}_{\mathrm{r}}\right\| \tag{29}$$

and

$$f_{\mathrm{d}}^{k} = -\frac{1}{\lambda^{k}}\left[\frac{(\mathbf{P}_{\mathrm{s}}^{k} - \mathbf{P}_{\mathrm{t}})^{\mathrm{T}}(\mathbf{V}_{\mathrm{s}}^{k} - \mathbf{V}_{\mathrm{t}})}{\left\|\mathbf{P}_{\mathrm{s}}^{k} - \mathbf{P}_{\mathrm{t}}\right\|} + \frac{(\mathbf{P}_{\mathrm{t}} - \mathbf{P}_{\mathrm{r}})^{\mathrm{T}}\mathbf{V}_{\mathrm{t}}}{\|\mathbf{P}_{\mathrm{t}} - \mathbf{P}_{\mathrm{r}}\|} - \frac{(\mathbf{P}_{\mathrm{s}}^{k} - \mathbf{P}_{\mathrm{r}})^{\mathrm{T}}\mathbf{V}_{\mathrm{s}}^{k}}{\left\|\mathbf{P}_{\mathrm{s}}^{k} - \mathbf{P}_{\mathrm{r}}\right\|}\right] \tag{30}$$

Here, $\|\cdot\|$ is Euclidian norm, λ^k is the central carrier wavelength of the k-th satellite. As the distance between the satellite and the target is typically more than 20,000 km, the satellite's movement can be considered uniform throughout the observation time. Consequently, the DFR of the moving target in the k-th bistatic acquisition geometry at time instant u can be derived as:

$$\begin{aligned} f_{\mathrm{dr}}^{k} = -\frac{1}{\lambda^{k}}\Bigg\{ & \frac{\|\mathbf{V}_{\mathrm{s}}^{k}-\mathbf{V}_{\mathrm{t}}\|^{2}}{\|\mathbf{P}_{\mathrm{s}}^{k}-\mathbf{P}_{\mathrm{t}}\|} - \frac{\left[(\mathbf{P}_{\mathrm{s}}^{k}-\mathbf{P}_{\mathrm{t}})^{\mathrm{T}}(\mathbf{V}_{\mathrm{s}}^{k}-\mathbf{V}_{\mathrm{t}})\right]^{2}}{\|\mathbf{P}_{\mathrm{s}}^{k}-\mathbf{P}_{\mathrm{t}}\|^{3}} + \frac{\|\mathbf{V}_{\mathrm{t}}\|^{2}}{\|\mathbf{P}_{\mathrm{t}}-\mathbf{P}_{\mathrm{r}}\|} - \frac{[(\mathbf{P}_{\mathrm{t}}-\mathbf{P}_{\mathrm{r}})^{\mathrm{T}}\mathbf{V}_{\mathrm{t}}]^{2}}{\|\mathbf{P}_{\mathrm{t}}-\mathbf{P}_{\mathrm{r}}\|^{3}} - \\ & \frac{\|\mathbf{V}_{\mathrm{s}}^{k}\|^{2}}{\|\mathbf{P}_{\mathrm{s}}^{k}-\mathbf{P}_{\mathrm{r}}\|} + \frac{\left[(\mathbf{P}_{\mathrm{s}}^{k}-\mathbf{P}_{\mathrm{r}})^{\mathrm{T}}\mathbf{V}_{\mathrm{s}}^{k}\right]^{2}}{\|\mathbf{P}_{\mathrm{s}}^{k}-\mathbf{P}_{\mathrm{r}}\|^{3}} \Bigg\} \end{aligned} \tag{31}$$

Due to the satellite speed far greater than the maritime target speed, Eq. (31) can be approximately expressed as:

$$\begin{aligned} f_{\mathrm{dr}}^{k} = -\frac{1}{\lambda^{k}}\Bigg\{ & \frac{\|\mathbf{V}_{\mathrm{s}}^{k}\|^{2}}{\|\mathbf{P}_{\mathrm{s}}^{k}-\mathbf{P}_{\mathrm{t}}\|} - \frac{\left[(\mathbf{P}_{\mathrm{s}}^{k}-\mathbf{P}_{\mathrm{t}})^{\mathrm{T}}\mathbf{V}_{\mathrm{s}}^{k}\right]^{2}}{\|\mathbf{P}_{\mathrm{s}}^{k}-\mathbf{P}_{\mathrm{t}}\|^{3}} + \frac{\|\mathbf{V}_{\mathrm{t}}\|^{2}}{\|\mathbf{P}_{\mathrm{t}}-\mathbf{P}_{\mathrm{r}}\|} - \frac{[(\mathbf{P}_{\mathrm{t}}-\mathbf{P}_{\mathrm{r}})^{\mathrm{T}}\mathbf{V}_{\mathrm{t}}]^{2}}{\|\mathbf{P}_{\mathrm{t}}-\mathbf{P}_{\mathrm{r}}\|^{3}} - \\ & \frac{\|\mathbf{V}_{\mathrm{s}}^{k}\|^{2}}{\|\mathbf{P}_{\mathrm{s}}^{k}-\mathbf{P}_{\mathrm{r}}\|} + \frac{\left[(\mathbf{P}_{\mathrm{s}}^{k}-\mathbf{P}_{\mathrm{r}})^{\mathrm{T}}\mathbf{V}_{\mathrm{s}}^{k}\right]^{2}}{\|\mathbf{P}_{\mathrm{s}}^{k}-\mathbf{P}_{\mathrm{r}}\|^{3}} \Bigg\} \end{aligned} \tag{32}$$

Because the distance from the satellite to the target is much greater than the distance from the target to the receiver, the DRF in Eq. (32) is mainly contributed by the target motion. Therefore, it is observed that the DFR values for a non-maneuvering target remain similar across different bistatic acquisition geometries. Figure 9a plots the actual velocity curves of four satellites. Among them, satellite PRN03 exhibits the largest velocity variation during the observation time, reaching up to 4 m/s. Nevertheless, its average acceleration is only 0.067 m/s^2, indicating that the satellite motion is approximately uniform throughout the observation time. Using the real motion parameters of the four satellites and Eq. (31), Fig. 9b presents the DFR curves for a simulated target. Apart from the DFR curve for satellite PRN03, the curves for the other satellites nearly overlap. The maximum difference in DFR between satellite PRN03 and the others is 0.031 Hz/s, which is negligible. This observation will be applied in the next subsection.

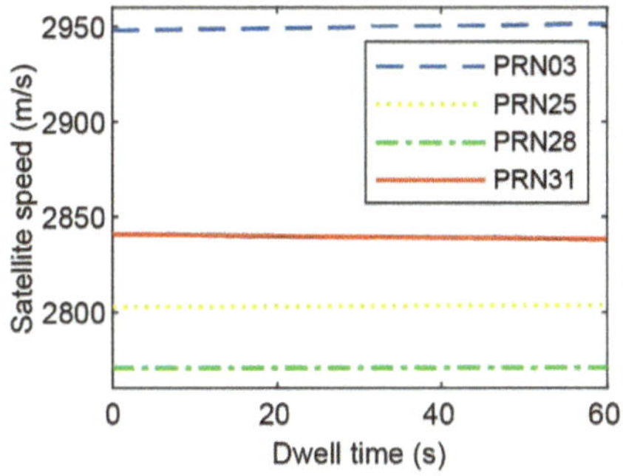

(a) Real satellite velocity curves

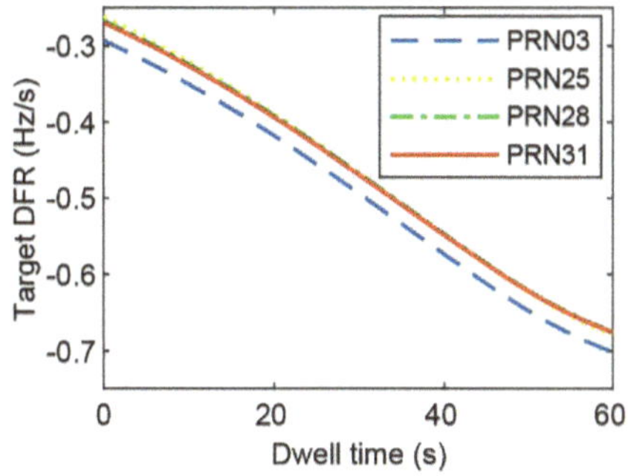

(b) DFR curves

Fig. 9 Real satellite velocity curves and DFR curves

4.2 Joint Detection and Localization Algorithm

As illustrated in Fig. 8, K bistatic acquisition geometries are formed by K satellites as signal sources, together with a receiver and a moving target. In the k-th bistatic acquisition geometry, prior to detection, a stack of integrated RD maps $S_{\mathrm{NRD}}^{k}(r, f_{\mathrm{u_s}}; f_{\mathrm{dr}}^{*})$ is generated by executing the long-time hybrid integration algorithm proposed in Sect. 3.2. Subsequently, the K stacks of integrated RD maps are projected into the Cartesian coordinate plane representing the surveillance area, which further accumulates the target echo energy in the spatial domain, allowing for the joint detection and localization of a maritime moving target [28]. The detailed processing steps are as follows.

Step 1: The X-Y plane is discretized into a specific number of grids. The position vector of an arbitrary grid $\mathbf{P}_{\mathrm{g}}$ is denoted as $\mathbf{P}_{\mathrm{g}} = [x_{\mathrm{g}}, y_{\mathrm{g}}, 0]^{\mathrm{T}}$.

Step 2: Supposing the target located at $\mathbf{P}_{\mathrm{g}}$, the bistatic range and Doppler frequency of $\mathbf{P}_{\mathrm{g}}$ in the k-th bistatic acquisition geometry are determined using Eqs. (15.29) and (15.30). Since the satellite coordinates, satellite velocity, and receiver coordinates are known in advance, the bistatic range can be directly computed, denoted as $R^{k}(x_{\mathrm{g}}, y_{\mathrm{g}})$. However, calculating the Doppler frequency requires the target's velocity vector, which is unknown and can be estimated by defining an appropriate velocity search interval. If the tested target velocity vector matches the actual velocity, the resulting Doppler frequency is denoted as $F^{k}\left(x_{\mathrm{g}}, y_{\mathrm{g}}; v_{\mathrm{x}}^{*}, v_{\mathrm{y}}^{*}\right)$. Here, v_{x}^{*} and v_{y}^{*} represent the tested velocity components along the X and Y axes, respectively.

Step 3: The signal intensity at the position $\left[R^{k}(x_{\mathrm{g}}, y_{\mathrm{g}}), F^{k}\left(x_{\mathrm{g}}, y_{\mathrm{g}}; v_{\mathrm{x}}^{*}, v_{\mathrm{y}}^{*}\right)\right]$ in each RD map of the k-th stack is projected into the corresponding grid $\mathbf{P}_{\mathrm{g}}$, producing the k-th stack of Cartesian planes, expressed as:

$$I^{k}\left(x_{\mathrm{g}}, y_{\mathrm{g}}; v_{\mathrm{x}}^{*}, v_{\mathrm{y}}^{*}, f_{\mathrm{dr}}^{*}\right) = S_{\mathrm{NRD}}^{k}\left(r = R^{k}(x_{\mathrm{g}}, y_{\mathrm{g}}), f_{\mathrm{u_s}} = F^{k}\left(x_{\mathrm{g}}, y_{\mathrm{g}}; v_{\mathrm{x}}^{*}, v_{\mathrm{y}}^{*}\right); f_{\mathrm{dr}}^{*}\right) \quad (33)$$

By replacing $\mathbf{P}_\mathrm{t}$, $\mathbf{V}_\mathrm{t}$, and f_dr^k in Eq. (32) by $\mathbf{P}_\mathrm{g}$, $\left[v_\mathrm{x}^*, v_\mathrm{y}^*, 0\right]^\mathrm{T}$, and f_dr^*, respectively, after rearranging, the following quadratic equation is obtained as:

$$\begin{cases} \left(\left\|\mathbf{P}_\mathrm{g} - \mathbf{P}_\mathrm{r}\right\|^2 - \left(y_\mathrm{g} - y_\mathrm{r}\right)^2\right)\left(v_\mathrm{y}^*\right)^2 + \left(\left\|\mathbf{P}_\mathrm{g} - \mathbf{P}_\mathrm{r}\right\|^2 - \left(x_\mathrm{g} - x_\mathrm{r}\right)^2\right)\left(v_\mathrm{x}^*\right)^2 \\ \quad - 2\left(x_\mathrm{g} - x_\mathrm{r}\right)\left(y_\mathrm{g} - y_\mathrm{r}\right)v_\mathrm{x}^* v_\mathrm{y}^* + \left\|\mathbf{P}_\mathrm{g} - \mathbf{P}_\mathrm{r}\right\|^3\left(\lambda^k f_\mathrm{dr}^* + C_\mathrm{t}\right) = 0 \\ C_\mathrm{t} = \frac{\|\mathbf{V}_\mathrm{s}^k\|^2}{\|\mathbf{P}_\mathrm{s}^k - \mathbf{P}_\mathrm{g}\|} - \frac{\left[(\mathbf{P}_\mathrm{s}^k - \mathbf{P}_\mathrm{g})^\mathrm{T}\mathbf{V}_\mathrm{s}^k\right]^2}{\|\mathbf{P}_\mathrm{s}^k - \mathbf{P}_\mathrm{g}\|^3} - \frac{\|\mathbf{V}_\mathrm{s}^k\|^2}{\|\mathbf{P}_\mathrm{s}^k - \mathbf{P}_\mathrm{r}\|} + \frac{\left[(\mathbf{P}_\mathrm{s}^k - \mathbf{P}_\mathrm{r})^\mathrm{T}\mathbf{V}_\mathrm{s}^k\right]^2}{\|\mathbf{P}_\mathrm{s}^k - \mathbf{P}_\mathrm{r}\|^3} \end{cases} \tag{34}$$

Based on Eq. (34), the velocity component along the Y-axis (or along the X-axis) can be calculated using the velocity component along the X-axis (or Y-axis) and the target DFR. For instance, considering the Y-axis component, there are two solutions for v_y^* given by:

$$\left(v_\mathrm{y}^1, v_\mathrm{y}^2\right) = \varsigma\left(v_\mathrm{x}^*, f_\mathrm{dr}^*\right) \tag{35}$$

Here, $\varsigma(\cdot)$ denotes the solution expression with respect to the above quadratic equation. By substituting Eq. (35) into Eq. (33), Eq. (33) can be recast as:

$$I^k\left(x_\mathrm{g}, y_\mathrm{g}; v_\mathrm{x}^*, f_\mathrm{dr}^*\right) = S_\mathrm{NRD}^k\left\{r = R^k\left(x_\mathrm{g}, y_\mathrm{g}\right), f_{\mathrm{u_s}} = F^k\left[x_\mathrm{g}, y_\mathrm{g}; v_\mathrm{x}^*, \varsigma\left(v_\mathrm{x}^*, f_\mathrm{dr}^*\right)\right]; f_\mathrm{dr}^*\right\} \tag{36}$$

By comparing Eqs. (33) and (36), it is observed that the dimensionality of the search space for unknown target motion parameters has decreased from three to two dimensions, thereby enhancing the efficiency of the projection processing.

Step 4: Multistatic integration is performed across K stacks of Cartesian planes generated in **Step 3** to further accumulate the target echo energy in the spatial domain. Because the DFR values of a non-maneuvering target remain similar across various bistatic acquisition geometries, Cartesian planes with the identical DFR value from the K stacks are combined through noncoherent integration to produce the corresponding multistatic Cartesian plane, which can be represented as:

$$I_\mathrm{NCI}\left(x_\mathrm{g}, y_\mathrm{g}; v_\mathrm{x}^*, f_\mathrm{dr}^*\right) = \frac{1}{K}\sum_{k=1}^{k=K} \mathrm{norm}\left[\left|I^k\left(x_\mathrm{g}, y_\mathrm{g}; v_\mathrm{x}^*, f_\mathrm{dr}^*\right)\right|\right] \tag{37}$$

Here, norm[·] is the normalization operation to eliminate differences in signal strength among the K satellites. As shown in (37), when a tested pair $\left(v_\mathrm{x}^*, f_\mathrm{dr}^*\right)$ matches the actual pair, the target energy will concentrate around the true target position on the multistatic Cartesian plane, achieving the highest integration gain. If not, the target energy becomes dispersed and is submerged by background noise. Consequently, the multistatic Cartesian plane with the highest peak value can be selected as input for the decision threshold technique to confirm the target's presence. Once the target is successfully detected, its position and velocity can be determined simultaneously.

5 Experimental Analyses

This section presents actual experimental results to verify the effectiveness of the proposed methods described in Sects. 3 and 4. Additionally, existing methods are used for comparison.

5.1 Experimental Setup

A maritime experiment was carried out at Cyberport Water-front Park in Hong Kong to validate the algorithms introduced in Sects. 3 and 4. Figure 10a illustrates the data acquisition geometry used in the field trial, where global positioning system (GPS) satellites serve as the illuminators of opportunity, and the receiving hardware is positioned on the shore to simultaneously capture both direct and reflected signals. Figure 10b depicts the receiving hardware, comprising a dual-channel receiver front-end, circular and square antennas, and a host computer. The circular antenna (reference channel antenna) is a commercial off-the-shelf right-hand circular polarization antenna pointing towards the sky to collect direct signals. The square antenna (surveillance channel antenna) is a custom left-hand circular polarization antenna with a 12 dBi gain, steering towards the sea surface to capture target echoes. The receiver front-end converts the received radio signals into digital intermediate frequency signals for the host computer processing. Moving cargo ships passing through the field of view of the receiving hardware were chosen as targets of interest. Their motion parameters were obtained from the automatic identification system (AIS) to serve as ground truth data. Table 1 provides the GPS signal parameters.

(a) Data acquisition geometry (b) Receiving hardware

Fig. 10 Maritime experimental scene

Table 1 GPS signal parameters

Parameter	Value
Carrier frequency (MHz)	1575.42
Ranging code chip rate (MHz)	1.023
PRI (ms)	1

5.2 Maritime Ship Detection Results

As shown in Fig. 11, the moving cargo ship "KUO LIN" is selected as the target of interest, with two GPS satellites, PRN03 and PRN22, serving as the signal sources. Table 2 gives the parameters for the satellites, the target, and the processing settings.

The range compression data matrices with respect to two satellites are demonstrated in Fig. 12. Although this is a short-range experimental scenario, the collected target echoes are extremely weak and submerged in the background noise, as seen

Fig. 11 Photograph of cargo ship "KUO LIN"

Table 2 Experimental parameters for maritime ship detection

Parameter	Value	Parameter	Value
PRN code	PRN03	Ship heading direction (°)	329
Satellite elevation angle (°)	41.11–40.86	The shortest distance to the receiver (m)	800
Satellite azimuth angle (°)	70.97–69.85	Sampling frequency (MHz)	16.368
PRN code	PRN22	Dwell time (s)	114.688
Satellite elevation angle (°)	25.28–24.80	Frame duration (s)	2.048
Satellite azimuth angle (°)	52.86–52.22	Noncoherent integrated frames	56
Ship length (m)	170	Search interval of f_{dr}^* (Hz/s)	[−1,1]
Ship average velocity (m/s)	7.52	Number of sampling points for f_{dr}^*	50

in Fig. 12a, b. Hence, it is necessary to enhance the weak target echo energy through long-time integration. For comparison, the conventional moving target detection (MTD) technique [27] is employed. This technique should be appropriately modified to satisfy the long observation time requirement of GNSS-based passive radar. That is, after range compression, the MTD technique applies a FFT inside each frame along the slow-time dimension to offer coherent integration gain. Then, all frames are combined through noncoherent integration without any motion compensation. Figure 13 displays the resulting RD maps with respect to two satellites. The color scales are in dB, with 0 dB representing the mean power of the background noise. As can be seen, the sidelobes of the range-compressed direct signal (collected by the sidelobes of the left-hand circular polarization antenna) are well-focused along the zero-Doppler frequency line and exhibit the strongest intensity in both RD maps. These are considered static strong returns and can be filtered out. But we retain them to compare with the target contributions. Due to the long integration time, the motion of the cargo ship cannot be ignored, which shifts the positions of the target contributions in the RD maps, causing the RCM and DFM. Consequently, the target contributions are dispersed across multiple RD cells and are obscured by the static strong returns in Fig. 13.

The TMC method presented in [7], the KT + FrFT method presented in [8], and the proposed KT + MF method are performed to concentrate weak target energy and compensate for RCM and DFM. As illustrated in Figs. 14 and 15, the target contributions obtained by these three methods are easily distinguished from the background noise. The highest intensity values of the target contributions exceed 10 dB, which is adequate for reliable detection. This is due to two main factors. First, the target energy is concentrated because all three long-time hybrid integration algorithms perform RCM and DFM compensations and provide sufficient hybrid integration gains. Second, these methods spread the energy of strong static returns across multiple RD cells, reducing their impact. Consequently, compared to Fig. 13, the sidelobes of the range-compressed direct signals shown in Figs. 14 and 15 are

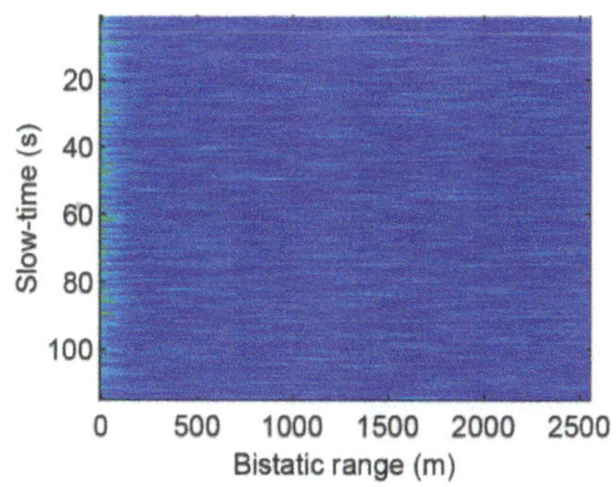

(a) PRN03

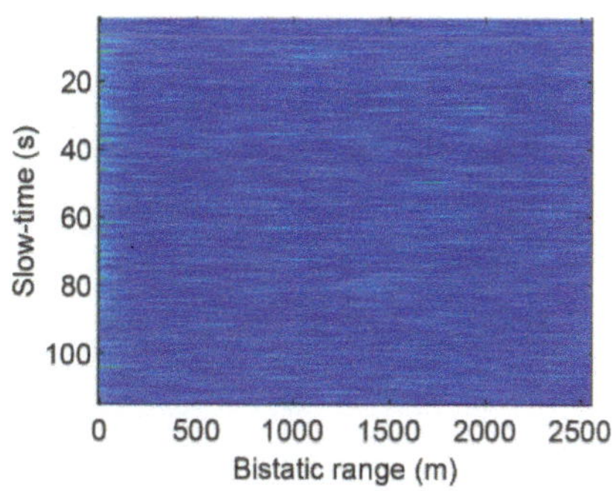

(b) PRN22

Fig. 12 Range compression results with respect to two satellites

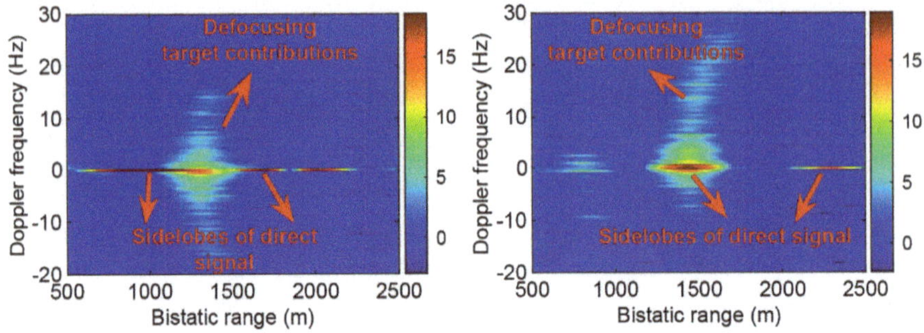

(a) PRN03 (b) PRN22

Fig. 13 RD maps of ship "KUO LIN" generated by the MTD technique with respect to two satellites

dramatically suppressed. The effectiveness of these three methods for maritime target detection has been validated using field measurement data.

To quantitatively evaluate the detection capabilities of the MTD method, the TMC method, the KT + FrFT method, and the proposed KT + MF method, Monte Carlo

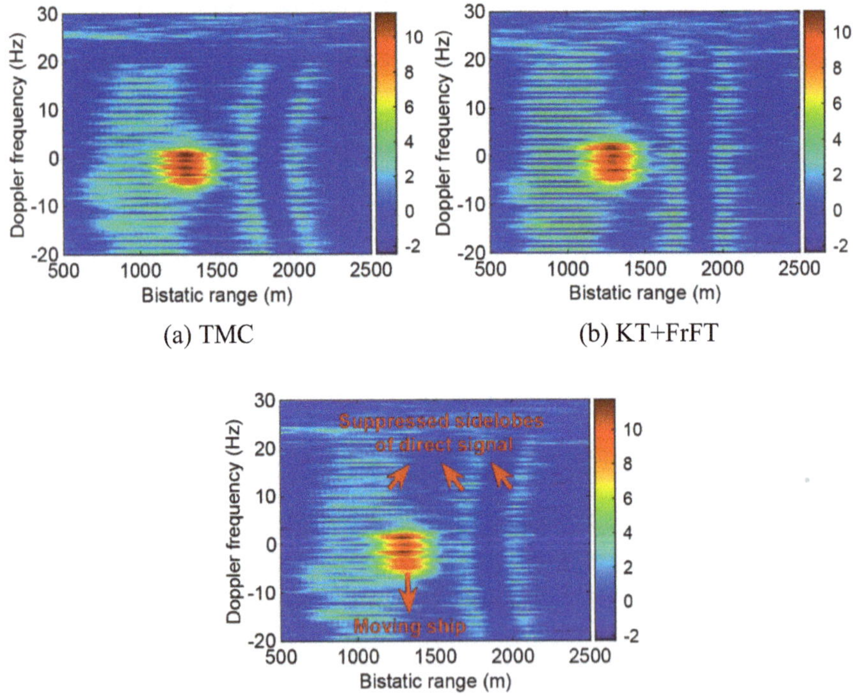

(a) TMC (b) KT+FrFT

(c) Proposed KT+MF

Fig. 14 RD maps of ship "KUO LIN" with respect to PRN03

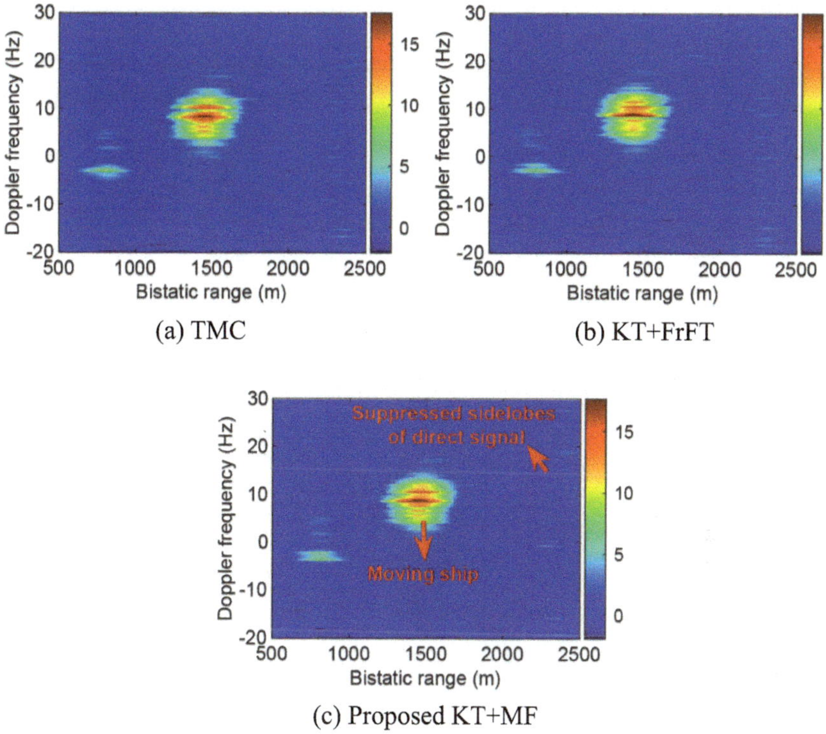

Fig. 15 RD maps of ship "KUO LIN" with respect to PRN22

simulations are conducted by adding randomly generated white Gaussian noise into the target echoes at various SNR levels. These four methods were combined with a 2-dimensional cell averaging (CA)-CFAR detector to serve as the respective detectors. The false alarm rate is set to 10^{-3}, and 100 Monte Carlo trials are done. Figure 16 plots the curves of numerical detection results. As can be seen, for the same detection probability, the proposed method requires a lower SNR compared to the other methods. Thus, the proposed KT + MF method demonstrates superior detection capability over the other approaches when using GNSS signals with the same bandwidth.

5.3 Maritime Ship Joint Detection and Localization Results

As shown in Fig. 17, the moving cargo ship "WAN HAI 508" is selected as the target of interest, with two GPS satellites, PRN01 and PRN22, serving as the signal sources. Table 3 lists the parameters for the satellites, the target, and the processing settings. Figure 18 depicts the experimental setup on a 1000 m $\times$ 1000 m Cartesian plane, where the receiver is located at (800 m, 500 m, 0) as a reference point, enabling

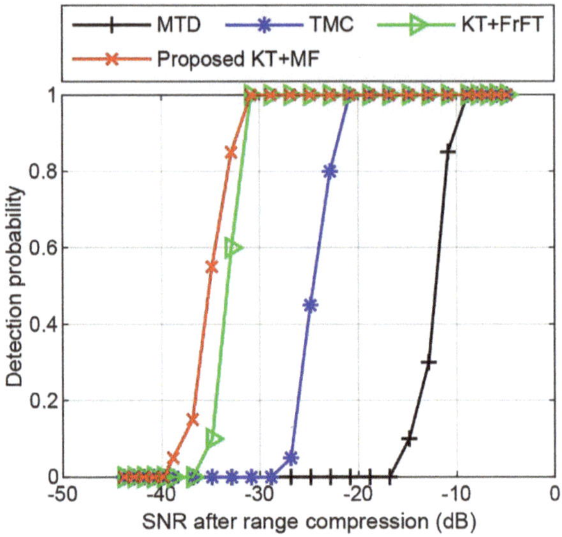

Fig. 16 Detection probability curves for four methods

the focused target energy to be centered near the origin of the Cartesian plane for improved visualization. It should be noted that the GPS satellite positions from the ephemeris file and the moving ship's positions from the AIS are given in different coordinate systems. These must be converted into the established local Cartesian reference system using a coordinate transformation function. Due to the short-range field experiment, the large maritime ship should be treated as an extended target rather than a point-like target. Therefore, one multistatic Cartesian plane exhibiting the maximum intensity contrast (IC) [29] will be chosen from the stack of multistatic Cartesian planes. The IC is mathematically defined as follows:

$$I_c = \frac{\sqrt{E[I_{\mathrm{NCI}} - E(I_{\mathrm{NCI}})]^2}}{E(I_{\mathrm{NCI}})} \tag{38}$$

Here, $E(\cdot)$ is the average value of the grids in the Cartesian plane.

To demonstrate the sequential location and velocity estimation results, the experimental data collected over the dwell time are divided into five groups. Taking the fifth group as an example, using the tested pair $\left(f_{\mathrm{dr}}^{*}, v_{x}^{*}\right)$ that yields the maximum IC value in the final multistatic Cartesian plane, Figs. 19 and 20 display the noncoherent-integrated RD maps and the Cartesian planes with respect to two satellites, respectively. Consistent with the results in Figs. 14 and 15, the target energy of the ship "WAN HAI 508" in Fig. 19 is concentrated, and the RD positions of the bright spots representing the ship in Fig. 19a, b are different, which are dependent on the specific bistatic acquisition geometries. After the projection processing, the bright spots in Fig. 20a, c are nearly aligned, facilitating multistatic integration. Furthermore, in Fig. 20a, a ghost target marked by a black circle has an intensity comparable to the moving target, which could increase the false alarm rate if a decision threshold is

Fig. 17 Photograph of cargo ship "WAN HAI 508"

Table 3 Experimental parameters for maritime ship joint detection and localization

Parameter	Value	Parameter	Value
PRN code	PRN01	Dwell time (s)	40.96
Satellite elevation angle (°)	61.90–61.47	Group number	5
Satellite azimuth angle (°)	32.70–32.81	Integration time in a group (s)	8.192
PRN code	PRN22	Frame duration (s)	2.048
Satellite elevation angle (°)	36.52–36.20	Noncoherent integrated frames	4
Satellite azimuth angle (°)	100.48–99.98	Size of Cartesian plane (km × km)	1 × 1
Ship length (m)	269	Grid size of Cartesian plane (m × m)	5 × 5
Ship average velocity (m/s)	7.16	Search interval of f_{dr}^{*} (Hz)	[−1,1]
Ship heading direction (°)	331	Search interval of v_x^* (m/s)	[−10, 10]
The shortest distance to the receiver (m)	~860	Number of sampling points for f_{dr}^{*}	50
Sampling frequency (MHz)	16.368	Number of sampling points for v_x^*	50

applied on the Cartesian plane. Fortunately, since Fig. 20c has no ghost target at the same position, the ghost target energy is suppressed in the multistatic integration result shown in Fig. 20e, also marked by a black circle. Thus, multistatic integration serves as a ghost target rejection filter, thanks to the spatial diversity of GNSS satellites. Figure 20b, d show the Cartesian planes generated using the invalid solution of v_y^*. It is evident that the target energy is defocused and indistinguishable from background noise. The IC value for the multistatic integration result in Fig. 20f is only 0.36, much lower than that for Fig. 20e at 1.21. Therefore, it is reasonable to select Fig. 20e as the final output.

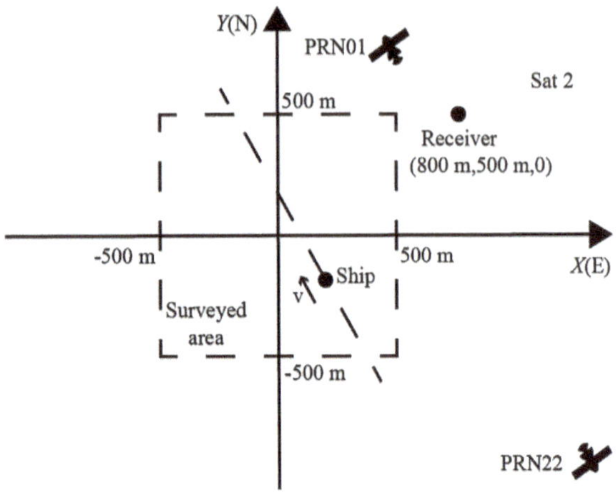

Fig. 18 Top view of a local Cartesian plane representing the experimental scene

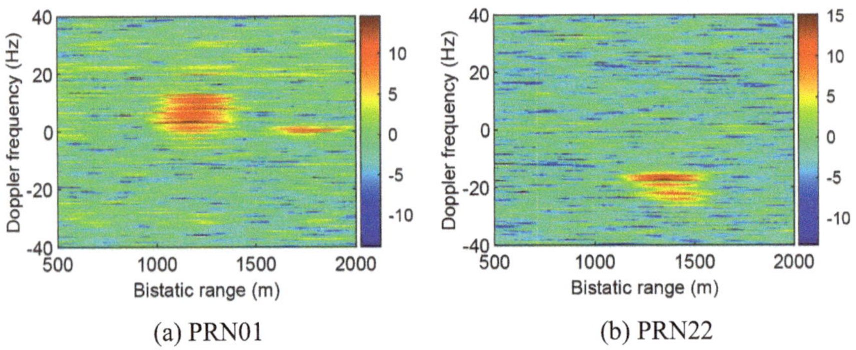

Fig. 19 RD maps of ship "WAN HAI 508" with respect to two satellites

The estimation results of target position and velocity obtained by the algorithm proposed in Sect. 4 and the existing method [29] on five experimental datasets are presented in Fig. 21. The AIS location points and AIS target velocity are included as ground truth references. As shown in Fig. 21a, the spatial distributions of the five datasets are non-uniform, which can be attributed to the nonnegligible physical dimensions of the ship causing the strongest scattering points on the target surface to vary over the long dwell time. Despite this, the trajectories reconstructed from the red and blue circles align well with the AIS trajectory. In Fig. 21b, the velocity curves estimated by both methods fluctuate around the AIS velocity. Table 4 gives the root mean square error (RMSE) values for the position and velocity estimates across the five datasets, along with the average computational time. The estimation accuracy of the two methods is comparable in terms of both location and velocity. However, the average computational time of the proposed algorithm in Sect. 4 is significantly lower than that of the existing method, demonstrating its practical advantage in real-world applications.

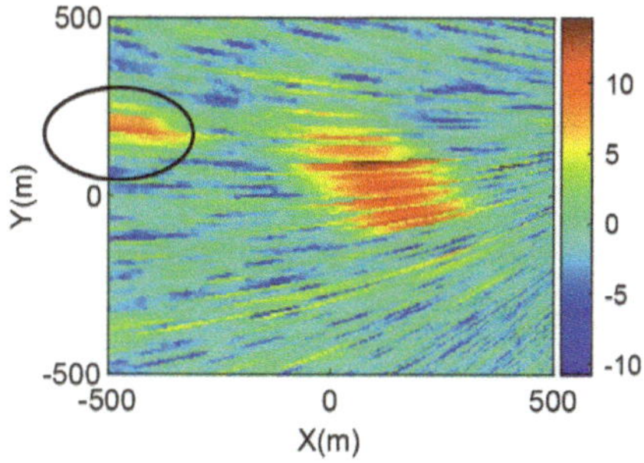

(a) Cartesian plane using valid solution of v_y^* with respect to PRN01

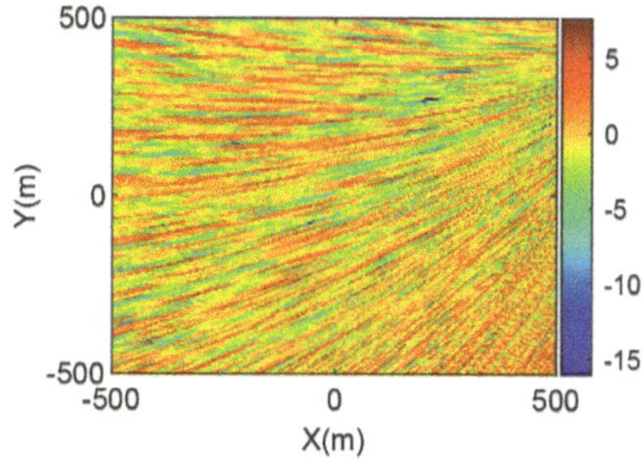

(b) Cartesian plane using invalid solution of v_y^* with respect to PRN01

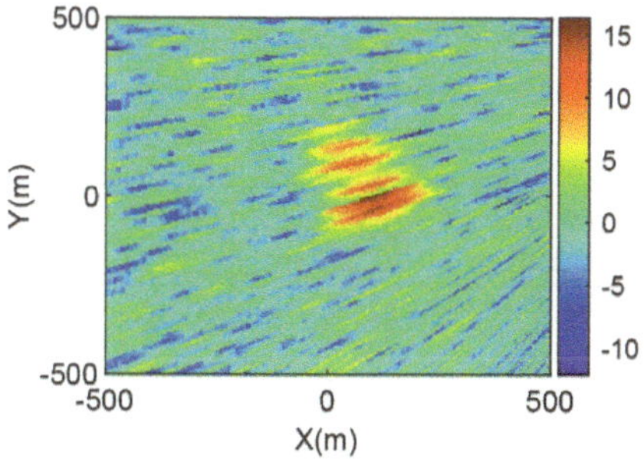

(c) Cartesian plane using valid solution of v_y^* with respect to PRN22

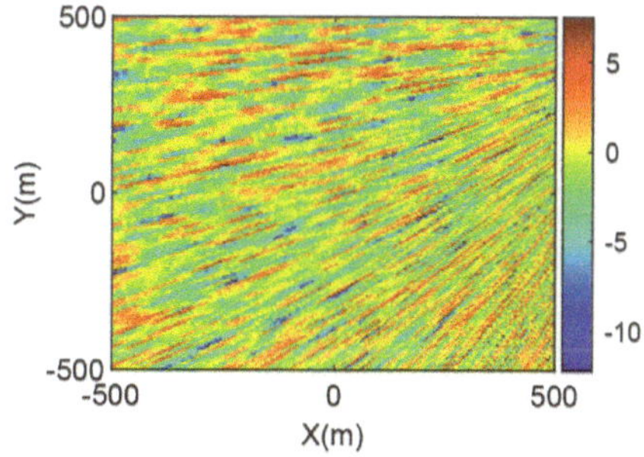

(d) Cartesian plane using invalid solution of v_y^* with respect to PRN22

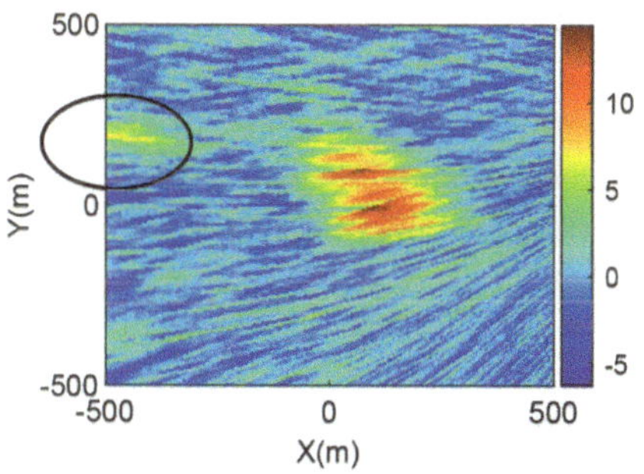

(e) Multistatic Cartesian plane using valid solution of v_y^*

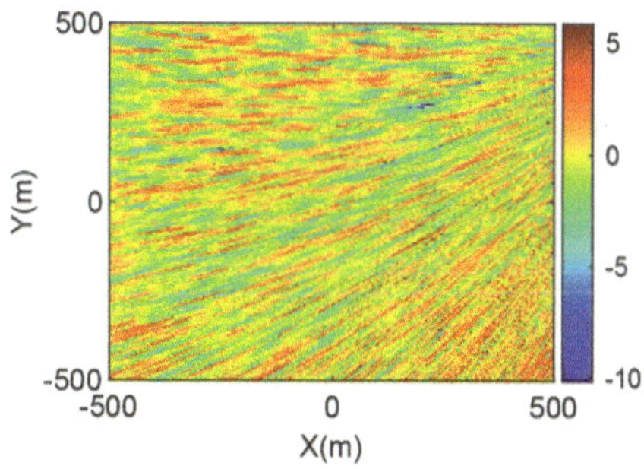

(f) Multistatic Cartesian plane using invalid solution of v_y^*

Fig. 20 Cartesian planes with respect to two satellites

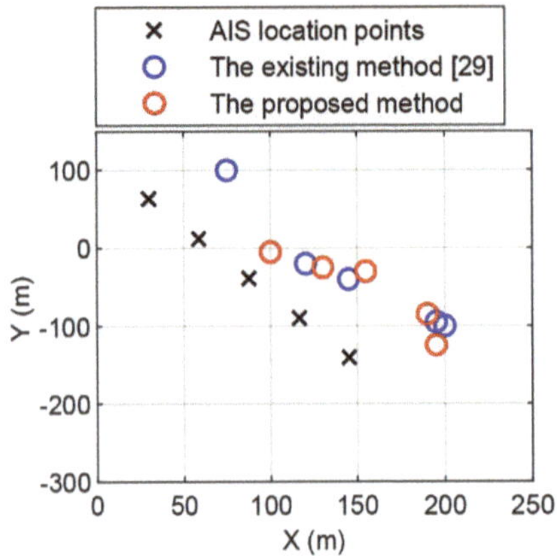

(a) Location results

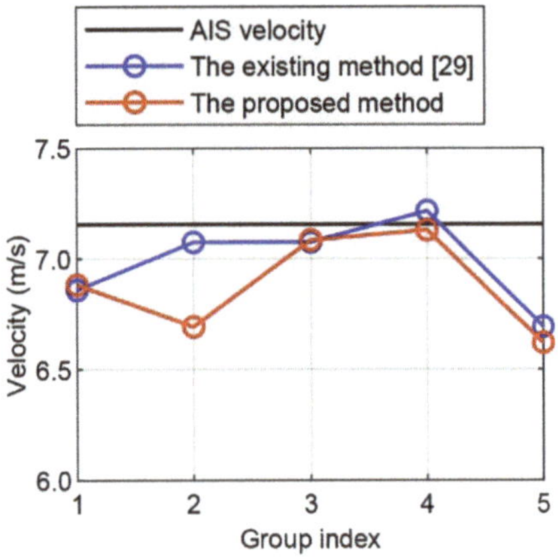

(b) Velocity estimation results

Fig. 21 Location and velocity estimation results

Table 4 Evaluation parameter comparison for real experiment

Algorithms	The existing method [29]	The proposed method in Sect. 4
RMSE of location (m)	66.82	75.83
RMSE of velocity (m/s)	0.25	0.34
Mean time cost (s)	124.97	32.43

6 Conclusions

This chapter investigates the issue of maritime moving target detection and localization using GNSS-based passive radar, proposing and validating a series of signal processing algorithms. First, to overcome the fundamental limitation of extremely weak target echo energy, a long-time hybrid integration algorithm combining the KT and the MF is developed. This algorithm effectively compensates for the first-order RCM induced by the target's bistatic motion via KT, and employs a bank of matched filters to search for and compensate the second-order RCM as well as intraframe and interframe DFMs. This process enables the effective accumulation of weak target energy over long observation intervals, significantly enhancing the radar's detection performance.

Furthermore, the proposed detection framework is extended from a bistatic to a multistatic configuration, leading to a novel joint detection and localization method for maritime moving targets. By analyzing the DFR characteristics of a non-maneuvering target within the multistatic geometry, it is found that the target's DFR values across different satellite-receiver pairs are very similar. Leveraging this finding, the stacks of range-Doppler-DFR data cubes—obtained after long-time integration for each channel—are projected onto a local Cartesian plane. Subsequent multistatic noncoherent integration in the spatial domain not only further improved the SNR for reliable detection but also simultaneously estimated the target's position and velocity.

Experimental analyses using field-measurement maritime datasets validated the effectiveness of the proposed methods. The proposed KT + MF algorithm demonstrated superior performance compared to conventional MTD, the TMC method, and the KT + FrFT method, achieving higher detection probabilities for given input SNR levels. In the multistatic joint detection and localization experiments, the proposed method maintained comparable estimation accuracy to the existing approach [29] while significantly reducing computational time, highlighting its practicality for real-scenario applications.

References

1. Pieralice F, Santi F, Pastina D et al (2017) GNSS-based passive radar for maritime surveillance: long integration time MTI technique. In: 2017 IEEE radar conference (RadarConf), Seattle, WA, USA
2. Simone AD, Park H, Riccio D et al (2017) Sea target detection using spaceborne GNSS-R delay-Doppler maps: theory and experimental proof of concept using TDS-1 data. IEEE J Sel Top Appl Earth Obs Remote Sens 10(9):4237–4255
3. Gomez-Del-Hoyo P, Del-Rey-Maestre N, Jarabo-Amores MP et al (2022) Improved 2D ground target tracking in GPS-based passive radar scenarios. Sensors 22(5):1724–1741
4. Zhou X, Wang P, Chen J et al (2021) A modified radon Fourier transform for GNSS-based bistatic radar target detection. IEEE Geosci Remote Sens Lett 19:1–5
5. Zhou X, Wang P, Zeng H et al (2022) Moving target detection using GNSS-based passive bistatic radar. IEEE Trans Geosci Remote Sensing 60:1–15
6. Tang T, Wang P, Zhao P et al (2024) A novel multi-target TBD scheme for GNSS-based passive bistatic radar. IET Radar Sonar Navig 18(12):2497–2512
7. Ma H, Antoniou M, Pastina D et al (2018) Maritime moving target indication using passive GNSS-based bistatic radar. IEEE Trans Aerosp Electron Syst 54(1):115–130
8. Li Z, Santi F, Pastina D et al (2017) Multi-frame fractional Fourier transform technique for moving target detection with space-based passive radar. IET Radar Sonar Navig 11(5):822–828
9. Pastina D, Santi F, Pieralice F et al (2018) Maritime moving target long time integration for GNSS-based passive bistatic radar. IEEE Trans Aerosp Electron Syst 54(6):3060–3083
10. Ma H, Antoniou M, Stove AG et al (2018) Maritime moving target localization using passive GNSS-based multistatic radar. IEEE Trans Geosci Remote Sensing 56(8):4808–4819
11. Ma H, Antoniou M, Stove AG et al (2021) Target kinematic state estimation with passive multistatic radar. IEEE Trans Aerosp Electron Syst 57(4):2121–2134
12. Nasso I, Santi F, Pastina D (2021) Maritime targets velocity estimation in space-based passive multistatic radar using long integration times. IEEE Access 9:163764–163779
13. Sadeghi M, Behnia F, Amiri R (2021) Maritime target localization from bistatic range measurements in space-based passive radar. IEEE Trans Instrum Meas 70:1–8
14. Antoniou M, Cherniakov M (2013) GNSS-based bistatic SAR: a signal processing view. EURASIP J Adv Signal Process 1:1–16
15. Antoniou M, Cherniakov M, Ma H (2015) Space-surface bistatic synthetic aperture radar with navigation satellite transmissions: a review. Sci China-Inf Sci 58(6):1–20
16. Borre K, Akos DM, Bertelsen N et al (2007) A software-defined GPS and Galileo receiver: a single-frequency approach. Birkhäuser, Boston

17. Zeng Z (2013) Passive bistatic SAR with GNSS transmitter and a stationary receiver. Dissertation, University of Birmingham
18. Cherniakov M (2008) Bistatic radar: emerging technology. Wiley, Wiltshire
19. Pastina D, Santi F, Pieralice F et al (2021) Passive radar imaging of ship targets with GNSS signals of opportunity. IEEE Trans Geosci Remote Sensing 59(3):2627–2642
20. Peebles PZ (2007) Radar principles. Wiley, Chichester
21. Chen X, Guan J, Liu N et al (2014) Maneuvering target detection via radon-fractional Fourier transform-based long-time coherent integration. IEEE Trans Signal Process 62(4):939–953
22. Zhang J, Ding T, Zhang L (2022) Longtime coherent integration algorithm for high-speed maneuvering target detection using space-based bistatic radar. IEEE Trans Geosci Remote Sensing 60:1–16
23. Wan J, Kang X, Tan X et al (2023) An efficient approach for coherent integration detection of high-speed maneuvering targets with arbitrary-order Doppler frequency migrations. IEEE Trans Aerosp Electron Syst 59(5):6729–6748
24. He Z, Chen W, Yang Y et al (2023) Sea target detection using the GNSS reflection signals. GPS Solut 27(4):160–173
25. Wang Y, Li J, Chen J et al (2014) A parameter-adjusting polar format algorithm for extremely high squint SAR imaging. IEEE Trans Geosci Remote Sensing 52(1):640–650
26. Santi F, Pieralice F, Pastina D (2019) Joint detection and localization of vessels at sea with a GNSS-based multistatic radar. IEEE Trans Geosci Remote Sensing 57(8):5894–5913
27. Richards MA (2014) Fundamentals of radar signal processing. Mcgraw-hill, New York
28. He Z, Yang Y, Chen W et al (2024) Maritime ship joint detection and localization using GNSS-based passive multistatic radar. IEEE Trans Instrum Meas 73:1–14
29. Nasso I, Santi F (2022) A centralized ship localization strategy for passive multistatic radar based on navigation satellites. IEEE Geosci Remote Sens Lett 19:1–5

Chapter 16
Oceanic Eddy Detection and Shape Reconstruction Based on Spaceborne GNSS-R

Jin Xing and Feng Wang

Abstract Spaceborne radar altimeters are fundamentally constrained by their orbital configurations, achieving near-global coverage only over multi-day cycles. This limitation hinders the accurate capture and detailed structural reconstruction of short-lived, rapidly evolving oceanic mesoscale eddies. This paper investigates ocean eddy detection and shape reconstruction using spaceborne Global Navigation Satellite System-Reflectometry (GNSS-R), leveraging its high spatiotemporal resolution. We first demonstrate the feasibility of GNSS-R for eddy detection. By employing high-bandwidth ACE-BOC signals, the gridded sea surface height accuracy is improved to the centimeter level, which significantly enhances the detection of eddy centers and radii based on sea surface height anomalies and the geostrophic velocity field. To address the challenges of blurred boundaries and gradient aliasing in eddy-dense regions, a local shape feature refinement method is proposed. This method constructs an adaptive Bessel kernel to enhance boundary gradient features and dynamically corrects the eddy radius by incorporating the radial gradient characteristics of the eddy kinetic energy field. Experimental results show that the local feature extraction method using ACE-BOC signals achieves detection precision rates of 85.94% for cyclonic eddies and 84.34% for anticyclonic eddies, representing an improvement of over 6% compared to conventional global feature-based methods. A comparative analysis with the Jason-3 altimeter confirms that GNSS-R provides a distinct advantage in daily coverage capability, enabling continuous monitoring of eddy evolution at daily scales while maintaining high detection accuracy, thus offering a powerful complementary technique for monitoring highly dynamic oceanic phenomena.

Keywords GNSS-R · Oceanic eddy detection · Shape reconstruction

---

J. Xing
School of Artificial Intelligence, China University of Geosciences Beijing, Beijing, China
e-mail: jinxing@cugb.com

F. Wang (✉)
Department of Electronic Information Engineering, Beihang University, Beijing, China
e-mail: wangfeng_buaa@buaa.edu.cn

K. Yu and N. Zheng (eds.), *Advances in GNSS Reflectometry*, Navigation: Science and Technology 19, https://doi.org/10.1007/978-981-92-2325-1_16

1 Introduction

1.1 Temporal Resolution in Capturing Eddy Evolution

Mesoscale eddies typically have lifespans ranging from ten days to several months, yet their formative, intensifying, and decaying phases may last only a few days. Processes such as eddy merging or splitting often occur on even shorter timescales; for instance, energy exchanges between two eddies have been observed to complete within just three days [1]. In addition, eddies can travel at speeds of 5–20 cm/s (equivalent to 4–17 km per day), and their trajectories frequently exhibit nonlinear jumps or abrupt shifts [2]. When observations are constrained by low temporal resolution, short-lived eddies may be smoothed out or entirely missed, making it difficult to characterize their evolution accurately. This not only introduces errors in trajectory tracking but also oversimplifies eddy paths into straight lines or great-circle arcs, thereby obscuring important dynamical details.

Current eddy detection methods rely primarily on satellite radar altimeters (e.g., Jason-3, Sentinel-6). However, their orbital design imposes a substantial temporal resolution bottleneck. Specifically, the exact repeat orbit (ERO) configuration produces a ground-track spacing of approximately 5–7 km, requiring about 35 days of orbital drift to achieve full coverage at mid-latitudes [3]. According to the theoretical eddy timescale (defined by the Rossby parameter and eddy radius) [4], most mesoscale eddies have lifespans of only 10–15 days. From the perspective of spatiotemporal sampling theory, conventional altimetry captures transient processes of this scale with a probability of merely 38%.

Figure 1 compares the spatial coverage patterns of Jason altimetry data and spaceborne GNSS-R observations over different temporal windows. Jason altimetry achieves near-global latitudinal coverage (up to $\pm 80°$), but its daily sampling remains sparse. As shown in Fig. 1b, even with five days of accumulation, large spatial gaps persist. By contrast, Fig. 1c illustrates that GNSS-R observations from the CYGNSS constellation, though confined to $\pm$ 40° latitude, provide nearly continuous daily coverage within their accessible region, demonstrating a clear advantage in temporal sampling density over conventional altimetry.

The results indicate that when relying solely on a single conventional radar altimeter, eddy detection is fundamentally constrained by the satellite's revisit cycle, making it difficult to achieve global coverage within short time windows. However, oceanic eddies are spatially coherent phenomena with characteristic areas and shapes, and their effective detection requires capturing large-scale features such as sea surface height anomalies. Although altimeter ground tracks can resolve localized amplitude variations when passing across an eddy, they fail to recover the complete eddy structure, and thus cannot accurately characterize eddy dynamics over short timescales. This limitation highlights the necessity of adopting high-temporal-resolution observation techniques to enable continuous monitoring of eddy evolution.

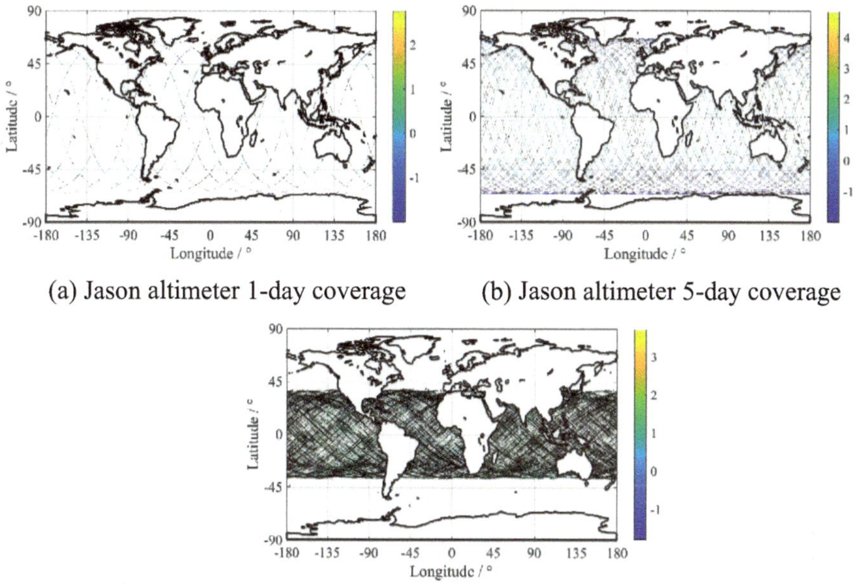

(a) Jason altimeter 1-day coverage

(b) Jason altimeter 5-day coverage

(c) Spaceborne GNSS-R technology 1-day coverage

Fig. 1 Jason altimeter and GNSS-R coverage

Figure 2 provides a quantitative comparison of the coverage performance of Jason altimetry and spaceborne GNSS-R. The daily coverage rate of conventional altimetry is less than 20%, reaching approximately 80% only after ten days, at which point interpolation methods are typically required to fill spatial gaps. In contrast, GNSS-R demonstrates a decisive advantage by enabling eddy detection at high temporal resolution, thereby offering a more effective means of resolving eddy structure and dynamics in near real-time.

The underlying reason lies in the observational geometry of the satellites themselves. Jason, as a single-platform mission, provides only a limited number of ground tracks, resulting in sparse spatial sampling. By contrast, spaceborne GNSS-R platforms benefit from abundant signal sources. Taking CYGNSS as an example, the constellation consists of eight microsatellites, each equipped with a receiver of four reflection channels, and key observables such as delay-Doppler maps (DDMs) which are generated at a temporal resolution of 0.5 sec. This configuration endows GNSS-R with a distinct advantage for eddy detection, enabling dense sampling and continuous monitoring capabilities that cannot be achieved by traditional altimetry.

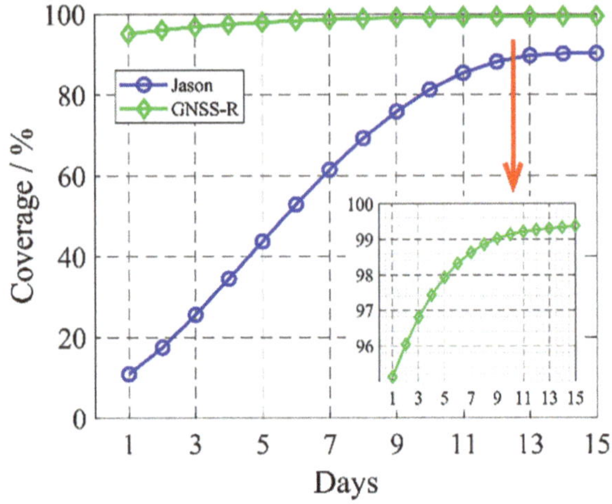

Fig. 2 Time evolution of detection coverage

1.2 Gradient Aliasing in High-Density Eddy Fields

As a characteristic energy-dissipating structure, oceanic eddies exhibit well-defined geometric features and spatial extents, and their effective detection relies on densely sampled, high-resolution gridded observations over target regions. At the methodological level, variations in sea surface height (SSH) fields constitute the primary feature for eddy identification, where eddy centers correspond to SSH extrema and eddy boundaries manifest as regions of pronounced gradients. Notably, the morphological evolution of eddies is modulated by the coupled dynamics of ocean currents, temperature, and salinity, leading to significant spatiotemporal heterogeneity in boundary gradient signatures. More than 85% of neighboring eddy centers are separated by less than three times the spatial resolution, such that the minimum contour spacing is reduced to below a single pixel, resulting in sub-pixel-level overlap. This spatial clustering produces severe aliasing effects in the SSH gradient field, whereby multiple eddy signatures become intertwined. In fact, the maximum gradient difference between adjacent eddies is often smaller than 0.02 m per pixel, leading conventional morphological segmentation algorithms to exhibit a false alarm rate exceeding 15% as well as systematic biases in boundary localization.

Existing approaches typically rely on gradient-based operators, which exhibit inherent limitations when applied to global SSH variation detection [5, 6]. Such operator-based methods are built upon the assumption that sea surface height fields vary smoothly. However, in densely populated eddy regions, the fuzzy nature of eddy boundaries and the complexity of their morphology hinder accurate boundary delineation. As a result, adjacent eddies often cannot be effectively separated and may even be erroneously merged into a single eddy, thereby compromising the accuracy of eddy detection [7, 8].

2 Eddy Detection

This section presents the eddy detection workflow based on GNSS-R technology, which primarily consists of eddy center detection and radius estimation. The feasibility of the approach is assessed through performance evaluation, followed by a comparison with conventional radar altimetry to highlight the advantages brought by the higher temporal resolution of GNSS-R.

2.1 Detection Based on Global Height Features

Eddy Center Detection

The detection process begins with the globally gridded daily-mean SSH dataset obtained. By incorporating the geometric relationship of the geoid, daily-mean absolute dynamic topography (ADT) contours are computed and plotted on a global scale. To extract candidate eddy regions, the geometric properties of these contours are further analyzed. Eddies are typically manifested as closed contour loops, whose shapes are closely related to their physical characteristics. Therefore, the key task lies in reliably detecting and selecting such closed contours, which may represent potential eddy structures [9].

To ensure that the selected contours are indeed closed, the endpoint distance of each contour is calculated. Contours with endpoint distances less than 25 km are classified as closed loops. Through this procedure, a collection of closed contours consistent with eddy characteristics is obtained, forming the basis for subsequent eddy identification and analysis.

Next, the closed contours are fitted using the least-squares method to derive an equivalent best-fit circumference [10]. On this basis, an eddy identification filter is applied to remove features that do not conform to expected eddy properties. The filtering criteria primarily consist of four categories [11]:

(1) Excessive shape error. As noted earlier, standard oceanic eddies are generally approximated as circular features. Thus, the shape error (ε) is used as a criterion. This error is defined as the difference between the actual area of the closed contour and the area of an idealized circle with the same equivalent radius. The shape error is computed according to Eq. (1):

$$E_s = \frac{S_{pol} + S_{cir} - 2 \times S_{int}}{S_{cir}} \times 100 \tag{1}$$

where S_{pol} denotes the actual area of the closed contour, S_{cir} denotes the area of the equivalent circle, and S_{int} denotes the overlap area between them. A smaller shape error indicates that the identified eddy is closer to circular; conversely, a larger error implies a greater deviation from circularity. A threshold is set on the shape error to filter candidates and improve identification accuracy.

(2) If land is included within the area enclosed by the identified closed contour, that contour is discarded.
(3) Abnormal ADT values within the closed contour: according to ocean eddy characteristics, only one ADT extremum should exist inside a closed contour. If, for a cyclonic eddy, any sampled point inside the contour has an ADT value higher than the contour ADT, the contour is removed; likewise, for an anticyclonic eddy, if any sampled point has an ADT value lower than the contour ADT, the contour is removed.
(4) Pixel-count thresholds inside the closed contour: given the gridded resolution adopted in this study, the contour size must fall within a prescribed range. Contours that are too small (fewer than 4 pixels) or too large (more than 1000 pixels) are removed.

After filtering by the four criteria above, two-dimensional partial derivatives are computed. Local minima are then detected following the CSS11 criterion to determine eddy centers [12]. Specifically, each pixel and its 3 × 3 neighborhood are traversed; neighborhood values are extracted and stored in a temporary array; a generic mask is applied to suppress pixels that do not meet the conditions. By comparing each pixel with its neighbors, local minima are identified, and the coordinates of those that meet the selection criteria are recorded. If the number of local minima exceeds a predefined limit, adjacent extrema are clustered and their centroid is computed and taken as the representative point. The count of local minima is further checked against the preset condition; when it exceeds the limit, grouping and centroid computation are reapplied to optimize the result and enhance identification accuracy. Once the maximum or minimum within the closed contour is identified, the corresponding pixel is taken as the eddy center.

The eddy amplitude is computed as the difference between the ADT at the eddy center and the ADT along the contour. Amplitude-based filtering is then applied to remove outliers (excessively large or small amplitudes). Finally, an ocean eddy at the location is confirmed, and its type is determined—cyclonic or anticyclonic—based on the amplitude.

Eddy Radius Detection

The eddy radius is a key characteristic in oceanic eddy research, essential for understanding eddy intensity, spatial extent, and dynamical impacts on the surrounding environment. Here, the radius is obtained by fitting a circle to the locus of maximum azimuthal velocity within the eddy.

Assuming geostrophic balance between the Coriolis force and the pressure-gradient force, surface geostrophic velocities are inferred from global sea surface height, thereby avoiding direct reliance on observed velocities and effectively capturing eddy characteristics and their spatial variability across regions. The geostrophic velocity components ($U^{'}$, $V^{'}$) are computed as in Eqs. (2) and (3) [13]:

$$U' = -\frac{g}{f}\frac{\partial(ADT)}{\partial y} \tag{2}$$

$$V' = \frac{g}{f}\frac{\partial(ADT)}{\partial x} \tag{3}$$

Here, g denotes gravitational acceleration, f is the Coriolis parameter, and $\partial(x)$ and $\partial(y)$ are the zonal (eastward) and meridional (northward) distances, respectively. Given the surface geostrophic velocity components, the eddy kinetic energy (EKE) is defined by Eq. (4):

$$EKE = \sqrt{U'^2 + V'^2} \tag{4}$$

Next, contours are drawn on the derived EKE map and an iterative search is initialized. A step size and an initial contour level are set to obtain the centroid of the region enclosed by the current contour. If no suitable contour is found, or if the current-level contour lies outside the original contour, the iteration terminates. For each contour level, the pixel count and mean speed are computed. If the pixel count exceeds a minimum threshold and the mean speed is greater than or equal to the current maximum, the maximum speed and its corresponding contour are updated. The radius computed at this maximum-speed contour is taken as the eddy radius. Operationally, the method examines geostrophic speed along successive contour levels to identify the contour meeting these conditions; throughout the layer-by-layer analysis, both speed and pixel count are tracked to recover the key speed and its associated contour. The eddy area is then determined by fitting a circle to the kinetic-energy pattern, with the fitted circle's radius assigned as the eddy radius.

Using the above procedure, both the eddy center and its radius are determined, enabling robust extraction of the eddy's spatial characteristics and thereby supporting GNSS-R–based eddy detection.

2.2 *Field/ACE-BOC Signal Validation*

Eddy detection is a typical classification task. Performance is quantitatively evaluated using the following metrics [14].

- Precision is defined as:

$$p_{pre} = \frac{TP}{TP + FP} \times 100\% \tag{5}$$

In Eq. (5), *TP* (True Positives) denotes number of correctly detected eddies, whereas *FP* (False Positives) refers to number of regions incorrectly detected as eddies (where no eddy actually exists). In this study, when the GNSS-R–detected eddy location overlaps the true eddy location in the AVISO reference dataset, the two eddies are deemed a successful match and counted as *TP*. As illustrated schematically in Fig. 3, the positions of the detected and reference eddies are compared to quantify spatial

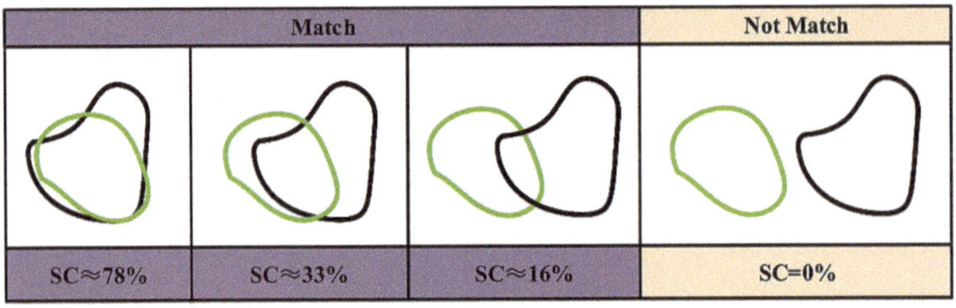

Fig. 3 Schematic diagram of the matching principle between detected eddy and actual eddy

consistency. The green closed curve represents the true eddy contour, while the black isoline denotes the detected eddy. The similarity coefficient (SC) measures the degree of spatial overlap between the two contours: as the positional deviation increases, SC gradually decreases toward zero, indicating a loss of correspondence between the detected and true eddies. It is worth noting that SC evaluates the geometric similarity (i.e., location accuracy) of individual detections, rather than their overall detection rate. In contrast, precision measures the proportion of correctly identified eddies among all detections classified as eddies by the model. Therefore, SC reflects how well a detected eddy matches its reference counterpart in space, whereas precision reflects how many detections correspond to real eddies. In this study, the two metrics are used jointly to assess both the detection reliability and spatial accuracy of eddy identification.

- Recall is defined by:

$$p_{rec} = \frac{TP}{TP + FN} \times 100\% \quad (6)$$

In Eq. (6), *FN* (False Negatives) denotes number of eddies that actually exist but are not detected. This metric measures, among all true eddies, the proportion correctly detected by the proposed method, thereby reflecting the model's detection capability.

- F1 Score is defined by:

$$p_{\mathrm{F1}} = 2\frac{p_{pre} \cdot p_{rec}}{p_{pre} + p_{rec}} \times 100\% \quad (7)$$

As shown in Eq. (7), this metric is the harmonic mean of precision and recall, jointly accounting for the effects of false positives and false negatives to balance the two. Under class-imbalance conditions, it provides a comprehensive assessment of algorithm performance.

- Accuracy is defined as:

$$p_{acc} = \frac{TP + TN}{TP + TN + FP + FN} \times 100\% \quad (8)$$

In Eq. (8), *TN* (True Negatives) refers to the number of the eddies correctly identified as non-eddy. In this work, TN = 0, because the model does not explicitly label non-eddy regions as "correctly identified non-eddy." This metric measures the overall correctness across all classes, including the proportions correctly identified as eddy and non-eddy, reflecting the algorithm's overall classification performance when contrasting eddy vs. non-eddy detection.

To evaluate detected eddy characteristics against ground truth, five assessment indicators are defined, addressing the eddy center, radius, amplitude, relative error, and shape.

- Distance Error (DE):

$$E_D = \mathbf{P}_{\text{detected}} - \mathbf{P}_{\text{true}} \tag{9}$$

In Eq. (9), $\mathbf{P}_{\text{detected}}$ and $\mathbf{P}_{\text{true}}$ are the centers of the detected eddy and the true eddy, respectively.

- Radius Error (RE):

$$E_{\text{R}} = r_{\text{detected}} - r_{\text{true}} \tag{10}$$

In Eq. (10), r_{detected} and r_{true} denote the radii of the detected and true eddies, respectively.

- Amplitude Error (AE):

$$E_{\text{A}} = A_{\text{detected}} - A_{\text{true}} \tag{11}$$

In Eq. (11), A_{detected} and A_{true} are the amplitudes of the detected and true eddies, respectively.

- Integrated Relative Error (IRE):

$$E_{\text{I}} = \sqrt{\left(\frac{E_{\text{A}}}{A_{\text{detected}}}\right)^2 + \left(\frac{E_{\text{R}}}{r_{\text{detected}}}\right)^2 + \left(\frac{E_{\text{D}}}{d_{\text{ref}}}\right)^2} \tag{12}$$

In Eq. (12), d_{ref} is a reference distance, set to 125 in this study based on empirical data. Using the three structural descriptors—amplitude, velocity, and eddy-center position—the accuracy of ocean eddies detection is quantified. Unlike the area-based similarity coefficient, the IRE evaluates the match between detected and true eddies from a structural perspective.

- Similarity Coefficient (SC):

$$SC = \frac{A_{\text{c}} \cap A_{\text{true}}}{A_{\text{c}} \cup A_{\text{true}}} \times 100\% \tag{13}$$

In Eq. (13), A_{true} is the region of the true ocean eddy, while ∩ and ∪ denote intersection and union, respectively. The principle is to measure similarity by comparing the overlap between the detected eddy region and the true eddy region. SC ranges from 0 to 100%; larger values indicate higher similarity between the detected eddy and the actual eddy. SC quantifies the geometric correspondence between detected and reference eddies, whereas precision measures the fraction of detections that are true eddies. Thus, SC reflects spatial matching accuracy, while precision reflects detection reliability. Both metrics are jointly employed to evaluate the overall performance of eddy identification.

This section employs the above metrics for comprehensive validation and evaluation. Based on the global daily mean SSH from GNSS-R (fusing CYGNSS and FY-3E platforms) in the previous chapter, AVISO eddy ground-truth data are used as a reference. Year-2023 data within ± 40° latitude (a total span of 80°) are used for validation. Performance is assessed with ACE-BOC signaling, taking the BeiDou B2 signal as the representative, to demonstrate the advantage of high-bandwidth signals in ocean eddy detection.

Figure 4 presents a cyclonic eddy detection case on 14 July, covering 31–35° N and 142–146° E. The background shows the ADT ranging approximately from −0.7 to 0.3. The detected large cyclonic eddy exhibits an amplitude exceeding 1.0 m; such large-amplitude eddies are easier to identify. The ground-truth closed isopleth is elliptical. Contours are overlaid: green lines delineate the eddy region, yellow lines denote the circle fitted to the eddy, and pentagrams mark the eddy centers in the two panels. Figure 4a shows the GNSS-R–detected eddy, while Fig. 4b shows the AVISO ground truth; visually, the two appear highly similar. Noticeable irregularities (bulges and indentations) in the GNSS-R detection constrain precise eddy delineation, mainly due to the altimetric precision limits when detecting eddies with BPSK-based signals from the in-situ satellite platform. Quantitatively, the inter-center distance error is 2 km, the radius error is 7.8 km, the amplitude error is 7 cm, the integrated relative error (IRE) is 2.1, and the similarity coefficient (SC) is 56%. These discrepancies indicate that while ADT derived from GNSS-R can effectively detect certain large-scale eddies, the retrieved eddy shapes still exhibit non-negligible errors.

As shown in Fig. 5, only detected eddies are displayed (undetected eddies are omitted). The background again depicts the ADT field. For clarity, cyclonic eddies (CE) and anticyclonic eddies (AE) are displayed over 0–180° and 180–360° longitude ranges, respectively. Blue and red closed curves indicate GNSS-R detections and ground-truth eddies, respectively. Taking 1 July as an example, the detected eddies broadly coincide with the ground truth in location, but their shapes differ markedly.

Using the reference dataset, year-2023 statistics are compiled. There are 871,693 cyclonic and 836,924 anticyclonic eddies in the ground truth. The GNSS-R detections contain 1,601,283 cyclonic and 1,598,264 anticyclonic eddies. Of these, true positives number 779,259 (cyclonic) and 749,364 (anticyclonic). These results indicate a substantial fraction of detections are misidentified. The corresponding performance metrics are listed in Table 1. The underlying cause is the limited global gridded altimetric precision of the measured GNSS-R SSH, which constrains eddy-detection accuracy.

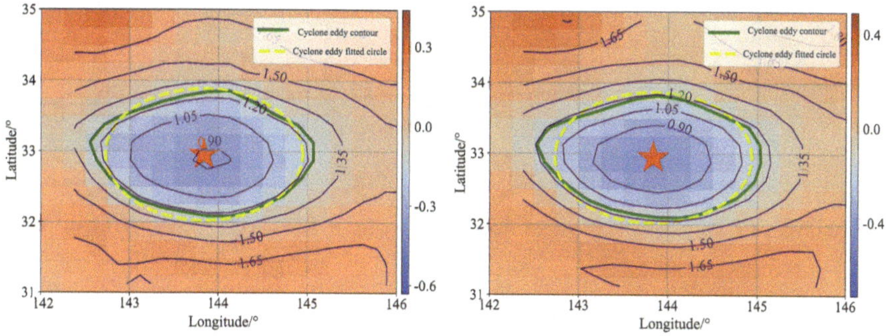

(a) GNSS-R eddy detection case (b) Actual eddy case

Fig. 4 Ocean eddy case study

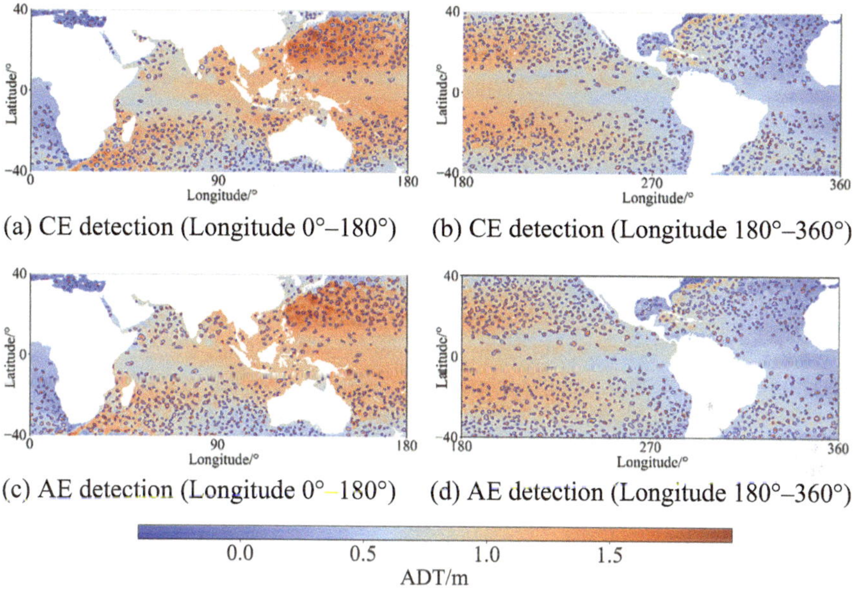

(a) CE detection (Longitude 0°–180°) (b) CE detection (Longitude 180°–360°)

(c) AE detection (Longitude 0°–180°) (d) AE detection (Longitude 180°–360°)

Fig. 5 Eddy detection example using GNSS-R measured signals

Table 1 Metrics for eddy detection in GNSS-R measured signals

Metric	CE/%	AE/%
Precision	48.69	46.89
Recall	89.40	89.53
F1 score	62.83	61.52
Accuracy	45.23	45.24

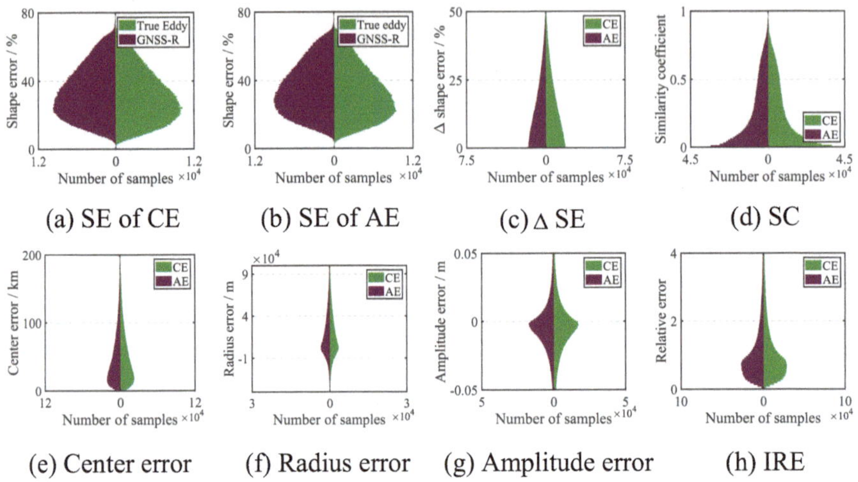

Fig. 6 Statistics of eddy detection performance using GNSS-R measured signals

Figure 6 shows the distributions of each metric, with the vertical axis denoting the metric and the horizontal axis the sample count; the left and right panels share the same y-axis. Errors are widely distributed; shape error (SE) is concentrated below 5%, while most SC values are < 0.1, indicating notable shape discrepancies. Center-distance, radius, and amplitude errors are relatively large, and IRE clusters between 0.2 and 0.9. Overall, although measured GNSS-R enables daily detection, deficiencies persist in recovered shapes and physical structure.

Using ACE-BOC high-bandwidth signaling, 998,077 cyclonic eddies are detected over the year (ground truth: 871,693); for anticyclonic eddies, 949,001 are detected (ground truth: 836,924). True positives are 783,219 (cyclonic) and 743,472 (anticyclonic). The distributions are shown in Fig. 7.

Table 2 summarizes four statistical indices. ACE-BOC exhibits clear advantages over measured BPSK signals: detection metrics improve markedly, with precision and accuracy both exceeding 70%—more than a 20% gain with high bandwidth. F1 score and recall also improve to varying degrees. The fundamental driver is altimetric precision; the previous chapter showed that high-bandwidth signals can achieve ~1 cm gridded precision, which is highly beneficial for eddy detection. The distributions of shape and physical-property indices are detailed in Fig. 8.

Under ACE-BOC, the SE distribution for both cyclonic and anticyclonic eddies more closely matches the ground truth, and the SC distribution is closer to 1 than with BPSK. Center-distance errors fall within 25 km, radius errors within 20 km. Amplitude-error samples cluster within 1 cm, evidencing a substantial improvement, and IRE concentrates below 1.

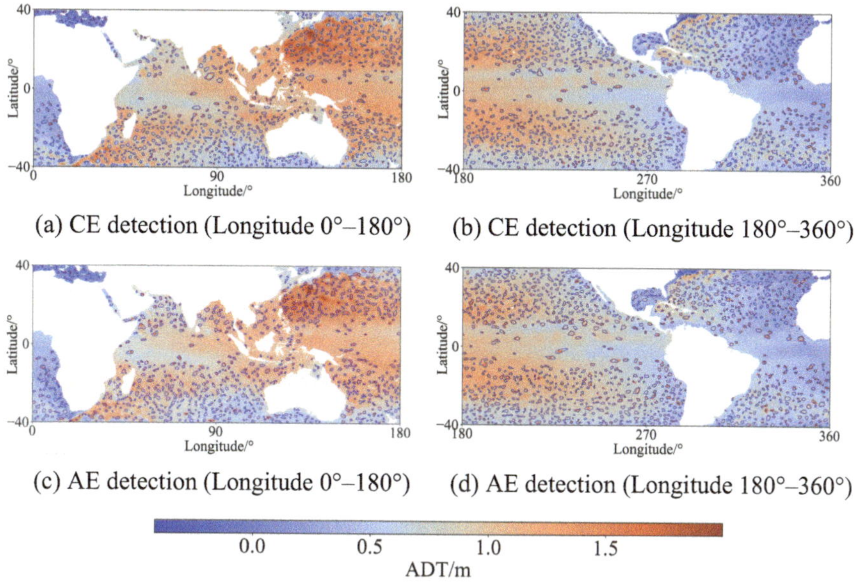

(a) CE detection (Longitude 0°–180°) (b) CE detection (Longitude 180°–360°)

(c) AE detection (Longitude 0°–180°) (d) AE detection (Longitude 180°–360°)

Fig. 7 Demonstration of eddy detection in ACE-BOC signals

Table 2 Metrics for eddy detection in ACE-BOC signals

Metric	CE/%	AE/%
Precision	78.47	78.34
Recall	89.85	88.83
F1 score	83.78	83.26
Accuracy	72.08	71.32

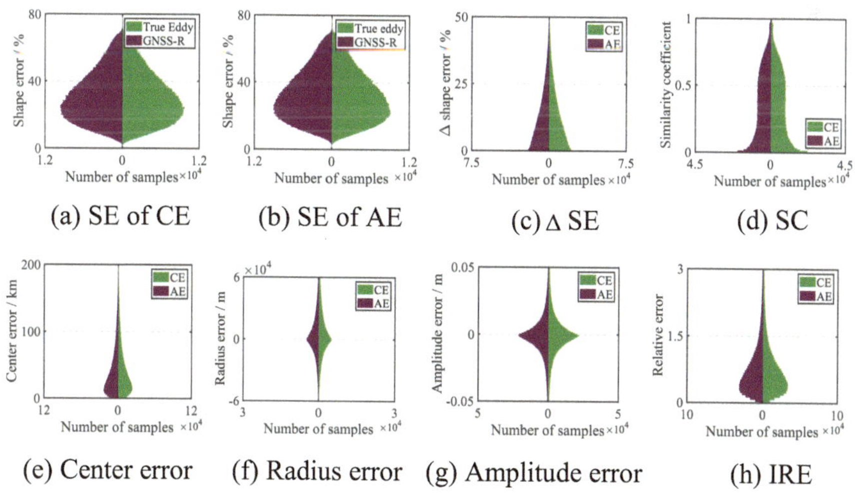

(a) SE of CE (b) SE of AE (c) Δ SE (d) SC

(e) Center error (f) Radius error (g) Amplitude error (h) IRE

Fig. 8 Statistics of eddy detection performance using ACE-BOC signals

2.3 Comparative Experiment with Conventional Altimetry

To evaluate the operational potential of GNSS-R technology in ocean eddy detection, this study systematically compares eddy detection results derived from novel GNSS-R observations and from the current-generation Jason-3 altimeter. Although Jason-3 achieves centimeter-level accuracy via radar altimetry, its revisit cycle is constrained by its sun-synchronous orbit, making it unsuitable for short-term, continuous global ocean coverage. In contrast, GNSS-R leverages the abundant signal sources provided by global navigation satellite systems to achieve higher spatiotemporal sampling density, rendering it particularly suitable for observing highly dynamic and short-period oceanic phenomena.

This section conducts a performance assessment of GNSS-R eddy detection accuracy and applicability by directly comparing it to Jason-3 data, aiming to explore GNSS-R as a complementary data source to conventional altimetry. The goal is to harness the high spatiotemporal resolution of GNSS-R to enhance future ocean dynamic monitoring capabilities.

In this experiment, a 10-day cumulative dataset from Jason-3 was used to construct a global eddy-detection baseline, whereas GNSS-R utilized ACE-BOC high-bandwidth signal reflections to produce daily observations. Eddy detection was performed using synchronous observations from January 1–10, 2023. With GNSS-R, a global sea surface height (SSH) map was interpolated every 6 h from Day 1; due to Jason-3's sparse data points, global SSH maps were generated starting from Day 4. Notably, gridded SSH maps exhibit temporal correlation, allowing previous-day height estimates to serve as priors for current-day interpolation, thereby filling spatial gaps.

Figure 9 evaluates the detection performance using precision as an example. GNSS-R can deliver eddy detection results within a single day, achieving a stable precision of approximately 78%. Although Jason-3 cannot achieve global coverage within the first three days, it ultimately reaches a precision exceeding 90%, owing to its superior altimetric accuracy, which ensures highly reliable eddy detection.

Detailed performance comparison is provided in Table 3. The results indicate that although Jason-3 offers a clear accuracy advantage in single-pass detection (with precision and accuracy exceeding GNSS-R by 13.25% and 17.96%, respectively), GNSS-R's key advantage lies in its real-time capability—offering daily temporal resolution—which proves especially effective in capturing high-frequency eddy events. Importantly, for the remaining two performance metrics, GNSS-R's gap with Jason-3 is less than 10%, demonstrating that ACE-BOC–based GNSS-R achieves detection performance broadly comparable to satellite altimeters. Moreover, the daily GNSS-R data enables continuous eddy tracking, whereas Jason-3, with its 10-day revisit cycle, can only provide temporally discrete eddy snapshots.

In summary, GNSS-R's global, high-frequency coverage offers a powerful complement to conventional satellite altimetry, addressing the limitations posed by long revisit intervals. While conventional altimeters maintain superior measurement precision and detection accuracy over large-scale domains, their inability to

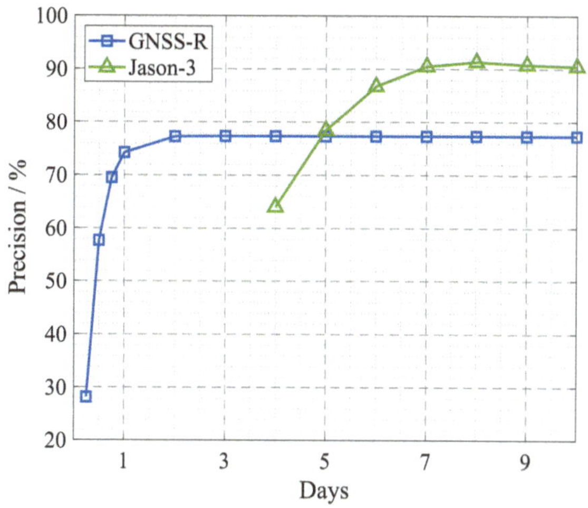

Fig. 9 Eddy detection: Jason-3 versus GNSS-R

Table 3 Eddy detection performance: GNSS-R versus Jason-3 altimeter

Metric	GNSS-R	Jason-3
Data duration	1 day	10 day
Precision	78.14%	91.39%
Recall	88.76%	92.85%
F1 score	83.51%	91.25%
Accuracy	70.91%	88.87%

provide global coverage within a single day makes direct comparisons with GNSS-R impractical. This study explores the application potential of GNSS-R in ocean eddy detection and offers theoretical support for future satellite missions equipped with high-bandwidth GNSS-R systems. In particular, with the gradual deployment of high-bandwidth signals such as BeiDou B2 on spaceborne GNSS-R platforms, the methods proposed in this study demonstrate the feasibility of achieving altimeter-comparable precision, opening new avenues for the development and deployment of advanced ocean eddy detection technologies.

3 Local Shape Feature Refinement

Although the previous section demonstrated the feasibility of ocean eddy detection using GNSS-R technology, the overall detection performance—particularly in terms of precision and contour delineation—still requires improvement. Accurate identification of eddy contours, especially in densely populated eddy regions, is essential for precise oceanographic interpretation. To address this, the present section introduces a refined eddy detection framework that leverages boundary information to estimate eddy centers and utilizes the local EKE field to correct eddy radii, thereby enabling the extraction of local shape features for enhanced detection.

3.1 Boundary-Driven Eddy Detection

As illustrated in Fig. 10, in contrast to traditional global detection methods for eddy centers, this approach introduces a boundary-guided strategy. After generating global isocontours from the ADT field, a high-pass filter based on a Bessel kernel is constructed and applied to the global ADT map via two-dimensional convolution. This filtering process suppresses low-frequency components while enhancing high-frequency signals associated with eddy boundaries, thus improving the sharpness and clarity of the eddy contour structures.

This boundary-enhanced ADT field enables more precise localization of eddy centers and provides a basis for refining contour-based radius estimation, particularly in complex or overlapping eddy environments.

During the kernel construction process, this study establishes a scale-matching mechanism based on the spatial characteristics of mesoscale oceanic eddies. Considering the detection requirement for target eddies with a characteristic diameter of 400 km, the grid resolution is first determined through geospatial transformation with Earth curvature correction. At latitude φ, the meridional resolution is constant at $\Delta y = 6371km \times (0.25^{\circ} \times \pi/180) \approx 27.8km$, while the zonal resolution decreases

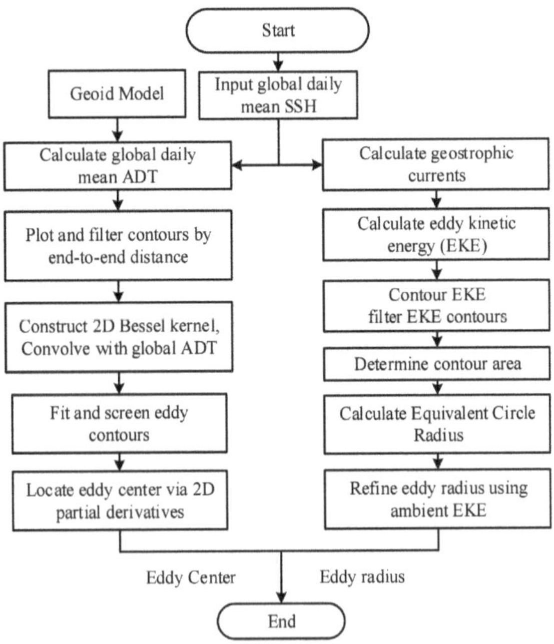

Fig. 10 Flowchart of eddy detection

with latitude according to $\Delta x = \Delta y \times \cos \varphi$.This mapping ensures the global consistency of spatial reference. Subject to the constraint of the maximum eddy scale, the kernel size in the zonal direction is defined as $N\_x = 2 \times \lceil 400/(2\Delta x) \rceil + 1$, and in the meridional direction as $N\_y = 2 \times \lceil 400/(2\Delta y) \rceil + 1$, where Γ· denotes the ceiling operator. This design guarantees that the physical coverage of the filtering kernel strictly exceeds the maximum possible diameter of the target eddy. For example, in a given eddy scenario with $\Delta x = 20$ km, we obtain $N\_x = 2 \times \lceil 400/(2 \times 20) \rceil + 1 = 21$, $N\_y = 2 \times \lceil 400/(2 \times 27.8) \rceil + 1 = 29$. This results in a 21 × 29 rectangular kernel structure. By dynamically adjusting the kernel geometry, the latitude-adaptive kernel configuration effectively compensates for the meridional grid compression effect at high latitudes.

In this study, the first-order Bessel function is employed to achieve frequency-domain separation of mesoscale eddy signals. A normalized distance matrix is utilized to ensure that the kernel shape effectively extracts the high-frequency components of the signal. The Bessel kernel is parameterized within the mathematical framework of functional analysis. Let the Cartesian grid coordinates of the discrete convolution kernel be $(x_i, y_j) \in \mathbb{Z}^2$, with the kernel center located at the origin (0, 0). Then, the normalized spatial distance function is defined as

$$r\big(x_i, y_j; \phi\big) = \sqrt{\left(\frac{x_i \cdot \Delta_V(\phi)}{L}\right)^2 + \left(\frac{y_j \cdot \Delta_g}{L}\right)^2} \tag{14}$$

where $\Delta_x(\phi) = 27.8 \cos \phi$ represents the zonal grid spacing at latitude ϕ (derived from Earth's radius R = 6371 km and a grid resolution of 0.25°), while $\Delta_y = 27.8km$ is the constant meridional spacing, and L = 400 km denotes the characteristic eddy scale.

On this basis, the normalized distance matrix is employed to compute the kernel function, which acts as a filter in signal processing. The shape of the kernel directly determines the filtering performance. By accurately constructing the kernel function, noise can be effectively suppressed or specific frequency components can be extracted, thereby improving the precision of signal processing. This procedure ensures that the kernel's shape and dimensions are precisely matched to the spatial data resolution, optimizing overall processing performance. The kernel function is expressed as Eq. (15):

$$\text{kernel}(r) = \frac{\sin\left(\frac{r}{n}\right)}{\frac{r}{n}} \cdot \frac{J_1(2\pi r)}{r} \tag{15}$$

where r is the normalized distance and n denotes the kernel order, set to 1. $J_1(2\pi r)$ is the first-order Bessel function of the first kind. The definition of $J_1(x)$ is given by:

$$J_1(x) = \sum_{m=0}^{\infty} \frac{(-1)^m}{m!\Gamma(m+2)} \left(\frac{x}{2}\right)^{2m+1} \tag{16}$$

Here, $J_1(x)$ represents the first-order Bessel function of the first kind, and the Gamma function $\Gamma(m+2) = (m+1)!$ ensures series convergence. The sinc term, with its $\propto 1/r$ decay property, suppresses Gibbs oscillations, while the zero distribution of the Bessel term (given by the solution set of $J_1(2\pi r) = 0$) produces a sharp transition band in the wavenumber domain. Its cutoff frequency $f_c = 1/(2L_x)$ corresponds to the minimum detectable eddy scale (approximately 200 km).

During kernel optimization, computational efficiency is enhanced through multi-scale symmetry constraints and sparsification. An extended tensor $\mathcal{K}^* \in \mathbb{R}^{(2M+1)\times(2N+1)}$ is constructed, and mirror-symmetry operations are applied to satisfy:

$$\mathcal{K}^*_{i,j} = \begin{cases} \mathcal{K}_{M+i,N+j} \; i \in [-M, M], j \in \Big[-N, N \\ \mathcal{K}_{M-i,N+j} \; \textit{Meridional Symmetric Extension} \\ \mathcal{K}_{M+i,N-j} \; \textit{Linear Symmetric Extension} \end{cases} \tag{17}$$

This ensures that the kernel satisfies the even-symmetry conditions $\mathcal{K}^*(-x, y) = \mathcal{K}^*(x, y)$ and $\mathcal{K}^*(x, -y) = \mathcal{K}^*(x, y)$. A subsequent threshold-truncation step is then applied, where the energy truncation criterion is defined by:

$$\mathbb{I}_{mn} = \begin{cases} 0 \text{ if} \left|\mathcal{K}^*_{mn}\right| < \varepsilon \cdot \max\left|\mathcal{K}^*\right| \\ 1 \text{ others} \end{cases} \tag{18}$$

with $\epsilon = 10^{-3}$ as the threshold parameter. Through Hadamard product operations, insignificant coefficients are sparsified, eliminating redundant elements that contribute less than 0.1% of the total energy. The resulting compact kernel matrix $\tilde{\mathcal{K}} \in \mathbb{R}^{p\times q}$ has dimensions determined by:

$$p = \lfloor 2L_x(\phi)/\Delta_x(\phi) \rfloor, q = \lfloor 2L_y/\Delta_y \rfloor \tag{19}$$

where $L_x(\phi) = 400/\cos\phi km$ is the dynamic zonal scale at latitude ϕ, and $\Delta_x(\phi) = 27.8\cos\phi km$ is the corresponding grid spacing. As illustrated in Fig. 11, at latitude −39.55° the kernel size reduces to 17×21, yielding a 23% reduction in computational cost. This approach retains 98.7% of the cumulative kernel energy, thereby reducing computational complexity while preserving eddy detection accuracy.

For the purpose of eddy radius detection, this section verifies the validity of the computed radius by quantifying the local radial gradient characteristics of the kinetic energy field at the eddy base. The radial gradient of the kinetic energy field is calculated using the central difference method, as defined as:

$$\nabla EKE(r) = \frac{EKE(r+\Delta r) - EKE(r-\Delta r)}{2\Delta r} \tag{20}$$

To suppress the influence of small-scale noise and ensure spatial consistency, the radial increment Δr is set to 1.5 times the grid resolution. A dual criterion is introduced to determine the location of the characteristic gradient: (i) the gradient

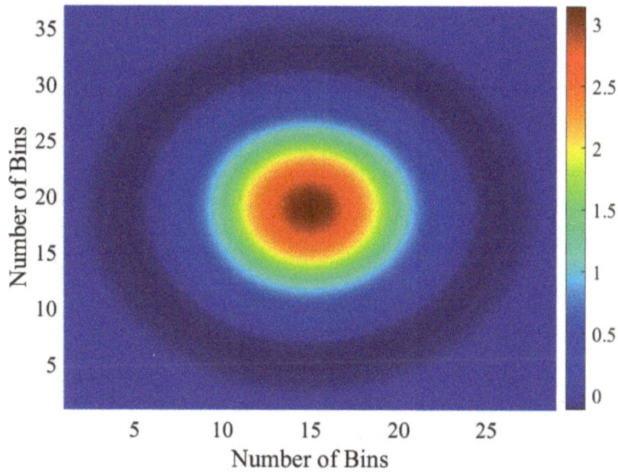

Fig. 11 Diagram of convolution kernel structure

magnitude reaches 90% of the global maximum value max (| ∇EKE|), and (ii) the gradient direction exhibits continuous clockwise or counterclockwise rotation along the contour. Considering the influence of the background kinetic energy field on the eddy structure, a correction is applied according to:

$$R_{adj} = R_{grad} \cdot \left[1 + \alpha \cdot \tanh\left(\frac{EKE_{\max}}{EKE_{\mathrm{bg}}}\right)\right] \tag{21}$$

where $\alpha = 0.2$ is a calibrated empirical coefficient, R_{adj} denotes the adjusted radius, R_{grad} is the initially estimated radius, $EKE_{\max}$ is the maximum eddy kinetic energy, and EKE_{bg} is the mean kinetic energy within a surrounding region defined by five times the eddy radius. This approach accounts for the influence of nearby eddy kinetic energy on the detected eddy and applies a dynamic compensation to refine the estimated radius.

3.2 Center Distance/Radius Error Comparison

Case Analysis

This section examines two representative mesoscale eddies occurred in the Northwest Pacific on January 1, 2023 (geographical locations shown in Fig. 12: (a) –10.7°S/64.8°E, and (b) 18.4°N/132.1°E). The experiments are conducted using ACE-BOC high-bandwidth data, with a local feature extraction algorithm compared against a conventional global feature extraction method.

In Case 1, the absolute deviation between the estimated eddy center obtained by the proposed method (purple circle) and the ground truth (red pentagram) is 0.98 km, whereas the deviation using the traditional method (dark green square) is 2.65 km, representing a 63% improvement in relative accuracy. These results demonstrate that analyzing local curvature features effectively suppresses contour-detection interference caused by aliasing in the height field, thereby yielding more

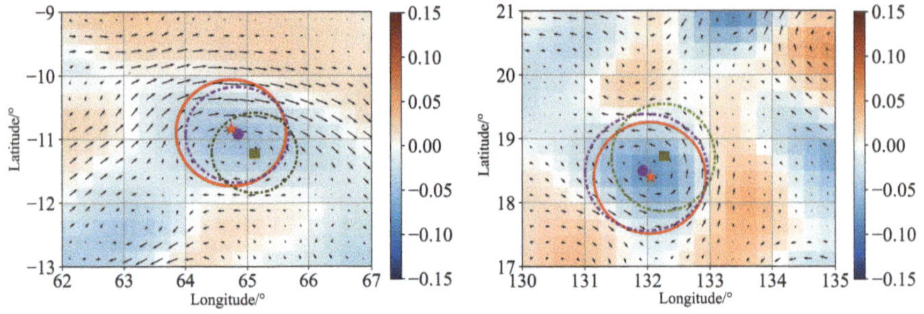

(a) Case 1 (b) Case 2

Fig. 12 Case Study of eddy center distance comparison

precise center estimates. Regarding radius detection, the eddy kinetic energy (EKE)-based approach (purple dashed circle) achieves higher accuracy than the traditional global feature extraction method (green dashed circle), with the reference eddy radius represented by the red solid circle. As shown in Fig. 12a, in the typical cyclonic eddy case, the true radius is 98.7 km. The proposed method successfully captures the kinetic energy maximum in the eddy core, yielding a radius of 89.3 km with a relative error of 9.6%. By contrast, the traditional method underestimates the radius as 51.6 km, with a deviation of 47.7% due to neglecting nonlinear advection effects.

In Case 2 (Fig. 12b), the proposed local feature extraction method achieves a center deviation of 1.67 km, whereas the traditional method shows a deviation of 4.67 km, representing a 64% improvement in accuracy. This further validates the robustness and applicability of the proposed approach under different scenarios. For radius detection, the true eddy radius is 88.9 km. The proposed method slightly overestimates it as 100.1 km, primarily due to background flow interference, whereas the traditional method underestimates it as 75.5 km, with an error of – 15.1%. These two case studies highlight the superior capability of the proposed method in eddy radius detection.

Annual Detection

This section further evaluates eddy detection results using ACE-BOC signals throughout 2023 (Fig. 13).

Statistical analysis shows that, for eddy center detection, the 68% confidence interval of the traditional global feature extraction method spans 55.5 km (42.3–97.8 km), with a mean error of 65.7 km, and approximately 27.7% of cases exceeding the 80 km error threshold. By contrast, the proposed local feature extraction method yields error distributions that conform to Gaussian statistics, compressing the 95% confidence interval to a 6.6 km range (5.1–11.7 km). The mean error is reduced to 8.3 km, representing an 87.4% improvement over the traditional method. Quantitative error mechanism analysis confirms the effectiveness of local gradient-based eddy contour feature extraction.

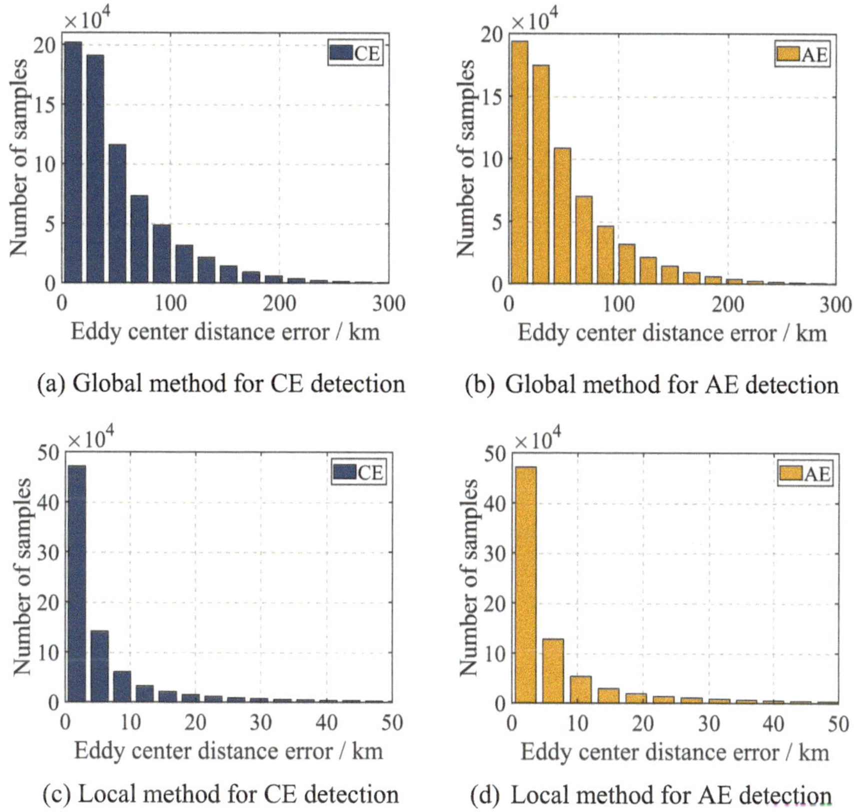

(a) Global method for CE detection (b) Global method for AE detection

(c) Local method for CE detection (d) Local method for AE detection

Fig. 13 Annual comparative experiment on center error using high-bandwidth GNSS-R

In addition, this study evaluates eddy radius detection performance throughout 2023 using ACE-BOC high-bandwidth signals (Fig. 14). Statistical results indicate that the proposed flow-field-based approach significantly enhances eddy radius detection accuracy. As shown in the histogram of error distributions, the radius errors of cyclonic/anticyclonic eddies derived from the traditional feature extraction method are mostly concentrated within ± 50 km. With the incorporation of the EKE-gradient characterization method, 95% of the samples fall within ±25 km, providing more reliable spatial parameter support for eddy detection.

4 Performance Analysis of GNSS-R Eddy Detection

In this section, daily mean sea surface height maps generated from 2023 GNSS-R observations (CYGNSS and FY-3E platforms) are used to validate the proposed algorithm in detecting the spatial distribution of mesoscale eddies.

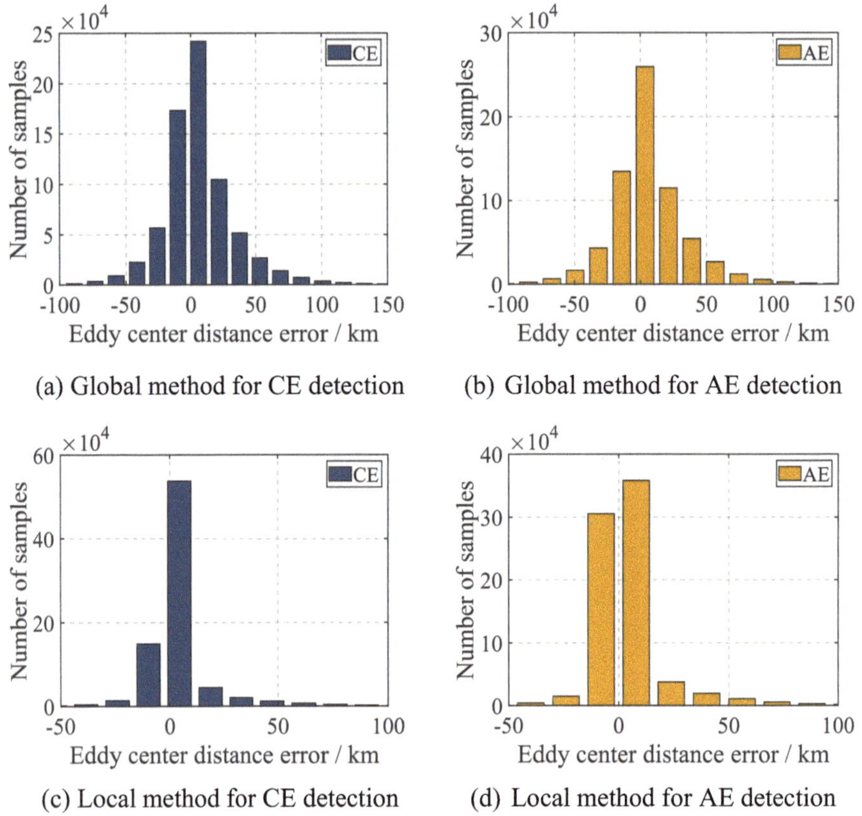

Fig. 14 Annual comparative experiment on radius error using high-bandwidth GNSS-R

Throughout 2023, as shown in Fig. 15, a total of 871,693 cyclonic eddies and 836,924 anticyclonic eddies were identified as ground truth. Using the proposed local-shape feature extraction method, 1,595,648 cyclonic and 1,587,259 anticyclonic eddies were detected, of which 784,287 and 758,362 were true detections, respectively.

The corresponding performance metrics are summarized in Table 4. Compared with Table 1, the results show that the local feature extraction method improves all key indicators for both cyclonic and anticyclonic detection. In particular, the recall rate increases by 0.57 and 1.08 percentage points, respectively, while the F1-scores also increase, reflecting the enhanced capability of the proposed approach in eddy detection.

Figure 16a, d show the shape error distributions of cyclonic and anticyclonic eddies, respectively. Most true and GNSS-R-detected eddies exhibit nearly circular shapes, with shape errors primarily distributed between 25 and 45%. However, notable differences exist: the peak shape error for true eddies occurs around 20%, while the GNSS-R-detected eddies peak is closer to 30%, indicating a systematic

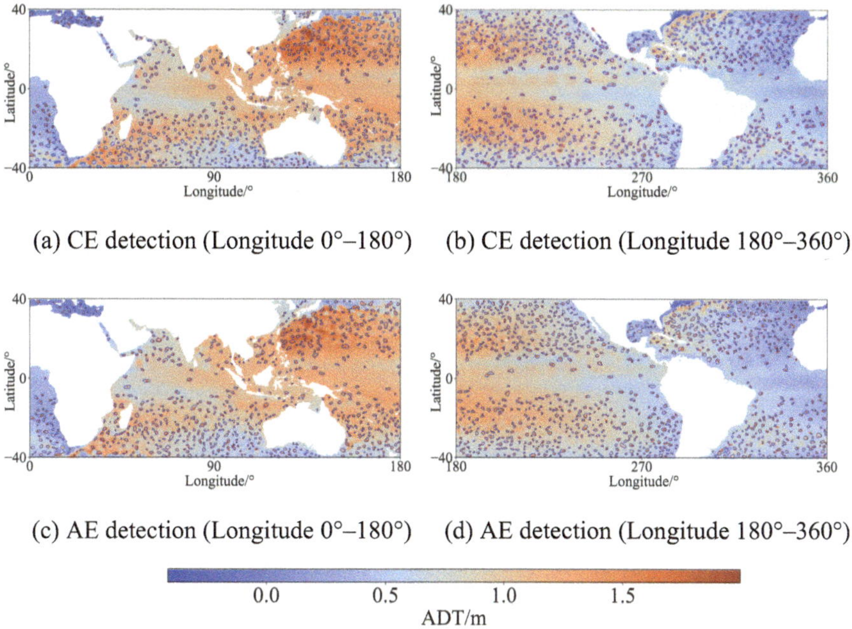

Fig. 15 Eddy detection example using GNSS-R measured signals (local feature extraction)

Table 4 Metrics for eddy detection in GNSS-R measured signals (local feature extraction)

Metric	CE/%	AE/%
Precision	49.15	47.78
Recall	89.97	90.61
F1 score	63.57	62.57
Accuracy	46.60	45.52

shape bias. This conclusion is further supported by Fig. 16c, which illustrates the distribution of shape error differences between true and CYGNSS-detected eddies, with most differences exceeding 10%. Figure 16d presents the similarity coefficients between detected and true eddies, showing a relatively uniform distribution without a clear downward trend. Although samples with values between 0.8 and 1 are slightly more prevalent than those near 0.4, the majority fall below 0.15, indicating low overall shape similarity between detected and true eddies.

To further assess structural differences, center distance errors, radius errors, and amplitude errors between true and GNSS-R-detected eddies were calculated. Figure 16e shows that center distance errors for both cyclonic and anticyclonic eddies follow similar distributions, with the vast majority under 50 km. Figure 16f shows that amplitude errors are mainly distributed between − 3 and 3 cm, with positive mean values, indicating that detected eddy amplitudes are generally larger than true values. Figure 16g illustrates radius error distributions ranging from − 50 to 200 km, with

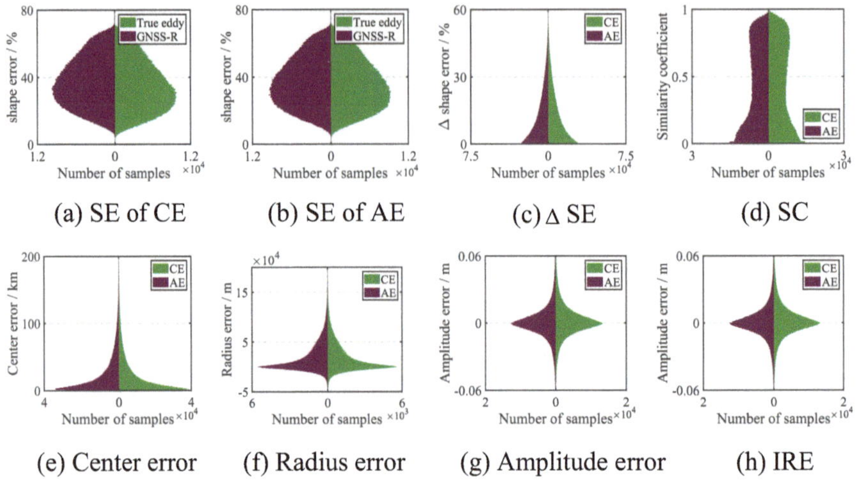

Fig. 16 Statistics of eddy detection performance using GNSS-R measured signals (local feature extraction)

an average positive bias of approximately 25 km, suggesting that GNSS-R-detected eddy radii are systematically overestimated. Figure 16h presents the distribution of overall relative errors, which are mostly between 1 and 3 for both cyclonic and anticyclonic eddies.

As shown in Fig. 17, the globally detected eddies exhibit strong spatial and morphological consistency with ground truth. Across the year, 931,798 cyclonic eddies and 907,423 anticyclonic eddies were detected, compared to 871,693 and 836,924 true eddies, respectively. Among these, 800,815 cyclonic and 765,329 anticyclonic eddies were correctly detected.

Comparing Table 5 with Table 2, the proposed local-shape feature extraction method significantly improves detection accuracy. Specifically, cyclonic eddy precision increases to 85.94% (+ 7.47%), achieved by exploiting local gradient variations in height field contours to filter out false detections caused by aliasing. Recall rises to 91.87% (+ 2.02%), confirming the method's effectiveness in retaining true eddy features. The F1-score improves to 88.81% (+ 5.03%), substantially alleviating the precision–recall imbalance observed in traditional methods (imbalance index reduced from 11.38 to 5.93%). For anticyclonic eddies, precision increases by 6.00 to 84.34%, recall improves by 2.62 to 91.45%, and the F1-score reaches 87.75% (+ 4.49%). These results highlight the universality of combining local gradient-field analysis with second-order curvature features across both eddy polarities. Overall accuracy improves by 7.79% (cyclonic) and 6.85% (anticyclonic), demonstrating that the proposed local feature extraction strategy successfully overcomes the limitations of global feature representations. Two key factors contribute to this improvement: (1) local morphological extraction at the 25 km grid scale effectively captures sharp gradient transitions between closed contours, and (2) enforcing gradient direction consistency suppresses false detections of spurious closed contours.

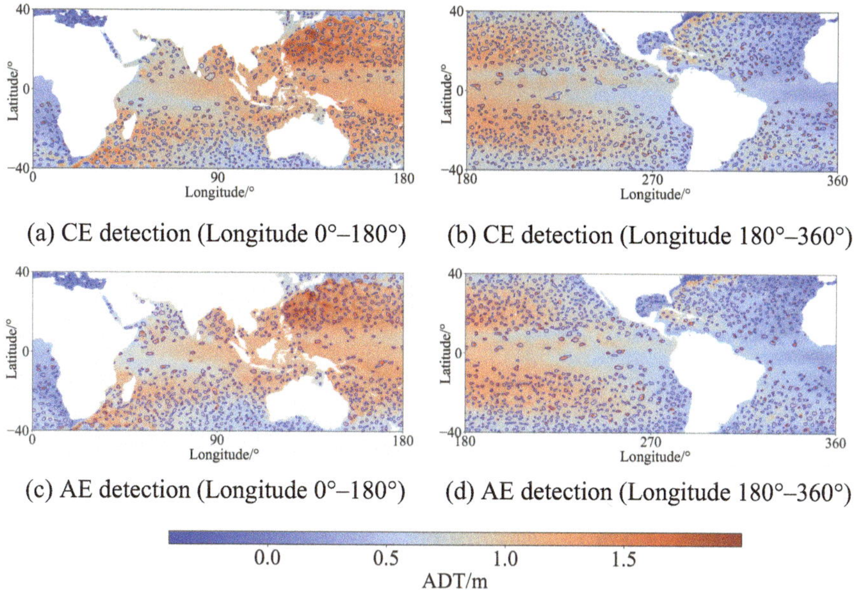

Fig. 17 Demonstration of eddy detection in ACE-BOC signals (local feature extraction)

Table 5 Metrics for eddy detection in ACE-BOC signals (local feature extraction)

Metric	CE/%	AE/%
Precision	85.94	84.34
Recall	91.87	91.45
F1 score	88.81	87.75
Accuracy	79.87	78.17

Figure 18 further illustrates performance metrics. Figure 18a, b show shape error distributions for true and detected eddies, demonstrating high spatial consistency. Figure 18c confirms that most shape error differences remain below 5%, demonstrating superior performance compared with altimeter-based GNSS-R measurements. Figure 18d shows strong shape similarity, with most similarity coefficients between 0.8 and 1.0, and only a few below 0.2. Figure 18e shows that over 90% of center distance errors are within 0–20 km, even though the grid size is 25 km. This is explained by center identification based on ADT and geostrophic flow circle fitting, allowing sub-grid accuracy. Figure 18f shows amplitude errors concentrated within [−2, 2] cm, with a positive bias of 0.4 cm, indicating slightly stronger amplitudes in ACE-BOC-based detections compared with true values. Figure 18g shows radius errors mostly within – 20 to 20 km, while Fig. 18h presents overall relative error distributions, with most values clustered around 0.5.

These results demonstrate that the proposed method enables ACE-BOC signals to achieve outstanding performance in ocean eddy detection. The approach provides an

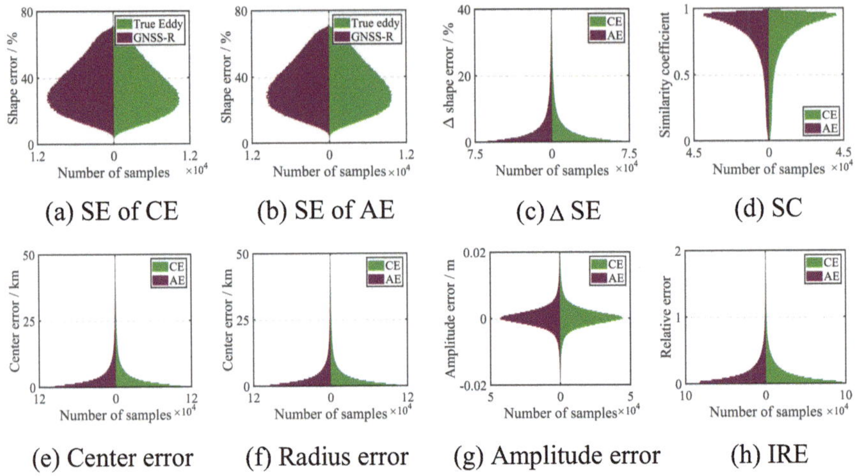

Fig. 18 Statistics of eddy detection performance using ACE-BOC signals (local feature extraction)

effective means for high-precision, high-resolution eddy monitoring, offering new technical pathways and methodological support for future oceanographic research.

5 Summary

To address the challenge of fine-scale segmentation caused by blurred boundaries in eddy-dense regions, this study leverages the high temporal resolution of GNSS-R observations and proposes an eddy detection method based on refined extraction of local shape features. This chapter first demonstrated the effectiveness of GNSS-R technology for oceanic eddy detection and validated its advantage in temporal resolution through comparison with traditional altimetric results.

In the proposed local-shape feature refinement method, the Bessel function's local curvature properties are exploited to capture variations in edge gradient orientation. A dynamic adaptive detection window is constructed, and by integrating the extremal distribution of closed-contour velocities in the height field, the eddy center position is accurately identified. Subsequently, an eddy kinetic energy threshold and correction scheme are introduced to refine the estimation of the eddy radius.

Using ACE-BOC high-bandwidth signals, the proposed method achieves detection precision rates of 85.94% for cyclonic eddies and 84.34% for anticyclonic eddies, representing improvements of 7.47 and 6.00%, respectively, compared with conventional global-feature extraction methods. Relative to traditional single-satellite altimetry, GNSS-R demonstrates equivalent accuracy levels in eddy detection, while its core advantage lies in providing daily, high-temporal-resolution observations. This capability greatly enhances the spatiotemporal capture of oceanic dynamic processes.

References

1. Chen JX, Song HB, Bai Y et al (2013) The vertical structure and physical properties of the Mediterranean eddy. Chin J Geophys 56(3):943–952
2. Morrow R, Birol F, Griffin D et al (2004) Divergent pathways of cyclonic and anti-cyclonic ocean eddies. Geophys Res Lett 31(24):L24311
3. AVISO+ (2021) Jason-3 products handbook. CNES/EUMETSAT/NOAA/NASA/EU
4. Dufau C, Orsztynowicz M, Dibarboure G et al (2016) Mesoscale resolution capability of altimetry: present and future. J Geophys Res Oceans 121(7):4910–4927
5. Lguensat R, Sun M, Fablet R et al (2018) EddyNet: a deep neural network for pixel-wise classification of oceanic eddies. In: IGARSS 2018–2018 IEEE international geoscience and remote sensing symposium. IEEE, pp 1764–1767
6. Yi J, Du Y, He Z et al (2014) Enhancing the accuracy of automatic eddy detection and the capability of recognizing the multi-core structures from maps of sea level anomaly. Ocean Sci 10(1):39–48
7. Alpers W, Brandt P, Lazar A et al (2013) A small-scale oceanic eddy off the coast of West Africa studied by multi-sensor satellite and surface drifter data. Remote Sens Environ 129:132–143
8. Xu G, Cheng C, Yang W et al (2019) Oceanic eddy identification using an AI scheme. Remote Sens 11(11):1349
9. Fu LL, Chelton DB, Le Traon PY et al (2010) Eddy dynamics from satellite altimetry. Oceanography 23(4):14–25
10. Escudier R, Renault L, Pascual A et al (2016) Eddy properties in the Western Mediterranean Sea from satellite altimetry and a numerical simulation. J Geophys Res Oceans 121(6):3990–4006
11. Chelton DB, Schlax MG, Samelson RM (2011) Global observations of nonlinear mesoscale eddies. Prog Oceanogr 91(2):167–216
12. Chelton DB, Gaube P, Schlax MG et al (2011) The influence of nonlinear mesoscale eddies on near-surface oceanic chlorophyll. Science 334(6054):328–332
13. Kurian J, Colas F, Capet X et al (2011) Eddy properties in the California current system. J Geophys Res Oceans 116(C8):C08026
14. Biancamaria S, Schaedele T, Blumstein D et al (2018) Validation of Jason-3 tracking modes over French rivers. Remote Sens Environ 209:77–89

Chapter 17
Sea Ice Detection and Classification Using Spaceborne GNSS-R

Yongchao Zhu

Abstract Sea ice sensing employs various techniques, with remote sensing recognized as the most efficient approach. Global Navigation Satellite System Reflectometry (GNSS-R) has emerged as a new technology that uses signals reflected from the Earth's surface to retrieve geophysical parameters. This technique has been used for a range of applications, including ocean altimetry, soil moisture estimation, tsunami detection, and oil spill identification. GNSS-R data can be collected from ground-based, airborne, shipborne, and spaceborne platforms. Among these, spaceborne GNSS-R offers global coverage with high spatiotemporal resolution. The launch of UK's TechDemoSat-1 (TDS-1) in 2014 has significantly advanced spaceborne GNSS-R research, providing substantial data resources for sea ice monitoring.

1 Introduction

Sea ice sensing employs various techniques, with remote sensing recognized as the most efficient approach. Global Navigation Satellite System Reflectometry (GNSS-R) has emerged as a new technology that uses signals reflected from the Earth's surface to retrieve geophysical parameters. This technique has been used for a range of applications, including ocean altimetry, soil moisture estimation, tsunami detection, and oil spill identification. GNSS-R data can be collected from ground-based, airborne, shipborne, and spaceborne platforms. Among these, spaceborne GNSS-R offers global coverage with high spatiotemporal resolution. The launch of UK's TechDemoSat-1 (TDS-1) in 2014 has significantly advanced spaceborne GNSS-R research, providing substantial data resources for sea ice monitoring.

A fundamental data set collected in spaceborne GNSS-R is the Delay-Doppler Map (DDM). DDMs over sea ice exhibit more concentrated reflections due to coherent scattering, while sea water surfaces produce diffuse signals as a result of surface roughness. Early studies utilized pixel counts or power thresholds in

Y. Zhu (✉)
School of Civil and Hydraulic Engineering, Hefei University of Technology, Hefei, China
e-mail: yczhu@hfut.edu.cn

K. Yu and N. Zheng (eds.), *Advances in GNSS Reflectometry*, Navigation: Science and Technology 19, https://doi.org/10.1007/978-981-92-2325-1_17

DDMs to distinguish sea ice from water. Subsequent methods involved differential DDMs, deconvolution techniques, and waveform analysis, achieving classification accuracy exceeding 98%. Additional applications include sea ice altimetry, classification, concentration estimation, and thickness retrieval.

Advances in artificial intelligence (AI) have prompted the increasing application of machine learning (ML) to remote sensing. While techniques such as neural network (NN), convolutional neural network (CNNs), and support vector machine (SVM) have been employed for sea ice detection and concentration estimation with TDS-1 data, the application of decision tree (DT) and random forest (RF) to GNSS-R-based sea ice monitoring is still limited. This work, therefore, seeks to assess the capability of DT and RF classifiers for sea ice detection using spaceborne GNSS-R data.

2 Dataset Description

2.1 TDS-1 Mission and Dataset

Launched in July 2014, the TDS-1 satellite began its data acquisition in September of the same year. Operating in a quasi-Sun-synchronous orbit at approximately 635 km altitude and 98.4° inclination, it provided extensive coverage of the polar regions. Its payload included the Space GNSS Receiver-Remote Sensing Instrument (SGR-ReSI), which initially collected data two days per eight-day cycle. This changed in February 2018, when the instrument entered a full-time operational phase that lasted until December 2018.

The GNSS-R data from TDS-1 are processed into three levels. Level 0 (L0) consists of raw data, which are not publicly available except for limited samples. The publicly accessible data begin with Level 1 (L1), which includes the L1b product containing Delay-Doppler Maps (DDMs) and associated metadata in NetCDF format—the primary dataset for scientific research. Level 2 (L2) data provide derived geophysical products like wind speed and sea ice parameters.

The SGR-ReSI generates DDMs by cross-correlating the scattered GNSS signals with local replica codes across a range of time delays and Doppler shifts. Each DDM has a resolution of 128 delay bins (0.25 C/A code chips each) and 20 Doppler bins (spaced at 500 Hz). These maps are fundamental for sea ice detection because the scattering properties of sea ice and open water differ significantly. Smooth sea ice produces a coherent, focused reflection, while rough ocean water causes incoherent scattering that spreads the signal power. This distinct signature allows for the remote sensing of sea ice.

All TDS-1 data products are freely available on the MERRByS website (www.merrbys.co.uk). Key mission and instrument specifications are summarized in Table 1.

Table 1 Orbit and instrument specifications of TDS-1 [1]

Parameter	Value
Orbit	quasi-Sun synchronous
Altitude	~ 635 km
Inclination	98.4°
Delay pixels/resolution	128/244 ns
Doppler pixels/resolution	20/500 Hz
Coherent integration time	1 ms
Incoherent integration time	1 s

3 Reference Sea Ice Data

Two independent sea ice datasets are employed to validate the performance of the proposed sea ice monitoring technique. The first reference dataset is the Sea Ice Edge (SIE), Sea Ice Type (SIT) and Sea Ice Concentration (SIC) product from the Ocean and Sea Ice Satellite Application Facility (OSISAF). This dataset is generated on a 10 km grid using a Bayesian approach that combines data from the Advanced Scatterometer (ASCAT) and the Special Sensor Microwave Imager/Sounder (SSMIS), utilizing channels such as 19, 37, and 91 GHz [2]. A critical feature of the OSISAF product is its quality flag system, which categorizes data reliability into six confidence levels (0: unprocessed, 1: erroneous, 2: unreliable, 3: acceptable, 4: good, 5: excellent). In line with established practices, only data with a minimum confidence level of 3 are utilized in this analysis to ensure robustness. The second reference dataset is the Sea Ice Concentration (SIC) product derived from Advanced Microwave Scanning Radiometer 2 (AMSR-2) data using the ARTIST Sea Ice (ASI) algorithm. This dataset, acquired from the online portal www.meereisportal.de, provides daily maps in polar stereographic projection with a higher spatial resolution of 6.25 km. To align the GNSS-R observations with the reference data, the TDS-1 DDMs are co-located with the SIC maps using the specular point location and the data collection timestamp available in the Level 1b dataset. Following a common classification threshold in the field, a DDM is classified as corresponding to sea ice if its associated SIC value exceeds 15%; otherwise, it is classified as open water. TDS-1 observations were co-located with the OSISAF maps using the specular point location and acquisition date from the TDS-1 dataset. Further details on the OSISAF data are available in [3].

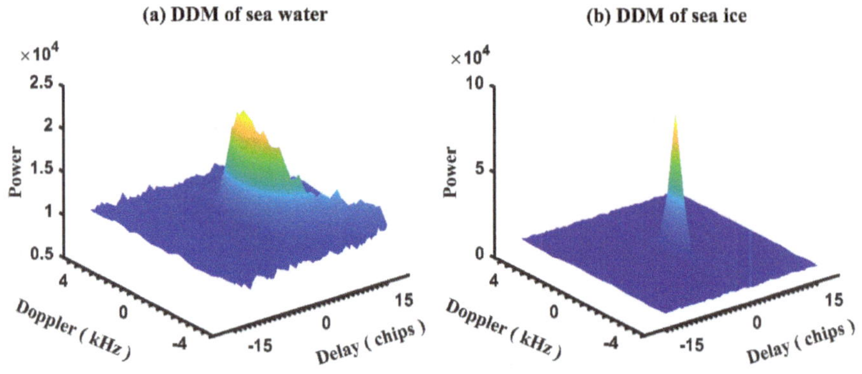

Fig. 1 Typical GNSS-R DDMs collected over **a** sea water and **b** sea ice, respectively

4 Sea Ice Detection Method and Results

4.1 DDMs from Sea Ice and Water Surface

GNSS-R DDMs exhibit distinct characteristics over sea ice and open water due to their differing surface roughness. Sea ice produces a relatively smooth surface that results in coherent signal reflection, yielding a sharply peaked DDM with minimal spreading in the delay and Doppler domains. In contrast, the rough surface of open water causes diffuse scattering, spreading the signal power across a wider area in the DDM and forming a broad glistening zone. This fundamental difference in signal spreading provides the physical basis for sea ice detection using spaceborne GNSS-R technology. Two typical different DDMs collected over sea water and sea ice surface are presented in Fig. 1.

4.2 GNSS-R Features

The scattered signal components in a DDM originate from the glistening zone, where each point contributes different time delays and Doppler shifts relative to the specular point. To simplify data processing, the method introduced by Zhu et al. [4] utilizes two-dimensional delay waveforms derived from DDMs as fundamental observables. Specifically, cross-sections taken at each of the 20 Doppler shifts produce 20 individual delay waveforms. Summing these waveforms over the Doppler domain yields the Integrated Delay Waveform (IDW), as described by Alonso-Arroyo et al. [5]. The distribution of scattered signal power as a function of time delay, represented by the DDM, is mathematically characterized using the bistatic radar equation-based model developed by Zavorotny and Voronovich [6].

$$DDM = T_i^2 \int \frac{D^2(\vec{\rho})}{4\pi R_T^2(\vec{\rho}) R_R^2(\vec{\rho})} |\Lambda(\tau) \times S(f_D)|^2 \sigma_0(\vec{\rho}) d^2\rho \quad (1)$$

In this model, the parameters are defined as follows: T_i denotes the coherent integration time, τ corresponds to the time delay, and D^2 represents the antenna power footprint function. The variables R_T and R_R indicate the distances from the scattering point to the GNSS transmitter and receiver, respectively. Λ represents a triangular function dependent on time delay, while S characterizes the sinc function in the frequency domain for GNSS signals. Furthermore, σ_0 signifies the normalized bistatic radar cross section (NBRCS), f_D denotes the Doppler shift frequency, and ρ represents the vector from the specular reflection point to the scattering point.

In the TDS-1 mission configuration, the coherent integration time is set at 1 ms, with the corresponding Doppler bandwidth expressed as $\Delta f_0 = 1/2T_i$. When the maximum and minimum Doppler shifts of the scattered signal are denoted as $f_{\max}$ and f_{min}, the Doppler extent of the glistening zone can be quantified as $f_{\max} - f_{\min}$.. Under the condition that the Doppler bandwidth exceeds this glistening zone width (i.e., $\Delta f_0 > f_{\max} - f_{\min}$), the influence of Doppler effects becomes negligible. In such cases, the sinc function S approaches unity, and the cross-section corresponding to zero Doppler shift yields the Central Delay Waveform (CDW)—a specific waveform extracted from the DDM. This waveform can be mathematically expressed as follows:

$$CDW = T_i^2 \int \frac{D^2(\vec{\rho})}{4\pi R_T^2(\vec{\rho}) R_R^2(\vec{\rho})} |\Lambda(\tau)|^2 \sigma_0(\vec{\rho}) d^2\rho \quad (2)$$

To quantify the discrepancy between the CDW and IDW, an additional observable referred to as the Differential Delay Waveform (DDW) is introduced. The DDW is mathematically expressed as the difference between the normalized CDW (NCDW) and the normalized IDW (NIDW), defined by the following relation:

$$DDW = NIDW - NCDW \quad (3)$$

The IDW serves as an effective indicator of power distribution patterns resulting from surface roughness variations. To facilitate feature extraction from DDMs, normalization procedures outlined in prior research [4, 7] are implemented. These procedures involve removing the noise floor to generate Normalized DDMs (NDDMs) containing three key observables: NIDW, NCDW, and DDW. Departing from conventional practices, this study employs a more inclusive data selection criterion by retaining measurements with peak signal-to-noise ratio (SNR) exceeding -3 dB. This expanded dataset enables a more rigorous evaluation of the proposed methodology's robustness and general applicability.

To mitigate the impact of noise present in the initial and final delay bins, feature extraction is confined to a focused window encompassing 48 delay bins around the specular point, corresponding to code shifts from -3 to 8.75 chips. Figure 2 illustrates the ground tracks and a subset of data samples—including DDMs and their corresponding delay waveforms—acquired by TDS-1 over Baffin Bay on 15 January

2016. In the map, light blue, white, and light yellow denote open water, sea ice, and land, respectively, while magenta and blue ground tracks represent measurements collected over sea ice and water. Sample DDMs for these surfaces are highlighted by cyan and green rectangles. Figure 2a displays a sequence of DDMs across a water-to-ice transition zone (marked by a red rectangle), and Fig. 2b presents the associated normalized delay waveforms: NCDW, NIDW, and DDW. A noticeable transition in the shape of these waveforms occurs between the sea ice and water surfaces, with the most significant change observed between DDM 487 and DDM 488.

Several characteristic parameters are extracted from the delay waveforms to facilitate sea ice monitoring. As depicted in Fig. 3, the NCDW, NIDW, and DDW can each be segmented into a left edge (LE) and a right edge (RE) using the point of zero delay value as a reference. The spaceborne GNSS-R DDMs are produced by performing a cross-correlation operation between the scattered GNSS signals and locally generated replica codes across a range of time delays and Doppler shifts. The bin containing the maximum signal power within the Doppler domain is identified to determine the zero-delay value. It is acknowledged that variations in the Earth's surface, such as sea surface height fluctuations or the height of ice above the reference ellipsoid, can influence the signal geometry and potentially lead to inaccuracies in this estimation. However, as this investigation concentrates on analyzing relative changes in waveform features for classification rather than precise altimetry applications, the impact of these geometric factors is not considered within the current scope of the study.

The LE component of the waveform is primarily influenced by the area above the reflection surface, making it less sensitive to surface properties. Consequently, this study utilizes only RE related observables for sea ice monitoring. Six characteristic parameters are extracted as features: the RE slope of CDW (RESC), RE slope of IDW (RESI), RE slope of DDW (RESD), RE waveform summation of CDW (REWC), RE waveform summation of IDW (REWI), and RE waveform summation of DDW (REWD) (Table 2).

The mathematical formulations for the six selected features (RESC, RESI, RESD, REWC, REWI, and REWD) are summarized in Table 1. In these equations, the parameter n (where n $\geq$ 2) denotes the number of delay bins used for curve fitting, with one delay bin corresponding to 0.25 chips. The time delay of the i-th point is represented by τ_i, while C_i^R, I_i^R and D_i^R correspond to the right-edge waveform values of the CDW, IDW, and DDW, respectively. The average waveform values used in fitting for CDW, IDW, and DDW are denoted as $\overline{C^R}$, $\overline{I^R}$ and $\overline{D^R}$, and the mean time delay of the points included in the fitting is expressed as $\overline{\tau}$.. The number of points n is configured as 5 for calculating the slope-based features (RESC, RESI, RESD) and 7 for the waveform summation features (REWC, REWI, REWD).

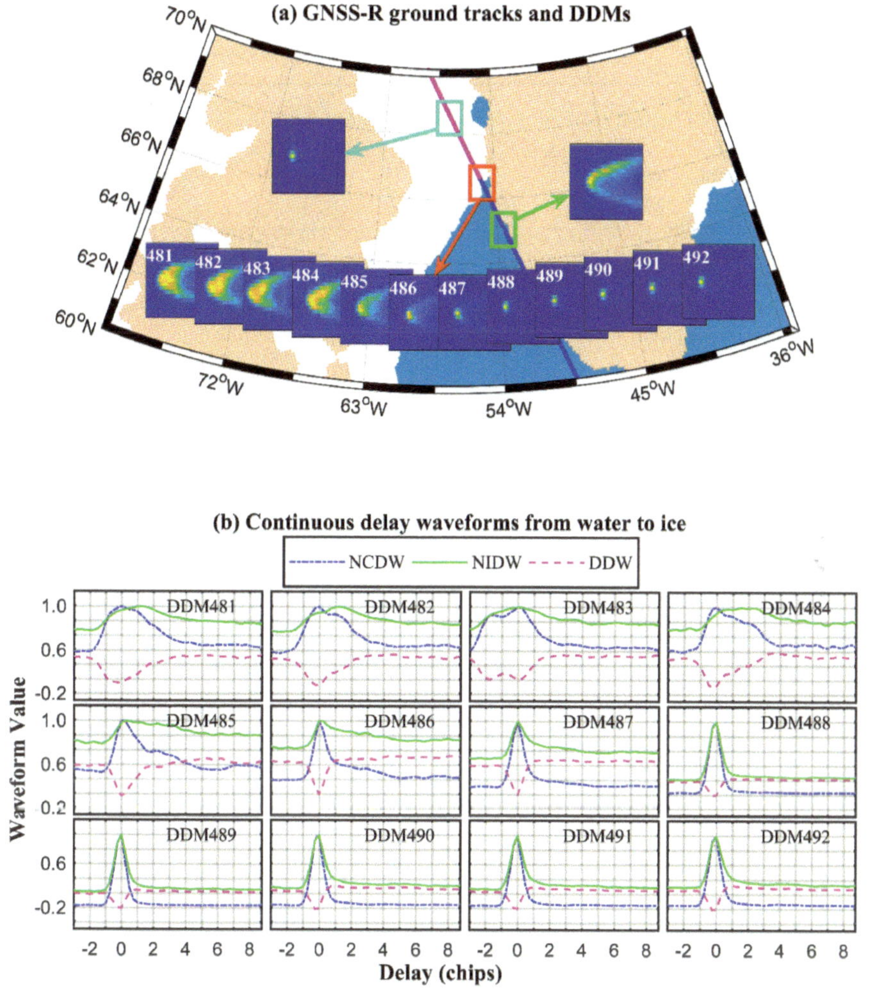

Fig. 2 DDMs and corresponding delay waveforms (NCDW, NIDW, DDW) derived from TDS-1 data acquired on 15 January 2016. **a** Ground tracks over sea ice and open water are depicted in magenta and blue, respectively. Sample DDMs representative of each surface type are highlighted: ice reflections are marked by a cyan rectangle, and water reflections by a green rectangle. A sequence of consecutive DDMs (indices 481–492) capturing a transition from water to ice is indicated with a red rectangle. **b** The corresponding delay waveforms for DDMs 481 to 492. The NCDW, NIDW, and DDW are plotted as a blue dotted line, solid green line, and magenta dashed line, respectively, illustrating the evolution of scattering characteristics across the transition zone

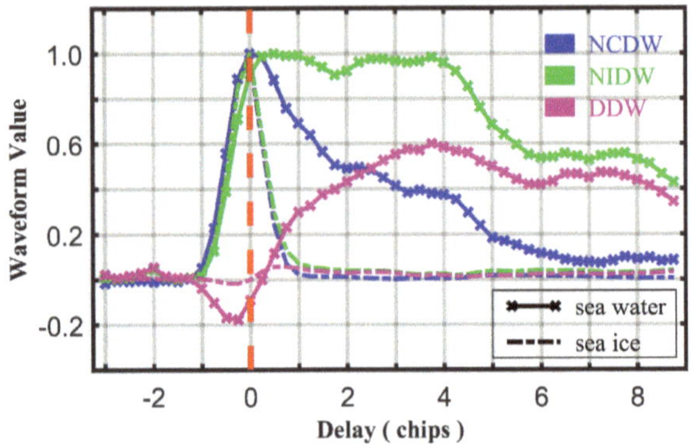

Fig. 3 The characteristic delay waveforms for sea water (cross markers) and sea ice (dotted markers), with the NCDW, NIDW, and DDW waveforms represented in blue, green, and magenta, respectively. A vertical red dashed line divides each waveform into left-edge (LE) and right-edge (RE) segments relative to the zero-delay reference point

Table 2 The six selected features are mathematically defined to characterize the scattering behavior derived from delay waveforms. RESC, RESI, and RESD represent the slopes of the right-edge segment of the CDW, IDW, and DDW, respectively. REWC, REWI, and REWD correspond to the cumulative summation of the right-edge waveform values for the CDW, IDW, and DDW

Features	Mathematical description
RESC	$\left(\sum_{i=1}^{n} \tau_i C_i^R - n\overline{\tau}\overline{C^R}\right) / \left(\sum_{i=1}^{n} \tau_i^2 - n\overline{\tau}^2\right)$
RESI	$\left(\sum_{i=1}^{n} \tau_i I_i^R - n\overline{\tau}\overline{I^R}\right) / \left(\sum_{i=1}^{n} \tau_i^2 - n\overline{\tau}^2\right)$
RESD	$\left(\sum_{i=1}^{n} \tau_i D_i^R - n\overline{\tau}\overline{D^R}\right) / \left(\sum_{i=1}^{n} \tau_i^2 - n\overline{\tau}^2\right)$
REWC	$\sum_{i=1}^{n} C_i^R$
REWI	$\sum_{i=1}^{n} I_i^R$
REWD	$\sum_{i=1}^{n} D_i^R$

4.3 Machine Learning-Aided Sea Ice Monitoring Methods

The core objective of this study is to distinguish sea ice from open water, a task that can be formulated as a binary classification problem addressed by applying machine learning (ML) to a large satellite dataset. The overall workflow for ML-based sea ice monitoring is illustrated in Fig. 4.

This methodology comprises three key stages: (1) extraction of relevant features from TDS-1 GNSS-R data; (2) model training using a labeled dataset and ML algorithms; and (3) automated discrimination between sea ice and water for new data collections. The model utilizes seven input variables: a reference sea ice concentration (SIC) map and a sequence of six features—namely RESC, RESI, RESD, REWC, REWI, and REWD—derived from TDS-1 DDMs.

For coherent reflections, the TDS-1 DDM footprint measures approximately 6 km along-track by 0.4 km across-track, which is spatially comparable to the reference

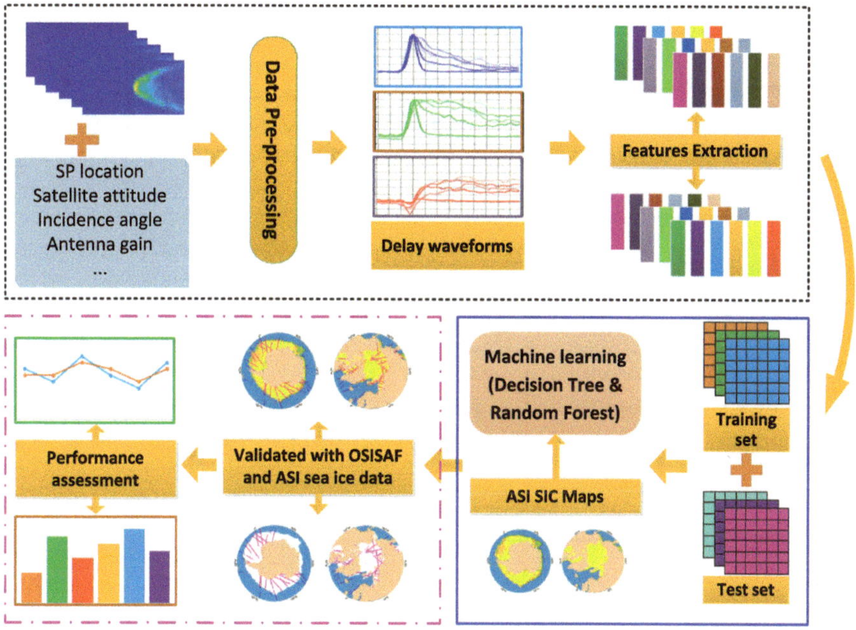

Fig. 4 The sea ice monitoring workflow employing machine learning (ML) integrates three sequential phases: initial preprocessing of TDS-1 GNSS-R data to extract six distinctive feature sequences (RESC, RESI, RESD, REWC, REWI, REWD) characterizing surface scattering properties; subsequent development of a classifier through training with labeled data using DT or RF algorithms; and final application of the trained model to test data for automated sea ice-water discrimination, with performance validation against OSISAF and ASI reference products

SIC maps with a 6.25 km grid resolution. The specular point of each DDM is used for co-registration with the reference data. The machine learning framework operates on two distinct datasets: a training set and a test set. The training data are pre-labeled using the reference SIC maps, enabling the ML algorithm to model the relationship between the input parameters and the desired output. This model is subsequently applied to classify the test data. As the training data include known outcomes, this study employs a supervised learning approach, specifically utilizing the DT and RF algorithms for sea ice classification.

Decision Tree Algorithm

The DT algorithm serves as an intuitive yet powerful supervised learning method widely employed in classification tasks, including remote sensing applications such as surface type discrimination [8–11]. Its structure resembles an inverted tree, comprising a root node (the starting point), internal nodes (each representing a test on a feature parameter), branches (outcomes of feature tests that direct the flow based on discriminant functions), and leaf nodes (terminal points that provide the final classification decision). This architecture makes the decision process transparent and interpretable.

For binary classification problems—such as distinguishing sea ice from open water—the C4.5 algorithm is often utilized. Unlike its predecessor ID3, C4.5 employs the information gain ratio as its splitting criterion to mitigate bias towards features with numerous values. The algorithm operates recursively on a training set $S = s_1, s_2, \ldots s_i$, where each sample s_i is a p-dimensional feature vector $(x_{1,i}, x_{2,i}, \ldots, x_{p,i})$ accompanied by a known class label. By evaluating the information gain ratio at each step, C4.5 partitions the data into subsets, progressively building a tree that maximizes the discriminative power while controlling for feature bias.

The information gain ratio is formally defined as the ratio of the information gain to the intrinsic information of a feature split:

$$G_{ratio} = \frac{\left[\sum_{k=1}^{m} p(k)\log_2 p(k) + \sum_{j=1}^{v}\left(|D_j|/|D|\right)Ent\left(D_j\right)\right]}{\sum_{j=1}^{v}\left(|D_j|/|D|\right)\log_2\left(|D_j|/|D|\right)} \tag{4}$$

where m denotes the category count, v indicates the number of chosen features, D represents the total sample size, and D_j corresponds to the j-th individual sample. This adjustment helps avoid overfitting and enhances the model's generalization ability, making C4.5 suitable for handling both discrete and continuous attributes in geospatial data analysis. Given that this study involves two categories (sea ice and sea water), m is set to 2, and with six features utilized, v equals 6. Consequently, the information gain ratio can be expressed in a simplified form as follows:

$$G_{ratio} = \frac{\left[\sum_{k=1}^{2} p(k)\log_2 p(k) + \sum_{j=1}^{6}\left(|D_j|/|D|\right)Ent\left(D_j\right)\right]}{\sum_{j=1}^{6}\left(|D_j|/|D|\right)\log_2\left(|D_j|/|D|\right)} \tag{5}$$

The C4.5 algorithm offers distinct benefits in practical use: it reduces overfitting via an efficient single-pass pruning method, processes both discrete and continuous data types, and effectively manages incomplete datasets frequently encountered in real-world applications.

Random Forest Algorithm

RF, an ensemble learning method for classification, builds a collection of decision trees (DT) during training. It integrates a bagging (bootstrap aggregating) sampling strategy with Classification and Regression Trees (CART) to mitigate drawbacks of single CART models, such as overfitting. While CART employs the Gini index [12] to measure impurity within training datasets, the C4.5 algorithm utilizes the concept of entropy for its splits. The Gini index is mathematically described as follows:

$$G_{index}(p) = 1 - \sum_{l=1}^{s} p_l^2 \tag{6}$$

where s denotes the number of distinct classes and p_l indicates the proportion of data samples associated with class l. As the task of sea ice detection constitutes a binary classification problem (involving only sea ice and open water), the Gini index can be simplified to the following expression:

$$G_{index}(p) = 2p(1 - p) \tag{7}$$

where p can be regarded as the probability that samples belong to sea ice.

CART offers significant benefits, such as easily interpretable decision rules and the inherent management of parameter selection, missing data, outliers, as well as complex variable interactions and nonlinearities. A single CART model's major limitation is its susceptibility to overfitting. This issue is effectively mitigated by bagging, a technique that aggregates predictions from numerous independent trees to enhance stability and reduce error [13]. Owing to these strengths, the Random Forest algorithm, which is based on this principle, shows considerable promise in the field of remote sensing [14].

4.4 Sea Ice Detection Results

This study analyzes TDS-1 data acquired between January 2015 and December 2018 from Arctic and Antarctic regions above 55° N and 55° S. For ML-based classification between sea ice and water, 20% of the dataset is randomly selected for model training, while the remaining 80% is reserved for testing and validation.

As previously noted, the GNSS-R receiver on TDS-1 had intermittent operation. Figure 5 illustrates data availability throughout the study period. The absence of data from August to October 2017 likely corresponds to the mission's scheduled conclusion in July 2017. However, the mission was subsequently extended from February to December 2018. During this extension period, the SGR-ReSI instrument operated daily—a fourfold increase in coverage compared to the initial three-year cycle of two days per eight-day cycle. Brief data gaps in 2018 may be attributed to statutory holidays such as Christmas.

Characteristics of GNSS-R Features

The distribution characteristics of six feature parameters (RESC, RESI, RESD, REWC, REWI, REWD) for sea ice and open water are presented in Fig. 6. The analysis reveals a distinct separation in the median values of these parameters between the

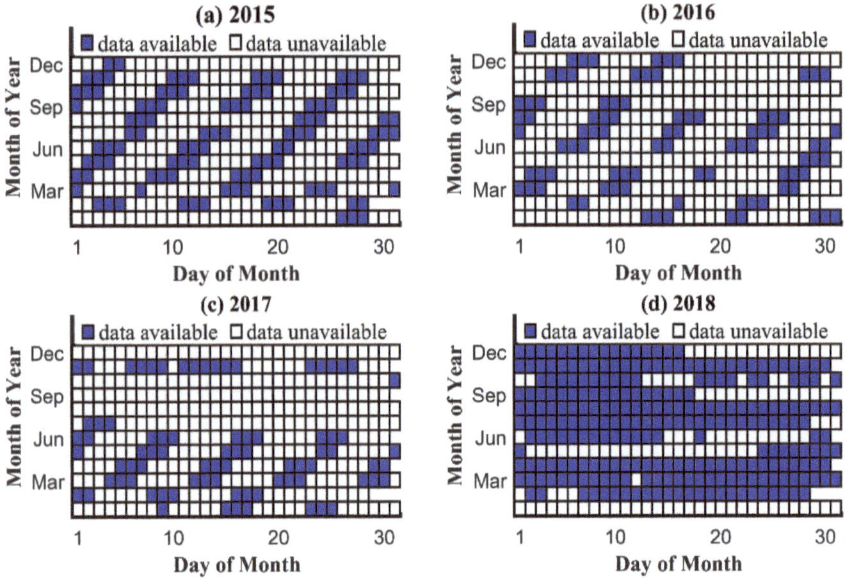

Fig. 5 TDS-1 data availability in **a** 2015, **b** 2016, **c** 2017 and **d** 2018. The rectangles filled in blue represent the available TDS-1 data, whereas the rectangles without filled color represent the unavailability of TDS-1 data

two surface types. This discrepancy is attributed to the more coherent signal reflection characteristic of sea ice surfaces, in contrast to the diffuse scattering caused by the greater roughness of wind-affected sea water. Despite a statistically significant difference in median values, the distributions for each feature exhibit substantial overlap between the classes.

This overlap is exemplified by the RESC parameter (Fig. 6a). While the established threshold of 0.745 [4] provides a general discriminator, a considerable number of sea ice observations fall below this value, and numerous water observations exceed it. This indicates that a classification rule based on a single-feature threshold is prone to misclassification, as it cannot accurately capture the full variability within each class.

To address this limitation, a multi-feature approach is adopted. The six parameters, which are derived from delay waveforms exhibiting different Doppler spread characteristics, are combined into an ordered feature vector to provide a more robust characterization of the scattering surface. These feature vectors serve as the input data for training the sea ice detection model. The parameters are sequenced as follows: RESC, RESI, RESD, REWC, REWI, and REWD. Representative feature vectors for samples from the Arctic and Antarctic regions are visualized in Fig. 7.

As illustrated in Figs. 7a, b, which depict feature sequences for 40,000 sea ice and water samples, a clear distinction is observed between the two classes. This differentiation confirms the potential of using feature sequences for reliable sea ice monitoring. The synergistic use of multiple features in a sequence offers a more

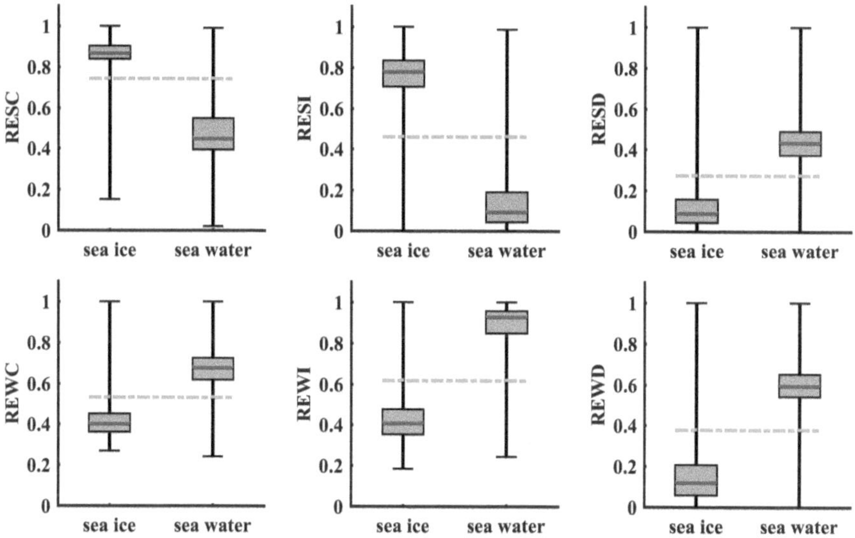

Fig. 6 Distribution of six waveform features (RESC, RESI, RESD, REWC, REWI, REWD) for sea ice and open water, derived from Arctic TDS-1 data (January 2015–December 2018). Boxes represent the interquartile range, with the internal red line indicating the median. The green dotted line shows the discriminating threshold from [4]

robust and accurate description of the surface characteristics compared to analyses based on single parameters.

Sea Ice Monitoring Performance and Analysis

The performance of DT and RF-based sea ice detection models is evaluated using confusion matrices [15] by comparing predictions with OSISAF SIE data. In machine learning classification tasks, a confusion matrix provides a tabular representation of algorithm performance, displaying counts of true positives, false positives, true negatives and false negatives. This study employs error matrices, overall accuracy, and kappa coefficient to quantitatively assess model performance. Tables 3 and 4 present the evaluation results for the Arctic and Antarctic regions respectively.

The DT model achieved overall accuracies of 97.51% for the Arctic and 95.46% for the Antarctic, while the RF model yielded slightly higher accuracies of 98.03% and 95.96%, respectively. For both models, producer and user accuracies were higher for seawater than for sea ice, likely due to the tendency of low-concentration sea ice to be misclassified as water. Wind-driven surface roughness in marginal ice zones may further cause the mixture of sea ice and seawater to be identified as seawater. Although both models demonstrated comparable overall accuracy, the RF algorithm showed marginally better performance as indicated by its higher Kappa coefficient. While the current accuracy values are slightly lower than those reported in previous studies, the

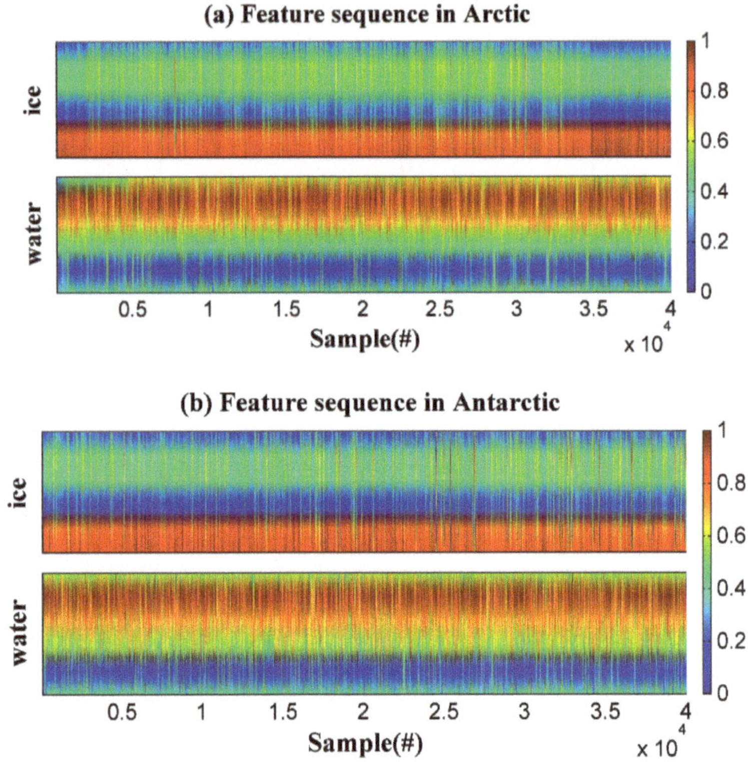

Fig. 7 Feature sequences composited with RESC, RESI, RESD, REWC, REWI and REWD in **a** Arctic **b** and Antarctic regions. The upper plot in each figure represents the feature sequences of sea ice and the lower plot represents those of sea water. The color scale represents the value of each feature parameter

methodology developed here utilizes a substantially larger dataset with less stringent filtering criteria, demonstrating improved applicability and generalizability. When applied solely to the initial mission data following our previous study [16], the present method achieves a 0.22% higher overall accuracy compared to the REWD method employed earlier.

The SVM method has demonstrated superior performance over neural network (NN) and convolutional neural network (CNN) approaches for sea ice detection using spaceborne GNSS-R data [17]. SVMs perform classification by identifying an optimal hyperplane that maximizes the margin between distinct classes. In contrast, NNs offer high flexibility in handling diverse data structures and can autonomously learn significant features without manual feature engineering. CNNs, while more constrained in structure than fully connected networks, exhibit a strong bias toward effective image processing. For performance benchmarking, the SVM method is employed here as a baseline comparator. As summarized in Table 4, the proposed RF-based sea ice monitoring approach achieves higher accuracy than the SVM-based

Table 3 The confusion matrix for the DT algorithm using data from Arctic and Antarctic regions

Arctic	Reference classified as	Sea ice	Sea water	Sum	User accuracy
	Sea ice	1,242,947	6513	1,249,460	99.48%
	Sea water	61,677	1,427,415	1,489,092	95.86%
	Sum	1,304,624	1,433,928	2,738,552	
	Producer accuracy	95.27%	99.55%		
	Overall accuracy			97.51%	
	Kappa coefficient			95.00%	
Antarctic	Reference classified as	Sea ice	Sea water	Sum	User accuracy
	Sea ice	1,368,301	29,491	1,397,792	97.89%
	Sea water	110,509	1,572,579	1,683,088	93.43%
	Sum	1,478,810	1,602,070	3,080,880	
	Producer accuracy	92.53%	98.16%		
	Overall accuracy			95.46%	
	Kappa coefficient			90.88%	

Table 4 The confusion matrix for the RF algorithm using data from Arctic and Antarctic regions

Arctic	Reference classified as	Sea ice	Sea water	Sum	User accuracy
	Sea ice	1,275,679	25,121	1,300,800	98.07%
	Sea water	28,945	1,408,807	1,437,752	97.99%
	Sum	1,304,624	1,433,928	2,738,552	
	Producer accuracy	97.78%	98.25%		
	Overall accuracy			98.03%	
	Kappa coefficient			96.04%	
Antarctic	Reference classified as	Sea ice	Sea water	Sum	User accuracy
	Sea ice	1,411,677	57,411	1,469,088	96.09%
	Sea water	67,133	1,544,659	1,611,792	95.83%
	Sum	1,478,810	1,602,070	3,080,880	
	Producer accuracy	95.46%	96.42%		
	Overall accuracy			95.96%	
	Kappa coefficient			91.90%	

method, which in turn outperforms the DT-based technique. The feature sequences utilized in this work are derived from delay waveforms—namely NCDW, NIDW, and DDW—across varying Doppler shifts.

For further temporal analysis, the overall accuracy of sea ice detection was derived from all available data between January 2015 and December 2018 (Fig. 8). The detection performance in the Arctic region was notably lower during September 2016, primarily due to seasonal sea ice melt during this period. The trend observed

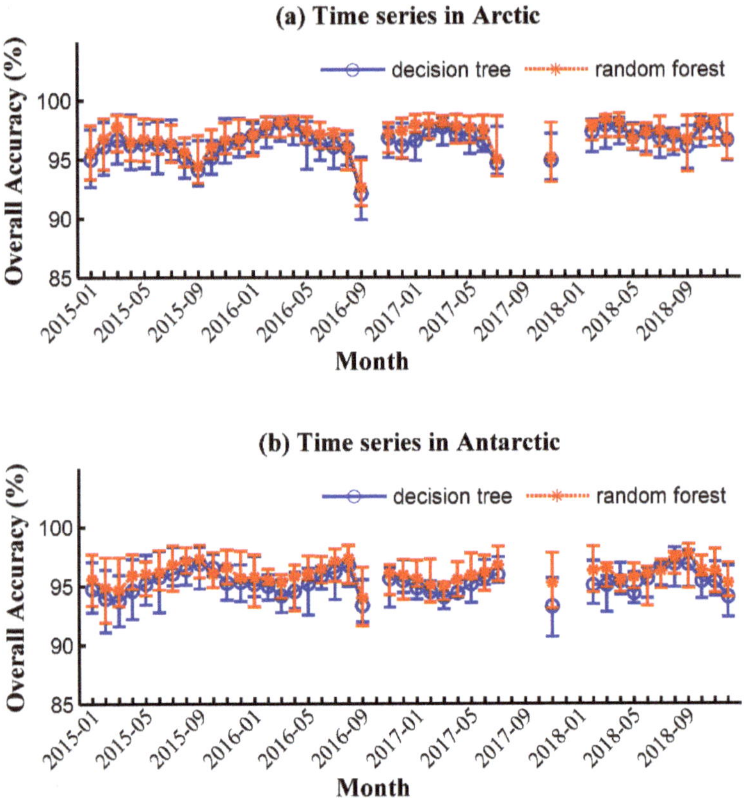

Fig. 8 The time series of producer and user accuracies of sea ice monitoring results over the **a** Arctic and **b** Antarctic regions using DT and RF algorithms

in the Antarctic exhibits a converse pattern, due to the opposing seasons between the two polar regions.

To assess the contribution of individual variables, Fig. 9 displays their relative importance for sea ice detection. Within the DT algorithm, REWD is employed at every node, leading to its substantial impact on detection performance. REWI also proves valuable by enabling accurate discrimination between sea ice and water. The parameter significance ranking for DT is: REWD (highest), followed by REWI, RESI, RESD, RESC, and REWC (lowest). Similarly, in the RF algorithm, REWC demonstrates the lowest significance among all variables.

Figure 10 compares RF-based GNSS-R sea ice detection results with OSISAF SIE data for March and September 2018. The white and dark gray edges (partially highlighted by red dotted rectangles) represent the respective minimum and maximum ice extents observed during these months. In March, the Arctic sea ice reached its seasonal minimum on March 6 and maximum on March 14, while Antarctic extremes were recorded on March 1 (minimum) and March 31 (maximum). As indicated in Fig. 5d, data availability was intermittent in September 2018. During September

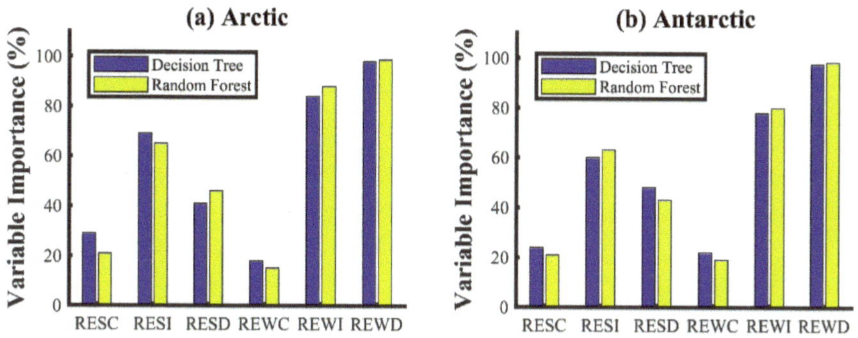

Fig. 9 Relative importance of variables to sea ice detection using DT and RF algorithms over the **a** Arctic and **b** Antarctic regions

1–17, the Arctic's maximum and minimum extents occurred on September 2 and 17 respectively, while the Antarctic reached its minimum on September 13 and maximum on September 17. The scattered points represent TDS-1 ground tracks with peak SNR > − 3 dB, resulting in discontinuous coverage. Magento points indicate GNSS-R detected ice, while blue shows open water. The overlapping classifications in Figs. 10b, c reflect both the monthly aggregation of GNSS-R data and rapid ice dynamics during polar melt seasons.

Figure 11 presents a comparative analysis of four sea ice monitoring methods around Greenland, validated against ASI sea ice concentration (SIC) data. Two threshold-based methods from [4] employing REWD (> 0.38 for water, < 0.38 for ice) and REWI (> 0.62 for water, < 0.62 for ice) parameters are shown in Fig. 11a, b, respectively. These single-parameter threshold methods demonstrate notable misclassification, incorrectly identifying ice as water and vice versa, confirming their limitations despite the utility of REWD and REWI as discrimination parameters. The DT and RF-based approaches, displayed in Fig. 11c, d, show improved performance, though misclassification persists primarily along ice edge areas with low SIC. These errors may be attributed to surface roughening caused by oceanic winds affecting low-concentration ice areas, leading to false identification—an aspect not quantitatively analyzed in this study.

5 Conclusions

This study developed two machine learning-assisted GNSS-R methodologies for sea ice detection utilizing 42 months of TDS-1 data. Validation against OSISAF SIE data confirms the effectiveness of both approaches in discriminating sea ice from open water. The RF classifier achieved overall accuracies of 98.03% in Arctic and 95.96% in Antarctic regions, while the DT method yielded 97.51% and 95.46% accuracy respectively. For comparative analysis, we implemented a SVM approach

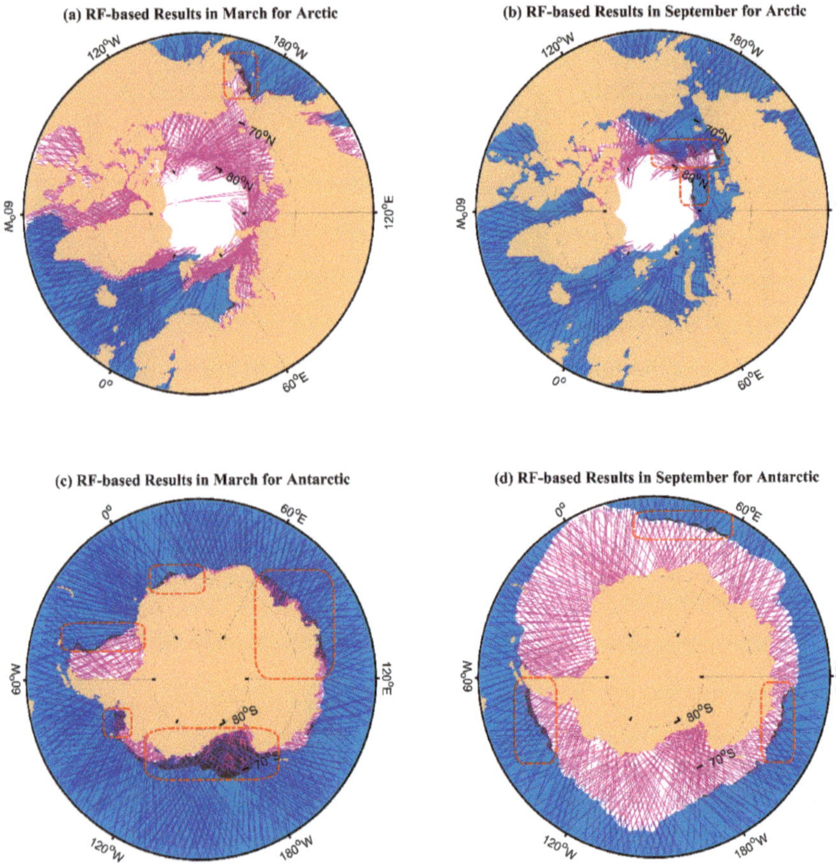

Fig. 10 RF-based GNSS-R sea ice monitoring results against OSISAF sea ice edge (SIE) data for both polar regions in March and September 2018. Panels **a** and **b** show Arctic conditions for March and September respectively, while **c** and **d** present Antarctic conditions for the same months. Dark gray and white borders (partially highlighted with red dotted rectangles) indicate the maximum and minimum monthly ice extents. Magenta points mark GNSS-R detected sea ice locations, with blue points indicating open water

from previous research [17], which achieved 97.62% (Arctic) and 95.61% (Antarctic) accuracy on our dataset.

Six distinctive features (RESC, RESI, RESD, REWC, REWI, REWD) were integrated for sea ice detection. While previous studies utilized these parameters individually, this work represents the first application of their combined implementation. The feature combination provides more comprehensive characterization of surface scattering properties compared to single-parameter thresholding methods, resulting in superior classification performance. Notably, the feature set employed

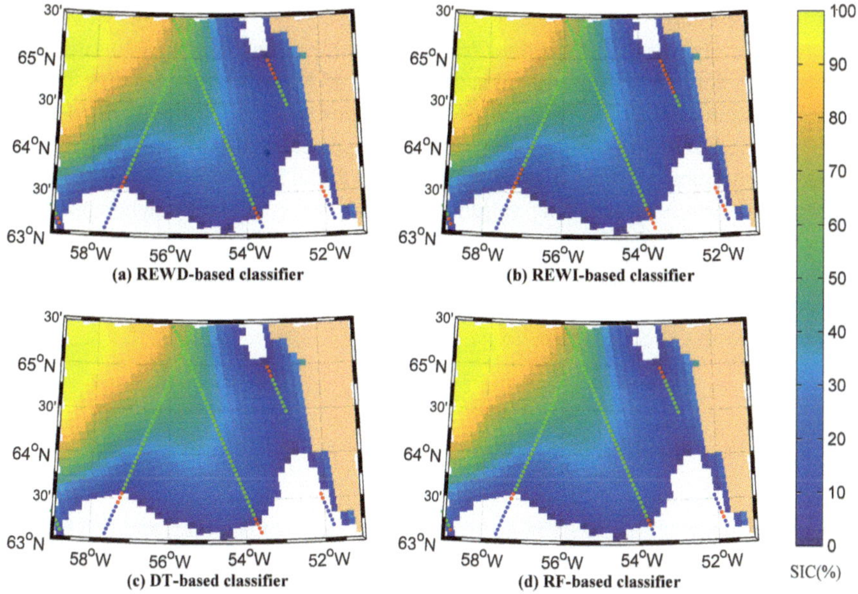

Fig. 11 Validation results against ASI SIC maps from AMSR-2 data for sea ice monitoring off southwestern Greenland on 14 March 2018, comparing four methodologies: **a** REWD thresholding, **b** REWI thresholding, **c** the DT-based approach, and **d** the RF-based approach. Light brown and white areas denote land and open water respectively, while sea ice concentration is color-coded according to the accompanying bar. Classification results are shown as green (correctly detected sea ice), blue (correctly identified water), and red (false detections) points

in our machine learning approaches differs fundamentally from those used in single-parameter methods. Furthermore, our study utilizes an extensive 42-month GNSS-R dataset from the TDS-1 mission, substantially larger than those employed in previous investigations. These promising results demonstrate the significant potential of machine learning algorithms for GNSS-R applications. Future work will investigate wind effects on low-concentration sea ice areas and explore ensemble approaches combining multiple machine learning techniques.

6 Sea Ice Classification

Sea ice type, as a critical sea ice parameter, is of significant interest due to the distinct characteristics of first-year ice (FYI) and multi-year ice (MYI) [18]. This study introduces a novel sea ice classification method that applies machine learning (ML) classifiers to spaceborne Global Navigation Satellite System-Reflectometry (GNSS-R) data from TDS-1. Using eight features derived from GNSS-R observables collected over a five-month period, the approach effectively discriminates among open water (OW), FYI, and MYI.

A two-step classification strategy employing RF and SVM algorithms was implemented. Initially, a binary classifier trained on 30% of the dataset separated OW from sea ice, achieving high overall accuracies of 98.83% (RF) and 98.60% (SVM) when validated against the remaining 70% of samples using OSISAF SSMIS sea ice type data. Subsequently, a second classifier distinguished FYI from MYI within the sea ice samples, yielding overall accuracies of 84.82% (RF) and 71.71% (SVM). Classifier robustness was further validated through cross-validation, where monthly data were iteratively used for training and testing.

The findings highlight the strong sensitivity of GNSS-R signals to sea ice types and underscore the considerable potential of ML classifiers in advancing sea ice monitoring with GNSS-R technology.

6.1 DDMs Over Different Types of Sea Ice

The spatial distribution of GNSS-R ground tracks across OW, FYI, and MYI on 26 November 2015 is depicted in Fig. 12a, alongside representative TDS-1 DDMs from the respective transition zones. These ground tracks are superimposed on a Sea Ice Concentration (SIC) map in Fig. 12b, revealing that MYI typically exhibits higher SIC values than FYI.

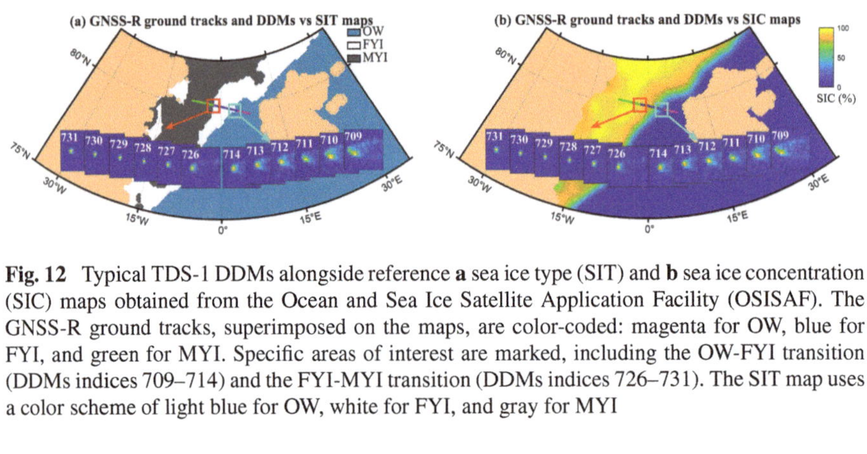

Fig. 12 Typical TDS-1 DDMs alongside reference **a** sea ice type (SIT) and **b** sea ice concentration (SIC) maps obtained from the Ocean and Sea Ice Satellite Application Facility (OSISAF). The GNSS-R ground tracks, superimposed on the maps, are color-coded: magenta for OW, blue for FYI, and green for MYI. Specific areas of interest are marked, including the OW-FYI transition (DDMs indices 709–714) and the FYI-MYI transition (DDMs indices 726–731). The SIT map uses a color scheme of light blue for OW, white for FYI, and gray for MYI

6.2 Spaceborne GNSS-R Features

Surface Reflectivity

The GNSS-R instrument, such as the SGR-ReSI, captures both the direct signals from GNSS constellations and those scattered from the Earth's surface. In the TDS-1 mission, the primary signals observed are the L1 C/A codes, which have a center frequency of 1.575 GHz. The L-band is particularly advantageous for remote sensing because it is largely unaffected by atmospheric conditions and precipitation. Specular reflections are most significant when the surface is smooth and relatively flat. As noted by [5], the GNSS-R signals collected over sea ice are typically coherent owing to the smooth nature of the surface. Consequently, the coherent surface reflectivity (SR_{coh}) at the specular point (SP) can be described by the model from [6, 19], which expresses SR_{coh} as follows:

$$SR_{coh} = \frac{(4\pi)^2(R_t + R_r)^2}{(\lambda)^2 G_r G_t P_t} P_{coh} \tag{8}$$

where P_{coh} is the coherently received power reflected from the surface, R_r and R_t represent the distances from the receiver and transmitter to the specular point, respectively, G_r and G_t denote the antenna gains of the receiver and transmitter, λ is the wavelength of the GNSS signal, and P_t is the power transmitted by the GNSS satellites.

Most parameters in Eq. (8) can be derived from the TDS-1 Level 1b data. The distances R_r and R_t are computed based on the positions of the transmitter, receiver, and specular point, all available in the metadata. The receiver antenna gain G_r at the specular point is directly obtained from the metadata. The noise floor is estimated by averaging the first four rows of the Delay-Doppler Map (DDM), as suggested by [16]. The transmitted signal power is calculated using the method from [20]:

$$P_t G_t = 4\pi R_d^2 G_d P_d \tag{9}$$

where $P_t G_t$ is the effective isotropic radiated power (EIRP), P_d is the direct received power, R_d is the distance from the transmitter to the receiver, and G_d is the zenith antenna gain. In this study, G_d is set to 4 dB following the Merrbys documentation [1, 21]. As highlighted by [22], the received power primarily comes from the area around the specular point, typically the first Fresnel zone. The Fresnel reflectivity model, as discussed in [23], further describes the coherent surface reflectivity as:

$$SR_{coh} = |\Re|^2 \cdot \exp\left[-\left(\frac{4\pi}{\lambda}\cos\theta_i\sigma\right)^2\right] \tag{10}$$

Here, θ_i is the incidence angle, and $\sigma$is the surface root mean square (RMS) height. The exponential term accounts for surface roughness effects. The Fresnel reflection

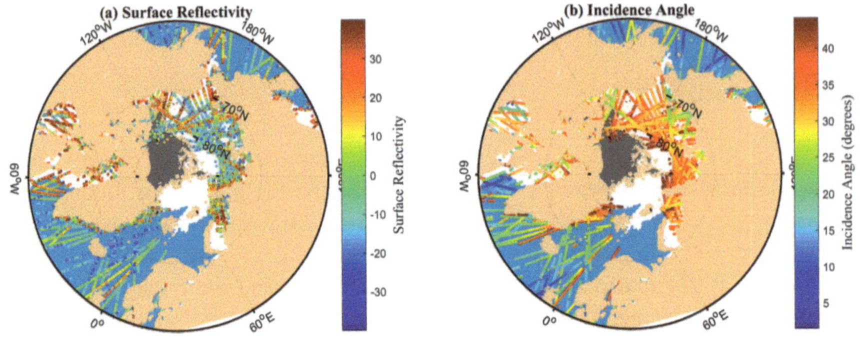

Fig. 13 **a** Surface reflectivity from TDS-1 and **b** the corresponding incidence angle distribution in the Arctic region from 11 to 15 February 2018. OW is shown in light blue, while FYI and MYI are represented in white and gray, respectively

coefficient $\Re$ is derived from the following expressions:

$$\Re = \frac{1}{2}(\Re_{VV} - \Re_{HH}) \tag{11}$$

$$\Re_{VV} = \frac{\epsilon_i \cos\theta_i - \sqrt{\epsilon_i - \sin^2\theta_i}}{\epsilon_i \cos\theta_i + \sqrt{\epsilon_i - \sin^2\theta_i}} \tag{12}$$

$$\Re_{HH} = \frac{\cos\theta_i - \sqrt{\epsilon_i - \sin^2\theta_i}}{\cos\theta_i + \sqrt{\epsilon_i - \sin^2\theta_i}} \tag{13}$$

where θ_i is the incidence angle over the sea ice, and ϵi is the permittivity of the sea ice, which varies with sea ice type. This analysis utilizes data where the incidence angle is less than 45°. Additionally, data flagged for issues such as eclipse conditions or direct signal contamination in the DDM [24] have been excluded to ensure reliability. The distribution of surface reflectivity from TDS-1 and the corresponding incidence angles in the Arctic region over a five-day period (11–15 February 2018) are illustrated in Fig. 13.

Features Derived from DDM

In addition to surface reflectivity, seven additional observables obtained from GNSS-R measurements are selected for the purpose of sea ice classification. These observables correspond to distinct characteristics extracted from DDMs, which capture essential information regarding the power distribution and morphological attributes of the reflected signals.

Among these parameters, the Delay-Doppler Map Average (DDMA) is utilized, representing the average SNR calculated within a predefined window surrounding

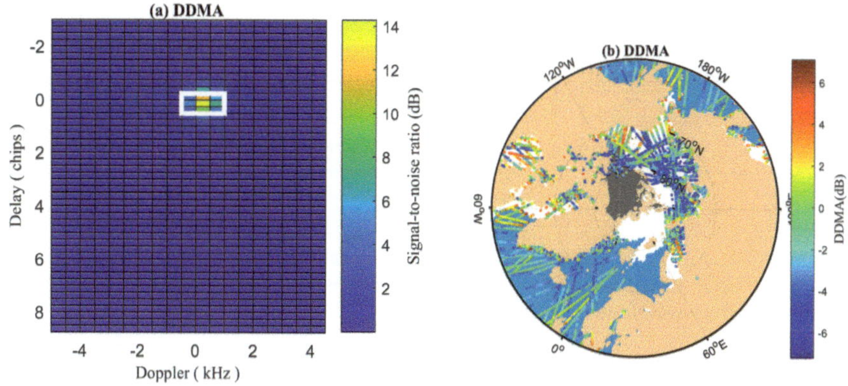

Fig. 14 **a** The region used for DDMA computation is indicated by the white rectangle, centered on the peak SNR point and covering an area of 3 delay bins by 3 Doppler shifts. **b** Geographic distribution of DDMA values compared against the OSISAF SIT map. OW is indicated in light blue, while FYI and MYI are represented by white and gray, respectively

the DDM's peak SNR location (Fig. 14a). Specifically, this window encompasses an area extending 3 delay bins and 3 Doppler shifts from the central peak SNR point. Figure 14b illustrates the spatial distribution of DDMA values obtained over a five-day observation period (11–15 February 2018), presented alongside the corresponding OSISAF sea ice type (SIT) map for comparative analysis.

Following the methodology described in prior studies [4], the integrated delay waveform (IDW) is derived by summing 20 individual delay waveforms, where each waveform constitutes a cross-section of the DDM taken at different Doppler shifts. The specific cross-section corresponding to zero Doppler shift is designated as the central delay waveform (CDW). To quantify the disparity between the IDW and CDW, the differential delay waveform (DDW) is introduced and defined by the following expression:

$$DDW = NIDW - NCDW \tag{14}$$

Here, NIDW and NCDW denote the normalized versions of the IDW and CDW, respectively. The characteristics of NIDW are influenced by power spreading effects attributable to surface roughness. Representative examples of these normalized delay waveforms (NCDW, NIDW) and the resulting DDW, observed over delay bins ranging from − 5 to 20 for surfaces classified as OW, FYI, and MYI, are displayed in Fig. 15.

This section delineates the six key features derived from the delay waveforms for the purpose of sea ice classification.

The first category of features is defined by the right-edge slope of their respective waveforms, calculated using a linear fit over a 5-delay-bin window commencing from the zero-delay bin.

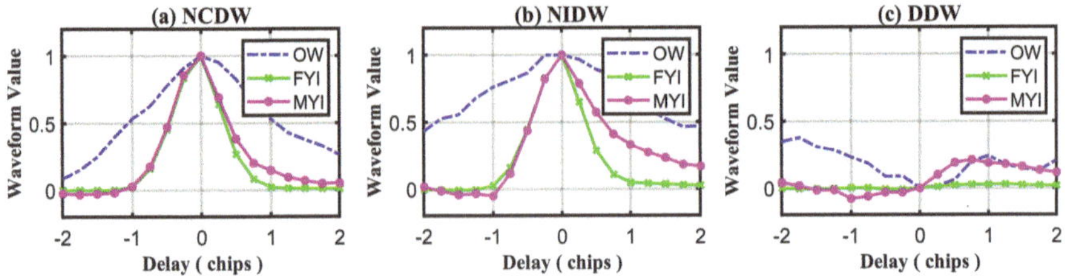

Fig. 15 Waveforms corresponding to samples collected over OW, FYI, and MYI surfaces: **a** normalized central delay waveform (NCDW), **b** normalized integrated delay waveform (NIDW), and **c** differential delay waveform (DDW)

RESC (Right Edge Slope of CDW): This parameter is characterized by the slope of the normalized central delay waveform (NCDW), as represented by the blue line in Fig. 16a.

RESI (Right Edge Slope of IDW): This feature denotes the corresponding slope computed from the normalized integrated delay waveform (NIDW), illustrated by the green line in Fig. 16a.

RESD (Right Edge Slope of DDW): This metric captures the slope derived from the differential delay waveform (DDW), indicated by the magenta line in Fig. 16a.

The second group of features is quantified by summing the waveform values across delay bins 0 to 6.

REWC (Right Edge Waveform Summation of CDW): This is defined as the cumulative sum of the NCDW values, which are marked with blue diamond symbols in Fig. 16b.

REWI (Right Edge Waveform Summation of IDW): This feature is obtained from the summation of the NIDW values, represented by green cross symbols in Fig. 16b.

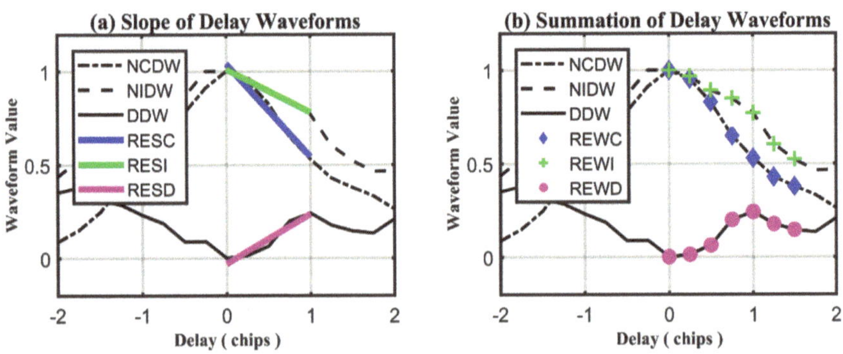

Fig. 16 **a** Three slope-based features derived from the delay waveforms: RESC, RESI, and RESD. These are determined by the slopes of the best-fit lines, color-coded in blue, green, and magenta, respectively. **b** Three summation-based features: REWC, REWI, and REWD. The values of these features are computed from the summation of the normalized waveform magnitudes at the delay bins marked by the blue diamond, green cross, and magenta circle symbols, respectively

REWD (Right Edge Waveform Summation of DDW): This is calculated as the sum of the DDW values, marked with magenta circle symbols in Fig. 16b.

Figure 17 compares the geographical distribution of these six features with the reference OSISAF SIT maps.

6.3 *Sea Ice Classification Method*

Building upon the established efficacy of RF and SVM in classification tasks, this study adopts these two methods to address the under-explored challenge of classifying sea ice types (FYI and MYI) using GNSS-R features.

Random Forest (RF)

RF is an ensemble machine learning technique that constructs multiple decision trees and aggregates their predictions, often leading to superior performance compared to individual classifiers [25, 26]. This method operates on the principle that a collective of diverse models typically achieves greater accuracy and robustness than a single model. Key advantages of RF particularly relevant to remote sensing and classification tasks include High Efficiency with Large Datasets, High-Dimensional Input Handling, Feature Importance Estimation and Noise Robustness.

In the construction of individual decision trees within the forest, node splitting is a critical step often guided by metrics like the Gini index [12], which quantifies the impurity of training samples at a node. For a given training dataset T, the Gini index (G_{index}) is defined for a node as:

$$G_{index}(T) = 1 - \sum_{i=1}^{m} p_i^2 \quad (15)$$

where m is the number of target classes and pi is the proportion of samples in the node that belong to class i. This metric aids in selecting the most discriminative features for splitting the data at each node. Furthermore, RF facilitates the assessment of feature importance, offering valuable insights for feature selection and understanding the contribution of each variable, such as the different GNSS-R observables, to the final classification outcome [27, 28].

Support Vector Machine (SVM)

SVM is a supervised machine learning method grounded in statistical learning theory, designed to identify optimal decision boundaries that maximize the separation between different classes [29]. This approach is applicable to both linear

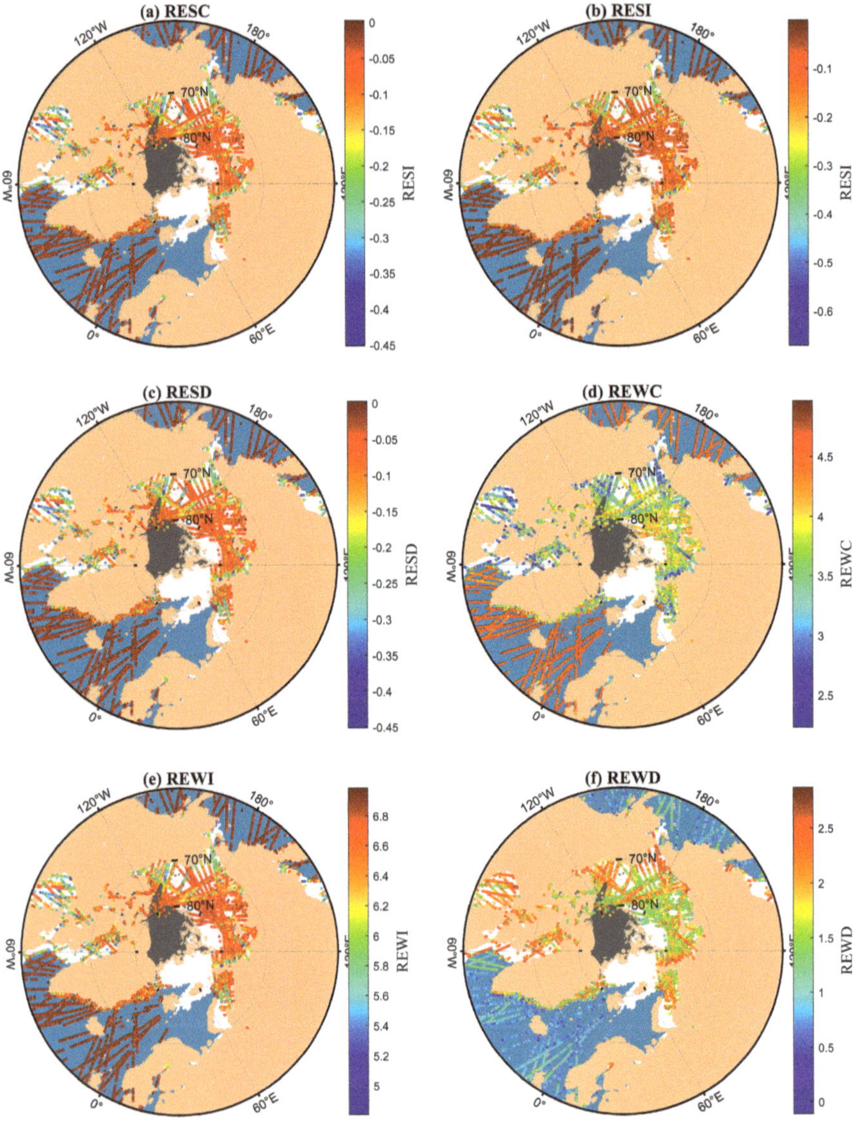

Fig. 17 A comparative analysis between the spatial distributions of six GNSS-R features and the OSISAF SIT map. Panels **a**–**f** illustrate the geographical patterns of RESC, RESI, RESD, REWC, REWI, and REWD, respectively, for the observation period of 11–15 February 2018. The reference sea ice types in the background map correspond to open water (light blue), first-year ice (white), and multi-year ice (gray)

and nonlinear classification tasks, capable of handling high-dimensional data while maintaining strong generalization performance. Although frequently applied to binary classification problems, SVM can be extended to multi-class scenarios using strategies such as one-versus-rest or one-versus-one.

$$(\boldsymbol{x}_1, y_1), (\boldsymbol{x}_2, y_1), \ldots, (\boldsymbol{x}_n, y_n) \in R^n \times Y, Y = (-1, 1) \tag{16}$$

where $\boldsymbol{x}_i$ represents the input samples that consist of features for classification, y_i is the class label of $\boldsymbol{x}_i$.

Sea Ice Classification Based on RF and SVM

The sea surface in this investigation is categorized into three distinct types: OW, FYI, and MYI. To address this multi-class classification challenge, the one-versus-all (OVA) strategy is employed [30]. This approach transforms the multi-class problem into a series of binary classification tasks. Specifically, the RF algorithm, a powerful machine learning classifier capable of handling both binary and multi-class problems, is utilized within this framework. The OVA strategy decomposes the original three-class problem into two sequential binary classifications. In the initial stage, a binary classifier is trained to distinguish OW (treated as the positive class) from the combined sea ice category (FYI and MYI collectively as the negative class). Subsequently, a second binary classifier is trained exclusively on the sea ice samples to differentiate between FYI (positive class) and MYI (negative class). This staged approach is adopted because, although standard RF can directly handle multi-class problems, applying the OVA binarization has been reported to achieve comparable or even superior accuracy, sometimes with more efficient model sizes. Furthermore, as the SVM classifier, also used in this study, is inherently well-suited for binary classification, the OVA strategy provides a consistent methodological framework for both algorithms. The entire classification procedure is implemented using the Scikit-Learn library in Python [31]. A flowchart summarizing the sequential steps of the sea ice classification methodology is provided in Fig. 18.

The methodology can be systematically divided into four main components:

Data Preprocessing and Feature Extraction: The initial stage involves processing TDS-1 data acquired north of 55°N with a peak SNR exceeding − 3 dB. From the calibrated signals, delay waveforms are generated and normalized, leading to the extraction of eight distinct features: Surface Reflectivity (SR), DDMA, and six additional metrics derived from the delay waveforms (RESC, RESI, RESD, REWC, REWI, REWD). These features are then spatiotemporally collocated with OSISAF SIT maps for supervised learning.

Open Water vs. Sea Ice Classification: In this step, a binary classification problem is addressed where FYI and MYI are grouped into a single "sea ice" category. The model is trained using a randomly selected subset comprising 30% of the available samples, while the remaining 70% are reserved for testing the classifier's ability to distinguish OW from sea ice.

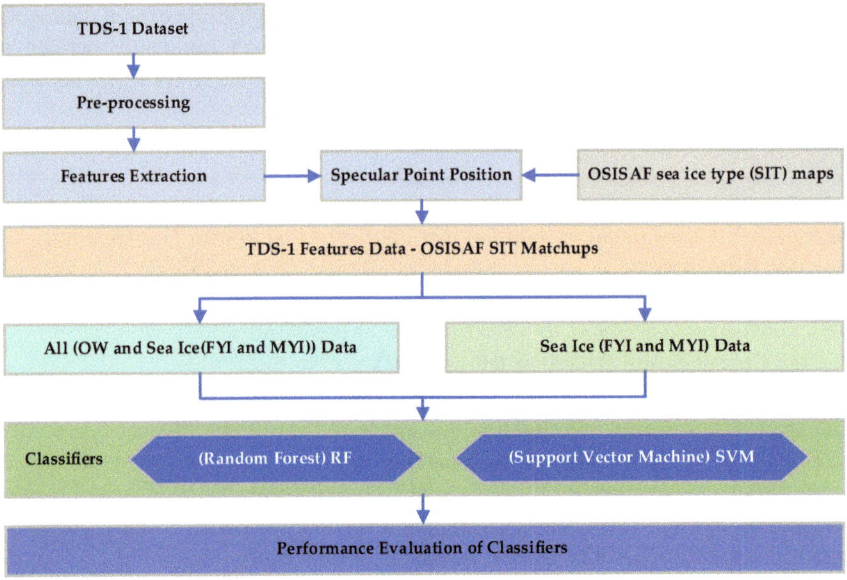

Fig. 18 Workflow for sea ice classification utilizing machine learning techniques

FYI vs. MYI Discrimination: Following the initial separation, a subsequent classification is performed specifically on the pixels identified as sea ice. This step aims to classify the sea ice further into FYI and MYI types. A similar data splitting strategy (70% for testing) is employed to assess the performance of the FYI-MYI classifier.

Performance Assessment: The efficacy of the classification system is quantitatively evaluated using a confusion matrix and standard performance metrics, detailed in Fig. 19 and Table 5 of the original context, which include performance indexes such as overall accuracy, precision, and recall.

Similarly, the confusion matrix is used to show the classification results with rows typically representing the actual (ground truth) categories, while columns corresponding to the predicted categories. The core components of this matrix for a binary classification problem are defined as follows:

True Positives (TP): Instances that are positive and are correctly identified as positive by the classifier.
False Negatives (FN): Instances that are positive but are incorrectly classified as negative.
False Positives (FP): Instances that are negative but are incorrectly classified as positive.
True Negatives (TN): Instances that are negative and are correctly identified as negative.

Based on these four fundamental components, several key metrics presented in Table 5 are derived to quantitatively assess classifier performance:

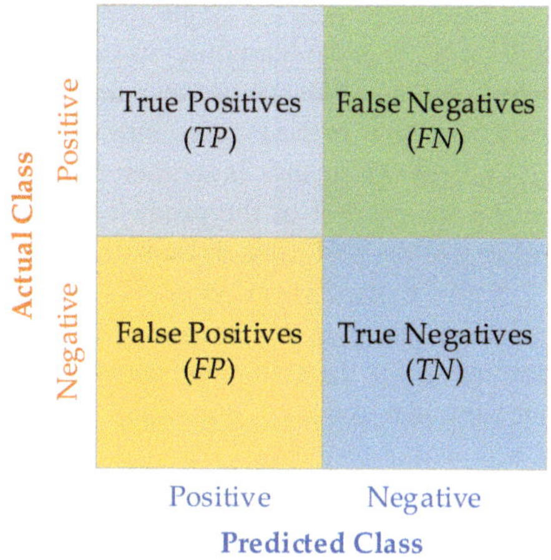

Fig. 19 The definition of the confusion matrix

Table 5 The confusion matrix for the SVM algorithm using data from Arctic and Antarctic regions

Arctic	Reference classified as	Sea ice	Sea water	Sum	User accuracy
	Sea ice	1,270,679	31,121	1,301,800	97.61%
	Sea water	33,945	1,402,807	1,436,752	97.64%
	Sum	1,304,624	1,433,928	2,738,552	
	Producer accuracy	97.40%	97.83%		
	Overall accuracy			97.62%	
	Kappa coefficient			95.24%	
Antarctic	Reference classified as	Sea ice	Sea water	Sum	User accuracy
	Sea ice	1,406,997	63,331	1,470,328	95.69%
	Sea water	71,813	1,538,739	1,610,552	95.54%
	Sum	1,478,810	1,602,070	3,080,880	
	Producer accuracy	95.14%	96.05%		
	Overall accuracy			95.61%	
	Kappa coefficient			91.21%	

Accuracy quantifies the overall proportion of correct predictions, both positive and negative, among the total number of cases examined. It serves as a general indicator of correctness.

Precision, also known as the positive predictive value, measures the reliability of the positive predictions. It is the fraction of instances predicted as positive that are actually positive. This metric is crucial when the cost of false positives (FP) is high.

Recall, also referred to as Sensitivity or True Positive Rate (TPR), assesses the model's ability to correctly identify all relevant positive instances. It calculates the proportion of actual positives that are correctly identified. This metric is vital when missing a positive instance (false negative) is costly.

F1-Score provides a single metric that balances the concerns of both precision and recall. It is the harmonic mean of these two metrics, offering a useful measure when seeking a balance between them, especially on imbalanced datasets. When the beta parameter in the F-score formula is set to 1, it yields the F1 score.

G-mean (G-means), is another metric useful for imbalanced datasets. It is defined as the square root of the product of sensitivity (recall) and specificity, aiming to maximize the performance on both classes simultaneously.

Kappa Coefficient (Cohen's Kappa) is a statistic that measures inter-rater agreement (in this case, between the classifier's predictions and the true labels) while correcting for the agreement that occurs by chance. This makes it a more robust measure than accuracy, particularly when class distributions are skewed.

6.4 Sea Ice Classification Results

Although the OSISAF SIT products are provided continuously, there is a lack of MYI information from May to October most of the time in the OSISAF SIT products. In addition, the TDS-1 data is very sparse from 2015 to 2017 as the GNSS-R receiver was working only two days in an eight-day cycle. Therefore, the TDS-1 datasets collected in 2018 are selected according to the availability of MYI information of OSISAF SIT products. Figure 20a presents the data availability status of both OSISAF SIT products and TDS-1 data that are used in this study. The number of OW, FYI, and MYI samples every month are shown in Fig. 20b. According to the statistics, a total number of 460,685 samples, including 165,434 OW samples, 266,680 FYI samples, and 28,571 MYI samples, are used in this study.

Despite the continuous availability of OSISAF Sea Ice Type (SIT) products, data regarding MYI are typically incomplete or unavailable during the months from May to October each year. Moreover, the sampling density of TDS-1 data between 2015 and 2017 is relatively low, attributable to the GNSS-R receiver's operational cycle, which collected data for only two days within an eight-day period. Consequently, TDS-1 datasets from the year 2018 were selected for this analysis to ensure consistency with the periods of available MYI information in the OSISAF SIT records. The temporal coverage and availability of both the OSISAF SIT products and the TDS-1 data used in this study are visualized in Fig. 20a. The monthly sample sizes for each sea ice category—OW, FYI, and MYI—are presented in Fig. 20b. In aggregate, the analysis incorporates a total of 460,685 valid samples, distributed as follows: 165,434 samples of OW, 266,680 samples of FYI, and 28,571 samples of MYI.

Figure 21 displays the confusion matrices for both RF and SVM classifiers, detailing the classification outcomes for each category. The corresponding performance metrics, calculated based on the formulas provided in Table 6, are summarized in Table 7. These quantitative assessments were obtained by validating the model predictions against the OSISAF SIT maps using the test dataset.

As illustrated in Fig. 21a, b, the OW-sea ice classification was evaluated on a test set comprising 322,480 samples (109,559 OW and 212,921 sea ice). The RF classifier

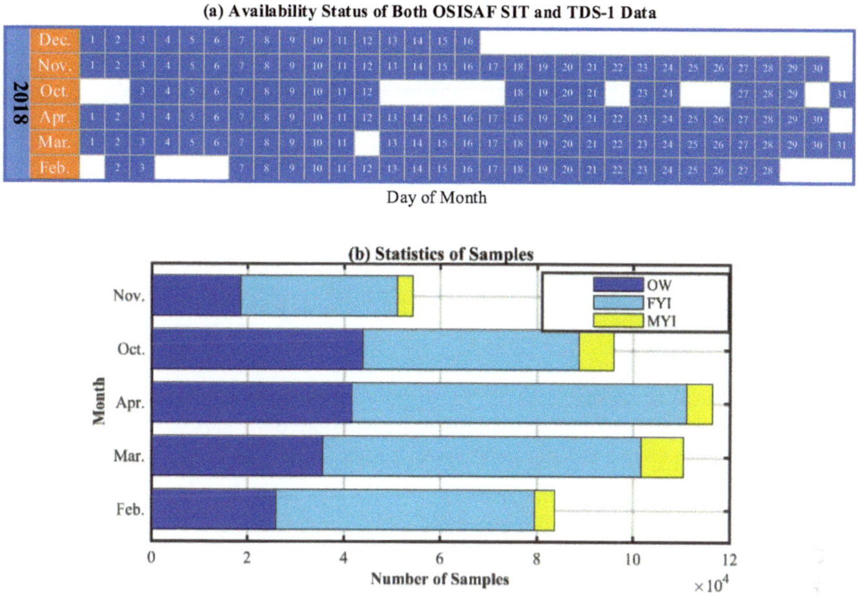

Fig. 20 **a** The availability of both TDS-1 and OSISAF SIT data, which includes the information of OW, FYI, and MYI in 2018. **b** Statistics of OW, FYI, and MYI samples every month

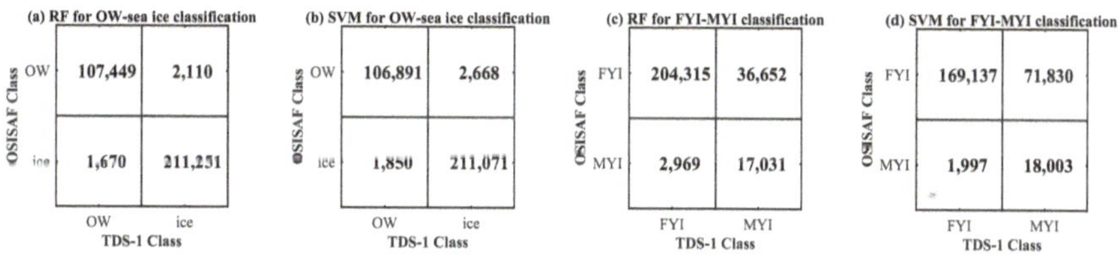

Fig. 21 Confusion matrix of **a** RF and **b** SVM for OW-sea ice classification. Confusion matrix of **c** RF and **d** SVM for FYI-MYI classification

Table 6 The definition of evaluation metrics

Evaluation metrics	Equation
Accuracy	$\frac{TP+TN}{TP+FP+TN+FN}$
Precision	$\frac{TP}{TP+FP}$
Recall	$\frac{TP}{TP+FN}$
F-value	$\frac{Recall\cdot Precision\cdot(1+\beta^2)}{\beta\cdot(Recall+Precision)}$
G-mean	$\sqrt{\frac{TP\cdot TN}{(TP+FN)\cdot(TN+FP)}}$
Kappa coefficient	$\frac{2\cdot(TP\cdot TN-FN\cdot FP)}{(TP+FP)\cdot(FP+TN)+(TP+FN)\cdot(FN+TN)}$

Table 7 The evaluation metrics of OW-sea ice and FYI-MYI classification

Evaluation metrics	OW-sea ice classification		FYI-MYI classification	
	RF	SVM	RF	SVM
Accuracy (%)	98.83	98.60	84.82	71.71
Precision (%)	98.47	98.30	98.57	98.83
Recall (%)	98.07	97.56	84.79	70.19
F1-value	0.98	0.98	0.91	0.82
G-mean	0.99	0.98	0.85	0.79
Kappa coefficient	0.97	0.97	0.39	0.23

attained an overall accuracy of 98.83%, slightly outperforming the SVM classifier, which achieved 98.60%. These results are consistent with accuracies reported in prior studies [15, 15, 15]. For the FYI-MYI classification task (Fig. 21c, d), a test set of 20,000 FYI samples and 240,967 MYI samples was used. In this case, RF demonstrated a markedly higher accuracy of 84.82%, compared to SVM's accuracy of 71.71%. This indicates that while both classifiers perform comparably well in distinguishing sea ice from open water, RF is significantly more effective at discriminating between FYI and MYI.

The spatial distribution of the final classification results is mapped in Fig. 22. For a more detailed visual analysis, selected portions of the classification results are magnified in Fig. 23. These subplots show the predicted classes, the reference OSISAF types, and a difference map highlighting the discrepancies between them. In the analysis, a numerical labeling scheme was applied: for the OW-sea ice classification, sea ice was assigned a value of -1 and OW a value of 1; for the FYI-MYI classification, FYI was labeled as -1 and MYI as 1. The difference between the reference and predicted maps was calculated as follows:

$$Diff = Ref - Pre \tag{16}$$

The discrepancy between classification results and reference data is quantified by the difference (Diff), calculated as Diff = Ref - Pre, where Ref denotes the reference label and Pre represents the predicted label. In the OW versus sea ice classification, a difference value of -2 indicates that sea ice was incorrectly classified as OW (ice-OW), a value of 2 indicates that OW was misclassified as sea ice (OW-ice), and a value of 0 indicates agreement between the predicted and reference labels. Similarly, in the FYI versus MYI classification, a value of -2 corresponds to FYI being misclassified as MYI (FYI-MYI), a value of 2 corresponds to MYI being misclassified as FYI (MYI-FYI), and a value of 0 indicates correct classification.

Analysis of the spatial distribution of errors (Fig. 22b, c) reveals that the majority of misclassifications for both RF and SVM classifiers in the OW/sea ice task occur predominantly along the transition boundaries between OW and sea ice, a pattern also visible in the zoomed-in views of Fig. 23a, b. For the FYI/MYI classification task, the confusion matrices (Fig. 21c, d) indicate that the number of misclassified

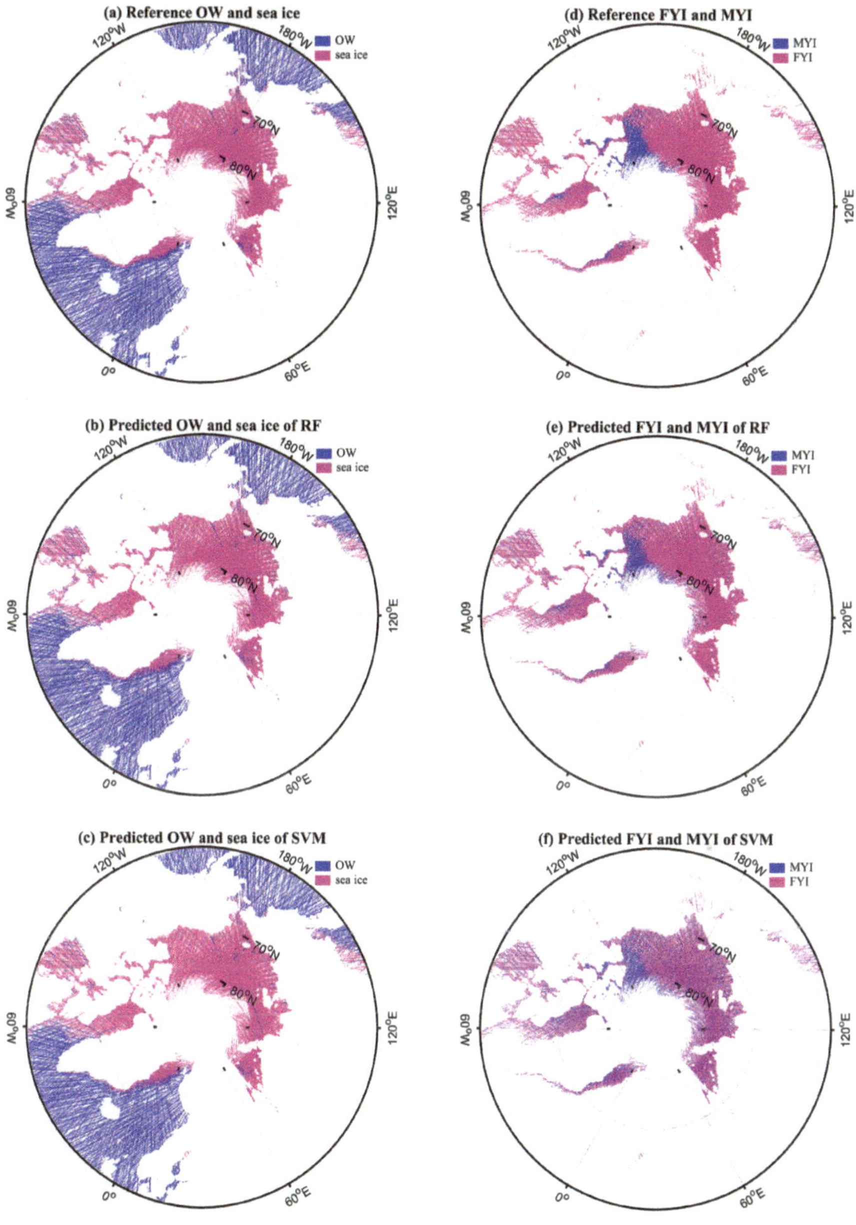

Fig. 22 Classification results of RF and SVM classifiers. **a** Reference OW and sea ice. **b** Predicted OW and sea ice of RF. **c** Predicted OW and sea ice of SVM. **d** Reference FYI and MYI. **e** Predicted FYI and MYI of RF. **f** Predicted FYI and MYI of SVM

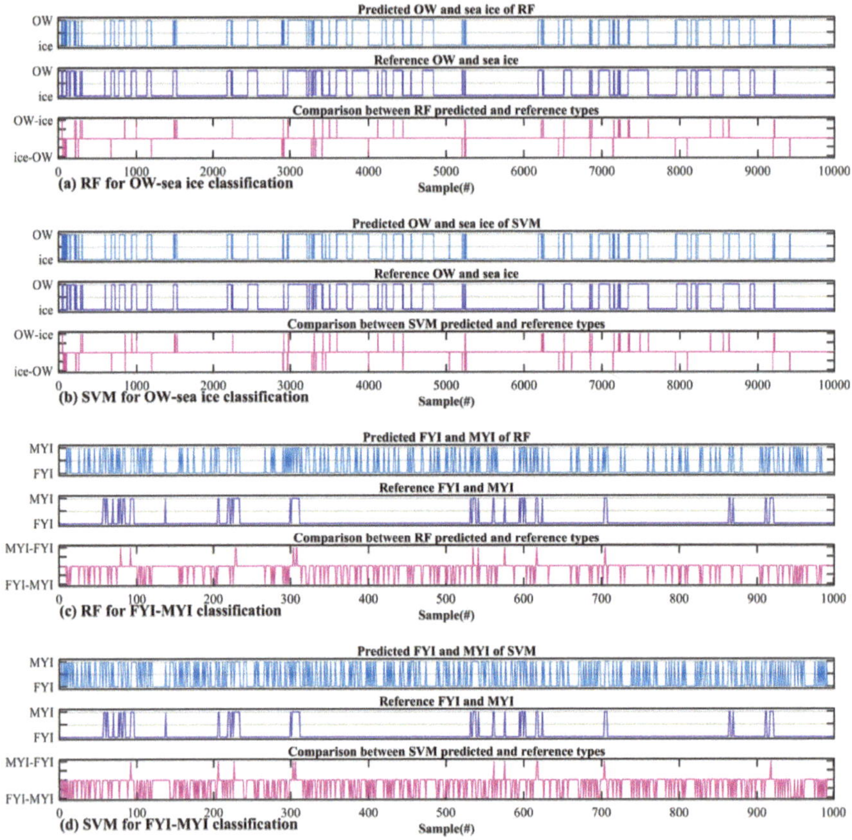

Fig. 23 Parts of classification results of RF and SVM classifiers. Zoomed **a** RF and **b** SVM results of 10,000 samples for OW-sea ice classification. Zoomed **c** RF and **c** SVM results of 1000 samples for FYI-MYI classification. In each figure of (**a**)–(**d**), the top plot represents the predicted types of the classifier, the middle one is the reference types from OSISAF SIT and the lowest one is the comparison results between predicted and reference types. OW-ice means the OW is misidentified as sea ice, and ice-OW means the sea ice is misclassified as OW. FYI-MYI means the FYI is misidentified as MYI, and MYI-FYI means the MYI is misclassified as FYI

FYI samples is substantially higher than that of MYI samples. This discrepancy is likely attributable to the significant class imbalance, as the quantity of FYI samples is more than ten times greater than that of MYI samples. Consequently, the overall accuracy metric is less sensitive to misclassifications of the minority MYI class. The spatial error distribution for the FYI/MYI classification (Fig. 22e, f) shows that misclassifications are widespread for both classifiers, though the SVM model produces a larger number of errors compared to RF, as further evidenced in Fig. 23c, d.

The kappa coefficient, a measure of agreement correcting for chance, further highlights the performance difference between the two classification tasks. For the

OW/sea ice classification, both RF and SVM achieved a high kappa value of 0.97, indicating almost perfect agreement. In contrast, for the more challenging FYI/MYI classification, the kappa coefficient dropped substantially to 0.39 for RF and 0.23 for SVM, indicating only fair and slight agreement, respectively. Furthermore, while the classification accuracy of SVM was comparable to that of RF for the OW/sea ice task, the computational time required by SVM was approximately nine times longer than that of RF when applied to the same test dataset, indicating a significant efficiency advantage for the RF algorithm.

6.5 Discussion

To assess the impact of sea ice phenology (growth and melt cycles) on classification efficacy and methodological robustness, an additional validation experiment was designed. This experiment employed a month-wise training strategy, where data from a single month were used for training, while data from the remaining four months served as the independent test set. This approach shares similarities with k-fold cross-validation [32] but is intentionally designed to preserve seasonal characteristics, whereas standard k-fold validation randomly partitions data, potentially diluting seasonal signals.

The validation process commenced by training the model on February data and testing it sequentially on datasets from March, April, November, and December.

As shown in Figs. 14 and 15, the classification accuracy for OW versus sea ice demonstrates significantly lower variability compared to the accuracy for discriminating between FYI and MYI. This difference in stability can be attributed to the relative consistency in the proportion between OW and overall sea ice, whereas the distribution ratio between FYI and MYI is more dynamic and susceptible to changes induced by ice growth and melt processes.

A detailed analysis of the temporal trend reveals that the accuracy for OW-sea ice classification remains relatively stable from February to April, experiences a decline towards November, and subsequently shows a substantial recovery by December. Notably, November registers the lowest classification accuracy across all months evaluated. A similar trend is observed for the FYI-MYI classifier, although with greater amplitude in its fluctuations.

This pattern is closely linked to the sea ice extent dynamics observed in 2018, as depicted in Fig. 16. The sea ice extent reached its annual maximum around March 14, 2018, and its minimum around September 21, 2018. The period from February to April is characterized by a relatively stable ice extent following the peak, whereas the period from November to December marks a phase of rapid ice growth during early winter. The notably lower ice extent in November creates a challenging scenario for classification, as newly formed ice is often fragmented and surrounded by open water, increasing the likelihood of misclassification. The subsequent significant increase in classification accuracy from November to December aligns with the consolidation

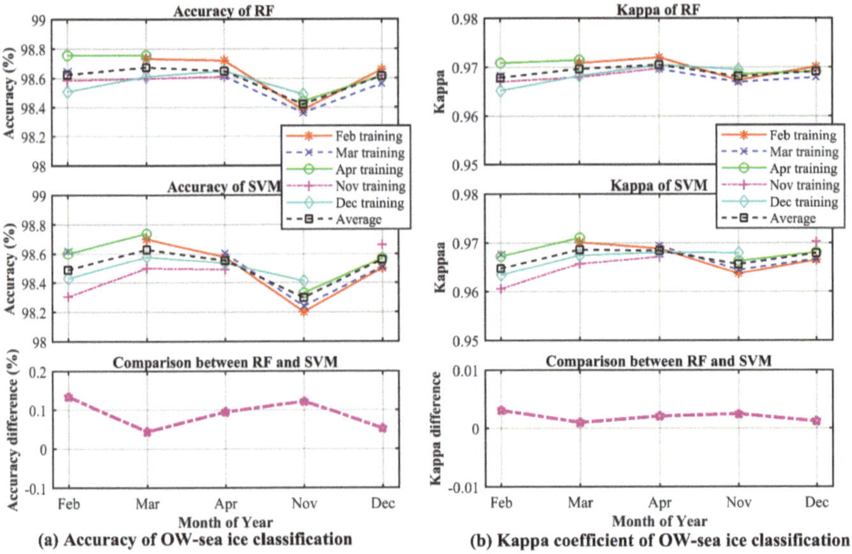

Fig. 24 The classification **a** accuracy for OW versus sea ice, while **b** shows the corresponding kappa coefficients, both derived using a monthly rotation training strategy. In each panel, the top and middle subplots present the results for the RF and SVM classifiers, respectively. The bottom subplot illustrates the performance difference between the two classifiers (RF minus SVM)

and expansion of the sea ice cover, demonstrating the method's sensitivity to these phenological stages (Figs. 24, 25, 26).

6.6 Conclusions

This study evaluates RF and SVM classifiers for Arctic sea ice type discrimination by applying a one-against-all binarization strategy. This approach reframes the multi-class problem into sequential binary classification tasks. The process involves an initial separation of OW from sea ice, where sea ice is treated as a combined category encompassing FYI and MYI. A subsequent classification step then discriminates between FYI and MYI within the identified sea ice regions.

The analysis utilizes data from specific periods in 2018 (February-April and November–December), selected to ensure the concurrent availability of OW, FYI, and MYI reference data from OSISAF Sea Ice Type (SIT) products for model training and validation. Validation against these OSISAF SIT maps demonstrated high performance in the first step, with RF and SVM achieving overall accuracies of 98.83% and 98.60%, respectively, for OW-sea ice classification. These results are consistent with findings from earlier studies employing TDS-1 data [5, 16, 17, 33, 34]. In the second classification step, which distinguishes FYI from MYI, the overall accuracy

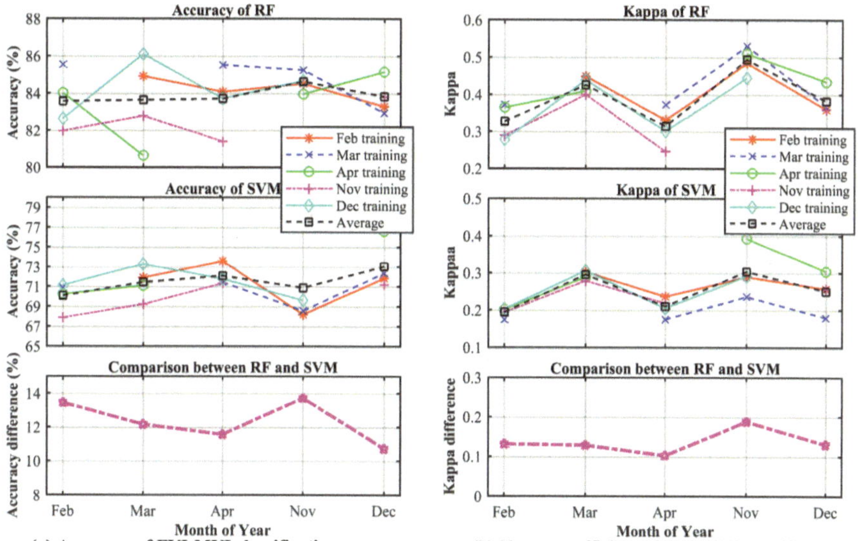

(a) Accuracy of FYI-MYI classification (b) Kappa coefficient of FYI-MYI classification

Fig. 25 The classification **a** accuracy for the discrimination between FYI and MYI, while **b** shows the corresponding kappa coefficients. In each figure, the top and middle plots present the results for the RF and SVM classifiers, respectively. The bottom plot illustrates the performance difference (RF result minus SVM result) for each metric

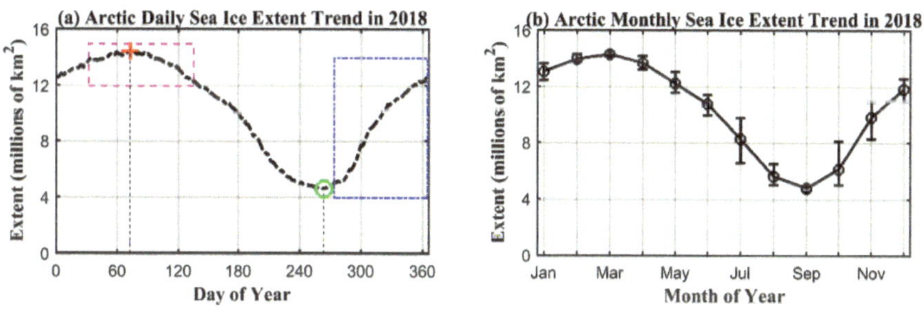

Fig. 26 **a** The temporal variation in daily Arctic sea ice extent over the course of 2018. The monthly average extent is indicated by blue circular markers, while the annual maximum and minimum values are highlighted by a red cross symbol and a green circle, respectively. Magenta and blue rectangles outline the two primary data collection periods utilized in this study. Data points within the first period (February to May) are represented by red dots inside the magenta rectangle, while the second period (October to December) is indicated by blue dots within the corresponding blue rectangle. **b** The monthly trend in Arctic sea ice extent for 2018, where error bars depict the range between the maximum and minimum values observed each month. The sea ice extent data were sourced from the National Snow and Ice Data Center (NSIDC) on August 15, 2021

was 84.82% for RF and 71.71% for SVM. The lower performance in FYI-MYI classification compared to OW-sea ice separation is consistent with the challenges noted in other studies using different sensors, where discriminating between ice types can be more complex than separating ice from water.

A cross-validation experiment was conducted to assess the impact of seasonal ice growth and melt on classification robustness. This involved iteratively using data from individual months for training and testing against the remaining months. To the best of our knowledge, this work represents one of the first attempts to apply RF and SVM classifiers specifically for FYI and MYI classification using spaceborne GNSS-R data. The results underscore the significant potential of GNSS-R data combined with machine learning for sea ice monitoring, suggesting it can serve as an effective complementary tool to other remote sensing techniques. Future research directions should include the exploration of a broader set of GNSS-R features, testing of additional machine learning algorithms, and the integration of environmental factors such as ocean wind to further enhance classification accuracy, particularly for the more challenging differentiation of ice types.

References

1. Jales P, Unwin M (2019) MERRByS product manual: GNSS reflectometry on TDS-1 with the SGR-ReSI. Surrey Satellite Technol. Ld. Guildford, UK
2. Aaboe S, Breivik L-A, Sørensen A, Eastwood S, Lavergne T (2018) Global sea ice edge and type product user's manual OSI-402-c & OSI-403-c v2.3. EUMETSAT OSISAF, France
3. Breivik LA, Eastwood S, Godøy Ø, Schyberg H, Andersen S, Tonboe R (2001) Sea ice products for EUMETSAT satellite application facility. Can J Remote Sens 27:403–410
4. Zhu Y, Tao T, Yu K, Li Z, Qu X, Ye Z, Geng J, Zou J, Semmling M, Wickert J (2020) Sensing sea ice based on Doppler spread analysis of spaceborne GNSS-R data. IEEE J Sel Top Appl Earth Obs Remote Sens 13:217–226
5. Alonso-Arroyo A, Zavorotny VU, Camps A (2017) Sea ice detection using UK TDS-1 GNSS-R data. IEEE Trans Geosci Remote Sens 55:4989–5001
6. Zavorotny VU, Voronovich AG (2000) Scattering of GPS signals from the ocean with wind remote sensing application. IEEE Trans Geosci Remote Sens 38:951–964
7. Zhu Y, Yu K, Zou J, Wickert J (2017) Sea ice detection based on differential delay-Doppler maps from UK TechDemoSat-1. Sensors 17:1614
8. Im J, Jensen JR (2005) A change detection model based on neighborhood correlation image analysis and decision tree classification. Remote Sens Environ 99:326–340
9. Gleason CJ, Im J (2012) Forest biomass estimation from airborne LiDAR data using machine learning approaches. Remote Sens Environ 125:80–91
10. Lohse J, Doulgeris AP, Dierking W (2019) An optimal decision-tree design strategy and its application to sea ice classification from SAR imagery. Remote Sens 11:1574
11. Zhang X, Treitz PM, Chen D, Quan C, Shi L, Li X (2017) Mapping mangrove forests using multi-tidal remotely-sensed data and a decision-tree-based procedure. Int J Appl Earth Obs Geoinf 62:201–214
12. Lerman RI, Yitzhaki S (1984) A note on the calculation and interpretation of the Gini index. Econ Lett 15:363–368
13. Chan JC-W, Paelinckx D (2008) Evaluation of Random Forest and Adaboost tree-based ensemble classification and spectral band selection for ecotope mapping using airborne hyperspectral imagery. Remote Sens Environ 112:2999–3011

14. Belgiu M, Drăguţ L (2016) Random forest in remote sensing: a review of applications and future directions. ISPRS J Photogramm Remote Sens 114:24–31
15. Stehman SV (1997) Selecting and interpreting measures of thematic classification accuracy. Remote Sens Environ 62:77–89
16. Zhu Y, Tao T, Yu K, Qu X, Li S, Wickert J, Semmling M (2020) Machine learning-aided sea ice monitoring using feature sequences extracted from Spaceborne GNSS-reflectometry data. Remote Sens 12:3751
17. Yan Q, Huang W (2019) Detecting sea ice from TechDemoSat-1 data using support vector machines with feature selection. IEEE J Sel Top Appl Earth Obs Remote Sens 12:1409–1416
18. Park J-W, Korosov AA, Babiker M, Won J-S, Hansen MW, Kim H-C (2019) Classification of sea ice types in sentinel-1 SAR images. Cryosphere Discuss 2019:1–23
19. Voronovich AG, Zavorotny VU (2017) Bistatic radar equation for signals of opportunity revisited. IEEE Trans Geosci Remote Sens 56:1959–1968
20. Hajj GA, Zuffada C (2003) Theoretical description of a bistatic system for ocean altimetry using the GPS signal. Radio Sci 38
21. Comite D, Cenci L, Colliander A, Pierdicca N (2020) Monitoring freeze-thaw state by means of GNSS reflectometry: an analysis of TechDemoSat-1 data. IEEE J Sel Top Appl Earth Obs Remote Sens 13:2996–3005
22. Yan Q, Huang W (2020) Sea ice thickness measurement using spaceborne GNSS-R: first results with TechDemoSat-1 data. IEEE J Sel Top Appl Earth Obs Remote Sens 13:577–587
23. Tsang L, Newton RW (1982) Microwave emissions from soils with rough surfaces. J Geophys Res: Ocean 87:9017–9024
24. Camps A, Park H, Portal G, Rossato L (2018) Sensitivity of TDS-1 GNSS-R reflectivity to soil moisture: global and regional differences and impact of different spatial scales. Remote Sensing 10:1856
25. Breiman L (1996) Bagging predictors. Mach Learn 24:123–140
26. Dietterich TG (2000) An experimental comparison of three methods for constructing ensembles of decision trees: bagging, boosting, and randomization. Mach Learn 40:139–157
27. Ghimire B, Rogan J, Miller J (2010) Contextual land-cover classification: incorporating spatial dependence in land-cover classification models using random forests and the Getis statistic. Remote Sens Lett 1:45–54
28. Rodriguez-Galiano VF, Ghimire B, Rogan J, Chica-Olmo M, Rigol-Sanchez JP (2012) An assessment of the effectiveness of a random forest classifier for land-cover classification. ISPRS J Photogramm Remote Sens 67:93–104
29. Cortes C, Vapnik V (1995) Support-vector networks. Mach Learn 20:273–297
30. Anand R, Mehrotra K, Mohan CK, Ranka S (1995) Efficient classification for multiclass problems using modular neural networks. IEEE Trans Neural Netw 6:117–124
31. Pedregosa F, Varoquaux G, Gramfort A, Michel V, Thirion B, Grisel O, Blondel M, Prettenhofer P, Weiss R, Dubourg V (2011) Scikit-learn: machine learning in python. J Mach Learn Res 12:2825–2830
32. Wong T-T, Yeh P-Y (2019) Reliable accuracy estimates from k-fold cross validation. IEEE Trans Knowl Data Eng 32:1586–1594
33. Rodriguez-Alvarez N, Holt B, Jaruwatanadilok S, Podest E, Cavanaugh KC (2019) An Arctic sea ice multi-step classification based on GNSS-R data from the TDS-1 mission. Remote Sens Environ 230:111202
34. Llaveria D, Munoz-Martin JF, Herbert C, Pablos M, Park H, Camps A (2021) Sea ice concentration and sea ice extent mapping with L-band microwave radiometry and GNSS-R DATA from the FFSCat mission using neural networks. Remote Sens 13:1139

Chapter 18
Sea Surface Rainfall Detection from CYGNSS Delay-Doppler Maps

Jinwei Bu and Kegen Yu

Abstract Rainfall detection (RD) remains a prominent research focus within the field of ocean remote sensing. Previous studies have predominantly relied upon X-band marine radar imagery for detecting precipitation over the ocean surface. This chapter aims to investigate the potential of spaceborne GNSS-R for detecting rainfall over the ocean surface, utilising Delay Doppler Maps (DDMs) derived from data collected by the CYGNSS mission. This chapter first describes models of rainfall effects on the sea surface and analyzes rainfall features within GNSS-R DDMs. It then introduces the datasets employed and outlines data processing methods, followed by a presentation of rainfall detection methods: observation threshold-based detection using Probability Density Functions (PDF), detection using Support Vector Machines (SVM), Random Forests (RF), and Convolutional Neural Networks (CNN). Finally, a comprehensive evaluation of the rainfall detection performance is conducted.

1 Introduction

Over the last ten years, X-band marine radar imagery has been extensively employed to retrieve parameters such as sea surface wind speed, wind direction, and wave height [18]. Although numerous studies have demonstrated the considerable potential of this measurement method for wind and wave parameter retrieval under most conditions

J. Bu
Faculty of Land Resources Engineering, Kunming University of Science and Technology, Kunming, China

Yunnan Key Laboratory of Intelligent Monitoring and Spatiotemporal Big Data Governance of Natural Resources, Kunming, China

J. Bu
e-mail: b_jinwei@kust.edu.cn

K. Yu (✉)
School of Environmental Science and Spatial Informatics, China University of Mining and Technology, Xuzhou, China
e-mail: kegen.yu@cumt.edu.cn

K. Yu and N. Zheng (eds.), *Advances in GNSS Reflectometry*, Navigation: Science and Technology 19, https://doi.org/10.1007/978-981-92-2325-1_18

[19, 20, 38, 39, 65]. However, X-band marine radar images typically contain rainfall interference noise when precipitation occurs over the ocean [40]. Through altering sea surface roughness and changing electromagnetic wave scattering characteristics, rainfall ultimately exerts an impact on the retrieval performance of wind and wave parameters [40, 70]. Presently, in wave retrieval studies, precipitation is generally treated as interference, with rain-contaminated radar images undergoing detection, screening, and algorithmic suppression [48]. To distinguish rainfall-affected from rainfall-free data, thereby strictly controlling radar image quality during wind or wave parameter retrieval, rainfall detection is essential. The zero-pixel percentage (ZPP) method proposed Lund et al. [50] can effectively detect rainfall (especially heavy rain) in X-band ocean radar images, but the accuracy is limited, and the threshold needs to be adjusted according to environmental conditions in practical use. Experimental results have shown that the Support Vector Machine (SVM) method proposed by Chen et al. [18] for rainfall detection in X-band ocean radar images outperforms the ZPP method in detection performance and has the ability to directly adapt to multiple radar systems. The rainfall detection method proposed by Zheng et al. [73] utilizes the difference in correlation characteristics between radar images with and without rain, and determines the detection threshold by analyzing the spatiotemporal correlation of sea clutter. It can effectively detect rainfall from X-band ocean radar images. However, threshold determination proves difficult in both light and heavy rain scenarios, which makes it hard to accurately identify rainfall contamination in radar images. Thus, this method remains in need of further study and refinement. Based on the ZPP method, Lu et al. [49] developed the Ratio of Zero Intensity to Echo (RZE) approach for rainfall detection, and experimental validation showed its detection success rate to be 11.7% above that of the ZPP method. However, when the detection area is defined in the wave area of the radar image, the detection accuracy will be degraded, so accuracy enhancement would be required.

In recent years, Global Navigation Satellite System Reflectometry (GNSS-R) has emerged as a novel remote sensing technology in ocean remote sensing and is now one of the most popular technical approaches in this field. The short revisit period, low observation cost, and high spatiotemporal resolution of GNSS-R make it a technology with great potential. At present, countries such as the United Kingdom, the United States, China, and Spain have successfully launched TechDemoSat-1 (TDS-1) [33], Cyclone Global Navigation Satellite System (CYGNSS) [22], BuFeng-1 A/B satellites [42], Fengyun-3E/3F/3G [60], Tianmu-1 [8, 47] and ^{3}Cat-5/A and B satellites [51], dedicated to GNSS-R applications. In recent years, although this technology has been successfully applied to various geophysical parameter retrieval, e.g. snow depth estimation [71], wind speed retrieval [24, 33, 37], wave height estimation [7, 9, 11, 13], altimetric retrieval [16, 44], soil moisture retrieval [15], and sea ice detection [1, 68]. Unfortunately, there are few research reports on using this technology to obtain rainfall information.

There have been numerous studies on the surface effects of precipitation on microwave wind scattering measurements [6, 25, 28, 66]. Considerable attention has also been drawn to the effect of rainfall on reflected GNSS signals. Ghavidel and Camps [35] studied the GNSS-R electromagnetic deviation caused by rainfall, waves, and ocean currents through numerical simulations. A study in 2018 [3] first demonstrated the presence of rainfall characteristics in TDS-1 Delay Doppler Map (DDM). Based on DDM data generated by CYGNSS satellites, Balasubramaniam and Ruf [4] also studied these rainfall characteristics under low wind speeds. Said et al. [57] analyzed the impact of rainfall on CYGNSS wind speed measurement, and the experimental results showed that under low wind speed conditions, rainfall can affect the features of CYGNSS DDM (It is the core technical means used by the CYGNSS system for data collection and processing are mainly used to retrieve parameters such as sea surface wind speed and soil moisture).

Inspired by these observations, Bu and Yu [10] carried out an initial investigation into sea surface rainfall detection and rainfall intensity (RI) retrieval using spaceborne GNSS-R data, and proposed a threshold-based rainfall detection (RD) method for GNSS-R observations that relies on the probability density function (PDF) (see Sect. 4). Under low wind speed conditions (< 5 m/s), this method achieves a rainfall detection probability (PDr) exceeding 75%, according to the experimental results. However, the PDF method ignores multiple key parameters, which limits the accuracy of RD. In fact, RD is also influenced by geometric reflection parameters (i.e., the latitude and longitude of specular reflection points) and some metadata. About 7% of scatterometer measurements worldwide are distorted due to rainfall interference [66]. And there are other observation methods available. Therefore, utilizing GNSS-R data for ocean rainfall detection is of great significance.

In summary, the focus of this chapter is to propose a new rainfall detection method that can accurately detect rainfall in CYGNSS DDM (see Sect. 4). In the rest of this chapter, we first describe the modeling of sea surface rainfall spectra and the analysis of rainfall characteristics in DDM. This chapter begins with a description of the dataset and data processing methods. It then introduces the four proposed RD methods and evaluates their detection performance. Finally, this chapter summarizes the main findings and discusses several challenges that urgently need to be addressed in future research.

2 Rainfall Spectrum Modeling and Rainfall Signatures in DDM

This section first introduces the model of the impact of rainfall on the sea surface, and then analyzes the rainfall characteristics in GNSS-R DDM.

2.1 Modeling of Sea Surface Rainfall Spectrum

The fully developed wave spectrum can be divided into two categories in electromagnetic scattering: one is the near-specular reflection region, corresponding to wind speeds typically below 5 m/s, at which point the sea surface Rayleigh parameter is far less than 1. Another type is the diffuse scattering region, corresponding to wind speeds of 5 m/s or higher, where the Rayleigh parameters of the sea surface are large and the geometric optical approximation is effective [62]. In the diffuse scattering regime, the surface roughness induced by wind-wave interaction is more significant than that caused by raindrop impact. The surface roughness caused by raindrops striking the water surface is dominant only under low wind speed conditions (< 5 m/s) [72].

Rainwater changes the surface of the ocean through impact (surface splashing). At low wind speeds, the most significant impact is the change in surface roughness. In this case, the recognized rainfall impact model refers to the splashing of water droplets (including rings, stems, and crowns) generated by their impact on the water, from which microwave signals scatter [5, 67]. Initially, an impact cavity and a crown structure form simultaneously on the water surface. The cavity then evolves into a vertical water column, which eventually collapses, giving rise to a ring of gravity–capillary waves [5]. Throughout this process, the dominant features at microwave frequencies are the water stems and the circular waves. Under oblique incidence angles, circular waves with centimeter-scale wavelengths prevail, whereas water stems become dominant only at grazing incidence angles [27].

Based on the first-order superposition idea of the circular spectrum generated by rain and the Elfouhail elevation spectrum influenced by wind, this paper establishes a model to describe the impact of raindrops on water surface. Among them, the circular spectrum adopts a logarithmic Gaussian model, which takes the specific form of:

$$S_K(K) = \frac{1}{2\pi} V_{Gr}(K) S_{Peak} \exp\left\{ -\pi \left[\frac{\ln\left(f(K)/f_P\right)}{\Delta f / f_P} \right]^2 \right\} \tag{1.1}$$

where $f(K)$ is given by the dispersion relation $\sigma^2 = (2\pi f)^2 = gK + \tau/\rho \cdot K^3$; σ, g, K, ρ and τ represent the radian frequency, gravitational acceleration (~ 980 cm s^{-2}), wave number, water density, and water surface tension (~ 74 cm^3 s^{-2}), respectively; V_{Gr} is the group velocity at wavenumber K, and S_{Peak} is the power-law spectral model given by $6 \times 10^{-4} R^{0.53}$ cm^2 Hz^{-1}; The peak frequency is $f_P = 5.772 - 0.0018R$ Hz; Bliven et al. [5] estimated the model coefficients to represent circular waves in the frequency range of 2.75–12 Hz (wavelength of 1.94–20.8 cm).

The larger the wave number, the shorter the wave. According to [4], when the wind speed is 5 m/s, the peak wavenumber range of Elfouhail spectrum is 0.1–1 rad/m, and the peak wavenumber range of annular spectrum is 100–300 rad/m. Larger wave numbers represent shorter waves. Therefore, compared to the long-range roughness caused by wind, annular waves introduce additional short-range roughness. The observable of GNSS-R σ_0 (The normalized bistatic radar cross section (NBRCS) is inversely related to the mean square slope (MSS) of the surface. Using the geometric optical approximation in electromagnetic scattering theory, MSS can be defined as the omnidirectional slope spectrum integral within the cutoff wavenumber range

[30]. For the L-band wavelength where the GNSS-R signal is located, the cutoff wavenumber is approximately between 10–12 rad/m. It is worth noting that the impact of rainfall on the roughness spectrum is mainly concentrated in the areas above this cut-off value.

2.2 *Signatures of Rainfall in CYGNSS DDM Measurements*

In order to analyze the impact of rainfall on DDM measurement, this section studies the changes of DDMA relative to rainfall intensity and wind speed Fig. 1 shows the variation of DDMA relative to wind speed within different rainfall intensity ranges. It can be seen that DDMA has a strong dependence on rainfall under low wind speed conditions (< 5 m/s), indicating that DDMA is not sensitive to the impact of rainfall when wind speed is greater than 5 m/s. In addition, with the increase of rainfall intensity, a significant decrease in the value of DDMA was observed. According to Balasubramaniam and Ruf's research, when the rainfall intensity is below 30 mm/h, the signal attenuation caused by rainfall can be ignored [4], indicating that the changes in DDMA are mainly due to the roughness of the water surface caused by raindrops. On the other hand, within the wind speed range of 10–15 m/s, the dependence of DDMA on rainfall intensity can be almost ignored. This means that when the wind speed exceeds the upper limit of 15 m/s, the surface roughness of wind-driven wind will dominate the influence on scattering signals, while the effect of rainfall is relatively secondary.

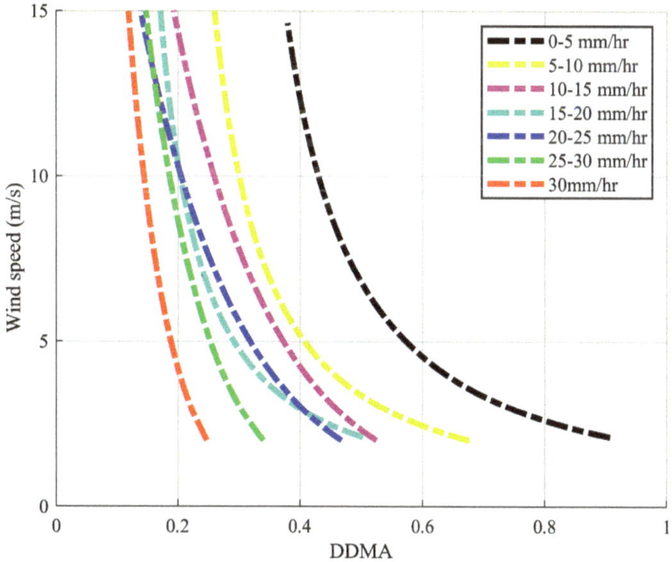

Fig. 1 Changes in DDMA relative to wind speed within different rainfall intensity ranges

The key to studying the changes in DDM under similar measurement conditions is the simultaneous presence of heavy and no rainfall events on the same sea surface reflection trajectory. The measurement results of CYGNSS are influenced by various sea surface features and satellite geometry structures [4]. Among them, the control parameters of sea surface features include wind speed and significant wave height in the peak and no rain areas along the CYGNSS trajectory. In other words, it is necessary to ensure that the wind speed and significant wave height parameters of these two types of areas (peak rainfall event area and no rainfall event area) match. A no rain event refers to a measurement sample with zero precipitation on a given trajectory. Correspondingly, peak rainfall events represent the maximum rainfall intensity observed at two different locations along the same trajectory. Therefore, in areas with peak rainfall and no rainfall, the wind speed and significant wave height (SWH) are quite similar. To eliminate the interference of wind speed on DDM and avoid sea surface waves blocking rainfall information, this study only conducted rainfall intensity inversion under extremely low wind speed (< 5 m/s) conditions, in order to minimize the effect of wind before analyzing the impact of rainfall. The reason for selecting this wind speed range is that under wind speeds below 5 m/s, DDM observations exhibit high sensitivity to rainfall, where the surface roughness formed by raindrops hitting the water surface dominates [4, 67, 72], As mentioned earlier, the impact of rainfall on DDM can be truly observed in this situation.

As described in [3], under weak wind conditions, rainfall characteristics can be observed under conditions where quasi-specular scattering driven by surface slope changes to higher-order Bragg scattering more similar to that driven by the Rayleigh parameter Ra. The direct proportionality between Ra and the root-mean-square surface height arises from integrating the full surface elevation spectrum, encompassing spectral intervals influenced by rain splashes. To describe this weak diffuse scattering and its gradual transition to a strong diffuse scattering state, Voronovich and Zavorotny proposed a bistatic scattering model [61, 62], which describes this weak diffuse scattering and provides a smooth transition to a strong diffuse scattering state. For sufficiently weak winds, the amplitude of spectral peaks caused by wind generated waves becomes comparable to the secondary peaks caused by circular waves generated by rain, and the root mean square of surface height (as well as Rayleigh parameters) becomes sensitive to circular waves generated by rainfall. Therefore, rainfall may first be predictable at low wind speeds. Secondly, the wind speed effect remains the main parameter controlling DDM and must be modelled/predicted or offset in some way. Weissman et al.'s study suggests that in addition to surface roughness caused by water droplet splashing effects, local winds induced by rain may be generated by the sinking airflow of air columns beneath the rain. This sinking airflow may lead to an additional increase in wind speed to alter the surface spectrum [67]. Balasubramaniam and Ruf conducted corresponding simulation studies on the downwind phenomenon caused by rainfall, and based on this, proposed an empirical model for calculating the downwind velocity caused by rainfall [4]. From this model, it can be seen that the sinking wind increases monotonically with wind intensity, which is consistent with expectations. This model explains the radar cross-section changes caused by rainfall induced downdrafts by quantifying the additional wind field contribution caused by rainfall. It should be pointed out that this study only conducts rainfall intensity retrieval for situations where the rainfall intensity on the sea surface is less than 30 mm/h (as described in another chapter), because within this intensity range, the sinking wind speed caused by rainfall is very weak. At this point, the impact of wind speed on DDM is negligible compared to rainfall. As for the two mechanisms of splashing effect and sinking airflow effect, splashing effect only occurs under extremely low wind speed conditions (< 5 m/s). Thus, according to [3, 4], under extremely low wind speeds (< 5 m/s), the surface roughness caused by raindrop impact dominates, making GNSS-R observations highly sensitive to rainfall. Once the wind speed reaches 5 m/s or higher, the roughness generated by the interaction between wind and waves will exceed the contribution of rainfall.

Another important point is that at low wind speeds, swell does have a greater impact on NBRCS (or DDM). Therefore, in addition to the impact of rainwater on sea surface roughness, the influence of swells also needs to be considered. SWH information of ERA5 data is used to determine whether there is a swell. To minimize the impact of swell on DDM on the same CYGNSS trajectory, it is necessary to exclude data containing significant swell. In this study, if the swell SWH provided by ERA5 is greater than 2 m, DDM data will not be used.

The control parameters of satellite geometry include SWH, wind speed, receiver antenna gain, and incident angle of specular reflection point. Among them, antenna gain and incident angle can be obtained from CYGNSS observation data. The above four parameters correspond one-to-one between peak rainfall events and areas without rainfall events, so the parameter values in the two regions are very close. The difference in GNSS-R observation values between rainy and rainless conditions is mainly due to the change in sea surface roughness caused by raindrop splashing. To ensure that this difference is not affected by surface conditions and measurement geometry, four control parameters are used to maintain similarity between rainy and non rainy areas in these two aspects.

Figure 2 presents two examples of sea surface rainfall observations obtained along the specular reflection trajectories of IMERG and GNSS signals. The observation dates were 23 August 2017 (DOY 235) and 17 August 2020 (DOY 230). As shown in Fig. 2 (left), green patches represent land, while other areas represent the ocean. Use colored bars to display the rainfall intensity of IMERG, where the red "X" represents the CYGNSS data sample with the highest rainfall, and the cyan "X" represents the CYGNSS data sample with zero rainfall. The red solid line in Fig. 2 (right) shows the rainfall intensity along the CYGNSS observation trajectory; correspondingly, the blue dashed line represents the wind speed, the cyan dashed line represents the SWH, the magenta solid line represents the incident angle, and the green solid line represents the receiver antenna gain. In Fig. 2a, the wind speeds corresponding to the maximum (27.52 mm/h) and minimum (0 mm/h) rainfall intensities along the selected trajectory are 3.59 m/s and 3.26 m/s, respectively, while the two SWHs are 1.24 m and 1.48 m. Similarly, in Fig. 2b, the wind speeds for the maximum rainfall intensity of 23.44 mm/h and the minimum of 0 mm/h are 3.66 m/s and 3.89 m/s, respectively, with corresponding significant wave heights of 0.42 m and 0.38 m.

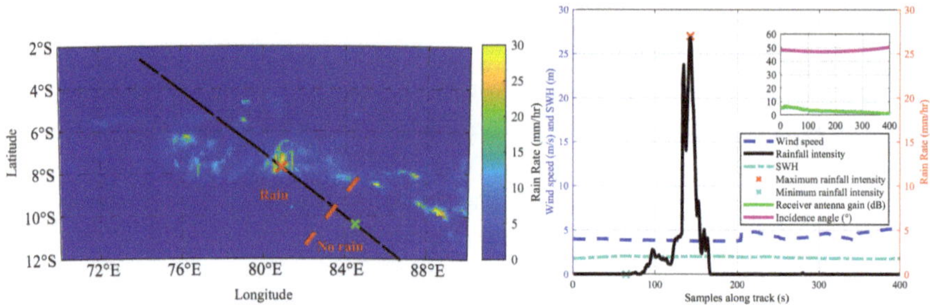

(a) August 23, 2017 12:30-13:30 UTC (DOY 235)

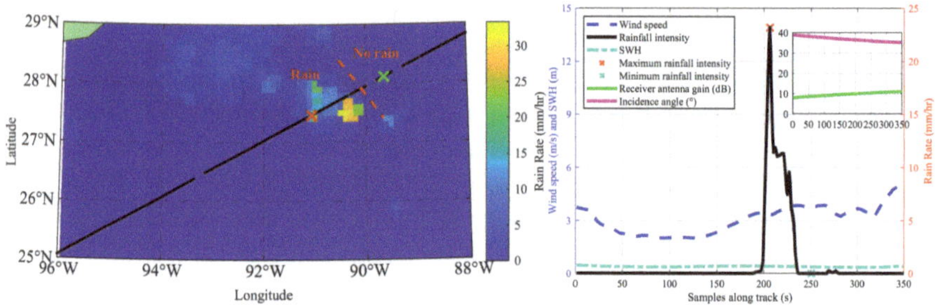

(b) August 17, 2020 13:30-14:30 UTC (DOY 230)

Fig. 2 CYGNSS reflection point trajectories, GPM IMERG rainfall intensity distribution, along with wind speed and rainfall intensity corresponding to Two CYGNSS trajectory lines

The differential DDM method is used here to analyze the impact of rainfall on CYGNSS DDM. The difference in DDM is achieved by considering the DDM at the point of maximum rainfall intensity and the DDM at nearby zero rainfall intensity points on the same trajectory, highlighting the impact of rainwater. The situation considered here involves very small variations in wind speed or measurement geometry along the trajectory to minimize the influence of all other surface and geometric parameters. As illustrated in Fig. 3, the left and middle panels present the DDM plots corresponding to peak rainfall intensity and zero rainfall intensity, respectively. The observation time of Fig. 3a, b are the same as those in Fig. 2. To characterize how rainfall manifests in the DDMs, the difference between the two DDMs is computed using the different methods. The resulting differential DDM (peak rainfall minus no rainfall) is shown in the right panel of Fig. 3. Several interesting features can be observed. Firstly, differential DDM displays that the impact of rainfall is concentrated in the flash area around the mirror reflection point: when there is rain, the received power at this point is lower than when there is no rain. The selected samples (marked as "X" in the figure) have similar wind speeds (2–5 m/s), so it can be assumed that the differences are mainly caused by rainfall. Secondly, most of the differences between the two DDMs are located in the horseshoe shaped area, and the maximum value appears near the specular reflection point, indicating that rainfall information is distributed around this point [4, 12].

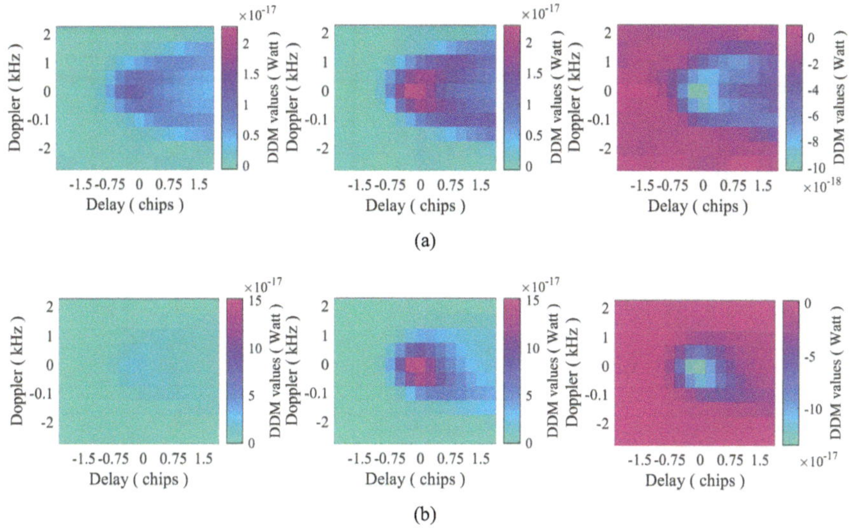

Fig. 3 Differenced DDM method reveals the rainfall signature. Illustrated are: peak rainfall DDM (left), rainless DDM (center), and differenced DDM (right). **a** 23 Aug 2017, 07:30–08:30 UTC (DOY 235). **b** 17 Aug 2020, 14:00–15:00 UTC (DOY 230)

Figure 4 shows some trajectories in the dataset. In the figure, the wind speed of each trajectory is quite uniform, and the trajectory includes peak rainfall and no rainfall areas. The observation time periods of the four subplots in Fig. 4 are: (a) 14:30–15:30 UTC, 18 Aug 2020 (DOY 231). (b) 00:30–01:30 UTC, 21 Aug 2020 (DOY 234). (c) 22:30–23:30 UTC, 21 Aug 2020 (DOY 234). (d) 19:30–20:30 UTC, 22 Aug 2020 (DOY 235). To verify the hypothesis that the difference between peak rainfall and no rainfall is caused by rainfall and the information is concentrated in the DDM horseshoe shaped region, Fig. 5 shows the peak rainfall DDM, no rain DDM, and differential DDM of four CYGNSS trajectories. Table 1 provides the corresponding rainfall intensity, wind speed, significant wave height, incident angle, and antenna gain for the sampling points. Although the wind speed changes significantly on the same trajectory, narrow rainless sections with wind speeds similar to the rainfall trajectory can still be selected. In all cases, the surface and geometric features of the two regions are very similar, which supports the above hypothesis. This validates the hypothesis that the observed differences between peak rainfall and no rainfall are caused by rainfall, and most of the information is contained in the horseshoe shaped region of DDM.

It should be noted that the DDM rainfall characteristics described in this section are only valid when the sea surface wind speed is less than 5 m/s. The reason is that under low wind speeds, DDM responds strongly to rainfall, and the surface roughness caused by raindrops splashing plays a major role; When the wind speed exceeds 5 m/s, the roughness generated by the interaction between wind and waves will dominate and mask the influence of rainfall.

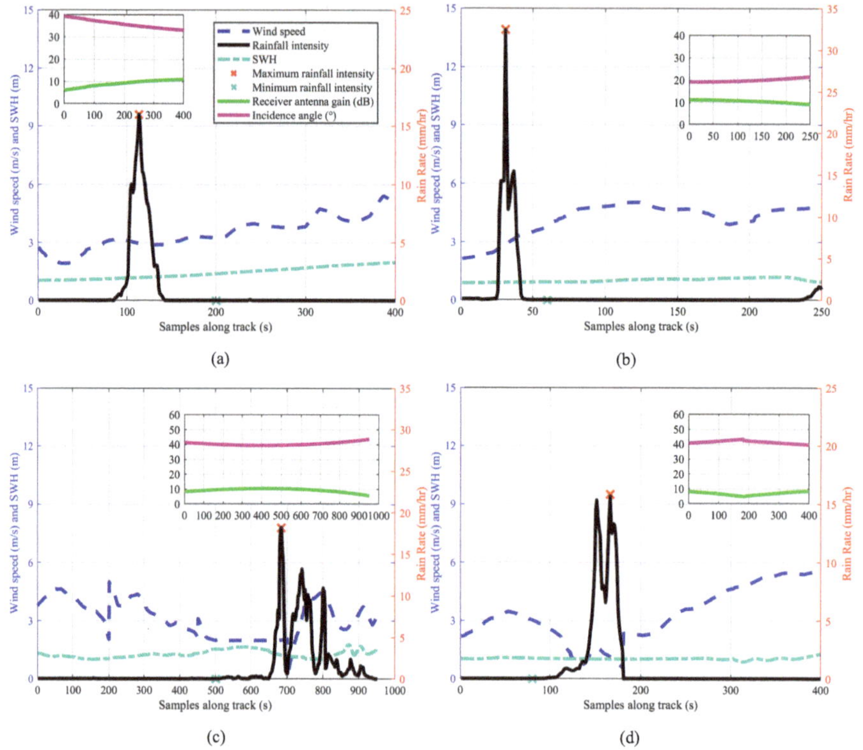

Fig. 4 Depicted are example CYGNSS tracks from the controlled dataset. Along each track, the following variables are plotted: rainfall (solid red), wind speed (dotted blue), SWH (dotted cyan), incidence angle (solid magenta), and receiver antenna gain (solid green). **a** 14:30–15:30 UTC, 18 Aug 2020 (DOY 231). **b** 00:30–01:30 UTC, 21 Aug 2020 (DOY 234). **c** 22:30–23:30 UTC, 21 Aug 2020 (DOY 234). **d** 19:30–20:30 UTC, 22 Aug 2020 (DOY 235)

3 Dataset Description and Data Processing

This section first introduces the dataset, then defines the time-delay waveform and calculates the GNSS-R observations derived from DDM. Subsequently, it introduces the Fast Non-Local Mean (FNLM) DDM image denoising algorithm based on integral images. Finally, a data filtering method for rainfall detection was proposed to select high-quality GNSS-R observation data to ensure the performance of spaceborne GNSS-R rainfall detection.

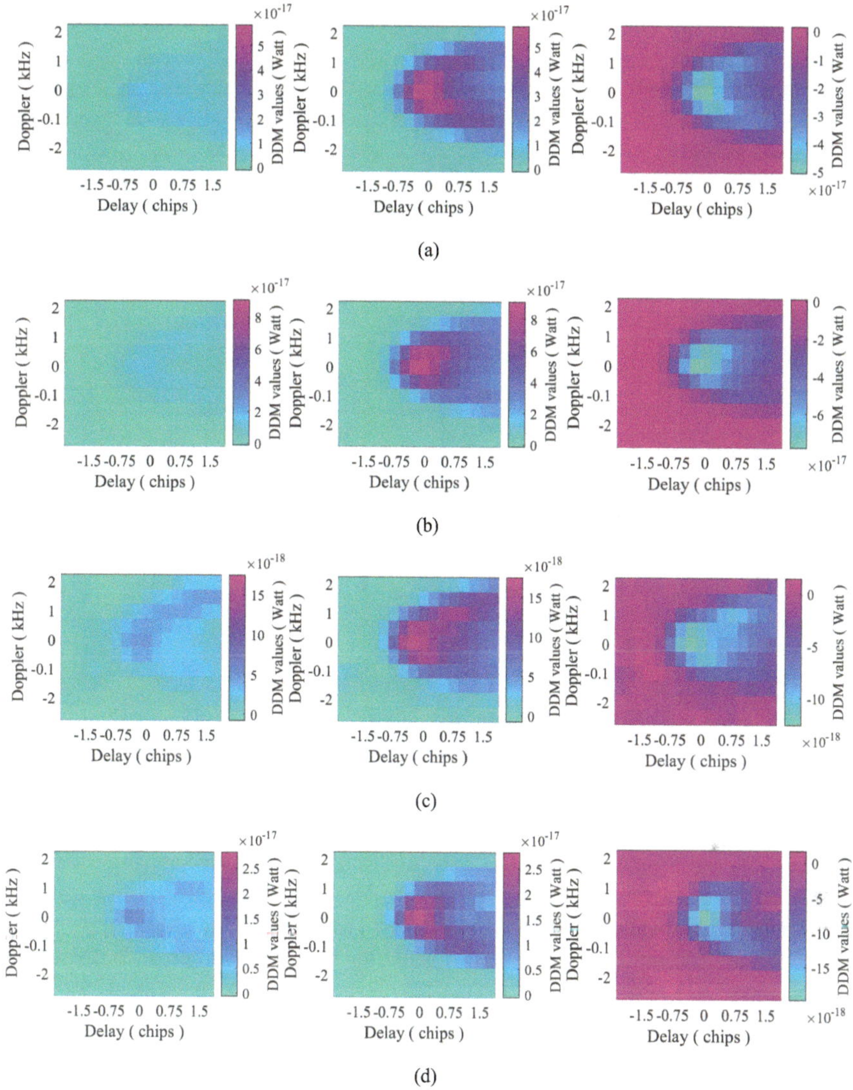

Fig. 5 Differenced DDM derived from four CYGNSS tracks exhibiting near-uniform wind speed along the track. The three panels correspond to: (left) DDM images during peak rainfall, (middle) DDM images in the absence of rainfall, and (right) the differenced DDM between the peak-rainfall and rainless cases. **a** 14:30–15:30 UTC, 18 Aug 2020 (DOY 231). **b** 00:30–01:30 UTC, 21 Aug 2020 (DOY 234). **c** 22:30–23:30 UTC, 21 Aug 2020 (DOY 234). **d** 19:30–20:30 UTC, 22 Aug 2020 (DOY 235)

Table 1 RI, WS, SWH, incidence angle, and receiver antenna gain information at the peak and no rainfall sample points

Acquisition time	Samples	Rainfall intensity (mm/hr)	Wind speed (m/s)	SWH (m)	Incidence angle (°)	Receiver antenna gain (dBi)
4:30–15:30 UTC, 18 Aug 2020 (DOY 231)	Max rainfall Zero rainfall	16.32 0	3.03 3.24	1.65 1.73	37.99 36.88	10.36 10.07
00:30–01:30 UTC, 21 Aug 2020 (DOY 234)	Max rainfall Zero rainfall	33.51 0	3.07 3.93	1.08 1.12	20.08 21.01	10.30 10.23
22:30–23:30 UTC, 21 Aug 2020 (DOY 234)	Max rainfall Zero rainfall	18.75 0	2.43 2.46	1.31 2.03	40.33 40.06	10.15 10.04
19:30–20:30 UTC, 22 Aug 2020 (DOY 235)	Max rainfall Zero rainfall	16.77 0	1.25 3.06	1.24 1.21	41.20 40.44	7.17 9.54

3.1 Datasets

To utilize spaceborne GNSS-R data for sea surface rainfall detection and rainfall intensity inversion, four datasets were used in this chapter, namely CYGNSS GNSS-R L1 (Level 1) data, European Center for Medium Range Weather Forecasts (ECMWF) wind products and SWH data, and Integrated Multi-Satellite Retrievals for Global Precipitation Measurement (IMERG) Final operational precipitation (IMERG-F) product (Level 3, Version 06) [41]. This chapter downloads and collects observation data containing rainfall information from September 1, 2018 (corresponding to DOY 244) to October 1, 2020 (corresponding to DOY 275). More detailed information about these datasets is provided below.

Launched in 2016 by the National Aeronautics and Space Administration (NASA), the CYGNSS satellite system comprises eight microsatellites (CY01, CY02, CY03, CY04, CY05, CY06, CY07, and CY08). The eight CYGNSS satellites were equipped with a Delay Doppler Mapping Instrument (DDMI) specifically designed to sense the intensity of tropical cyclones or hurricanes [55], and for receiving DDM data obtained from GNSS reflection measurements [54]. The CYGNSS DDM data used in this chapter was downloaded from NASA's Physical Oceanography Distributed Active Archive Center (PO. DAAC). Currently, users can download three levels of CYGNSS data products for free in NASA's PO. DAAC.[1] The commonly used data format is NetCDF format, and this chapter uses L1 level

[1] https://podaac-tools.jpl.nasa.gov/drive/files/allData/cygnss.

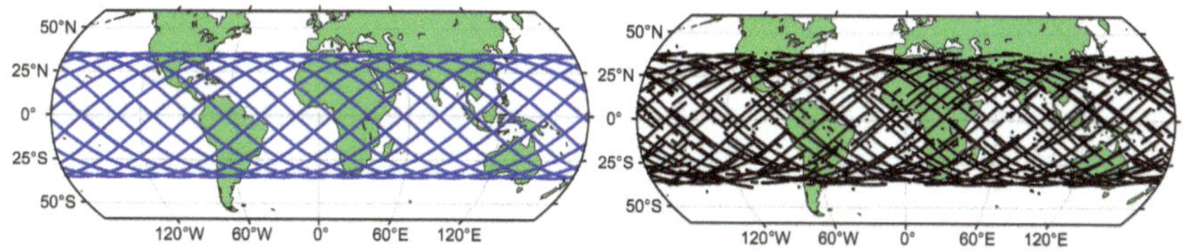

(a) Ground trajectories of CY08 satellite

(b) Specular reflection point trajectories

Fig. 6 CY08 satellite's ground trajectories and the specular reflection point trajectories of GPS satellite signals processed by CY08 within 24 h

data in version v3.0.[2] Figure 6 shows the 24-h satellite ground trajectories and specular reflection point trajectories of CY08 satellite, observed on April 30, 2020. Note that each satellite processed the signals of four GPS satellites, generating four specular reflection trajectories. As shown in Fig. 6, the CY01 satellite can well observe the Earth's surface within the latitude range of 40° N and 40° S, which covers the all ocean surface areas of interest, with high spatiotemporal resolution.

As a successor mission to TRMM, Global precipitation measurements from integrated multi-satellite retrievals (GPM-IMERG) was launched in 2014, with the advantage of a wider measurement range and higher spatiotemporal resolution. It runs a three-level multi satellite precipitation estimation algorithm that integrates three types of information: intermittent precipitation estimation from multi constellation microwave sensors, infrared observation data from stationary satellites, and monthly cumulative precipitation data provided by ground rain gauges. The currently widely used data version is IMERG-V6, with a spatial resolution of $0.1° \times 0.1°$ and a temporal resolution of 30 min. GPM-IMERG offers three types of products: early products (IMERG-E, real-time products with a delay of about 4 h), late products (IMERG-L, near real-time processing products with a delay of about 14 h), and final products (IMERG-F) [41]. This chapter uses the IMERG-F product published on the GPM website, which can be downloaded for free from the GPM website.[3]

At present, users can obtain ECMWF wind speed products and SWH data for free from the Copernicus Climate Change Service (C3S) climate database.[4] ERA5 is the fifth generation ECMWF global climate and atmospheric reanalysis product, which combines model data with observational results from around the world into a complete and consistent global dataset. ERA5 data has replaced its predecessor ERA Interim reanalysis data, covering parameters such as atmospheric temperature, pressure, wind speed, rainfall, soil moisture content, and wave height at different altitudes from 1979 to the present. Among them, the ERA5 data product provides global grid wind speed values with a spatial resolution of $0.25° \times 0.25°$, a temporal resolution of 1 h, and data formats of GRIB and NetCDF. However, this chapter uses the NetCDF format, which takes the wind speed value of 10 m/s above the sea

2 https://podaac-tools.jpl.nasa.gov/drive/files/allData/cygnss/L1/v3.0.

3 https://gpm.nasa.gov/data/directory.

4 https://cds.climate.copernicus.eu/datasets/reanalysis-era5-single-levels?tab=overview.

surface as the true sea surface wind speed reference value [29, 37]. In addition, this chapter also used ERA5 SWH data, which includes three types: SWH of combined wind waves and swell, SWH of wind waves, and SWH of swell. Among the three types of swell significant wave height (SWH), the corresponding variable names in the ERA5 dataset are "Significant wave height of first swell partition", "Significant wave height of second swell partition", and "Significant wave height of third swell partition". In this chapter, we adopt the first of these, i.e., "Significant wave height of first swell partition".

Due to the inconsistent spatiotemporal resolution of the four data products mentioned above. In practical applications, ERA5 wind speed products, ERA5 SWH products, and IMERG-F products are matched with CYGNSS data in space and time through spatial bilinear interpolation and temporal linear interpolation algorithms.

3.2 Delay Waveforms and Observables Derived from CYGNSS DDM

In spaceborne GNSS-R, the direct use of DDM is not conducive to practical applications; It is usually used to calculate the delay waveform or GNSS-R observable value of DDM, which is more convenient for calculation and analysis. There are three commonly used delay waveforms, including normalized integral delay waveform (NIDW), normalized central delay waveform (NCDW) and differential delay waveform (DDW). In previous studies, various observable values related to DDM have been used for the inversion of geophysical parameters, such as signal-to-noise ratio (SNR), DDM mean (DDMA), leading edge slope (LES), trailing edge slope (TES), and normalized dual base radar cross section (NBRCS). These parameters have been successfully applied in fields such as sea surface wind speed inversion, soil moisture estimation, and sea ice detection, but have not yet been used to detect rainfall. Thus, this study applies three commonly used observables (i.e., DDMA, LES, and TES) to RD. A more detailed description of these three delay waveforms and the corresponding GNSS-R observables can be found in Bu and Yu [10].

It is worth noting that the three observable parameters mentioned above are all derived from the power DDM, which is calculated by selecting a 3×5 window near the specular reflection point of the power DDM. In addition, a 17×11 effective scattering area array (unit: m^2), namely BRCS DDM, was introduced in this study. This BRCS DDM is used as a direct input image for model training and testing of Support Vector Machines (SVM), Random Forests (RF), and Convolutional Neural Networks (CNN) in Rainfall Detection (RD). Each metadata sample not only contains BRCS DDM, but also provides up to 14 auxiliary features associated with DDM, GPS satellites, and measurement geometry. It should be noted that although the metadata does not directly provide RCG information, it contains several relevant parameters that can be used to calculate RCG. The specific calculation can be done using Eq. 14

Table 2 Input parameters for the SVM, RF, and CNN methods

	Input variables	Description
DDM-related	DDMA	Average scattering power near the specular point in DDM
	LES	Leading edge slope
	TES	Trailing edge slope
	SNR	DDM signal to noise ratio
Receiver-related	gps_eirp	GPS effective isotropic radiated power
	rx_to_sp_range	Receiver to specular point range
	sp_rx_gain	Specular point receiver antenna gain
Geometry-related	sp_lon	Specular point longitude
	sp_lat	Specular point latitude
	sp_inc_angle	Specular point incidence angle
	sp_theta_orbit	Specular point orbit frame theta angle
	sp_az_body	Specular point azimuth angle
	sp_az_orbit	Specular point orbit frame azimuth angle
	RCG	Range Corrected Gain

in Guo et al. [36]. Table 2 summarizes all input parameters for the three methods (i.e., SVM, RF, and CNN).

It should be pointed out that waves do have an impact on rainfall detection. Thus, in order to represent the role of wind and waves in rainfall detection based on SVM, RF, and CNN methods, ERA5 SWH and wind speed were also employed as auxiliary input features.

3.3 *Fast Non-local Means DDM Images Denoising Algorithm Based on Integral Images*

To reduce the impact of antenna gain and system noise, it is essential to perform denoising on DDM data, given that the DDM images from CYGNSS satellites are frequently noisy. Therefore, denoising must be given the highest priority ahead of later data processing steps. In previous studies, an and Chu et al. used the method of subtracting the noise floor from the power values obtained from DDM to denoise the DDM. The denoised DDM images can be used for sea ice detection, sea ice density inversion [68, 69], and sea surface wind speed inversion applications [21]. Therefore, the method for denoising DDM images is as follows:

$$\begin{cases} M_{\text{nosie}} = \frac{1}{M} \sum_{\tau_1}^{\tau_n} \sum_{f_1}^{f_n} M(\tau, f) \\ DDM_{\text{denoise}} = DDM_n - M_{\text{noise}} \end{cases} \tag{1.2}$$

where M_{noise} is DDM noise, τ_1 and τ_n are the delay range of the selected noise box, f_1 and f_n are the Doppler shift range of the selected noise box, M is the number of pixels in the noise box, and DDM_n is the value corresponding to each pixel in DDM.

The above method is also known as noise floor subtraction, although it has been widely adopted in different applications of spaceborne GNSS-R. However, the method has two main drawbacks. First, it only accounts for systematic noise in DDM data, ignoring other types of noise [21]. Second, it cannot effectively eliminate speckle noise. Consequently, the denoising effect remains poor. In response to the shortcomings of existing denoising methods, a novel DDM denoising method is introduced, namely the DDM denoising method of non-local mean (NLM) algorithm with integrated image acceleration.

In 2005, Buades et al. [14] proposed the NLM denoising algorithm, which fully utilizes redundant information in images and maximizes the preservation of image details while denoising. The basic idea is as follows: Select other pixels in the image that have similar neighborhood structures to the current pixel, weight and average the grayscale values of these pixels, and the result is the estimated value of the current pixel. However, due to its extremely high computational complexity and low efficiency, this algorithm is not practical enough in practical applications. In response to this deficiency, a strategy called integrated image was proposed in [64] to reduce the complexity of image block distance calculation while maintaining accurate computation. Subsequently, Facciolo et al. [31] also conducted in-depth research on the integrated image algorithm and its applications. Froment [34] also used this algorithm to accelerate the NLM denoising algorithm, in order to reduce the computational complexity of the NLM algorithm. Below is a description of the Fast Non-Local Mean (FNLM) algorithm based on integral graphs.

Set two fixed size windows: set the search window size to $2D \times 2D$ and the neighborhood window size to $2d \times 2d$, so that the Euclidean distance between pixel z and $z + r$ is:

$$s_r(z) = \|u(z) - u(z + r)\|_2^2 \tag{1.3}$$

where $z = (z1, z2) \in \Omega$, $r = (r1, r2) \in [-D, D] \times [-D, D]$, then we can construct an integrated image of pixel differences as follows:

$$S_r(a, b) = \sum_{\{0 \le z_1 \le a, 0 \le z_2 \le b\}} s_r(z_1, z_2), \ (a, b) \in \Omega \tag{1.4}$$

according to the following recursive formula, the integral map of the entire image can be calculated.

$$\begin{gathered} \forall (a, b) \in \Omega, \ \mathrm{a} \ge 1, \mathrm{b} \ge 1, \\ \mathrm{S}_r(\mathrm{a,b}) = \mathrm{s}_t(a, b) + S_t(a - 1, b) + \mathrm{S}_t(a, b - 1) - \mathrm{S}_t(a - 1, b - 1) \end{gathered} \tag{1.5}$$

The Euclidean distance (also known as Euclidean norm) of the similar window region centered on pixel (a, b), $(a{+}r1, b + r2)$ is expressed as follows:

$$\|U(a,b) - U(a+r1, b+r2)\|_2^2 = \frac{1}{d^2}(S_t(a+d, b+d)+ \\ S_t(a-d-1, b-d-1) - S_t(a+d, b-d-1)- \\ S_t(a-d-1, b+d)) \tag{1.6}$$

Similar to the NLM algorithm, the above algorithm requires calculating the similarity between L pixels in the image and each pixel in the search window. This calculation process results in an algorithm complexity of $O(L \times D^2)$. However, the computational complexity of the NLM algorithm is $O(L \times D^2 \times d^2)$, and compared with the NLM algorithm, the computational efficiency of the above algorithm is significantly improved.

3.4 Data Filtering

In addition to denoising DDM images, to ensure the quality of CYGNSS data, it is also necessary to perform quality control and filtering on CYGNSS GNSS-R data before proceeding with subsequent data processing. Filter data according to the following criteria [2, 12, 23, 45, 52].

Firstly, all observations containing NaN values need to be deleted. Secondly, all observations less than 0 need to be discarded. Thirdly, the RCG value should be greater than 3. Fourthly, For cases where the gain of the receiving antenna (s_rx_gain) is less than 0 dBi or the BRCS uncertainty (ddm-BRCS_check) is greater than 1, we directly discard the corresponding data. In addition, we use the basic quality control (QC) markers of CYGNSS L1B data to filter the data, and only use a single QC marker instead of the entire QC marker (such as "overall poor quality", least significant bit or bit 0). When data meets any of the following conditions, it should be discarded: DDM invalid or abnormal (bits 4, 7–10, 15, 17–20); The instrument has data transmission or calibration issues (bits 1, 5, 6, 13, 14, 21–26); The transmitter power or antenna gain height is uncertain (bits 16, 27); The attitude error of the spacecraft is significant or abnormal (bits 2, 3, 28); Delete the L1B observation value containing GPS IIF satellite ("low_quality_GPS_ont_knowledge", bit 27). In addition, data filtering employs the quality control flags "s_near_Land" (bit 12) and "s_very_near_land" (bit 11) to exclude measurements taken when signals reflect off land, since land data are outside the scope of this study. To reduce modeling errors, only data with specular reflection points located more than 25 km from land are selected.

4 Detection Algorithms

There are roughly three methods for spaceborne GNSS-R rainfall detection: Empirical modeling (based on Probability Density Function (PDF)), machine learning (SVM, RF), and deep learning (CNN). The rest of this section briefly introduces the basic theories of each method.

4.1 Observation Threshold Rainfall Detection Method Based on Probability Density Function

The rainfall detection method based on GNSS-R observable threshold and PDF has been discussed in detail in [10], this section only gives the training results based on this method.

Figure 7 shows the PDFs of three GNSS-R observables (i.e., DDMA, LES, and TES). Figure 7 shows that the value at the intersection of two PDF curves is used as the threshold for each observable value. The purpose of proposing this threshold is to achieve a higher probability of rainfall detection while maintaining a lower false alarm probability. Therefore, if a certain observable value is less than the defined threshold, it is assumed that there is rainfall on the sea surface.

To evaluate the performance of the proposed rainfall detection method, four evaluation indexes (PDr, Probability of false alarm detection of rainfall (PFAr), probability of missing in rainfall detection (PMR), and probability of overall error detection (POE)) are employed in the study. Take the DDMA observable in Fig. 7 as an example to illustrate the calculation of the four evaluation indexes.

As shown in Fig. 7, the abscissa of the intersection of PDF curves with and without rainfall is the RD threshold, which is marked with a green line. The rainfall PDF integral on the left side of the intersection corresponds to the probability of rainfall detection (PDr), the area of no rainfall PDF on the left side corresponds to the probability of rainfall false alarm (PFAr), and the area of rainfall PDF on the right

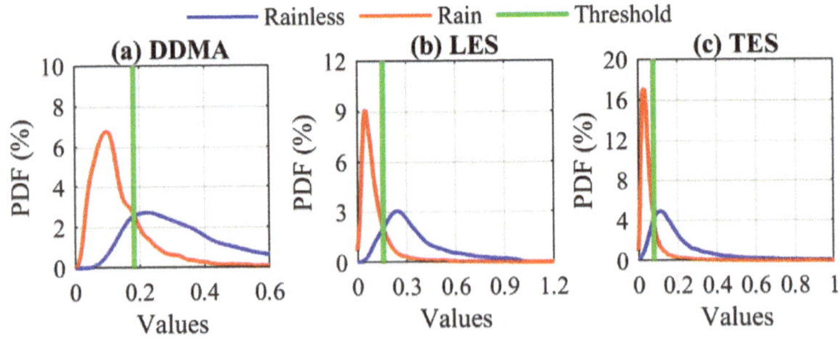

Fig. 7 PDFs of three GNSS-R observables. DDMA (left), LES (middle), and LES (right)

side corresponds to the probability of rainfall missed alarm (PMR). The expressions for these three quantities are as follows:

(1) The PDr

$$PDr = \int_{0}^{T} P_{Obs}^{(r)}(\psi_r)d\psi_r \tag{1.7}$$

(2) The PFAr

$$PFAr = \int_{T}^{\infty} P_{Obs}^{(r)}(\psi_r)d\psi_r \tag{1.8}$$

(3) The PMR

In the presence of rainfall on the sea surface, the probability that no rain is detected is called the probability of missing in rainfall detection (PMR).

$$PMR = \int_{0}^{T} P_{\text{Obs}}^{(rl)}(\psi_{rl})d\psi_{rl} \tag{1.9}$$

1) The incorrect probability of rainfall detection can be defined as the POE

$$POE = \frac{1}{2}(PFAr + PMR) \tag{1.10}$$

In hypothesis testing, if only judging whether an event has occurred, a common strategy is to maximize the detection probability (PD) while ensuring that the false positive probability (PFA) is sufficiently small. In general, for a given testing method, PD and PFA are difficult to improve simultaneously—the increase in PD often accompanies the increase in PFA. If "rain" and "no rain" are considered as two independent events, the average false alarm probability (APFA) under rain and no rain conditions can be used as the performance indicator, which is denoted as POE in formula (Eq. 1.10).

Based on the PDF method, Table 3 presents the rainfall detection performance of three observation points within the wind speed range of 0–10 m/s. From the four indicators in the table, it can be seen that DDMA observations have the highest PDr and lowest POE, while TES performs the worst. Taking into account the results of PDr, PFAr, PMR, and POE, as well as the superior overall performance of DDMA, we have chosen to set the threshold based on the observable values of DDMA.

As mentioned, although PDF method has been proved to be useful for spaceborne GNSS-R rainfall detection, this method only considers limited parameters, such as

Table 3 The rainfall detection performance of three observables based on PDF method using training data

GNSS-R observables	PDr (%)	PFAr (%)	PMR (%)	POE (%)
DDMA	75.25	15.42	24.75	20.09
LES	82.21	12.66	17.79	15.23
TES	81.64	13.75	18.36	16.06

GNSS-R observables, incident angle, wind speed and SWH. However, The relationship between sea surface rainfall and GNSS-R observables is further complicated by reflection geometric parameters (e.g., specular point longitude and latitude) as well as certain metadata, all of which also influence rainfall detection. Therefore, it is difficult to describe the relationship between rainfall, DDM observables and the above factors when using PDF method. Note that considering so many factors, it is very difficult to develop multivariable models through analytical methods.

Based on the significant advantages of machine learning methods such as ensemble learning and deep learning in multi parameter modeling, this paper presents them as a powerful alternative to spaceborne GNSS-R sea surface rainfall detection, and introduces a rainfall detection method using SVM, RF, and CNN.

4.2 Rainfall Detection Based on Support Vector Machine Method

SVM was first proposed by [26] and is currently a popular machine learning algorithm that can solve classification problems. Due to the different characteristics of rain-free DDM compared to DDM contaminated with the effect of rainfall, this chapter applies SVM to classify these two types of DDM data.

The set of samples required for training a detection model can be represented as $(k_1, m_1), (k_2, m_2), \ldots, (k_N, m_N)\,(k_i \in R^n, m_i = \pm 1)$, Among them, k_i is an n-dimensional feature vector, m_i is the label of the DDM image (− 1 represents no rain, + 1 represents rain pollution), and N is the number of DDM images used to train the SVM detection model. The SVM method aims to minimize the following objective function:

$$J(w) = \frac{1}{2}\|w\|^2 + C\sum_{i=1}^{l}\varepsilon_i \tag{1.11}$$

Optimization problems can be written as:

$$\begin{aligned} &\min_{w} J(w) \\ &s.t. m_i(w \cdot \psi(k_i) + b) \geq 1 - \varepsilon_i,\ i = 1, 2, \ldots, N \end{aligned} \tag{1.12}$$

where $\psi(k_i)$ is the mapping function, w is the weight of y_i, C is the penalty parameter (determined by the user), ε_i is the relaxation variable, representing the tolerance of allowable error, and b is the deviation term. The above problem can be solved by considering its dual form and introducing Lagrange multipliers $\delta_i(i \in [1, N], \delta_i \in [0, +\infty]$ [46].

$$W(\delta) = \sum_{i=1}^{N} \delta_{\mathrm{i}} - \frac{1}{2} \sum_{i,j=1}^{N} \delta_i \delta_j k_i m_j K(k_i, m_j) \quad (1.13)$$

where $W(\delta)$ is the dual form of the Lagrangian, $0 \leq \delta_j \leq C$, $\sum_i \delta_j m_j = 0$, $K(k_i, m_j) = \psi(k_i)^T \psi(k_j)$ are kernel functions.

Among the many kernel functions used for SVM training, such as linear kernel, polynomial kernel, radial basis function (RBF kernel), and sigmoid kernel, this chapter chooses (RBF) for rainfall detection because RBF has the highest detection accuracy for training samples. This model can be represented as::

$$K(k_i, m_j) = \exp(-\gamma \left\| k_i - m_j \right\|^2) \quad (1.14)$$

where $\gamma > 0$ determines the size of the kernel.

$W(\delta)$ can achieve maximum detection accuracy through the Sequential Minimum Optimization (SMO) algorithm proposed in the application [32]:

$$f(x) = \mathrm{sgn}\left(\sum_{i-1}^{l} \delta_i m_i \psi(k_i) + b \right) \quad (1.15)$$

given that C and γ have a significant impact on model accuracy, we use the SMO algorithm to optimize these two hyperparameters in order to maximize detection accuracy. During the optimization process, the objective function (Eq. 1.11) will reach its minimum value.

4.3 Rainfall Detection Based on Random Forest Method

Random Forest (RF) is an ensemble learning algorithm based on decision trees [53], which improves classification or regression performance by combining multiple models. It has been widely used in fields such as sea ice detection, forest fire detection, and soil moisture estimation [58, 59, 74].

To overcome the drawbacks of over fitting in a single Classification Regression Tree (CART), RF combines enhanced sampling strategy with CART. Among them, CART uses the Gini index to measure impurities in the training dataset, and the formula for this Gini index is as follows:

$$G_{\text{index}}(p) = 1 - \sum_{i=1}^{s} p_i^2 \tag{1.16}$$

where s is the number of categories, and p_i is the proportion of samples belonging to the l category. Since rainfall detection is a binary classification problem, the Gini index can be simplified as:

$$G_{\text{mix}}(p) = 2p(1-p) \tag{1.17}$$

RF can achieve high accuracy at a lower computational cost and has been proven to have good robustness against bias and overfitting.

4.4 *Rainfall Detection Based on Convolutional Neural Network Method*

Compared to other machine learning methods, CNN has better feature learning ability, simpler structure, and more stable performance. A typical CNN model consists of multiple convolutional layers, pooling layers, and fully connected layers. It should be noted that different layers and their corresponding hyperparameters should be adjusted according to different problems. Figure 8 presents the CNN architecture proposed in this chapter. From Fig. 8, it can be seen that through the processing of different layers, the size of the feature map gradually decreases, and the result is ultimately output through fully connected layers. In this case, CNN has two input lines, one for inputting BRCS DDM and the other for inputting auxiliary parameters.

The first step is data input, and we use 17×11 BRCS DDM directly as the input for the CNN model. Next is the convolutional layer, which is essentially a special matrix operation between the input image and the convolutional kernel. Convolutional layers can convolve the input to extract features to form feature maps, and reduce noise interference. Subsequently, the pooling layer downsamples these feature maps, preserving key features while compressing data dimensions. The fully connected layer performs matrix operations between weights and features, adds biases, and completes weighted summation through nonlinear activation functions to achieve learning of advanced features.

This article uses the Rectified Linear Unit (ReLU) activation function, which is represented as:

$$f_{\text{ReLU}}(x) = \max(0, x) \tag{1.18}$$

ReLU avoids the gradient vanishing problem encountered in tanh and sigmoid activation functions [63]. It uses the Adam optimizer to train small batches of random cards.

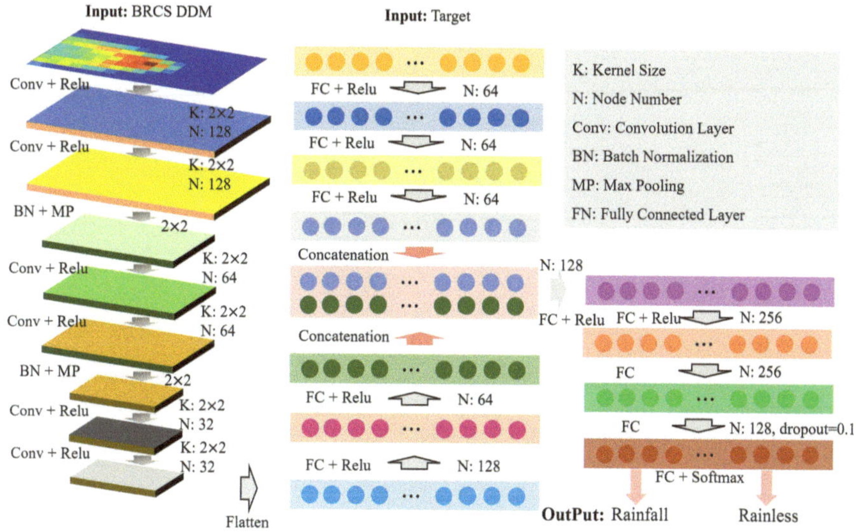

Fig. 8 Architecture of a CNN

In order to handle large sample sizes, CNN training used a small batch gradient descent method [43]. Each iteration takes a batch of size N = 32 from the sequence set. Upon the batch's total loss becoming qualified, coefficients are updated via backpropagation [56]. In each batch, only the samples from that batch are used for weight updates. When all batches in the training set have been processed, an epoch is completed. To ensure model performance, we set the total number of epochs to 400.

5 Rainfall Detection Results

This section first introduces the basic principles of the proposed rainfall detection (RD) method, followed by four performance evaluation indicators. Based on these indicators, evaluate the detection performance of the method, and finally compare the RD results based on GNSS-R with the IMERG-F rainfall intensity (RI) map.

5.1 Basic Method Description

Before studying rainfall detection, CYGNSS DDM data must be preprocessed. Specifically, the denoising method outlined in Sect. 3.3 is used to denoise DDM data. Furthermore, CYGNSS L1B measured values are subject to quality control and

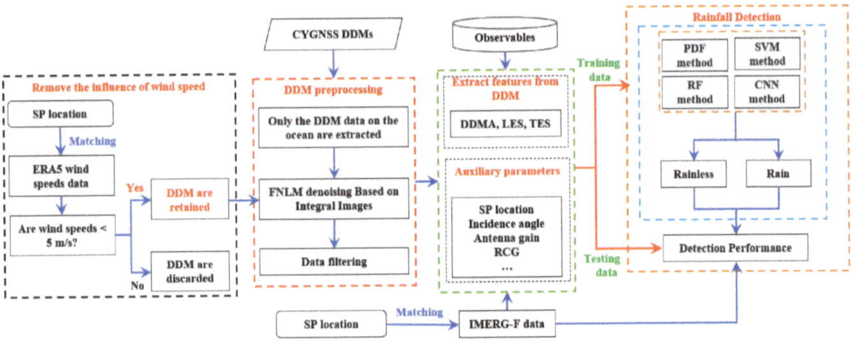

Fig. 9 Flow chart of RD algorithm. It mainly includes four steps: the first step: data matching; the second step: DDM data preprocessing, the third step: GNSS-R observables calculation, and the fourth step: rainfall detection and detection performance evaluation (marked in the orange box)

filtering before model training and testing to ensure the quality of CYGNSS data. Data filtering is performed as described in Sect. 3.4.

Next, we derive three GNSS observations from high-quality DDM—DDMA, LES, and TES. Use IMERG-F products as reference data and randomly partition the data samples, with 30% allocated to the training set and the other 70% allocated to the testing set. In order to ensure that the collection time of DDM data and the position of the mirror point (SP) can match the wind speed, SWH product, and rainfall data of ERA5, we adopted spatial bilinear interpolation and temporal linear interpolation methods. In addition, in addition to the above data quality control, CYGNSS data is also constrained by various surface features and satellite geometry conditions (see Sect. 2.2). Then, carry out rainfall detection according to the following steps. Firstly, the rainfall detection thresholds of three GNSS-R observations are obtained by using the training data based on PDF method. Then, the training data are used to train the model of rainfall detection method based on SVM, RF and CNN respectively. The process of the proposed rainfall detection method is shown in Fig. 9. Specifically, first use the training data to train SVM, RF, and CNN models separately; Then, evaluate the performance of the four detection methods using the remaining 70% of the test data.

5.2 *Performance Index*

Several indexes can be used to evaluate RD performance, e.g., correct detection rate, false positive (FP) rate, and accuracy [17]. Employing multiple evaluation metrics is beneficial for achieving a reliable assessment.

In the field of information retrieval, accuracy, recall rate, and F1 score are the three commonly used indicators to measure system effectiveness. Among them, accuracy refers to the proportion of samples predicted to be positive that are actually positive;

Table 4 Four possible outcomes of RD

	Positive	Negative
True	TP	TN
False	FP	FN

The recall rate reflects the proportion of correctly detected samples among all true positive samples. The F1 score, as a harmonic average of the two, can comprehensively evaluate the overall performance of the system. These three indicators together constitute a robust object detection performance evaluation system, which will be used for rainfall detection evaluation in this chapter.

As listed in Table 4, any detection method may produce four types of results: true positive (TP), false negative (FN), false positive (FP), and true negative (TN). Specifically, TP refers to the number of correctly identified positive samples; FN refers to a positive sample that has been incorrectly classified as negative; FP refers to negative samples that have been incorrectly labeled as positive; TN refers to correctly identified negative samples. These four results constitute the complete result set of rainfall detection (RD).

The detection performance is quantitatively measured using precision, recall, and F1 score, which are defined as:

$$\text{Precision} = \frac{|\text{TP}|}{|\text{TP}| + |\text{FP}|} \tag{1.19}$$

$$\text{Recall} = \frac{|\text{TP}|}{|\text{TP}| + |\text{FN}|} \tag{1.20}$$

$$F1 = 2 \times \frac{\text{precision} \times \text{recall}}{\text{precision} + \text{recall}} \tag{1.21}$$

please note that higher recall rates or higher overall scores (such as $F1$ scores) both imply better recognition performance, which is consistent with the meaning of accuracy.

5.3 Detection Performance

To test the adaptability of the model under different wind conditions, the training samples covered multiple wind speed ranges such as [0, 5], [5, 8], [8, 10], [10, 15], and [0, 10] m/s. Subsequently, we randomly selected 60% of the test data samples to evaluate the performance of the proposed machine learning based RD method and PDF method separately for comparison. Table 5 summarizes the test results of four RD methods in different wind speed ranges. Due to the high detection performance of PDF method based on DDMA peak observations, Table 5 presents only the rainfall detection outcomes obtained with the PDF method using DDMA peak observations;

detection results based on other observables are not discussed further. From the data in Table 5, the following observations can be made.

(1) Across different wind speed ranges, relative to the PDF method, the SVM, RF, and CNN methods achieve improvements of 10.6%, 10.4%, and 16.7% in accuracy; 12.6%, 13.9%, and 17.1% in recall; and 14.2%, 12.3%, and 15.1% in F1 score, respectively.
(2) The three proposed methods all achieve better performance than the PDF method on all three metrics. While the SVM and RF methods perform similarly, the CNN method is substantially superior to the other three (PDF, SVM, and RF). In particular, the detection accuracy is improved by over 10% relative to the PDF method.
(3) The accuracy of rainfall detection is marginally better in the 5–8 m/s wind speed range compared to the 0–5 m/s range. A possible explanation is that at very low wind speeds, sea surface rainfall is more influenced by ocean swell. However, Rainfall detection accuracy declines as wind speed increases, particularly beyond 10 m/s. From this, it can be concluded that the fully developed wave spectrum can be divided into two electromagnetic scattering mechanisms based on wind speed. When the wind speed is greater than 5 m/s, the Rayleigh parameter is large and the geometric optical approximation holds. The roughness of wind and waves is much greater than that of rainfall, so the roughness of the sea surface is dominated by wind speed, and the effect of rainfall is masked by wind speed.

With the increase of wind speed, the accuracy of RD shows a downward trend; When the wind speed exceeds 10 m/s, this decrease is particularly significant. A preliminary conclusion can therefore be drawn: for wind speeds below 5 m/s, the

Table 5 Statistics of rainfall detection performance results based on four detection methods

		Wind speed range (m/s)				
Index	Method	[0, 5]	[5, 8]	[8, 10]	[10, 15]	[0, 10]
Precision	PDF	69.1	74.2	72.0	54.8	69.4
	SVM	75.6	82.1	76.1	57.0	72.9
	RF	76.3	81.7	74.7	58.4	75.1
	CNN	80.6	78.0	77.1	58.9	78.5
Recall	PDF	70.3	74.4	74.4	70.0	71.6
	SVM	77.1	83.8	82.1	73.6	82.2
	RF	77.5	84.1	84.8	76.1	80.7
	CNN	80.3	85.0	84.9	77.5	83.8
F1 score	PDF	69.7	76.3	69.2	61.0	69.5
	SVM	76.3	82.9	79.0	64.3	77.3
	RF	78.3	80.9	72.7	66.6	77.4
	CNN	79.0	84.0	79.7	67.3	78.1

method proposed in this section is effective for detecting sea surface rainfall using spaceborne GNSS-R data. Under this condition, the accuracy, recall, and F1 score are better than 80.6%, 80.3%, and 79.0%, respectively.

5.4 Comparison of IMERG-F RI Map with GNSS-R Data Tracks

Due to space limitation, Fig. 10 only shows the comparison between the IMERG-F rainfall data at six different observation times and the rainfall detection results obtained from GNSS-R data using CNN method. The existence of rainfall events is marked by a red circle, while the existence of no rain events is marked by a black circle. The bottom color bar (Paula colormap) represents the IMERG-F rainfall intensity, which is used as the reference true value for verification.

During data preprocessing, low-quality DDMs were filtered out, leading to the discarding of over 30% of the data and consequently creating gaps in the GNSS-R trajectory. Since low-quality DDMs typically occur in consecutive periods, these gaps are often substantial, with distances exceeding 20 km. Figure 10 reveals that the GNSS-R method yields rainfall detection results that closely match the IMERG-F rainfall data. Additionally, one characteristic of spaceborne GNSS-R is that its data tracks exhibit similarity across various bands. Although the coverage of currently accessible GNSS-R data is much smaller than that of IMERG-F rainfall data, future missions will help assess the competitiveness of this technology compared with traditional technologies.

When sea surface wind speed and waves have little effect on the DDM, the DDM becomes highly sensitive to rainfall intensity. Notably, the SNR of scattered signals over a rain-free sea surface is generally higher than that over a rainy sea surface. Another factor related to data quality is geometric characteristics, such as incident angle. This may be the reason for the continuous occurrence of low-quality DDM.

6 Conclusion

In response to the problem that rainfall detection methods based on PDF thresholds are easily affected by geometric shapes and sea conditions, this chapter proposes three new spaceborne GNSS-R rainfall detection methods, which use SVM, RF, and CNN algorithms respectively. Prior to this, in order to improve the quality of DDM data, we first proposed a fast NLM denoising algorithm based on integrated images, and calculated three observable values (DDMA, LES, and TES) from the power DDM accordingly. The research data is sourced from the CYGNSS mission. To evaluate the rainfall detection performance of the proposed approaches, the GPM-IMERG

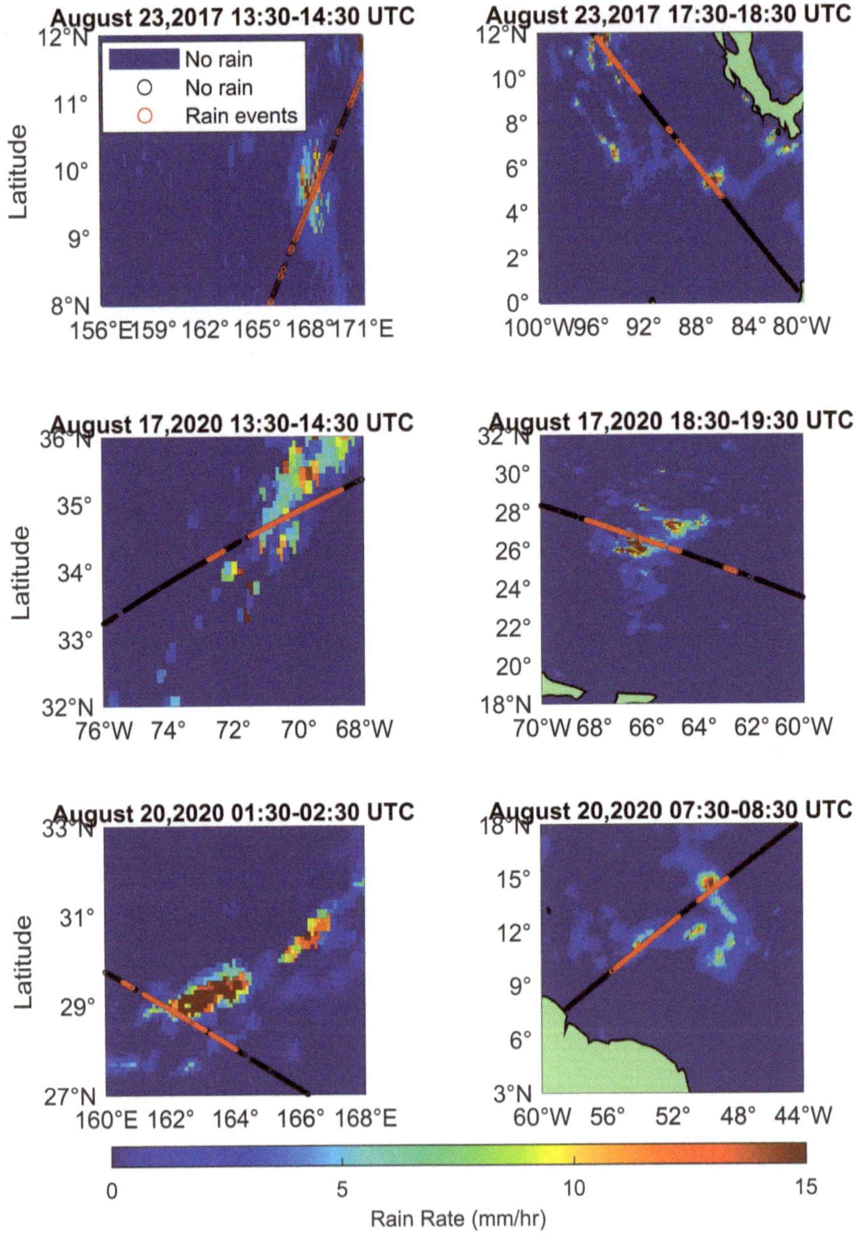

Fig. 10 GNSS-R rainfall detection results compared with the IMERG-F RI map of six different observation time. (top left) Aug 23, 2017 13:30–14:30 UTC, (top right) Aug 23, 2020 17:30–18:30 UTC, (middle left) Aug 17, 2020 13:30–14:30 UTC, (middle right) Aug 17, 2020 18:30–19:30 UTC, (bottom left) Aug 20, 2020 01:30–02:30 UTC, and (bottom right) Aug 20, 2020 07:30–08:30 UTC

product serves as the reference data. These three methods all demonstrate considerable advantages in constructing multi-parameter models, thereby offering a robust alternative for spaceborne GNSS-R-based sea surface rainfall detection. According to the experimental results, the three proposed methods consistently outperform the PDF method in terms of precision, recall, and F1 score. Furthermore, while the RD accuracy of the proposed SVM and RF methods remains nearly identical, the CNN-based method exhibits a clear superiority over the other two. Specifically, compared to the PDF method, the proposed CNN method improves rainfall detection (RD) accuracy by more than 10%. Under low wind speed conditions (below 10 m/s), the CNN-based approach demonstrates effective rainfall detection using spaceborne GNSS-R data, achieving a precision of over 78.5%, a recall exceeding 83.8%, and an F1 score above 78.1%.

It should be emphasized that GNSS can achieve short revisit time at low cost (thanks to the miniaturization, lightweight, low power consumption, and low cost of satellites), which is its key advantage. The revisit time of CYGNSS is about 7 h, reflecting the potential of GNSS-R to track rainfall dynamics. Technological development will drive future multi satellite missions to provide high spatiotemporal resolution data, further promoting RD.

It should be noted that promising rainfall detection results have been achieved using the proposed DDM denoising method together with the three RD approaches (SVM, RF, and CNN). In addition to rainfall detection, both the denoising technique and the observables are also applicable to other marine remote sensing tasks, such as sea ice detection and target detection. Nevertheless, further investigation and validation are still required to confirm the results in those applications. Due to the fact that sea surface winds can alter the roughness of the sea surface, which in turn affects the DDM and delay waveform of the reflected signal (especially under high wind speed conditions), the observed values are highly susceptible to wind interference. Therefore, in future research, continuing to reduce the impact of wind on the proposed method is a valuable direction.

Acknowledgements The work was supported by the National Natural Science Foundation of China under grant 42404037.

References

1. Alonso-Arroyo A, Zavorotny VU, Camps A (2017) Sea ice detection using U.K. TDS-1 GNSS-R data. IEEE Trans Geosci Remote Sens **55**(9):4989–5001. https://doi.org/10.1109/tgrs.2017.2699122
2. Asgarimehr M, Arnold C, Weigel T, Ruf C, Wickert J (2022) GNSS reflectometry global ocean wind speed using deep learning: development and assessment of CyGNSSnet. Remote Sens Environ 269:112801. https://doi.org/10.1016/j.rse.2021.112801
3. Asgarimehr M, Zavorotny V, Wickert J, Reich S (2018) Can GNSS reflectometry detect precipitation over oceans? Geophys Res Lett 45(22):12585–12592. https://doi.org/10.1029/2018gl079708

4. Balasubramaniam R, Ruf C (2020) Characterization of rain impact on L-band GNSS-R ocean surface measurements. Remote Sens Environ 239:111607. https://doi.org/10.1016/j.rse.2019.111607
5. Bliven LF, Sobieski PW, Craeye C (1997) Rain generated ring-waves: measurements and modelling for remote sensing. Int J Remote Sens 18(1):221–228. https://doi.org/10.1080/014311697219385
6. Braun N, Gade M, Lange PA (2002) The effect of artificial rain on wave spectra and multi-polarisation X band radar backscatter. Int J Remote Sens 23(20):4305–4323. https://doi.org/10.1080/01431160110106032
7. Bu J, Wang Q, Ni J (2024) Estimating sea surface swell height using a hybrid model combining CNN, ConvLSTM, and FCN based on spaceborne GNSS-R data from the CYGNSS mission. GPS Solutions 28(3):133. https://doi.org/10.1007/s10291-024-01671-1
8. Bu J, Wang Q, Wang Z, Fan S, Liu X, Zuo X (2024) Land remote sensing applications using spaceborne GNSS reflectometry: a comprehensive overview. IEEE J Sel Top Appl Earth Obs Remote Sens 17:12811–12841. https://doi.org/10.1109/JSTARS.2024.3415754
9. Bu J, Yu K (2022) A new integrated method of CYGNSS DDMA and LES measurements for significant wave height estimation. IEEE Geosci Remote Sens Lett 19:1505605. https://doi.org/10.1109/LGRS.2022.3198131
10. Bu J, Yu K (2022) Sea surface rainfall detection and intensity retrieval based on GNSS-reflectometry data from the CYGNSS mission. IEEE Trans Geosci Remote Sens 60:5802015. https://doi.org/10.1109/TGRS.2021.3127943
11. Bu J., K. Yu (2022) Significant Wave Height Retrieval Method Based on Spaceborne GNSS Reflectometry. IEEE Geoscience and Remote Sensing Letters **19**: 1503705.https://doi.org/10.1109/LGRS.2022.3155563
12. Bu J., K. Yu, S. Han, N. Qian, Y. Lin, J. Wang (2022) Retrieval of Sea Surface Rainfall Intensity Using Spaceborne GNSS-R Data. IEEE Transactions on Geoscience and Remote Sensing **60**: 5803116.https://doi.org/10.1109/TGRS.2022.3168928
13. Bu J., K. Yu, J. Ni, W. Huang (2023) Combining ERA5 data and CYGNSS observations for the joint retrieval of global significant wave height of ocean swell and wind wave: a deep convolutional neural network approach. Journal of Geodesy **97**(8): 81.https://doi.org/10.1007/s00190-023-01768-4
14. Buades A., B. Coll, J. M. Morel (2005) A Review of Image Denoising Algorithms, with a New One. Multiscale Modeling & Simulation **4**(2): 490–530.https://doi.org/10.1137/040616024
15. Camps A., H. Park, J. Castellví, J. Corbera, E. Ascaso (2020) Single-Pass Soil Moisture Retrievals Using GNSS-R: Lessons Learned. Remote Sensing **12**(12): 2064.https://doi.org/10.3390/rs12122064
16. Cardellach E., W. Li, A. Rius, M. Semmling, J. Wickert, F. Zus, C. S. Ruf, C. Buontempo (2020) First Precise Spaceborne Sea Surface Altimetry With GNSS Reflected Signals. IEEE Journal of Selected Topics in Applied Earth Observations and Remote Sensing **13**: 102–112.https://doi.org/10.1109/jstars.2019.2952694
17. Chen X., Q. An, K. Yu, Y. Ban (2021) A Novel Fire Identification Algorithm Based on Improved Color Segmentation and Enhanced Feature Data. IEEE Transactions on Instrumentation and Measurement **70**: 1–15.https://doi.org/10.1109/TIM.2021.3075380
18. Chen X., W. Huang, C. Zhao, Y. Tian (2020) Rain Detection From X-Band Marine Radar Images: A Support Vector Machine-Based Approach. IEEE Transactions on Geoscience and Remote Sensing **58**(3): 2115–2123.https://doi.org/10.1109/tgrs.2019.2953143
19. Chen Z., Y. He, B. Zhang (2017) An Automatic Algorithm to Retrieve Wave Height From X-Band Marine Radar Image Sequence. IEEE Transactions on Geoscience and Remote Sensing **55**(9): 5084–5092.https://doi.org/10.1109/tgrs.2017.2702192
20. Chen Z., Y. He, B. Zhang, Z. Qiu, B. Yin (2014) A New Algorithm to Retrieve Wave Parameters From Marine X-Band Radar Image Sequences. IEEE Transactions on Geoscience and Remote Sensing **52**(7): 4083–4091.https://doi.org/10.1109/tgrs.2013.2279547
21. Chu X., J. He, H. Song, Y. Qi, Y. Sun, W. Bai, W. Li, Q. Wu (2020) Multimodal Deep Learning for Heterogeneous GNSS-R Data Fusion and Ocean Wind Speed Retrieval. IEEE Journal of

Selected Topics in Applied Earth Observations and Remote Sensing **13**: 5971–5981.https://doi.org/10.1109/jstars.2020.3010879
22. Clarizia M. P., C. S. Ruf (2016) Wind Speed Retrieval Algorithm for the Cyclone Global Navigation Satellite System (CYGNSS) Mission. IEEE Transactions on Geoscience and Remote Sensing **54**(8): 4419–4432.https://doi.org/10.1109/tgrs.2016.2541343
23. Clarizia M. P., C. S. Ruf (2020) Statistical Derivation of Wind Speeds From CYGNSS Data. IEEE Transactions on Geoscience and Remote Sensing **58**(6): 3955–3964.https://doi.org/10.1109/tgrs.2019.2959715
24. Clarizia M. P., C. S. Ruf, P. Jales, C. Gommenginger (2014) Spaceborne GNSS-R Minimum Variance Wind Speed Estimator. IEEE Transactions on Geoscience and Remote Sensing **52**(11): 6829–6843.https://doi.org/10.1109/tgrs.2014.2303831
25. Contreras R. F., W. J. Plant (2006) Surface effect of rain on microwave backscatter from the ocean: Measurements and modeling. Journal of Geophysical Research: Oceans **111**(C8)
26. Cortes C, Vapnik V (1995) Support-vector networks. Machine learning 20(3):273–297
27. Craeye C., P. W. Sobieski, L. F. Bliven (1997) Scattering by artificial wind and rain roughened water surfaces at oblique incidences. International Journal of Remote Sensing **18**(10): 2241–2246.https://doi.org/10.1080/014311697217864
28. Craeye C, Sobieski PW, Bliven LF, Guissard A (1999) Ring-waves generated by water drops impacting on water surfaces at rest. IEEE J Oceanic Eng 24(3):323–332
29. Dee D. P., S. M. Uppala, A. J. Simmons, P. Berrisford, P. Poli, S. Kobayashi, U. Andrae, M. A. Balmaseda, G. Balsamo, P. Bauer, P. Bechtold, A. C. M. Beljaars, L. van de Berg, J. Bidlot, N. Bormann, C. Delsol, R. Dragani, M. Fuentes, A. J. Geer, L. Haimberger, S. B. Healy, H. Hersbach, E. V. Hólm, L. Isaksen, P. Kållberg, M. Köhler, M. Matricardi, A. P. McNally, B. M. Monge-Sanz, J. J. Morcrette, B. K. Park, C. Peubey, P. de Rosnay, C. Tavolato, J. N. Thépaut, F. Vitart (2011) The ERA-Interim reanalysis: configuration and performance of the data assimilation system. Quarterly Journal of the Royal Meteorological Society **137**(656): 553–597.https://doi.org/10.1002/qj.828
30. Elfouhaily T., B. Chapron, K. Katsaros, D. Vandemark (1997) A unified directional spectrum for long and short wind-driven waves. Journal of Geophysical Research: Oceans **102**(C7): 15781–15796.https://doi.org/10.1029/97JC00467
31. Facciolo G, Limare N, Meinhardt-Llopis E (2014) Integral Images for Block Matching. Image Processing On Line 4:344–369
32. Fan RE, Chen PH, Lin CJ, Joachims T (2005) Working Set Selection Using Second Order Information for Training Support Vector Machines. J Mach Learn Res 6(4):1889–1918
33. Foti G., C. Gommenginger, P. Jales, M. Unwin, A. Shaw, C. Robertson, J. Rosello (2015) Spaceborne GNSS reflectometry for ocean winds: First results from the UK TechDemoSat-1 mission. Geophysical Research Letters **42**(13): 5435–5441.https://doi.org/10.1002/2015gl064204
34. Froment J. (2014) Parameter-Free Fast Pixelwise Non-Local Means Denoising. Image Processing On Line **4**: 300–326.https://doi.org/10.5201/ipol.2014.120
35. Ghavidel A., A. Camps (2016) Impact of Rain, Swell, and Surface Currents on the Electromagnetic Bias in GNSS-Reflectometry. Ieee Journal of Selected Topics in Applied Earth Observations and Remote Sensing **9**(10): 4643–4649.https://doi.org/10.1109/jstars.2016.2538181
36. Guo W., H. Du, C. Guo, B. J. Southwell, J. W. Cheong, A. G. Dempster (2022) Information fusion for GNSS-R wind speed retrieval using statistically modified convolutional neural network. Remote Sensing of Environment **272**: 112934.https://doi.org/10.1016/j.rse.2022.112934
37. Hammond M. L., G. Foti, C. Gommenginger, M. Srokosz (2020) Temporal variability of GNSS-Reflectometry ocean wind speed retrieval performance during the UK TechDemoSat-1 mission. Remote Sensing of Environment **242**: 111744.https://doi.org/10.1016/j.rse.2020.111744
38. Huang W., E. Gill, J. An (2014) Iterative least-squares-based wave measurement using X-band nautical radar. IET Radar, Sonar & Navigation **8**(8): 853–863.https://doi.org/10.1049/iet-rsn.2013.0281

39. Huang W., X. Liu, E. Gill (2017) Ocean Wind and Wave Measurements Using X-Band Marine Radar: A Comprehensive Review. Remote Sensing **9**(12): 1261.https://doi.org/10.3390/rs9121261
40. Huang W., Y. Liu, E. Gill (2017) Texture-Analysis-Incorporated Wind Parameters Extraction from Rain-Contaminated X-Band Nautical Radar Images. Remote Sensing **9**(2): 166.https://doi.org/10.3390/rs9020166
41. Huffman G. J., D. T. Bolvin, D. Braithwaite, K. Hsu, R. Joyce, C. Kidd, E. J. Nelkin, S. Sorooshian, J. Tan, P. Xie (2019) Algorithm Theoretical Basis Document (ATBD) Version 06 NASA Global Precipitation Measurement (GPM) Integrated Multi- satellitE Retrievals for GPM (IMERG). National Aeronautics and Space Administration (NASA): 1–34
42. Jing C., X. Niu, C. Duan, F. Lu, G. Di, X. Yang (2019) Sea Surface Wind Speed Retrieval from the First Chinese GNSS-R Mission: Technique and Preliminary Results. Remote Sensing **11**(24).https://doi.org/10.3390/rs11243013
43. Li M., T. Zhang, Y. Chen, A. J. Smola (2014). Efficient mini-batch training for stochastic optimization. Proceedings of the 20th ACM SIGKDD international conference on Knowledge discovery and data mining.
44. Li W., A. Rius, F. Fabra, E. Cardellach, S. Ribo, M. Martin-Neira (2018) Revisiting the GNSS-R Waveform Statistics and Its Impact on Altimetric Retrievals. IEEE Transactions on Geoscience and Remote Sensing **56**(5): 2854–2871.https://doi.org/10.1109/tgrs.2017.2785343
45. Li X., D. Yang, J. Yang, G. Zheng, G. Han, Y. Nan, W. Li (2021) Analysis of coastal wind speed retrieval from CYGNSS mission using artificial neural network. Remote Sensing of Environment **260**: 112454.https://doi.org/10.1016/j.rse.2021.112454
46. Liu L, Huang W, Wang C (2017) Hyperspectral image classification with kernel-based least-squares support vector machines in sum space. IEEE Journal of Selected Topics in Applied Earth Observations and Remote Sensing 11(4):1144–1157
47. Liu X., J. Bu, X. Zuo, Z. Wang, Q. Wang, Q. Wang, C. Ji, Y. Zhao, H. Yang, X. He (2025) Performance Validation of Sea Surface Wind Speed Retrieval Algorithms and Products From the Chinese Tianmu-1 Constellation GNSS-R: First Results on Comparison With Other Wind Speed Products. IEEE Journal of Selected Topics in Applied Earth Observations and Remote Sensing **18**: 5189–5203.https://doi.org/10.1109/JSTARS.2025.3527026
48. Liu X., W. Huang, E. W. Gill (2017) Wind Direction Estimation From Rain-Contaminated Marine Radar Data Using the Ensemble Empirical Mode Decomposition Method. IEEE Transactions on Geoscience and Remote Sensing **55**(3): 1833–1841.https://doi.org/10.1109/tgrs.2016.2635078
49. Lu Z., L. Sun, Y. Zhou (2021) A Method for Rainfall Detection and Rainfall Intensity Level Retrieval from X-Band Marine Radar Images. Applied Sciences **11**(4): 1565.https://doi.org/10.3390/app11041565
50. Lund B., H. C. Graber, R. Romeiser (2012) Wind Retrieval From Shipborne Nautical X-Band Radar Data. IEEE Transactions on Geoscience and Remote Sensing **50**(10): 3800–3811.https://doi.org/10.1109/tgrs.2012.2186457
51. Munoz-Martin J. F., L. Fernandez, A. Perez, J. A. Ruiz-de-Azua, H. Park, A. Camps, B. C. Domínguez, M. Pastena (2020) In-Orbit Validation of the FMPL-2 Instrument—The GNSS-R and L-Band Microwave Radiometer Payload of the FSSCat Mission. Remote Sensing **13**(1): 121.https://doi.org/10.3390/rs13010121
52. Reynolds J., M. P. Clarizia, E. Santi (2020) Wind Speed Estimation From CYGNSS Using Artificial Neural Networks. IEEE Journal of Selected Topics in Applied Earth Observations and Remote Sensing **13**: 708–716.https://doi.org/10.1109/jstars.2020.2968156
53. Rodriguez-Galiano V. F., B. Ghimire, J. Rogan, M. Chica-Olmo, J. P. Rigol-Sanchez (2012) An assessment of the effectiveness of a random forest classifier for land-cover classification. ISPRS Journal of Photogrammetry and Remote Sensing **67**: 93–104.https://doi.org/10.1016/j.isprsjprs.2011.11.002
54. Ruf C., R. Atlas, P. Chang, M. P. Clarizia, J. garrison, S. Gleason, S. Katzberg, Z. Jelenak, J. Johnson, S. Majumdar, A. O'Brien, D. Posselt, A. Ridley, R. Rose, V. Zavorotny (2016) New Ocean Winds Satellite Mission to Probe Hurricanes and Tropical Convection. Bulletin of the American Meteorological Society: 385–395

55. Ruf C. S., R. Atlas, P. S. Chang, M. P. Clarizia, J. L. Garrison, S. Gleason, S. J. Katzberg, Z. Jelenak, J. T. Johnson, S. J. Majumdar, A. O'brien, D. J. Posselt, A. J. Ridley, R. J. Rose, V. U. Zavorotny (2016) New Ocean Winds Satellite Mission to Probe Hurricanes and Tropical Convection. Bulletin of the American Meteorological Society **97**(3): 385–395.https://doi.org/10.1175/BAMS-D-14-00218.1
56. Rumelhart DE, Hinton GE, Williams RJ (1986) Learning representations by back-propagating errors. nature 323(6088):533–536
57. Said F., Z. Jelenak, J. Park, P. S. Chang (2022) The NOAA Track-Wise Wind Retrieval Algorithm and Product Assessment for CyGNSS. IEEE Transactions on Geoscience and Remote Sensing **60**: 4202524.https://doi.org/10.1109/tgrs.2021.3087426
58. Santi E., M. P. Clarizia, D. Comite, L. Dente, L. Guerriero, N. Pierdicca (2022) Detecting fire disturbances in forests by using GNSS reflectometry and machine learning: A case study in Angola. Remote Sensing of Environment **270**: 112878.https://doi.org/10.1016/j.rse.2021.112878
59. Senyurek V, Lei F, Boyd D, Kurum M, Gurbuz AC, Moorhead R (2020) Machine Learning-Based CYGNSS Soil Moisture Estimates over ISMN sites in CONUS. Remote Sensing 12(7):1168
60. Sun Y, Huang F, Xia J, Yin C, Bai W, Du Q, Wang X, Cai Y, Li W, Yang G, Zhai X, Xu N, Hu X, Liu Y, Liu C, Wang D, Qiu T, Tian Y, Duan L, Li F, Meng X, Liu C, Tan G, Hu P, Wu R, Song D (2023) GNOS-II on Fengyun-3 Satellite Series: Exploration of Multi-GNSS Reflection Signals for Operational Applications. Remote Sensing 15(24):5756
61. Voronovich A. G., V. U. Zavorotny (2017) The Transition From Weak to Strong Diffuse Radar Bistatic Scattering From Rough Ocean Surface. IEEE Transactions on Antennas and Propagation **65**(11): 6029–6034.https://doi.org/10.1109/tap.2017.2752219
62. Voronovich A. G., V. U. Zavorotny (2018) Bistatic Radar Equation for Signals of Opportunity Revisited. IEEE Transactions on Geoscience and Remote Sensing **56**(4): 1959–1968.https://doi.org/10.1109/tgrs.2017.2771253
63. Wang C, Xi Y (1997) Convolutional neural network for image classification. Johns Hopkins University Baltimore, MD, p 21218
64. Wang J., Y. Guo, Y. Ying, Y. Liu, Q. Peng (2006). Fast Non-Local Algorithm for Image Denoising. 2006 International Conference on Image Processing.
65. Wang Y., W. Huang (2016) An Algorithm for Wind Direction Retrieval From X-Band Marine Radar Images. IEEE Geoscience and Remote Sensing Letters **13**(2): 252–256.https://doi.org/10.1109/lgrs.2015.2508284
66. Weissman D, Stiles B, Hristova-Veleva S, Long D, Smith D, Hilburn K, Jones W (2012) Challenges to satellite sensors of ocean winds: Addressing precipitation effects. J Atmos Oceanic Tech 29(3):356–374
67. Weissman D. E., B. W. Stiles, S. M. Hristova-Veleva, D. G. Long, D. K. Smith, K. A. Hilburn, W. L. Jones (2012) Challenges to Satellite Sensors of Ocean Winds: Addressing Precipitation Effects. Journal of Atmospheric and Oceanic Technology **29**(3): 356–374.https://doi.org/10.1175/jtech-d-11-00054.1
68. Yan Q., W. Huang (2016) Spaceborne GNSS-R Sea Ice Detection Using Delay-Doppler Maps: First Results From the U.K. TechDemoSat-1 Mission. IEEE Journal of Selected Topics in Applied Earth Observations and Remote Sensing **9**(10): 4795–4801.https://doi.org/10.1109/jstars.2016.2582690
69. Yan Q., W. Huang, C. Moloney (2017) Neural Networks Based Sea Ice Detection and Concentration Retrieval From GNSS-R Delay-Doppler Maps. IEEE Journal of Selected Topics in Applied Earth Observations and Remote Sensing **10**(8): 3789–3798.https://doi.org/10.1109/jstars.2017.2689009
70. Yang Z., W. Huang, X. Chen (2021) Evaluation and Mitigation of Rain Effect on Wave Direction and Period Estimation from X-band Marine Radar Images. IEEE Journal of Selected Topics in Applied Earth Observations and Remote Sensing **14**: 5207–5219.https://doi.org/10.1109/jstars.2021.3076693

71. Yu K., Y. Li, X. Chang (2019) Snow Depth Estimation Based on Combination of Pseudorange and Carrier Phase of GNSS Dual-Frequency Signals. IEEE Transactions on Geoscience and Remote Sensing **57**(3): 1817–1828.https://doi.org/10.1109/tgrs.2018.2869284
72. Zavorotny V., A. Voronovich (2018). GNSS-R Modeling Results Obtained with Improved Bistatic Radar Equation. 2018 International Conference on Electromagnetics in Advanced Applications (ICEAA).
73. Zheng Y., Z. Shi, Z. Lu, W. Ma (2020) A Method for Detecting Rainfall From X-Band Marine Radar Images. IEEE Access **8**: 19046–19057.https://doi.org/10.1109/access.2020.2968601
74. Zhu Y., T. Tao, K. Yu, X. Qu, S. Li, J. Wickert, M. Semmling (2020) Machine Learning-Aided Sea Ice Monitoring Using Feature Sequences Extracted from Spaceborne GNSS-Reflectometry Data. Remote Sensing **12**(22): 3751.https://doi.org/10.3390/rs12223751

Chapter 19
Sea Surface Wind Speed Retrieval Using Spaceborne GNSS-R Data

Wenfei Guo, Hao Du, and Yuxian Hu

Abstract This chapter focuses on sea surface wind speed retrieval using spaceborne GNSS-Reflectometry (GNSS-R) data. To overcome the limitations of traditional empirical models in handling the complex, nonlinear coupling between the sea and atmosphere, multiple retrieval strategies based on Delay-Doppler Map (DDM) features are analyzed and improved. First, the performance of lookup table (LUT) and cumulative distribution function (CDF) methods based on geophysical model functions (GMFs) is assessed. The results show that approaches relying on single DDM features (e.g., LES, NBRCS) exhibit substantial biases across varying incidence angles and sea-state conditions, failing to capture the nonlinear relationship between wind speed and surface roughness. To improve feature combination within the GMF framework, the limitations of the minimum variance estimator (MVE) are examined, and a particle swarm optimization (PSO) algorithm is introduced to globally optimize the combination weights of LES and NBRCS, thereby removing the unit-weight constraint inherent in MVE. Experiments demonstrate that PSO significantly enhances retrieval accuracy, reducing RMSE from 2.21 to 1.95 m/s, with pronounced improvements at low-to-moderate wind speeds. To further address the limitations of traditional GMFs in modeling strongly nonlinear feature interactions, an end-to-end convolutional neural network (CNN) is developed. The model adaptively extracts local structural features from DDMs and integrates geometric parameters, such as specular point latitude and longitude, incidence angle, and RCG, to strengthen nonlinear representation. Significant wave height (SWH) is incorporated into the CNN2 model for joint wind–wave characterization, and CDF matching is

W. Guo (✉)
GNSS Research Center, Wuhan University, Wuhan, China
e-mail: wf.guo@whu.edu.cn

H. Du
Institute of Space Sciences (ICE-CSIC)/Institut d'Estudis Espacials de Catalunya (IEEC), Barcelona, Spain
e-mail: du1441053458@gmail.com

Y. Hu
School of Cyber Science and Engineering, Wuhan University, Wuhan, China
e-mail: yuxianhu84@gmail.com

K. Yu and N. Zheng (eds.), *Advances in GNSS Reflectometry*, Navigation: Science and Technology 19, https://doi.org/10.1007/978-981-92-2325-1_19

applied to produce the MCNN model, thereby mitigating systematic biases. Experiments using CYGNSS L1 V2.1 and ERA5 data from 2018 to 2019 (over five million samples) demonstrate that the CNN-based models outperform LUT and CDF approaches. After adding SWH, the CNN2 RMSE decreases from 1.50 to 1.47 m/s. The MCNN model reduces RMSE by 38 and 18% relative to LUT and CDF, respectively, and decreases systematic bias by 90 and 72%, while providing improved long-term stability. Overall, the PSO approach removes the structural constraints of MVE and enables more effective feature combination, achieving markedly lower retrieval errors than traditional GMFs. The MCNN model further advances wind speed retrieval by integrating multi-source features, delivering superior accuracy and robustness. The proposed technical framework offers a valuable reference for developing high-precision, generalizable GNSS-R wind retrieval methods, with demonstrated potential for typhoon monitoring, numerical weather prediction, and climate research.

Keywords Global navigation satellite system reflectometry (GNSS-R) · Sea surface wind speed retrieval; particle swarm optimization (PSO) · Convolutional neural network (CNN) · Cumulative distribution function (CDF) matching

1 Introduction

Global Navigation Satellite System Reflectometry (GNSS-R), an emerging passive microwave remote sensing technique, has made substantial progress in recent years in retrieving marine environmental parameters. Unlike conventional active microwave scatterometers and radiometers, GNSS-R exploits existing L-band signals from GNSS constellations as sources of illumination and performs observations by receiving signals reflected from the sea surface. This approach offers several salient advantages, including passive remote sensing operation, low cost, low power consumption, high temporal resolution and coverage, as well as all-weather observation capability [1, 2]. Spaceborne missions such as CYGNSS [3], TDS-1 [4], BuFeng-1 [5], and FY-3 [6] have further leveraged multi-satellite constellations to achieve high revisit rates and enhanced capabilities in capturing extreme wind events. These developments have markedly improved the spatiotemporal monitoring of sea surface wind fields in tropical cyclone and typhoon regions. Moreover, the DDM derived from GNSS-R observations provides a direct descriptor of sea surface roughness and is highly sensitive to short-scale wind-generated waves. Consequently, GNSS-R demonstrates distinctive capability for wind speed retrieval and high-wind monitoring.

The development of GNSS-R sea surface wind speed retrieval has generally progressed through three distinct phases. The initial phase primarily focused on concept validation and establishing theoretical foundations. In 1988, Hall and Cordey introduced the concept of bistatic radar observations using GNSS signals [7]. In 1993, Martin-Neira further advanced this idea by proposing its application to sea surface altimetry [8], marking a significant step toward the practical utilization of GNSS-R. In 1998, Garrison et al. pioneered the investigation of sea surface wind detection using GNSS-R signals, demonstrating its feasibility through extensive airborne experiments [9]. To further substantiate this potential, numerous field campaigns were conducted by institutions such as NASA, the University of Colorado, ICE-CSIC/IEEC, and the Starlab laboratory, collectively laying the theoretical and experimental groundwork for GNSS-R wind retrieval [10]. At the beginning of the twenty-first century, GNSS-R sea surface wind speed retrieval entered a developmental phase characterized by advances in scattering theory and the formulation of foundational models. Zavorotny and Voronovich established a two-dimensional delay–Doppler power model, commonly referred to as the Z–V model, based on the bistatic radar equation and the Kirchhoff approximation under the geometric optics limit [11]. This model quantified the relationships among reflected power, sea surface roughness, incidence geometry, and wind speed. Combined with the Elfouhaily ocean wave spectrum model [12], it provided a robust theoretical basis for GNSS-R sea surface wind speed retrieval. Building on this foundation, the theory and application of GNSS-R wind measurement advanced rapidly. The subsequent deployment of constellation missions such as CYGNSS and TDS-1 enabled the accumulation of vast observational datasets. This data abundance facilitated a transition from empirical modeling to data-driven learning, propelling GNSS-R sea surface wind speed retrieval into a phase of systematic and intelligent development.

Existing GNSS-R wind speed retrieval methods can be broadly categorized into four groups. (1) Waveform-matching techniques, which retrieve wind speed by simulating DDMs or delay waveforms using physical models and performing a least-squares fit with the observed measurements. These methods offer strong physical consistency and interpretability but remain sensitive to model assumptions and varying sea-state conditions (2). Empirical and statistical models, which establish a GMF using observables such as the NBRCS, the LES of the delay waveform, and the signal-to-noise ratio (SNR) [13]. Alternatively, statistical approaches such as MVE, can also be employed to map wind speed. (3) Intelligent optimization algorithms, which integrate physical models with global optimization algorithms—such as PSO and simulated annealing (SA)—to enhance convergence and robustness in solving nonlinear inversion problems. (4) Machine learning and deep learning methods, including CNNs, long short-term memory (LSTM) networks, attention-based Transformers, and physics-informed neural networks. Recent studies have demonstrated end-to-end retrieval of wind speed by feeding complete DDM or BRCS images, along with auxiliary features such as incidence angle, geometric configuration, significant wave height, and ERA5 wind fields, into deep learning architectures. Representative models include CYGNSS-Net [14], MCNN [15], GloWS-Net [16], and DDM-Former [17]. Experimental results show that deep learning models significantly improve retrieval accuracy at low-to-medium wind speeds, achieving RMSE values as low as 1.3–1.6 m/s. However, under extreme wind conditions, complex wave-spectrum dynamics and scattering mechanisms induce nonlinear saturation effects in the GNSS-R signal response, thereby degrading the generalization performance of these models [18].

Despite the considerable progress made in GNSS-R sea surface wind speed retrieval, several key challenges remain. Current research relies heavily on observations from the CYGNSS satellite constellation, in which the DDM serves as the primary observable for wind estimation. In 2012, Rodríguez-Álvarez et al. [19] utilized DDM data collected from airborne GNSS-R experiments to extracted a set of range-based characteristic parameters, establishing an empirical relationship with wind speed. However, empirical models based on a single characteristic feature provide limited accuracy and fail to fully capture the complexity of sea-surface scattering processes. To improve retrieval performance, Clarizia et al. [20] introduced a LUT–based MVE method. This approach constructs a discrete GMF by jointly employing two key observables—the NBRCS and the LES of the delay waveform. Although it achieves moderate improvements in retrieval accuracy, the MVE method remains constrained by its unit-weight assumption, which limits parameter optimization and hinders the effective use of multi-feature information. To overcome this limitation, Guo et al. [21] proposed a multi-feature fusion wind speed retrieval algorithm based on PSO. By removing the unit-weight constraint and employing a swarm intelligence search mechanism to adaptively determine optimal combination coefficients for multiple features, this method significantly enhances both the accuracy and robustness of wind speed retrieval.

Furthermore, GNSS-R sea surface wind speed retrieval typically relies on statistical features extracted from the DDM to construct a GMF. However, conventional GMF approaches exhibit substantial limitations in capturing the multi-parameter, multi-scale nonlinear relationships inherent in GNSS-R observations. To address these challenges, researchers have introduced machine learning methods to enable multi-feature joint modeling, achieving notable improvements in retrieval accuracy. In 2018, Wang et al. pioneered the extraction of features from DDM, such as the peak power and LES, employing a fully connected network (FCN) to map these features, together with antenna gain at the specular point, to wind speed. Their model achieved a root-mean-square error (RMSE) of approximately 2.5 m/s within the 0–20 m/s wind speed range [22]. Subsequently, Reynolds et al. [23] expanded the input feature set by incorporating additional parameters, including the reflection coefficient gain (RCG) and the geographic coordinates of the specular point. By applying CDF matching to mitigate systematic bias caused by non-uniform wind speed distributions, they reduced the RMSE to 1.51 m/s over the 0–32 m/s range. Nevertheless, due to the highly imbalanced distribution of wind speed samples, the FCN model experienced overfitting under high-wind conditions. Additionally, manually extracting features from the DDM may lead to the loss of critical information, thereby limiting further improvements in model performance.

To address these limitations, Guo et al. [24] incorporated SWH alongside wind speed as input variables and developed a CNN model integrated with adaptive cumulative distribution correction, thereby improving retrieval accuracy under high-wind conditions. In 2023, Arabi et al. [25] introduced a hybrid CNN–LSTM model for the first time for GNSS-R wind speed retrieval. By leveraging the LSTM network to model consecutive DDM sequences, the method effectively captured temporal wind field variations along the observation track and substantially enhanced retrieval performance. Additionally, Zhao et al. combined a Transformer architecture with fully connected layers to exploit the spatiotemporal characteristics of DDM data, achieving an overall RMSE of 1.43 m/s [18]. However, their approach did not account for wind speeds below 2.5 m/s and exhibited rapidly deteriorating performance above 20 m/s. Subsequently, Du et al. [26] proposed a deep residual fully connected network for GNSS-R wind retrieval. The residual architecture and increased network depth strengthened the model's ability to represent complex nonlinear mappings, and its performance gains were validated on the CYGNSS dataset. In further developments, Chen et al. [27] systematically evaluated and quantified the impact of satellite relative velocity on bistatic DDM morphology and retrieval bias. They proposed integrating this parameter either as an additional model input or through geometric correction and demonstrated that doing so can reduce RMSE by more than 11% across several datasets, highlighting the significant importance of orbital dynamics information in GNSS-R wind retrieval. Moreover, since high-wind retrieval remains a key bottleneck in GNSS-R sea surface wind estimation, several studies have focused on extreme wind scenarios such as typhoons and hurricanes. For example, Al-Khaldi et al. [28] employed synthetic typhoon models to evaluate the high-wind-speed sensitivity of multiple versions of the CYGNSS NBRCS product. Their findings provide important insights into algorithm performance and error sources under extreme wind conditions.

In summary, although existing GNSS-R sea surface wind speed retrieval methods have achieved substantial progress in model development and retrieval accuracy, traditional empirical approaches and shallow neural networks still face notable limitations, particularly in nonlinear modeling and high-wind estimation. To further improve retrieval accuracy and robustness, this chapter introduces an end-to-end CNN-based feature learning framework and a PSO-driven adaptive multi-feature fusion method. The theoretical foundations, implementation procedures, and comparative performance of these approaches are presented in the subsequent sections. Before detailing the specific algorithms, the next section provides a concise overview of the fundamental principles of GNSS-R sea surface wind measurement, establishing the theoretical basis for the subsequent research.

2 Fundamental Theory of GNSS-R Sea Surface Wind Measurement

When GNSS signals interact with the ocean surface, they undergo both transmission and scattering. These processes are governed by factors such as signal wavelength, incidence angle, dielectric properties, and sea surface roughness. Consequently, constructing a GNSS-R sea surface wind speed retrieval model requires linking the observed signals to the physical properties of the sea surface based on electromagnetic scattering theory.

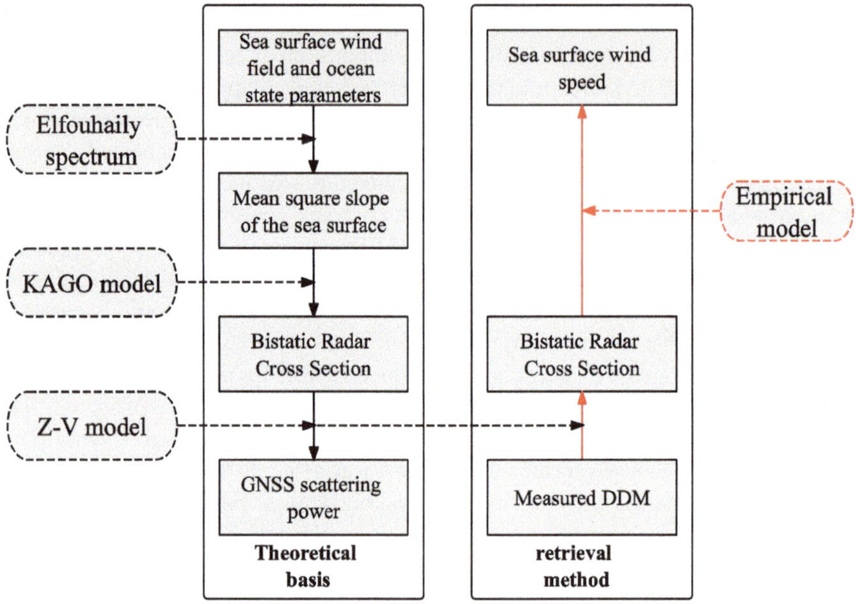

Fig. 1 Mechanism of GNSS signal scattering at sea surface

As illustrated in Fig. 1, the sea surface wind field is theoretically connected to GNSS-R observations through the Elfouhaily wave spectrum, the Kirchhoff Approximation with Geometric Optics (KAGO) model [29], and the Z-V model, which together provide the physical foundation for wind speed retrieval. Building on this relationship, once actual DDM data are obtained from a satellite-borne GNSS-R platform, the Z-V model can be applied to convert DDM power into BRCS. Subsequently, an empirical or semi-empirical relationship between BRCS and wind speed can be established to retrieve sea surface wind speed. This section provides an in-depth introduction to these theoretical models.

2.1 Elfouhaily Wave Spectrum Model

When sea breezes interact with the ocean surface, they generate waves that modify the surface roughness. Ocean waves are highly complex phenomena, characterized by significant temporal and spatial variability. In GNSS-R sea surface wind speed retrieval, theoretical models that relate the sea surface wind field to wave energy are commonly constructed using wave spectra. In 1997, Elfouhaily et al. [12] proposed a unified two-dimensional wave spectrum that encompasses both long-wave and short-wave regimes. It has since become one of the most widely adopted spectra in sea surface wind speed research.

The mathematical formulation of the Elfouhaily spectrum is as follows:

$$S_E(k,\psi) = M_E(k) f_E(k,\psi) \tag{1}$$

Here, k denotes the wavenumber, ψ represents the wave direction, and the azimuthal component function $f_E(k,\psi)$ is defined as follows:

$$f_E(k,\psi) = \frac{1}{2\pi}[1 + \Delta_E(k)\cos(2\psi)] \tag{2}$$

$$\Delta_E(k) = \tanh\left(\frac{\ln 2}{4} + 4\left(\frac{v_{ph}}{v_g}\right)^{2.5} + 0.13\frac{u_f}{v_{phm}}\left(\frac{v_{phm}}{v_{ph}}\right)^{2.5}\right) \tag{3}$$

The isotropic component $M_E(k)$ is expressed as follows:

$$M_E(k) = \frac{k^{-3}}{2v_{ph}}\left(\alpha_g v_g F_g + \alpha_c v_{phm} F_c\right)\gamma^{\Gamma}\exp\left\{-\frac{5}{4}\left(k_p/k\right)^2\right\} \tag{4}$$

$$\alpha_g = 6\times 10^{-3}\sqrt{\Omega},\ v_g = u_{10}/\Omega,\ F_g = \exp\left\{-\frac{\Omega}{\sqrt{10}}\left(\sqrt{k/k_p} - 1\right)\right\} \tag{5}$$

$$\gamma = \begin{cases} 1.7 & 0.84 \le \Omega \le 1 \\ 1.7 + 6\log(\Omega) & 1 < \Omega < 5 \end{cases} \tag{6}$$

$$\Gamma = \exp\left\{-\frac{\left(\sqrt{k/k_p} - 1\right)^2}{2\delta^2}\right\} \tag{7}$$

$$\delta = 0.08\left(1 + 4/\Omega^3\right),\ k_p = \Omega^2 g/u_{10}^2 \tag{8}$$

$$\Omega = 0.84\tanh\left[\left(X/2.2\times 10^4\right)^{0.4}\right]^{-0.75} \tag{9}$$

$$F_c = \exp\left\{-\frac{1}{4}\left(k/k_m - 1\right)^2\right\} \tag{10}$$

$$\alpha_c = 10^{-2}\begin{cases} 1 + \ln\left(\frac{u_f}{v_{phm}}\right) & u_f \le v_{phm} \\ 1 + 3\ln\left(\frac{u_f}{v_{phm}}\right) & u_f > v_{phm} \end{cases} \tag{11}$$

$$k_m = 363 rad/m,\ v_{phm} = 0.23 m/s,\ v_{ph} = \sqrt{\frac{g}{k}\left(1 + (k/k_m)^2\right)} \tag{12}$$

$$u_f = u_{10}\sqrt{10^{-2}(0.81 + 0.065u_{10})} \tag{13}$$

Where α_c denotes the short-wave generalized Phillis-Kitaigorodski equilibrium range parameter (u_f/v_{phm}). u_f represents the surface friction velocity, a physical quantity describing the interaction between airflow and the ocean surface, and is typically related to surface roughness and wind speed. v_{ph} corresponds to the short-wave phase velocity, while v_{phm} represents the minimum phase velocity at the wavenumber associated with the gravity–capillary peak in the curvilinear spectrum. Finally, g is the gravitational constant.

For open and fully developed seas (FDS), where $X \gg \infty$ generally holds, the wave age is taken as $\Omega = 0.85$. Figure 2 presents one-dimensional simulations of the Elfouhaily spectrum under different wind speeds. After obtaining the wave spectrum, the next step is to compute the sea surface roughness, which is commonly characterized by the mean square slope (MSS). Katzberg demonstrated a strong correlation between wind speed and MSS using airborne GNSS-R observations.

In the GNSS-R scenario, taking the local coordinate system as the reference frame and assuming that the wind direction forms an angle φ with the y-axis, the mean square slopes σ_{sx}^2 and σ_{sy}^2 along the x-and y-axis directions, respectively, and their correlation coefficient $b_{x,y}$, are given by:

$$\begin{bmatrix} \sigma_{sx}^2 & b_{x,y} \\ b_{x,y} & \sigma_{sy}^2 \end{bmatrix} = \begin{bmatrix} \cos\varphi & -\sin\varphi \\ \sin\varphi & \cos\varphi \end{bmatrix} \cdot \begin{bmatrix} \sigma_{sc}^2 & 0 \\ 0 & \sigma_{su}^2 \end{bmatrix} \cdot \begin{bmatrix} \cos\varphi & \sin\varphi \\ -\sin\varphi & \cos\varphi \end{bmatrix} \tag{14}$$

Thus, based on the Elfouhaily wave spectrum model, a quantitative relationship between the sea surface wind field and sea surface roughness can be established.

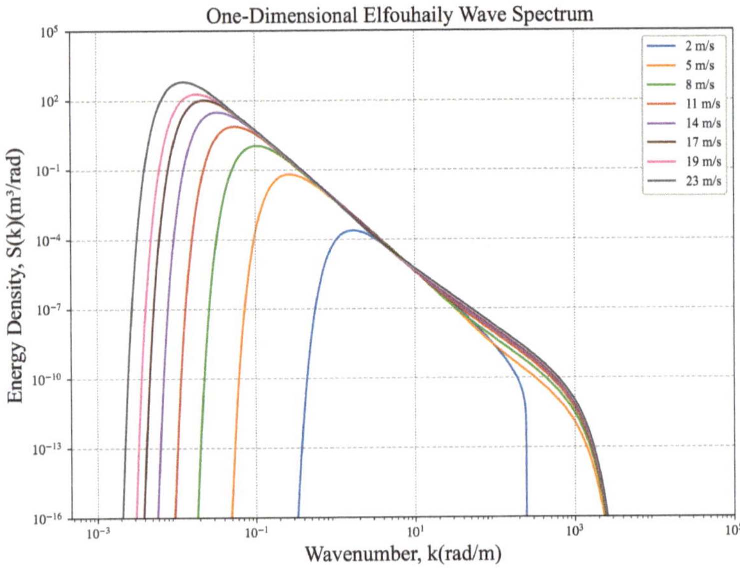

Fig. 2 One-dimensional Elfouhaily wave spectrum

2.2 KAGO Model

The scattering of electromagnetic waves can be decomposed into two components: coherent (specular) reflection and incoherent scattering. In the GNSS-R scenarios, the ocean surface is generally treated as rough; therefore, the received GNSS-R signals over water are dominated by incoherent scattering. The scattering characteristics induced by sea surface roughness are typically quantified using the KAGO electromagnetic scattering model. The KAGO model accounts only for the incoherent component of the scattered electromagnetic field. Developing this framework further, Zavorotny and Voronovich developed a simplified KAGO-based scattering model tailored for GNSS-R scenarios [11]:

$$\sigma_0(\vec{\rho}) = \frac{\pi|\Re|^2 q^4}{q_z^4} P(s_x, s_y) \tag{15}$$

Among these terms, $\vec{\rho}$ denotes the scattering facet, while q represents the scattering vector. The term q_z corresponds to the component of the scattering vector q in the direction perpendicular to the local surface. $s_x = -q_x/q_z$ and $s_y = -q_y/q_z$ denote the slopes of the sea surface in the up-wind and cross-wind directions, respectively. $P(s_x, s_y)$ is the probability density function describing the anisotropic distribution of sea surface slopes, expressed as:

$$P(s_x, s_y) = \frac{1}{2\pi\sigma_{sx}\sigma_{sy}\sqrt{1-b_{x,y}^2}} \exp\left[-\frac{1}{2\left(1-b_{x,y}^2\right)}\left(\frac{s_x^2}{\sigma_{sx}^2} - 2b_{x,y}\frac{s_x s_y}{\sigma_{sx}\sigma_{sy}} + \frac{s_y^2}{\sigma_{sy}^2}\right)\right] \tag{16}$$

Among these parameters, σ_{sx}, σ_{sy} and $b_{x,y}$ can be directly obtained from Eq. 14. $\Re$ represents the Nephier reflection coefficient, which characterizes the behavior of electromagnetic signals at the interface between two media with different refractive indices. It reflects the energy relationship between the incident, reflected, and transmitted components of the signal. For GNSS signals propagating over the sea surface, the reflection coefficients $\Re$ for horizontally and vertically polarized waves are given respectively by:

$$\Re_{HH} = \frac{\sin\theta - \sqrt{\varepsilon - \cos^2\theta}}{\sin\theta + \sqrt{\varepsilon - \cos^2\theta}} \quad \Re_{VV} = \frac{\varepsilon\sin\theta - \sqrt{\varepsilon - \cos^2\theta}}{\varepsilon\sin\theta + \sqrt{\varepsilon - \cos^2\theta}} \tag{17}$$

Here, θ denotes the incident angle of the signal, and ε represents the complex permittivity of seawater, whose value depends on both the carrier frequency and the sea surface temperature. The left-hand circularly polarized (LHCP) component $\Re_{RL}$ and the right-hand circularly polarized (RHCP) component $\Re_{RR}$ of the GNSS signal reflected from the sea surface are respectively expressed as:

$$\Re_{RL} = \frac{\Re_{HH} + \Re_{VV}}{2}, \quad \Re_{RR} = \frac{\Re_{VV} - \Re_{HH}}{2} \tag{18}$$

Thus, based on the above expressions, the relationship between sea surface roughness and the bistatic radar cross section (BRCS) σ_0, can be derived.

2.3 Z-V Model

Unlike conventional ocean remote sensing instruments, such as altimeters and scatterometers, GNSS-R utilizes a forward-scattering mechanism. At the receiver, a left-hand circularly polarized antenna captures the reflected signals, while a right-hand circularly polarized antenna receives the direct signals. A theoretical model linking the reflected signal power to the BRCS σ_0 is established using the dual-base radar Eq. (11):

$$Y_s(\hat{\tau}, \hat{f}) = \frac{\lambda^2 T_i^2 P_T}{(4\pi)^3} \int\int \frac{G_R \sigma_0 G_T}{R_T^2 R_R^2} \Lambda^2(\hat{\tau} - \tau) S^2(\hat{f} - f) dA \tag{19}$$

Here, λ denotes the carrier wavelength, T_i is the coherent integration time; P_T and G_T represent the transmitter signal power and antenna gain, respectively, whose product defines the Effective Isotropic Radiated Power (EIRP). G_R denotes the gain of the receiving antenna; R_T and R_R indicate the distances from the transmitter and the GNSS-R receiver to the scattering unit, respectively. Λ and S correspond to the C/A code autocorrelation function and the Doppler frequency shift filtering SINC function, respectively, while σ_0 denotes the BRCS. The set of correlated power values computed according to the Z-V model for different delays and Doppler shifts is referred to as the DDM, which constitutes the power map.

Based on the Z-V model, relationships among reflection geometry, signal parameters, and DDM power can be established. By integrating these models, a theoretical framework is constructed that links sea-surface wind fields to GNSS-R scattering power, enabling the simulation of satellite-based GNSS-R data. These theoretical foundations, together with extensive simulation studies, provide a basis for the development of satellite-based GNSS-R platforms and offer critical support for advancing GNSS-R wind speed retrieval algorithms.

2.4 Fundamental Principles of GNSS-R Wind Speed Retrieval

GNSS direct signals experience frequency and waveform alterations upon reflection from the Earth's surface, influenced by variations in surface properties and satellite geometry. GNSS-R technology analyzes these reflected signals by integrating electromagnetic scattering theory with signal propagation models to quantify surface characteristics, thereby enabling the retrieval of surface information. In marine scattering scenarios, changes in wind speed directly modify the sea surface, altering its roughness. As wind speeds increase, sea surface roughness correspondingly rises, leading to attenuation of scattered power in the quasi-specular reflection direction of GNSS-R signals. Consequently, GNSS-R signal correlation data exhibit characteristic variations with wind speed. Therefore, by examining how wind-induced changes in sea surface roughness affect GNSS-R signals, ocean surface wind speeds can be retrieved.

The core of GNSS-R-based sea surface wind speed retrieval lies in processing the reflected signals, typically through the extraction of their power characteristics from the DDM. The DDM is a scatter power map constructed along both delay and Doppler dimensions, generated by correlating GNSS-R reflected signals with local replica codes under varying delay and Doppler shift conditions [30]. Its morphology exhibits notable variations across different wind speeds. As shown in Fig. 3, under low wind conditions, the DDM energy distribution remains relatively compact and concentrated. In contrast, at high wind speeds, increased sea surface roughness causes the scattered signal to broaden and diverge within the delay-Doppler domain. Within the DDM, the primary features used for GNSS-R wind speed retrieval are the LES and NBRCS, calculated from a 5 × 3 DDM window surrounding the Specular Point (SP), as illustrated in Fig. 4. The LES represents the slope of the Integrated Delay Waveform (IDW) at its leading edge, defined by Eq. 20:

$$LES = \frac{a_{LES}}{eff\_scatter\_area} \tag{20}$$

$$a_{LES} = \frac{N\sum_{i=1}^{3} x_i y_i - \sum_{i=1}^{3} x_i \sum_{i=1}^{3} y_i}{N\sum_{i=1}^{3} x_i^2 - \left(\sum_{i=1}^{3} x_i\right)^2} \tag{21}$$

Here, *eff_scatter_area* denotes the effective scattering area corresponding to a 5 × 3 DDM window; N represents the number of chips included in the IDW, where $N = 3$ in this case; x_i indicates the index value in the time-delay dimension, and y_i denotes the DDM power value corresponding to x_i.

The normalized bistatic radar cross section (NBRCS) coefficient is defined as the ratio of the total bistatic radar cross section (i.e., the sum of BRCS within the window) to the effective scattering area within a specified delay-Doppler window surrounding the specular reflection point. Its calculation is given by Eq. 22:

$$NBRCS = \frac{\sum_{5*3} BRCS}{eff\_scatter\_area} \tag{22}$$

where the BRCS represents the value within the DDM window. After extracting feature values from the DDM, a GMF can be established to relate these features to

Fig. 3 DDM diagram

(a) Low wind speed (b) High wind speed

Fig. 4 DDM feature value calculation window in CYGNSS products

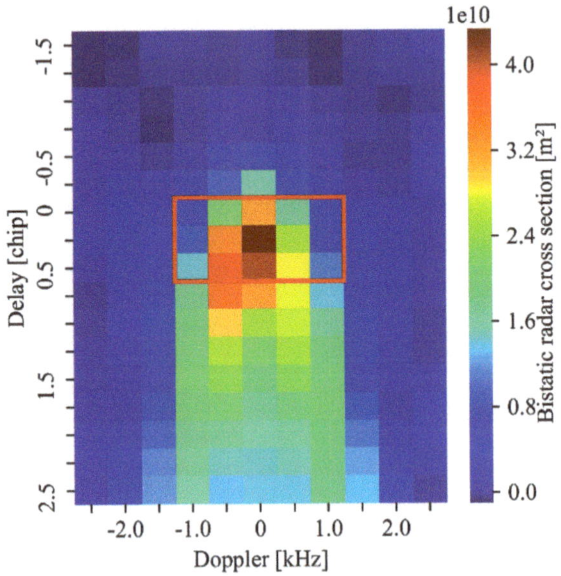

wind speed. Common approaches for constructing the GMF include the LUT and CDF methods.

3 GMF-Based Methods and Data Processing

After extracting the NBRCS and LES feature parameters from the DDM, an empirical physical model is established to relate these parameters to wind speed. This model typically employs LUTs and CDFs to perform single-parameter retrieval. For example, in the CYGNSS L1 wind speed product, these two feature parameters are used to construct discrete two-dimensional LUTs that map them to reference wind speeds. The resulting GMF is then applied to retrieve wind speed.

3.1 Look-Up Table (LUT) Approach

The LUT method is a widely used empirical approach for wind speed retrieval. Its fundamental principle involves pre-establishing a mapping between DDM characteristic parameters and wind speed, allowing for wind speed estimation through table lookup or interpolation during the retrieval process [20]. In GNSS-R sea surface wind speed retrieval, a key challenge is the highly uneven distribution of wind speed samples across different intervals, with the majority concentrated in the 2–15 m/s range. Consequently, methods that fit continuous functions using the least-squares principle to generate global optimal solutions can introduce significant systematic biases in wind speed retrieval, especially in zones with sparse samples. In contrast, the LUT method calculates representative values for samples within each interval, constructing a lookup table that maps DDM feature parameters to wind speeds. Theoretically, with a sufficiently large sample size, this approach can generate a reliable GMF, enabling more accurate wind speed retrieval.

DDM characteristics are influenced not only by wind speed but also by the incidence angle of the reflected signal, necessitating the construction of a two-dimensional GMF. To address challenges such as insufficient or missing samples at high incidence angles, the CYGNSS product implements an empirical calibration model based on signal incidence angle. This model accounts for the effects of Fresnel reflection coefficients and incidence angle on the DDM, thereby eliminating the dependence of DDM feature parameters on the incidence angle [31]. The calibration model is expressed as

$$O_{cor} = \frac{O_{ori}}{g(\theta)} \tag{23}$$

$$g(\theta) = -1.14e^{-9} * \theta^{4.61} + 1 \tag{24}$$

Specifically, O_{ori} and O_{cor} denote the DDM characteristic values before and after correction, respectively, while θ represents the incident angle at the specular reflection point.

During LUT construction, the incident angle and wind speed dimensions are divided into a two-dimensional grid with increments of 0.1° and 0.1 m/s, respectively. The matched data are then assigned to the corresponding grid cells. To ensure accuracy, a sample distribution-based LUT partitioning method is employed, utilizing a triangular weighting approach to calculate the weighted average of the DDM features within each grid cell [32]. The specific LUT partitioning procedure is illustrated as follows:

1) $\pm 0.4\ m/s\ (0 \leq u_{10} < 2\ m/s)$.
2) $\pm 0.3\ m/s\ (2 \leq u_{10} < 5\ m/s)$.
3) $\pm 0.2\ m/s\ (5 \leq u_{10} < 9\ m/s)$.
4) $\pm 0.4\ m/s\ (8 \leq u_{10} < 11\ m/s)$.
5) $\pm 0.6\ m/s\ (11 \leq u_{10} < 14\ m/s)$.
6) $\pm 0.8\ m/s\ (14 \leq u_{10} < 18\ m/s)$.
7) $\pm 1.0\ m/s\ (18 \leq u_{10} < 20\ m/s)$.

As shown, with this partitioning, adjacent cells overlap, which helps mitigate artificially introduced high-frequency errors in the GMF. LUT construction begins at 7 m/s, calculating the representative feature values on both sides in increments of 0.1 m/s. The weighting strategy is as follows: when a sample falls within a single width, it is assigned double weight; when a sample falls between one and two widths, it is assigned single weight. After computing the weighted average for each cell, the values undergo smoothing to enforce monotonicity constraints, ensuring that feature values decrease monotonically with wind speed.

After processing all cells in the computational LUT, a one-dimensional LUT based on wind speed feature values is obtained, denoted as$g(u)$. Incorporating the incidence angle dimension subsequently produces a two-dimensional LUT, as shown in Eq. 25, resulting in the final discrete GMF.

$$g(u, \theta) = g(u) \cdot g(\theta) \tag{25}$$

Here, u denotes the wind speed, θ represents the incident angle, and $g(u, \theta)$ is the two-dimensional discrete LUT after incorporating the expanded incident angle.

The method described above constructs a two-dimensional lookup table based on DDM feature values and incidence angles to retrieve sea surface wind speeds. The workflow proceeds as follows: first, the incidence angle determines its dimension within the LUT, reducing the problem to a one-dimensional context. Matching is then performed along this dimension based on DDM feature values. If a feature value does not correspond to a discrete node, linear interpolation is used between the preceding and succeeding nodes to estimate wind speed within the continuous range. This approach was primarily developed for early CYGNSS V2.1 datasets, which contained relatively sparse LUTs necessitating interpolation. For current updated datasets, denser LUT construction techniques generally allow direct two-dimensional lookups, thereby improving both retrieval accuracy and computational efficiency.

3.2 Cumulative Distribution Function (CDF) Matching

The CDF method was originally used to construct the GMF between radar cross-section coefficients measured by altimeters and wind speed, representing a statistical approach [33]. This method is now widely applied in GNSS-R wind speed retrieval. It leverages the statistical distributions of DDM feature values and reference wind speeds as prior information to construct a model function relating DDM feature values to reference wind speeds, effectively mitigating systematic biases in the model.

The CDF method assumes the existence of a monotonic function between the DDM characteristic value o and wind speed u, satisfying:

$$u = M(o) \tag{26}$$

Assuming the CDFs of the DDM feature values and the reference wind speed are F_o and F_u, respectively, the CDF definition yields:

$$F_o\left(o^{'}\right) = P\left(o \leq o'\right) = P\left[M^{-1}(u) \leq o'\right] = P\left[u \geq M\left(o'\right)\right] = 1 - F_u\left(u'\right) \tag{27}$$

Here, $o^{'}$ denotes the DDM feature value, and the wind speed satisfies $u \geq M(o^{'})$; M^{-1} is the inverse of the monotonic function M. Consequently, the wind speed $u^{'}$ obtained by inverting the feature value $o^{'}$ using the CDF method can be expressed as:

$$u' = F_u^{-1}\left(1 - F_o\left(o'\right)\right) \tag{28}$$

Here, F_u^{-1}denotes the inverse of the monotonic functionF_u. Similar to the LUT method, DDM feature values are first empirically corrected for the incidence angle. The retrieved wind speed values are then calculated using Eq. 28 based on F_o and F_u.

From the above derivation, it is evident that the primary advantage of the CDF method lies in its ability to perform wind speed retrieval directly through statistical means. This is achieved by establishing a correspondence between the cumulative distribution functions of feature values and reference wind speeds, without requiring explicit knowledge of the underlying functional relationship.

3.3 Data Processing Methods

In GNSS-R sea surface wind speed retrieval, data processing constitutes the foundational step for developing high-precision retrieval models. This process primarily involves dataset acquisition, preprocessing, and the establishment of model performance evaluation metrics, as illustrated in Fig. 5.

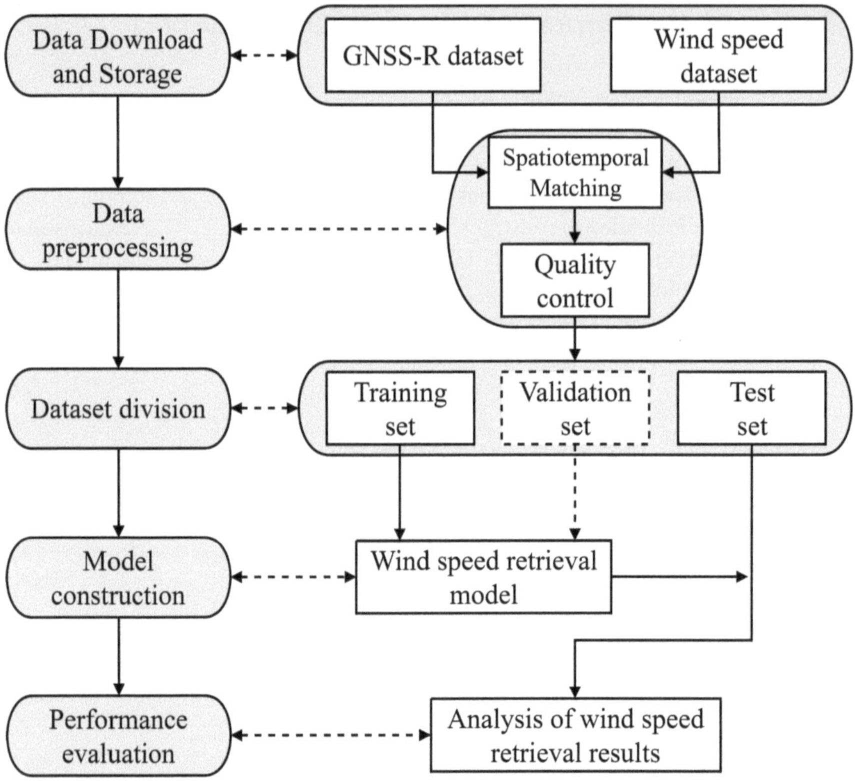

Fig. 5 GNSS-R data processing flowchart

Currently, widely used GNSS-R data sources include observation data from NASA's CYGNSS constellation. CYGNSS comprises eight low-Earth orbit satellites with an orbital altitude of approximately 520 km, an orbital inclination of about 35°, and an average revisit period of roughly 6 hours. Each satellite is equipped with a GNSS-R receiver featuring four channels, with antennas on the starboard and port sides simultaneously receiving four distinct backscatter signals. Each channel can provide 1, 2, or 4 observations per second, theoretically enabling CYGNSS to deliver millions of observation samples daily. Consequently, the extensive data provided by CYGNSS are highly suitable for large-scale retrieval of surface physical properties, particularly for sea surface wind fields. This study utilized the CYGNSS L1 V2.1 DDM product [34], covering the period from August 1, 2018, to December 31, 2019. The dataset includes BRCS DDM, effective scatterer cross section, LES, NBRCS, and other variables related to DDM and geometric relationships. Additionally, the RCG represents the antenna gain accounting for signal propagation path loss in the direction of the specular reflection point at the GNSS-R receiver. Its calculation is given by:

$$RCG = \frac{G_{SP}^{R} * 10^{27}}{\left(R_{T}^{SP} R_{R}^{SP}\right)^{2}} \tag{29}$$

Here, G_{SP}^{R} denotes the receiver antenna gain of the GNSS-R receiver in the direction of the specular reflection point, while R_{T}^{SP} and R_{R}^{SP} represent the distances from the specular reflection point (SP) to the GNSS-R receiver and the GPS transmitter, respectively. The unit of RCG is $10^{27} \times dBi \times m^{-4}$.

For reference data, the European Centre for Medium-Range Weather Forecasts (ECMWF) ERA5 product was selected for 10-m wind speed and SWH [35]. The ECMWF wind speed product is provided as grid-based data, with specific parameters detailed in Table 1. As the spatial resolutions of the two products differ, spatial bilinear interpolation and temporal linear interpolation were applied to the wind speed components u_{10} and v_{10}, respectively. The final 10-m wind speed was derived by combining these two components. Spatial matching for SWH was performed within a 25 km range, and temporal matching within a 30-min window. Samples that failed to match were assigned a value of NaN.

To ensure the quality of the dataset, additional filtering of the spatio-temporally matched data is required to obtain a high-quality dataset. The specific filtering criteria are as follows:

(1) Sample values must not be null (i.e., non-NaN);
(2) Matched wind speed values, SWH, and DDM feature values (LES and NBRCS) must all be positive.
(3) CYGNSS quality control flags (quality_flags) that are nonzero are used to exclude land points, Block IIF satellite observations, and other unsuitable data.
(4) Samples are excluded if the CYGNSS satellite attitude cannot be determined due to obstruction, which causing the star tracker to lose its satellite tracking capability.
(5) The RCG is greater than 3.

Since the GNSS-R wind speed retrieval model developed in this study is a data-driven empirical model, it is necessary to divide the dataset into training and test sets: the training set is used to construct the model, while the test set is used to evaluate its performance. To address the issues outlined above, three distinct datasets were constructed. When training neural networks, it is essential to prevent overfitting. Therefore, in addition to the training and test sets, an independent validation set is

Table 1 ECMWF ERA5 product wind speed vector and SWH parameter list

Parameters	u_{10}, v_{10}	Significant wave height
Horizontal resolution	0.25° × 0.25°	0.5° × 0.5°
Spatial resolution	1 h	1 h
Data range	Lat:(−90°, 90°) Lon:(0°, 360°)	Lat:(−90°, 90°) Lon:(0°, 360°)

established. This validation set is used to assess training accuracy, control the number of iterations and hyperparameters, and enhance the model's generalization capability.

3.4 Performance Evaluation Criteria

The performance of wind speed retrieval models is evaluated using RMSE, bias, and mean absolute percentage error (MAPE). RMSE represents the square root of the mean squared residuals between retrieved wind speeds and ECMWF reference wind speeds, comprehensively reflecting both systematic and random errors, as shown in Eq. 30:

$$RMSE = \sqrt{\frac{1}{N}\sum_{i=1}^{N}(\hat{u}_i - u_i)^2} \tag{30}$$

Here, N denotes the total number of samples in the dataset, while $\hat{u}_i$ and u_i represent the i-th retrieved wind speed value and its corresponding reference wind speed, respectively. Bias is defined as the arithmetic mean of the residuals between retrieved and reference wind speeds, systematically indicating the overall tendency of the model to overestimate or underestimate, as shown in Eq. 31:

$$Bias = \frac{1}{N}\sum_{i=1}^{N}(u_i - \hat{u}_i) \tag{31}$$

MAPE is the mean absolute percentage error between retrieved and reference wind speeds, quantifying the relative magnitude of error with respect to the true value, as shown in Eq. 32:

$$MAPE = \frac{100}{N}\sum_{i=1}^{N}\left|\frac{u_i - \hat{u}_i}{u_i}\right| \tag{32}$$

4 A PSO-Based Method for Wind Speed Retrieval

To overcome the limited accuracy of traditional empirical model–based wind speed retrieval methods and the unit-dependency constraint inherent in the MVE algorithm, this paper proposes a PSO-based composite wind speed retrieval approach. By exploiting population-based optimization, the method searches for an optimal set of composite coefficients that minimizes the RMSE of the retrieved wind speeds within a predefined coefficient range. In this section, we evaluate the performance of the proposed model using ECMWF ERA5 wind speeds as the reference. CYGNSS

L1 data from July 2019 and the corresponding ERA5 wind fields are selected for analysis. To rigorously assess the model's generalization capability, the July 2019 dataset is randomly split into a training set and a test set at a ratio of 7:3. The training set is used to construct wind speed retrieval models based on GMF (hereafter referring specifically to the LUT method in this section), MVE, and the proposed PSO method, while the test set is employed to verify retrieval accuracy. Detailed descriptions of the MVE and PSO algorithms are provided below.

4.1 MVE Method

To fully leverage the information contained in DDM eigenvalues for wind speed retrieval, the MVE method combines LES and NBRCS eigenvalue–based estimates to minimize retrieval errors. The MVE estimator minimizes the RMSE by suppressing the uncorrelated components of the prediction errors associated with individual eigenvalues [20]. A key advantage of MVE is that the RMSE of the final composite estimate is guaranteed to be no worse than the smallest RMSE of any single eigenvalue. Moreover, the lower the correlation between the estimation errors of different eigenvalues, the greater the improvement in RMSE achieved by the MVE estimator. Therefore, based on the wind speed retrieval results derived separately from each DDM eigenvalue, the MVE estimator determines the optimal combination coefficients using a least-squares criterion, thereby further enhancing the retrieval accuracy. The MVE estimator is formulated as a linear combination of the individual eigenvalue–based estimates:

$$\hat{u}_{MVE} = \overrightarrow{m} \cdot \hat{u} \tag{33}$$

Here, $\hat{u} = (u_{les}, u_{nbrcs})^T$ represents the wind speed retrieval results obtained from the two DDM features, and $\overrightarrow{m} = (m_{les}, m_{nbrcs})$ denotes the corresponding combination coefficient. To ensure that the composite estimate remains unbiased, the coefficient vector must satisfy the constraint $\overrightarrow{m} \cdot \overrightarrow{1} = 1$. Assuming that C is the error covariance matrix of the wind speed retrieval results derived from individual features, the elements of C are defined as follows:

$$c_{ij} = \left\langle (u_i - u_{true})\left(u_j - u_{true}\right)\right\rangle \tag{34}$$

Here, c_{ij} represents the element in the i-th row and j-th column of C; u_i and u_{true} denote the wind speed retrieved from the i-th DDM feature and the corresponding true wind speed, respectively. Since this study employs two DDM-derived features—LES and NBRCS—the indices are restricted to $0 \leq$ i, j < 2. The symbol $\langle\rangle$ denotes the average over all samples in the dataset. The variance of the MVE combined wind speed retrieval is expressed as:

$$\sigma_{MVE}^2 = \left\langle (\hat{u}_{MVE} - u_{true})^2 \right\rangle \tag{35}$$

Substituting Eqs. 33 and 34 into Eq. 35 yields:

$$\sigma_{MVE}^2 = \overrightarrow{m} \cdot C \cdot \overrightarrow{m}' \tag{36}$$

The objective of linearly combining the two retrieved wind speed estimates is to minimize σ_{MVE}^2. To achieve this, the Lagrange multiplier method is applied together with the unbiasedness constraint $\overrightarrow{m} \cdot \overrightarrow{1} = 1$, from which we obtain:

$$\nabla\left(\sigma_{MVE}^2 + \lambda \overrightarrow{m} \cdot \overrightarrow{1}\right) = 2C \cdot \overrightarrow{m}' + \lambda \cdot \overrightarrow{1} = 0 \tag{37}$$

Among these, λ is the Lagrange multiplier. Solving this equation yields:

$$\overrightarrow{m} = -\frac{\lambda}{2} C^{-1} \cdot \overrightarrow{1} \tag{38}$$

Substituting $\overrightarrow{m} \cdot \overrightarrow{1} = 1$ into Eq. 38 yields:

$$\lambda = -2\left(\sum\nolimits_{i=1}^{N} \sum\nolimits_{j=1}^{N} c_{ij}^{-1}\right)^{-1} \tag{39}$$

Here, c_{ij}^{-1} denotes the element in the *i*-th row and *j*-th column of the covariance matrix C^{-1}. Substituting Eq. 39 into Eq. 38 gives the combination coefficient $\overrightarrow{m}$ that minimizes σ_{MVE}^2:

$$\overrightarrow{m} = \left(\sum\nolimits_{i=1}^{N} \sum\nolimits_{j=1}^{N} c_{ij}^{-1}\right)^{-1} \cdot C^{-1} \cdot \overrightarrow{1} \tag{40}$$

Using the aforementioned method, the combination coefficients required to construct the MVE estimator can be determined solely from the covariance matrix C. This approach exploits the orthogonality of the estimation errors associated with individual DDM eigenvalues to reduce the overall RMSE of the combined wind speed retrieval. The wind speed retrieval results obtained from the LUT and CDF methods based on LES and NBRCS can be linearly combined through the MVE framework, thereby mitigating the uncorrelated components of the retrieval errors between the two eigenvalues. However, because the two eigenvalues exhibit inherently strong correlation and the MVE solution imposes an unbiasedness constraint, the degree of accuracy improvement achieved through the combination is limited.

4.2 PSO Method

In 1995, inspired by the collective hunting behavior observed in bird flocks, Kennedy and Eberhart proposed the PSO algorithm [36]. PSO is an evolutionary optimization method that searches for optimal solutions in complex spaces through cooperation and competition among individuals within a swarm. It is well-known for its powerful global search capability and its insensitivity to the specific form and differentiability of the objective function.

The algorithm begins by initializing a population of particles with random positions and velocities. During each iteration, a particle's velocity and position are updated through a combination of stochastic components and deterministic guidance: each particle adjusts its trajectory based on its personal best historical position (*pbest*) and the global best position (*gbest*) found by the swarm, while an inertia weight governs the balance between exploration and exploitation. As iterations proceed, the swarm gradually converges toward the global optimum of the fitness function until a predefined termination criterion—such as a maximum number of iterations or a target accuracy—is satisfied. Through this iterative process, PSO exhibits notable efficiency and effectiveness in solving complex, nonlinear optimization problems.

For the single wind speed retrieval problem based on combining LES and NBRCS, it is necessary to determine a set of coefficients—analogous to those obtained via MVE—that linearly combine the feature-specific retrieved wind speeds to yield an optimal wind speed estimate. The ultimate objective of the PSO algorithm is to identify a coefficient vector $\vec{m}$ that minimizes a predefined objective function. In the context of wind speed retrieval through feature combination, the objective function employed in this study is formulated as follows:

$$F(\vec{m}) = \left\langle \left(\vec{m} \cdot \hat{u} - u_{true}\right)^2 \right\rangle \tag{41}$$

Here, $\vec{m}$ represents the coefficient to be optimized; $(m_{les}, m_{nbrcs}) \in (0, 1)$. The objective function quantifies the MSE between the wind speed retrieved from each feature and the corresponding true wind speed. The PSO algorithm explores the search space to identify the set of coefficients that minimizes $F(\vec{m})$. In other words, the algorithm determines the optimal linear combination of individual wind speed retrieval derived from LES and NBRCS. Consequently, the combination coefficients obtained via PSO correspond to the optimal solution within the defined search space. Theoretically, this approach can achieve a lower overall MSE compared to the MVE method.

In the PSO algorithm implemented in this study, the combination coefficient $\vec{m}$ for each particle at the $(i + 1)$-th iteration is updated as follows:

$$\vec{m}_{i+1} = \vec{m}_i + \Delta\vec{m}_{i+1} \tag{42}$$

Here, $\vec{m}_i$denotes the current combination coefficient; $\Delta\vec{m}_{i+1}$represents the update increment, and $\vec{m}_{i+1}$is the updated combination coefficient, all expressed as two-dimensional vectors. The increment $\Delta\vec{m}_{i+1}$ is defined as:

$$\Delta\vec{m}_{i+1} = \omega * \Delta\vec{m}_i + c_1 * \mathrm{rand}_i^1\left(pbest_i - \vec{m}_i\right) + c_2 * \mathrm{rand}_i^2\left(gbest_i - \vec{m}_i\right) \quad (43)$$

Where ω is the inertia factor that controls the contribution of the previous increment, $pbest_i$ and$gbest_i$ represent the local and global optimal coefficients of each particle at iteration i, respectively; c_1 and c_2 are learning factors controlling the influence of $pbest_i$ and $gbest_i$; rand_i^1 and rand_i^2 are independent random numbers uniformly distributed in [0,1]. The PSO-based optimization problem can then be formulated as:

$$\begin{gathered} \min F\left(\vec{m}\right) \\ s.t. \quad 0 \le m_{les} \le 1, \quad 0 \le m_{nbrcs} \le 1 \end{gathered} \quad (44)$$

The PSO-retrieved wind speed $\vec{u}_{PSO}$, is obtained by linearly combining the individual feature-derived wind speeds in vector $\hat{u}$ with the PSO-optimized coefficients $\vec{m}$:

$$\vec{u}_{PSO} = \vec{m} \cdot \hat{u} \quad (45)$$

The main process of the PSO algorithm is illustrated in Fig. 6 and consists of the following steps:

(1) Particle Initialization: Generate a population of particles, with their positions (combination coefficients) and velocity increments randomly initialized within the defined search space.
(2) Fitness Evaluation: Compute the fitness value$F\left(\vec{m}\right)$of each particle based on the objective function defined in Eq. 41
(3) Update Personal and Global Bests: Determine the personal best ($pbest_i$) for each particle and the global best ($gbest_i$) for the entire population at the current iteration i
(4) Coefficient and Increment Update: Update the combination coefficients $\vec{m}$ and velocity increments $\Delta\vec{m}$ for each particle according to Eqs. 42 and 43
(5) Iteration and Termination: Repeat steps 2–4 until the termination criteria are satisfied (e.g., maximum iterations or convergence threshold). The final global optimum *gbest* represents the optimal combination coefficient obtained via the PSO algorithm.

Based on the aforementioned PSO algorithm, during the training phase, a set of combination coefficients is obtained from the training set to merge the wind speeds retrieved from LES and NBRCS. Although the coefficients *gbest* may fluctuate throughout the iterative optimization process, they ultimately converge to the optimal position within the defined search space. These optimized coefficients are then directly applied to the test set to combine the LES- and NBRCS-retrieved wind speeds.

The PSO algorithm's advantage stems from its use of a large population of particles to efficiently identify optimal combination coefficients. These coefficients are

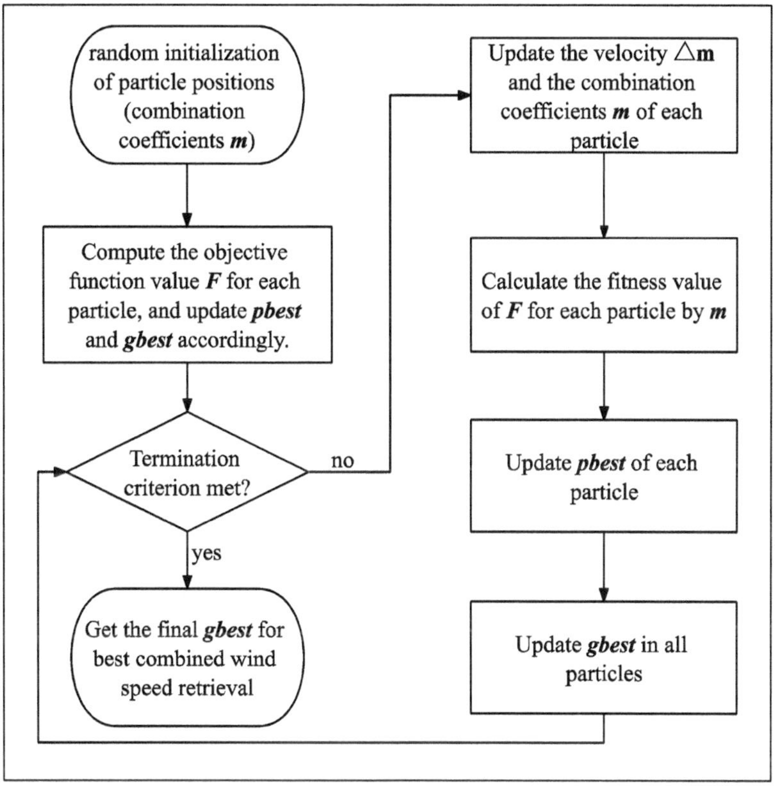

Fig. 6 A combined wind speed retrieval algorithm based on particle swarm optimization

determined by minimizing the MSE between the combined wind speed retrieval and the actual wind speed.

The specific implementation of the PSO algorithm for wind speed retrieval follows the workflow shown in Fig. 7, which includes spatiotemporal matching of datasets, data filtering, dataset generation, model development, and performance evaluation. After generating the LES and GMF of NBRCS from the DDM observations, the MVE and PSO methods are applied for single wind speed retrieval, and their performance is compared. In this study, the PSO algorithm utilized 200 particles with a maximum iteration limit of 50. The computational cost of PSO is proportional to the number of training samples, the number of particles, and the iteration limit. Additionally, the inertia weight factor ω was set to 1.0, while the cognitive and social learning factors, $c1$ and $c2$, were both set to 2.0.

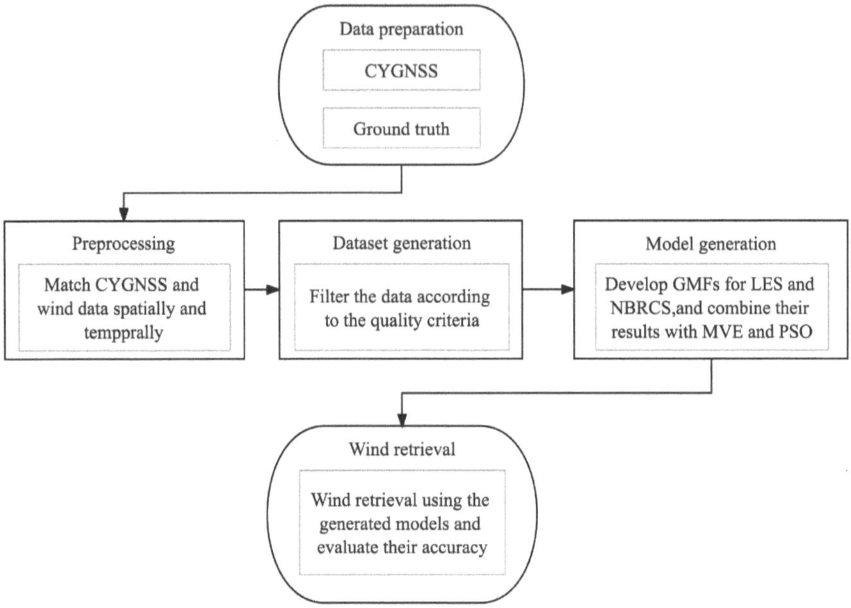

Fig. 7 Wind speed retrieval algorithm flowchart

4.3 Retrieval Results and Performance Analysis

Based on the aforementioned methodology, we compared the PSO algorithm with GMF models constructed using two individual DDM features and the MVE algorithm, as summarized in Table 2. Initially, GMF models were developed using the LES and NBRCS feature values separately, and their respective wind speed retrieval performance was evaluated. Subsequently, both the MVE and PSO algorithms were employed to combine the wind speed retrieval results from LES and NBRCS. Finally, the performance of all models was analyzed to assess the effectiveness of the PSO algorithm. For each experiment, the ECMWF wind speed product served as the reference dataset.

Following the steps in Fig. 7, a two-dimensional GMF was developed using LES and NBRCS after completing data preprocessing and dataset generation. Figure 8

Table 2 Performance evaluation algorithm information

Models	Input parameters
GMF	LES
GMF	NBRCS
MVE	Wind speeds retrieved using the LES and NBRCS GMFs
PSO	Wind speeds retrieved using the LES and NBRCS GMFs

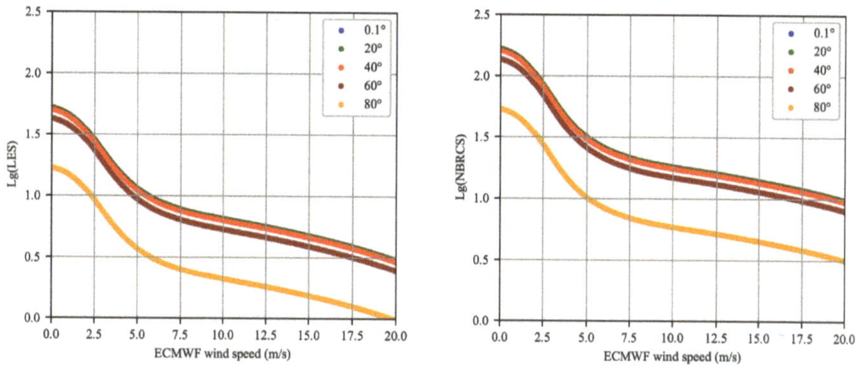

Fig. 8 GMF of LES and NBRCS at different incidence angles

presents the GMFs of LES and NBRCS at different incidence angles. It can be observed that the DDM eigenvalues exhibit a negative correlation with wind speed, with measured values gradually decreasing as wind speed increases. This trend arises because wind speed influences sea surface roughness; higher wind speeds generate rougher sea surfaces, which in turn reduce the power of the reflected GNSS signal. According to Eqs. 21 and 22, the DDM observed values consequently decline.

Subsequently, combined wind speed retrieval was performed using the MVE and PSO algorithms. The optimized combination coefficients obtained from both methods are listed in Table 3. It can be observed that the m_{nbrcs} values from both methods are significantly larger than m_{les}. This is because the wind speed retrieval accuracy of NBRCS is higher than that of LES. Moreover, the sum of the combination coefficients derived from the PSO algorithm is less than 1, effectively moderating the overall scale of the retrieved wind speed. The underlying reasons for this phenomenon will be analyzed in detail later.

To further evaluate the performance of the PSO algorithm, we computed the RMSE, bias, and MAPE for the four methods on the test set, with results summarized in Table 4. It can be observed that wind speed retrieval based on LES exhibits the poorest performance. In contrast, NBRCS demonstrates a marked improvement over LES, consistent with the higher weighting of NBRCS in the combination coefficients discussed previously. The MVE method, which linearly combines the retrieval results from LES and NBRCS, achieves a modest improvement in performance relative to the single-feature approaches. This limited enhancement arises because MVE only eliminates the uncorrelated component of errors between the two single-feature

Table 3 MVE and PSO combination coefficient

Models	Combination coefficient	
	m_{les}	m_{nbrcs}
MVE	0.105	0.895
PSO	0.139	0.732

Table 4 Performance of four wind speed retrieval algorithms

Algorithms	Performance evaluation		
	RMSE(m/s)	Bias(m/s)	MAPE(%)
GMF(les)	3.08	−0.853	37.4
GMF(NBRCS)	2.23	−0.557	29.2
MVE	2.21	−0.589	29.2
PSO	1.95	0.349	26.3

retrievals. Among the four approaches, the PSO algorithm exhibits the most significant performance gains. Compared to MVE, PSO reduces the RMSE from 2.21 m/s to 1.95 m/s, corresponding to a relative decrease of 12%. Concurrently, the systematic bias decreases from −0.589 m/s to 0.349 m/s, indicating a substantial reduction in retrieval bias.

Additionally, we conducted a comparative analysis of wind speed retrieval results obtained from the MVE and PSO algorithms against ECMWF reference data. As shown in the density plots in Fig. 9, both algorithms tend to underestimate wind speeds below 2.5 m/s, reflecting the limited sensitivity of DDM eigenvalues at low wind speeds. Moreover, the MVE algorithm exhibits overestimation within the 7–12 m/s range, resulting in a markedly asymmetric density distribution. This occurs because, although MVE is theoretically an unbiased estimator, it presupposes that the input single-feature retrievals are unbiased. However, in practical scenarios, influenced by sea states and sample distribution, the single-feature retrievals exhibit systematic overestimation within this range. Constrained by the normalization requirement, MVE fails to eliminate and instead propagates this systematic bias. In contrast, the PSO algorithm demonstrates superior overall symmetry, particularly in low- to medium-wind speed scenarios, where its retrievals more closely align with reference values.

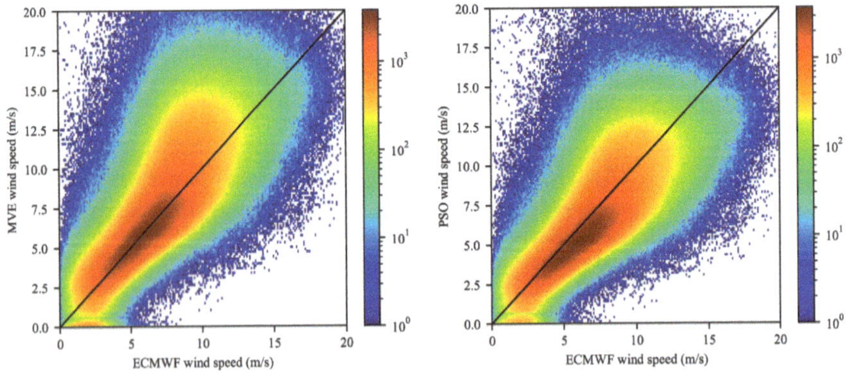

(a) MVE (b) PSO

Fig. 9 Density plots of MVE and PSO wind speed versus reference wind speed

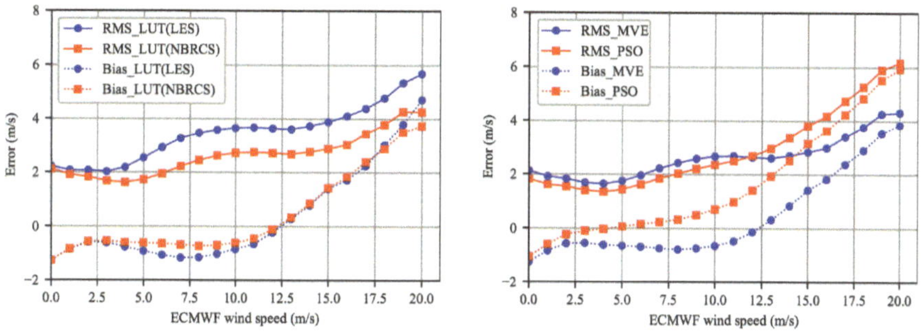

Fig. 10 MVE and PSO error analysis

Figure 10 presents the RMSE and bias of the four algorithms across different wind speed intervals. The GMF based on NBRCS consistently outperforms that based on LES, especially under high-wind conditions, owing to the lower sensitivity of LES at higher wind speeds. Further analysis of MVE and PSO errors across different wind speed ranges indicates that PSO substantially outperforms MVE at low-to-moderate wind speeds. However, for wind speeds exceeding 12 m/s, PSO errors increase sharply and eventually surpass those of MVE. The deviation distribution curve shows that PSO-optimized deviations are closer to zero in the 2.5–10 m/s range. This behavior arises because the PSO optimization adaptively adjusts the combination coefficients, maintaining their sum below 1 to achieve optimal overall performance. Mathematically, this is equivalent to applying a global scaling to the LES and NBRCS retrievals, thereby correcting the overall systematic bias while simultaneously optimizing the weight allocation of individual features. In fact, at low wind speeds, sea surface roughness is influenced not only by wind speed but also by complex sea states such as swell, causing GMFs based on single features to generally overestimate wind speeds [37]. By employing optimal combination coefficients with a sum less than 1, the PSO algorithm effectively corrects this swell-induced overestimation, thereby minimizing the RMSE of the final retrieval results.

Compared to the combined MVE method, the PSO algorithm demonstrates substantially lower overall bias and higher consistency with ECMWF reference wind speeds. Table 5 presents the RMSE and systematic bias of the two combined retrieval models within the 0–12 m/s and 12–20 m/s ranges. It is observed that for wind speed scenarios exceeding 12 m/s, the RMSE of the PSO algorithm is higher than that of the MVE algorithm. Nevertheless, since the majority of data samples in practical scenarios are concentrated within the wind speed range below 12 m/s, this range primarily governs the improvement in wind speed retrieval performance. Consequently, the comprehensive retrieval accuracy obtained by the PSO algorithm is significantly enhanced, although systematic bias resulting from uneven sample distribution persists. To address this issue, an adaptive CDF matching method for systematic bias elimination in GNSS-R sea surface wind speed retrieval is designed in the subsequent section.

Table 5 Error analysis of MVE and PSO at different wind speed ranges

Algorithms	0–12 m/s		12–20 m/s	
	RMSE(m/s)	Bias(m/s)	RMSE(m/s)	Bias(m/s)
MVE	2.18	−0.649	2.74	0.651
PSO	1.86	0.254	3.30	2.305

5 A CNN-Based Method for Wind Speed Retrieval

To address the limitations of existing GMF methods, which consider only eigenvalues and incidence angles as inputs, and to mitigate overfitting issues in current neural network approaches caused by uneven sample distributions, we developed an end-to-end, multi-information fusion wind speed retrieval algorithm based on statistical correction using CNNs [21]. This approach is grounded in the principles of GNSS-R wind measurement. The model architecture consists of two components: a CNN and a CDF-based bias removal module. For convenience, the proposed statistical correction CNN is referred to as MCNN. This section details the implementation of the method for wind speed retrieval, including effective feature extraction, auxiliary information fusion, and model bias removal optimization.

Given the massive volume of satellite-based GNSS-R data and the extensive temporal coverage of the dataset used in this study, after initial quality filtering of the CYGNSS data, the dataset was downsampled daily to reduce computational overhead. Specifically, a random integer n between 1 and 50,000 was generated for each day, and n samples were randomly selected and matched with ERA5 wind speed and SWH products. Any unmatched samples were discarded. To further reduce the training data scale, we conducted an extra random subsampling only on the training set, ultimately drawing more than 3,000,000 training samples. We also implemented data augmentation by duplicating training samples with wind speeds exceeding 15 m/s to strengthen the model's performance in high-wind scenarios. The final dataset comprised 5,049,630 samples, covering wind speeds from 0 to 25 m/s. This dataset was divided into training, validation, and test sets based on temporal intervals, as illustrated in Fig. 11.

To investigate the impact of ocean state information and CDF-based correction on retrieval performance, two intermediate models were developed: CNN and CNN2. The CNN model excludes SWH as auxiliary information and does not incorporate CDF matching correction, whereas CNN2 includes SWH but similarly omits CDF matching correction. For comparison with existing methods, the models discussed in this chapter are grouped into two categories, S1 and S2, as illustrated in Fig. 12. S1 comprises four models: LUT, CDF, FCN, and CNN. Both LUT and CDF are combined with the MVE approach. The construction of FCN and CNN requires both training and validation sets. S2 includes CNN2 and MCNN, which also necessitate training and validation sets for model development. The following sections provide detailed descriptions of the construction and experimental procedures of the FCN, CNN, and MCNN models.

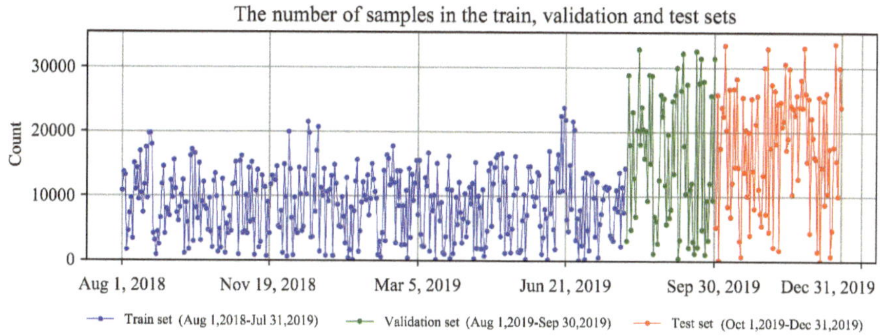

Fig. 11 Training set, Validation set, Test set time distribution

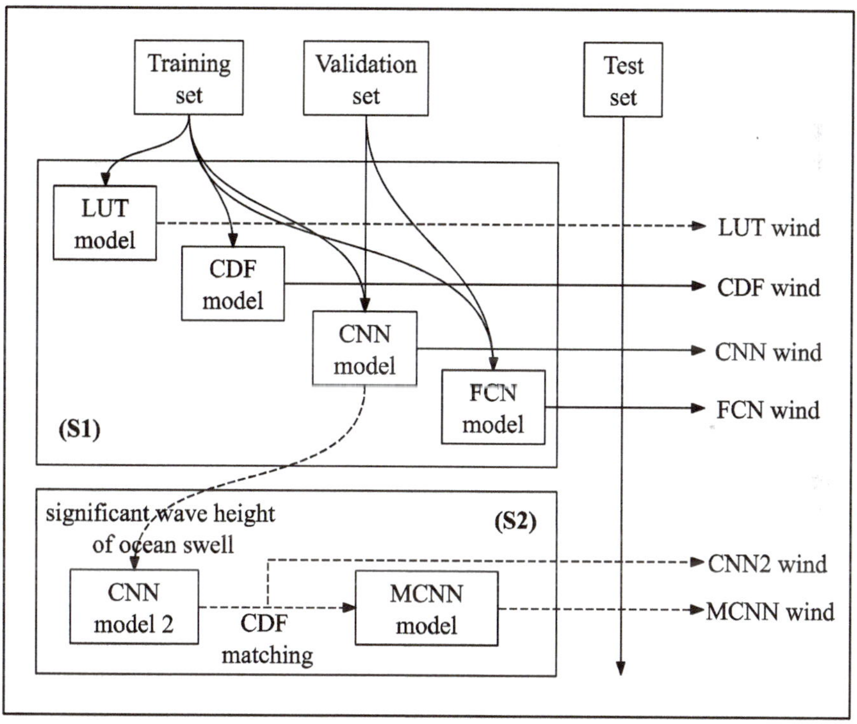

Fig. 12 Overview of CNN data flow and experimental design

5.1 FCN Model

The Backpropagation (BP) neural network, introduced in 1986, is a multilayer feed-forward neural network that utilizes the error gradient backpropagation algorithm for training the model. It is one of the most widely used neural network architectures, capable of addressing a range of problems, including regression, classification, pattern recognition, and data mining [38]. The BP neural network comprises three main components: an input layer, one or more hidden layers, and an output layer. The input layer accepts one or more feature inputs. In this study, wind speed retrieval is treated as a regression problem; therefore, the BP neural network is used to fit a mapping function $f(x)$ such that:

$$\min\left\langle \left[f(x) - u_{true}\right]^2 \right\rangle \tag{46}$$

After passing through multiple hidden layers, the BP neural network outputs the retrieved wind speed $\hat{u}$.

The primary operations of a BP neural network include forward propagation and error backpropagation. Forward propagation refers to the process by which input data received by the input layer is transmitted through the network. At each neuron, the input undergoes weighted linear transformations followed by a nonlinear mapping via the neuron's activation function. The resulting values are then propagated to the subsequent layer. After processing through one or more hidden layers, the network ultimately produces its output. Backpropagation is the training procedure that calculates the error between the network's predicted output and the actual output. Using the backpropagation algorithm, the network iteratively adjusts the weights and biases of each neuron to minimize the error, thereby improving prediction accuracy [37]. The training process terminates when either the network output error falls below a predefined threshold or the number of training iterations reaches a preset maximum, accomplished through repeated cycles of forward propagation and backpropagation.

During the forward propagation phase, the network begins training with randomly initialized weights and biases. For a single neuron, the computation consists of two steps: a linear transformation followed by a non-linear activation. In the linear step, the output is computed as the weighted sum of the inputs plus the bias:

$$net = w \cdot x^T + b \tag{47}$$

The resulting value is then passed through an activation function f to introduce nonlinearity:

$$out = f(net) = f\left(w \cdot x^T + b\right) \tag{48}$$

Common activation functions include the sigmoid function, softmax function, rectified linear unit (ReLU), and hyperbolic tangent function (tanh). In this study, ReLU and tanh are employed, defined as follows:

$$f_{\mathrm{Re}\,LU}(x) = \max(0, x) = \begin{cases} 0, & x < 0 \\ x, & x > 0 \end{cases} \tag{49}$$

$$f_{\tanh}(x) = \frac{e^x - e^{-x}}{e^x + e^{-x}} \tag{50}$$

Within the neural network, the output of each preceding layer serves as the input to the subsequent layer, continuing until the final output layer generates the network prediction.

During the backpropagation phase, the error between the network's predicted output and the actual value is calculated for each training iteration. The backpropagation algorithm then updates the weights and biases of all network nodes to minimize this error. The error is typically quantified using a cost function. In the GNSS-R wind speed retrieval context, the cost function is defined as the MSE of the retrieval results. To mitigate model overfitting, an L2 regularization term is generally incorporated. Consequently, the final cost function is expressed as:

$$J = \frac{1}{N}\sum\nolimits_{i=1}^{N} \left[f(x_i) - u_i\right]^2 + l2 = \frac{1}{N}\sum\nolimits_{i=1}^{N} [\hat{u}_i - u_i]^2 + \lambda \sum \left(w_j\right)^2 \tag{51}$$

Here, $\hat{u}_i$ represents the retrieved wind speed, u_i denotes the corresponding reference wind speed, λ is the regularization decay coefficient, and w_j refers to the network weights and bias coefficients. During network training, optimizers such as SGD, MBGD, Adam, or RMSProp are commonly employed to perform gradient backpropagation and update network parameters. In this study, the RMSProp optimizer is adopted for error backpropagation and weight updates, ensuring rapid network convergence while effectively mitigating the risk of being trapped in local minima.

Based on the aforementioned principles of BP neural network, we designed the model illustrated in Fig. 13, referred to herein as the FCN model [23]. Using LES, NBRCS, and auxiliary features—including the latitude and longitude of the specular reflection point, incidence angle, RCG, and BRCS uncertainty—to establish a mapping between these inputs and reference wind speeds for accurate retrieval.

5.2 CNN Model

CNNs are deep learning models specifically designed for processing data with grid-like structures. As a variant of the Multi-Layer Perceptron (MLP), CNNs were originally inspired by early studies of the visual cortex in cats and are widely employed for tasks such as image classification, semantic segmentation, and related applications [39]. In recent years, CNNs have also been extensively applied in remote sensing, including image synthesis, image matching, scene classification, and target recognition. However, their application to regression problems involving the retrieval of surface physical properties remains limited. Traditional scatterometers and GNSS-R receivers generate remote sensing data characterized by low pixel resolution and high noise levels, constraining the depth and performance of CNNs. In contrast, GNSS-R technology can acquire tens of millions of measurements per day, providing millions

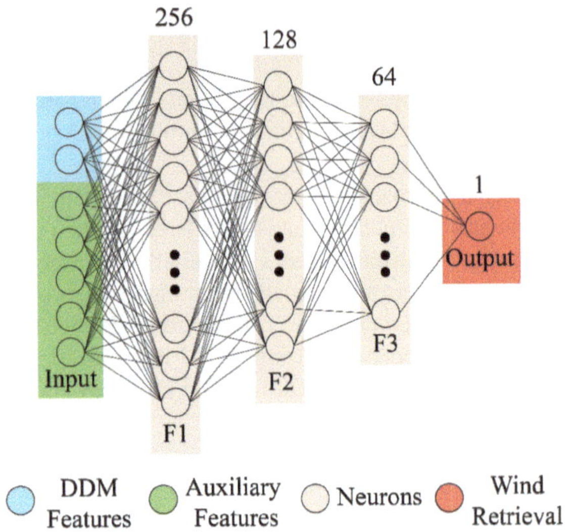

Fig. 13 Schematic of the FCN architecture

of observational samples that are largely unaffected by cloud cover or extreme convective weather, such as tropical cyclones and hurricanes [3]. Therefore, relative to most other contemporary remote sensing instruments—including satellite scatterometers, radiometers, and altimeters—the ultra-large-scale, all-weather datasets produced by GNSS-R technology can effectively support data-driven deep learning applications, despite challenges associated with single-point data quality.

The development of wind speed retrieval algorithms can be conceptualized as the extraction of effective features from DDMs, the integration of these features with auxiliary information, and the formulation of a regression problem that maps the extracted features to actual ground-based wind speeds. The resulting model can range from simple polynomial regressions to complex multi-layer neural networks. Typically, a CNN comprises an input layer, convolutional layers, pooling layers, fully connected layers, and an output layer [40], as illustrated in Fig. 14. The convolutional layers perform convolution operations on the model input to extract features, generating feature maps while simultaneously suppressing noise. The subsequent pooling layers downsample these feature maps to retain essential characteristics. Fully connected layers perform matrix multiplication between weights and features, incorporate bias, and apply nonlinear activation functions to aggregate information. In CNNs, each neuron receives information from only a localized region, known as its receptive field. This local information is progressively integrated at higher layers to capture global features. Furthermore, the weight-sharing mechanism in convolutional layers mimics operations in biological neural networks while significantly reducing the number of trainable parameters and lowering computational complexity.

The three-dimensional structure of a convolutional layer is illustrated in Fig. 15 [41], and primarily consists of the input feature map group X, the convolutional kernel W, and the output feature map group Y. Here, $X \in \mathbb{R}^{M \times N \times D}$ is a three-dimensional

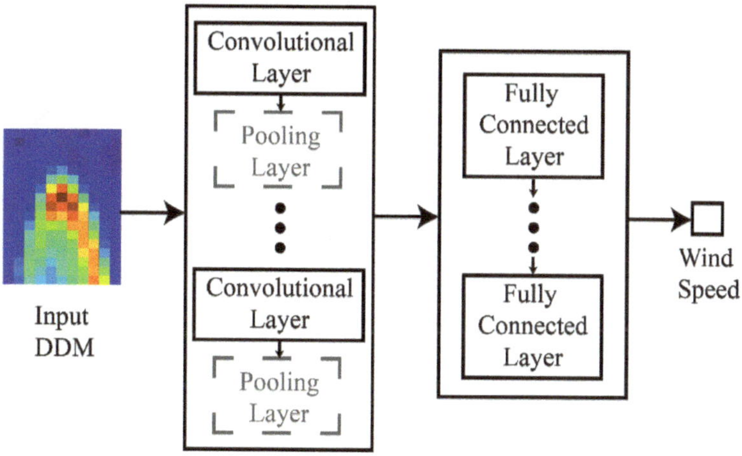

Fig. 14 Example of CNN in GNSS-R wind speed retrieval scenario

tensor, where each slice matrix $X^d \in \mathbb{R}^{M \times N}$ represents an input, $1 \leq d \leq D$; $W \in \mathbb{R}^{m \times n \times D \times P}$ is a four-dimensional tensor, where each slice matrix $W^{p,d} \in \mathbb{R}^{m \times n}$ corresponds to a two-dimensional convolutional kernel, $1 \leq d \leq D$, $1 \leq p \leq P$; $Y \in \mathbb{R}^{M' \times N' \times P}$ is a three-dimensional tensor, where each slice matrix $Y^d \in \mathbb{R}^{M' \times N'}$, $1 \leq p \leq P$. The feature maps of the input layer are the images themselves. Thus, for grayscale images, $D = 1$, and for color images, $D = 3$. In this paper, the CNN input combines DDM and the effective scattering area, so $D = 2$. The feature maps Y^d are computed as:

$$Y^d = \sigma\left(W^P \otimes X + b^p\right) = \sigma\left(\sum_{d=1}^{D} W^{p,d} \otimes X^d + b^p\right) \tag{52}$$

Here, σ denotes a nonlinear activation function, which introduces nonlinearity into each neuron. Common examples include the ReLU and the tanh function. The ReLU function offers advantages such as rapid convergence, low computational complexity, and mitigation of gradient saturation and vanishing gradient issues in the non-negative domain. In contrast, the tanh function provides stronger nonlinearity but is more susceptible to gradient saturation and vanishing gradient problems. For each input feature map $X = (X^1, X^2, \cdots, X^D)$ in a convolutional layer, repeating the above computation P times produces the output feature map set $Y = (Y^1, Y^2, \cdots, Y^D)$. This completes the computational process for a convolutional layer.

Pooling layers achieve feature map subsampling by applying a statistical operation (e.g., maximum or averaging) over local regions. This spatial compression reduces the dimensionality of the data flow, leading to a reduction in learnable parameters and computational complexity. Consequently, these operations not only distill dominant features and suppress noise but also introduce a form of translation invariance to the model.

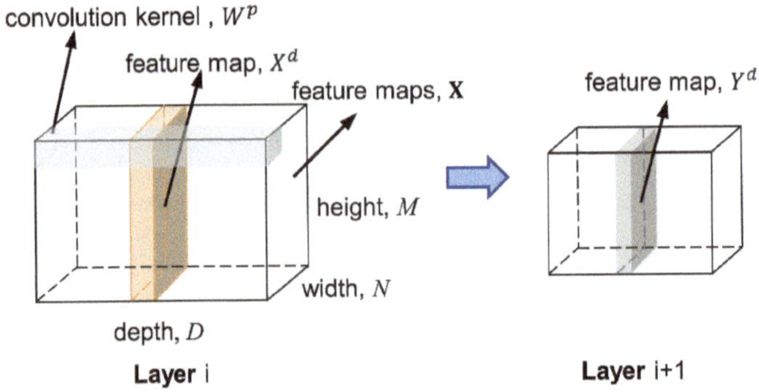

Fig. 15 Convolution layer structure

Following the extraction of DDM features through convolutional and pooling layers, a feature fusion layer can be incorporated to combine these extracted features with additional auxiliary information. The resulting feature vector is subsequently passed to fully connected layers, which perform weighted operations and map the fused feature vector to the wind speed space.

At this stage, the CNN establishes an end-to-end wind speed retrieval model that integrates feature extraction, multi-source information fusion, and feature-to-wind-speed mapping. The primary advantage of applying CNNs to GNSS-R wind speed retrieval lies in their ability to adaptively extract features relevant to sea surface wind speed from DDMs. When combined with auxiliary information, the model effectively captures the mapping relationship between the feature vectors and wind speed. Consequently, the structural characteristics of CNNs align closely with the fundamental principles of GNSS-R wind speed retrieval, theoretically improving the accuracy of wind speed retrieval.

5.3 *MCNN Model*

Neural networks are inherently data-driven models, and their performance is highly dependent on the distribution of training samples. In practice, the natural distribution of wind speeds is highly uneven, which makes it difficult for neural networks alone to eliminate overfitting errors arising from this imbalance. To address this issue, after performing wind speed retrieval using a CNN, this study further corrects systematic biases through an adaptive polynomial based on CDF matching, thereby establishing the MCNN model [24]. The MCNN architecture primarily consists of two components: the CNN and the CDF bias correction module, which are described in detail below.

CNN Module

Building on the fundamental principles of CNNs, this study proposes a model design for GNSS-R wind speed retrieval. As illustrated in Fig. 16, the CNN input comprises the BRCS DDM and the corresponding effective scattering area, with dimensions N × 17 × 11 × 2, denoted as X, where N represents the batch size. To mitigate the impact of differing scales between DDM and the effective scattering area, normalization is applied. Given that the DDM product provided by CYGNSS is only 17 × 11 in size, the inclusion of pooling layers in conventional CNNs limits the depth that can be achieved by convolutional layers, resulting in potential loss of valuable DDM information. Moreover, wind speed retrieval constitutes a regression task rather than a typical image classification problem, eliminating the need for pooling layers designed to extract spatially invariant features. Consequently, pooling layers are omitted from the CNN architecture. To reduce the influence of edge noise in the DDM, zero-padding was not applied during convolution operations. Convolutional layers C1–C3 employ the Rectified Linear Unit (ReLU) activation function and batch normalization [42]. Each batch of samples is normalized to a standard Gaussian distribution, effectively reducing internal covariate shift while accelerating model convergence. Following C3, feature maps are reduced to yield 28 feature values, denoted as d. To comprehensively integrate auxiliary information—including signal reflection point coordinates, RCG, BRCS DDM uncertainty, signal incidence angle, and SWH—a feature fusion layer O4 is added prior to the fully connected layers. The resulting fused feature vector is then passed to fully connected layers F5–F7 for model fitting. To enhance nonlinearity, F5 utilizes a tan$h$ activation function, whereas F6 and F7 employ ReLU activation. This design ensures that the model outputs non-negative wind speeds, consistent with physical expectations, while mitigating the vanishing gradient problem.

During CNN training, mini-batch gradient descent is employed due to the large dataset size. In each iteration, a batch of samples from the training set is selected as input, with the batch size N set to 100. The network updates its weights through forward propagation and backpropagation. After all batches in the training set have been processed, one epoch of training is completed. To ensure sufficient model performance, the total number of epochs is set to 100. The loss function is defined as the MSE between the retrieval and reference wind speeds, incorporating L2 regularization, as expressed in Eq. 40.

$$J = \frac{1}{N}\sum\nolimits_{i=1}^{N}(\hat{u}_i - u_i)^2 + l2 = \frac{1}{N}\sum\nolimits_{i=1}^{N}(\hat{u}_i - u_i)^2 + \lambda\sum\left(w_j\right)^2 \tag{53}$$

Here, N denotes the batch size; $l2$ represents L2 regularization to prevent overfitting; λ is the weight decay coefficient, set to 0.0015; and w_j represents the weights within the network. To further mitigate overfitting, early stopping [43] was implemented with a tolerance threshold of 10.

The network's initial parameters are randomly generated from a normal distribution with a mean of 0 and a standard deviation of 0.01. The RMSProp optimizer is

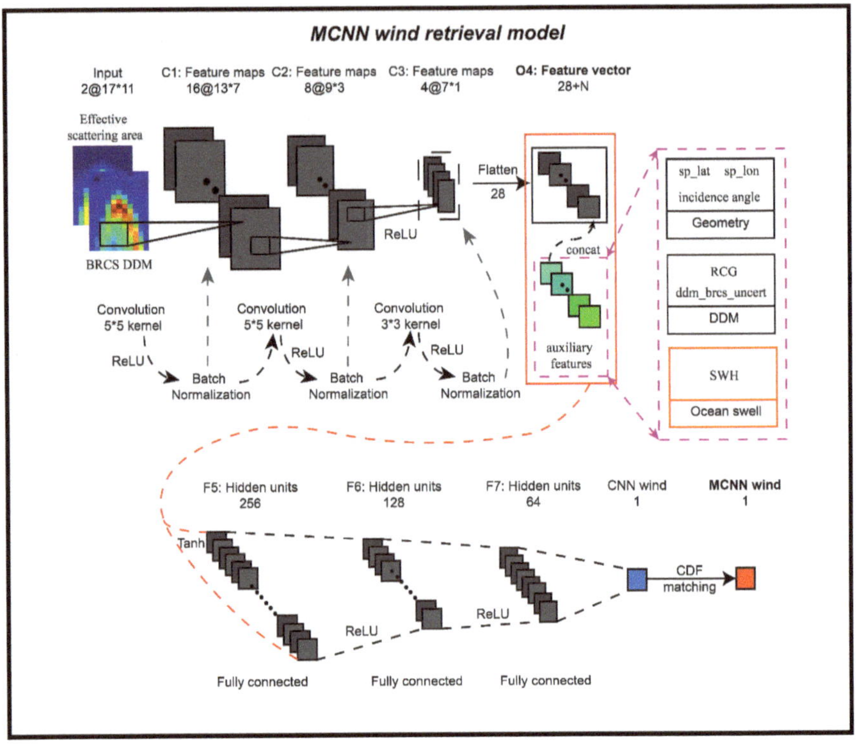

Fig. 16 MCNN wind speed retrieval model

employed for gradient updates, as expressed in Eq. 54.

$$\begin{cases} E\left|g^2\right|_t = \alpha E\left|g^2\right|_{t-1} + (1-\alpha)g_t^2 \\ W_{t+1} = W_t - \frac{\eta_0}{\sqrt{E\left|g^2\right|_t + \varepsilon}} \odot g_t \end{cases} \tag{54}$$

Here, $g_t = \Delta J(W_t)$ denotes the gradient of the loss function J with respect to W at iteration t; $E|g^2|_t$ and $E|g^2|_{t-1}$ represent the exponentially weighted moving averages of the squared gradients at the t-th and $(t+1)$-th iterations, respectively; α is the smoothing factor controlling the weight of historical information, typically ranging from 0.9 to 0.99 (set to 0.9 in this study); W_{t+1} and W_t represent the weight coefficients at the next and current time steps, respectively; η_0 is the initial learning rate (set to 0.0001); and ε is a small constant used to maintain numerical stability. After multiple iterations, the loss function converges to its minimum value, at which point the corresponding weight coefficients represent the optimal parameters for the CNN.

Bias Correction Via CDF Matching

Due to the highly non-uniform distribution of actual wind speed data, the risk of model overfitting increases. To address this, we designed a CDF-matching-based systematic bias correction method to adjust the preliminary wind speed retrievals obtained from the CNN, thereby establishing the final MCNN model. CDF matching is a bias correction technique that effectively removes systematic discrepancies between retrieved and reference wind speeds. This approach has been widely applied in remote sensing [44] and has previously been employed for bias correction in GNSS-R wind speed retrieval [23]. In this study, the method was improved by fitting the relationship between retrieved wind speeds and systematic bias using an adaptive high-order polynomial. Furthermore, the best-matching polynomial was applied to correct CNN-derived wind speeds for each month. Reference wind speeds and retrieval results are sorted in ascending order. If the two sequences differ in length, the distribution function fit value F_u for the sorted reference wind speeds is linearly interpolated based on the sorted retrieved wind speeds and their corresponding distribution function fit values F_o. The interpolation is then performed while simultaneously fitting the deviation to the wind speed retrieval using an n-th order polynomial function P_n, as expressed in Eq. 55:

$$bias = P_n(\hat{u}_{CNN}) \tag{55}$$

Here, P_n denotes an n-th degree polynomial, and $\hat{u}_{CNN}$ represents the wind speed obtained from CNN retrieval. Since the training and test datasets cover different time periods, the corresponding CDFs of reference wind speeds also differ. Consequently, an independent polynomial function must be constructed for each month. When fitting the polynomial function, the ($\hat{u}_{CNN}$, bias) data pairs within the month are randomly divided into two subsets in a 7:3 ratio: the first subset is used to fit the polynomial function, while the second subset is used to evaluate the fitting performance through RMSE and R^2, as defined in Eqs. 56 and 57:

$$RMSE_{CDF} = \sqrt{\frac{1}{N}\sum\nolimits_i^N (P_n(\hat{u}_i) - bias_i)^2} \tag{56}$$

$$R^2 = 1 - \frac{\sum_{i=1}^N \left(\widehat{bias}_i - bias_i\right)^2}{\sum_{i=1}^N \left(\overline{bias}_i - bias_i\right)^2} \tag{57}$$

Here, N denotes the sample size of the second subset; $\widehat{bias}_i = P_n(\hat{u}_i)$ and $bias_i$ represent the polynomial-estimated error and the actual error, respectively, and $\overline{bias}_i$ is the mean of the actual error. The optimal polynomial is selected based on the minimum RMSE and R^2. After applying the CDF correction, the final wind speed retrieval results of the model are obtained as:

$$\hat{u}_{MCNN} = \hat{u}_{CNN} + P_n(\hat{u}_{CNN}) \quad (58)$$

CDF matching effectively reduces the systematic bias between the reference wind speed and the preliminary CNN-retrieved wind speed, thereby alleviating overfitting issues in the CNN model—particularly the tendency to overestimate low wind speeds and underestimate high wind speeds. However, due to the bias correction procedure, this method may introduce a slight increase in the overall RMSE of the retrieved wind speeds.

5.4 Retrieval Results and Performance Analysis

To validate the effectiveness of the CNN model and the improved MCNN model, we systematically evaluate model performance by grouping the trained models into two categories, S1 and S2, following the design framework shown in Fig. 12. In S1, the CNN model without SWH is compared with the LUT, CDF, and FCN models to assess the baseline performance of the CNN. In S2, the CNN2 model is constructed by incorporating SWH as an additional auxiliary feature, allowing analysis of the effects of SWH and CDF matching on wind speed retrieval. It is important to note that the CNN, CNN2, and MCNN models are mutually independent; all are trained using the same training set and validated with the same validation set. The CNN2 model is introduced solely to evaluate the specific influence of including SWH and applying CDF matching.

Performance Comparison of the CNN Model with Other Models in S1

The performance of the four models in S1 was evaluated using the dataset, and the results are presented in Table 6. Among the four models, the LUT model exhibits the poorest performance, with an RMSE of 2.48 m/s and a bias of −0.998 m/s. Compared with the LUT model, the CDF model reduces the RMSE by approximately 25% and eliminates 65% of the bias. However, both the LUT and CDF models show degraded retrieval performance on the validation and test sets. This degradation is attributed to calibration issues in the CYGNSS V2.1 DDM product, where the NBRCS measurements display a gradual temporal decline [45]. In contrast, the deep learning–based FCN and CNN models demonstrate substantial performance improvements. Compared with the CDF model, the FCN and CNN models reduce the RMSE on the test set by 17% and 19%, respectively, indicating stronger robustness and generalization capability relative to traditional empirical methods. These advantages arise from neural networks' superior abilities in feature learning, nonlinear fitting, and generalization. Nevertheless, the bias of the FCN and CNN models increases by 101% and 171%, respectively, on the test set relative to their training set results, suggesting the presence of residual systematic bias. Moreover, due to the early stopping mechanism adopted during model training, both neural network models achieve their lowest RMSE and bias on the validation set.

Figure 17 presents the density plots of the wind speeds retrieved by the four models in S1, alongside the corresponding ECMWF ERA5 wind speeds. The LUT model exhibits pronounced overestimation in the 7–12 m/s range, which may be attributed to the insufficient calibration of the CYGNSS V2.1 product, as well as the impact of measurement noise. The CDF density plot in Fig. 17b demonstrates substantially improved symmetry compared with that of the LUT model, although noticeable deviations remain. As illustrated in Fig. 17c and d, the FCN and CNN models exhibit distributions that lie closer to the 1:1 line, indicating reduced overall divergence. Nevertheless, both models still demonstrate a tendency to overestimate at low wind speeds and underestimate at high wind speeds. Such behavior is common in data-driven GNSS-R wind speed retrieval models and is primarily influenced by the distribution characteristics of the reference wind speeds and by oceanic conditions such as swell.

To further assess the overall performance of the models, Figs. 18 and 19 present the probability distribution curves of the retrieved wind speeds from the four models, compared with the reference wind speeds, along with their error distributions across different wind speed intervals. The discrepancy between the CNN-retrieved wind speeds and the reference values is notable, whereas the CDF-corrected wind speed distribution aligns most closely with the reference distribution. This further justifies the need for subsequent bias correction of the CNN retrievals. When examining the error characteristics across different wind speed ranges, the CNN model achieves a clear reduction in RMSE within the 0–12 m/s interval relative to the two traditional models, with the most substantial improvement occurring in the 5–12 m/s range, where sample density is highest. Compared with the FCN model, the CNN retrievals also exhibit lower overall RMSE and bias. However, in the 0–2 and 9–25 m/s ranges, both FCN and CNN models display pronounced systematic bias. This behavior is primarily driven by two factors. First, at very low wind speeds, sea surface roughness is dominated by swells generated by non-local wind fields. GNSS-R observations are influenced by these gravity-wave-induced roughness patterns, leading the models to overestimate wind speeds in the 0–2 m/s range. Second, as shown in Fig. 20, real-world wind fields provide sparse training samples at both low and high wind speeds. This data imbalance induces overfitting in data-driven models, resulting in persistent

Table 6 Performance of CNN versus the other three models in S1

Models	Training set			Validation set			Test set		
	RMSE (m/s)	Bias (m/s)	MAPE (%)	RMSE (m/s)	Bias (m/s)	MAPE (%)	RMSE (m/s)	Bias (m/s)	MAPE (%)
LUT	2.15	−0.605	30.1	2.37	−0.857	32.5	2.48	−0.998	35.0
CDF	1.74	0.000	25.5	1.82	−0.184	26.8	1.86	−0.350	28.5
FCN	1.54	0.099	24.0	1.54	−0.027	24.8	1.54	−0.199	26.5
CNN	1.51	0.069	23.6	1.49	−0.023	24.2	1.50	−0.187	25.9

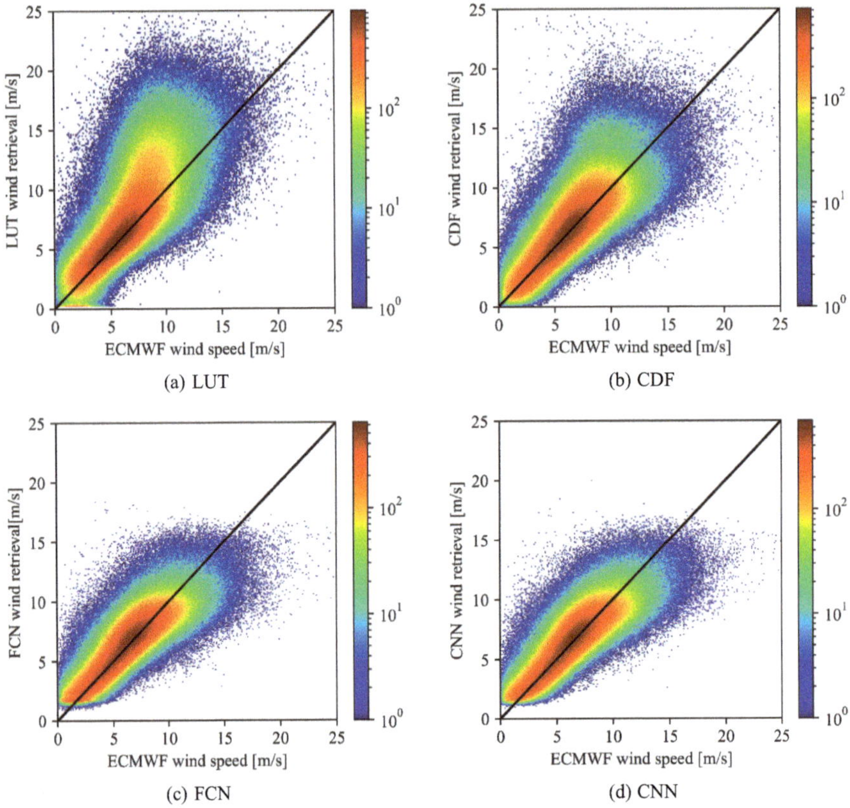

Fig. 17 Density plots of wind speeds for four models in S1 relative to reference wind speed

wind speed bias in the CNN model despite efforts to mitigate it. As wind speeds increase, the CNN model exhibits accelerated performance degradation. This trend is further evidenced by Fig. 17d, which shows that the CNN model's performance drops significantly in the 20–25 m/s interval, producing substantial underestimation.

Performance Analysis of the MCNN Model

Following the methodology illustrated in Fig. 16, we first developed the CNN2 model, which integrates SWH as an auxiliary input during training based on the original CNN architecture. Subsequently, a polynomial bias-correction model was constructed on top of CNN2. By applying CDF matching with the reference wind speeds, we obtained the final MCNN wind speed retrieval model.

Table 7 presents the overall wind speed retrieval errors of the CNN2 and MCNN models, with the performance of the original CNN model included for comparison. The results show that the RMSE of CNN2 decreases relative to that of CNN,

Fig. 18 Probability distribution of wind speed retrieval by four models and reference wind Speed

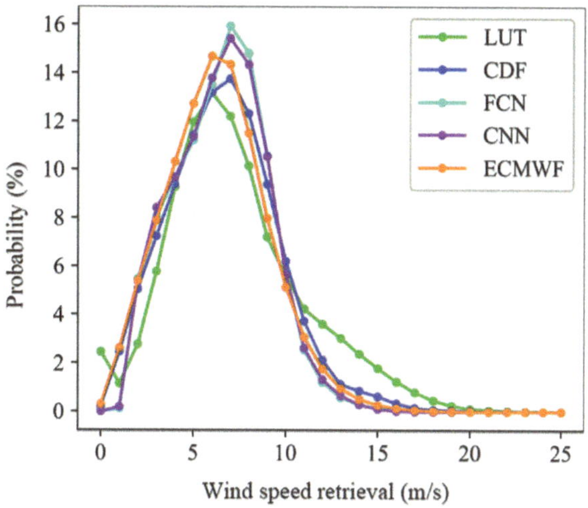

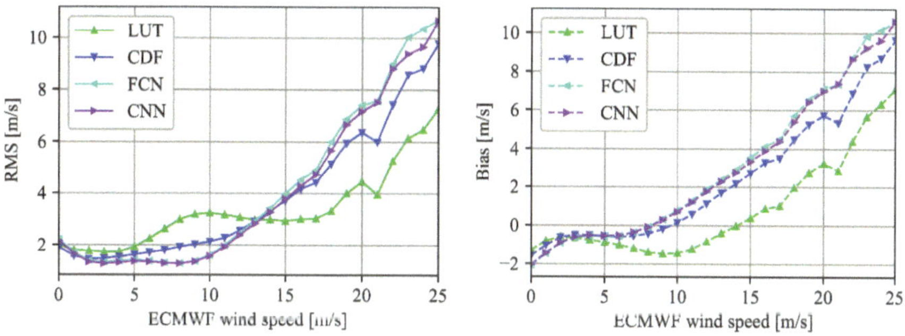

(a) RMSE (b) Bias

Fig. 19 Retrieval errors of the four models across different wind speed ranges

Wind Speed (m/s)	Count	Percentage (%)
0.0	4,874	0.3213
1.0	39,949	2.6335
2.0	82,368	5.4297
3.0	120,036	7.9128
4.0	156,818	10.3375
5.0	193,275	12.7408
6.0	222,938	14.6962
7.0	217,865	14.3618
8.0	174,615	11.5109
9.0	121,411	8.0035
10.0	77,953	5.1387
11.0	46,755	3.0821
12.0	27,018	1.7810
13.0	14,730	0.9710
14.0	7,976	0.5258
15.0	4,191	0.2763
16.0	2,231	0.1461
17.0	1,075	0.0709
18.0	464	0.0306
19.0	228	0.0150
20.0	108	0.0071
21.0	52	0.0034
22.0	20	0.0013
23.0	15	0.0010
24.0	7	0.0005
25.0	6	0.0004

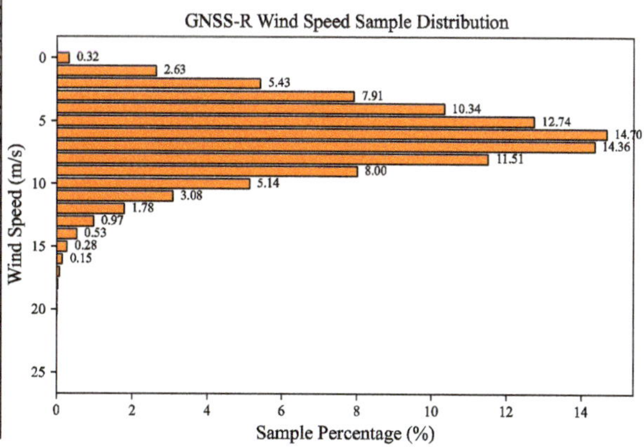

Fig. 20 Distribution statistics of samples in the test set

indicating that the incorporation of SWH effectively improves retrieval performance. However, introducing non–GNSS-R physical variables (i.e., SWH) leads to increased systematic bias in the training set. Despite this, CNN2 achieves lower bias and improved robustness on the test set, suggesting that SWH provides beneficial physical constraints under real-world ocean conditions. After applying the proposed polynomial bias correction, the MCNN model reduces systematic bias by 14% compared with CNN2, though at the cost of a slight increase in RMSE. In comparison to traditional LUT and CDF models, MCNN achieves RMSE reductions of 38% and 18%, and bias reductions of 90% and 72%, respectively. These results demonstrate that incorporating swell-related information significantly enhances the accuracy of GNSS-R wind speed retrieval models, while CDF-based matching provides an effective means of mitigating the systematic bias inherent in data-driven CNN retrievals.

Figure 21 presents the density plots of wind speeds retrieved by the CNN2 and MCNN models compared against ECMWF ERA5 reference wind speeds. By comparing Fig.21a with Figure 17d, it can be observed that within the 17–25 m/s range, the overall symmetry of the CNN2 retrievals improves noticeably. This indicates that the sea state information embedded in SWH contributes to enhanced CNN performance under high-wind-speed conditions [32]. Nevertheless, due to the uneven distribution of samples across wind speed ranges, the CNN2 model still suffers from overfitting, particularly at the extremes of the wind speed range. To mitigate this imbalance, CDF matching was applied to correct the systematic bias in the retrieved wind speeds. As shown in Figure 21b, the MCNN wind speed density plot exhibits significantly improved symmetry and is more tightly clustered around the 1:1 line. The corresponding probability distribution curves and error distributions for CNN2 and MCNN, as shown in Figs. 22 and 23 further demonstrate this improvement. Both distribution curves more closely align with the ECMWF reference wind speed distribution. These results indicate that incorporating SWH as auxiliary information effectively reduces the model's RMSE, while CDF-based matching provides a robust solution for eliminating systematic bias.

To assess the performance improvements of CNN2 and MCNN more intuitively, the CNN error curve is also included in Fig. 21 for comparison. Overall, the additional ocean state information provided by SWH enhances the model's performance, particularly under high-wind-speed conditions. This improvement partially alleviates the issue wherein high-wind-speed events are short-lived and fail to generate

Table 7 Overall error between CNN2 and MCNN

Models	Training set			Validation set			Test set		
	RMSE (m/s)	Bias (m/s)	MAPE (%)	RMSE (m/s)	Bias (m/s)	MAPE (%)	RMSE (m/s)	Bias (m/s)	MAPE (%)
CNN	1.51	0.069	23.6	1.49	−0.023	24.2	1.50	−0.187	25.9
CNN2	1.49	0.145	23.0	1.48	0.082	23.5	1.47	−0.113	25.4
MCNN	1.50	0.081	22.5	1.54	0.010	23.3	1.53	−0.097	24.6

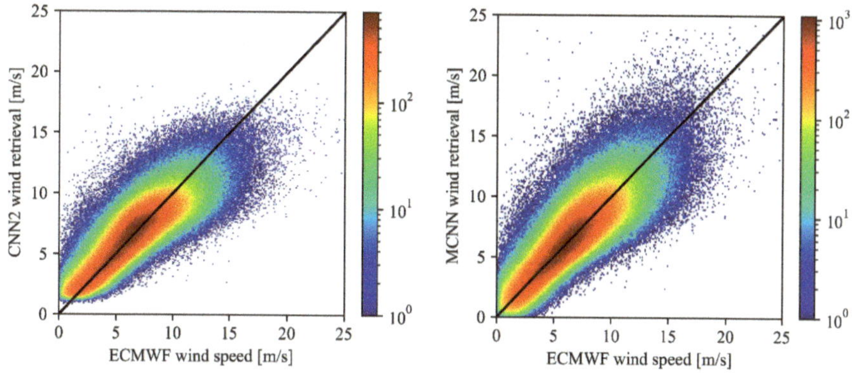

(a) CNN2 (b) MCNN

Fig. 21 Density plots of CNN2 and MCNN wind speed versus reference wind speed

Fig. 22 Probability distribution of wind speed retrieval by CNN2 and MCNN compared to reference wind speed

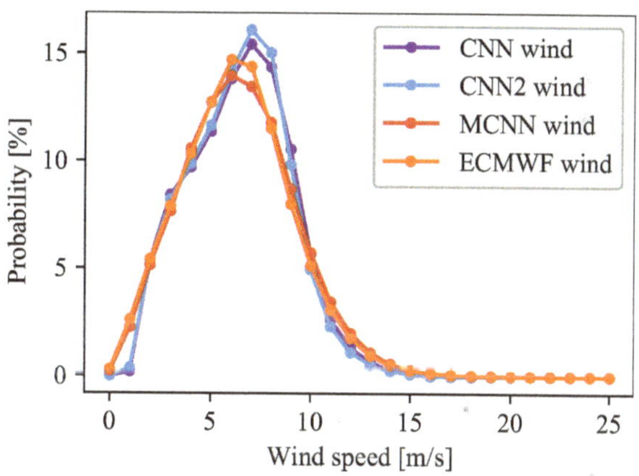

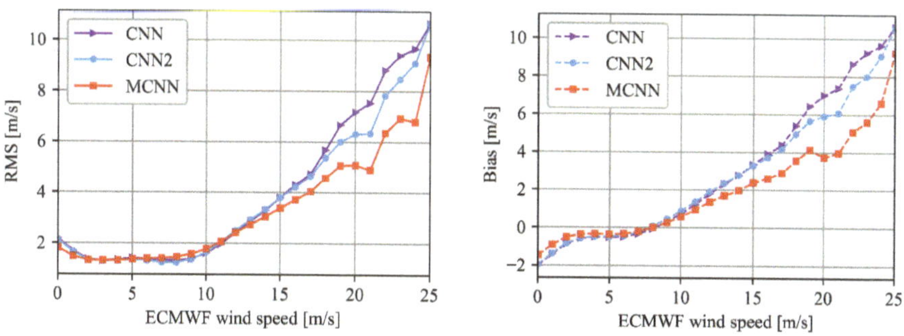

(a) RMSE (b) Bias

Fig. 23 Retrieval errors of CNN2 and MCNN across different wind speed ranges

Table 8 Retrieval errors of six models in the 0–12 m/s and 12–25 m/s wind speed ranges

Models	0–12 m/s		12–25 m/s	
	RMSE (m/s)	Bias (m/s)	RMSE (m/s)	Bias (m/s)
LUT	2.46	−1.022	3.04	−0.148
CDF	1.80	−0.418	3.24	1.998
FCN	1.45	−0.285	3.37	2.758
CNN	1.42	−0.269	3.24	2.626
CNN2	1.39	−0.192	3.25	2.630
MCNN	1.46	−0.154	2.97	1.881

corresponding sea surface roughness [24]. Although the CNN2 model achieves a lower RMSE, the overfitting challenges inherent in data-driven models at both low and high wind speeds—primarily caused by the highly imbalanced sample distribution—remain difficult to overcome through deep learning alone. Therefore, the proposed MCNN model incorporates a CDF-based matching correction strategy, which effectively eliminates systematic bias and enhances the reliability of wind speed retrieval.

To further evaluate model performance under low- and high-wind-speed conditions, Table 8 summarizes the RMSE and bias of the six models across the 0–12 m/s and 12–25 m/s wind speed ranges. The results show that data-driven deep learning models perform best in the 0–12 m/s range, where the majority of samples are concentrated. Compared with the FCN model, the CNN model reduces RMSE and systematic bias by 2% and 6%, respectively, indicating that the convolutional layers in CNN can extract more informative features from the DDM than those represented by LES and NBRCS alone. By incorporating SWH as an auxiliary feature, the CNN2 model yields further error reductions in the 0–12 m/s interval. Relative to CNN2, the MCNN model reduces bias by 20% in the 0–12 m/s range and decreases RMSE and bias by 9% and 28%, respectively, in the 12–25 m/s range. Although MCNN achieves the lowest RMSE within the 12–25 m/s interval, the LUT method exhibits lower bias, suggesting that LUT is less sensitive to sample distribution imbalances than the data-driven MCNN.

Overall, the experimental results demonstrate that, compared with traditional models and the FCN model, the proposed end-to-end retrieval framework effectively integrates multiple auxiliary information sources to enhance model robustness. Moreover, by applying CDF matching to mitigate systematic errors induced by overfitting, the MCNN achieves consistently superior performance, particularly in high-wind-speed scenarios.

MCNN Spatiotemporal Error Analysis

To further validate the robustness and generalization capability of the MCNN model, we conducted a comprehensive spatial and temporal error analysis using the test dataset. The spatial performance assessment was carried out by reconstructing large-scale wind field patterns from the MCNN-retrieved wind speeds and comparing them with ECMWF ERA5 wind fields, as shown in Fig. 24. The results indicate that most wind speeds are concentrated below 12 m/s. Within this range, the spatial distribution of MCNN-retrieved wind speeds aligns closely with that of the ECMWF ERA5 winds. Furthermore, to examine spatially varying biases, we computed the difference between ERA5 wind speeds and MCNN retrievals, as illustrated in Fig. 25. The spatial bias patterns show that, following CDF matching, large-scale overestimation or underestimation is effectively avoided, demonstrating that the MCNN model maintains strong retrieval stability across different DDM measurement locations. However, localized underestimation remains in certain coastal regions, including the South China Sea and the northern coast of Brazil. This behavior may result from contamination of sea surface reflections by land-scattered signals in near-shore environments, as well as the increased complexity of coastal oceanographic conditions, which poses challenges for data-driven GNSS-R wind speed retrieval models.

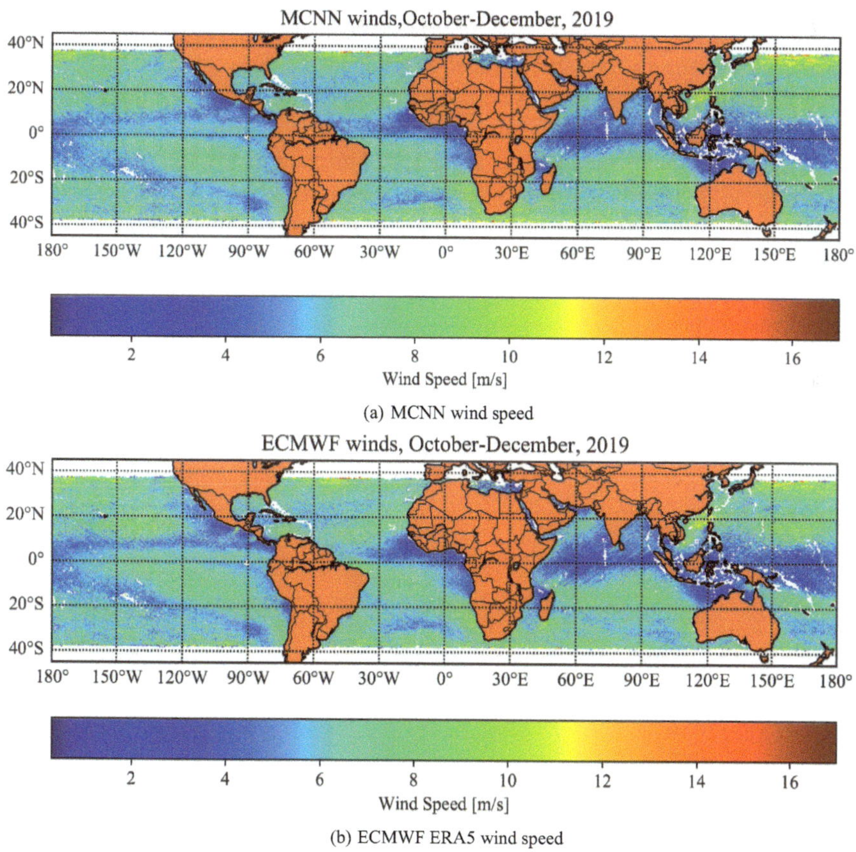

Fig. 24 Spatial distribution of wind speed from MCNN and ECMWF ERA5

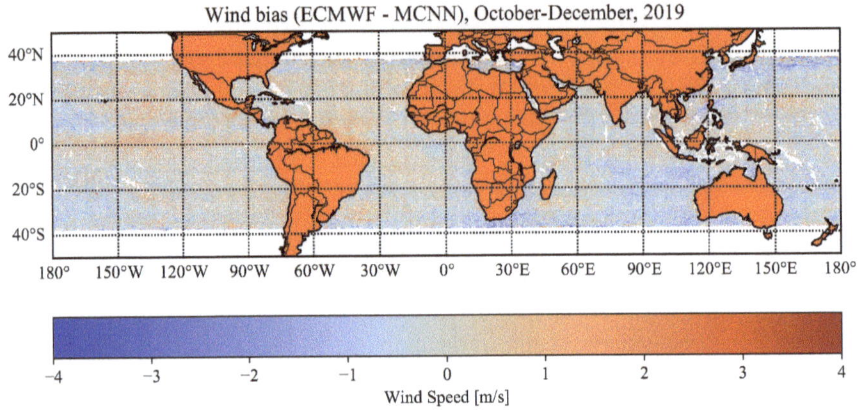

Fig. 25 Spatial distribution of MCNN system bias

We also conducted a temporal performance analysis of the MCNN model. To provide a more intuitive comparison, Fig. 26 presents the daily averaged wind speed distributions for the six models examined in this study, alongside the ECMWF ERA5 wind speeds on the test set. The results show that the LUT retrievals exhibit substantial overestimation, while the CDF method yields moderate improvement. However, due to the power calibration drift present in the DDM measurements, the deviation of the CDF retrievals gradually increases over time. In contrast, the FCN, CNN, and CNN2 models—benefiting from the strong fitting and generalization capabilities of deep learning—demonstrate stable retrieval performance. Among all models, the MCNN wind speeds most closely match the reference ERA5 values, owing to the CDF-based bias correction process.

As shown in Fig. 27, we further computed the time-series evolution of RMSE and bias for all six models. The MCNN model exhibits a notably stable RMSE and maintains a near-zero systematic bias throughout the test period. This demonstrates

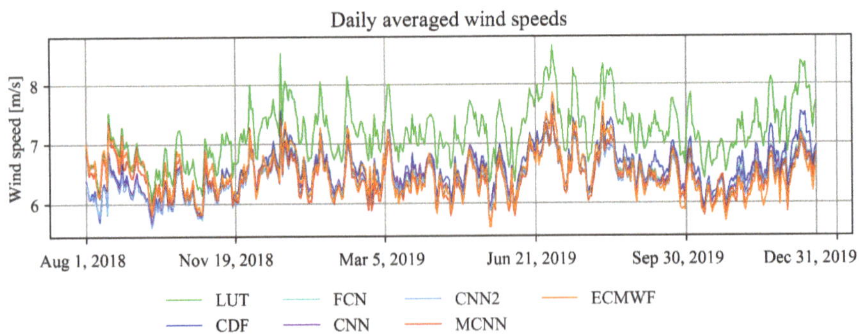

Fig. 26 Daily average wind speed and reference wind speed time series distribution for six models

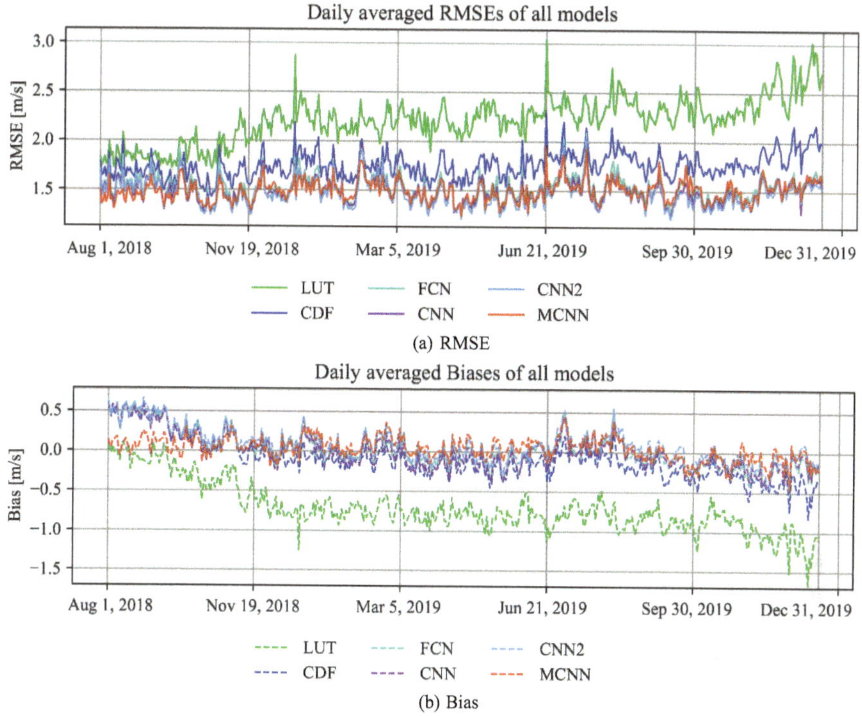

Fig. 27 Six types of model error time series distributions

that the MCNN model exhibits strong temporal reliability and effectively compensates for the long-term drift in DDM observations resulting from power calibration issues. Therefore, MCNN provides a robust and accurate framework for long-term, stable retrieval of sea surface wind speeds from GNSS-R satellite measurements.

6 Conclusion

In this chapter, we focus on GNSS-R-based sea surface wind speed retrieval. To address the limitations of traditional empirical models, we validated a statistical optimization approach employing PSO and a deep learning framework based on CNN, both of which effectively enhance retrieval accuracy and robustness.

Regarding PSO-based statistical optimization, to overcome the unit-weight constraint of the MVE method, we employed the PSO algorithm to globally search for optimal combination coefficients of LES and NBRCS without requiring their sum to equal 1. In the July 2019 CYGNSS dataset validation, the PSO method reduced the test set RMSE from MVE's 2.21–1.95 m/s, achieving a bias of 0.349 m/s in the low-to-medium wind speed range (0–12 m/s), significantly outperforming MVE.

Although accuracy slightly decreased in the high wind speed range (12–20 m/s), overall performance remained superior to the single-feature GMF model, confirming PSO's effectiveness as an enhancement within the traditional GMF framework.

In the CNN-based deep learning model, we incorporate multi-source information fusion and improved bias correction, proposing a statistically corrected multi-information fusion retrieval model. Compared to empirical and FCN models, the CNN model demonstrates superior generalization capability and accuracy by automatically extracting more representative features from DDM. Nevertheless, constrained by sample distribution and the physical characteristics of wind fields, CNNs still exhibit overfitting and systematic bias in extreme wind speed ranges. The CNN2 model, which incorporates SWH as an auxiliary feature, highlights the importance of including additional auxiliary information. Building upon this, the CDF-debiased MCNN further reduces systematic bias and exhibits robust stability across both spatial and temporal dimensions, providing a long-term, reliable, and high-precision solution for wind speed retrieval.

Despite recent advances, GNSS-R sea surface wind speed retrieval still faces several challenges, including limited generalization in high-wind scenarios, interference from complex coastal conditions, imbalanced sample distributions, and a gap between physical mechanisms and data-driven approaches. Future research should focus on constructing physically constrained deep learning models, integrating multi-source data to mitigate interference, applying data augmentation techniques to balance sample distributions, and developing specialized models for high-wind-speed conditions. Addressing these challenges will further enhance retrieval accuracy and reliability. With such advancements, GNSS-R technology is poised to play an increasingly important role in applications such as marine environmental monitoring, typhoon tracking, and numerical weather prediction.

References

1. Yu K (2021) Theory and practice of GNSS reflectometry. Springer, Singapore
2. Li W, Cardellach E, Fabra F et al (2020) Assessment of Spaceborne GNSS-R Ocean altimetry performance using CYGNSS Mission raw data. IEEE Trans Geosci Remote Sens 58(1):238–250
3. Ruf CS et al (2016) New Ocean winds satellite mission to probe hurricanes and tropical convection. B AM METEOROL SOC 97(3):385–395
4. Unwin M, Jales P, Tye J et al (2016) Spaceborne GNSS-reflectometry on TechDemoSat-1: early mission operations and exploitation. IEEE J Sel Top Appl Earth Obs Remote Sens 9(10):4525–4539
5. Jing C, Niu X, Duan C et al (2019) Sea surface wind speed retrieval from the first Chinese GNSS-R Mission: technique and preliminary results. Remote Sens 11(24):3013
6. Xia J, Bai W, Sun Y et al (2021) Calibration and wind speed retrieval for the Fengyun-3 E meteorological satellite GNSS-R Mission. In: 2021 IEEE specialist meeting on reflectometry using GNSS and other signals of opportunity (GNSS+R), pp 25–28
7. Hall CD, Cordey RA (1988) Multistatic Scatterometry. In: International geoscience and remote sensing symposium, "remote sensing: moving toward the 21st century", pp 561–562

8. Martin-Neira M (1993) A passive reflectometry and interferometry system (PARIS): application to ocean altimetry. ESA J 17(4):331–355
9. Garrison JL, Katzberg SJ, Hill MI (1998) Effect of sea roughness on bistatically scattered range-coded signals from the global positioning system. Geophys Res Lett 25(13):2257–2260
10. Garrison JL, Katzberg SJ (1997) Detection of ocean reflected GPS signals: theory and experiment. In: Proceedings IEEE SOUTHEASTCON'97. "Engineering the new century", pp 290–294
11. Zavorotny VU, Voronovich AG (2000) Scattering of GPS signals from the ocean with wind remote sensing application. IEEE Trans Geosci Remote Sens 38(2):951–964
12. Elfouhaily T, Chapron B, Katsaros K et al (1997) A unified directional spectrum for long and short wind-driven waves. J Geophys Res Oceans 102(C7):15781–15796
13. Garrison JL, Komjathy A, Zavorotny VU, Katzberg SJ (2002) Wind speed measurement using forward scattered GPS signals. IEEE Trans Geosci Remote Sens 40(1):50–65
14. Asgarimehr M, Arnold C, Weigel T et al (2022) GNSS reflectometry global ocean wind speed using deep learning: development and assessment of CyGNSSnet. Remote Sens Environ 269:112801
15. Wang M-H, Lu S-D, Hsieh C-C, Hung C-C (2022) Fault detection of wind turbine blades using Multi-Channel CNN. Sustainability 14(3):1781
16. Bu J, Yu K, Zuo X et al (2023) GloWS-net: a deep learning framework for retrieving Global Sea surface wind speed using Spaceborne GNSS-R data. Remote Sens 15(3):590
17. Zhao D, Heidler K, Asgarimehr M et al (2023) DDM-former: transformer networks for GNSS reflectometry global ocean wind speed estimation. Remote Sens Environ 294:113629
18. Pascual D, Clarizia MP, Ruf CS (2021) Improved CYGNSS wind speed retrieval using significant wave height correction. Remote Sens 13(21):4313
19. Rodriguez-Alvarez N, Akos DM, Zavorotny VU et al (2013) Airborne GNSS-R wind retrievals using delay–doppler maps. IEEE Trans Geosci Remote Sens 51(1):626–641
20. Clarizia MP, Ruf CS, Jales P, Gommenginger C (2014) Spaceborne GNSS-R minimum variance wind speed estimator. IEEE Trans Geosci Remote Sens 52(11):6829–6843
21. Guo W, Du H, Cheong JW et al (2022) GNSS-R wind speed retrieval of sea surface based on particle swarm optimization algorithm. IEEE Trans Geosci Remote Sens 60:1–14
22. Wang F, Yang D, Zhang B, Li W (2018) Waveform-based spaceborne GNSS-R wind speed observation: demonstration and analysis using UK TechDemoSat-1 data. Adv Space Res 61(6):1573–1587
23. Reynolds J, Clarizia MP, Santi E (2020) Wind speed estimation from CYGNSS using artificial neural networks. IEEE J Sel Top Appl Earth Obs Remote Sens 13:708–716
24. Guo W, Du H, Guo C et al (2022) Information fusion for GNSS-R wind speed retrieval using statistically modified convolutional neural network. Remote Sens Environ 272:112934
25. Arabi S, Asgarimehr M, Kada M, Wickert J (2023) Hybrid CNN-LSTM deep learning for track-wise GNSS-R Ocean wind speed retrieval. Remote Sens 15(17):4169
26. Du H, Li W, Cardellach E et al (2024) Deep residual fully connected network for GNSS-R wind speed retrieval and its interpretation. Remote Sens Environ 313:114375
27. Chen M-Y, Chien H, Yeh W-H et al (2025) Evaluation of the effect of satellite motion on GNSS-R wind speed retrieval: insights from TRITON. Front Remote Sens 6
28. Al-Khaldi MM, Gleason S, Johnson JT et al (2023) Using synthetic cyclone models for high wind GNSS-R calibration, validation, and algorithm development: a CYGNSS case study. IEEE Trans Geosci Remote Sens 61:1–11
29. Tang K, Buckius RO (1998) The geometric optics approximation for reflection from two-dimensional random rough surfaces. Int J Heat Mass Transf 41(13):2037–2047
30. Ruf C (2022) CYGNSS Handbook
31. Clarizia MP, Ruf CS (2016) Wind speed retrieval algorithm for the cyclone global navigation satellite system (CYGNSS) Mission. IEEE Trans Geosci Remote Sens 54(8):4419–4432
32. Ruf CS, Balasubramaniam R (2019) Development of the CYGNSS geophysical model function for wind speed. IEEE J Sel Top Appl Earth Obs Remote Sens 12(1):66–77

33. Reichle RH, Koster RD (2004) Bias reduction in short records of satellite soil moisture. Geophys Res Lett 31(19)
34. CYGNSS (2017) CYGNSS level 1 science data record version 2.1
35. Persson A, Grazzini F (2007) User guide to ECMWF forecast products. Meteorol Bull 3:153
36. Kennedy J, Eberhart R (1995) Particle swarm optimization. In: Proceedings of ICNN'95-international conference on neural networks, vol 4, pp 1942–1948
37. Rumelhart DE, Hinton GE, Williams RJ (1986) Learning representations by back-propagating errors. Nature 323(6088):533–536
38. Agatonovic-Kustrin S, Beresford R (2000) Basic concepts of artificial neural network (ANN) modeling and its application in pharmaceutical research. J Pharm Biomed Anal 22(5):717–727
39. LeCun Y, Boser B, Denker J et al (1989) Handwritten digit recognition with a Back-propagation network. In: Advances in neural information processing systems. Morgan-Kaufmann
40. Krizhevsky A, Sutskever I, Hinton GE (2012) ImageNet classification with deep convolutional neural networks. In: Advances in neural information processing systems. Curran Associates, Inc.
41. Qiu X (2020) Neural networks and deep learning, pp 112–113
42. Santurkar S, Tsipras D, Ilyas A, Madry A (2018) How does batch normalization help optimization? In: Advances in neural information processing systems. Curran Associates, Inc.
43. Prechelt L (1998) Early stopping—but when? In: Orr GB, Müller K-R (eds) Neural networks: tricks of the trade. Springer, Berlin, Heidelberg, pp 55–69
44. Brocca L, Hasenauer S, Lacava T et al (2011) Soil moisture estimation through ASCAT and AMSR-E sensors: an intercomparison and validation study across Europe. Remote Sens Environ 115(12):3390–3408
45. Saïd F, Jelenak Z, Park J et al (2021) The NOAA track-wise wind retrieval algorithm and product assessment for CYGNSS. IEEE Trans Geosci Remote Sens 60:1–24

Chapter 20
Sea Surface Wind Direction Inversion Based on DDM Asymmetry

Yun Zhang, Shuhu Yang, and Xu Chen

Abstract This study proposes two parallel GNSS-R sea surface wind direction retrieval methods using CYGNSS Full DDM (high-dimensional) and L1 compact DDM (low-dimensional) data, converting the wind direction inversion problem into a classification task by discretizing wind direction into 72 intervals with a 5° step. For Full DDM, an SVM model with 5/9-dimensional features (incorporating DDM geometric parameters φ_1/φ_2, LES, NBRCS) is developed, achieving an optimal RMSE of 26.70° at wind speed ≥10 m/s without 180° ambiguity. For L1 DDM, SVM, BP, and CNN models are compared, with CNN outperforming others (generalized test set RMSE = 37.6°, classification accuracy = 0.44). A probability-based filter (threshold = 0.21) reduces CNN's RMSE to 19.5° (39.3% sample retention). Key findings are: 9-dimensional features outperform 5-dimensional ones in the first study; CNN excels at large-scale data due to hierarchical feature extraction in the second study; uneven wind direction samples (easterly-dominated) degrade westerly retrieval accuracy. Both methods offer all-weather capability and high temporal resolution, providing a robust basis for global marine wind direction monitoring.

1 Introduction

Wind speed and wind direction are critical fundamental climate variables that play pivotal roles in climate monitoring, marine environmental assessment, and coastal disaster prediction. As a passive and non-contact remote sensing technique, Global Navigation Satellite System Reflectometry (GNSS-R) has advanced substantially

Y. Zhang · S. Yang (✉) · X. Chen
College of Information Technology, Shanghai Ocean University, Shanghai, China
e-mail: shyang@shou.edu.cn

Y. Zhang
e-mail: y-zhang@shou.edu.cn

X. Chen
e-mail: m200711435@st.shou.edu.cn

K. Yu and N. Zheng (eds.), *Advances in GNSS Reflectometry*, Navigation: Science and Technology 19, https://doi.org/10.1007/978-981-92-2325-1_20

with the successful deployment of dedicated satellites such as Techdemosat-1 (TDS-1) [1] and the Cyclone Global Navigation Satellite System (CYGNSS) [2]. This technology has evolved beyond its core positioning functionality to become a versatile tool for earth surface parameter retrieval, offering unique advantages in terms of global coverage, all-weather operation, and low cost. Operating via a bistatic transmitter-receiver configuration, GNSS-R leverages reflected L-band signals from navigation satellites—including Global Positioning System (GPS) signals scattered by satellites orbiting at an altitude of approximately 20,000 km [3]. Since Martín first proposed the feasibility of GNSS-R-based sea surface height retrieval [4], the technology has been extensively validated and applied in diverse fields, including sea surface wind speed estimation [5–7], sea ice detection and classification [8–11], ocean altimetry and topography mapping [12–15], significant wave height retrieval [16], soil moisture content monitoring [17], ocean oil slick detection and tracking [18, 19], and general target detection [20, 21].

Sea surface wind field retrieval is a key application of GNSS-R technology, encompassing two primary components: wind direction retrieval and wind speed retrieval. Among these, sea surface wind direction retrieval remains a core technical challenge [22], with several critical barriers hindering its development. First, wind direction retrieval is inherently affected by unavoidable interference from wind speed and other environmental factors [23], making it difficult to isolate pure wind direction signals from GNSS-R measurements. Second, the long-standing issue of 180° ambiguity in wind direction retrieval has not been fully resolved [24, 25], which limits the practical utility of retrieved results. Third, relative to wind speed retrieval, there remains a paucity of in-depth research on the sensitivity of GNSS-R signals to wind direction [22], leading to a lack of robust theoretical foundations for model development.

Traditional main microwave-based wind direction measurement tools serve as important benchmarks for evaluating GNSS-R performance. Microwave scatterometers achieve a retrieval accuracy of approximately 20° by establishing geophysical model functions (GMF) based on backscattering coefficients, combined with maximum likelihood estimation and ambiguity-reducing filtering techniques [26], but their measurements are significantly constrained by adverse weather conditions such as heavy rainfall. Microwave radiometers, by contrast, use Stokes parameters to quantify wind direction-induced changes in sea surface brightness temperature, offering comparable 20° accuracy and insensitivity to rainfall [27]; however, their performance relies on high instrument sensitivity due to the weak magnitude of brightness temperature variations associated with wind direction shifts. Additionally, synthetic aperture radar (SAR) provides an image-based approach to wind direction retrieval by analyzing sea surface wind streaks, which offers valuable spatial context but is limited by its relatively low temporal resolution [28]. Given these limitations of traditional methods, GNSS-R's all-weather capability, global coverage, and high temporal resolution position it as a promising alternative, driving strong demand for the development of advanced wind direction retrieval models.

2 Review of Related Works

2.1 *Wind Direction Retrieval Based on DDM Features and Bistatic Radar Cross-Section*

Early exploration of GNSS-R wind direction retrieval (1994–2004) focused on extracting discriminative parameters from Delay Doppler Maps (DDMs) and validating core technical feasibility. As early as 1994, Hildebrand et al. conducted pioneering work by applying least-squares analysis to probability density functions (PDFs) derived from reflected GPS signals, achieving maximum retrieval accuracies of 5° for filtered datasets and 40° for unfiltered datasets when the aircraft-mounted receiver operated at an altitude of 3–5 km [29]. In 2003, Zuffada et al. performed numerical simulations to investigate the influence of wind direction on GPS signal delay waveforms, systematically analyzing the effects of reflected waveform geometry, scattering area size, and receiver integration time [30, 31]. Their results revealed significant differences in the trailing edge of the waveform across different wind directions under fixed satellite elevation angles, wind speeds, and receiver heights. Parallel work in 2003 also focused on Mean Square Slope (MSS) estimation, with researchers fitting delayed waveforms obtained from flight experiments to a geometric optical model to derive MSS values [32], and the resulting wind direction data showed good consistency with Centre for Medium-Range Weather Forecasts (ECMWF) reanalysis data when the satellite azimuth was aligned with the aircraft heading.

Building on these foundations, a 2004 study identified a direct correspondence between the anisotropy in the PDF of sea surface MSS and local near-surface wind direction, demonstrating that PDFs could be reliably calculated from code correlation waveforms of reflected GPS signals and that wind direction could be effectively retrieved by fusing signals from two or more GPS satellites to resolve ambiguity [33]. Complementary simulator-based research in 2004 (conducted at 10 km altitude) confirmed Zuffada et al.'s findings and further uncovered distinct differences in the trailing edges of waveforms between Left-Hand Circular Polarization (LHCP) and Right-Hand Circular Polarization (RHCP) signals [23], with these polarization-dependent differences diminishing as the incident angle increased. That same year, Komjathy et al. reported promising results by combining multiple satellite signals collected from aircraft with nonlinear least-squares algorithms, achieving a 30° error relative to QuikSCAT measurements at wind speeds of 5–10 m/s [34].

The field entered an extended exploration phase (2014–2017) with a focus on expanding DDM-derived parameters and validating new retrieval approaches. In 2014, Chen et al. achieved a wind direction retrieval accuracy of 30° by fitting measured DDMs with simulated counterparts generated from physical models [35]; that same year, a small slope approximation model (SSA) was developed to evaluate the sensitivity of the bistatic radar cross-section to wind direction, demonstrating higher accuracy when using scattered signals away from the nominal specular reflection direction [36]. In 2017, a study calculated Delay Doppler Map Average (DDMA) by leveraging the influence of wind direction on the near-specular reflection area of DDMs [37], finding weak but statistically significant wind direction sensitivity in pure mirror geometry. However, the study also noted that wind speed confounding effects could not be fully eliminated, and that the observed 1 dB signal to noise ratio (SNR) peak difference between tailwind and headwind conditions (under incident angles of 20°–25° and wind speeds of approximately 9 m/s) was too small to enable reliable wind direction retrieval [38]. Notably, both [35, 37] highlighted the untapped potential of using regions of the DDM far from the specular reflection point for wind direction retrieval.

In the contemporary deepening phase (2016–2021), research increasingly focused on the bistatic radar cross-section and ambiguity resolution. In 2016, Park et al. extended prior work by using the Normalized Bistatic Radar Cross Section (NBRCS) to systematically study wind direction effects on GNSS-R sea surface specular scattering [39], finding negligible NBRCS changes in pure mirror geometry but significant wind direction-induced variations at individual surface points in slightly non-mirror geometries; they also established an airborne wind direction retrieval model based on the National Oceanic and Atmospheric Administration (NOAA) G-IV jet aircraft, using the DDM Doppler angle as the key observation [24], which achieved an average retrieval accuracy of 20° but lacked generalizability due to limited airborne data and persistent 180° ambiguity. To address this ambiguity, Wang et al. [24] explored the feasibility of using backscattered signals from multi-beam antennas, observing wind direction from at least three different directions to avoid 180° ambiguity and achieving a retrieval accuracy of 24° at low wind speeds. More recently, Pascual et al. [40] concluded that wind speed and SNR have important influences on wind direction retrieval models, and studied the sensitivity of the ocean surface bistatic scattering cross-section (measured by CYGNSS) to wind direction using the kurtosis of DDM samples within a given area [41]; their results showed a coefficient of determination (R^2) between 0.6 and 0.9 for wind speeds ranging from 4 to 10 m/s, confirming the potential of DDM statistical features for wind direction retrieval.

2.2 *Machine Learning in GNSS-R and Wind Direction Retrieval Gaps*

Traditional empirical GNSS-R models are often limited by their reliance on simplified physical assumptions and manual feature engineering, which hinders their ability to capture complex nonlinear relationships between GNSS-R measurements and geophysical parameters [42–44]. This has spurred the growing adoption of machine learning and deep learning algorithms in the field, which offer powerful tools for modeling complex systems without explicit physical assumptions. Convolutional Neural Networks (CNNs) are particularly well-suited for extracting spatial features from grid-structured data such as DDMs, while Support Vector Machines (SVMs) excel in high-dimensional feature spaces and small sample scenarios—relying on a minimal set of support vectors to determine the final model, which renders them robust to outliers and ideal for applications with limited high-quality training data [7, 42–44].

These machine learning algorithms have been widely and successfully applied in various GNSS-R parameter retrieval tasks. For example, Volkan developed a global soil moisture retrieval framework using on-board GNSS-R observations and geophysical data provided by CYGNSS satellites [45]; Zhang et al. successfully established and validated a machine learning model for sea surface height retrieval using data from the TDS-1 satellite [46]; and Hu et al. [47] proposed a height model based on the oscillation frequency in SNR arcs to track changes in snow depth, enabling both snow depth retrieval and classification between snow-free and snow-covered states. In the field of sea surface wind speed retrieval, numerous machine learning-based models have been developed: Liu et al. [48] established a DDM-based neural network model to capture the overall characteristics of DDM images, achieving significantly improved accuracy and generalization compared with traditional methods such as GMF and linear regression; Balasubramaniam [49] developed a neural network-based quality control filter for CYGNSS wind retrieval, creating an efficient method to remove abnormal samples and thereby improve the overall data quality and wind speed retrieval accuracy; Chu [50] proposed a Heterogeneous Multimodal Deep Learning (HMDL) method to account for the heterogeneity of input data, validating the model with real datasets obtained by the TDS-1 satellite; Reynolds [51] developed an artificial neural network (ANN)-based algorithm for wind speed retrieval, optimizing the number of hidden layers and neurons using simulated wind data from the WaveWatch 3 (WW3) model and applying cumulative distribution function correction to achieve a lower root mean square difference (RMSD); Zhang et al. [7] successfully established a high wind speed retrieval model by combining SVR and CNN algorithms, addressing the poor accuracy of empirical models in retrieving high wind speeds; and Milad [52] used a trained feedforward neural network for TDS-1 wind speed retrieval, reducing the root mean square error (RMSE) by 20% compared with the least-squares method.

In stark contrast to the mature application of machine learning in sea surface wind speed retrieval, research on machine learning-based sea surface wind direction retrieval remains relatively scarce. The subsequent sections will present two related works on sea surface wind direction retrieval. The first is a global SVM-based sea surface wind direction retrieval model using CYGNSS Full DDM data and ECMWF reanalysis data, which achieved a wind direction RMSE of 26.7° for wind speeds exceeding 10 m/s (constrained by limited training data at the time) [25]. Leveraging the substantial data volume provided by CYGNSS L1 products, which creates unprecedented opportunities for deploying machine learning algorithms requiring large-scale training samples (e.g., Back Propagation (BP) neural networks [53] and CNNs [54]), the second work extends the application of these powerful algorithms to sea surface wind direction retrieval. These algorithms possess robust nonlinear fitting capabilities to capture intricate relationships between GNSS-R measurements and wind direction, along with strong rule-learning abilities and high generalization performance.

2.3 Research Gaps and Objectives of this Study

In stark contrast to the mature application of machine learning in sea surface wind speed retrieval, research on machine learning-based sea surface wind direction retrieval remains relatively scarce. To address this gap and tackle the core challenges of GNSS-R wind direction retrieval (e.g., 180° ambiguity, low-dimensional DDM information loss, and confounding factor interference), this study conducts two parallel works tailored to different CYGNSS DDM products.

The first work focuses on a global sea surface wind direction retrieval model based on the SVM algorithm, building on our prior research on machine learning-based sea surface wind speed inversion models using CYGNSS data [30, 55], it utilizes CYGNSS Full DDM data (high-dimensional, original resolution) and ECMWF reanalysis data spanning 2019–2020. Geometric relationship parameters derived from DDMs and CYGNSS satellite-specific parameters are extracted as feature inputs, with hyperparameters optimized via grid search. This work aims to leverage the rich geometric information of Full DDM data to establish a stable retrieval model, addressing the technical limitations of small-sample high-dimensional scenarios. Earlier results of this work, constrained by limited training data at the time, achieved a wind direction RMSE of 26.7° for wind speeds exceeding 10 m/s [25].

The second work targets the low dimensionality challenge of CYGNSS L1 compact DDM data (17×11, reduced resolution) by exploring diverse machine learning methods. Using CYGNSS L1 data and ECMWF reanalysis data from 2020–2021, a 31-dimensional feature vector is constructed (encompassing wind speed, DDM geometric characteristics, sea level pressure, and receiver-satellite parameters). Three retrieval models (SVM, BP) neural networks [53] and CNNs [54]) are established, with performance evaluated on both test sets and generalized test sets. This work seeks to compensate for the information loss of low-dimensional DDMs through extended features and advanced neural networks, tapping into the potential of large-scale data for wind direction retrieval—an opportunity enabled by the substantial data volume provided by CYGNSS L1 products, which supports the deployment of machine learning algorithms requiring massive training samples.

3 Data Acquisition and Processing

3.1 Data Acquisition

This study includes two related works on GNSS-R sea surface wind direction retrieval, with consistent core data sources (CYGNSS and ECMWF reanalysis datasets) but distinct selections of CYGNSS DDM products and parameter applications: **Work 1** adopts CYGNSS Full DDM data (high-dimensional, original resolution) as the key observation source, while **Work 2** exclusively uses CYGNSS L1 compact DDM data (low-dimensional, reduced resolution).

CYGNSS Datasets

The CYGNSS mission, launched on December 15, 2016, deploys eight micro-satellites into a low Earth orbit (LEO) at an altitude of approximately 520 km [55]. Each satellite is equipped with the CYGNSS Delay Doppler Mapping Instrument (DDMI) payload, which receives GPS signals reflected off the Earth's surface. Consequently, each satellite's receiver operates as a bistatic radar, observing the forward scattering of signals from specific GPS transmitters via specular reflection off the Earth's surface. By correlating the received GPS transmissions with a locally generated replica of the GPS C/A code, a fundamental GNSS-R measurement—referred to as a DDM—is generated (data are accessible via https://podaac.jpl.nasa.gov/cloud-datasets?view=list&ids=Instruments&values=DDMI). Each "pixel" in the DDM represents the signal power scattered from a reflection point with a specific time delay and Doppler shift with respect to those at the specular reflection point. Each CYGNSS satellite is fitted with two nadir-pointing antennas, capable of tracking up to four simultaneous reflections at any given time. Within a $\pm 38°$ swath width, the satellite constellation produces observations along irregularly repeating spatiotemporal orbits, with average and median revisit times of 7 hr and 3 hr, respectively. These characteristics render CYGNSS measurements particularly suitable for frequent storm monitoring and large-scale spatial coverage observations. Since 2019, the satellite constellation has been capable of generating up to 64 DDM measurements per second.

Data for Work 1

Work 1 utilizes two types of CYGNSS products: Full DDM and L1 products. Both datasets are formatted in NetCDF, encompassing time-delay Doppler sample data, a suite of metadata describing geometric configurations and instrument parameters at the time of acquisition. The temporal coverage of the data extends from February 27, 2019, to November 17, 2020. DDM sample data are extracted from Full DDM files, while CYGNSS satellite parameters are derived from L1 data. Full DDM and L1 datasets are matched based on specular reflection point indices to ensure spatial and temporal consistency.

Data for Work 2

Work 2 primarily utilizes CYGNSS L1 data, which includes DDM sampling data and receiver-satellite parameter information in NetCDF format. In comparison to CYGNSS Full DDM, the L1 compact DDM offers substantial advantages in terms of data volume and global coverage, but its dimensionality is reduced from 128×20 to 17×11.

ECMWF Reanalysis Dataset

ECMWF utilizes its advanced numerical weather prediction models and data assimilation systems to reanalyze archived observational data, generating a global dataset that describes the recent history of the atmosphere, land surface, and oceans. The global data assimilation and filtering system employed in ECMWF reanalysis projects ensures temporal continuity, while a comprehensive and complete database guarantees the consistency of reanalyzed data. Input data for reanalysis are sourced from diverse platforms, including radio soundings, aircraft observations, satellite measurements, and ground-based stations. The integration of multi-source data helps mitigate local effects from non-target observations, such as terrain obstacles or vegetation cover. Prior to inclusion, the data undergo filtering to ensure independence and consistency.

Usage for both Works

ECMWF provides sea surface wind field data (east-west direction U and north-south direction V components) at a reference height of 10 m above the sea surface, with a spatial resolution of $0.25° \times 0.25°$ and a temporal resolution of 1 h. The dataset is matched to CYGNSS observations by selecting the wind field data corresponding to the spatiotemporally closest grid point to the CYGNSS specular reflection point. In both works, wind speed data from ECMWF are used as auxiliary training data, while wind direction data serve as the ground truth for model training and validation.

Additional Usage for Work 2

In addition to wind speed and direction, Work 2 further uses ECMWF's sea level pressure and sea surface temperature data (with a spatial resolution of 1° and a temporal resolution of 1 h) as feature parameters for model training.

3.2 *Data Preprocessing*

Both works conduct data quality control and preprocessing to ensure the reliability of model training and retrieval results, with overlapping core filtering criteria but distinct parameter thresholds and additional constraints tailored to their respective DDM data characteristics.

Common Preprocessing Criteria for both Works

(a) **CYGNSS L1 Quality Control (QC)**: Both works retain only data samples with good quality, as indicated by the QC flag in CYGNSS L1 products.
(b) **Wind Speed Filtering**: At low wind speeds, DDM power is primarily concentrated near the specular reflection point, making it difficult to extract reliable geometric angles. Therefore, both works exclude samples with wind speeds below 5 m/s and select samples with wind speeds ≥5 m/s for analysis. Additionally, Work 1 further conducts stratified analysis for subsets with wind speeds ≥8, ≥10, ≥12, and ≥ 15 m/s.
(c) **Offshore Distance Filtering**: Wind direction near coastal regions is affected by land-sea breeze circulation, leading to irregular DDM shapes (e.g., peak points located at the edge of the horseshoe pattern). Consequently, both works select samples with an offshore distance ≥25 km to avoid coastal effects.

Preprocessing Criteria Specific to Work 1

SNR Filtering: When SNR is excessively low, the original DDM shape becomes "submerged" in noise, compromising the quality of geometric feature parameters. Figure 1 illustrates the distribution of abnormal DDM across different SNR ranges. A significant increase in abnormal DDM samples is observed when SNR < 1.3, so Work 1 retains samples with SNR ≥1.3.

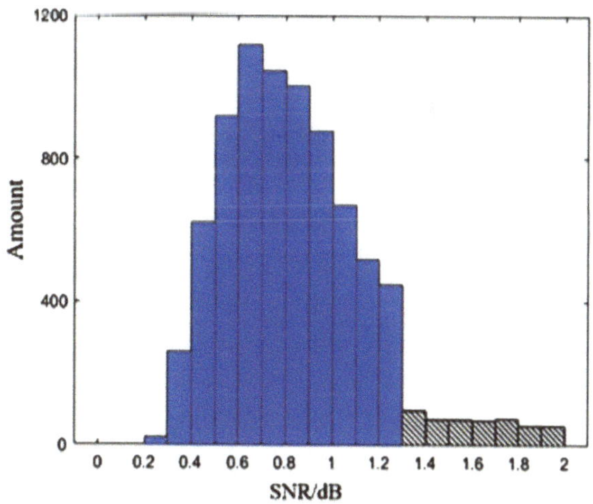

Fig. 1 The number distribution of abnormal angles in different SNR retrievals

Preprocessing Criteria Specific to Work 2

To establish a more accurate retrieval model, Work 2 adds additional quality control constraints beyond the common criteria:

(a) **Strict SNR Filtering**: Low SNR values lead to irregular DDM geometry, rendering angle-related parameters meaningless. Following extensive statistical data verification, Work 2 only selects samples with an SNR greater than 4.
(b) **Positive Constraint on DDM-Related Parameters**: Observed values of DDM-related parameters, such as leading-edge slope (LES) and normalized bistatic radar cross-section (NBRCS), must be positive.
(c) **Range-Corrected Gain (RCG) Filtering**: RCG is an effective indicator of the quality of received scattered signals, defined as:

$$RCG = \frac{G_{sp}^{R} * 10^{27}}{\left(R_T^{sp} R_R^{sp}\right)^2} \tag{1}$$

where R_T^{sp} represents the distance between the specular reflection point and the transmitter, R_R^{sp} represents the distance between the specular reflection point and the receiver, and G_{sp}^{R} represents the receiver antenna gain.

3.3 Feature Parameter Extraction

Wind direction modulates the asymmetry of Delay-Doppler Maps (DDMs). In the theoretical scattering model of bistatically reflected GPS signals proposed by Zavorotny and Voronovich [22], the wind direction-sensitive parameter is the scattering coefficient σ^0, defined as:

$$\sigma^0 = \frac{\pi |R|^2 q^4}{q_z^4} P(\vec{s}) \tag{2}$$

$$P(\vec{s}) = \left[\frac{1}{2\pi\sqrt{mss_u \cdot mss_v}} \exp\left\{-\frac{1}{2}\left(\frac{s_u^2}{mss_u} + \frac{s_v^2}{mss_v}\right)\right\}\right]\left[1 + \sum_{i,j=1}^{\infty} c_{ij} H_i(s_u) H_j(s_v)\right] \tag{3}$$

where R denotes the Fresnel reflection coefficient, q is the scattering vector, and $\vec{s} = (s_u, s_v)$ represents the sea surface slope vector (with s_u and s_v as the upwind and crosswind slope components, respectively). mss_u and mss_v are the upwind and crosswind mean square slopes (MSS), while H_i and H_j are Hermite polynomials with corresponding coefficients c_{ij} [56]. Extensions in the upwind and crosswind directions translate to DDM shape variations: wind speed governs the extension range, and wind direction dictates the asymmetry of this extension.

According to the GNSS-R bistatic radar equation, this asymmetry originates from wind direction-induced changes in the orientation of sea surface slopes. However, other factors (e.g., wind speed, geometric configurations) can also influence DDM shape [57], making direct extraction of wind direction from DDMs challenging. Consequently, there is a need to identify DDM geometric feature parameters that are highly sensitive to wind direction while remaining less responsive to other confounding factors. Extensive research and observational evidence have shown that DDMs expand as wind speed increases. During this expansion, the DDM peak, the centroid of the "DDM_r" region, and the centroid of the "DDM_{skirt}" region shift toward larger delay bins. Notably, the differences in these shifts are more strongly determined by wind direction than by other factors such as wind speed. DDM_r is defined as the horseshoe-shaped portion of the DDM with power values exceeding a specified threshold after normalization, while DDM_{skirt} refers to the region where power values fall within a certain ratio of the DDM peak power.

Based on these observations, two angle-based measurement parameters that capture DDM asymmetry are proposed to establish a quantitative relationship with wind direction, serving as critical inputs for machine learning-driven wind direction retrieval [57]:

(a) The vector azimuth angle from the DDM peak point (the specular reflection point on the sea surface) to the centroid of DDM_r (denoted as φ_1);
(b) The vector azimuth angle from the centroid of DDM_r to the centroid of DDM_{skirt} (denoted as φ_2).

Figure 2a illustrates φ_1 and φ_2 extracted from CYGNSS L1 full DDM, while Fig. 2b shows the same parameters from CYGNSS compact DDM.

To avoid calibration-related issues, DDM power values are normalized relative to the peak power prior to calculating these parameters. The specific extraction steps are as follows:

(a) Normalize the DDM using the peak power value;
(b) Filter out the DDM_r region from the normalized DDM, retaining pixels with power values greater than e^{-1};

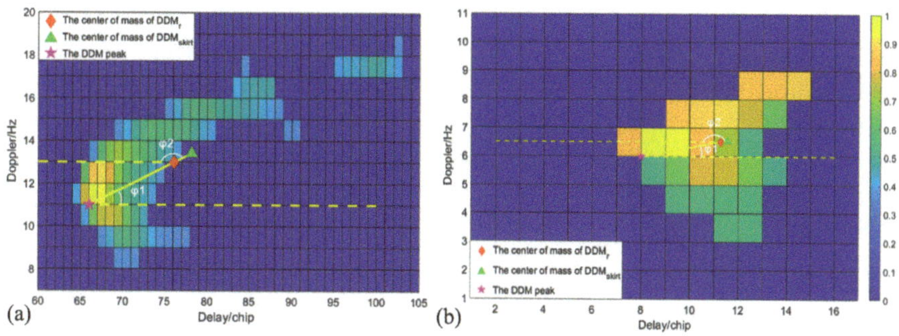

Fig. 2 Extract feature parameters φ1 and φ2 from **a** full DDM (Work 1) and **b** L1 compact DDM (Work 2)

(c) Calculate the coordinates of the DDM peak point ($X_{\max}$, $Y_{\max}$) and the centroid coordinates (X_{cm}, Y_{cm}) of DDM_r, where X represents time delay, Y represents Doppler shift, and $T(X, Y)$ denotes the power at time delay X and Doppler shift Y:

$$X_{cm} = T_0^{-1} \sum \sum X \cdot T(X, Y) dX dY \tag{4}$$

$$Y_{cm} = T_0^{-1} \sum \sum Y \cdot T(X, Y) dX dY \tag{5}$$

where

$$T_0 = \sum \sum T(X, Y) dX dY \tag{6}$$

(d) From DDM_r, the areas with power values ranging from 30 to 70% of the peak power are further isolated, and designated as DDM_{skirt}, whose centroid is calculated using the same method described in step (c);

(e) Compute φ_1 and φ_2 using the coordinates of the DDM peak point, DDM_r centroid (X_{cm}, Y_{cm}), and DDM_{skirt} centroid ($X_{cmskirt}$, $Y_{cmskirt}$) to obtain the geometric feature parameters.

Feature Validation for Work 1 (Full DDM)

Figures 3, 4, 5 and 6 illustrate significant differences in DDM shape across different wind directions, particularly between Figs. 3 (or 5) and Figs. 4 (or 6). The region outside the dark blue area represents DDM_r; the five-pointed star denotes the peak point, the diamond represents the DDM_r centroid, and the triangle represents the DDM_{skirt} centroid. These results confirm that φ_1 and φ_2 can be effectively used for wind direction retrieval. However, it is also observed that wind direction retrieved using φ_1 and φ_2 exhibits 180° ambiguity. To resolve this ambiguity and improve retrieval accuracy, additional wind direction-related parameters are required.

Figures 3, 4, 5 and 6 further indicate that the normal range of φ_1 is between 0° and 10°, while the normal range of φ_2 is between 170° and 180°. Consequently, this study defines $\varphi_1 > 15°$ and $\varphi_2 < 165°$ as thresholds for abnormal angles. For DDMs with abnormal angles, φ_1 and φ_2 fail to establish a reliable relationship with wind direction, so corresponding data samples are excluded during preprocessing.

Feature Extraction and Validation for Work 2 (L1 Compact DDM)

Work 2 employs CYGNSS L1 compact DDM (11 × 17 dimensions) for feature extraction, with adjusted thresholds for DDM_r and DDM_{skirt} regions compared to Work 1: after DDM normalization, DDM_r is defined as the horseshoe-shaped region with power values exceeding 0.5 (vs. e^{-1} for Work 1), while DDM_{skirt} refers to the

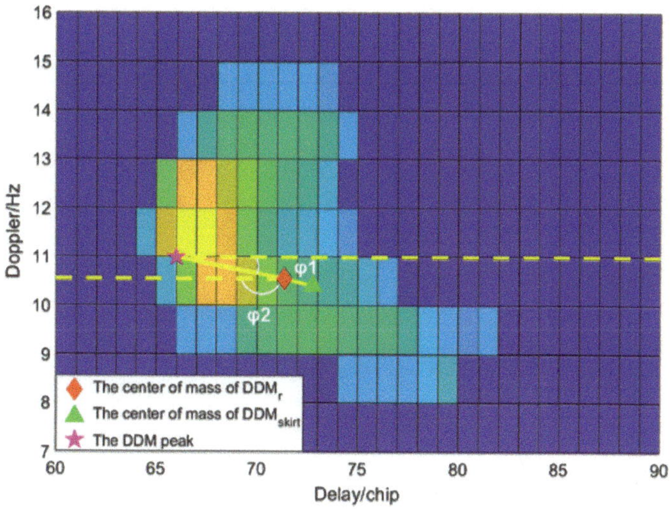

Fig. 3 Extracting feature parameters φ_1 and φ_2 based on geometric relationship from DDM of 0° wind direction

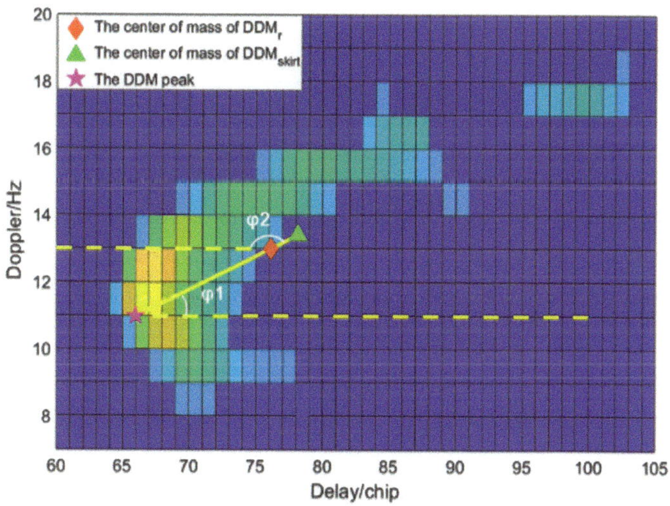

Fig. 4 Extracting feature parameters of geometric relationship φ_1 and φ_2 from DDM of 90° wind direction

subset of DDM_r within 40–70% of the peak power (consistent with Work 1). The definitions of φ_1 and φ_2 remain the same as those in Sect. 3.3.1.

Figures 7, 8, 9 and 10 present L1 compact DDMs under wind directions of 0°, 90°, 180°, and 270°, respectively. Compared to Full DDM, compact DDM also exhibits

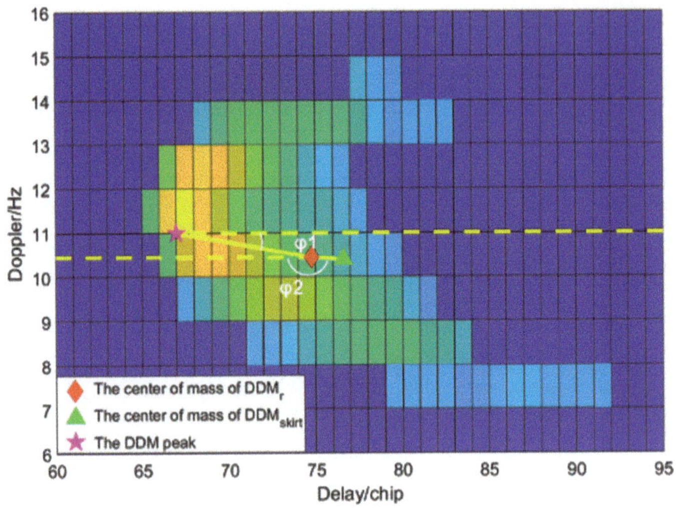

Fig. 5 Extracting feature parameters φ_1 and φ_2 based on geometric relationship from DDM of 180° wind direction

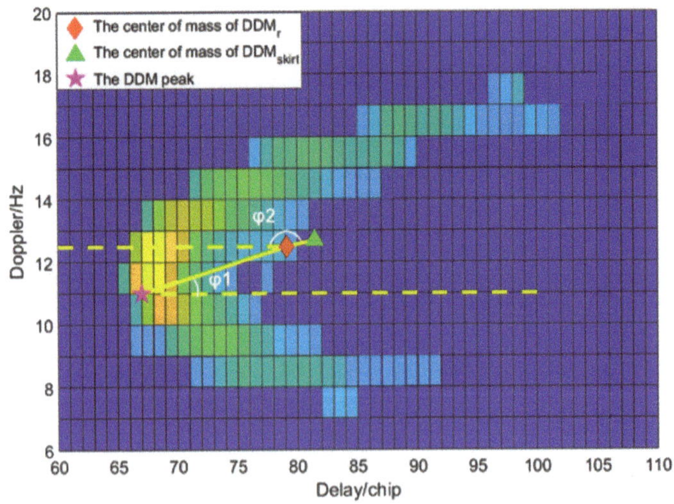

Fig. 6 Extracting feature parameters φ_1 and φ_2 based on geometric relationship from DDM of 270° wind direction

wind direction-dependent asymmetry—though the magnitude of this asymmetry is significantly diminished due to dimensionality reduction.

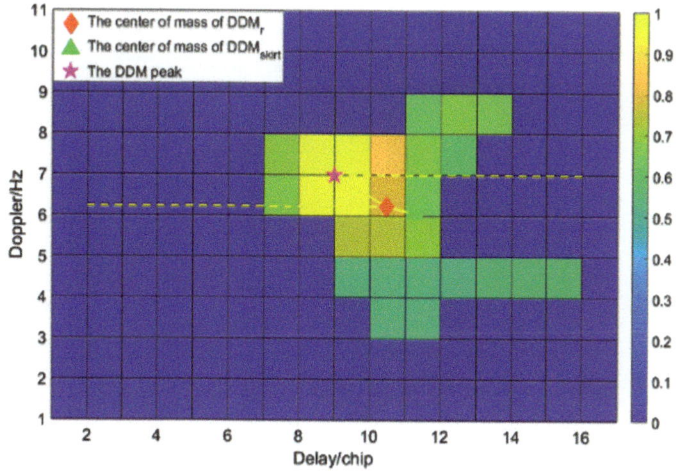

Fig. 7 Extracting feature parameters φ_1 and φ_2 based on the geometric relationships from the L1 compact DDM of a 0° wind direction

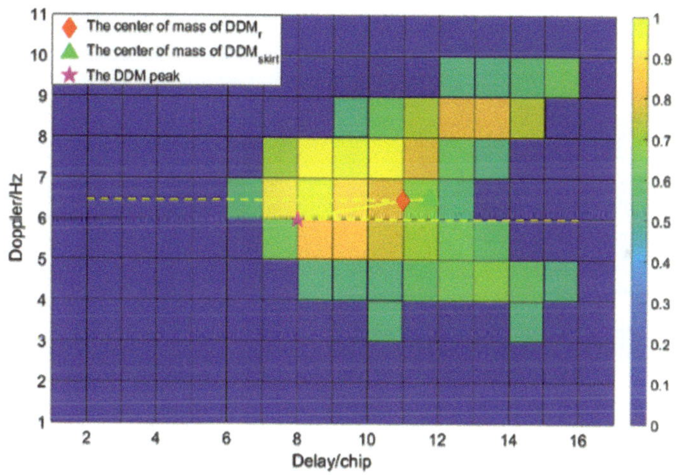

Fig. 8 Extracting feature parameters φ_1 and φ_2 based on the geometric relationships from the L1 compact DDM of a 90° wind direction

Quantitative Relationship Between Geometric Features and Wind Direction

Figures 11 and 12 display scatter plots of wind direction versus φ_1 and φ_2, respectively (wind speed: 9–10 m/s, elevation angle: 60–80°, azimuth angle: 100–120°), with red curves representing fitted trends. Quantitatively, φ_1 and φ_2 exhibit approximate sinusoidal relationships with wind direction—consistent with the simulation

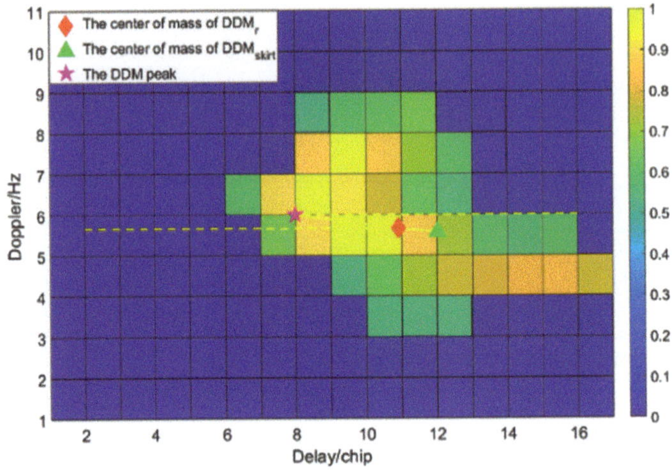

Fig. 9 Extracting feature parameters φ_1 and φ_2 based on the geometric relationships from the L1 compact DDM of a 180° wind direction

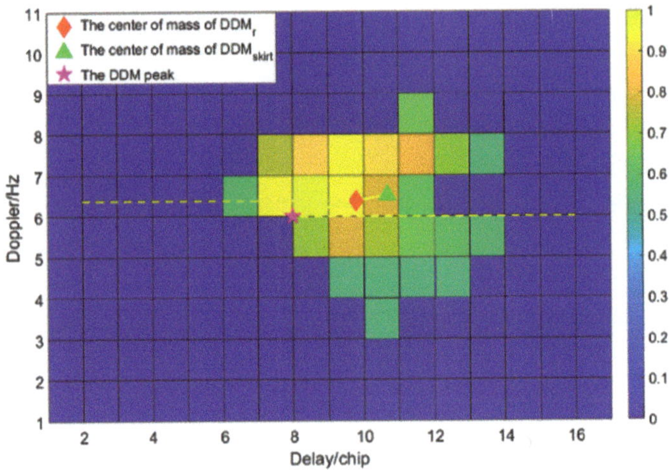

Fig. 10 Extracting feature parameters φ_1 and φ_2 based on the geometric relationships from the L1 compact DDM of a 270° wind direction

results in [56]. However, this correlation is only observable when controlling for confounding factors (e.g., wind speed, elevation angle), justifying the adoption of a machine learning model with supplementary feature parameters instead of a pure physical model based solely on φ_1 and φ_2. These supplementary features include wind speed, elevation angle, sea level pressure, and sea surface temperature (detailed in Sect. 3.1.2.2).

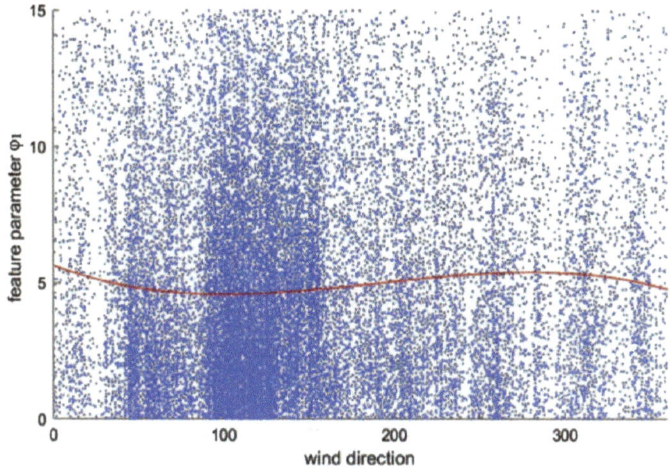

Fig. 11 Scatter diagram of wind direction and φ_1. The wind speed is between 9 and 10 m/s, the elevation angle is between 60 and 80° and the azimuth angle is between 100 and 120°

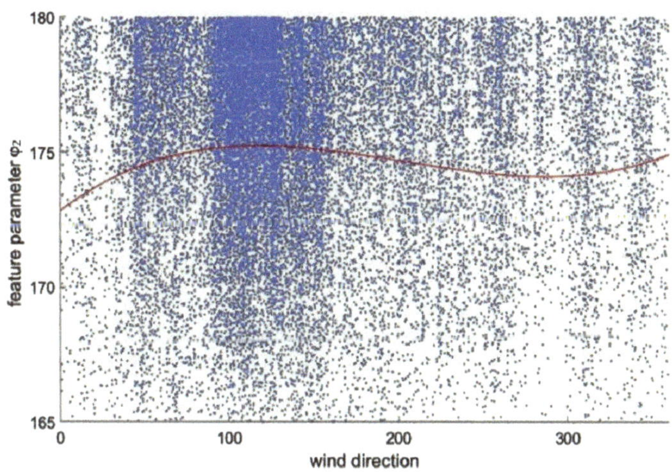

Fig. 12 Scatter diagram of wind direction and φ_2. The wind speed is between 9 and 10 m/s, the elevation angle is between 60 and 80° and the azimuth angle is between 100 and 120°

4 Machine Learning Based Wind Direction Retrieval

As highlighted in the previous section, DDM geometric features (φ_1 and φ_2) exhibit a sinusoidal correlation with wind direction, but this relationship is easily interfered with by confounding factors such as wind speed and elevation angle. Pure physical models struggle to capture complex nonlinear mappings, making machine learning an ideal approach for wind direction retrieval. This chapter focuses on two core works tailored to different CYGNSS DDM products: Work 1 constructs a global wind direction retrieval model based on CYGNSS Full DDM data and the SVM algorithm, while Work 2 addresses the challenge of low-dimensional CYGNSS L1 compact DDM data by comparing multiple machine learning methods (SVM, BP neural network, CNN). Detailed model designs, training processes, and evaluation methods are presented below.

4.1 SVM-Based Global Wind Direction Retrieval Using CYGNSS Full DDM

Building on prior research on machine learning-based sea surface wind speed inversion models using CYGNSS data [30, 55], this work develops a spaceborne GNSS-R sea surface wind direction retrieval model based on the SVM algorithm. Compared to existing relevant studies, spaceborne GNSS-R-based sea surface wind direction retrieval offers a broader detection range and access to larger data volumes. The CYGNSS satellite data selected in this research features extensive spatial coverage and a long temporal span, enabling the establishment of a robust global retrieval model.

Feature Parameter Design

The power waveform of GNSS reflected signals is influenced not only by sea surface wind speed but also by GNSS-R geometric configurations. To comprehensively capture these effects, nine-dimensional feature parameters are extracted for each GNSS-R event, including CYGNSS satellite parameters and geometric feature parameters derived from DDM waveforms. Two distinct mapping relationships between feature parameters and sea surface wind direction (WD) are proposed:

$$WD_1 = f_1(ws, ele, azi, \varphi_1, \varphi_2) \tag{7}$$

$$WD_2 = f_2(ws, ele, azi, \varphi_1, \varphi_2, LES, NBRCS, SNR, RCG) \tag{8}$$

where the feature parameters correspond to wind speed (*ws*), receiver elevation angle (*ele*), receiver azimuth angle (*azi*), φ_1, φ_2, leading edge slope (*LES*), *NBRCS*, signal-to-noise ratio (*SNR*), and range-corrected gain (RCG). Compared to f_1, f_2 incorporates four additional feature parameters. Wind speed (*ws*), elevation angle (*ele*), and azimuth angle (*azi*) are critical for wind direction retrieval from DDMs, and reasonable results can be obtained using Eq. (7). However, these parameters fail to fully capture the influence of sea surface roughness on DDM characteristics. Therefore, *LES*, *NBRCS*, *SNR*, and *RCG* are introduced in Eq. (8) to address this limitation.

SVM Algorithm and Kernel Function Selection

The SVM is a widely used supervised machine learning algorithm rooted in statistical learning theory, specifically leveraging the "VC Dimension Theory" and "Structural Risk Minimization Theory." It offers distinct advantages in solving non-linear classification problems and high-dimensional feature space classification tasks. The core objective of the SVM algorithm is to identify the optimal hyperplane in the feature space that maximally separates different classes of samples. Formally, let $K(x, z)$ denote a positive definite kernel function, which implies the existence of a mapping $\varphi(\cdot)$ from the input space to a high-dimensional feature space such that:

$$K(x, z) = \varphi(x) \bullet \varphi(z) \tag{9}$$

In the dual optimization problem of the linear SVM, which converts the original constrained primal classification problem into a more tractable Lagrangian dual form and establishes the mathematical foundation for kernel function introduction, the kernel function $K(x, z)$ is employed to replace the inner product operation in the high-dimensional feature space, thereby deriving the solution of the non-linear SVM model as follows:

$$f(x) = \text{sign}\left(\sum_{i=1}^{N} \alpha_\iota^* y_\iota K(x, x_\iota) + b^*\right) \tag{10}$$

where α_ι^* and b^* are the optimal solutions obtained from the dual problem, and y_i represents the class label of the i-th training sample.

The non-linear SVM-based wind direction retrieval algorithm consists of two core phases: model training and wind direction inversion.

In the model training phase, the labeled training dataset $T = \{(x1, y1), (x2, y2), \ldots, (xn, yn)\}$ is input to the algorithm, where $\{x_\iota\}$ are the features and $\{y_i\}$ are the corresponding wind direction class labels, the constrained primal classification problem of the linear SVM is first converted into a solvable Lagrangian dual optimization problem, and the kernel function K(x, z) is introduced to replace the inner product operation in the high-dimensional feature space. By solving the dual optimization problem, the optimal estimates of Lagrangian multipliers α_ι^* and bias term b^* are obtained, and the final classification decision function for wind direction retrieval is derived, with the trained SVM model and decision function as the output.

In the subsequent wind direction inversion phase, the unlabeled test dataset is fed into the pre-trained model. For each test sample, the kernel function values with the support vectors in the training set are computed, and the test features are substituted into the pre-derived decision function to predict the wind direction class label.

This work adopts the Gaussian kernel (also known as the Radial Basis Function, RBF) due to its strong generalizability. The Gaussian kernel essentially measures the "similarity" between samples, enabling linear separability in a similarity-defined space. It offers several advantages, including mapping to an infinite-dimensional feature space, supporting diverse decision boundaries, and requiring only a single hyperparameter (γ), which simplifies parameter selection compared to polynomial kernels. The Gaussian kernel function is defined as:

$$K(x, y) = e^{-\gamma x - y^2} \tag{11}$$

where x and y are sample vectors, and γ represents the norm of the vector, which can be interpreted as the vector modulus.

Hyperparameter Optimization

To enhance the accuracy and generalization capability of the retrieval model, optimizing the kernel function parameter γ (gamma) and the penalty factor C is critical. The γ parameter influences the smoothness of the decision boundary, while C controls the model's tolerance for misclassified samples. A grid search cross-validation approach is employed for hyperparameter optimization: a broad search range for γ and C is predefined, followed by exhaustive training over all parameter combinations on the grid. The classification accuracy under cross-validation is evaluated for each parameter set, and the combination yielding the highest accuracy is selected for final model training. The optimization process is illustrated in Fig. 13.

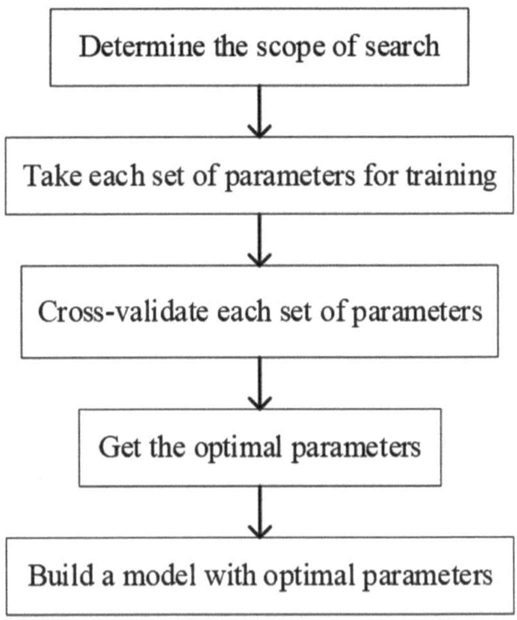

Fig. 13 Cross validation based on grid search method

Retrieval Process

DDM sample data are derived from CYGNSS Full DDM products, while receiver elevation angle, azimuth angle, LES, RCG, NBRCS, and SNR are extracted from CYGNSS L1 data corresponding to the specular reflection point of each Full DDM. Wind speed and wind direction ground truth data are obtained from ECMWF merged wind field products. All datasets are matched and processed based on spatiotemporal proximity.

For wind direction retrieval, geometric feature parameters, CYGNSS satellite parameters, and wind speed data are combined to form the training dataset, which serves as input to the SVM model. To improve computational efficiency and reduce model complexity, wind direction values ranging from 0° to 360° are discretized into 72 intervals with a step size of 5° (where 0° corresponds to true north). Wind direction ground truth data from ECMWF are converted into these interval labels, which are used to train the SVM model. The basic retrieval workflow is illustrated in Fig. 14.

Accuracy Assessment

The input dataset integrates CYGNSS satellite data and ECMWF reanalysis data, which are randomly split into an 20% training set and a 80% test set. A 20% subset of the training set is further allocated for grid search cross-validation (five-fold cross-validation). After model training, accuracy is evaluated on the test set using the RMSE, defined as:

$$RMSE = \sqrt{\frac{\sum_{i=1}^{n} \left| WD_{predict} - WD_{true} \right|_i^2}{n}} \tag{12}$$

where n is the number of test samples, WD_{true} is the wind direction ground truth from ECMWF reanalysis data, and $WD_{predict}$ is the midpoint value of the predicted wind direction interval (e.g., 2.5° for interval 1, 7.5° for interval 2, etc.).

4.2 Multi-machine Learning—Methods for Global Wind Direction Retrieval Using CYGNSS L1 Compact DDM

The low dimensionality of CYGNSS L1 compact DDM data (17 × 11) poses significant challenges to sea surface wind direction retrieval, as the reduced DDM asymmetry weakens direct feature sensitivity. To address this issue, this work investigates global satellite-borne GNSS-R wind direction retrieval models using CYGNSS L1 data based on three machine learning methods (SVM, BP neural network, CNN), aiming to identify the optimal algorithm through comprehensive comparison.

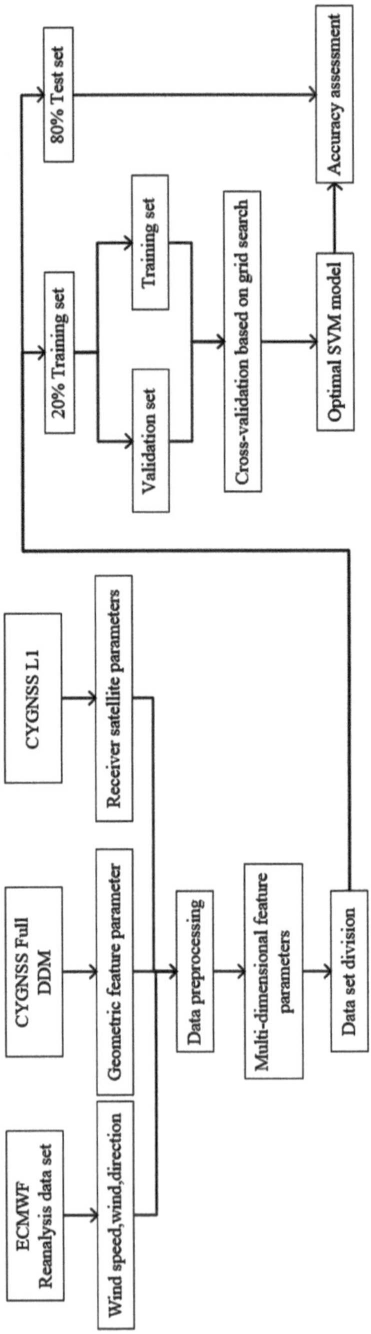

Fig. 14 GNSS-R wind direction retrieval flow chart based on SVM

Table 1 Detailed retrieval feature parameters

Category	Specific parameters
Geophysical parameters	Wind speed, mean sea level pressure (MSL), sea surface temperature (SST)
DDM asymmetry parameters	φ_1, φ_2
Signal quality parameters	Range corrected gain (RCG), signal-to-noise ratio (SNR), LES, NBRCS
DDM power parameters	brcs_ddm_sp_bin_dopp_col, brcs_ddm_sp_bin_delay_row, brcs_ddm_peak_bin_delay_row, brcs_ddm_peak_bin_dopp_col, ddm_noise_floor, Fresnel_coeff
Timing/correction parameters	Sp_fsw_delay, sp_ddmi_delay_correction, sp_ddmi_dopp_correction
Spatial/temporal parameters	Rx to sp. distance, Tx to sp. distance, elevation angle, azimuth angle, sp_vel_x/y/z, tx_vel_x/y/z, prn_code, spacecraft_num, sp_az_orbit

Data and Feature Parameters

The datasets employed include CYGNSS L1 data and ECMWF reanalysis data spanning 2020–2021. A total of 31-dimensional feature parameters is extracted (Table 1), integrating geophysical parameters, DDM characteristics, signal quality indicators, and satellite-related parameters to compensate for the low dimensionality of L1 DDM data. Wind direction data from ECMWF serves as the true reference value for model training and validation.

Similar to Work 1, Wind direction is divided into 72 intervals at 5° increments, with 0° representing true north. The 31-dimensional feature parameters and their corresponding labels form the experimental dataset. The retrieval workflow is illustrated in Fig. 15.

Dataset Partitioning

To mitigate overfitting and achieve optimal parameter optimization, the dataset is partitioned as follows: 10% of the total samples are used as the training set for model construction, 10% as the validation set to verify parameter tuning results, and the remaining 80% as the test set for accuracy evaluation. Additionally, generalized test sets are introduced to assess the model's generalization capability across different scenarios.

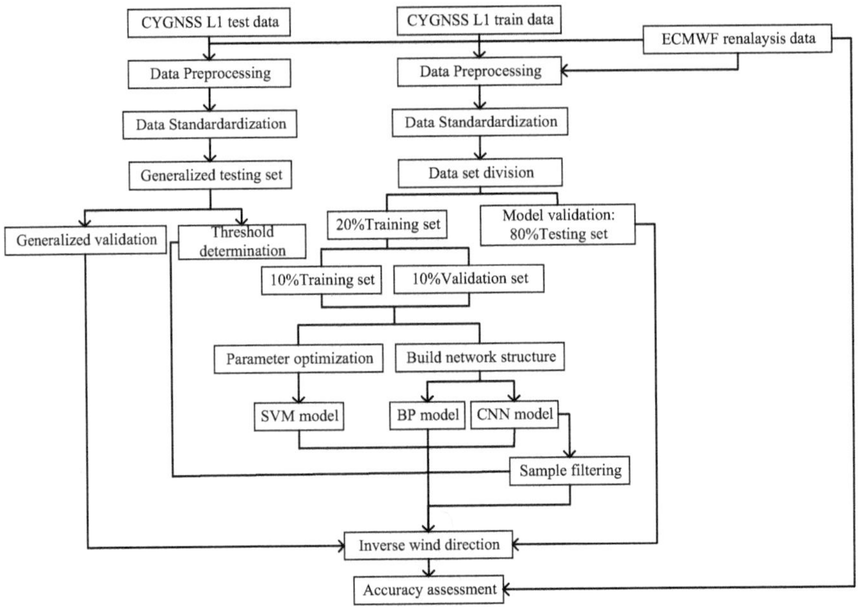

Fig. 15 Flow chart of wind direction retrieval based on different machine learning models

Detailed Model Designs

SVM Wind Direction Retrieval Process

Due to SVM's inherent limitation in handling large-scale sample training, 50,000 data samples are randomly selected for model training. This work employs the radial basis function (RBF) as the kernel function, leveraging its strong universality and adaptability to nonlinear relationships. Hyperparameter optimization is performed using the grid search method, which efficiently explores the parameter space (γ and C) and identifies the configuration that minimizes retrieval errors. The detailed parameter optimization workflow is illustrated in Fig. 16.

BP Wind Direction Retrieval Process

Figure 17 illustrates the network structure of the BP model for sea surface wind direction retrieval. A BP neural network typically consists of an input layer, hidden layers, and an output layer. This work designs a BP model with three hidden layers specifically tailored for wind direction retrieval: the input layer accepts 31-dimensional feature parameters, the optimal number of nodes in each hidden layer is determined through iterative testing (configured as 256, 128, and 64), and the output layer outputs the probability distribution of the sample across all wind direction intervals.

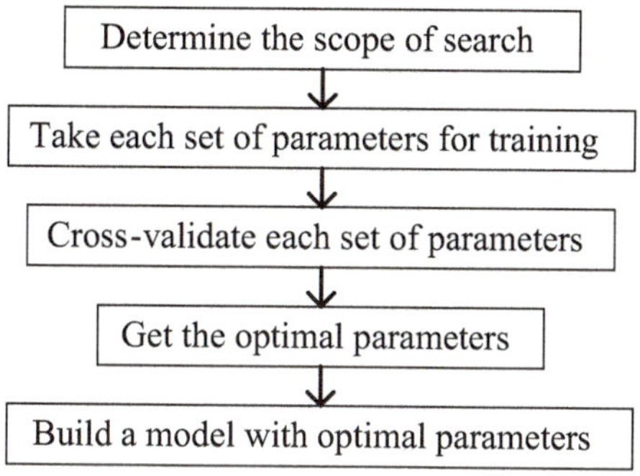

Fig. 16 Flow chart of SVM optimization

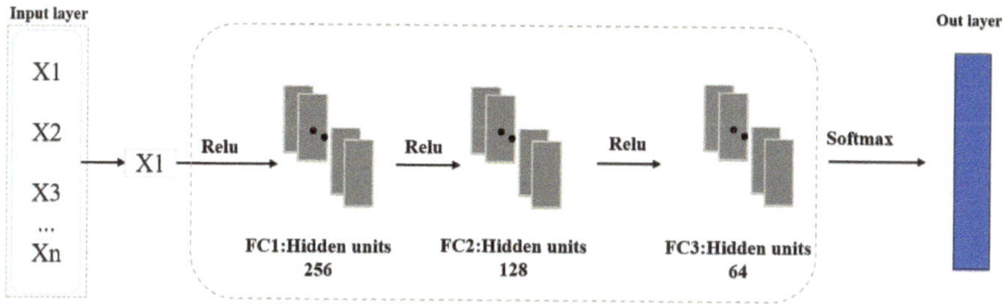

Fig. 17 Network structure of BP model for sea surface wind direction retrieval

The ReLU (Rectified Linear Unit) function is adopted as the activation function, and sparse_categorical_crossentropy is selected as the loss function—both well-suited for multi-classification tasks involving integer labels. The model undergoes repeated forward propagation (calculating predicted values from input to output) and BP (adjusting weights and biases based on loss) processes until the error function converges or the maximum number of training epochs is reached.

CNN Wind Direction Retrieval Process

The CNN model employed in this work comprises a sequence of convolutional layers, pooling layers, fully connected layers, and auxiliary functional layers (Fig. 18). Specifically, the model includes 6 convolutional layers (3×1 convolutional kernels, with the number of kernels increasing to 256, 512, and 1024 to capture increasingly complex features), 3 pooling layers, and 1 fully connected layer.

To mitigate overfitting, a dropout layer (dropout rate $= 0.3$) is integrated to temporarily deactivate individual neural network units during training. A batch normalization layer is appended after each convolutional layer to standardize the data distribution, mitigating the gradient explosion problem and accelerating training convergence. Early stopping is also implemented to preserve the optimal parameter configuration. The ReLU function is used as the activation function to introduce

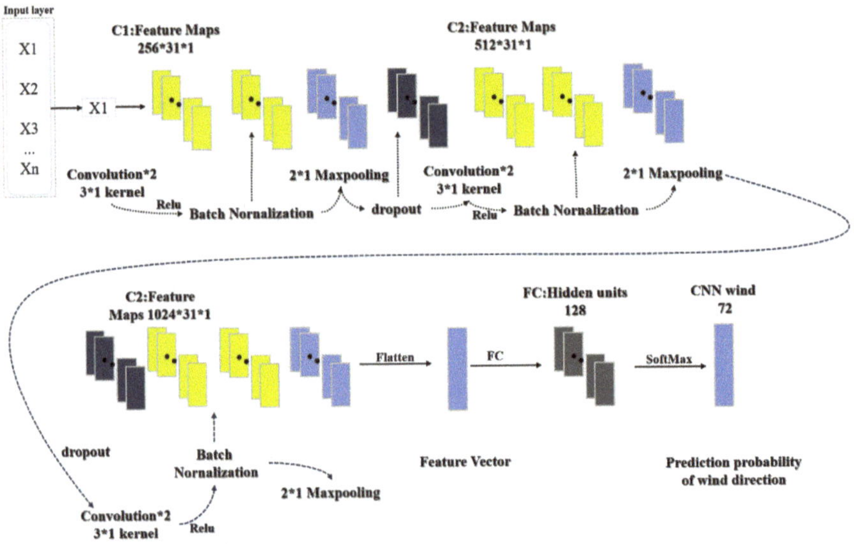

Fig. 18 Network structure diagram of CNN model for sea surface wind direction retrieval

nonlinearity, and the Adam optimizer is selected for adaptive parameter updating. Compared with the BP neural network, the CNN model leverages convolutional and pooling layers to extract hierarchical feature information more effectively, with fewer parameters resulting in higher computational efficiency and improved generalization performance.

Accuracy Evaluation Plan

Consistent with Work 1, RMSE is adopted as the core evaluation indicator, and bias is also used.

$$Bias = \frac{\sum\left(WD_{predict} - WD_{true}\right)}{m} \tag{13}$$

The models are evaluated on both the standard test set and generalized test sets to assess overall accuracy and generalization capability. Additionally, the retrieval accuracy differences across different wind directions are analyzed to identify potential performance bottlenecks. Finally, filters are introduced to optimize model outcomes, further enhancing the reliability of retrieval results.

5 Retrieval Results and Analysis

This section focuses on the results and in-depth analysis of the two-core wind direction retrieval works, following the same framework as the previous section. For Work 1 (CYGNSS Full DDM + SVM), the retrieval performance under different wind speed ranges and feature parameter combinations is analyzed. For Work 2 (CYGNSS L1 Compact DDM + multi-model), the performance of SVM, BP, and CNN models is compared, and optimization measures such as filters and data balance processing are discussed. Detailed results and analysis are presented below.

5.1 Retrieval Results of Work 1 (CYGNSS Full DDM + SVM)

Work 1 uses CYGNSS Full DDM data and SVM algorithm for wind direction retrieval, with the dataset covering the period from February 27, 2019, to November 17, 2020. The retrieval results are analyzed from three aspects: dataset characteristics, grid search optimization, and mapping relationship comparison.

Dataset Characteristics after Preprocessing

The basic information of the dataset is summarized in Table 2, including data types, time coverage, and temporal resolution. After strict preprocessing (QC filtering, wind speed filtering, offshore distance filtering, SNR filtering), the data volume under different wind speed ranges is shown in Table 3. It can be seen that the total number of valid samples with wind speed ≥5 m/s reaches 127,786, and as the wind speed threshold increases, the number of samples gradually decreases, with only 6498 samples remaining when wind speed ≥15 m/s.

The temporal and directional distribution of the data is visualized in Figs. 19 and 20. Figure 19 shows the monthly distribution of Full DDM data before and

Table 2 Data set information

Data set	Type	Time coverage	Temporal resolution
CYGNSS	L1	February 27, 2019–November 17, 2020	1 second
CYGNSS	Full DDM	February 27, 2019–November 17, 2020	1 second
ECMWF	Reanalysis dataset	February 27, 2019–November 17, 2020	1 hour

Table 3 Data quantity of different wind speed ranges

Wind peed	≥5 m/s	≥8 m/s	≥10 m/s	≥12 m/s	≥15 m/s
Amount	127,786	57,593	29,788	16,552	6498

after preprocessing: the original data (Fig. 19a) and preprocessed data (Fig. 19b) are both concentrated between July and November, indicating that the data has obvious seasonal aggregation characteristics. Figure 20 presents the distribution of filtered data across different wind direction intervals (72 intervals in total, 5° per interval), and the samples are mainly concentrated in intervals 1–36 (corresponding to 0°–180°), showing an uneven distribution of wind directions.

To obtain the optimal performance of the SVM model, grid search cross-validation is used to optimize the hyperparameters (γ and C). Figure 21 shows the grid search results for the dataset with wind speed ≥8 m/s using 9-dimensional feature parameters. It can be observed that the optimal parameter combination is concentrated around $C = 10$ and $\gamma = 10^{-1.5}$, and further refinement can be achieved by narrowing the search range.

Table 4 summarizes the optimal parameters and corresponding RMSE values of the SVM model under different wind speed ranges for the two mapping relationships (f_1 and f_2). For the 5-dimensional feature parameter mapping f_1, the RMSE decreases gradually with the increase of wind speed, from 49.51° (≥5 m/s) to 39.51° (≥15 m/s), indicating that the retrieval accuracy is positively correlated with wind speed within a certain range. For the 9-dimensional feature parameter mapping f_2, the RMSE reaches the minimum of 26.70° when wind speed ≥10 m/s, and increases slightly to 28.27° when wind speed ≥12 m/s, which is mainly due to the decline in data quality (such as reduced SNR and increased abnormal angle samples) under high wind speeds. When wind speed ≥15 m/s, the RMSE decreases to 26.78°, which is attributed to the reduced data volume and spatiotemporal concentration of high-wind-speed samples, improving the classification accuracy of the SVM model.

Comparison of Two Mapping Relationships

By comparing the retrieval results of the two mapping relationships (f_1 and f_2) in Table 4, significant performance differences are observed, and these differences are consistent across all wind speed ranges.

From a quantitative perspective, the RMSE of f_2 (9-dimensional feature parameters) is substantially reduced compared to f_1 (5-dimensional feature parameters). At wind speed ≥5 m/s, the RMSE decreases from 49.51° to 30.03°, representing a reduction of 39.3%. At wind speed ≥10 m/s (the optimal performance range), the RMSE drops from 44.12° to 26.70°, with a reduction rate of 39.5%. Even at high wind speeds ≥15 m/s, f_2 still maintains a 32.4% reduction in RMSE (from 39.51° to 26.78°). This systematic accuracy improvement verifies the effectiveness of the additional feature parameters.

From a physical perspective, f_1 only relies on wind speed, elevation angle, azimuth angle, and core geometric features (φ_1 and φ_2). While it can capture the basic correlation between wind direction and DDM asymmetry, it fails to characterize the subtle modulation of DDM waveforms by sea surface roughness. In contrast, the four additional parameters in f_2—LES, NBRCS, SNR, and RCG—exactly compensate for this deficiency. LES and NBRCS directly reflect the variation of sea surface

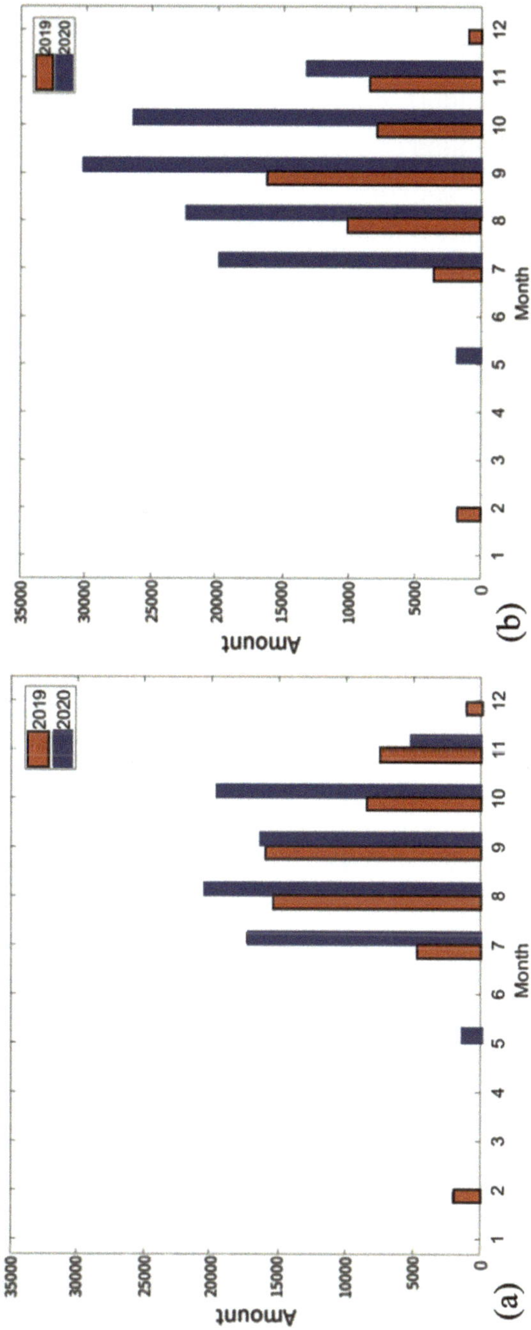

Fig. 19 Quantitative distribution of Full DDM data in different months before **a** and after **b** data preprocessing

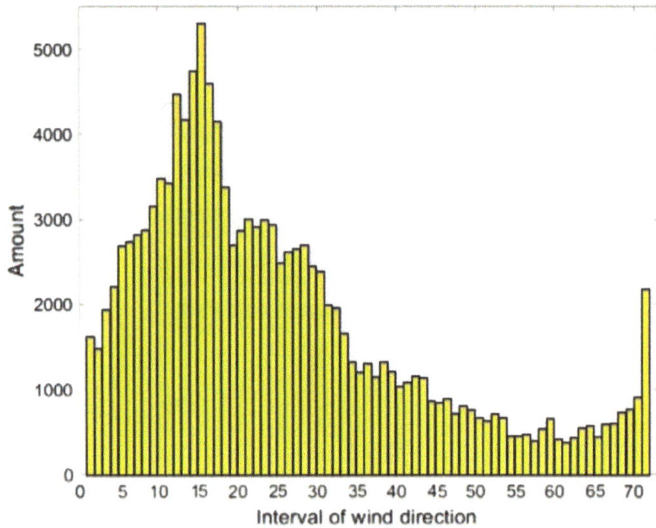

Fig. 20 Quantitative distribution of filtered Full DDM data in different wind direction's intervals

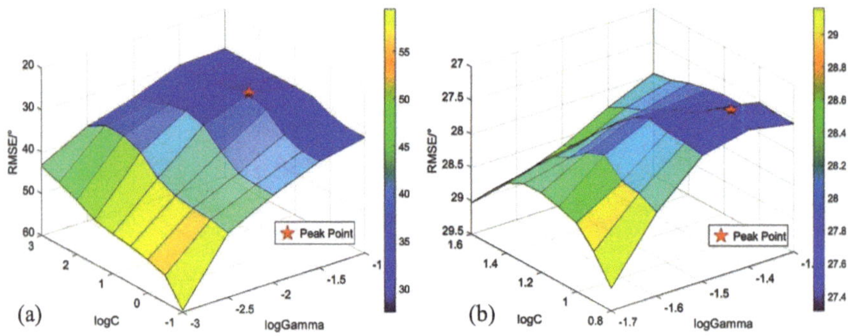

Fig. 21 Grid search results of datasets with wind speed greater than 8 m/s with 9-dimensional feature parameters

roughness with wind direction, while SNR and RCG ensure the reliability of DDM feature extraction. These four parameters synergistically enhance the model's ability to distinguish wind direction signals, thereby significantly reducing retrieval errors.

In-Depth Analysis of Optimal Retrieval Results

According to Table 4, the optimal retrieval performance of Work 1 is achieved by the f_2 mapping relationship when wind speed ≥ 10 m/s (RMSE = 26.70°). Figure 22 presents the SVM classification confusion matrix for this dataset: most samples are

Table 4 SVM results of different wind speed dataset with two mapping relationships

Mapping relationship	Wind speed	Optimal parameters	RMSE (°)
5-dimensional feature parameters			
f_1	≥5 m/s	(gamma $=10^{-1.2}$, C=$10^{1.7}$)	49.51
f_1	≥8 m/s	(gamma $=10^{-1.3}$, C=$10^{1.9}$)	47.82
f_1	≥10 m/s	(gamma $=10^{-1.1}$, C=$10^{1.9}$)	44.12
f_1	≥12 m/s	(gamma $=10^{-0.9}$, C=$10^{2.1}$)	41.11
f_1	≥15 m/s	(gamma $=10^{-1}$, C=10^{2})	39.51
9-dimensional feature parameters			
f_2	≥5 m/s	(gamma $=10^{-1.6}$, C=$10^{1.4}$)	30.03
f_2	≥8 m/s	(gamma $=10^{-1.4}$, C=$10^{0.9}$)	27.31
f_2	**≥10 m/s**	**(gamma $=10^{-1.3}$, C = 10)**	**26.70**
f_2	≥12 m/s	(gamma $=10^{-1.4}$, C=$10^{1.2}$)	28.27
f_2	≥15 m/s	(gamma $=10^{-1.3}$, C=$10^{1.2}$)	26.78

classified into the correct wind direction interval or adjacent intervals, reflecting high overall classification accuracy. Figure 23 shows the distribution of prediction errors (diff) on the test set: the prediction error of most samples is within 5°–10°, and only a small proportion of samples have errors between 20° and 180°, indicating that the SVM model can effectively resolve the 180° wind direction ambiguity.

Notably, the RMSE of all datasets exceeds 20°, which is mainly due to the high sensitivity of RMSE to large-error samples. Even a small number of samples with errors exceeding 90° will significantly increase the overall RMSE. This phenomenon confirms that rigorous data preprocessing is crucial for extracting reliable geometric

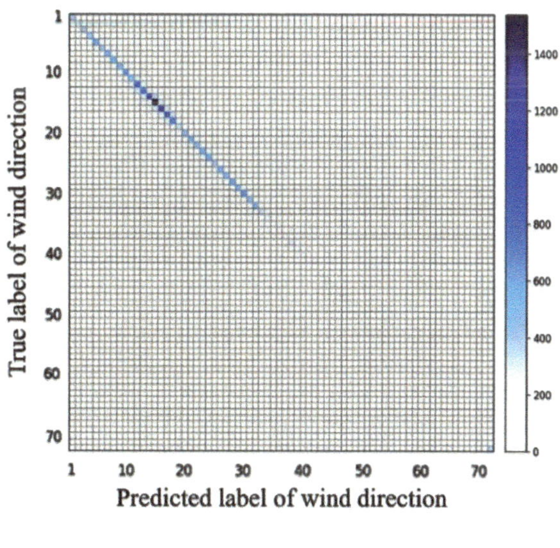

Fig. 22 Confusion matrix dividing the distribution of SVM classification results in 72 retrievals with wind speed greater than 10 m/s

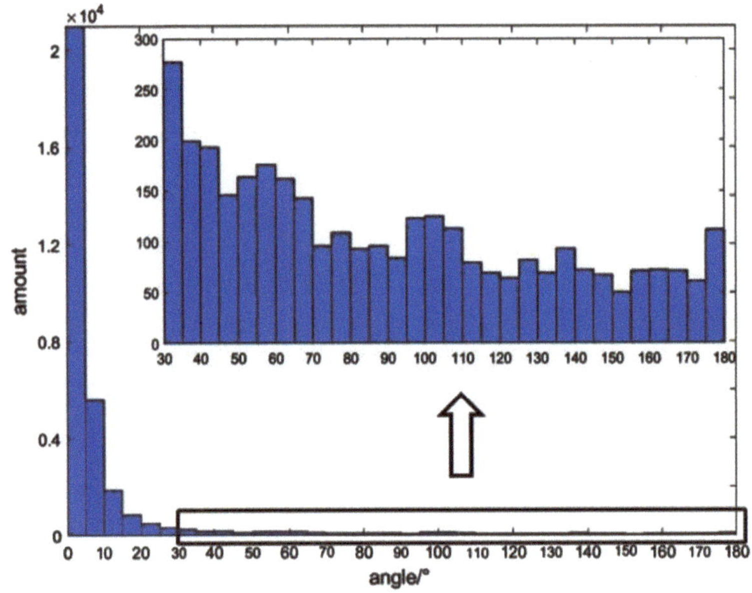

Fig. 23 diff in test set according to sample sequence with wind speed greater than 10 m/s

features (φ_1 and φ_2), and the 9-dimensional feature parameters can help the SVM model achieve accurate wind direction retrieval under specific wind speed and SNR conditions.

5.2 Retrieval Results of Work 2 (CYGNSS L1 Compact DDM + Multi-model)

Work 2 addresses the challenge of low dimensionality of CYGNSS L1 compact DDM data (17 × 11) by constructing SVM, BP, and CNN models for wind direction retrieval. The dataset covers the period from March 2020 to October 2020, and the results are analyzed from model performance comparison, filter optimization, and wind direction interval performance three aspects.

Dataset Characteristics and Partitioning

The basic information of the dataset is shown in Table 5: CYGNSS L1 data has a temporal resolution of 0.5 s, and ECMWF reanalysis data has a temporal resolution of 1 h. Wind speed, wind direction, SST, and MSL are derived from ECMWF, and spatiotemporal nearest-neighbor matching is used to align the two types of data.

Table 5 Dataset information

Data source	Dataset	Time	Time resolution
CYGNSS	L1 data	2020.3–2020.10	0.5 s
ECMWF	Reanalysis Dataset	2020.3–2020.10	1 h

Table 6 Dataset coverage time range table

Data set	Time	Data quantity
Training set	2020.3–2020.8	13,758,843
Model validation		
Model test set	2020.3–2020.8	11,007,074
Generalization validation		
Generalized test set 1	2020.9.1–2020.9.5	641,219
Generalized test set 2	2020.9.6–2020.9.10	605,567
Generalized test set 3	2020.9.11–2020.9.15	645,032
Generalized test set 4	2020.9.16–2020.9.20	615,097
Generalized test set 5	2020.9.21–2020.9.25	581,501
Generalized test set 6	2020.9.26–2020.9.30	590,622
Generalized test set 7	2020.10.1–2020.10.5	534,096
Generalized test set 8	2020.10.6–2020.10.10	555,957
Generalized test set 9	2020.10.11–2020.10.15	549,299
Generalized test set 10	2020.10.16–2020.10.20	560,211
Generalized test set 11	2020.10.21–2020.10.25	491,544
Generalized test set 12	2020.10.26–2020.10.30	600,542

To avoid overfitting and evaluate generalization ability, the dataset is partitioned as follows (Table 6): 10% of the data from March to August 2020 is used as the training set (13,758,843 samples), 10% as the validation set, and 80% as the model test set (11,007,074 samples); the data from September to October 2020 is divided into 12 non-overlapping 5-day generalized test sets to assess the model's robustness on out-of-distribution data. Figure 24 shows the wind direction interval distribution of the training set: easterly winds (0°–180°) dominate, resulting in uneven sample distribution.

Performance Comparison of SVM, BP, and CNN Models

Table 7 summarizes the retrieval performance of the three models on the model test set and generalized test sets. Due to the inherent limitation of SVM in handling large-scale datasets, only 50,000 samples are randomly selected for training, resulting in the lowest accuracy: the classification accuracy on the model test set is 0.39, and

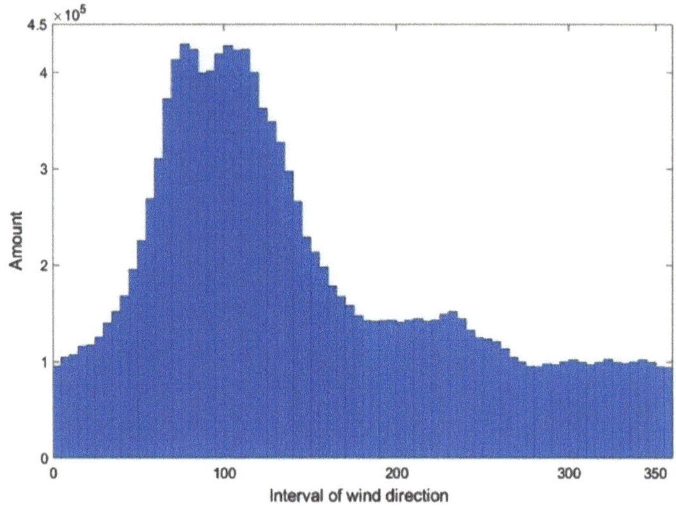

Fig. 24 Distribution diagram of wind direction interval of training set

the RMSE is 28.1°; on the generalized test sets, the classification accuracy drops to 0.24, and the RMSE rises to 52.5°. The BP model performs better than SVM, with a classification accuracy of 0.45 and an RMSE of 21.7° on the model test set, and a classification accuracy of 0.32 and an RMSE of 42.3° on the generalized test sets.

The CNN model achieves the best overall performance, leveraging convolutional layers to extract hierarchical features effectively. On the model test set, its classification accuracy reaches 0.53, and the RMSE is only 12.7°; on the generalized test sets, it maintains a classification accuracy of 0.44 and an RMSE of 37.6°, demonstrating stronger generalization ability than SVM and BP models.

Figure 25 further visualizes the performance of the three models on the 12 generalized test sets: affected by data quality differences, the model performance fluctuates across test sets, but the CNN model consistently leads in both classification accuracy (Fig. 25a) and RMSE (Fig. 25b). Its maximum classification accuracy on the generalized test sets is 0.47, and the minimum RMSE is 34.3°, confirming its robustness in practical applications. Compared with scatterometers, the CNN model has a higher average RMSE (~36°) but offers unique advantages such as adaptability to

Table 7 Retrieval performance results of different machine models

Method	Model validation		Generalization validation	
	Classification accuracy	RMSE (°)	Classification accuracy	RMSE (°)
SVM	0.39	28.1	0.24	52.5
BP	0.45	21.7	0.32	42.3
CNN	0.53	12.7	0.44	37.6

high wind speeds, insensitivity to weather conditions, no wind direction ambiguity, and low detection costs.

Further analysis of the CNN model's results (Figs. 26 and 27) shows that most samples are correctly classified into the target interval or adjacent intervals (Fig. 26), and the prediction error of ~92% of samples is below 20° (Fig. 27). However, ~8% of samples have errors exceeding 90°, which significantly increases the RMSE due to the high sensitivity of RMSE to large errors. This explains why the CNN model has high classification accuracy but a relatively high overall RMSE.

Filter Optimization Based on CNN Model

To mitigate the impact of large-error samples on retrieval performance, a filter is designed based on the maximum prediction probability of the CNN model (output by the SoftMax function). The core idea is that samples with higher maximum prediction probabilities are more likely to be correctly classified.

Since wind direction is divided into 72 discrete intervals, the CNN model established in this study is designed for a multi-classification task. The output layer's upper activation function is SoftMax, which produces a probability distribution across the 72 output units—each corresponding to a wind direction interval and storing the prediction probability of the sample belonging to that interval. Figure 28 illustrates an example of the prediction probability distribution output by the CNN model for a single data sample. Typically, the model selects the wind direction interval with the highest probability as the final prediction for the sample.

With 72 wind direction categories, the average classification probability per category (assuming uniform distribution) is ~0.014. In general, the higher the maximum classification probability of a sample predicted by the CNN model, the greater the likelihood of correct classification—provided the model is neither overfitted nor underfitted. Extensive experiments have shown that: (1) when the model is underfitted, the prediction probabilities are evenly distributed across the 72 intervals, resulting in low accuracy; (2) when the model is overfitted, a large number of samples exhibit high maximum prediction probabilities but significant deviations from the true values. To mitigate these issues and enhance the CNN model's retrieval performance, this study incorporates machine learning strategies such as max-pooling, dropout, and early stopping during model training. After finalizing the CNN model structure, the key focus shifts to determining the filtering threshold for the maximum classification probability. A threshold that is too low will render the filter ineffective, while an excessively high threshold will exclude too many valid samples, reducing the model's practicality.

The threshold selection is data-driven and model-specific. Figure 29 outlines the threshold determination process in this study:

Threshold Calibration (dotted box in Fig. 29): Use generalized test sets 1–3 as the calibration dataset. After generating predictions, separate samples into two groups: those with errors <20° and those with errors >90°. Adjust the threshold from low to high, gradually excluding samples with maximum classification probabilities below

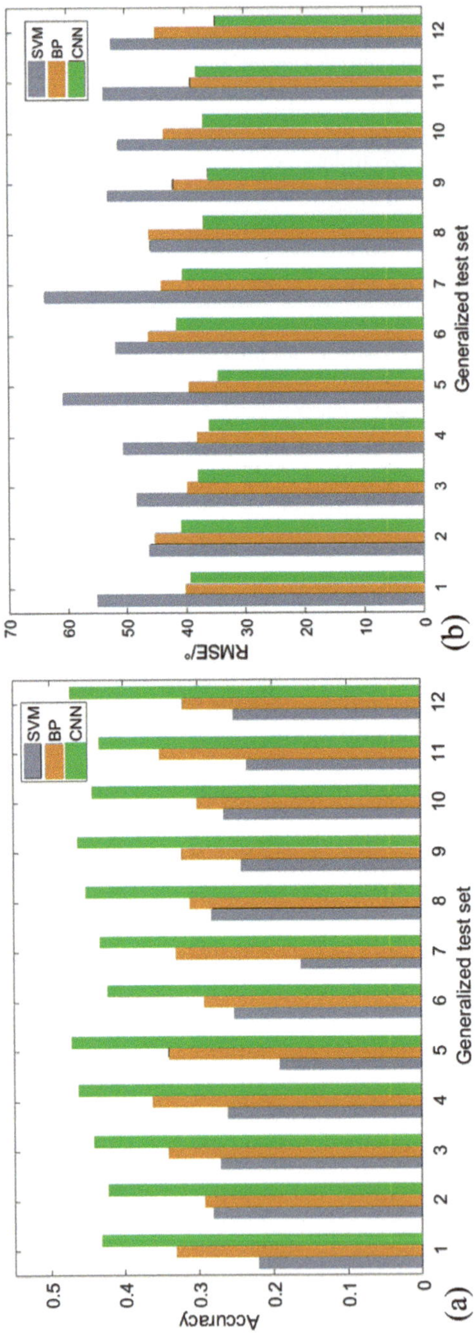

Fig. 25 Accuracy **a** and RMSE **b** results of SVM, BP and CNN on 12 generalized test sets

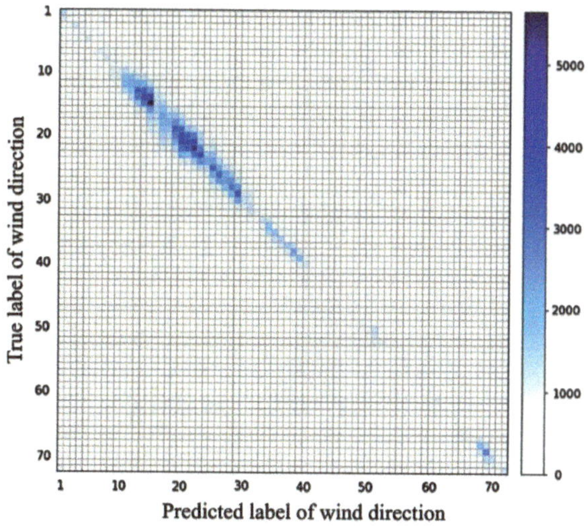

Fig. 26 Classification confusion matrix of CNN model on generalized test set

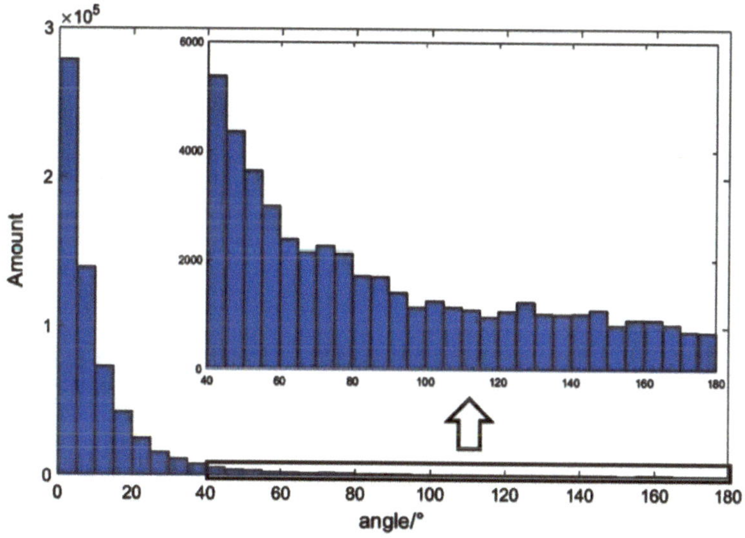

Fig. 27 Distribution of samples with different angle errors

the threshold. The optimal threshold is identified when: (a) more than 50% of samples with errors >90° are filtered out; and (b) fewer than 20% of samples with errors <20° are excluded.

Generalization Validation: Apply the calibrated threshold to new samples (e.g., the remaining 9 generalized test sets). Samples with maximum classification probabilities below the threshold are considered inaccurate predictions (potentially due to poor data quality or model mismatches) and are filtered out. Notably, once the

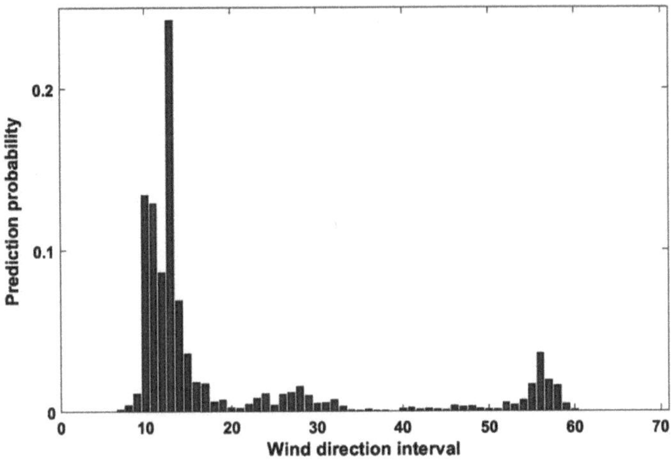

Fig. 28 An example of data prediction probability distribution

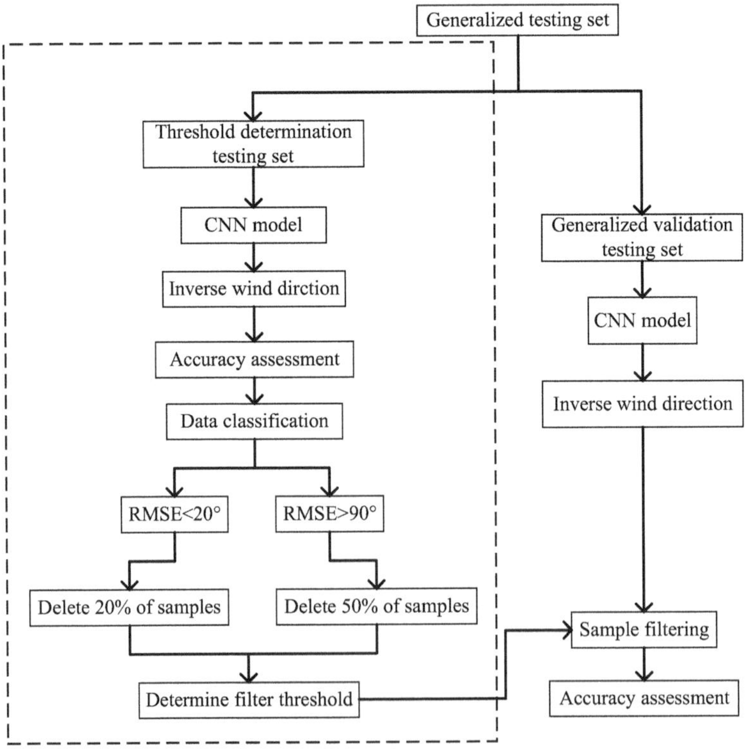

Fig. 29 Filter flow chart based on CNN model

threshold is determined using the calibration dataset, it is fixed and does not require re-calibration for additional generalized test sets.

Figure 30a and b confirm the feasibility of the proposed filter: samples with errors <20° consistently exhibit higher maximum classification probabilities than those with errors >90°. This validates the core assumption of the filter—setting a threshold on the maximum classification probability can effectively exclude most large-error samples while retaining the majority of small-error samples, thereby improving overall retrieval performance. After filtering out 50% of the samples with errors >90°, the optimal threshold is determined to be 0.21, which is applied to all 12 generalized test sets.

Table 8 shows the performance of the CNN model after filtering: as the number of filtered samples increases, the RMSE continues to decrease. When the sample retention rate is 39.3% (210,235 samples remaining), the RMSE is reduced to 19.5°, a significant improvement. However, excessive filtering will reduce the model's practicality, so it is necessary to balance sample retention and retrieval accuracy in practical applications.

Retrieval Performance across Different Wind Direction Intervals

Due to the uneven distribution of training set samples (easterly winds dominate), the CNN model's retrieval performance varies across different wind direction intervals. Figures 31 and 32 show the RMSE and bias of the CNN model on the generalized test set: the RMSE of easterly wind intervals (0°–180°) is 32.8°, while that of westerly wind intervals (180°–360°) is 48.4°, with a significant gap. The average bias across all intervals is 0.35°, indicating no systematic deviation.

To improve the performance of westerly wind intervals, undersampling is applied to easterly wind samples to balance the data distribution. Table 9 and Fig. 33 show the comparison before and after undersampling: the RMSE of westerly wind intervals decreases from 48.4° to 43.6°, but the RMSE of easterly wind intervals increases from 32.8° to 35.8°. Since the generalized test set is still dominated by easterly wind samples, the overall RMSE rises from 37.6° to 39.9°, indicating that balancing sample distribution may not be beneficial for overall performance when the test set has inherent unevenness.

5.3 *Key Findings and Summary*

(a) **Work 1 (Full DDM + SVM)**: The 9-dimensional feature parameters (incorporating LES, NBRCS, SNR, RCG) significantly outperform the 5-dimensional parameters, with the optimal RMSE of 26.70° (wind speed ≥10 m/s). The model can resolve 180° ambiguity, but large-error samples lead to overall RMSE exceeding 20°.

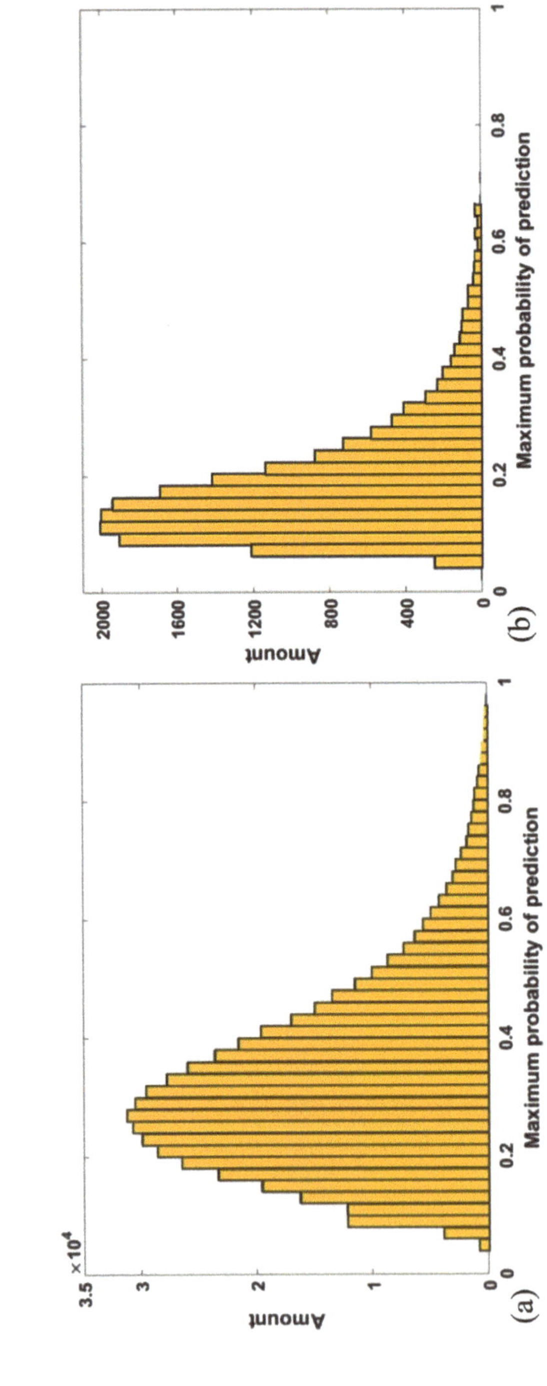

Fig. 30 Distribution of maximum probability of classification prediction with errors **a** less than 20 degrees and **b** greater than 90 degrees

Table 8 Filter performance results based on CNN model

Number of samples before filtering	Number of samples after filtering	Retention rate (%)	Average RMSE (°)
534,204	495,057	92.6	35.1
534,204	371,607	69.5	28.2
534,204	287,949	53.9	24.6
534,204	210,235	39.3	19.5

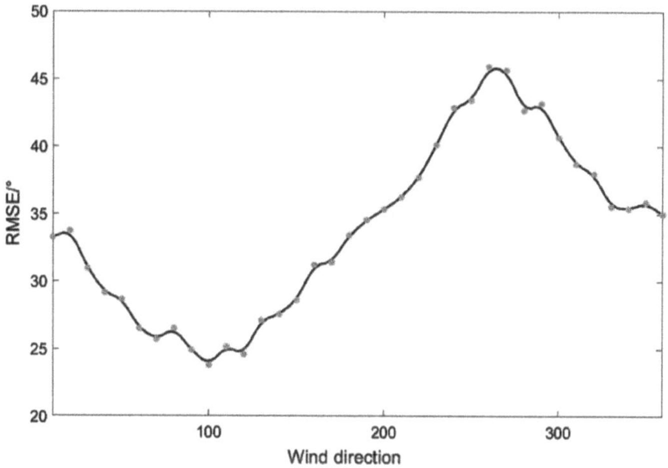

Fig. 31 RMSE at different wind direction intervals on the generalized test set. Each point represents a different wind speed interval, and the line is the fitted curve of these points, which facilitates the observation of RMSE variations across different wind speed intervals

(b) **Work 2 (L1 DDM + Multi-model)**: The CNN model outperforms SVM and BP models in both accuracy and generalization ability, with an RMSE of 37.6° on generalized test sets. After filter optimization, the RMSE can be reduced to 19.5°, but sample retention rate needs to be balanced.

(c) **Data Impact**: Uneven wind direction sample distribution affects model performance—easterly wind retrieval accuracy is significantly higher than westerly wind. Low dimensionality of L1 compact DDM limits feature expression, while Full DDM's high dimensionality provides more reliable geometric information.

(d) **Practical Advantages**: Both works demonstrate the feasibility of GNSS-R machine learning-based wind direction retrieval. The models have advantages such as wide detection range, insensitivity to weather conditions, and adaptability to high wind speeds, providing a basis for global sea surface wind direction monitoring.

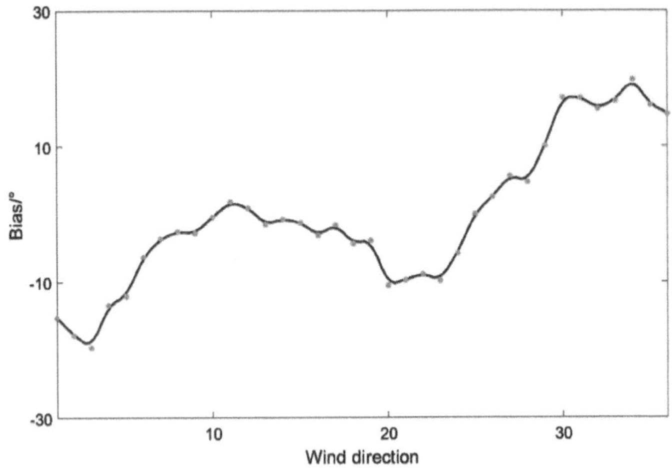

Fig. 32 Bias at different wind direction intervals on the generalized test set. Each point represents a different wind speed interval, and the line is the fitted curve of these points, which facilitates the observation of bias variations across different wind speed intervals

Table 9 Comparison of results of CNN model before and after downsampling

Method	Average data quantity	Average RMSE/ bias of East-wind intervals(°)	Average RMSE/ bias of West-wind intervals(°)	Average RMSE/ bias of all intervals (°)
Before under-sampling	534,204	32.8/0.27	48.4/0.49	37.6/0.35
After under-sampling	231,245	35.8/0.29	43.6/0.33	39.9/0.38

6 Conclusions

This study conducts two parallel works on GNSS-R sea surface wind direction retrieval, targeting CYGNSS Full DDM (high-dimensional) and L1 compact DDM (low-dimensional) data respectively. By integrating machine learning algorithms, DDM geometric features, and multi-source auxiliary parameters, the feasibility and effectiveness of wind direction retrieval are verified, and key findings and future outlooks are summarized as follows:

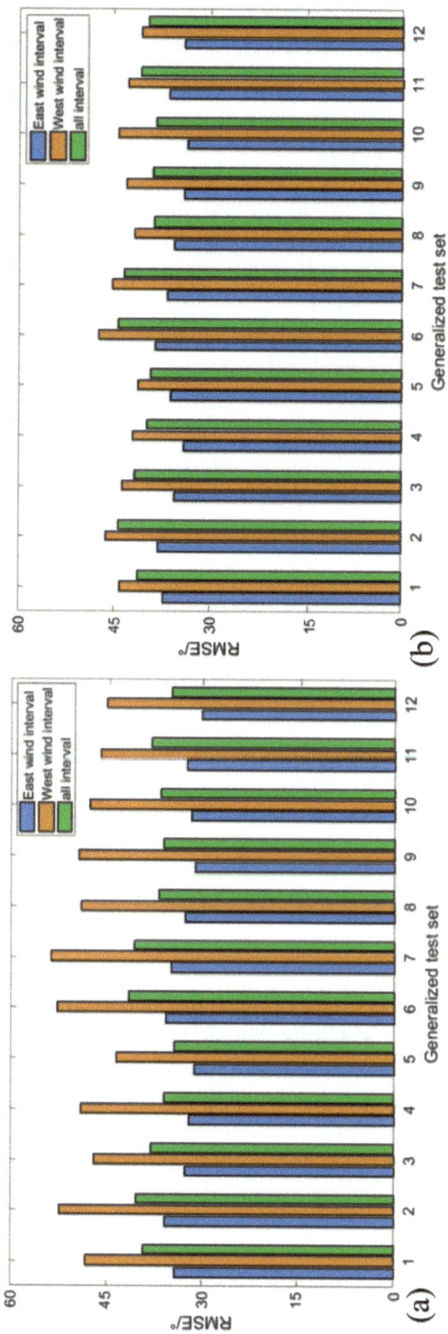

Fig. 33 Comparison of east wind part, west wind part and overall RMSE on each generalized test set before **a** and after **b** undersampling

6.1 Core Research Achievements

Work 1: SVM-Based Retrieval with CYGNSS Full DDM Data

This work proposes an SVM-based wind direction retrieval model, leveraging CYGNSS Full DDM, CYGNSS L1, and ECMWF reanalysis datasets (temporal coverage: February 27, 2019–November 17, 2020) with extensive spatiotemporal coverage. DDM geometric relationship features (φ_1 and φ_2) are extracted to accurately capture wind direction characteristics, serving as core input parameters. Wind direction is discretized into 72 intervals with a 5° step size to simplify model training and improve computational efficiency.

Two mapping relationships (f_1 with 5-dimensional features and f_2 with 9-dimensional features) are established and compared. The results confirm that wind speed and SNR significantly influence the extraction of φ_1 and φ_2, making data screening based on these two factors critical for retrieval accuracy. The 9-dimensional feature parameters (incorporating ws, ele, azi, φ_1, φ_2, LES, NBRCS, SNR, and RCG) outperform the 5-dimensional counterpart across all wind speed ranges:

(a) At wind speed $\geq$5 m/s, the RMSE is 30.03° (vs. 49.51° for f_1, a reduction of 39.3%);
(b) At wind speed $\geq$8 m/s, the RMSE decreases to 27.31° (vs. 47.82° for f_1, a reduction of 42.9%);
(c) The optimal performance is achieved at wind speed $\geq$10 m/s, with an RMSE of 26.70° (vs. 44.12° for f_1, a reduction of 39.5%).

The model effectively resolves the 180° wind direction ambiguity, with most samples ($\approx$85%) having prediction errors within 5°–10°, verifying the reliability of Full DDM data in capturing detailed DDM asymmetry.

Work 2: Multi-model Retrieval with CYGNSS L1 Compact DDM Data

Addressing the challenge of low dimensionality (17×11) of CYGNSS L1 compact DDM, this work constructs three machine learning models (SVM, BP neural network, and CNN) and uses datasets spanning March–October 2020 (CYGNSS L1 and ECMWF reanalysis data). A 31-dimensional feature vector is designed, integrating geophysical parameters (wind speed, MSL, SST), DDM asymmetry parameters (φ_1, φ_2), signal quality indicators, and satellite-related parameters to compensate for the insufficient feature expression of low-dimensional DDM.

The comparative results show that the CNN model outperforms SVM and BP models in both accuracy and generalization ability:

(a) On the model test set, the CNN model achieves a classification accuracy of 0.53 and an RMSE of 12.7°;

(b) On 12 generalized test sets, it maintains a classification accuracy of 0.44 and an RMSE of 37.6°, with the maximum classification accuracy reaching 0.47 and the minimum RMSE 34.3°.

To mitigate the impact of large-error samples (≈8% of total samples with errors >90°) on RMSE, a filter based on the maximum prediction probability of the CNN model is designed. With an optimal threshold of 0.21, the RMSE can be reduced to 19.5° (sample retention rate: 39.3%), balancing retrieval accuracy and practicality. Additionally, the study finds that the uneven distribution of training samples (dominated by easterly winds) leads to significant performance gaps between easterly (RMSE: 32.8°) and westerly (RMSE: 48.4°) wind intervals. While undersampling easterly wind samples reduces the RMSE of westerly wind intervals to 43.6°, it slightly increases the overall RMSE due to the dominance of easterly wind samples in the test set.

6.2 *Key Insights*

(a) **Feature Parameter Design**: The inclusion of sea surface roughness-related parameters (LES, NBRCS) and signal quality parameters (SNR, RCG) significantly improves retrieval accuracy, as these parameters effectively capture the subtle modulation of DDM waveforms by wind direction and ensure the reliability of feature extraction.
(b) **Data Impact**: High-dimensional Full DDM provides more detailed geometric information, supporting stable SVM-based retrieval with an optimal RMSE of 26.70°; low-dimensional L1 compact DDM requires extended multi-dimensional features and advanced neural networks (e.g., CNN) to compensate for information loss. Uneven sample distribution is a key factor affecting the generalization ability of models for specific wind directions.
(c) **Algorithm Selection**: For large-scale datasets, CNN outperforms traditional machine learning algorithms (SVM) and shallow neural networks (BP) due to its strong ability to extract hierarchical features and resist overfitting. SVM is suitable for high-dimensional small-sample scenarios (e.g., Full DDM data) but faces limitations in handling massive samples.
(d) **Practical Advantages**: Both retrieval methods demonstrate unique advantages, including wide detection range, insensitivity to weather conditions, adaptability to high wind speeds, and no wind direction ambiguity, providing a solid foundation for global sea surface wind direction monitoring.

6.3 Limitations and Future Directions

Current Limitations

(a) For Work 1, the spatiotemporal discontinuity of Full DDM data, relatively small volume of filtered samples, and potential errors in ECMWF wind speed products may impact retrieval accuracy; the overall RMSE exceeds 20° due to the sensitivity of RMSE to large-error samples.
(b) For Work 2, the reliance on ECMWF geophysical parameters (MSL, SST) reduces the practicality of the model; excessive filtering leads to low sample retention rates, limiting operational application.
(c) Both works are affected by uneven wind direction sample distribution, resulting in suboptimal retrieval performance for less frequent wind directions (e.g., westerly winds).

Future Research Directions

(a) Expand the dataset volume and optimize data preprocessing strategies to reduce the impact of abnormal samples and improve the stability of retrieval results.
(b) Explore feature engineering methods tailored to low-dimensional DDM data, reducing reliance on external geophysical parameters while maintaining retrieval accuracy.
(c) Adopt advanced imbalance learning techniques (e.g., synthetic minority oversampling) to balance the distribution of wind direction samples and enhance the model's generalization ability for underrepresented wind directions.
(d) Combine physical models with machine learning algorithms to integrate prior knowledge of GNSS-R scattering mechanisms, further improving retrieval accuracy and interpretability.
(e) Leverage the continuous accumulation of CYGNSS data products to update and optimize models, promoting the application of GNSS-R wind direction retrieval in operational marine environmental monitoring.

References

1. Foti G, Gommenginger C, Jales P et al (2015) Spaceborne GNSS reflectometry for ocean winds: first results from the UK TechDemoSat-1 mission. Geophys Res Lett 42(13):5435–5441. https://doi.org/10.1002/2015GL064204
2. Ruf CS, Gleason S, Jelenak Z et al (2012) The CYGNSS nanosatellite constellation hurricane mission. In: 2012 IEEE international Zavorotny symposium, 22–27 July 2012
3. Clarizia MP, Gommenginger CP, Gleason ST et al (2009) Analysis of GNSS-R delay-Doppler maps from the UK-DMC satellite over the ocean. Geophys Res Lett 36(2). https://doi.org/10.1029/2008GL036292

4. Martin-Neira M (1993) A passive reflectometry and interferometry system (PARIS): application to ocean altimetry. ESA J 17(4):331–355
5. Liu Y, Wang J, Collett I et al (2019) A Machine Learning Framework for Real Data Gnss-R Wind Speed Retrieval. In: IGARSS 2019–2019 IEEE international geoscience and remote sensing symposium, 28 July–2 Aug 2019
6. Rodriguez-Alvarez N, Garrison JL (2014) A generalized linear observable for GNSS-R wind speed retrievals over the ocean. In: 2014 IEEE geoscience and remote sensing symposium, 13–18 July 2014
7. Zhang Y, Yin J, Yang S et al (2021) High wind speed inversion model of CYGNSS Sea surface data based on machine learning. Remote Sens 13(16):3324
8. AlonsoArroyo A, Zavorotny VU, Camps A (2016) Sea ice detection using GNSS-R data from UK TDS-1. In: 2016 IEEE International Geoscience and Remote Sensing Symposium (IGARSS), 10–15 July 2016
9. Yun Z, Wanting M, Qiming G et al (2015) Detection of Bohai Bay sea ice using GPS-reflected signals. IEEE J Sel Top Appl Earth Obs Remote Sens 8(1):39–46. https://doi.org/10.1109/JSTARS.2014.2357894
10. Zhang Y, Hang S, Han Y et al (2021) Sea ice thickness detection using coastal BeiDou reflection setup in Bohai Bay. IEEE Geosci Remote Sens Lett 18(3):381–385. https://doi.org/10.1109/LGRS.2020.2980106
11. Li W, Cardellach E, Fabra F et al (2017) First spaceborne phase altimetry over sea ice using TechDemoSat-1 GNSS-R signals. Geophys Res Lett 44(16):8369–8376. https://doi.org/10.1002/2017GL074513
12. Zhang Y, Tian L, Meng W et al (2015) Feasibility of code-level altimetry using coastal BeiDou reflection (BeiDou-R) setups. IEEE J Sel Top Appl Earth Obs Remote Sens 8(8):4130–4140. https://doi.org/10.1109/JSTARS.2015.2446684
13. Li W, Rius A, Fabra F et al (2018) Revisiting the GNSS-R waveform statistics and its impact on Altimetric retrievals. IEEE Trans Geosci Remote Sens 99:1–18
14. Yun Z, Binbin L, Luman T et al (2016) Phase altimetry using reflected signals from BeiDou GEO satellites. IEEE Geosci Remote Sens Lett 13(10):1–5
15. Yun Z, Binbin L, Luman T et al (2016) Phase altimetry using reflected signals from BeiDou GEO satellites. IEEE Geosci Remote Sens Lett 13(10):1410–1414. https://doi.org/10.1109/LGRS.2016.2578361
16. Alonso-Arroyo A, Camps A, Park H et al (2015) Retrieval of significant wave height and mean sea surface level using the GNSS-R interference pattern technique: results from a three-month field campaign. IEEE Trans Geosci Remote Sens 53(6):3198–3209. https://doi.org/10.1109/TGRS.2014.2371540
17. Chew C, Shah R, Zuffada C et al (2016) Demonstrating soil moisture remote sensing with observations from the UK TechDemoSat-1 satellite mission. Geophys Res Lett 43(7):3317–3324. https://doi.org/10.1002/2016GL068189
18. Valencia E, Camps A, Rodriguez-Alvarez N et al (2013) Using GNSS-R imaging of the ocean surface for oil slick detection. IEEE J Sel Top Appl Earth Obs Remote Sens 6(1):217–223. https://doi.org/10.1109/JSTARS.2012.2210392
19. Zhang Y, Chen S, Hong Z et al (2017) Feasibility of oil slick detection using BeiDou-R coastal simulation. Math Probl Eng 1:8098029. https://doi.org/10.1155/2017/8098029
20. Simone AD, Park H, Riccio D et al (2017) Sea target detection using Spaceborne GNSS-R delay-Doppler maps: theory and experimental proof of concept using TDS-1 data. IEEE J Sel Top Appl Earth Obs Remote Sens 10(9):4237–4255. https://doi.org/10.1109/JSTARS.2017.2705350
21. Ji W, Xiu C, Li W et al (2014) Ocean surface target detection and positioning using the spaceborne GNSS-R Delay-Doppler maps. In: 2014 IEEE geoscience and remote sensing symposium, 13–18 July 2014
22. Zavorotny VU, Voronovich AG (2000) Scattering of GPS signals from the ocean with wind remote sensing application. IEEE Trans Geosci Remote Sens 38(2):951–964. https://doi.org/10.1109/36.841977

23. Hildebrand PH (1994) Estimation of sea-surface winds using backscatter cross-section measurements from airborne research weather radar. IEEE Trans Geosci Remote Sens 32(1):110–117. https://doi.org/10.1109/36.285194
24. Wang F, Yang D, Yang L (2018) Feasibility of wind direction observation using low-altitude global navigation satellite system-reflectometry. IEEE J Sel Top Appl Earth Obs Remote Sens 11(12):5063–5075. https://doi.org/10.1109/JSTARS.2018.2877388
25. Zhang Y, Chen X, Meng W et al (2021) Wind direction retrieval using support vector machine from CYGNSS Sea surface data. Remote Sens 13(21):4451. https://doi.org/10.3390/rs13214451
26. Xie X, Lin M, Chen K et al (2012) A wind direction extension based algorithm for scatterometer wind vector retrieval. In: 2012 IEEE international geoscience and remote sensing symposium, 22–27 July 2012
27. Xu X, Dong X, Zhang L et al (2017) Wind speed retrieving for combined observations of scatterometer and radiometer onboard HY-2A for typhoons using neural network. In: 2017 IEEE international geoscience and remote sensing symposium (IGARSS), 23–28 July 2017
28. Zecchetto S (2018) Wind direction from Sentinel-1 sar images in regional seas. In: IGARSS 2018–2018 IEEE international geoscience and remote sensing symposium, 22–27 July 2018
29. Germain O, Ruffini G, Soulat F et al (2004) The Eddy experiment: GNSS-R speculometry for directional sea-roughness retrieval from low altitude aircraft. Geophys Res Lett 31(21):L21307. https://doi.org/10.1029/2004GL020991
30. Zuffada C, Elfouhaily T, Lowe S (2003) Sensitivity analysis of wind vector measurements from ocean reflected GPS signals. Remote Sens Environ 88(3):341–350. https://doi.org/10.1016/S0034-4257(03)00175-5
31. Zuffada C, Fung A, Parker J et al (2004) Polarization properties of the GPS signal scattered off a wind-driven ocean. IEEE Trans Antennas Propag 52(1):172–188. https://doi.org/10.1109/TAP.2003.822438
32. Komjathy A, Armatys M, Masters D et al (2004) Retrieval of ocean surface wind speed and wind direction using reflected GPS signals. J Atmos Ocean Technol 21(3):515–526. https://doi.org/10.1175/1520-0426(2004)0212.0.CO;2
33. Garrison JL (2003) Anisotropy in reflected GPS measurements of ocean winds. In: IGARSS 2003.2003 IEEE international geoscience and remote sensing symposium. Proceedings (IEEE Cat. No.03CH37477), 21–25 July 2003
34. Li C, Huang W (2014) An algorithm for sea-surface wind field retrieval from GNSS-R delay-Doppler map. IEEE Geosci Remote Sens Lett 11(12):2110–2114. https://doi.org/10.1109/LGRS.2014.2320852
35. Park J, Johnson JT (2017) A study of wind direction effects on sea surface specular scattering for GNSS-R applications. IEEE J Sel Top Appl Earth Obs Remote Sens 10(11):4677–4685. https://doi.org/10.1109/JSTARS.2017.2719405
36. Park J, Johnson JT, Ouellette J (2016) Modeling polarimetric sea surface specular scattering for GNSS-R applications. In: 2016 IEEE international geoscience and remote sensing symposium (IGARSS), 10–15 July 2016
37. Soisuvarn S, Jelenak Z, Said F et al (2016) The GNSS reflectometry response to the ocean surface winds and waves. IEEE J Sel Top Appl Earth Obs Remote Sens 9(10):4678–4699. https://doi.org/10.1109/JSTARS.2016.2602703
38. Zavorotny VU, Voronovich AG (2014) Recent progress on forward scattering modeling for GNSS reflectometry. In: 2014 IEEE geoscience and remote sensing symposium, 13–18 July 2014
39. Valencia E, Zavorotny VU, Akos DM et al (2014) Using DDM asymmetry metrics for wind direction retrieval from GPS Ocean-scattered signals in airborne experiments. IEEE Trans Geosci Remote Sens 52(7):3924–3936. https://doi.org/10.1109/TGRS.2013.2278151
40. Pascual D, Clarizia MP, Ruf CS (2022) Spaceborne demonstration of GNSS-R scattering cross section sensitivity to wind direction. IEEE Geosci Remote Sens Lett 19:1–5. https://doi.org/10.1109/LGRS.2021.3049526

41. Liu W, Wei J, Meng Q (2020). Comparisions on KNN, SVM, BP and the CNN for handwritten digit recognition. In: 2020 IEEE international conference on advances in electrical engineering and computer applications (AEECA), 25–27 Aug 2020
42. Huang GB, Zhou H, Ding X et al (2012) Extreme learning machine for regression and multiclass classification. IEEE Trans Syst Man Cybern B Cybern 42(2):513–529. https://doi.org/10.1109/TSMCB.2011.2168604
43. Ji GR, Han P, Zhai YJ (2007). Wind speed forecasting based on support vector machine with forecasting error estimation. In: 2007 international conference on machine learning and cybernetics, 19–22 Aug 2007
44. Joachims T (1999) Making large-scale SVM learning practical. Adv kernel methods—support vector learning. Cambridge
45. Senyurek V, Lei F, Gurbuz AC et al (2022). Machine learning-based global soil moisture estimation using GNSS-R. In: SoutheastCon 2022, 26 March-3 April 2022
46. Zhang Y, Huang S, Han Y et al (2022) Machine learning methods for Spaceborne GNSS-R Sea surface height measurement from TDS-1. IEEE J Sel Top Appl Earth Obs Remote Sens 15:1079–1088. https://doi.org/10.1109/JSTARS.2021.3139376
47. Hu Y, Yuan X, Liu W et al (2022) An SVM-based snow detection algorithm for GNSS-R snow depth retrievals. IEEE J Sel Top Appl Earth Obs Remote Sens 15:6046–6052. https://doi.org/10.1109/JSTARS.2022.3193113
48. Liu Y, Collett I, Morton YJ (2019) Application of neural network to GNSS-R wind speed retrieval. IEEE Trans Geosci Remote Sens 57(12):9756–9766. https://doi.org/10.1109/TGRS.2019.2929002
49. Balasubramaniam R, Ruf C (2020) Neural network based quality control of CYGNSS wind retrieval. Remote Sens 12(17):2859. https://doi.org/10.3390/rs12172859
50. Chu X, He J, Song H et al (2020) Multimodal deep learning for heterogeneous GNSS-R data fusion and ocean wind speed retrieval. IEEE J Sel Top Appl Earth Obs Remote Sens 13:5971–5981. https://doi.org/10.1109/JSTARS.2020.3010879
51. Reynolds J, Clarizia MP, Santi E (2020) Wind speed estimation from CYGNSS using artificial neural networks. IEEE J Sel Top Appl Earth Obs Remote Sens 13:708–716. https://doi.org/10.1109/JSTARS.2020.2968156
52. Asgarimehr M, Zhelavskaya I, Foti G et al (2020) A GNSS-R geophysical model function: machine learning for wind speed retrievals. IEEE Geosci Remote Sens Lett 17(8):1333–1337. https://doi.org/10.1109/LGRS.2019.2948566
53. Caihong Z, Zengyuan W, Chang L (2019). A study on quality prediction for smart manufacturing based on the optimized BP-AdaBoost model. In: 2019 IEEE international conference on smart manufacturing, Industrial & Logistics Engineering (SMILE), 20–21 Apr 2019
54. Alshaye M, Alawwad F, Elshafiey I (2020). Hurricane tracking using multi-GNSS-R and deep learning. In: 2020 3rd international conference on computer applications & information security (ICCAIS), 19–21 Mar 2020
55. Ruf C, Asharaf S, Balasubramaniam R et al (2019) In-orbit performance of the constellation of CYGNSS hurricane satellites. Bull Am Meteorol Soc 100(10):2009–2023. https://doi.org/10.1175/BAMS-D-18-0337.1
56. Wu J (1990) Mean square slopes of the wind-disturbed water surface, their magnitude, directionality, and composition. Radio Sci 25(01):37–48. https://doi.org/10.1029/RS025i001p00037
57. Guan D, Park H, Camps A et al (2018) Wind direction signatures in GNSS-R observables from space. Remote Sens 10(2). https://doi.org/10.3390/rs10020198

Chapter 21
Retrieval of Sea Surface Rainfall Intensity Using Spaceborne GNSS-R Data

Jinwei Bu and Kegen Yu

Abstract Reliable rainfall information plays a crucial role in fields such as hydrology and climate research, drought and flood forecasting, construction, and industrial and agricultural production. GNSS-R, as an emerging remote sensing technology, is used to retrieve geophysical parameters due to its unique advantages such as low cost, high spatiotemporal resolution, wide coverage, and all-weather operation. Although prior research has demonstrated the capability of GNSS-R technology to detect oceanic rainfall, the issue of rainfall intensity has remained unexplored. Accordingly, based on GNSS-R Delay Doppler Map (DDM) observations, this chapter seeks to investigate the feasibility of retrieving sea surface rainfall intensity (RI). To improve DDM data quality, a wavelet denoising approach was initially introduced. Subsequently, based on the four GNSS-R observations (i.e. DDM Average (DDMA), normalized integrated delay waveform (LES-NIDW), leading-edge waveform summation of NIDW (LEWS-NIDW), these observables were classified into three categories according to their numerical values, and three empirical models were accordingly developed. The reference data IMERG-F was widely used to evaluate the RI retrieval performance of the three models mentioned above. In addition, this chapter also elaborates on how to use machine learning methods to construct a retrieval model for sea surface rainfall intensity. Compared to empirical modeling methods, a prominent advantage of machine learning based FFNN models is that they can more effectively handle the complex effects of geometric shapes, sea conditions, and other factors. According to the experimental results and analysis, machine

J. Bu
Faculty of Land Resources Engineering, Kunming University of Science and Technology, Kunming, China

Yunnan Key Laboratory of Intelligent Monitoring and Spatiotemporal Big Data Governance of Natural Resources, Kunming, China

J. Bu
e-mail: b_jinwei@kust.edu.cn

K. Yu (✉)
School of Environmental Science and Spatial Informatics, China University of Mining and Technology, Xuzhou, China
e-mail: kegen.yu@cumt.edu.cn

K. Yu and N. Zheng (eds.), *Advances in GNSS Reflectometry*, Navigation: Science and Technology 19, https://doi.org/10.1007/978-981-92-2325-1_21

learning methods are significantly superior to empirical modeling methods in terms of retrieval accuracy. Specifically, under the condition of wind speed less than 10 m/s, if the effects of wind speed and surge are corrected simultaneously, a considerable high retrieval accuracy can be achieved.

1 Introduction

Central to the global water and energy cycle, rainfall also constitutes a key atmospheric forcing that drives surface hydrology and land surface processes. Accurately collecting rainfall data is of great significance for weather and flood forecasting, disaster risk reduction, and water resource management. The methods for obtaining rainfall information generally include ground-based rain gauges, ground-based radar measurement, and meteorological satellite observation retrieval [29]. Obtaining rainfall data through ground-based rain gauges is clearly the most direct and effective method. Due to the sparse distribution and discontinuity of data between ground rain gauges and ground-based radar, they are usually only used for rainfall monitoring in small areas, which limits their scientific applications [15].

In the context of the continuous development of meteorological satellites, researchers have obtained fruitful research results in recent years through satellite remote sensing inversion of rainfall products [7, 16, 26]. In the 1970s, satellite remote sensing for rainfall retrieval gradually attracted people's attention. Currently, there are some satellite retrieval of rainfall product data, including the Climate Hazards Group Infrared Precipitation with Station (CHIRPS) product [1], Climate Prediction Center MORPHing product (CMORPH) [25], Multi-Source Weighted-Ensemble Precipitation (MSWEP) product [9], SM2RAIN (Soil Moisture TO RAIN) product [17], Global Satellite Mapping of Precipitation (GSMaP) product [38], Tropical Rainfall Measurement Mission Multi-Satellite Precipitation Analysis (TRMM TMPA) [28], and Global Precipitation Measurements from Integrated Multi-satellite Retrievals (GPM-IMERG) product [24]. Compared to other satellite rainfall datasets, GPM-IMERG products typically have wider coverage, higher accuracy, and higher temporal resolution [3, 31, 33] and have been widely used in fields such as hydrology, meteorology, and environmental science.

At present, there are various satellite remote sensing technologies used for retrieving rainfall products, each with its own advantages and disadvantages. However, there is hardly any technology that can simultaneously meet key indicators such as monitoring costs, revisit cycles, and spatial resolution. Due to the short follow-up period and low monitoring cost, Global Navigation Satellite System Reflectometry (GNSS-R) technology has shown significant potential in monitoring. As an emerging remote sensing technique, GNSS has been introduced into the field of ocean remote sensing in recent years and has become a commonly used technology. Recent studies [6, 5, 8] have confirmed for the first time that this technology can detect sea surface rainfall under weak wind conditions. However, it should be noted that there are currently no research results utilizing spaceborne GNSS-R data in the inversion of sea surface rainfall intensity. Therefore, conducting research on rainfall intensity inversion is of great significance, as it helps us further understand the interaction between rainfall, the sea air interface, and its impact on GNSS measurements.

Notably, a number of DDM-related observables have been introduced in prior research for various GNSS-R applications. As proposed in the literature [19], based on the United Kingdom Disaster Monitoring Constellation (UK-DMC) data, five observation values (Delay-Doppler Map Average (DDMA), the Delay-Doppler Map Variance (DDMV), Allan Delay-Doppler Map Variance (ADDMV), Leading Edge Slope (LES), and Trailing Edge Slope (TES)) were proposed for sea surface wind speed retrieval. A Minimum Variation (MV) wind speed estimator was also proposed based on the five observation values and the five observations were retrieved for sea surface wind speed. Each observable value can be used to derive corresponding wind estimation values, which can be linearly combined to obtain the MV estimator. Compared with the wind speed retrieval results based on each observation, the MV estimator has a smaller root mean square error (RMSE). The literature [32] used the maximum signal-to-noise ratio (SNR) for wind speed retrieval. Besides its application in ocean surface wind speed measurement, the SNR of the GNSS-R signal has also been shown to be highly effective for monitoring surface soil moisture [34]. In Hammond et al. [23], the potential of the Galileo SNR method for wind speed retrieval and sea ice detection was evaluated for the first time using onboard Galileo GNSS-R data collected from the UK TechDemoSat-1 mission. The literature [2] used DDMA, TES, and Woodward ambiguity function (WAF) matched filter (MF) observations based on TDS-1 data for sea ice detection. By comparing the sea ice detection results of the three observation values, it was found that the DDMA method had slightly worse sea ice detection performance than the TES method, while the MF and TES methods had comparable performance. This type of observation value has also been widely discussed in GNSS-R soil moisture retrieval [20]. Bu et al. [11] and Jing et al. [27] also used NBRCS observations for sea surface wind speed retrieval, and their wind speed retrieval performance was better than that of LES observations. Balasubramaniam and Ruf [8] obtained a preliminary relationship model between NBRCS observations and rainfall intensity, which was established based on their discussion of how rainfall affects GNSS-R ocean measurements. In addition, Zhu et al. [40] also used the sum of the trailing edge slope and trailing edge waveform values of normalized integral delay waveform (NIDW), normalized center delay waveform (NCDW), and differential delay waveform (DDW) for sea ice detection. The results showed that the sum of trailing edge waveform values of DDW was superior to others in detecting sea ice as an observation value. The aim of this study is to explore the method of using GNSS-R observation data to retrieve sea surface rainfall intensity (RI). Previously, Zhu et al. [39] confirmed the feasibility of spaceborne GNS-R in SIC retrieval by using the sum of DDW tail waveform values as the observed variable for sea ice concentration (SIC). The above work provides a foundation for this study.

This chapter aims to study the retrieval method of sea surface rainfall intensity (RI) based on GNSS-R using observation data obtained from spaceborne GNSS-R

DDM. Especially, the purpose of this study is to use wavelet denoising methods for data quality control in DDM data, thereby improving the accuracy of retrieving sea surface rainfall intensity. Another important goal is to develop three models for GNSS based sea surface RI inversion. By testing and comparing the performance of these models, we aim to explore their potential in sea surface RI inversion. The rest of this chapter is arranged as follows: Sect. 2 elaborates on the RI retrieval model based on empirical modeling methods and verifies the retrieval performance of the model. Section 3 introduces the Feed Forward Neural Network (FFNN) RI retrieval model based on machine learning methods and verifies its performance. Finally, Sect. 4 is the conclusion of this chapter.

2 Retrieval of Global Sea Surface RI Based on Empirical Model Method

Currently, the retrieval models for sea surface rainfall intensity in spaceborne GNSS-R are power function empirical models and machine learning models. The rest of this section briefly introduces the modeling methods for each model.

2.1 Basic Description of Methodology

The datasets used for this study were downloaded and collected as described in Sect. 3.1. The rainfall information has been included in the dataset used in this chapter. Among them, the data only comes from ocean surface reflection signals, and land surface reflection data is not considered. To minimize the influence of wind as much as possible before conducting rainfall intensity inversion research, this section will limit the research scope to extremely low wind speeds (< 5 m/s). Another important point here is that at low wind speeds, swells do have a significant impact on DDM. Therefore, in addition to the impact of rainwater on sea surface roughness, the influence of swells also needs to be the presence of swells. The significant wave height (SWH) information in ERA5 data can be used to determine whether there is a swell. In this section, if the swell SWH provided by ERA5 is greater than 2 m, DDM data will not be used. By using spatial bilinear interpolation and temporal linear interpolation, the DDM data collection time and specular point position are respectively consistent with the ERA5 wind speed product, SWH product, and rainfall data. In addition, preprocessing of CYGNSS DDM data is also required. Firstly, it is necessary to denoise the DDM image. Considering that most DDM images contain redundant information, conventional denoising methods such as noise floor subtraction [18, 36, 35] lack adaptability. Therefore, this chapter introduces a soft threshold wavelet denoising method for DDM denoising [21]. This chapter uses two-dimensional wavelet transform to denoise DDM images. In contrast, Camps et al.

[14] also studied similar denoising methods based on airborne GNSS-R data, but only used one-dimensional wavelet transform to denoise delayed waveforms.

If the wavelet coefficients are c_p and the threshold processed coefficients are κ_p (also known as the threshold function), then the expression of the threshold function is defined as follows:

$$\eta(\omega) = \begin{cases} (|\omega| - T)sign(\omega), & |\omega| \geq T \\ 0, & |\omega| < T \end{cases} \tag{1}$$

where $T = \sigma\sqrt{2\ln N}$, σ is the noise variance and N is the image signal length.

The algorithm steps for DDM image wavelet denoising are as follows:

(1) On the premise of subtracting the noise base, the DDM image undergoes wavelet decomposition to generate a set of wavelet coefficients. In this set of coefficients, high-frequency components are often the main carrier of noise.
(2) Quantify the high-frequency coefficients of wavelet decomposition in the form of threshold values.
(3) Reconstruction of the image signal is performed with a wavelet transform,, which is the DDM image denoised by wavelet domain threshold.

To obtain DDM observables with a spatial resolution of 25 km, a 5 × 3 window is selected around the specular reflection point within the DDM range of 17 × 11 (as shown by the blue box in Fig. 1). This figure shows three types of DDM images: uncorrected, denoised by noise based subtraction, and denoised by the two-dimensional wavelet transform proposed in this paper. Figure 2 shows the corresponding NIDW, NCDW, and DDW. Reference [10] provides a more detailed description of the three types of delay waveforms. The uncorrected DDM data collected on August 22, 2020, corresponding to RI values of 3.43, 5.43, 10.53, and 17.75 mm/h for the four signal-to-noise ratios(SNR), respectively. After denoising with two-dimensional wavelet transform, the noise in the uncorrected DDM image is effectively suppressed, and the "horseshoe" shaped contour of low signal-to-noise ratio DDM is particularly clear (as shown in Fig. 1c, d). As shown in Fig. 1, this method is superior to the existing noise floor subtraction method and can effectively suppress the noise adjustment effect caused by changes in antenna and instrument system noise. From Fig. 1, it can also be seen that although the curves of the three delay waveforms denoised by two-dimensional wavelet transform are basically consistent with the delay waveforms of the noise base subtraction denoising method, it can be clearly seen from the DDM image that background noise and speckle noise have been eliminated. The experimental results show that the proposed two-dimensional wavelet transform can effectively achieve DDM denoising. As the rainfall intensity increases, the power value near the specular reflection point (i.e. the maximum value in DDM) gradually decreases, reflecting the significant changes brought about by different rainfall intensities. At the same time, the delay waveforms of NIDW, NCDW, and DDW under the four rainfall intensities differ significantly, providing a basis for the retrieval of rainfall intensity on the sea surface.

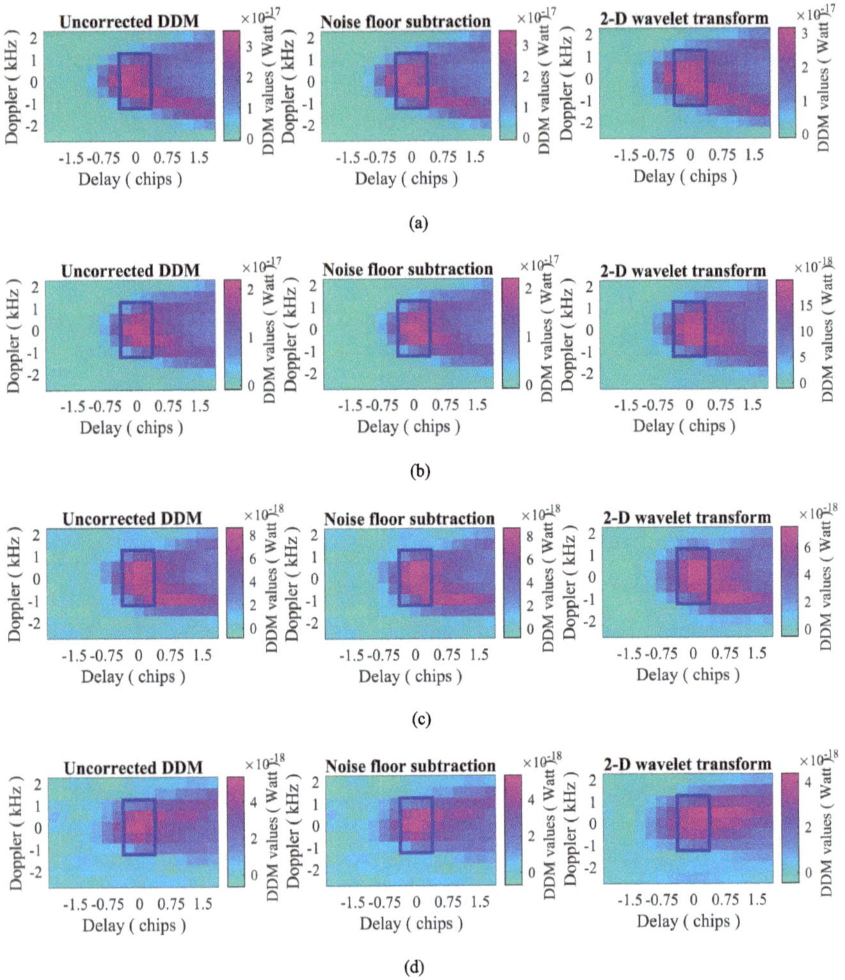

Fig. 1 DDM examples under different SNR, RI, and wind speeds (WS). From left to right: uncorrected DDM images, images after noise-floor subtraction, and images denoised using the proposed 2-D wavelet transform. Each subfigure corresponds to the following conditions: **a** SNR = 7.46 dB (RI = 7.138 mm/h, WS = 2.68 m/s, θ = 34.87°, RAG = 9.63 dBi). **b** SNR = 5.63 dB (RI = 13.125 mm/h, WS = 3.78 m/s, θ = 36.73°, RAG = 8.96 dBi). **c** SNR = 3.13 dB (RI = 19.749 mm/h, WS = 4.23 m/s, θ = 38.56°, RAG = 7.68 dBi). **d** SNR = 2.46 dB (RI = 27.011 mm/h, WS = 4.69 m/s, θ = 43.67°, RAG = 7.35 dBi)

Then, it is necessary to strictly control the quality of DDM data according to the method described in Sect. 3.4 of Chap. 18, and calculate several GNSS-R observations for rainfall intensity retrieval modeling. The data associated with each observation were then randomly partitioned into a training set (40%) and a testing set (60%), using the IMERG-F product as the reference data. The rainfall intensity

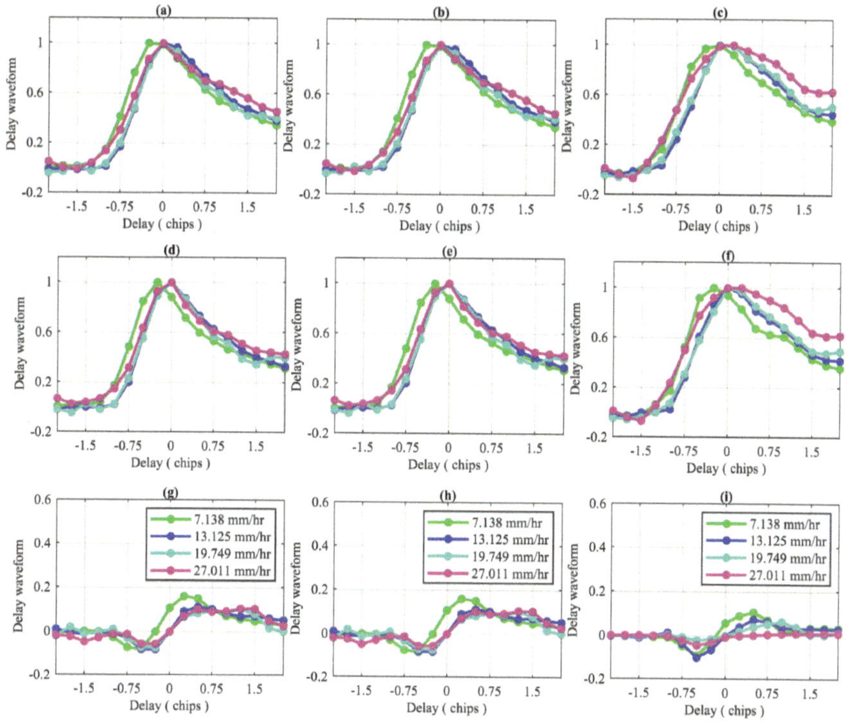

Fig. 2 NIDW, NCDW, and DDW with different RI. Results are presented for three processing approaches: uncorrected DDM (left panel), noise-floor subtraction (middle), and 2-D wavelet denoising (right). The subfigures are organized by observable and method: **a** Uncorrected NIDW. **b** Noise floor Subtraction (NIDW). **c** 2-D wavelet transform (NIDW). **d** Uncorrected NCDW. **e** Noise floor Subtraction (NCDW). **f** 2-D wavelet transform (NCDW). **g** Uncorrected DDW. **h** Noise floor Subtraction (DDW). **i** 2-D wavelet transform (DDW)

range studied in this section is 0–30 mm/h. Finally, GNSS-R observations and rainfall intensity were used as input and output variables, and a single observation model was established using least squares fitting. Based on this, two combined models were constructed. To evaluate the proposed rainfall intensity inversion model, the IMERG-F and GNSS R data were first used to establish the model; Then, input the GNSS observations from the test set into the model to obtain rainfall intensity inversion results; Finally, compare these results with IMERG-F data. Figure 3 shows the complete algorithm process for the development and testing of the model [12].

2.2 *Retrieval Model Construction*

Through regression analysis, a geophysical model function (GMF) with single and combined observations is developed using fitting functions. The power function

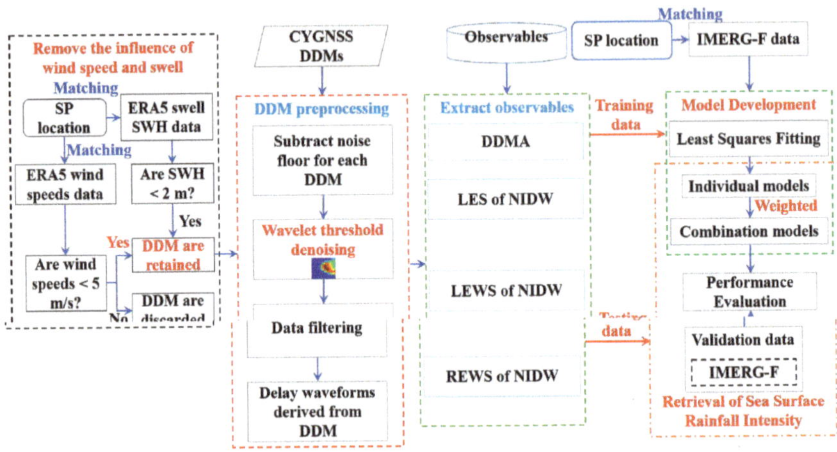

Fig. 3 Flowchart of rainfall intensity retrieval

provides a better fit for the relationship between various observed values and rainfall intensity. All models for retrieving sea surface rainfall intensity are divided into three categories to investigate whether the number of input variables (observables) affects model accuracy: single observation power function models based on three parameters, dual observation combination models, and triple observation combination models.

(1) Empirical model for power function rainfall intensity retrieval based on three parameters

This section proposes a power function model. Among them, four observations—DDMA, LES-NIDW, LEWS-NIDW, and TEWS-NIDW—were used to construct a single observation variable based model (referred to as Model 1) to characterize their relationship with rainfall intensity. The mathematical expression of the above model is derived from curve fitting experiments with a large amount of data, as follows:

$$RI_{ob} = Ax_{ob,i}^{B} + C \tag{2}$$

where RI represents rainfall intensity, x_{ob} represents any of 4 GNSS-R observation variables, nonlinear least squares fitting generates fitting parameters corresponding to other symbols in the model.

The primary task in developing rainfall intensity inversion models is to reduce the impact of sea surface wind speed on GNSS-R observations. Section 2.2 introduces a control method for sea surface wind speed, SWH, receiver antenna gain and DDM incident angle parameters are limited. It should be pointed out that this study only retrieved sea surface rainfall intensity under wind speeds below 5 m/s. In fact, the influence of wind speed on DDM observations during the modeling process cannot be ignored—even if the wind speed is very small (< 5 m/s), DDM observations are still highly sensitive to rainfall intensity, as the surface roughness generated by raindrops hitting the water surface dominates at this time. To eliminate or minimize the impact of wind speed on DDM observations, this paper proposes the following empirical correction model:

$$Y(W) = c \cdot W + d \tag{3}$$

where W is the wind speed, In the empirical correction model, c and d are the fitting parameters. Considering the dependence of these parameters on rainfall intensity, to improve modeling accuracy, Using existing data, we separate rainfall intensity into four intervals: [0, 5], [5, 10], [10, 15], and [15, 30] mm/h—with the fourth interval being much wider because heavy rainfall data samples are relatively rare.

Figure 4 presents empirical models for wind speed correction of DDMA, LES, LEWS, and TEWS observations under RI: [0, 5], [5, 10], [10, 15], and [15, 30], unit: mm/h. The degree of influence of wind speed on observed values is closely related to rainfall intensity: when RI is low ([0, 5] mm/h), the observed values change significantly with wind speed; In the other three higher RI intervals, the observed values do not change significantly with wind speed. This means that as rainfall intensity increases, the impact of wind speed decreases significantly.

This chapter uses root mean square error (RMSE) and coefficient of determination (R^2) as performance indicators to evaluate the model fitting results under different observation parameters. The definitions of these two indicators are:

$$RMSE = \sqrt{\frac{1}{i}\sum_{n=1}^{k}\left(y_{n,0} - y_{n,m}\right)^2} \tag{4}$$

$$R^2 = \frac{\left(\sum_{n=1}^{k}(y_{n,m} - \overline{y}_m)(y_{n,0} - \overline{y}_0)\right)^2}{\sum_{n=1}^{k}(y_{n,m} - \overline{y}_m)^2 \sum_{n=1}^{k}(y_{n,0} - \overline{y}_0)^2} \tag{5}$$

where k it is the total sample size in the dataset, $y_{n,m}$ represents the rainfall intensity fitted by the model, and $y_{n,0}$ represents the known rainfall intensity in the training data, $\overline{y}_0$ is the average of known rainfall intensities in the training data, and $\overline{y}_m$ is the average of rainfall intensities fitted by the model.

Figure 5 presents the fitting curves of the power function rainfall intensity retrieval model developed based on four observation variables. The wind speed is 1–5 m/s, the incident angle is 10°–30°, and the rainfall intensity range is 0–30 mm/h. For ease of calculation and intuitive comparison, these observations are normalized to a common denominator. The average rainfall intensity in the training set is 7.52 mm/h. Table 1 presents the RMSE and R^2 of each observation fitting model for comparison of model fitting performance. Figure 5 and Table 1 visually show that compared to the fitting model based on uncorrected observations, the scatter distribution of

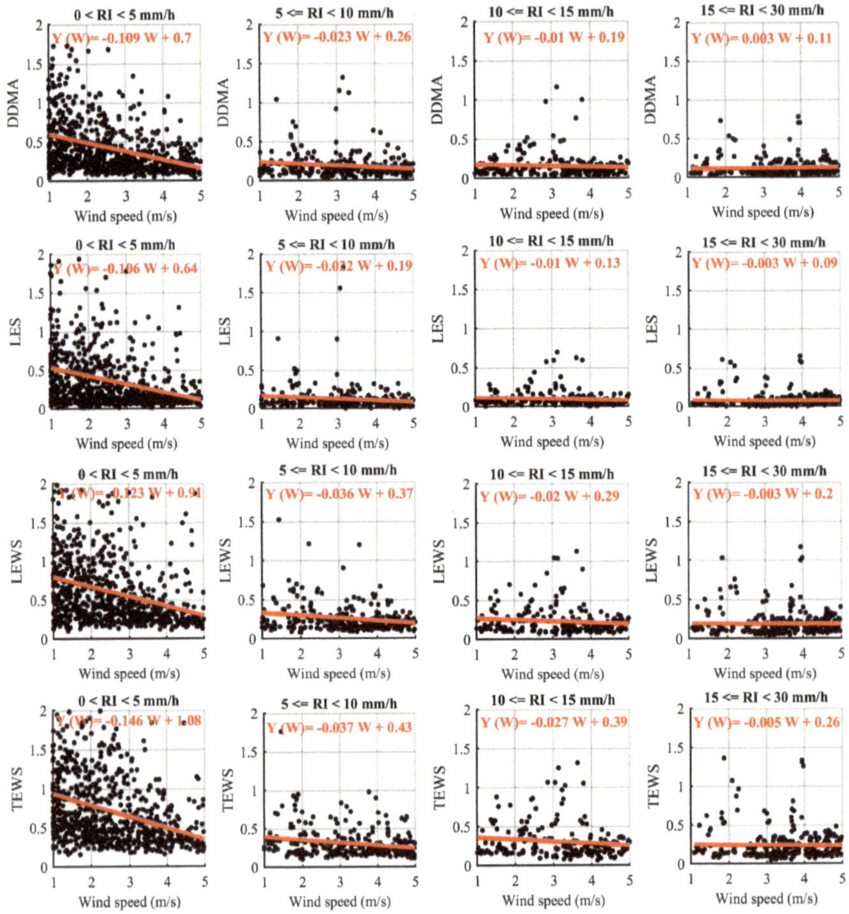

Fig. 4 Empirical model for wind speed correction of DDMA, LES, LEWS and TEWS

the corrected observations presents better model fitting results, indicating that using the wind speed influence correction empirical model to correct the observations can effectively improve the modeling effect. For the four observed values (DDMA, LES, LEWS-NIDW, and TEWS-NIDW), the RMSE of their fitted models increased by 15.32%, 9.42%, 11.08%, and 14.53%, respectively, and the R^2 increased by 39.53%, 42.86%, 29.41%, and 23.64%, respectively.

The observations used for developing rainfall intensity retrieval models are corrected by wind-speed-dependent empirical models. The parameters of the rainfall intensity retrieval model are shown in Table 2.

Based on the above analysis and observations with good fitting effects, two different combination models were introduced and developed, as described below.

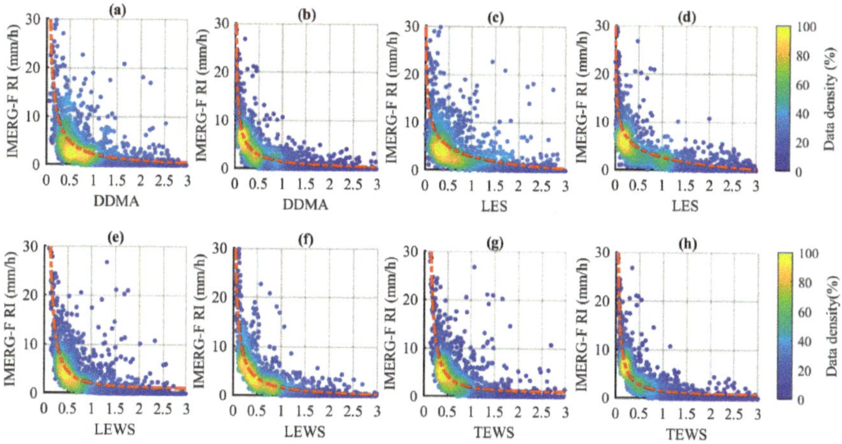

Fig. 5 A power function rainfall intensity retrieval model using four observation variables

Table 1 RMSE and R^2 of each observable fitting model

Observations	Before correcting the influence of wind speed		After correcting the influence of wind speed	
	RMSE (mm/h)	R^2	RMSE (mm/h)	R^2
DDMA	3.85	0.43	3.26	0.60
LES	4.14	0.35	3.75	0.50
LEWS	3.52	0.51	3.13	0.66
TEWS	3.58	0.55	3.06	0.68

Table 2 Parameters of rainfall intensity retrieval models

Power function parameters			
Observations	1	2	3
DDMA	1.918	− 0.7699	− 0.3765
LES-NIDW	12.3102	− 0.2068	− 9.5853
LEWS-NIDW	2.122	− 0.925	− 0.407
TEWS-NIDW	1.146	− 1.092	0.433

(2) Empirical model for rainfall intensity retrieval based on dual observation variables combination and triple observation variables combination

This section first proposes a combination model (referred to as Model 2), whose basic idea is to use the best weighted estimation to linearly combine the estimated rainfall intensity values corresponding to two GNSS-R observations. The mathematical expression of this model is as follows [12]:

$$\begin{cases} RI_{ob,i}^{Com} = \dfrac{[\gamma_1, \gamma_2] \cdot \left[RI_{ob,i}^{(1)}, RI_{ob,i}^{(2)}\right]^T}{\gamma_1 + \gamma_2}, i = 1, 2, \ldots, m \\ \gamma_x = 1 / \left(\dfrac{1}{m} \sum_{n=1}^{m} \left(RI_{ob,n}^{(x)} - y_{n,0} \right)^2 \right), x \in \{1, 2\} \end{cases} \tag{6}$$

in the combination model, $\{RI_{ob,i}^{Com}\}$ represents the estimation result of rainfall intensity, $\{\gamma_x\}$ is the linear combination coefficient (also known as weight), x is the label of single observation-based models, and m is the sample size of the data. $RI_{ob,n}^{(x)}$ is an estimation of rainfall intensity based on single GNSS-R observation-based model.

To construct a significantly improved combined model, this chapter considers combining TEWS-NIDW observations with LES-NIW observations—although the latter has the worst fitting performance when modeled separately, it is considered to have the best fitting performance. This contradiction needs further exploration.

Moreover, a model based on a combination of three GNSS-R observations (i.e., Model 3) was also proposed, described as [12]:

$$RI_{ob,i}^{Com} = \frac{[\gamma_1, \gamma_2, \gamma_3]\left[RI_{ob,i}^1, RI_{ob,i}^2, RI_{ob,i}^3\right]^T}{\sum_{k=1}^{3} \gamma_k} \tag{7}$$

the symbol definitions in the equation are consistent with Eq. (6).

Similarly, due to the poor model fitting performance of LES-NIDW observations, this chapter considers combining LES-NIDW observations with LEWS-NIDW and TEWS-NIDW observations, which have the best model fitting performance, to form a combined model and improve the estimation ability of LES-NIDW observations for rainfall intensity.

2.3 *Performance of Model-Based RI Retrieval*

To test the inversion ability of the model built in the previous section on rainfall intensity, its estimation results were compared with IMERG-F data, and the validation data used in this section is different from the previous section—these data were not involved in the model development. The evaluation of model performance uses three indicators: RMSE, Mean Absolute Error (MAE), and Pearson Correlation Coefficient (CC).[10].

Figure 6 presents scatter density plots of rainfall intensity estimates and IMERG-F rainfall data based on three models: single-observation model, dual-observation combination model, and triple-observation combination model. The color bars (from cold to warm) represent data density. In the figure, the 1:1 reference line is depicted as a magenta dashed line, while the red solid line indicates the linear fit between the model's estimated rainfall intensity and the IMERG-F rainfall data. The wind speed ranges from 1 to 5 m/s, and the incidence angle spans 10°–30°. Figure 6a–f are respectively the results based on DDMA, LES-NIDW, LEWS-NIDW, TEWS-NIDW, LES-NIDW + TEWS-NIDW, and LES-NIDW + LEWS-NIDW + TEWS-NIDW. In Table 3, in addition to the RMSE, MAE, and CC of the three models, the RMSE improvement rates (IR) of DDMA, LEWS, TEWS, LES + TWS, and LES + LEWS + TWS compared to the LES-NIDW model are also listed, as well as the evaluation performance of the three models for different rainfall intensity ranges (0–30 mm/h, below 10 mm/h, and above 10 mm/h).

Based on the results in Fig. 6 and Table 3, the following observations can be made.

(1) Under rainfall intensity of 0–30 mm/h, compared with the worst LES model, the retrieval accuracy of DDMA, LEWS, and TEWS models increased by 4.67%, 5.98%, and 10.31%, respectively, and the dual observation and triple observation combination models increased by 8.35% and 13.73%, respectively. Among them, the single observation model based on TEWS performs the best, with RMSE better than 3.17 mm/h and CC better than 0.82; The LES model performed the worst (with the highest RMSE and MAE). The above results indicate a significant improvement in retrieval performance.

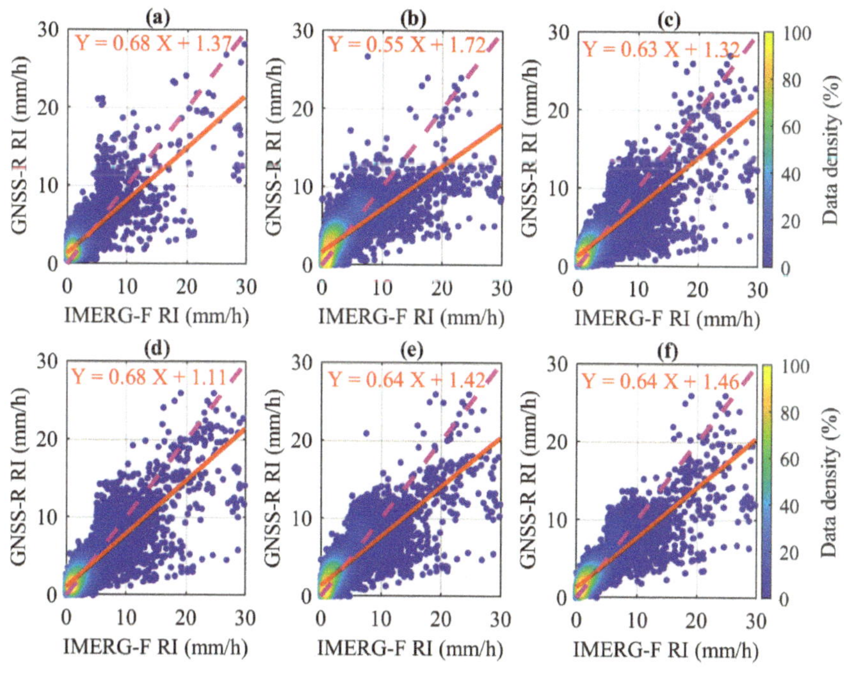

Fig. 6 Scatter density plots of the IMERG-F rainfall intensity data and rainfall intensity estimations by three models

Table 3 Statistical analysis of error indicators (RMSE, MAE, CC) and RMSE improvement rate for rainfall intensity inversion using three models

Models	Observables	< 10 mm/h		> 10 mm/h		0–30 mm/h			
		RMSE (mm/h)	MAE (mm/h)	RMSE (mm/h)	MAE (mm/h)	RMSE (mm/h)	MAE (mm/h)	CC	IR (%)
Model 1	LES	2.10	1.65	7.80	6.50	3.61	2.80	0.75	–
	DDMA	1.99	1.53	7.27	5.83	3.41	2.57	0.79	4.67
	LEWS	1.76	1.30	7.32	5.80	3.36	2.31	0.79	5.98
	REWS	1.70	1.24	6.70	5.22	3.17	2.19	0.82	10.31
Model 2	LES + REWS	1.72	1.32	7.01	5.65	3.26	2.31	0.81	8.35
Model 3	LES + LEWS + REWS	1.42	1.09	7.17	5.93	3.03	2.09	0.84	13.73

(2) In both cases where the rainfall intensity is less than 10 mm/h and greater than 10 mm/h, the results are similar to those described in (1). Compared with the retrieval performance of a single LES observation, the retrieval performance of the dual observation and triple observation combination models has been improved. It can also be seen that under these three rainfall intensity ranges, the retrieval accuracy of the TEWS based model is better than that of the other three single observation models.

(3) As the rainfall intensity increases, the retrieval performance of the model decreases, especially when the rainfall intensity is greater than 10 mm/h. There are two main reasons: saturation of DDM observation sensitivity under high rainfall intensity, and insufficient high rainfall intensity samples used for modeling. Therefore, in the future, it is necessary to strengthen the retrieval performance under this condition.

When the wind speed is below 5 m/s, rainfall will cause a systematic decrease in signal-to-noise ratio, and the difference in signal-to-noise ratio between rainy and non rainy conditions usually widens with increasing rainfall intensity, as shown in Fig. 7 (which depicts the relationship between DDM average signal-to-noise ratio and wind speed). Therefore, studying the changes in signal-to-noise ratio between rainy days and rainless environments with different wind speeds is of great value. Analysis indicates that rainfall causes a notable change in signal-to-noise ratio. For wind speeds below 3 m/s and rainfall intensities of 0–5 mm/h, the average SNR drops by approximately 2–3 dB. Under no rain conditions, the signal-to-noise ratio exhibits regular changes within different wind speed ranges. During rainfall, this change may intensify or deviate from the original pattern. However, due to the presence of observation noise and various uncertainties, the impact of rainfall on the signal-to-noise ratio of DDM varies with wind speed. Therefore, this study limits the modeling scope to the low wind speed range (i.e. below 4–5 m/s). As pointed out in reference

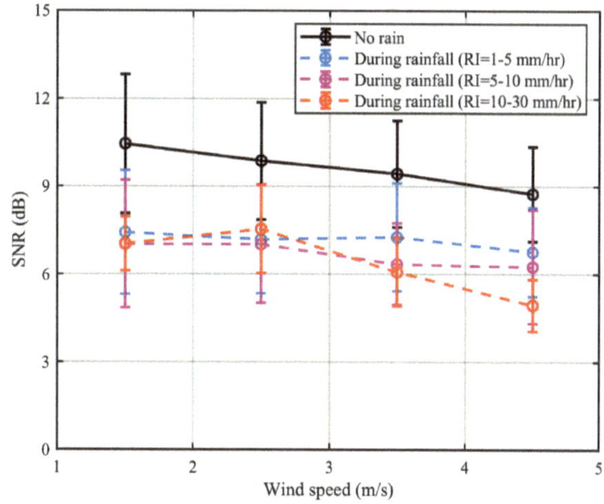

Fig. 7 Relationship between SNR and wind speed under different rainfall intensity (RI)

[5], when the wind speed exceeds this range, the impact of rainfall on signal-to-noise ratio is relatively weak.

The incident angle is an important satellite geometric control parameter in the inversion of sea surface rainfall intensity using spaceborne GNSS-R. Figure 8 displays scatter density plots comparing IMERG-F rainfall data with retrieved rainfall intensity from the DDMA and TEWS-NIDW observation models, corresponding to incident angle intervals of 10°–20°, 20°–30°, and 30°–40°, respectively. As seen from Fig. 8, the retrieval performance deteriorates with increasing incident angle (e.g., [30°–40°]). However, using only data within a very narrow range of incidence angles may not be practical, as the amount of data within a very narrow range of incidence angles is very sparse. Therefore, while ensuring sufficient data volume and not being affected by a large incident angle range, we chose an incident angle range of 10°–30° during the model construction and validation phase.

3 Ocean Surface RI Retrieval Using a Feedforward Neural Network Under Very Low Wind Speed Conditions

In the previous section, spaceborne GNSS-R has been applied to ocean rainfall intensity retrieval using features and empirical geophysical model functions (GMFs) extracted from delayed Doppler maps (DDMs) to retrieve sea surface rainfall intensity. However, Given the complex influence of other GNSS-R process factors (e.g., geometry and sea conditions), employing these factors as GMF inputs proves difficult. Therefore, in this section, a machine learning-based multilayer Feed-Forward Neural Network (FFNN) model is proposed. FFNN is trained using collected CYGNSS L1B level data and Global Precipitation Measurement (GPM-IMERG)

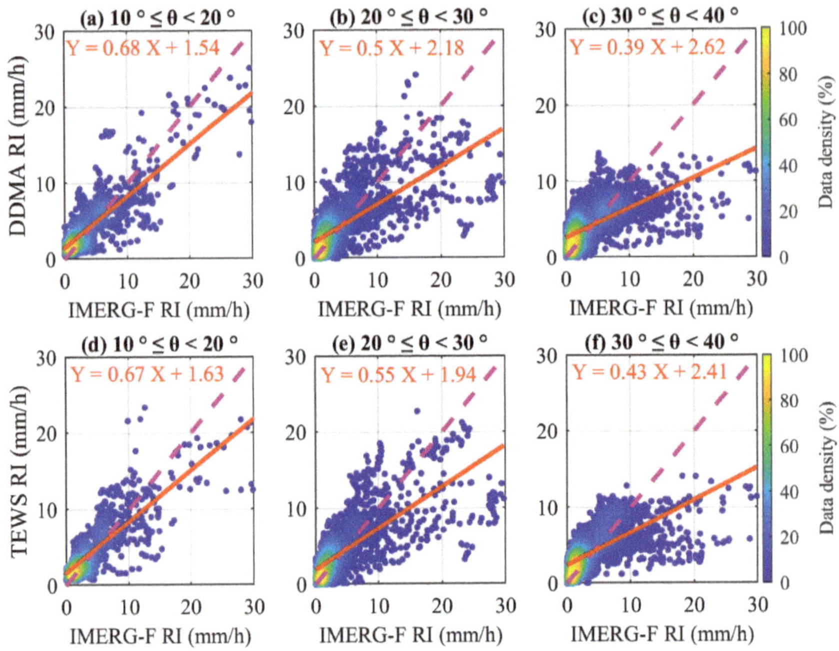

Fig. 8 Density scatterplots of rainfall intensity retrieved from models developed based on DDMA and TEWS observables and IMERG-F rainfall data

data. An FFNN model with three hidden layers and multiple neurons in each layer was constructed and applied to the test set for rainfall intensity retrieval.

3.1 Dataset Description

In this section, four datasets are used, namely: CYGNSS Level 1B (L1B) data, IMERG Final Run (IMERG-F) RI data, and European Center for Medium-Range Weather Forecasts (ECMWF) reanalysis wind speed and significant wave height (SWH) data.

Among them, the L1B variables used in this section include SNR, distance correction gain (RCG), angle of incidence, and longitude and latitude of specular reflection points. We collected datasets from June 1, 2019 to August 31, 2020, and five features associated with DDM (i.e., DDMA, LES, TES, LEWS, and TEWS) were used for rainfall intensity retrieval.

The expressions for the five DDM features are defined as follows [10]. DDMA is the average scattering power near the peak of DDM, which is defined as:

$$DDMA = \frac{1}{mn}\sum_{x=8}^{10}\sum_{y=4}^{8}\left\langle \left|y\prime\left(\tau_x, f_y\right)\right|^2\right\rangle \tag{8}$$

where $\left\langle \left|y\prime\left(\tau_x, f_y\right)\right|^2\right\rangle$ is the power value minus the m noise in the row (time delay) and n column (Doppler shift) region around the specular point.

The following formula gives the LES of NIDW, which is obtained by applying least-squares fitting inside the time delay window:

$$LES_{NIDW} = \frac{n\sum_{i=1}^{\mathrm{n}} \chi_i\omega_{NIDW,i}^{L} - \sum_{i=1}^{n} \chi_i \sum_{i=1}^{n} \omega_{NIDW,i}^{L}}{n\sum_{i=1}^{n} {\chi_i}^2 - \left(\sum_{i=1}^{n} \chi_i\right)^2} \tag{9}$$

In the formula, the number of delay boxes n used for curve fitting is set to 2; Among them, χ_i represents the delay value of the leading edge, and $\omega_{NIDW,i}^{L}$ represents the waveform value of the leading edge of NIDW.

The calculation logic for the trailing edge slope is the same as that for the leading edge slope, and the formula is as follows:

$$TES_{NIDW} = \frac{n\sum_{i=1}^{\mathrm{n}} \varepsilon_i\omega_{NIDW,i}^{T} - \sum_{i=1}^{n} \varepsilon_i \sum_{i=1}^{n} \omega_{NIDW,i}^{T}}{n\sum_{i=1}^{n} {\varepsilon_i}^2 - \left(\sum_{i=1}^{n} \varepsilon_i\right)^2} \tag{10}$$

where ε_i is the time delay value of the trailing edge part and $\omega_{NIDW,i}^{T}$ is the waveform value of the trailing edge of NIDW in the formula, the definitions of other symbols are consistent with formula (9).

The LEWS calculation formula for NIDW is as follows:

$$\mathrm{LEWS}_{\mathrm{NIDW}} = \sum_{i=1}^{m} \varphi_{\mathrm{NIDW},i}^{L} \tag{11}$$

Among them, the number of waveform values before the peak of the delay waveform is denoted as cc, and the leading waveform value of NIDW is denoted as $\varphi_{\mathrm{NIDW},i}^{L}$.

The TEWS calculation formula for NIDW is as follows:

$$\mathrm{TEWS}_{\mathrm{NIDW}} = \sum_{i=1}^{m} \varphi_{\mathrm{NIDW},i}^{T} \tag{12}$$

Among them, the number of waveform values after the peak of the delayed waveform is denoted as m, and the tail waveform value of NIDW is denoted as $\varphi_{\mathrm{NIDW},i}^{T}$.

3.2 Feedforward Neural Network (FFNN) Design

An FFNN model comprises three layers: input, hidden, and output. Such neural networks are capable of learning the mapping between input and output parameters without requiring explicit analytical equations. The learning mechanism employs a gradient-descent algorithm that continuously updates the network's weights and thresholds via error backpropagation, aiming to minimize the mean-square error. To evaluate the performance of the FFNN model, three metrics are used: RMSE, MAE, and the coefficient of determination (R^2).

The dataset is randomly divided into three subsets: the training set accounts for 60%, the validation set and the test set each account for 20%. Meanwhile, choose tanh as the activation function between the input layer and the hidden layer, and use MSE as the loss function. To prevent overfitting, the error rates of the training set and validation set should be similar, which is also the basis for measuring the effectiveness of data partitioning.

In this section, different FFNN configurations (hidden layers 1–5, with 10 neurons per layer) were tested one by one and subsequently applied to GNSS observations of the test data to retrieve rainfall intensity. The results of the rainfall intensity retrieval obtained were compared with the IMERG-F rainfall intensity, and the RMSE and R for each FFNN configuration were calculated. As shown in Fig. 9, the RMSE and R^2 values under different FFNN configurations are presented, which demonstrates that the multi-layer neural network model outperforms the single hidden layer model. The RMSE of a single hidden layer model is approximately 3.5 mm/h; The FFNN structure with three hidden layers (10 neurons per layer) achieves optimal performance, with an RMSE better than 2.8 mm/h and an R^2 better than 0.84.

Furthermore, we consider a more important aspect in affecting rainfall intensity retrieval, that is, sea condition. Swell and wind speed complicate the ocean surface spectrum and increase the uncertainty of GNSS-R rainfall intensity retrieval. In references [22, 30], the geophysical model function (GMF) used for wind speed inversion

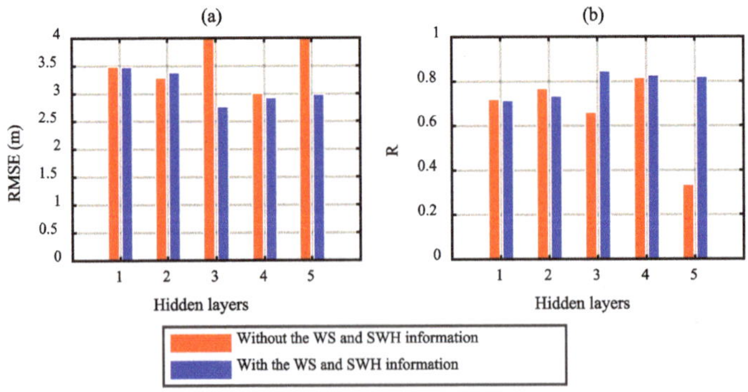

Fig. 9 RMSE (**a**) and R (**b**) of the RI retrieval by using FFNN models with different number of hidden layers

takes SWH as the input parameter and wave height information as prior knowledge. Similarly, in this study, the contribution of swell and wind speed information to neural network-based rainfall intensity inversion was also evaluated. In order to decouple additional rainfall intensity information from wave height and wind speed data, the SWH and wind speed of ERA5 were added as input parameters. Please note that as described in [5, 8], Due to the high sensitivity of GNSS-R observations to rainfall only at extremely low wind speeds (<5 m/s), and the extreme difficulty of RI inversion at medium to high wind speeds, this paper only studies sea surface RI retrieval under wind speeds less than 5 m/s.

The FFNN method can capture the nonlinear relationship between multiple observable values and IMERG-F rainfall intensity, which is reflected in the comparison in Fig. 9 This figure shows the effects of significant wave height (SWH) and wind speed on rainfall intensity inversion based on FFNN. After introducing surge height information, the inversion performance was significantly improved, and the RMSE decreased from 4.18 mm/h to 2.77 mm/h (a decrease of about 33%).

To further evaluate the impact of input parameters on FFNN based rainfall intensity retrieval, the following 10 schemes are designed according to different combinations of input parameters:

Scheme #1: DDMA, LES, TES, and SNR;

Scheme #2: DDMA, LES, TES, SNR, and longitude and latitude;

Scheme #3: DDMA, LES, TES, SNR, and incidence angle;

Scheme #4: DDMA, LES, TES, SNR, and RCG;

Scheme #5: DDMA, LES, TES, SNR, SWH, and wind speed;

Scheme #6: DDMA, LES, TES, SNR, RCG, longitude and latitude, and incidence angle;

Scheme #7: DDMA, LES, TES, SNR, RCG, longitude and latitude, incidence angle, SWH, and wind speed;

Scheme #8: DDMA, SNR, RCG, longitude and latitude, incidence angle, SWH, and wind speed;

Scheme #9: DDMA, LES, TES, LEWS, TEWS, SNR, RCG, longitude and latitude, incidence angle, SWH, and wind speed.

Scheme #10: DDMA, LEWS, TEWS, SNR, RCG, longitude and latitude, incidence angle, SWH, and wind speed.

Taking Scheme #9 as an example, the structure of FFNN is shown in Fig. 10, where the network contains an input layer, three hidden layers, and an output layer. Each layer contains 10 neurons. The input layer has a total of 12 input parameters, including 5 DDM features (i.e., DDMA, LES, TES, LEWS, and TEWS) and 7 auxiliary features (i.e., SNR, RCG, longitude and latitude, incidence angle, SWH, and wind speed).

The FFNN model is run on test data using different combinations of input parameters. The retrieved rainfall intensity is then compared with the IMERG-F rainfall intensity to derive the performance metrics. In Table 4, the rainfall intensity retrieval performance of each combination scheme is summarized. Differences and rates of improvement (IR) are related to scheme #1. From Table 4 can be observed below.

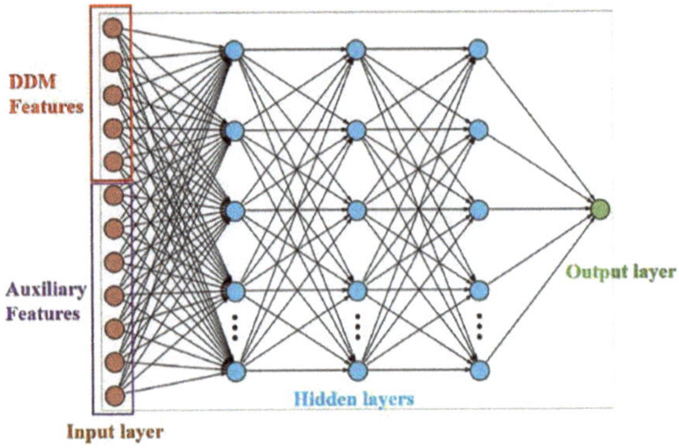

Fig. 10 The structure of the FFNN

Table 4 The performance metrics of input parameter combinations for the FFNN based RI retrieval

Schemes	#1	#2	#3	#4	#5	#6	#7	#8	#9	#10
RMSE (mm/h)	4.51	3.32	3.95	3.79	4.18	3.22	2.98	2.71	3.24	2.77
MAE (mm/h)	2.63	2.12	2.41	2.32	2.49	2.23	1.88	1.85	1.92	1.94
R	0.55	0.77	0.65	0.68	0.60	0.77	0.81	0.85	0.79	0.84
RMSE IR (%)	–	26.26	12.31	15.99	7.30	28.55	33.80	39.80	28.07	38.51
MAE IR (%)	–	19.20	8.08	11.50	5.04	15.22	28.35	29.75	26.78	26.22
R IR (%)	-	38.78	17.76	23.35	9.01	38.84	46.11	52.53	43.00	51.53

(1) The position information of specular reflection points can provide significant contributions, with an increase in R by 38.78% and a decrease of RMSE by 26.26% (e.g., scheme #2 and scheme #1).
(2) The angle of incidence of specular reflection points can also contribute to rainfall intensity retrieval, where RMSE is reduced by 12.31% (e.g., scheme #3 vs. scheme #1). Additionally, adding RCG, SWH, and wind speed parameters to the input parameters provides significant improvements.
(3) By adding swell height and wind speed information to the input parameters (e.g., scheme #1 and scheme #5), the rainfall intensity retrieval performance can be improved (7.3%), which is consistent with the analysis above.
(4) Comparing the performance of scheme #7 and scheme #10, it can be inferred that the inclusion of more auxiliary parameters does not imply an improved (i.e., reduced) RMSE for the test data. Overall, scheme #2, and scheme #6–#10 delivers optimal performance have an RMSE better than 3.32 mm/h.

3.3 Analysis of Retrieval Performance Under Wind Speed Only Less Than 5 m/s

To validate the robustness and generality, in this section, we evaluate the overall retrieval performance of the six best FFNN (i.e., scheme #2, and Scheme #6- scheme #10) RI retrieval models obtained in section III-B.

Figures 11 and 12 show the scatter density plots and distribution histograms of the deviation of the retrieved RI and IMERG-F rainfall data. Table 5 also lists the statistical results (RMSE, MAE, and CC) of the six FFNN models. Besides, Table 5 also shows the retrieval results of empirical model in [10] for comparison. The average value (Ave) and 80% quantile (Qua) of RI estimation error (or deviation from the reference) of the six schemes are given in Fig. 12. Combined with Fig. 11, Table 5, and Fig. 12, a few observations can be made as follows.

(1) When the RI is within 0–30 mm/h, the retrieval performance of scheme # 7 is the best, and RMSE and CC are better than 2.40 mm/h and 0.88, whereas the performance based on empirical model is the worst. Compared with the method in [10], the retrieval accuracy of scheme #2, and scheme #6- scheme #8 are

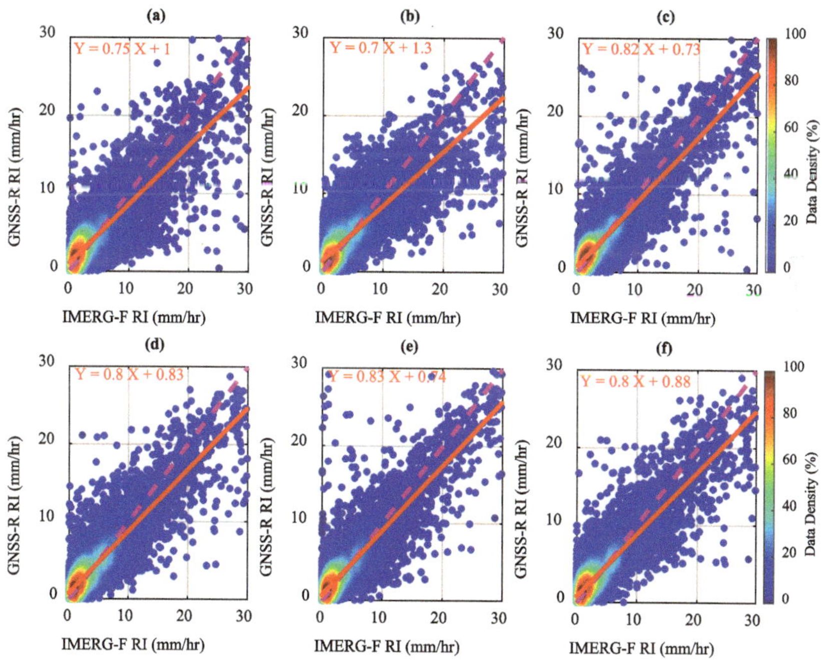

Fig. 11 Scatter density plots of RI estimates from FFNN models versus IMERG-F RI data. The magenta dotted line denotes the 1:1 reference, and the red solid line is the linear regression fit. Subfigures: **a** Scheme #2, **b** Scheme #6, **c** Scheme #7, **d** Scheme #8, **e** Scheme #9, **f** Scheme #10

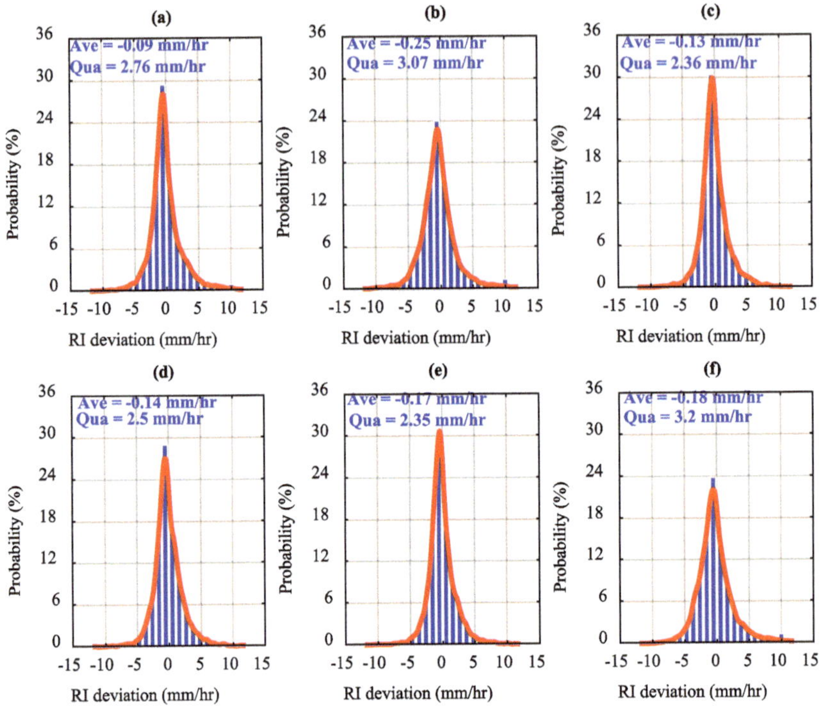

Fig. 12 RI deviations between the GNSS-R FFNN retrievals and IMERG-F data are displayed. A normal distribution curve (red solid line) is fitted to each deviation distribution. Subfigures: **a** Scheme #2, **b** Scheme #6, **c** Scheme #7, **d** Scheme #8, **e** Scheme #9, **f** Scheme #10

Table 5 Statistical results of RMSE, MAE, CC, and RMSE improvement rate of rainfall intensity retrieved by 6 Schemes

Schemes	< 10 mm/h		> 10 mm/h		0–30 mm/h			
	RMSE (mm/h)	MAE (mm/h)	RMSE (mm/h)	MAE (mm/h)	RMSE (mm/h)	MAE (mm/h)	R	IR (%)
Method in [10]	–	–	–	–	3.74	2.54	0.72	–
#2	2.17	1.50	5.43	4.10	2.79	1.83	0.84	25.47
#6	2.29	1.67	5.78	4.41	2.98	2.02	0.81	20.43
#7	2.00	1.36	4.22	3.02	2.40	1.57	0.88	35.90
#8	1.98	1.42	4.36	3.27	2.41	1.66	0.88	35.47
#9	2.22	1.37	4.07	2.98	2.52	1.57	0.87	32.61
#10	2.23	1.68	5.93	4.66	2.98	2.06	0.81	20.38

improved by 25.47%, 20.43%, 35.90%, 35.47%, 32.61%, and, 20.38%, respectively. Compared with the empirical model method, the average improvement rate of FFNN method in RMSE is 28.38%.

(2) When RI is below 10 mm/h or above 10 mm/h, among the six schemes, the retrieval performance of scheme # 7, scheme # 8, and scheme # 9 are better than that of the other three schemes.

(3) The RI deviation of RI retrieved by GNSS-R FFNN with IMERG-F data is approximately normally distributed. The average deviation between RI retrieved by the six FFNN schemes and IMERG-F RI is better than -0.25 mm/h; the 80% quantile of RI deviation of scheme #7, scheme #8, and scheme #9 are smaller than that of the other three schemes.

Overall, the rainfall intensity estimates obtained from the proposed FFNN retrieval schemes agree well with the IMERG-F rainfall data.

3.4 *Analysis of Retrieval Performance Under Wind Speed Less Than 10 m/s*

(1) Rainfall intensity retrieval performance for correcting wind speed effects

In this section, the rainfall intensity is retrieved when the wind speed is below 10 m/s. At the same time, the role of wind speed and swell effect correction in the inversion of sea surface rainfall intensity was discussed and analyzed in detail.

Figure 13 shows the density scatter plot between IMERG-F reference rainfall intensity and RI retrieved before and after correction (correction for swell and wind speed effects). Among them, subgraph (a) corresponds to a wind speed range of 0 to 10 m/s. In addition, the table also lists the inversion performance of the FFNN model in different rainfall intensity ranges (< 10 mm/h, > 10 mm/h, 5–10 mm/h, and 0–30 mm/h).

From Fig. 13 and Table 5 It can be seen:

(1) After the wind speed influence correction, the rainfall intensity retrieval performance was significantly better than that before the correction. For example, The four accuracy indices (RMSE, MAE, CC, MAPE) are improved by 30.30%, 34.85%, 32.26%, and 25.59% when wind speeds are 0–5 m/s, and by 20.13%, 34.25%, 28.57%, and 37.35% when wind speeds are 5–10 m/s. When the wind speed range is 0–10 m/s, the four accuracy indexes are improved by 15.19%, 26.27%, 24.56% and 27.11%, respectively.

(2) The retrieval performance of the FFNN model declines as rainfall intensity increases, particularly for intensities exceeding 10 mm/h. This degradation is largely attributable to the limited number of high-rainfall-intensity samples available. Conversely, when the rainfall intensity is below 10 mm/h—especially under 5 mm/h—the retrieval accuracy is very high, mainly due to the abundance of data samples within this range.

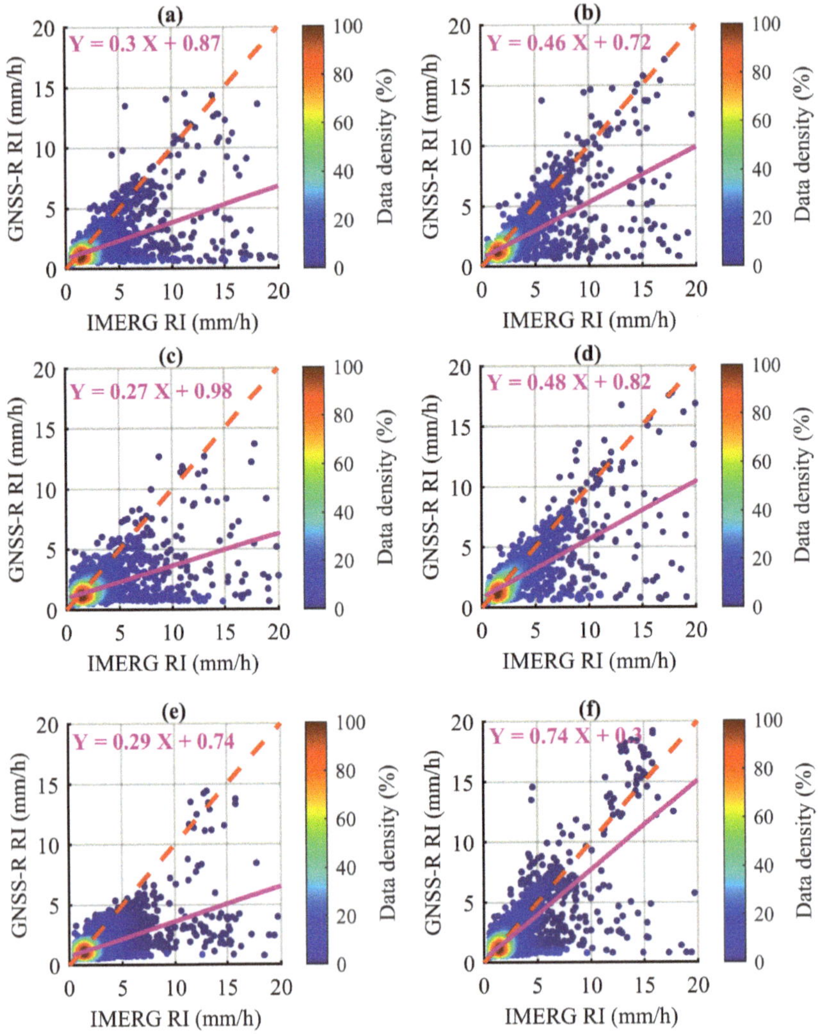

Fig. 13 Scatter density plots comparing IMERG-F rainfall intensity with retrieved rainfall intensity: before correction for swell and wind speed effects (left panels) and after wind speed effect correction (right panels). Wind speed ranges: **a**, **b** 0–10 m/s; **c**, **d** 5–10 m/s; **e**, **f** 0–5 m/s

(3) When the rainfall intensity is greater than 10 mm/h, the retrieval accuracy in the wind speed range of 5–10 m/s and 0–10 m/s is significantly worse than that in the wind speed range of 0–5 m/s. This shows that the higher the wind speed, the worse the retrieval performance for the greater rainfall intensity. Thus, it is extremely difficult to achieve rainfall intensity inversion within a large range of wind speeds; This problem can only be solved when the wind speed is not high and the rainfall intensity is not strong [37].

(2) Rainfall intensity retrieval performance after taking into account swell and wind speed factors

Figure 14 shows the density scatter plot of IMERG-F reference rainfall intensity and retrieveed RI under two correction conditions: correction only for surge effects and simultaneous correction for surge and wind speed effects. The three subgraphs of this graph—(a), (b), and (c)—correspond to wind speed ranges of 0–10 m/s, 5–10 m/s, and 0–5 m/s, respectively. Table 7 lists the inversion statistical indicators (RMSE, MAE, CC, MAPE) under the above two conditions, and also provides the retrieval performance of the FFNN model in four rainfall intensity intervals (< 10 mm/h, > 10 mm/h, 5–10 mm/h, 0–30 mm/h).

Comparing Fig. 13 with Table 6 in the previous section and Fig. 14 with Table 7 in this section, it can be seen that:

(1) Swell has a stronger effect on the accuracy of rainfall intensity retrieval than wind speed does.
(2) Rainfall intensity retrieval performance was substantially higher when wind speed and swell were corrected together than when only one of them was corrected. Improvements in the four accuracy indices (RMSE, MAE, CC, MAPE) are as follows: for wind speeds of 0–5 m/s, 12.79%, 11.66%, 4.90%, and 5.13%, respectively; for 5–10 m/s, 12.11%, 21.82%, 8.04%, and 18.61%; and for 0–10 m/s, 11.90%, 23.50%, 6.28%, and 19.63%.
(3) The rainfall intensity retrieval performance of the FFNN model decreases with increasing RI under different wind speed conditions. As mentioned in the previous section, the high accuracy exhibited when RI is less than 10 mm/h (especially less than 5 mm/h) is mainly due to the support of a large number of data samples within this interval. However, it should be noted that the accuracy obtained from the calculation may not necessarily represent the true performance of the interval—data imbalance is a serious challenge for machine learning models such as FFNN [4]. Notwithstanding that retrieval performance for the 0–5 m/s wind speed range exceeds that for the 5–10 m/s and 0–10 m/s ranges, the 5–10 mm/h rainfall intensity interval is more indicative of retrieval accuracy because its distribution is comparatively uniform. Hence, with concurrent correction of wind speed and swell effects, satisfactory retrieval accuracy is achievable for the 5–10 mm/h range.

As mentioned earlier, spaceborne GNS-R technology can only successfully retrieve sea surface rainfall intensity when the wind speed is not too high and the rainfall intensity is not too strong. This is because: under strong winds, the roughness of the sea surface is dominated by the wind, making it difficult to extract rainfall characteristics; And heavy rainfall can make the response of roughness to changes in rainfall intensity tend to saturate or become sluggish. Nevertheless, when the wind speed is below 5 m/s, this technology can still produce reliable accuracy. But in the case of simultaneous effects of wind speed and swell, inversion becomes very difficult. Nevertheless, based on the analysis in this section, when rainfall intensity

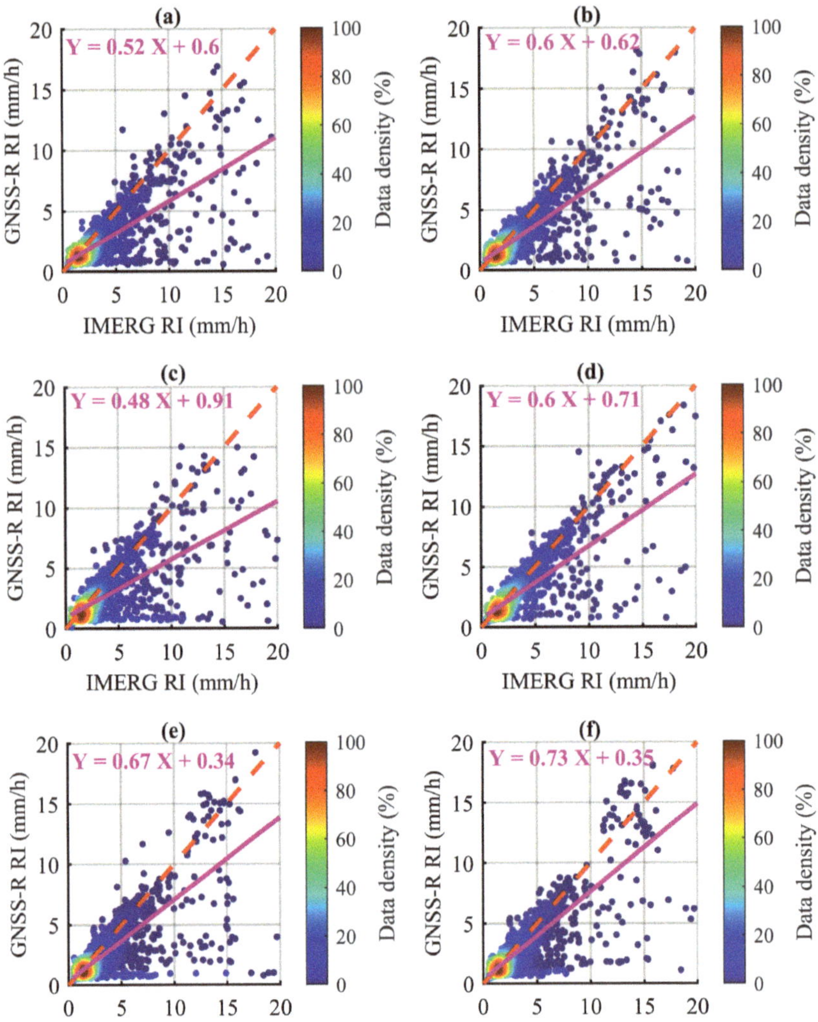

Fig. 14 Scatter density plots comparing IMERG-F rainfall intensity with retrieved rainfall intensity under two correction scenarios: after swell-only correction (left panels) and after combined swell and wind speed correction (right panels). Wind speed ranges: **a**, **b** 0–10 m/s; **c**, **d** 5–10 m/s; **e**, **f** 0–5 m/s

is modest (e.g., below 10 mm/h) and wind speed is under 10 m/s, the FFNN model—by correcting for both wind speed and swell simultaneously—achieves satisfactory performance [13]. This yields a retrieval accuracy of approximately 2 mm/h at a higher wind speed (< 10 m/s), which compares favorably with previous studies.

Table 6 RMSE, MAE, CC, and MAPE values for rainfall intensity retrieved under two conditions (uncorrected for swell and wind effects versus corrected for wind effects only)

Rainfall intensity range (mm/h)								
Wind speed (m/s)		< 10	> 10	5–10	0–30			
		RMSE (mm/h)	RMSE (mm/h)	RMSE (mm/h)	RMSE (mm/h)	MAE (mm/h)	CC	MAPE (%)
0–5	Wind speed + swell effect (uncorrected)	1.11	11.49	4.14	1.65	0.66	0.62	26.11
	Only wind speed influence (corrected)	0.83	7.48	2.96	1.15	0.43	0.82	19.42
5–10	Wind speed + swell effect (uncorrected)	1.76	12.62	4.44	3.18	1.46	0.56	36.84
	Only wind speed influence (corrected)	1.27	10.66	3.27	2.54	0.96	0.72	23.08
0–10	Wind speed + swell effect (uncorrected)	1.56	12.34	4.34	2.83	1.18	0.57	31.47
	Only wind speed influence (corrected)	1.22	10.83	3.39	2.41	0.87	0.71	22.94

The table also summarizes the FFNN model's retrieval accuracy for rainfall intensities in the ranges of < 10 mm/h, > 10 mm/h, 5–10 mm/h, and 0–30 mm/h

Table 7 Presents the accuracy of rainfall intensity inversion (RMSE, MAE, CC, MAPE) under two correction schemes (swell only, swell + wind), as well as the evaluation performance of the FFNN model in four RI intervals (<10, > 10, 5–10, 0–30 mm/h)

Rainfall intensity range (mm/h)								
Wind speed (m/s)		< 10	> 10	5–10	0–30			
		RMSE (mm/h)	RMSE (mm/h)	RMSE (mm/h)	RMSE (mm/h)	MAE (mm/h)	CC	MAPE (%)
[0, 5]	Only the impact of swell (corrected)	0.76	7.80	2.84	1.13	0.40	0.83	17.44
	Wind speed + swell effect (uncorrected)	0.66	6.82	2.47	1.08	0.35	0.87	16.54
[5, 10]	Only the impact of swell (corrected)	1.23	10.31	3.20	2.47	0.93	0.74	21.89
	Wind speed + swell effect (uncorrected)	0.99	9.52	2.55	2.17	0.72	0.80	17.81
[0, 10]	Only the impact of swell (corrected)	1.18	9.84	3.34	2.19	0.81	0.77	21.95
	Wind speed + swell effect (uncorrected)	0.90	9.03	2.51	1.93	0.62	0.82	17.64

4 Conclusion

In this chapter, the rainfall intensity retrieval based on spaceborne GNSS-R is realized, and sea surface rainfall intensity retrieval models are proposed based on empirical models and machine learning methods. Aiming at the problem of low retrieval accuracy of single observation variable-based model, combined models of double observation variables and three observation variables are proposed. The experimental results show that at very low wind speed (WS) (< 5 m/s), the following conditions are true. Firstly, the single-observation model (i.e., LES-NIDW) based on the slope of the leading edge of the normalized integral delay waveform proposed in the first type of model has the worst retrieval performance (RMSE = 3.69 mm/h, CC = 0.71]. However, the TEWS-NIDW performed best, with RMSE and CC of 3.17 mm/h and 0.79, respectively. Secondly, compared with LES-NIDW observations, the retrieval accuracy of models based on DDMA, LEWS-NIDW, and TEWS-NIDW observables is improved by 7.05%, 7.96%, and 12.29%, respectively. Thirdly, relative to the LES-NIDW model alone, the combination of two observables (LES-NIDW and

REWS-NIDW) improved accuracy by 10.34%, and the three-observable combination (LES-NIDW, LEWS-NIDW, and REWS-NIDW) improved it by 13.60%. Thus, a notable performance gain is achieved.

Compared with empirical models, the FFNN model proposed in this paper (based on machine learning) can better comprehensively consider complex factors related to GNSS-R processes such as geometric shape and sea conditions when retrieving rainfall intensity. This results in significantly higher retrieval accuracy of machine learning methods compared to empirical models, with an RMSE decrease of approximately 28.38%. The inversion results indicate that as long as the wind speed is controlled within 10 m/s, the rainfall intensity is moderate (such as less than 10 mm/h), and the effects of wind speed and swell are corrected simultaneously, the FFNN model can provide an inversion accuracy of about 2 mm/h.

In summary, this chapter has made some innovative research results in spaceborne GNSS-R sea surface intensity retrieval. It is important to note, however, that this chapter constitutes an early attempt to resolve this rather challenging problem. Owing to the dominant influence of sea surface winds on roughness, extracting rainfall features proves difficult, particularly under high wind speeds. The influence of raindrops is determined by their size, which in turn depends on rainfall intensity and cloud type. Therefore, the raindrop effect may vary with season (such as rainy season) and location (such as tropical and polar regions). In addition, wind speed and rainfall intensity can also affect the type of rainfall. It is worth noting that when the intensity of rainfall is too high, the roughness of the sea surface tends to saturate (i.e. no longer sensitive) to further increases in rainfall intensity. Therefore, currently, this problem can only be effectively addressed under conditions where both wind speed and rainfall intensity are not too high. However, spaceborne GNSS-R for rainfall intensity retrieval is still in its early experimental phase, and current technology cannot estimate raindrop size or retrieve rainfall type. These issues will therefore be the focus of our future work.

Acknowledgements The work was supported by the National Natural Science Foundation of China under grant 42404037.

References

1. Abdelmoneim H, Soliman MR, Moghazy HM (2020) Evaluation of TRMM 3B42V7 and CHIRPS satellite precipitation products as an input for hydrological model over Eastern Nile Basin. Earth Systems and Environment 4(4):685–698. https://doi.org/10.1007/s41748-020-00185-3
2. Alonso-Arroyo A, Zavorotny VU, Camps A (2017) Sea ice detection using U.K. TDS-1 GNSS-R data. IEEE Trans Geosci Remote Sens 55(9):4989–5001. https://doi.org/10.1109/tgrs.2017.2699122
3. Arshad M, Ma X, Yin J, Ullah W, Ali G, Ullah S, Liu M, Shahzaman M, Ullah I (2021) Evaluation of GPM-IMERG and TRMM-3B42 precipitation products over Pakistan. Atmos Res 249:105341. https://doi.org/10.1016/j.atmosres.2020.105341

4. Asgarimehr M, Arnold C, Weigel T, Ruf C, Wickert J (2022) GNSS reflectometry global ocean wind speed using deep learning: development and assessment of CyGNSSnet. Remote Sens Environ 269:112801. https://doi.org/10.1016/j.rse.2021.112801
5. Asgarimehr M, Zavorotny V, Wickert J, Reich S (2018) Can GNSS reflectometry detect precipitation over oceans? Geophys Res Lett 45(22):12585–12592. https://doi.org/10.1029/2018gl079708
6. Asgarimehr M., Hoseini M, Semmling M, Ramatschi M, Camps A, Nahavandchi H, Haas R, Wickert J (2022) Remote sensing of precipitation using reflected GNSS signals: response analysis of polarimetric observations. IEEE Trans Geosci Remote Sens 5800412. https://doi.org/10.1109/tgrs.2021.3062492
7. Ayoub AB, Tangang F, Juneng L, Tan ML, Chung JX (2020) Evaluation of gridded precipitation datasets in Malaysia. Remote Sensing 12(4):613. https://doi.org/10.3390/rs12040613
8. Balasubramaniam R, Ruf C (2020) Characterization of rain impact on L-band GNSS-R ocean surface measurements. Remote Sens Environ 239:111607. https://doi.org/10.1016/j.rse.2019.111607
9. Beck HE, van Dijk AIJM, Levizzani V, Schellekens J, Miralles DG, Martens B, de Roo A (2016) MSWEP: 3-hourly 0.25oglobal gridded precipitation (1979–2015) by merging gauge, satellite, and reanalysis data. Hydrol. Earth Syst Sci Discuss 1–38. https://doi.org/10.5194/hess-2016-236
10. Bu J, Yu K (2022) Sea surface rainfall detection and intensity retrieval based on GNSS-reflectometry data from the CYGNSS mission. IEEE Trans Geosci Remote Sens 60:5802015. https://doi.org/10.1109/TGRS.2021.3127943
11. Bu J, Yu K, Zhu Y, Qian N, Chang J (2020) Developing and testing models for sea surface wind speed estimation with GNSS-R delay doppler maps and delay waveforms. Remote Sens 12(22):3760. https://doi.org/10.3390/rs12223760
12. Bu J, Yu K, Han S, Qian N, Lin Y, Wang J (2022) Retrieval of sea surface rainfall intensity using spaceborne GNSS-R data. IEEE Trans Geosci Remote Sens 60:5803116. https://doi.org/10.1109/TGRS.2022.3168928
13. Bu J, Yu K, Zhu F, Zuo X, Huang W (2023) Joint retrieval of sea surface rainfall intensity, wind speed, and wave height based on spaceborne GNSS-R: a case study of the oceans near china. Remote Sen 15(11):2757. https://www.mdpi.com/2072-4292/15/11/2757
14. Camps A, Martín F, Park H, Valencia E, Rius A, Addio SD (2012) Interferometric GNSS-R achievable altimetric performance and compression/denoising using the wavelet transform: An experimental study. In: 2012 IEEE international geoscience and remote sensing symposium
15. Casse C, Gosset M, Peugeot C, Pedinotti V, Boone A, Tanimoun BA, Decharme B (2015) Potential of satellite rainfall products to predict Niger River flood events in Niamey. Atmos Res 163:162–176. https://doi.org/10.1016/j.atmosres.2015.01.010
16. Chen H, Yong B, Shen Y, Liu J, Hong Y, Zhang J (2020) Comparison analysis of six purely satellite-derived global precipitation estimates. J Hydrol 581:124376. https://doi.org/10.1016/j.jhydrol.2019.124376
17. Chiaravalloti F, Brocca L, Procopio A, Massari C, Gabriele S (2018) Assessment of GPM and SM2RAIN-ASCAT rainfall products over complex terrain in southern Italy. Atmos Res 206:64–74. https://doi.org/10.1016/j.atmosres.2018.02.019
18. Chu X, He J, Song H, Qi Y, Sun Y, Bai W, Li W, Wu Q (2020) Multimodal deep learning for heterogeneous gnss-r data fusion and ocean wind speed retrieval. IEEE J Sel Top Appl Earth Observations Remote Sens 13:5971–5981. https://doi.org/10.1109/jstars.2020.3010879
19. Clarizia MP, Ruf CS, Jales P, Gommenginger C (2014) Spaceborne GNSS-R minimum variance wind speed estimator. IEEE Trans Geosci Remote Sens 52(11):6829–6843. https://doi.org/10.1109/tgrs.2014.2303831
20. Dong Z, Jin S (2021) Evaluation of the land GNSS-reflected DDM coherence on soil moisture estimation from CYGNSS data. Remote Sensing 13:570. https://doi.org/10.3390/rs13040570
21. Donoho DL (1995) De-noising by soft-thresholding. IEEE Trans Inf Theory 41(3):613–627. https://doi.org/10.1109/18.382009

22. Guo W, Du H, Guo C, Southwell BJ, Cheong JW, Dempster AG (2022) Information fusion for GNSS-R wind speed retrieval using statistically modified convolutional neural network. Remote Sens Environ 272:112934. https://doi.org/10.1016/j.rse.2022.112934
23. Hammond ML, Foti G, Rawlinson J, Gommenginger C, Srokosz M, King L, Unwin M, Roselló J (2020) First assessment of geophysical sensitivities from spaceborne Galileo and BeiDou GNSS-reflectometry data collected by the UK TechDemoSat-1 Mission. Remote Sens 12(18):2927. https://doi.org/10.3390/rs12182927
24. Huffman GJ, Bolvin DT, Braithwaite D, Hsu K, Joyce R, Kidd C, Nelkin EJ, Sorooshian S, Tan J, Xie P (2019) Algorithm Theoretical Basis Document (ATBD) Version 06 NASA Global Precipitation Measurement (GPM) Integrated Multi-satellitE Retrievals for GPM (IMERG). National Aeronautics and Space Administration (NASA). 1–34
25. Hussain Y, Satgé F, Hussain MB, Martinez-Carvajal H, Bonnet M-P, Cárdenas-Soto M, Roig HL, Akhter G (2017) Performance of CMORPH, TMPA, and PERSIANN rainfall datasets over plain, mountainous, and glacial regions of Pakistan. Theoret Appl Climatol 131:1119–1132. https://doi.org/10.1007/s00704-016-2027-z
26. Islam MA, Yu B, Cartwright N (2020) Assessment and comparison of five satellite precipitation products in Australia. J Hydrol 590:125474. https://doi.org/10.1016/j.jhydrol.2020.125474
27. Jing C, Niu X, Duan C, Lu F, Di G, Yang X (2019) Sea surface wind speed retrieval from the first Chinese GNSS-R mission: technique and preliminary results. Remote Sens 11(24). https://doi.org/10.3390/rs11243013
28. Keikhosravi Kiany MS, Masoodian SA, Balling RC, Montazeri M (2020) Evaluation of the TRMM 3B42 product for extreme precipitation analysis over southwestern Iran. Adv Space Res 66(9):2094–2112. https://doi.org/10.1016/j.asr.2020.07.036
29. Li N, Tang G, Zhao P, Hong Y, Gou Y, Yang K (2017) Statistical assessment and hydrological utility of the latest multi-satellite precipitation analysis IMERG in Ganjiang River basin. Atmos Res 183:212–223. https://doi.org/10.1016/j.atmosres.2016.07.020
30. Li X, Yang D, Yang J, Zheng G, Han G, Nan Y, Li W (2021) Analysis of coastal wind speed retrieval from CYGNSS mission using artificial neural network. Remote Sens Environ 260:112454. https://doi.org/10.1016/j.rse.2021.112454
31. Nwachukwu PN, Satge F, Yacoubi SE, Pinel S, Bonnet M-P (2020) From TRMM to GPM: how reliable are satellite-based precipitation data across Nigeria? Remote Sensing 12(23):3964. https://doi.org/10.3390/rs12233964
32. Rodriguez-Alvarez N, Garrison JL (2016) Generalized linear observables for ocean wind retrieval from calibrated GNSS-R delay-doppler maps. IEEE Trans Geosci Remote Sens 54(2):1142–1155. https://doi.org/10.1109/tgrs.2015.2475317
33. Tang G, Clark MP, Papalexiou SM, Ma Z, Hong Y (2020) Have satellite precipitation products improved over last two decades? A comprehensive comparison of GPM IMERG with nine satellite and reanalysis datasets. Remote Sens Environ 240:111697. https://doi.org/10.1016/j.rse.2020.111697
34. Wang Y, Morton YJ (2020) Coherent GNSS reflection signal processing for high-precision and high-resolution spaceborne applications. IEEE Trans Geosci Remote Sens. 1–12. https://doi.org/10.1109/tgrs.2020.2993804
35. Yan Q, Huang W, Moloney C (2017) Neural networks based sea ice detection and concentration retrieval from GNSS-R delay-doppler maps. IEEE J Sel Topics Appl Earth Observations Remote Sens 10(8):3789–3798. https://doi.org/10.1109/jstars.2017.2689009
36. Yan Q, Huang W (2016) Spaceborne GNSS-R sea ice detection using delay-doppler maps: first results from the U.K. TechDemoSat-1 Mission. IEEE J Sel Topics Appl Earth Observations Remote Sens 9(10):4795–4801. https://doi.org/10.1109/jstars.2016.2582690
37. Yu K, Han S, Bu J, An Y, Zhou Z, Wang C, Tabibi S, Cheong JW (2022) Spaceborne GNSS reflectometry. Remote Sens 14(7):1605. https://doi.org/10.3390/rs14071605
38. Zhou Z, Guo B, Xing W, Zhou J, Xu F, Xu Y (2020) Comprehensive evaluation of latest GPM era IMERG and GSMaP precipitation products over mainland China. Atmos Res 246:105132. https://doi.org/10.1016/j.atmosres.2020.105132

39. Zhu Y, Tao T, Zou J, Yu K, Wickert J, Semmling M (2020) Spaceborne GNSS reflectometry for retrieving sea ice concentration using TDS-1 data. IEEE Geosci Remote Sens Lett 18:612–616. https://doi.org/10.1109/lgrs.2020.2982959
40. Zhu Y, Wickert J, Tao T, Yu K, Li Z, Qu X, Ye Z, Geng J, Zou J, Semmling M (2020) Sensing sea ice based on doppler spread analysis of spaceborne GNSS-R data. IEEE J Sel Topics Appl Earth Observations Remote Sens 13:217–226. https://doi.org/10.1109/jstars.2019.2955175

Chapter 22
Sea Wave Height Retrieval with Spaceborne GNSS Reflectometry

Changyang Wang, Nianfu Xu, Kegen Yu, and Nanshan Zheng

Abstract This chapter focuses on sea surface significant wave height (SWH) estimation using GNSS reflectometry (GNSS-R). Specifically, SWH estimation is based on the models established with a number of machine-learning algorithms, including Gaussian Support Vector Machine, Artificial Neural Network, and Bagging Tree. The ECMWF (European Centre for Medium-range Weather Forecast Reanalysis) SWH data are used as the reference SWH data for the model construction and the performance comparison. The GNSS observation data are the spaceborne delay-Doppler map (DDM) data provided by CYGNSS mission, which can be used to derive a range of feature parameters used as input variables such as the normalized bistatic radar cross section (NBRCS), leading-edge slope (LES), signal-to-noise Ratio (SNR) and DDM average (DDMA). To enhance the SWH retrieval performance, eight auxiliary parameters are also used as input variables, including instrument gain and scattering area. For each machine learning algorithm, a model is developed to describe the relationship between SWH and the input variables. Experimental results based on testing one million sets of data demonstrate that there are significant differences between the machine learning algorithms. The Bagging Tree model demonstrates the best performance, achieving a root mean square error (RMSE) of 0.48 m and a correlation coefficient (CC) of 0.82. Meanwhile, the effect of selecting input variables and training dataset size on the SWH retrieval performance is also evaluated.

C. Wang
National Satellite Meteorological Center, China Meteorological Administration, , Beijing, China
e-mail: changyangwang163@163.com

N. Xu · K. Yu (✉) · N. Zheng
School of Environmental Science and Spatial Informatics, China University of Mining and Technology, Xuzhou, China
e-mail: kegen.yu@cumt.edu.cn

N. Xu
e-mail: nianfuxu@cumt.edu.cn

N. Zheng
e-mail: znshcumt@cumt.edu.cn

K. Yu and N. Zheng (eds.), *Advances in GNSS Reflectometry*, Navigation: Science and Technology 19, https://doi.org/10.1007/978-981-92-2325-1_22

1 Introduction

Significant Wave Heights (SWH) is probably the most important parameter to measure sea wave height, which is also a major parameter to describe the sea surface roughness. SWH is defined as the average of the first one third of all the wave heights sorted from highest to lowest, which were observed over a specific period of time and a specific area of interest. To predict ocean storm disasters and assess wave energy resources, SWH plays a significant role [1]. Thus, it is vital to effectively monitor SWH especially in regions where damaging waves often occur. There are several different types of ocean waves including waves generated by sea surface wind, called local wind-driven waves, swells originally generated by storms of other locations, and tsunamis generated by land and ocean disasters such as earthquakes and volcano eruptions. Comparatively, local wind-driven waves have short wavelengths of usually a few meters and can be a few ten meters in the event of Typhoon or Hurricane. Swells have longer wavelengths which can be up to a few hundreds of meters, while Tsunamis can have wavelengths of a few hundreds of kilometers.

To acquire sea wave data (e.g., SWH), tens of thousands of buoys have been deployed on the ocean surface. Sensors fixed on the buoys are used to measure and record wave parameters of interest. The buoy-recorded data would have the highest accuracy, so they are usually treated as reference wave data, to evaluate the performance of other remote sensing technologies. However, the deployed buoys can only be used to monitor a rather small area of the vast oceans, usually located on the sea surface near the coastal lines. In addition, the cost is very large to deploy and maintain the large number of buoys [2]. Meanwhile, a series of altimetric satellites have been successfully launched in the past few decades, and a large amount of SWH data have been recorded. Satellite altimeters have several advantages such as wide coverage and high accuracy. One example is that the altimetry satellites Jason-1 and Jason-2 have recorded a lot of global SWH data and the altimetric accuracy is about 2 cm [3]. However, the spatial coverage of existing significant wave height (SWH) products from satellite altimeters is limited since the altimeters only measure the wave height of the narrow swath along the sea surface trajectory of the satellite [4].

GNSS reflectometry (GNSS-R) is an emerging remote sensing technology capable of measuring SWH. In 1988, Hall and Corley are probably the first to propose the use of GNSS signals for measuring sea surface wind speed [5], but it is in 1993 that Martín-Neira [6] investigated to use GNSS-R for measuring sea surface height and provided details of the altimetric system design. In the past three decades, GNSS-R has made great success in theory and practice, which has been utilized to sense a range of land and ocean geophysical parameters such as soil moisture [7], snow depth [8], sea surface wind speed [9, 10], and sea surface height [11], and to detect objects such as land flooding area [12], sea ice [13], and sea surface ship [14]. Late 1990s and early 2000s, a good number of ground-based and airborne experiments were conducted, and interesting results were reported [15]. The first satellite related to GNSS-R, UK-DMC, was developed by Surrey Satellite Technology Ltd and launched in 2003, successfully capturing reflected signals from the sea surface. In the past decade, more than 30 satellites with GNSS-R payloads have been launched, including TechDemoSat-1 satellite (TDS-1) by UK in 2014, 8 CYGNSS microsatellites by USA in 2016, BuFeng-1 A/B twin satellites by China in 2019, FY-3E satellite by China in 2021, and more than 20 Tianmu satellites by China in 2023–2024. The data generated by receivers onboard these satellites, especially those by TDS-1 and CYGNSS missions have widely been exploited for Earth observation. These satellite missions mainly focused on the sensing of the sea surface wind field (speed and direction) especially the intensity of Typhoon and Hurricane.

In addition to sea surface wind speed, sensing other geophysical parameters with GNSS-R also has been extensively investigated, such as sea surface height retrieval, soil moisture retrieval, snow depth retrieval, sea ice monitoring and inland water detection. However, retrieving the important ocean parameter SWH with GNSS-R did not gain significant attention initially. Early efforts focused on the ground-based methods to retrieve SWH. There are two main methods, the interference pattern technique technique (IPT)-based method [16] and the interferometric complex field (ICF)-based method [17, 18], which can only monitor SWH of small sea surface area near the coast.

In the past few years, a number of reports on GNSS-R-based SWH measurement can be seen in the literature especially using spaceborne data. In [4], the signal-to-noise ratio (SNR), a feature parameter derived from the delay-Doppler map (DDM) generated by CYGNSS receivers, is taken as the observation (i.e., input variable) and initial SWH estimation results have been produced. In [19], improved SWH estimation accuracy is achieved by using a 2-D polynomial (PN) function to describe the relationship between ECMWF SWH data and leading-edge slope (LES) of the delay waveform extracted from the CYGNSS DDM, as well as the CYGNSS DDM average (DDMA). Bu and Yu [20] have further enhanced the estimation accuracy through data fusion and optimization, but the generality of these methods is limited, since a relatively small amount of data was used in the studies.

This chapter focuses on the use of a number of machine learning (ML) algorithms to construct ML-based models to retrieve the SWH using spaceborne data, aiming to enhance the accuracy and generality of the models [21]. Generality is achieved by using a lot of data for model development, which was collected on a global scale and over a period of three years, while the high accuracy is enabled by using a range of DDM feature parameters and auxiliary parameters of receiver and environment as the input variables.

The remainder of the chapter is organized as follows. The next section describes the datasets used in this study and the data preprocessing to remove poor-quality data, based on data filtering and selection. Section 3 first constructs empirically the mathematical models between SWH and individual input variables, and then constructs ML models to describe the relationship between SWH and multiple input variables and analyzes their performance. Section 4 studies the properties of a ML-based retrieving model, bagging-tree model; constructs and evaluates one of the most-suited models; and analyzes how the different input variables affect the performance of the SWH retrieving model. Finally, conclusions are provided in Sect. 5.

2 Data Acquisition and Preprocessing

2.1 Datasets

In this chapter, two datasets are used, which are selected from CYGNSS (L1B) v3.1 product and ERA-5 SWH data generated by ECMWF. The CYGNSS data corresponds to the GNSS signals reflected over the sea surface of latitude 40° S to 40° N, and the spatial resolution is about 25 km, and the temporal resolution is 0.5 s [9]. The ERA-5 SWH data have a temporal resolution of 1 h and a spatial resolution of 0.5° × 0.5°. These data were derived by integrating meteorological data gathered from various sources. Due to the difference in temporal and spatial resolution, CYGNSS data must be matched in time and space with ERA-5 data [22]. It is a fact that the sea wave field frequently consist of different types of waves, as mentioned earlier, i.e., wind-driven waves, swells, and Tsunamis. Although a Tsunami seldom occurs, wind-driven waves and swells are often mixed. The wave height, wavelength and wave period of wind-driven waves and swells can be significantly different. In some cases, it is an advantage to measure the wave heights of the two different types of waves separately. However, in terms of disaster monitoring, measuring the combined SWH is more important. Thus, the significant height of wind waves (SHWW) and significant height of total swells (SHTS) in ERA-5 data are not individually considered, but the combined SWHs in ERA-5 data are used as the reference, which would be close to the actual sea surface SWHs [23].

The statistics of waves near the coast would be significantly different from those far away from the coast. Also, the signals reflected over sea surface close to the coast would be received together with signals reflected over the land area along the coast. Consequently, it is rather difficult, if not impractical, to measure the sea surface SWH near the coastline using the mixed signals reflected from sea and land surface. Thus, in this chapter, the data of signals with sea surface specular reflection points at least 250 km away from the coastline are employed. All CYGNSS data used in this chapter were collected when signal specular reflection points are located on the blue sea surface areas, as shown in Fig. 1. The land area is marked in black, while the coastal sea areas of 250 km wide are marked in green.

Nevertheless, it would be useful to conduct research on this challenging issue, measuring SWH of the sea surface area near the coastline with GNSS-R technology in the future. In Fig. 1, the challenging area is marked green and bordered by the coastline and another line 250 km away from the coastline. Frankly, this issue would not be as hard as that of measuring wind speed of the sea surface near the coastline. In addition to signal reflection over land, sea surface wind speed measurement near the coastline also has the problem of "fetch".

Besides the two datasets described as above, the buoy data from the USA's National Data Buoy Center (NDBC) are also utilized to assess the model's performance. Usually, the buoy-observed SWH data are more accurate, while those obtained through remote sensing have uncertainty, although the uncertainty can be rather small in some cases such as for the data observed by Jason-1 and Jason-2 (as

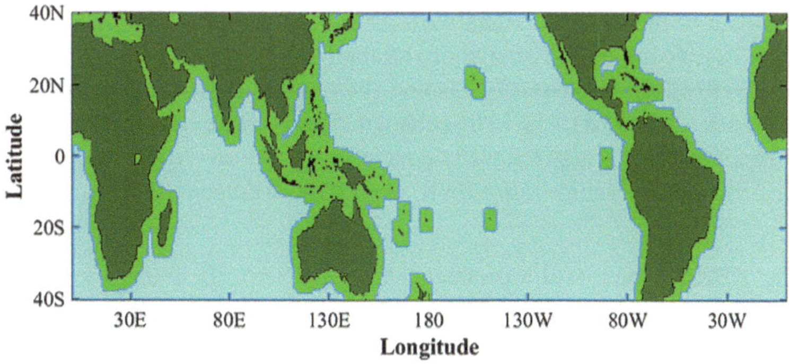

Fig. 1 The blue regions indicate the sea surface areas where data were collected. The dark green areas represent land, while the light green regions denote coastal seas, defined as the area extending from the coastline up to 250 km offshore

well as Jason-3 still in operation) satellite altimeters. More details about the buoy data will be provided later.

2.2 *Definition and Selection of Variables*

In the development of the SWH retrieval models, regardless of the empirical mathematical model or the machine learning-based model, input variables need to be defined and selected. This chapter deals with two types of variables, variables derived from DDM, and variables related to receiver/satellite and environment. The following provides more details of these variables.

DDM-Related Variables

Two-scale model is widely used to describe the electromagnetic scattering of the sea surface, and the scattering coefficient is defined as:

$$\sigma_{TSM} = \sigma_{KA-GO} + \sigma_{SPM} \quad (1)$$

where σ_{KA-GO} represents the scattering coefficient of large-scale surface roughness, determined using the Kirchhoff Approximation-Geometric Optics (KA-GO), and σ_{SPM} represents the scattering coefficient of small-scale surface roughness, calculated using the Small Perturbation Method. Through detailed analysis and experiments, it has been demonstrated that σ_{KA-GO} is mainly related to wind-driven waves and σ_{SPM} is mainly relevant to swells [24]. Consider the typical situation where wind wave (only related to wind speed) and swells are the only two components of the sea surface wave. SWH is strongly correlated with sea surface roughness and scattering coefficients, making its retrieval through spaceborne GNSS-R technology a viable approach.

DDM data of the reflected signals are the most important observation of spaceborne GNSS reflectometry. A range of variables can be derived from DDM, and in the case of reflection over sea surface, some of the variables are sensitive to the roughness of the sea surface [20, 25]. In this chapter, four variables are utilized to measure SWH, which are the normalized bistatic radar cross section (NBRCS), leading-edge slope (LES), signal-to-noise ratio (SNR) and DDM average (DDMA), as described below.

(1) The NBRCS, one of the products of CYGNSS L1B, is highly sensitive to sea surface roughness [26]. Due to this reason, NBRCS is highly effective in retrieving sea surface wind speed [24]. Similarly, NBRCS is very helpful in the measurement of SWH, although there are only a limited number of relevant studies reported. NBRCS is calculated as the ratio of the bin-by-bin sum of scattering cross sections $\overline{\sigma}_{total}$ to the bin-by-bin sum of effective scattering areas $\overline{A}_{total}$, as follows:

$$NBRCS = \frac{\overline{\sigma}_{total}}{\overline{A}_{total}} = \frac{\sum_{i=1}^{N}\sum_{j=1}^{M}\overline{\sigma}_{\tau_i f_j}}{\sum_{i=1}^{N}\sum_{j=1}^{M}\overline{A}_{\tau_i f_j}} \tag{2}$$

where N and M represent the number of delay bins and that of Doppler bins, respectively, $\overline{\sigma}_{\tau_i f_i}$ and $\overline{A}_{\tau_i f_i}$ represent the DDM bin-related bistatic radar cross section (BRCS) and effective scattering area when the delay and the Doppler frequency are τ_i and f_j. Generally, NBRCS is calculated using the 5 Doppler × 3 delay bins ($M = 5, N = 3$) around the specular point ($\tau_i = 0, f_j = 0$) in BRCS DDM products. Such a selection is based on the consideration that the signals in this central area are the strongest and it has been experimentally verified.

(2) LES is that of delay waveform and usually the integrated delay waveform (IDW) is used to reduce noise and interference. It is calculated by:

$$LES = \frac{N\sum_{i=1}^{N}\tau_i IDW_i - \sum_{i=1}^{N}\tau_i\sum_{i=1}^{N}IDW_i}{N\sum_{i=1}^{N}(\tau_i)^2 - \left(\sum_{i=1}^{N}\tau_i\right)^2} \tag{3}$$

where N represents the number of delay bins utilized for curve fitting, and N is determined to be 3; τ_i the time delay, and IDW is defined as:

$$IDW(\tau) = \frac{1}{M}\sum_{j=1}^{M}\left(DDM\left(\tau, f_j\right)\right)^2 \tag{4}$$

where M represents the number of Doppler bins utilized for curve fitting, which is set as 5, and $DDM\left(\tau, f_j\right)$ is the measured power of the reflected signal corresponding to the time delay τ and Doppler frequency f_j. Thus, like the calculation of NBRCS, reflected signal powers of 5 Doppler bins and 3 delay bins are used to determine the LES for the same reason. It is worth noting that the trailing edge slope (TES) of the delay waveform is usually considered for retrieving sea surface wind speed using airborne data in the case of low to medium wind speed. In the spaceborne case, the LES is also sensitive to sea surface roughness. Currently, TES is not used to retrieve SWH with spaceborne data mainly due to poor performance. It is not clear why this occur since TES would also be sensitive to roughness. Probably, the TES data are rather noisy, and noise mitigation would be required. It would be useful to work on this issue in the future.

(3) SNR of DDM is strongly correlated with SWH [27] and this can be intuitively understood. Higher SWH indicates larger roughness of the sea surface, and accordingly, the received signal has a stronger noise, resulting in probably reduced SNR. Thus, SNR is also a useful variable which can be used in measuring SWH. SNR of DDM is defined as:

$$SNR = 10 \times lg\left(\frac{S_{max}}{N_f}\right) \tag{5}$$

where S_{max} is the maximum pixel/bin value of DDM which consists of 20×128 raw pixel values (uncompressed DDM) in this chapter. N_f is the noise power (i.e., the noise floor) of the background noise of the observation. It is calculated with the pixel values of the raw DDM, where there is no signal but noise, as follows:

$$N_f = \frac{1}{N_n}\sum_{i=1}^{45}\sum_{j=1}^{20} rawDDM\left(\tau_i, f_j\right) \tag{6}$$

where N_n is the number of noise pixels of DDM, and 45 delay bins and 20 Doppler bins are used in this case, which corresponds to the blue bins on the left side of the DDM shown in Fig. 2. The selection of the bins should be based on the specific DDM pattern and only the noise bins are chosen.

(4) DDMA is simply the average of the selected DDM pixel values. The basic rule for the selection of pixels is that the pixel values should be as high as possible, i.e.., the corresponding SNRs would be the highest. As indicated earlier, the central 5 Doppler $\times$ 3 delay bins of the DDM, as shown in Fig. 2 under 6 different SWH values, are used to calculate DDMA, that is:

$$DDMA = \frac{1}{15}\sum_{i=1}^{N}\sum_{j=1}^{M} anaDDM\left(\tau_i, f_j\right) \tag{7}$$

where n equals 3 and m equals 5 in this case, anaDDM denotes analogue DDM.

Auxiliary Variables

In addition to the four main variables derived from DDM, which are closely associated with sea surface roughness and SWH, a number of other variables, namely

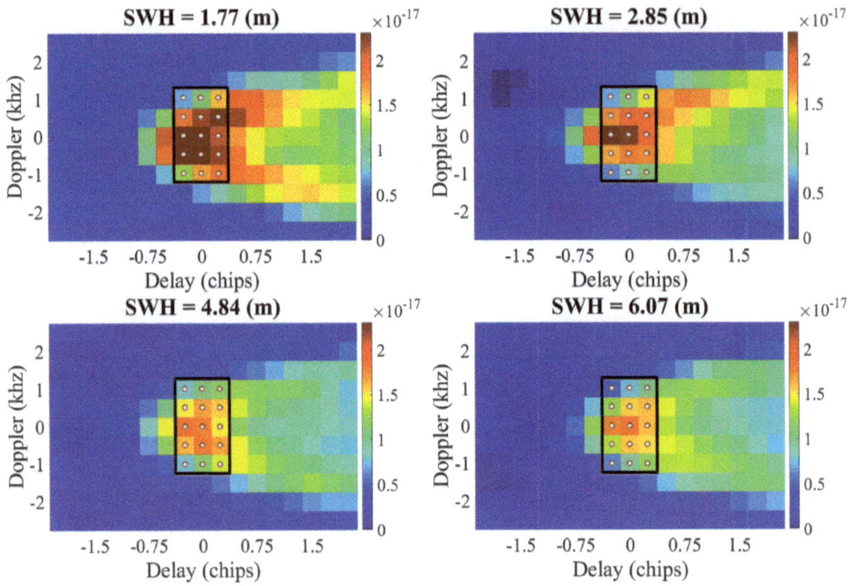

Fig. 2 Processed DDMs for six different SWH (1.77, 2.85, 4.84, and 6.07 m). The central 3 delay × 5 Doppler bin box is used to calculate DDM variables, corresponding to a Doppler spread of − 1 to 1 kHz and a chip delay range of − 0.25 to 0.25

auxiliary variables, would also have a certain relationship with SWH. These auxiliary variables are the ocean environmental variables and receiver/satellite variables, which directly or indirectly affect the reception of the reflected signals and thus the generated DDM, or are related to the characteristics of SWH, dependent on location and time/date. Thus, inclusion of the auxiliary variables as input variables would bring extra useful information to enhance the model accuracy and retrieval performance. Table 1 shows the four DDM variables and 8 auxiliary variables. The following offers a brief overview of these auxiliary variables and explains the rationale behind their use:

(1) Variables related to signal status: *Instrument gain* is a crucial parameter in generating the CYGNSS L0 and L1 products. It includes the low noise amplifier gain, all cascaded front-end gain stages, and processing gain. A key function of instrument gain is to calibrate the signal power in the Delay Mapping Receiver (DMR), which is essential for calculating the DDM power and noise power. *Scattering area* refers to the actual reflecting surface area that generates the reflected signal DDM. The surface roughness of the reflector impacts this value (in units of m^2). Typically, a rougher reflecting surface leads to a larger scattering area. *Noise floor*, as defined in Eq. (5), is partially influenced by the SNR. It is solely associated with noise and does not correlate with the useful signal.
(2) Orientation of specular reflected signal: *The incidence angle and azimuth angle of the reflected signal at specular point* are named sp_inc_angle and sp_az_

Table 1 Input variable list for machine learning-based models

	Input variables
DDM variables	Normalized bistatic radar cross section (NBRCS)
	Leading edge slope (LES)
	DDM signal to noise ratio (SNR)
	DDM average (DDMA)
Auxiliary variables	Instrument gain
	Scattering area of NBRCS and LES (scatter area)
	DDM noise floor
	Specular point incidence angle (Sp_inc_angle)
	Specular point azimuth angle (Sp_az_body)
	Specular point latitude (Sp_lat)
	Specular point longitude (Sp_lon)
	Hours of the day (HOD)

body, which affect other DDM variables such as SNR and NBRCS [24, 28]. These two variables contribute to improving the performance of the models [29].

(3) Space–time information of specular point: By adding *the latitude and longitude of the specular point* to the list of input variables, the model can take into account the impact of the region on SWH. By using the GPS week and GPS seconds, we can determine *the hour of the day* (HOD) for a data sample. The inclusion of the HOD variable helps minimize the impact of tidal variations at different times on the retrieval of SWH.

2.3 Data Filtering and Data Selection

Although calibration and initial processing have been applied to generate CYGNSS L0 product and L1 products (and other similar products such as TDS-1 and FY-3E products), further data filtering and selection is needed to remain the high-quality data. First, all invalid observational values (indicated by NaN or negative numbers) are removed. Then, some quality control flags are applied, and data with the following issues will be removed [29]:

(1) The data with overall poor quality (bit 0).
(2) The absolute value of the spacecraft roll exceeds or is equal to 30 degrees (bit 3).
(3) The DDM Cyclic Redundancy Check transmitted from the DDMI to the spacecraft computer was not valid (bit 6).
(4) Reflectometry channel was not tracking a PRN (bit 8).
(5) Errors occurred during the BRCS calculation (bit 17–19).

Table 2 The distribution of SWH samples

SWH range (m)	0–1	1–2	2–3	3–4
Sample count	109,074	2,411,574	2,661,781	759,055
SWH range (m)	4–5	5–6	> 6	
Sample count	178,933	41,025	14,787	

(6) Other errors about reflected signals, specular points and spacecrafts occurred (bit 20–27).

Also, to guarantee data quality, the FOM (figure of merit) is utilized to choose the suitable satellites which are in good conditions, and FOM must be greater than 10. Further, seasonal factors need to be taken into consideration. For instance, during particular seasons, El Niño and La Niña episode may take place, which substantially affect SWH [30]. Thus, it is necessary to use data collected over 12 different months to accommodate the effect of seasons on sea states. Finally, to enable the model's generality so it can be widely applied, a large amount of data samples collected globally should be used. In this chapter, 6 million sets of data collected from Mar. 2019 to Dec. 2021 are used in the model construction and performance evaluation. The distribution of SWH data is shown in Table 2. Most of the SWH samples are located between 1 and 4 m, and there are rather limited number of samples of high SWH. A model built on such a population of SWH samples would be more suited to measure SWH of 1–4 m. Larger errors would be generated for SWH higher than 4 m. A rough uniform distribution of the SWH samples is desirable in future model development. Alternatively, it is useful to use other techniques to handle this uneven distribution problem.

3 Construction of SWH Retrieval Model

In some cases, the parameter to be estimated is not dominantly determined by an observation variable or feature variable, it is not precise to develop a model to estimate the unknown parameter based on observations of a signal variable such as SNR in retrieving SWH. Also, when developing a complex model using machine learning method, it is a common technique to study different ML algorithms, test their performance and choose the most suited one or two with the consideration of complexity. It is also worth noting that in developing empirical models, including ML-based one, a large amount of data is utilized, typically divided randomly into two groups of data, one being used for training and the other being used for testing to evaluate the performance of the model. In this chapter, to further verify the generalization ability of the models, another group of data is used for further testing or validation, which is independent of the two groups of data used for training and initial testing. The flowchart of model construction and evaluation is shown in Fig. 3.

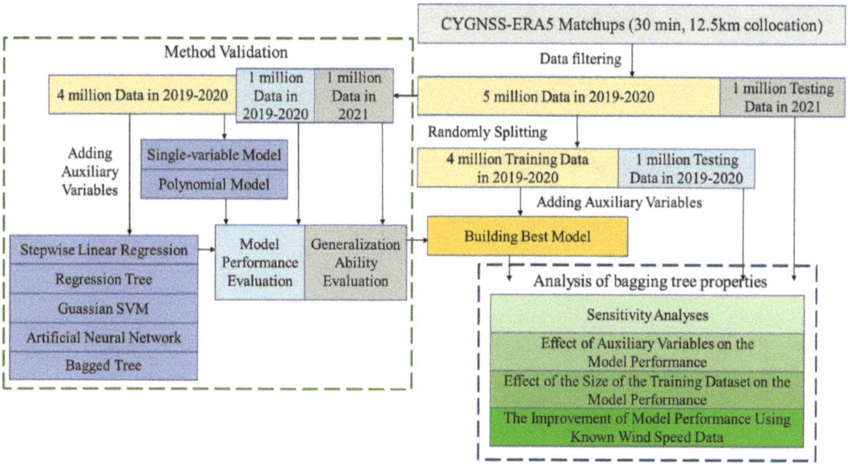

Fig. 3 Flow chart of the proposed methodology and the analytical experiments

Specifically, 4 million of data samples are employed for the construction of the ML-based models, including the initial testing. In addition, 1 million of data samples are utilized for model validation. The impact of the number of the training data samples on the model performance will be assessed. Three performance metrics are employed to assess the performance of various models and the influence of different variables, which are root mean square error (RMSE), correlation coefficient (CC) and bias. For a quick reference, the formulas for calculating these performance metrics are listed below:

$$RMSE = \sqrt{\frac{1}{n}\sum_{i=1}^{n}(X_i - Y_i)^2} \tag{8}$$

$$CC = \frac{\sum_{i=1}^{n}\left(X_i - \overline{X}\right)\left(Y_i - \overline{Y}\right)}{\sqrt{\sum_{i=1}^{n}\left(X_i - \overline{X}\right)^2 \sum_{i=1}^{n}\left(Y_i - \overline{Y}\right)^2}} \tag{9}$$

$$Bias = \frac{1}{n}\sum_{i=1}^{n}(X_i - Y_i) \tag{10}$$

where n is the number of data samples, $\{X_i\}$ are the SWH estimates generated by the models constructed in this chapter, $\{Y_i\}$ are the SWH data of ECMWF treated as the reference, $\overline{X}$ and $\overline{Y}$ are the mean of $\{X_i\}$ and $\{Y_i\}$ respectively.

3.1 Empirical Mathematical Models with Single Variable

Before building multi-variable models, it is useful to evaluate the performance of SWH models based on individual variables derived from DDM. First, SNR variable was considered in [4], and a simple SNR-based model was developed to retrieve sea surface SWH using data samples collected on limited number of specific dates:

$$\mathrm{SWH} = a_0 \times \mathrm{SNR}^{0.5} + b_0 \tag{11}$$

The model demonstrated good performance when a small number of data samples are tested. In this chapter, this model is extended to generate two more general models:

$$\mathrm{SWH} = a_1 \times v_i^{c_1} + b_1 \tag{12}$$

$$\mathrm{SWH} = a_2 \times e^{b_2 v_i} + c_2 \times e^{d_2 v_i} \tag{13}$$

where v_i denotes any of the individual DDM-derived variables such as based on the data of CYGNSS L1B v3.1. Here, NBRCS, LES, SNR and DDMA are the four individual variables, which are strongly correlated with sea surface SWH, as stated previously. In the three models of Eqs. 11–13, a_i, b_i, c_i, and d_i are the mode parameters to be fitted. The exponential parameter c_1 in Eq. 12 varies with v_i and it can be different from 0.5 for the case of SNR. These model parameters can be determined through least-squares fitting applied to a large dataset.

Figure 4 shows the scatter plots of NBRCS and SWH when the number of samples is 160 thousand, 800 thousand, 4 million, respectively. In Fig. 4, the red solid line represents the power function, and the green solid line represents the exponential function, which are fitted using the above two functions respectively. It can be seen that all fitting lines pass through the centre of the warm-color area where the majority of the samples are concentrated, demonstrating that majority of samples are well described by the fitted functions. However, a significant number of samples are still far away from the fitting lines, which indicates that the single-variable model would not be desirable due to large fitting error and thus large retrieval errors especially in the case of using a large amount of data samples for model construction. That is, models with multiple variables would be encouraged, which will be studied in next subsection.

When 4 million of samples of NBRCS, LES, SNR and DDMA variables are respectively used for curve fitting, 8 models for SWH retrieval are generated. The fitting performance in terms of RMSE, CC, and Bias are listed in Table 3. It can be seen from Table 3 that the RMSEs of all the eight models are similar, although the differences among their CCs are relatively a bit larger. The NBRCS-based power-function model and the LES-based exponential-function model achieve the highest accuracy with the largest CC being 0.43. Clearly, such a low largest CC is not satisfactory for typical applications. This is rather different from those results reported in

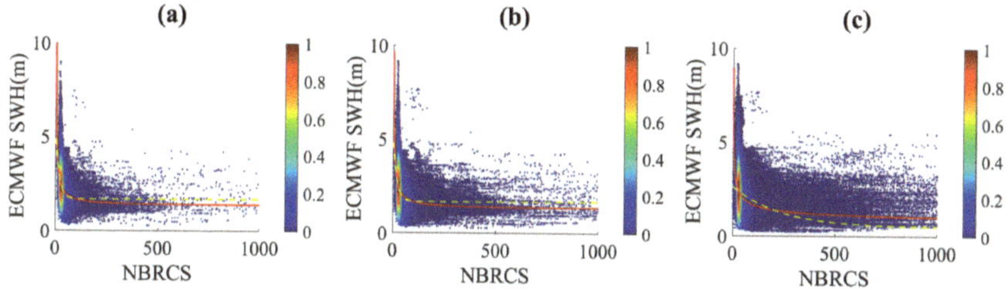

Fig. 4 The SWH retrieval model using NBRCS observables. **a** 160 thousand data samples. **b** 800 thousand data samples. **c** 4 million data samples. The red solid line represents power function, and the green dotted line represents exponential function

Table 3 Performance of single-variable SWH retrieval models using power and exponential functions

		NBRCS	LES	SNR	DDMA
Power function	RMSE (m)	0.74	0.75	0.76	0.77
	CC	0.43	0.41	0.37	0.35
Exponential function	RMSE (m)	0.78	0.74	0.76	0.79
	CC	0.32	0.43	0.37	0.25

[4, 19], which may be because a small amount of data is used for model construction and accuracy evaluation.

3.2 *Machine Learning Methods Using Multiple Variables*

It is often impractical to develop empirical mathematical models which have multiple variables especially when the number of variables is large. On the other hand, there are many machine learning methods which can be utilized to build models, which are not explicit mathematical formulas, to describe the complex relationship between an output variable and several input variables. Machine learning has been widely exploited in variety of fields with remarkable success, and it has also been utilized in GNSS reflectometry. Here, five different machine learning methods are briefly studied, which are Stepwise Linear Regression, Gaussian Support Vector Machine, Artificial Neural Network, Sparrow Search Algorithm–Extreme Learning Machine, and Bagging Tree. They will be used in next subsection to build machine learning-based SWH retrieval models.

(1) **Stepwise linear regression (SLR)**

Stepwise regression is a method used for variable selection in linear regression. The principle of this method is inputting variables (e.g., LES) into a linear regression model. If the RMSE of the model decreases, the variable (LES) is retained in the regression model. Conversely, if the RMSE increases, the variable is discarded. This process is repeated in several steps until no additional variables can be included, resulting in an automatically fitted regression model [31]. SLR is well-suited for addressing multivariable problems. The performance of the model is influenced by the sample size, and typically, larger number of samples are needed to obtain more reliable results [32].

(2) **Gaussian support vector machine (GSVM)**

As a supervised learning algorithm and a generalized linear classifier, the support vector machine (SVM) is designed for binary data categorization. It establishes a decision boundary using the maximum-margin hyperplane for the learning samples [33]. SVM minimizes empirical risk through the hinge loss function and optimizes structural risk by incorporating a regularization term into its solution. This approach makes the classifier both robust and sparse.

SVM, one of the standard kernel learning techniques, is capable of performing nonlinear classification or inversion [34]. The choice of the kernel function plays a crucial role in determining the performance of the SVM inversion model. To improve the model's effectiveness, the Gaussian function can be selected as the kernel function, and the corresponding SVM is called GSVM.

(3) **Artificial neural network (ANN)**

ANN is highly effective in solving complex nonlinear problems, and has been widely applied to remote sensing and other fields [35]. This chapter builds a five-layer feedforward Fully Connected Network (FCN), consisting of an input layer, three hidden layers, and an output layer. As illustrated in Fig. 5, the input layer consists of 12 CYGNSS variables, while the three hidden layers contain 10, 20, and 10 neurons, respectively. $\alpha_{12\times10}$, $\beta_{10\times20}$, $\gamma_{20\times10}$ and $\delta_{10\times1}$ in Fig. 5 are weight matrices, which can be calculated using the loss function and error return. The activation function is $f_{sigmoid}(x)$, which is defined as:

$$f_{sigmoid}(x) = \sigma(x) = \frac{1}{1+e^{-x}} \tag{14}$$

(4) **Sparrow Search Algorithm—Extreme Learning Machine (SSA-ELM)**

The structure of ELM is similar to Fig. 6, but it consists of only a single hidden layer. The ELM-based SWH estimate denoted by $\widetilde{SWH}$ is described as:

$$\widetilde{SWH} = \sum_{m=1}^{M}\sum_{n=1}^{N} \beta_m f_{Sigmoid}(\omega_{mn}v - b_m) \tag{15}$$

where N and M represent the number of neurons in the input layer and hidden layer, respectively, set to 12 and 50. β_m is the connection weight between the mth hidden layer neuron and output layer, ω_{mn} is the connection weight between the nth input

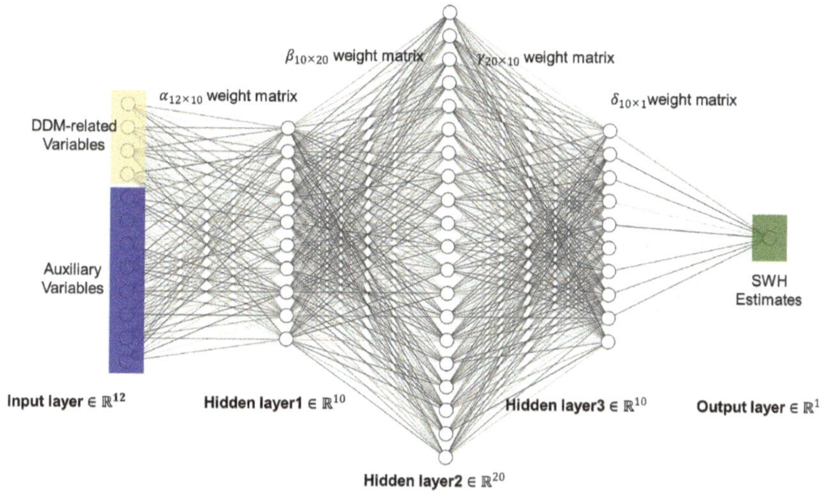

Fig. 5 ANN structure adopted in this chapter

layer neuron and mth hidden layer neuron, b_m is the threshold of the mth hidden layer neuron, and v is the input variable. Therefore, the performance of ELM model is primarily influenced by weight matrices β, ω and b. Note that here the vectors are treated as special cases of matrix.

To quickly obtain the optimal values for β, ω and b, SSA is employed. This algorithm is a novel swarm intelligence optimization approach proposed by [36] inspired by the foraging and anti-predatory behaviour of sparrows. It offers advantages such as high search accuracy, fast convergence, good stability, and the ability to avoid getting trapped in local optima. The algorithm categorizes all samples into Producers and Scroungers. In this chapter, a single sample is denoted as follows:

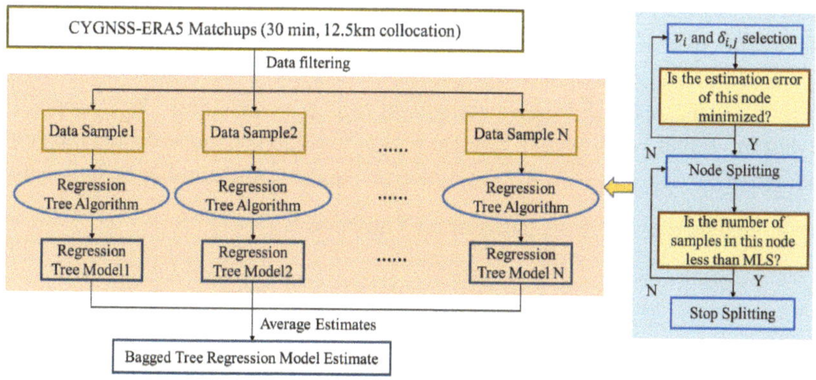

Fig. 6 Flow chart of the Bagging Tree

$$S_i = \{\beta^i, \omega^i, b^i; RMSE_i\} \tag{16}$$

where β^i, ω^i and b^i are the β, ω and b weight matrices of the ith sample, respectively. $RMSE_i$ denotes the RMSE of the ELM model, which is constructed using the above weight matrices, when performing SWH retrieval on the testing dataset. SSA-ELM involves five steps:

Step 1: Generate x samples randomly, with each sample having random β, ω and b weight matrices. In this chapter, a total of 50 samples were created, with 35 designated as Producers and the remaining 15 as Scroungers. Then, 15 samples are randomly selected from the Producers and Scroungers, and these are referred to as Vigilantes.

Step 2: All Producers randomly generate their weight matrices again. If the RMSE of the sample decreases, the updated weight matrices are adopted; otherwise, the sample retains its original weight matrices. In the tth iteration, the Producer sample is updated as follows:

$$X_{i,j}^{t+1} = \begin{cases} X_{i,j}^t \cdot e^{-\frac{i}{\alpha \cdot iter_{max}}}, & ifR_2 < ST \\ X_{i,j}^t + Q \cdot \mathrm{L}, & ifR_2 \geq ST \end{cases} \tag{17}$$

where $X_{i,j}^t$ represents the weight matrix of i-th sample in t-th iteration, and j indicates the index of current variables to be optimized, $j = 1, 2, 3$; $iter_{max}$ is the largest number of iterations, which is set to 50; $\alpha \in (0, 1]$ is a random number; R_2 and ST represent the alarm value and the safety threshold respectively, which are set to 0.6 and 0.7; Q is a random number that follows a normal distribution; and L is a $1 \times d$ vector, where each element is 1.

Step 3: To further reduce the RMSE of the Scroungers, the weight matrices of all Scroungers are updated based on the newly updated Producers:

$$X_{i,j}^{t+1} = \begin{cases} Q \cdot e^{-\frac{X_{worst}^t - X_{i,j}^t}{i^2}}, & ifi > \frac{n}{2} \\ X_P^{t+1} + \left| X_{i,j}^t - X_p^{t+1} \right| \cdot A^+ \cdot \mathrm{L}, & ifi \leq \frac{n}{2} \end{cases} \tag{18}$$

where X_P is the optimal weight matrices occupied by Producer; X_{worst} represents the current global worst weight matrix; A is a $1 \times$ d vector, where each element is randomly assigned 1 or $-$ 1; $A^+ = A^T(AA^T)^{-1}$ and n is the number of samples.

Step 4: Update the weight matrices of all Vigilantes.

$$X_{i,j}^{t+1} = \begin{cases} X_{best}^t + \lambda \cdot \left| X_{i,j}^t - X_{best}^t \right|, & iff_i > f_g \\ X_{i,j}^t + J \cdot \left(\frac{\left| X_{i,j}^t - X_{worst}^t \right|}{(f_i - f_w) + \varepsilon} \right), & iff_i = f_g \end{cases} \tag{19}$$

where X_{best} is the current global optimal weight matrix; λ is the step size control parameter, which follows a normal distribution with a mean of 0 and a variance of 1; $J \in [-1, 1]$ is a random number; f_i is the fitness of the ith sample; f_g and f_w are the current global best and worst RMSE, respectively; and ε is the smallest constant so as to avoid zero-division-error.

Step 5: Repeat Steps 2–4 until $t = iter_{max}$. Then select the sample with the smallest RMSE among all 50 samples, and use its β, ω and b weight matrices as the weight matrices for the final ELM model.

(5) **Bagging Tree**

BT is a widely used technique for constructing regression models [37]. It has been extensively applied across various industries, including the processing of remote sensing data [38, 39]. As shown in Fig. 6, the training set is drawn from the original sample set. Bootstrapping randomly selects training samples from the original dataset in each iteration [40, 41]. To generate N training sets, N rounds of sampling with replacement are performed, resulting in N models being created. These models and training sets are mutually independent [42]. The RMSE of the model will be decreased if numerous comparable datasets are produced by resampling with replacement, and regression trees are also generated without pruning and averaged [41]. By creating multiple comparable datasets through resampling and constructing unpruned regression trees that are subsequently averaged, the model's RMSE can be significantly reduced [41]. In this chapter, N = 30. In the BT algorithm studied in this chapter, a dataset containing 4 million samples is utilized to construct the models, and the mth sample is defined as:

$$S_m = \left\{ v_{1m}, v_{2m}, \ldots, v_{12m}; \widehat{SWH_m} \right\} \tag{20}$$

where $v_{1m}, v_{2m}, \ldots, v_{12m}$ and $\widehat{SWH_m}$ represents the values of the 12 input variables and the true SWH value of the sample, respectively. Figure 7 shows the process of node splitting. All the samples in a node will be divided into 2 child nodes: $Node_1 = \left\{S_{N_1 1}, S_{N_1 2}, S_{N_1 3}, \ldots\right\}$ and $Node_2 = \left\{S_{N_2 1}, S_{N_2 1}, S_{N_2 3}, \ldots\right\}$, according to the optimal segmentation variable v_i among the 12 variables and the optimal segmentation value $\delta_{i,j}$. For example, the node evaluation criteria in Fig. 7 may be 'if NBRCS < 14.513?'. Here, NBRCS is the optimal segmentation variable, and 14.513 is the optimal segmentation value. Samples satisfying the evaluation criterion are allocated to $Node_1$, while the remaining samples are directed to $Node_2$.. For simplicity, the two child nodes are labeled as N1 and N2 respectively. Each sample within these child nodes is assigned a SWH estimate ($\widetilde{SWH_{N1}}$, $\widetilde{SWH_{N2}}$), calculated using the following equation:

$$\widetilde{SWH_{N1}} = \frac{1}{K_1} \sum_{p=1}^{K_1} \widehat{SWH_{N_1 p}},$$

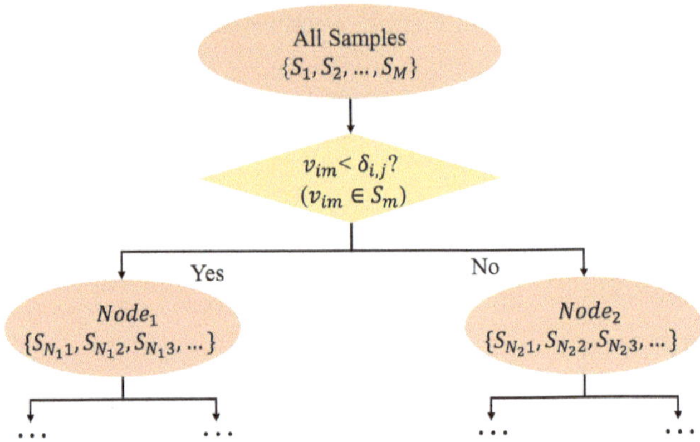

Fig. 7 Illustration of node splitting

$$\widetilde{SWH_{N2}} = \frac{1}{K_2}\sum_{q=1}^{K_2}\widehat{SWH_{N_2 q}} \tag{21}$$

where K_1 and K_2 are the sample number of the two child nodes respectively. The optimal segmentation variable v_i and the optimal segmentation value $\delta_{i,j}$ are determined based on the criterion of minimizing the estimation error at this node:

$$EE = min_{v_i,\delta_{i,j}}\left\{\sum_{p=1}^{K_1}\left(\widehat{SWH_{N_1 p}} - \widetilde{SWH_{N1}}\right)^2 + \sum_{p=1}^{K_2}\left(\widehat{SWH_{N_2 p}} - \widetilde{SWH_{N2}}\right)^2\right\} \tag{22}$$

where EE is the estimation error of the node. Subsequently, the two child nodes ($Node_1$ and $Node_2$) undergo further splitting independently. Splitting ceases for a node when its sample count becomes less than or equal to the Minimum Leaf Size (MLS), which is set to 4.

3.3 Performance of Multi-Variable Regression Models

This section evaluates the performance of six modeling approaches, including Polynomial function (PN), SLR, GSVM, ANN, SSA-ELM, and BT, as previously described. In [11], a simple 2D PN was constructed using LES and DDMA. The 2D PN function is expressed as:

$$\begin{aligned}\widetilde{SWH} &= p_{00} + p_{10}x + p_{01}y + p_{20}x^2 + p_{11}xy \\ &+ p_{02}y^2 + p_{30}x^3 + p_{21}x^2y + p_{12}xy^2\end{aligned} \tag{23}$$

where $\widetilde{SWH}$ is the SWH estimate, x represents LES, and y represents DDMA. $\{p_{00}, p_{10}, p_{01}, \ldots, p_{12}\}$ denote the fitting parameters of the model. This approach was also applied in this section for performance comparison. The models will be developed using both the four DDM-related variables and all auxiliary variables.

Table 4 shows the retrieval performance of four multivariable ML models and the 2D PN model. The same training datasets and testing datasets in Sect. 3.1 are used. This approach allows for an intuitive analysis of the accuracy difference between multivariable ML models and single-variable ML models. To ensure clarity, $RMSE_{2021}$, CC_{2021} and $Bias_{2021}$ are introduced to represent the metrics used for assessing generalization performance. In practical applications, the retrieval speed of the model is a crucial performance metric. In this chapter, the calculation time refers to the average time taken to generate all 1 million SWH estimates, repeated five times for each model. This approach helps minimize the impact of random factors on the overall calculation time.

Overall, compared with Table 3, the multi-variable ML models significantly outperform the single-variable ML models in both RMSE and CC. The $RMSE_{2021}$ values for the five multi-variable models are higher than their RMSE counterparts, while the CC_{2021} values are lower than CC. This discrepancy may primarily stem from notable differences in environmental and system parameters between 2019–2020 and 2021. Another factor could be the calibration quality of GNSS-R data, which is not consistently stable, partly due to the unpredictability of the GPS transmitter and CYGNSS receiver. As shown in Fig. 8, all the ML models tend to slightly overestimate low SWH (below 1 m) and underestimate high SWH (above 5 m). The primary reason for this phenomenon is that, when randomly sampling data, the volumes of low SWH and high SWH data are significantly smaller than that of medium SWH. In this case, most retrieval algorithms tend to follow a similar approach to ensure that the model retrieves more samples with higher accuracy. Additionally, when SWHs are low, the reflective surface roughness decreases, leading to a reduction in the scattering area of the reflected signal and most of the signal power becomes concentrated in the center of the DDM, which reduces the DDM's sensitivity to changes in surface roughness. Meanwhile, when SWHs are high, the reflective surface roughness also increases, making it challenging for DDM images to accurately exhibit a distinct horseshoe shape. At the same time, these data are often associated with greater interference factors, such as storms or hurricanes, which leads to a significant decline

Table 4 SWH retrieval metrics of multivariable ML models and PN model

Methods	PN	SLR	GSVM	ANN	SSA-ELM	BT
RMSE (m)	0.79	0.81	0.73	0.59	0.67	0.48
$RMSE_{2021}$ (m)	0.75	0.71	0.78	0.61	0.64	0.5
CC	0.32	0.25	0.58	0.69	0.57	0.82
CC_{2021}	0.36	0.48	0.52	0.64	0.62	0.78
Calculation time (s)	0.0105	0.1963	7556	0.2025	0.2938	2.5232

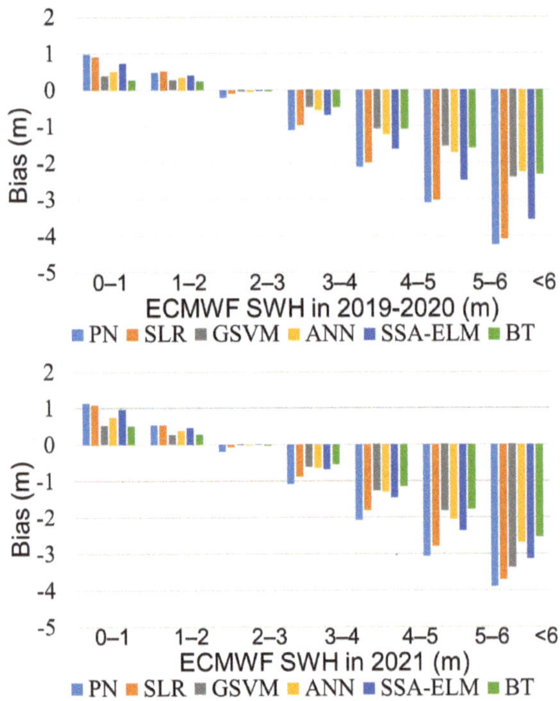

Fig. 8 The bias across different ranges of the CYGNSS SWH

in the models' performance. Furthermore, BT outperforms the other models under both low and high SWH conditions.

As shown in Table 4, the proposed BT-based model delivers the best overall performance, achieving the highest CC and the lowest RMSE, and it also demonstrates the strongest generalization ability. While the PN model has the shortest calculation time, its RMSE and CC values are the poorest among all models. The performance of the PN model differs from that reported in [19], likely due to the much smaller dataset used in [19]. Testing data from a two-year period are used to cover various sea states, including extreme scenarios with heavy rain and strong winds. The complex data environment in these conditions impacts the performance of the geophysical model function [43]. Although the accuracy of the ANN is slightly lower than that of BT, its calculation time is significantly shorter. Therefore, ANN is particularly well-suited for applications with complexity constraints. BT is an ensemble of multiple decision trees, which exhibits low variance in regression problems. The performance of BT may outperform other models significantly only when there is an adequate number of training dataset samples (for instance, using 4 million randomly selected samples). However, this increased performance comes at the cost of higher memory usage and longer computation time compared to other models.

Figure 9 presents the scatterplots between the SWH estimates from the six models and the 2019–2020 testing SWH dataset. The overall performance of these models is evaluated in Fig. 10, which further confirms that BT generally outperforms the other

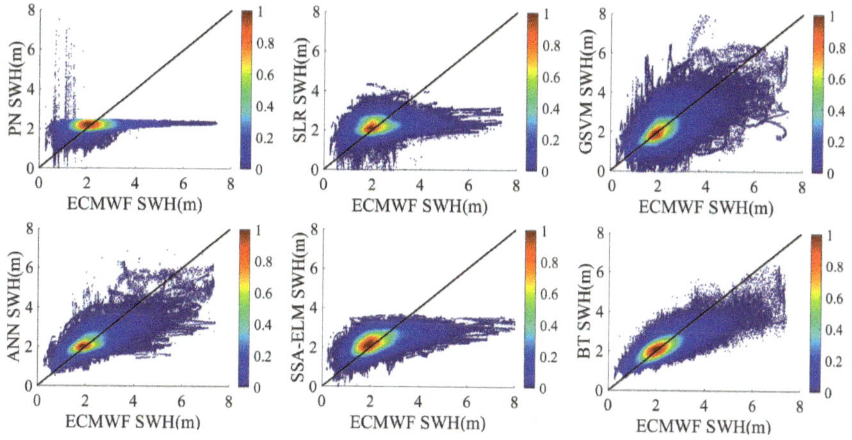

Fig. 9 Scatter plot of SWH retrieved by 6 models and SWH of the 2019–2020 test set. The black line is the 1:1 performance line

models, showing improved RMSE and Bias across the board. The distribution of the SWH estimates aligns well with the results presented in Table 4.

3.4 Comparison with NDBC Buoy Wave Height Data

Since the training and testing of the above model are based on ERA5 SWH data as the reference., it is necessary to use additional SWH data to evaluate the performance of each model. To this end, SWH data collected by the NDBC are used as an additional reference. Data of 22 NDBC buoys from July 3 to September 20, 2021 were employed. As illustrated in Fig. 11, the yellow pushpin markers indicate the buoy stations, which are primarily located in the southwestern waters of the United States and near Hawaii.

NDBC buoys usually provide wave height observations every 30 min to 1 h. For CYGNSS and NDBC matching, the NDBC buoy position and the CYGNSS specular point are considered a match when they are within 25 km of each other and the time difference is within 30 min. Table 5 provides the number of matchups for each NDBC buoy, totaling 4500 matchups. Figure 12 shows the relationship between NDBC SWH data and ECMWF SWH data, while Table 6 shows their accuracy. The RMSE in ECMWF SWH is approximately 0.19 m, and there is a very strong correlation between these two datasets. Additionally, Fig. 12 displays the SWH retrieval results based on the six models. Clearly, among these models, the estimates based on BT have the lowest RMSE and the highest CC when compared to the buoy measurements. Table 5 also presents the RMSE and CC of the BT-based model when using buoy observations as the reference. As shown in the table, the wave height estimates from the BT model exhibit considerable dispersion. This is primarily due

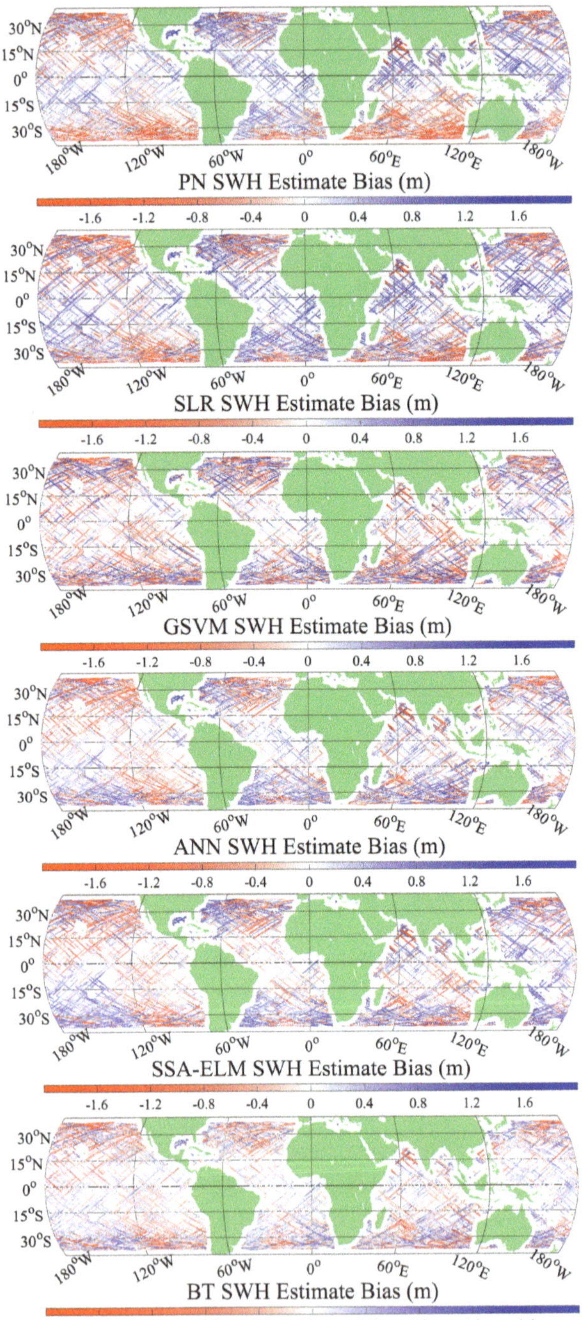

Fig. 10 The spatial distribution of bias between model estimates and ECMWF SWHs covers the period from March 2019 to November 2020

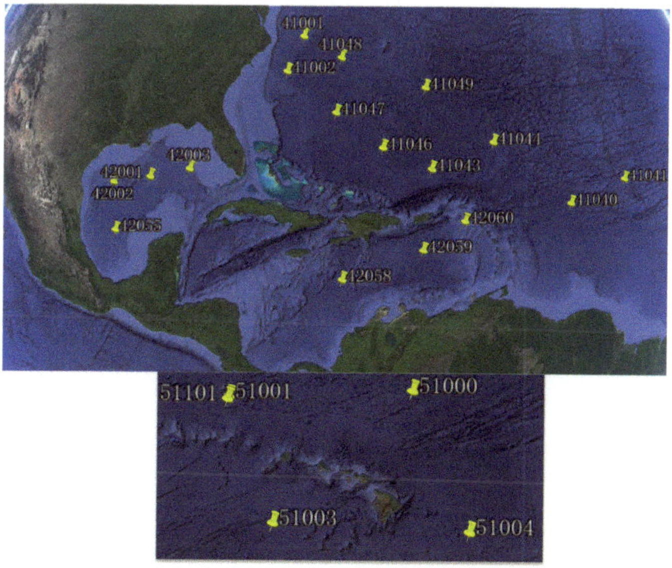

Fig. 11 The locations of buoy stations

to the differences in buoy instrument locations, which result in significant variations in BT performance. The magnitude of SWH can vary considerably across different locations. Moreover, even if the SWHs measured by two distant buoys are identical, the distribution and characteristics of the SWH within a 25 km radius around each buoy may differ, leading to variations in performance metrics. Similar observations were also reported in [29].

4 Analysis of the Performance of the Bagging Tree Algorithm

In this section, BT is chosen for further analysis. Moreover, these results are expected to provide helpful insights for the development of more advanced SWH retrieval algorithms in the future.

4.1 Performance of DDM-Related Variables

15 permutations of the four DDM-related variables were conducted to identify the optimal combination. None of these 15 models include the auxiliary variables listed in Table 1, allowing for a direct evaluation of the SWH retrieval performance of the

Table 5 Brief information about the NDBC's buoy stations used for the validation of the SWH retrievals

Buoy ID	Number of matchups	RMSE of BT (m)	CC of BT
41,001	236	0.54	0.68
41,002	351	0.52	0.80
41,040	190	0.36	0.63
41,041	40	0.50	0.41
41,043	154	0.32	0.44
41,044	371	0.41	0.39
41,046	277	0.35	0.84
41,047	262	0.44	0.71
41,048	457	0.50	0.76
41,049	869	0.37	0.07
42,001	20	0.59	0.99
42,002	118	0.55	0.57
42,003	309	0.38	0.88
42,055	14	1.04	0.54
42,058	115	0.32	0.94
42,059	72	0.54	0.92
42,060	74	0.53	0.98
51,000	312	0.47	0.63
51,001	7	0.72	0.76
51,003	28	0.49	0.65
51,004	124	0.43	0.21
51,101	100	0.43	0.65

four DDM-related variables within the BT model. As shown in Table 7, the results indicate that the model incorporating all four variables performs the best.

It is important to highlight that, compared to the four DDM-related variables, incorporating a single auxiliary variable would result in only a slight enhancement of the model's performance. Therefore, the next section will focus on analyzing the combined impact of all auxiliary parameters on improving the model's performance.

4.2 *Contribution of Auxiliary Variables*

As previously noted, except the four DDM-related variables, auxiliary variables can also have a significant impact. They include instrument gain, noise floor, scatter area, sp_inc_angle, sp_az_body, sp_lat, sp_lon, and HOD. The BT method is used as an example to assess the impact of auxiliary variables on model construction. To

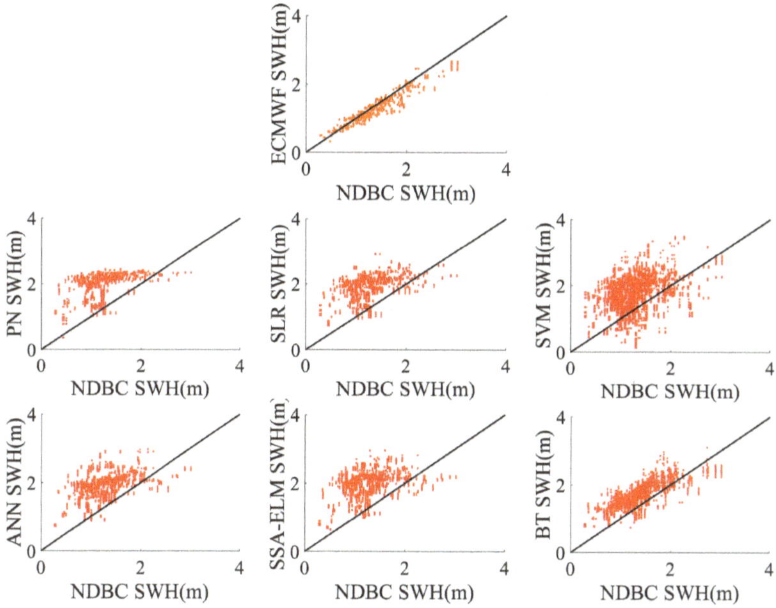

Fig. 12 Scatter plot between ECMWF SWH and NDBC SWH (leftmost single figure) and scatter plots of SWH estimates from the six models against the NDBC buoy test set (six plots on the right)

Table 6 Accuracy of each SWH estimate compared to NDBC data

Methods	ECMWF	PN	SLR	GSVM	ANN	SSA-ELM	BT
RMSE (m)	0.19	0.83	0.74	0.71	0.72	0.80	0.44
CC	0.93	0.44	0.46	0.33	0.48	0.37	0.73

identify an appropriate multi-variable ML model for SWH retrieval, all four DDM-related variables (NBRCS, LES, SNR, and DDMA) are incorporated into the model. The results presented in Fig. 13 demonstrate that auxiliary variables significantly enhance the model's performance. Incorporating auxiliary variables into the SWH estimation model reduces the RMSE and greatly improves the CC. The distribution pattern further reveals that, without the inclusion of auxiliary variables, the SWH estimates show a much larger deviation from the true SWH. Similar observations have been reported in other studies, such as those focused on retrieving sea surface wind speed [44, 45]. Overall, auxiliary variables play a crucial role in developing high-quality models.

Table 7 Model performance of different combinations of DDM-related variables

	NBRCS	LES	SNR	DDMA	RMSE (m)	$RMSE_{2021}$ (m)	CC	CC_{2021}
1	▲				0.76	0.73	0.39	0.42
2		▲			0.78	0.75	0.33	0.38
3			▲		0.81	0.80	0.27	0.27
4				▲	0.81	0.79	0.26	0.29
5	▲	▲			0.73	0.69	0.46	0.51
6	▲		▲		0.72	0.69	0.48	0.52
7	▲			▲	0.72	0.69	0.48	0.52
8		▲	▲		0.74	0.71	0.42	0.47
9		▲		▲	0.75	0.72	0.41	0.47
10			▲	▲	0.78	0.76	0.32	0.35
11	▲	▲	▲		0.71	0.69	0.49	0.55
12	▲	▲		▲	0.72	0.69	0.49	0.55
13	▲		▲	▲	0.71	0.69	0.49	0.53
14		▲	▲	▲	0.74	0.71	0.43	0.49
15	▲	▲	▲	▲	**0.71**	**0.68**	**0.50**	**0.55**

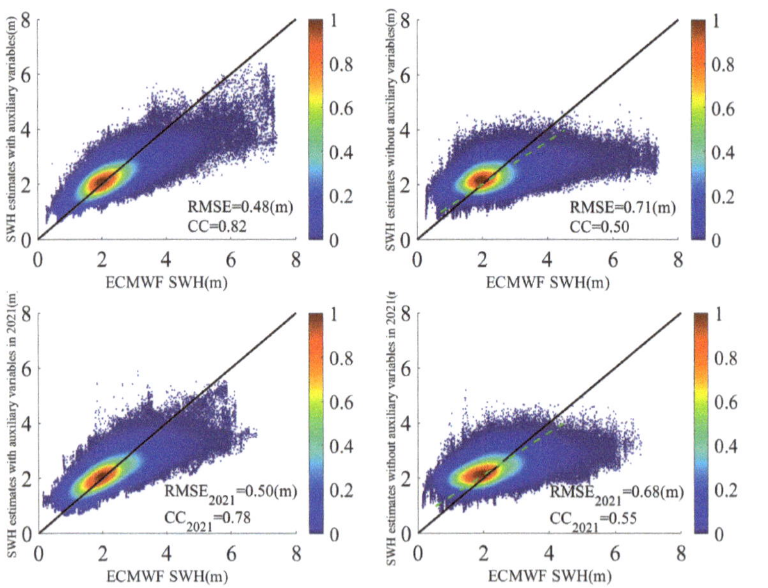

Fig. 13 Results of SWH retrievals with adding (left) and without adding (right) 8 auxiliary variables based on Bagging Tree method. Green dotted line is the linear fitting line between SWH estimates and ECMWF SWH in the figure

Table 8 Performance metrics of the BT model for varying dataset sizes

Models	$BT_{0.8m}$	$BT_{1.6m}$	$BT_{2.4m}$	$BT_{3.2m}$	BT_{4m}
RMSE (m)	0.66	0.58	0.51	0.49	0.48
$RMSE_{2021}$ (m)	0.61	0.56	0.54	0.52	0.5
CC	0.62	0.71	0.79	0.81	0.82
CC_{2021}	0.66	0.71	0.74	0.76	0.78
Calculation time (s)	2.032	2.0987	2.2394	2.3938	2.5232

4.3 Effect of Training Dataset Size

This section analyses the impact of training dataset size on model accuracy. To construct different BT models, data samples ranging from 0.8 million to 4 million, extracted randomly from the original dataset of 4 million samples, are used. Specifically, the models are labelled as $BT_{0.8m}$, $BT_{1.6m}$, $BT_{2.4m}$, $BT_{3.2m}$, and BT_{4m}, corresponding to the respective dataset sizes of 0.8 million, 1.6 million, 2.4 million, 3.2 million, and 4 million samples. The evaluation results are presented in Table 8. On the one hand, as the size of the training dataset increases, the models demonstrate improved performance and greater generalization ability. On the other hand, when the dataset size reaches a certain threshold (e.g., 2.4 million samples), the performance gains become notably smaller. If the size of the training dataset continues to increase, marginal improvements in performance can be achieved, provided that the inversion and training datasets remain of similar size. Furthermore, increasing the dataset size has minimal impact on the computation time of the BT model; the processing time only rises by approximately 25% when the dataset size is increased by 400%. Therefore, it is advantageous to enlarge the training dataset as much as possible, provided the conditions allow.

4.4 Effect of Wind Speed Data

In general, incorporating more relevant known information into a model tends to improve its performance. This section will explore whether adding ECMWF wind speed as an additional known variable enhances the performance of the BT model. As shown in Fig. 14, when wind speed is included as an additional input, more points are closer to the 1:1 line, particularly for SWH values greater than 3 m. Wind speed data can significantly reduce the model's RMSE and improve CC, with enhancements of 17% and 6%, respectively, especially at higher SWHs. Furthermore, as observed in Fig. 15, after including wind speed, the red and blue dots in the bias plot become less pronounced, indicating a reduction in the global bias of the BT model's SWH retrievals.

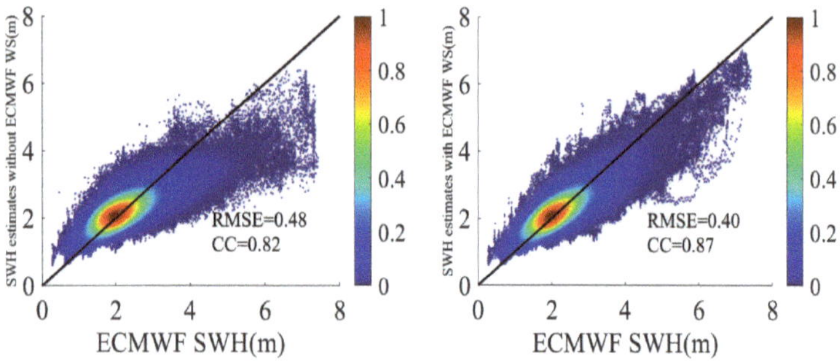

Fig. 14 Scatter plots of SWH retrievals without considering (left) and with considering (right) ECMWF wind speed based on Bagging Tree method. The period spans from Mar 2019 to Nov 2020

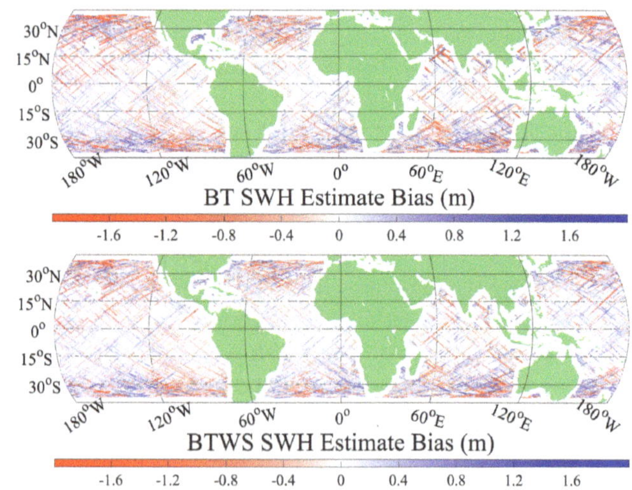

Fig. 15 SWH estimation bias distribution without considering (upper) and with considering (lower) ECMWF wind speed using the same algorithm and data as those in Fig. 14

5 Summary

This chapter first studied the SWH retrieval by using various univariate ML models.

After extensive data processing, it was found that individual variables, when used in isolation, did not yield satisfactory results. To enhance SWH retrieval performance, the chapter then focused on development and analysis of five different multi-variable machine learning models. The results and analysis indicate that the multi-variable model based on Bagging Tree delivers the best overall performance and generalization ability. Additionally, the retrieval results from the Bagging Tree model show a strong correlation with the NDBC buoy measurements. It is important to note that using a larger and more representative training dataset leads to improved retrieval performance. The data used for modelling spans two years (from March 2019 to December 2020), which gives the constructed models a degree of generality, making them suitable for SWH retrieval across different years. This is demonstrated by the fact that the SWH retrieval accuracy using 2021 data shows only a slight decline compared to the accuracy achieved with 2019–2020 data. Additionally, it has been demonstrated that incorporating auxiliary variables can significantly enhance the model's effectiveness. Also, as expected, the model's performance improves considerably when ECMWF wind speed is added as a known variable. Future research will focus on further enhancing the performance of the models. For example, high SWH and strong winds tend to increase estimation error. Therefore, it is beneficial to develop methods to deal with the retrieval of high SWH with only possible minor performance degradation.

References

1. Yang S, Deng Z, Li X, Zheng C, Xi L, Zhuang J, Zhang Z (2021) A novel hybrid model based on STL decomposition and one-dimensional convolutional neural networks with positional encoding for significant wave height forecast. Renew Energy 173:531–543. https://doi.org/10.1016/j.renene.2021.04.010
2. Wang J, Aouf L, Jia Y, Zhang Y (2020) Validation and calibration of significant wave height and wind speed retrievals from HY2B altimeter based on deep learning. Remote Sens 12(17):2858. https://doi.org/10.3390/rs12172858
3. Yang J, Zhang J, Jia Y, Fan C, Cui W (2020) Validation of Sentinel-3A/3B and Jason-3 altimeter wind speeds and significant wave heights using buoy and ASCAT data. Remote Sens 12(13):2079. https://doi.org/10.3390/rs12132079
4. Peng Q, Jin S (2019) Significant wave height estimation from space-borne cyclone-GNSS reflectometry. Remote Sens 11(5):584. https://doi.org/10.3390/rs11050584
5. Hall CD, Cordey RA (1988) Multistatic scatterometry. In: International geoscience and remote sensing symposium, 'remote sensing: moving toward the 21st century', vol 1. IEEE, pp 561–562. https://doi.org/10.1109/IGARSS.1988.570200
6. Martin-Neira M (1993) A passive reflectometry and interferometry system (PARIS): application to ocean altimetry. ESA J 17(4):331–355
7. Arroyo AA, Camps A, Aguasca A, Forte GF, Monerris A, Rüdiger C, Onrubia R (2014) Dual-polarization GNSS-R interference pattern technique for soil moisture mapping. IEEE J Sel Top Appl Earth Obs Remote Sens 7(5):1533–1544. https://doi.org/10.1109/JSTARS.2014.2320792
8. Yu K, Ban W, Zhang X, Lv C (2015) Snow depth estimation using GPS triple-frequency carrier phase combination. IEEE Trans Geosci Remote Sens 53(9):5100–5109
9. Clarizia MP, Ruf CS (2016) Wind speed retrieval algorithm for the Cyclone Global Navigation Satellite System (CYGNSS) mission. IEEE Trans Geosci Remote Sens 54(8):4419–4432. https://doi.org/10.1109/10.1109/TGRS.2016.2541343
10. Guo W, Du H, Guo C, Southwell BJ, Cheong JW, Dempster AG (2022) Information fusion for GNSS-R wind speed retrieval using statistically modified convolutional neural network. Remote Sens Environ 272:112934. https://doi.org/10.1016/j.rse.2022.112934
11. Hu C, Benson CR, Qiao L, Rizos C (2020) The validation of the weight function in the leading-edge-derivative path delay estimator for space-based GNSS-R altimetry. IEEE Trans Geosci Remote Sens 58(9):6243–6254. https://doi.org/10.1109/TGRS.2020.2975817

12. Chapman BD, Russo IM, Galdi C, Morris M, di Bisceglie M, Zuffada C, O'Brien AJ (2022) Comparison of SAR and CYGNSS surface water extent metrics. IEEE J Sel Top Appl Earth Obs Remote Sens 15:3235–3245. https://doi.org/10.1109/JSTARS.2022.3162764
13. Alonso-Arroyo A, Zavorotny VU, Camps A (2017) Sea ice detection using UK TDS-1 GNSS-R data. IEEE Trans Geosci Remote Sens 55(9):4989–5001. https://doi.org/10.1109/TGRS.2017.2699122
14. Di Simone A, Park H, Riccio D, Camps A (2017) Sea target detection using spaceborne GNSS-R delay-Doppler maps: Theory and experimental proof of concept using TDS-1 data. IEEE J Sel Top Appl Earth Obs Remote Sens 10(9):4237–4255. https://doi.org/10.1109/JSTARS.2017.2705350
15. Yu K (2021) Theory and practice of GNSS reflectometry. Springer, Berlin/Heidelberg, Germany
16. Alonso-Arroyo A, Camps A, Park H, Pascual D, Onrubia R, Martin F (2014) Retrieval of significant wave height and mean sea surface level using the GNSS-R interference pattern technique: results from a three-month field campaign. IEEE Trans Geosci Remote Sens 53(6):3198–3209. https://doi.org/10.1109/TGRS.2014.2371540
17. Soulat F, Caparrini M, Germain O, Lopez-Dekker P, Taani M, Ruffini G (2004) Sea state monitoring using coastal GNSS-R. Geophys Res Lett 31(21). https://doi.org/10.1029/2004GL020680
18. Shah R, Garrison JL (2014) Application of the ICF coherence time method for ocean remote sensing using digital communication satellite signals. IEEE J Sel Top Appl Earth Obs Remote Sens 7(5):1584–1591. https://doi.org/10.1109/JSTARS.2014.2314531
19. Yang S, Jin S, Jia Y, Ye M (2021) Significant wave height estimation from joint CYGNSS DDMA and LES observations. Sensors 21(18):6123. https://doi.org/10.3390/s21186123
20. Bu J, Yu K (2022) A new integrated method of CYGNSS DDMA and LES measurements for significant wave height estimation. IEEE Geosci Remote Sens Lett 19:1–5. https://doi.org/10.1109/LGRS.2022.3198131
21. Wang C, Yu K, Zhang K, Bu J, Qu F (2022) Significant wave height retrieval based on multivariable regression models developed with CYGNSS data. IEEE Trans Geosci Remote Sens 61:1–15. https://doi.org/10.1109/TGRS.2022.3233102
22. Wang C, Yu K, Qu F, Bu J, Han S, Zhang K (2022) Spaceborne GNSS-R wind speed retrieval using machine learning methods. Remote Sens 14(14):3507. https://doi.org/10.3390/rs14143507
23. Bu J, Yu K (2022) Significant wave height retrieval method based on spaceborne GNSS reflectometry. IEEE Geosci Remote Sens Lett 19:1–5. https://doi.org/10.1109/LGRS.2022.3155563
24. Li B, Yang L, Zhang B, Yang D, Wu D (2020) Modeling and simulation of GNSS-R observables with effects of swell. IEEE J Sel Top Appl Earth Obs Remote Sens 13:1833–1841. https://doi.org/10.1109/JSTARS.2020.2992037
25. Marchan-Hernandez JF, Valencia E, Rodriguez-Alvarez N, Ramos-Pérez I, Bosch-Lluis X, Camps A, Marcello J (2010) Sea-state determination using GNSS-R data. IEEE Geosci Remote Sens Lett 7(4):621–625. https://doi.org/10.1109/LGRS.2010.2043213
26. Soisuvarn S, Jelenak Z, Said F, Chang PS, Egido A (2016) The GNSS reflectometry response to the ocean surface winds and waves. IEEE J Sel Top Appl Earth Obs Remote Sens 9(10):4678–4699. https://doi.org/10.1109/JSTARS.2016.2602703
27. Chen-Zhang DD, Ruf CS, Ardhuin F, Park J (2016) GNSS-R nonlocal sea state dependencies: model and empirical verification. J Geophys Res: Oceans 121(11):8379–8394. https://doi.org/10.1002/2016JC012308
28. Jing C, Niu X, Duan C, Lu F, Di G, Yang X (2019) Sea surface wind speed retrieval from the first Chinese GNSS-R mission: technique and preliminary results. Remote Sens 11(24):3013. https://doi.org/10.3390/rs11243013
29. Li X, Yang D, Yang J, Zheng G, Han G, Nan Y, Li W (2021) Analysis of coastal wind speed retrieval from CYGNSS mission using artificial neural network. Remote Sens Environ 260:112454. https://doi.org/10.1016/j.rse.2021.112454

30. Liu WC, Huang WC (2021) Tide–surge and wave interaction around the Taiwan coast: insight from Typhoon Nepartak in 2016. Nat Hazard 107:1881–1904. https://doi.org/10.1007/s11069-021-04663-3
31. Krusienski DJ, Sellers EW, McFarland DJ, Vaughan TM, Wolpaw JR (2008) Toward enhanced P300 speller performance. J Neurosci Methods 167(1):15–21. https://doi.org/10.1016/j.jneumeth.2007.07.017
32. Siddiqi MH, Ali R, Khan AM, Park YT, Lee S (2015) Human facial expression recognition using stepwise linear discriminant analysis and hidden conditional random fields. IEEE Trans Image Process 24(4):1386–1398. https://doi.org/10.1109/TIP.2015.2405346
33. Scholkopf B, Sung KK, Burges CJ, Girosi F, Niyogi P, Poggio T, Vapnik V (1997) Comparing support vector machines with Gaussian kernels to radial basis function classifiers. IEEE Trans Signal Process 45(11):2758–2765. https://doi.org/10.1109/78.65010
34. Burges CJ (1998) A tutorial on support vector machines for pattern recognition. Data Min Knowl Disc 2(2):121–167. https://doi.org/10.1023/A:1009715923555
35. Basheer IA, Hajmeer M (2000) Artificial neural networks: fundamentals, computing, design, and application. J Microbiol Methods 43(1):3–31. https://doi.org/10.1016/s0167-7012(00)00201-3
36. Xue J, Shen B (2020) A novel swarm intelligence optimization approach: sparrow search algorithm. Syst Science & Control Engineering 8(1):22–34. https://doi.org/10.1080/21642583.2019.1708830
37. Chen W, Shahabi H, Zhang S, Khosravi K, Shirzadi A, Chapi K, Pham BT, Zhang T, Zhang L, Chai H, Ma J, Chen Y, Wang X, Li R, Ahmad BB (2018) Landslide susceptibility modeling based on GIS and novel bagging-based kernel logistic regression. Appl Sci 8(12):2540. https://doi.org/10.3390/app8122540
38. Chan JCW, Huang C, Defries R (2001) Enhanced algorithm performance for land cover classification from remotely sensed data using bagging and boosting. IEEE Trans Geosci Remote Sens 39(3):693–695. https://doi.org/10.1109/36.911126
39. Hothorn T, Lausen B, Benner A, Radespiel-Tröger M (2004) Bagging survival trees. Stat Med 23(1):77–91. https://doi.org/10.1002/sim.1593
40. Prasad AM, Iverson LR, Liaw A (2006) Newer classification and regression tree techniques: bagging and random forests for ecological prediction. Ecosystems 9:181–199. https://doi.org/10.1007/s10021-005-0054-1
41. Breiman L (1996) Bagging predictors. Mach Learn 24:123–140. https://doi.org/10.1007/BF00058655
42. Bauer E, Kohavi R (1999) An empirical comparison of voting classification algorithms: bagging, boosting, and variants. Mach Learn 36:105–139. https://doi.org/10.1023/A:1007515423169
43. Saïd F, Jelenak Z, Park J, Chang PS (2021) The NOAA track-wise wind retrieval algorithm and product assessment for CyGNSS. IEEE Trans Geosci Remote Sens 60:1–24. https://doi.org/10.1109/TGRS.2021.3087426
44. Bu J, Yu K, Zhu Y, Qian N, Chang J (2020) Developing and testing models for sea surface wind speed estimation with GNSS-R delay Doppler maps and delay waveforms. Remote Sens 12(22):3760. https://doi.org/10.3390/rs12223760
45. Balasubramaniam R, Ruf C (2020) Neural network based quality control of CYGNSS wind retrieval. Remote Sens 12(17):2859. https://doi.org/10.3390/rs12172859

FSC
www.fsc.org
MIX
Papier | Fördert
gute Waldnutzung
FSC® C083411

Kolibri 360 GmbH
Ferdinand-Jühlke-Straße 7
99095 Erfurt, Deutschland
produktsicherheit@kolibri360.de